GRADUATE STUDIES
IN MATHEMATICS **188**

Introduction to Algebraic Geometry

Steven Dale Cutkosky

American Mathematical Society
Providence, Rhode Island

EDITORIAL COMMITTEE

Dan Abramovich
Daniel S. Freed (Chair)
Gigliola Staffilani
Jeff A. Viaclovsky

2010 *Mathematics Subject Classification.* Primary 14-01.

For additional information and updates on this book, visit
www.ams.org/bookpages/gsm-188

Library of Congress Cataloging-in-Publication Data
Names: Cutkosky, Steven Dale, author.
Title: Introduction to algebraic geometry / Steven Dale Cutkosky.
Other titles: Algebraic geometry
Description: Providence, Rhode Island : American Mathematical Society, [2018] | Series: Graduate studies in mathematics ; volume 188 | Includes bibliographical references and index.
Identifiers: LCCN 2017045552 | ISBN 9781470435189 (alk. paper)
Subjects: LCSH: Geometry, Algebraic. | AMS: Algebraic geometry – Instructional exposition (textbooks, tutorial papers, etc.). msc
Classification: LCC QA564 .C8794 2018 | DDC 516.3/5–dc23
LC record available at https://lccn.loc.gov/2017045552

Copying and reprinting. Individual readers of this publication, and nonprofit libraries acting for them, are permitted to make fair use of the material, such as to copy select pages for use in teaching or research. Permission is granted to quote brief passages from this publication in reviews, provided the customary acknowledgment of the source is given.

Republication, systematic copying, or multiple reproduction of any material in this publication is permitted only under license from the American Mathematical Society. Requests for permission to reuse portions of AMS publication content are handled by the Copyright Clearance Center. For more information, please visit **www.ams.org/publications/pubpermissions**.

Send requests for translation rights and licensed reprints to **reprint-permission@ams.org**.

© 2018 by the author. All rights reserved.
Printed in the United States of America.

∞ The paper used in this book is acid-free and falls within the guidelines
established to ensure permanence and durability.
Visit the AMS home page at **http://www.ams.org/**

10 9 8 7 6 5 4 3 2 1   23 22 21 20 19 18

To Hema, Ashok, and Maya

Contents

Preface	xi
Chapter 1. A Crash Course in Commutative Algebra	1
§1.1. Basic algebra	1
§1.2. Field extensions	6
§1.3. Modules	8
§1.4. Localization	9
§1.5. Noetherian rings and factorization	10
§1.6. Primary decomposition	13
§1.7. Integral extensions	16
§1.8. Dimension	19
§1.9. Depth	20
§1.10. Normal rings and regular rings	22
Chapter 2. Affine Varieties	27
§2.1. Affine space and algebraic sets	27
§2.2. Regular functions and regular maps of affine algebraic sets	33
§2.3. Finite maps	40
§2.4. Dimension of algebraic sets	42
§2.5. Regular functions and regular maps of quasi-affine varieties	48
§2.6. Rational maps of affine varieties	58
Chapter 3. Projective Varieties	63
§3.1. Standard graded algebras	63

§3.2.	Projective varieties	67
§3.3.	Grassmann varieties	73
§3.4.	Regular functions and regular maps of quasi-projective varieties	74

Chapter 4. Regular and Rational Maps of Quasi-projective Varieties — 87

§4.1.	Criteria for regular maps	87
§4.2.	Linear isomorphisms of projective space	90
§4.3.	The Veronese embedding	91
§4.4.	Rational maps of quasi-projective varieties	93
§4.5.	Projection from a linear subspace	95

Chapter 5. Products — 99

§5.1.	Tensor products	99
§5.2.	Products of varieties	101
§5.3.	The Segre embedding	105
§5.4.	Graphs of regular and rational maps	106

Chapter 6. The Blow-up of an Ideal — 111

§6.1.	The blow-up of an ideal in an affine variety	111
§6.2.	The blow-up of an ideal in a projective variety	120

Chapter 7. Finite Maps of Quasi-projective Varieties — 127

§7.1.	Affine and finite maps	127
§7.2.	Finite maps	131
§7.3.	Construction of the normalization	135

Chapter 8. Dimension of Quasi-projective Algebraic Sets — 139

§8.1.	Properties of dimension	139
§8.2.	The theorem on dimension of fibers	141

Chapter 9. Zariski's Main Theorem — 147

Chapter 10. Nonsingularity — 153

§10.1.	Regular parameters	153
§10.2.	Local equations	155
§10.3.	The tangent space	156
§10.4.	Nonsingularity and the singular locus	159
§10.5.	Applications to rational maps	165

§10.6.	Factorization of birational regular maps of nonsingular surfaces	168
§10.7.	Projective embedding of nonsingular varieties	170
§10.8.	Complex manifolds	175

Chapter 11.	Sheaves	181
§11.1.	Limits	181
§11.2.	Presheaves and sheaves	185
§11.3.	Some sheaves associated to modules	196
§11.4.	Quasi-coherent and coherent sheaves	200
§11.5.	Constructions of sheaves from sheaves of modules	204
§11.6.	Some theorems about coherent sheaves	209

Chapter 12.	Applications to Regular and Rational Maps	221
§12.1.	Blow-ups of ideal sheaves	221
§12.2.	Resolution of singularities	225
§12.3.	Valuations in algebraic geometry	228
§12.4.	Factorization of birational maps	232
§12.5.	Monomialization of maps	236

Chapter 13.	Divisors	239
§13.1.	Divisors and the class group	240
§13.2.	The sheaf associated to a divisor	242
§13.3.	Divisors associated to forms	249
§13.4.	Calculation of some class groups	249
§13.5.	The class group of a curve	254
§13.6.	Divisors, rational maps, and linear systems	259
§13.7.	Criteria for closed embeddings	264
§13.8.	Invertible sheaves	269
§13.9.	Transition functions	271

Chapter 14.	Differential Forms and the Canonical Divisor	279
§14.1.	Derivations and Kähler differentials	279
§14.2.	Differentials on varieties	283
§14.3.	n-forms and canonical divisors	286

Chapter 15.	Schemes	289
§15.1.	Subschemes of varieties, schemes, and Cartier divisors	289
§15.2.	Blow-ups of ideals and associated graded rings of ideals	293

§15.3.	Abstract algebraic varieties	295
§15.4.	Varieties over nonclosed fields	296
§15.5.	General schemes	296

Chapter 16. The Degree of a Projective Variety — 299

Chapter 17. Cohomology — 307

§17.1.	Complexes	307
§17.2.	Sheaf cohomology	308
§17.3.	Čech cohomology	310
§17.4.	Applications	312
§17.5.	Higher direct images of sheaves	320
§17.6.	Local cohomology and regularity	325

Chapter 18. Curves — 333

§18.1.	The Riemann-Roch inequality	334
§18.2.	Serre duality	335
§18.3.	The Riemann-Roch theorem	340
§18.4.	The Riemann-Roch problem on varieties	343
§18.5.	The Hurwitz theorem	345
§18.6.	Inseparable maps of curves	348
§18.7.	Elliptic curves	351
§18.8.	Complex curves	358
§18.9.	Abelian varieties and Jacobians of curves	360

Chapter 19. An Introduction to Intersection Theory — 365

§19.1.	Definition, properties, and some examples of intersection numbers	366
§19.2.	Applications to degree and multiplicity	375

Chapter 20. Surfaces — 379

§20.1.	The Riemann-Roch theorem and the Hodge index theorem on a surface	379
§20.2.	Contractions and linear systems	383

Chapter 21. Ramification and Étale Maps — 391

§21.1.	Norms and Traces	392
§21.2.	Integral extensions	393
§21.3.	Discriminants and ramification	398

§21.4.	Ramification of regular maps of varieties	406
§21.5.	Completion	408
§21.6.	Zariski's main theorem and Zariski's subspace theorem	413
§21.7.	Galois theory of varieties	421
§21.8.	Derivations and Kähler differentials redux	424
§21.9.	Étale maps and uniformizing parameters	426
§21.10.	Purity of the branch locus and the Abhyankar-Jung theorem	433
§21.11.	Galois theory of local rings	438
§21.12.	A proof of the Abhyankar-Jung theorem	441
Chapter 22.	Bertini's Theorems and General Fibers of Maps	451
§22.1.	Geometric integrality	452
§22.2.	Nonsingularity of the general fiber	454
§22.3.	Bertini's second theorem	457
§22.4.	Bertini's first theorem	458
Bibliography		469
Index		477

Preface

This book is an introductory course in algebraic geometry, proving most of the fundamental classical results of algebraic geometry.

Algebraic geometry combines the intuition of geometry with the precision of algebra. Starting with geometric concepts, we introduce machinery as necessary to model important ideas from algebraic geometry and to prove fundamental results. Emphasis is put on developing facility with connecting geometric and algebraic concepts. Examples are constructed or cited illustrating the scope of these results. The theory in this book is developed in increasing sophistication, giving (and refining) definitions as required to accommodate new geometric ideas.

We work as much as possible with quasi-projective varieties over an algebraically closed field of arbitrary characteristic. This allows us to interpret varieties through their function fields. This approach and the use of methods of algebraic number theory in algebraic geometry have been central to algebraic geometry at least since the time of Dedekind and Weber (*Theorie der algebraischen Functionen einer Veränderlichen* [46], translated in [47]). By interpreting the geometric concept of varieties through their regular functions, we are able to use the techniques of commutative algebra.

Differences between the theory in characteristic 0 and positive characteristic are emphasized in this book. We extend our view to schemes, allowing rings with nilpotents, to study fibers of regular maps and to develop intersection theory. We discuss the cases of nonclosed ground fields and nonseparated schemes and some of the extra considerations which appear in these situations. A list of exercises is given at the end of many sections and chapters.

The classic textbooks *Basic Algebraic Geometry* [**136**] by Shafarevich, *Introduction to Algebraic Geometry* [**116**] by Mumford, and *Algebraic Geometry* [**73**] by Hartshorne, as well as the works of Zariski, Abhyankar, Serre, and Grothendieck, have been major influences on this book.

The necessary commutative algebra is introduced and reviewed, beginning with Chapter 1, "A Crash Course in Commutative Algebra". We state definitions and theorems, explain concepts, and give examples from commutative algebra for everything that we will need, proving some results and giving a few examples, but mostly giving references to books on commutative algebra for proofs. As such, this book is intended to be self-contained, although a reader may be curious about the proofs for some cited results in commutative algebra and will want to either derive them or look up the references. We give references to several books, mostly depending on which book has the exact statement we require.

A reader should be familiar with the material through Section 1.6 on primary decomposition before beginning Chapter 2 on affine varieties. Depending on the background of students, the material in Chapter 1 can be skipped, quickly reviewed at the beginning of a course, or be used as an outline of a semester-long course in commutative algebra before beginning the study of geometry in Chapter 2.

Chapters 2–10 give a one-semester introduction to algebraic geometry, through affine and projective varieties. The sections on integral extensions, dimension, depth, and normal and regular local rings in Chapter 1 can be referred to as necessary as these concepts are encountered within a geometric context. Chapters 11–20 provide a second-semester course, which includes sheaves, schemes, cohomology, divisors, intersection theory, and the application of these concepts to curves and surfaces.

Chapter 21 (on ramification and étale maps) and Chapter 22 (on Bertini's theorems and general fibers of maps) could be the subject of a topics course for a third-semester course. The distinctions between characteristic 0 and positive characteristic are especially explored in these last chapters. These two chapters could be read any time after the completion of Chapter 14.

I thank the students of my classes, especially Razieh Ahmadian, Navaneeth Chenicheri Chathath, Suprajo Das, Arpan Dutta, Melissa Emory, Kyle Maddox, Smita Praharaj, Thomas Polstra, Soumya Sanyal, Pham An Vinh, and particularly Roberto Núñez, for their feedback and helpful comments on preliminary versions of this book. I thank Maya Cutkosky for help with the figures.

Chapter 1

A Crash Course in Commutative Algebra

In this chapter we review some basics of commutative algebra which will be assumed in this book.

All rings will be commutative (with identity 1). The natural numbers, $\{0, 1, 2, \ldots\}$, will be denoted by $\mathbb{N}$. The positive integers, $\{1, 2, \ldots\}$, will be denoted by $\mathbb{Z}_+$. Throughout this book, k will be an algebraically closed field (of arbitrary characteristic) unless specified otherwise.

1.1. Basic algebra

The starting point of commutative algebra is the fact that every ring R has a maximal ideal and thus has at least one prime ideal [**13**, Theorem 1.3].

We will say that a ring R is a local ring if R has a unique maximal ideal. We will denote the maximal ideal of the local ring R by m_R. If $\phi : R \to S$ is a ring homomorphism and I is an ideal in S, then $\phi^{-1}(I)$ is an ideal in R. If P is a prime ideal in S, then $\phi^{-1}(P)$ is a prime ideal in R. Suppose that R, S are local domains with maximal ideals m_R, m_S, respectively. We will say that S dominates R if $R \subset S$ and $m_S \cap R = m_R$. We will write $\mathrm{QF}(R)$ for the quotient field of a domain R.

A fundamental fact is the following theorem.

Lemma 1.1. *Let $\pi : R \to S$ be a surjective ring homomorphism, with kernel K.*

1) *Suppose that I is an ideal in S. Then $\pi^{-1}(I)$ is an ideal in R containing K.*

1

2) *Suppose that J is an ideal in R such that J contains K. Then $\pi(J)$ is an ideal in S.*

3) *The map $I \mapsto \pi^{-1}(I)$ is a 1-1 correspondence between the set of ideals in S and the set of ideals in R which contain K. The inverse map is $J \mapsto \pi(J)$.*

4) *The correspondence is order preserving: for ideals I_1, I_2 in S, $I_1 \subset I_2$ if and only if $\pi^{-1}(I_1) \subset \pi^{-1}(I_2)$.*

5) *For an ideal I in S, I is a prime ideal if and only if $\pi^{-1}(I)$ is a prime ideal in R.*

6) *For an ideal I in S, I is a maximal ideal if and only if $\pi^{-1}(I)$ is a maximal ideal in R.*

In the case when $S = R/K$ and $\pi : R \to R/K$ is the map $\pi(x) = x + K$ for $x \in R$, we have that $\pi(J) = J/K$ for J an ideal of R containing K.

Proof. [84, Theorem 2.6]. □

A ring S is an R-algebra if there is a given ring homomorphism $\phi : R \to S$. This gives us a multiplication $rs = \phi(r)s$ for $r \in R$ and $s \in S$. Suppose that S is an R-algebra by a homomorphism $\phi : R \to S$ and T is an R-algebra by a homomorphism $\psi : R \to T$. Then a ring homomorphism $\sigma : S \to T$ is an R-algebra homomorphism if $\sigma(\phi(r)) = \psi(r)$ for all $r \in R$.

A proof of the following universal property of polynomial rings can be found in [84, Theorem 2.11].

Theorem 1.2. *Suppose that R and S are rings and $R[x_1, \ldots, x_n]$ is a polynomial ring over R. Suppose that $\overline{\phi} : R \to S$ is a ring homomorphism and $t_1, \ldots, t_n \in S$. Then there exists a unique ring homomorphism $\Phi : R[x_1, \ldots, x_n] \to S$ such that $\Phi(r) = \overline{\phi}(r)$ for $r \in R$ and $\Phi(x_i) = t_i$ for $1 \leq i \leq n$.*

A polynomial ring over a ring R is naturally an R-algebra. With the notation of the previous theorem, S is an R-algebra by the homomorphism $\overline{\phi}$, making ϕ an R-algebra homomorphism.

Suppose that $R \subset S$ is a subring and $\Lambda \subset S$ is a subset. Then $R[\Lambda]$ is defined to be the smallest subring of S containing R and Λ. For $n \in \mathbb{N}$, letting $R[x_1, \ldots, x_n]$ be a polynomial ring, we have that

$$R[\Lambda] = \{f(t_1, \ldots, t_n) \mid t_1, \ldots, t_n \in \Lambda \text{ and } f \in R[x_1, \ldots, x_n]\}.$$

If $\Lambda = \{t_1, \ldots, t_n\}$, write $R[\Lambda] = R[t_1, \ldots, t_n]$.

Suppose that we have a surjective ring homomorphism $\Phi : R[x_1, \ldots, x_n] \to S$ from the polynomial ring $R[x_1, \ldots, x_n]$. Letting I be the kernel of Φ,

1.1. Basic algebra

we have an induced isomorphism $R[x_1, \ldots, x_n]/I \cong S$. Letting $\overline{R} = \Phi(R) \cong R/I \cap R$, we have that $S = \overline{R}[t_1, \ldots, t_n]$ where $t_i = \Phi(x_i)$. More abstractly, if I is an ideal in the polynomial ring $R[x_1, \ldots, x_n]$, let $S = R[x_1, \ldots, x_n]/I$. Let $\overline{R} = R/(I \cap R) \subset S$ and $\overline{x}_i = x_i + I$ in S. Then $S = \overline{R}[\overline{x}_1, \ldots, \overline{x}_n]$.

An element $x \in R$ is a zero divisor if $x \neq 0$ and there exists $0 \neq y \in R$ such that $xy = 0$. An element $x \in R$ is nilpotent if $x \neq 0$ and there exists $n \in \mathbb{N}$ such that $x^n = 0$. The radical of an ideal I in R is

$$\sqrt{I} = \{f \in R \mid f^n \in I \text{ for some } n \in \mathbb{N}\}.$$

A ring R is reduced if whenever $f \in R$ is such that $f^n = 0$ for some positive integer n, we have that $f = 0$. Suppose that I is an ideal in a ring R. The ring R/I is reduced if and only if $\sqrt{I} = I$.

An R-algebra A is finitely generated if A is generated by a finite number of elements as an R-algebra, so that A is a quotient of a polynomial ring over R in finitely many variables.

If A is nonzero and is generated by $u_1, \ldots, u_n$ as a R-algebra, then $\overline{R} = R1_A$ is a subring of A and $A = \overline{R}[u_1, \ldots, u_n]$. In particular, A is a quotient of a polynomial ring over R.

If K is a field and A is a nonzero K-algebra, then we can view K as a subring of A by identifying K with $K1_A$.

The following lemma will be useful in some of the problems in Chapter 2.

Lemma 1.3. *Suppose that K is a field, $K[x_1, \ldots, x_n, z]$ is a polynomial ring over K, and $f_1, \ldots, f_r, g \in K[x_1, \ldots, x_n]$. Then*

$$A = K[x_1, \ldots, x_n, z]/(f_1(x_1, \ldots, x_n), \ldots, f_r(x_1, \ldots, x_n), z - g(x_1, \ldots, x_n))$$
$$\cong K[x_1, \ldots, x_n]/(f_1(x_1, \ldots, x_n), \ldots, f_r(x_1, \ldots, x_n)).$$

Proof. Let $\overline{x}_1, \ldots, \overline{x}_n, \overline{z}$ be the classes of $x_1, \ldots, x_n, z$ in A. We have that A is generated by $\overline{x}_1, \ldots, \overline{x}_n$ and $\overline{z}$ as a K-algebra, and $\overline{z} = g(\overline{x}_1, \ldots, \overline{x}_n)$, so $A = K[\overline{x}_1, \ldots, \overline{x}_n]$ is generated by $\overline{x}_1, \ldots, \overline{x}_n$ as a K-algebra.

By the universal property of polynomial rings, we have a K-algebra homomorphism $\Phi : K[x_1, \ldots, x_n] \to A$ defined by $\Phi(x_i) = \overline{x}_i$ for $1 \leq i \leq n$. Since $A = K[\overline{x}_1, \ldots, \overline{x}_n]$, Φ is surjective.

We now compute the kernel of Φ. The elements $f_i(x_1, \ldots, x_n)$ are in Kernel(Φ) since $\Phi(f_i) = f_i(\overline{x}_1, \ldots, \overline{x}_n) = 0$. Suppose

$$h(x_1, \ldots, x_n) \in \text{Kernel}(\Phi).$$

Then $\Phi(h(x_1, \ldots, x_n)) = h(\overline{x}_1, \ldots, \overline{x}_n) = 0$ in A, and so $h(x_1, \ldots, x_n)$ is in the ideal

$$(f_1(x_1, \ldots, x_n), \ldots, f_r(x_1, \ldots, x_n), z - g(x_1, \ldots, x_n))$$

of $K[x_1, \ldots, x_n, z]$ and so

$h(x_1, \ldots, x_n)$
$= a_1(x_1, \ldots, x_n, z) f_1(x_1, \ldots, x_n) + \cdots + a_r(x_1, \ldots, x_n, z) f_r(x_1, \ldots, x_n)$
$+ b(x_1, \ldots, x_n, z)(z - g(x_1, \ldots, x_n))$

for some $a_i, b \in K[x_1, \ldots, x_n, z]$. Setting $z = g(x_1, \ldots, x_n)$, we have

$$h(x_1, \ldots, x_n) = a_1(x_1, \ldots, x_n, g(x_1, \ldots, x_n)) f_1(x_1, \ldots, x_n)$$
$$+ \cdots + a_r(x_1, \ldots, x_n, g(x_1, \ldots, x_n)) f_r(x_1, \ldots, x_n).$$

Thus h is in the ideal $(f_1, \ldots, f_r)$ in $K[x_1, \ldots, x_n]$, and so Kernel$(\Phi) = (f_1, \ldots, f_r)$. Thus $A \cong K[x_1, \ldots, x_n]/(f_1, \ldots, f_r)$. □

The following theorem justifies the common identification of polynomials and polynomial functions over an infinite field.

Theorem 1.4. *Suppose that L is an infinite field and $f \in L[x_1, \ldots, x_n]$ is a nonzero polynomial. Then there exist elements $a_1, \ldots, a_n \in L$ such that $f(a_1, \ldots, a_n) \neq 0$.*

Proof. We prove the theorem by induction on n. A nonzero polynomial $f(x) \in L[x]$ has at most finitely many roots so, since L is infinite, there exists $a \in L$ such that $f(a) \neq 0$.

Assume that $n > 1$ and the theorem is true for $n - 1$ indeterminates. Expand

$$f(x_1, \ldots, x_n) = B_0 + B_1 x_n + \cdots + B_d x_n^d$$

where $B_i \in L[x_1, \ldots, x_{n-1}]$ for all i and $B_d \neq 0$. By induction, there exist $a_i \in L$ such that $B_d(a_1, \ldots, a_{n-1}) \neq 0$. Thus

$f(a_1, \ldots, a_{n-1}, x_n)$
$= B_0(a_1, \ldots, a_{n-1}) + B_1(a_1, \ldots, a_{n-1}) x_n + \cdots + B_d(a_1, \ldots, a_{n-1}) x_n^d$

is a nonzero polyomial in $L[x_n]$. Hence we can choose $a_n \in L$ such that $f(a_1, \ldots, a_{n-1}, a_n) \neq 0$. □

Theorem 1.5 (Chinese remainder theorem). *Let A be a ring and $I_1, \ldots, I_n$ be ideals in A such that $I_i + I_j = A$ for $i \neq j$ (I_i and I_j are coprime). Given elements $x_1, \ldots, x_n \in A$, there exists $x \in A$ such that $x \equiv x_i \bmod I_i$ for all i.*

Proof. [95, page 94]. □

Corollary 1.6. *Let A be a ring and $I_1, \ldots, I_n$ be ideals in A. Assume that $I_i + I_j = A$ for $i \neq j$. Let*

$$f : A \to \bigoplus_{i=1}^n A/I_i$$

be homomorphism induced by the canonical maps of A onto each factor A/I_i. Then the kernel of f is $I_1 \cap I_2 \cap \cdots \cap I_n = I_1 I_2 \cdots I_n$ and f is surjective, so we have an isomorphism $A/\bigcap I_i \cong \prod A/I_i$.

Proof. [95, page 95]. □

Exercise 1.7. Suppose that R is a domain and $0 \neq f \in R$. Show that $R[x]/(xf-1) \cong R[\frac{1}{f}]$. Hint: Start by using the universal property of polynomial rings to get an R-algebra homomorphism $\phi : R[x] \to K$ where K is the quotient field of R and $\phi(x) = \frac{1}{f}$.

Exercise 1.8. Let K be a field and $R = K[x,y]/(y^2 - x^3) = K[\overline{x}, \overline{y}]$ where $\overline{x}, \overline{y}$ are the classes of x and y in R. Show that R is a domain. Let R_1 be the subring $R_1 = R[\frac{\overline{y}}{\overline{x}}]$ of the quotient field of R. Show that $R_2 = R[t]/(\overline{x}t - \overline{y})$ is not a domain, so that the K-algebras R_1 and R_2 are not isomorphic.

Exercise 1.9. Let K be a field and $R = K[x,y]$ be a polynomial ring in the variables x and y. Let R_1 be the subring $R_1 = R[\frac{y}{x}]$ of the quotient field of R. Let $R_2 = R[t]/(xt - y)$. Show that the K-algebras R_1 and R_2 are isomorphic and that $R_1 = K[x, \frac{y}{x}]$ is a polynomial ring in the variables x and $\frac{y}{x}$.

Exercise 1.10. Suppose that R is a domain and $f, g \in R$ with $g \neq 0$. Show that $R[\frac{f}{g}] \cong R[t]/(tg - f)$ if and only if $(tg - f)$ is a prime ideal in $R[t]$.

Exercise 1.11. Let A be a ring and X be the set of all prime ideals in A. For each subset E of A, let $V(E)$ be the set of all prime ideals in A which contain E. Prove that:

a) If I is the ideal generated by E, then $V(E) = V(I) = V(\sqrt{I})$.

b) $V(0) = X$ and $V(1) = \emptyset$.

c) If $\{E_s\}_{s \in S}$ is any family of subsets of A, then

$$V\left(\bigcup_{s \in S} E_s\right) = \bigcap_{s \in S} V(E_s).$$

d) $V(I \cap J) = V(IJ) = V(I) \cup V(J)$ for any ideals I, J of A

This exercise shows that the sets $V(E)$ satisfy the axioms for closed sets in a topological space. We call this topology on X the Zariski topology and write $\text{Spec}(A)$ for this topological space.

Exercise 1.12. Suppose that R is a local ring and $f_1, \ldots, f_r \in R$ generate an ideal I of R. Suppose that I is a principal ideal. Show that there exists an index i such that $I = (f_i)$. Give an example to show that this is false in a polynomial ring $k[x]$.

Exercise 1.13. Let κ be a field, and define $\mathrm{Map}(\kappa^n, \kappa)$ to be the set of maps (of sets) from κ^n to κ. Since κ is a κ-algebra, $\mathrm{Map}(\kappa^n, \kappa)$ is a κ-algebra, with the operations $(\phi + \psi)(\alpha) = \phi(\alpha) + \psi(\alpha)$, $(\phi\psi)(\alpha) = \phi(\alpha)\psi(\alpha)$, and $(c\phi)(\alpha) = c\phi(\alpha)$ for $\phi, \psi \in \mathrm{Map}(\kappa^n, \kappa)$, $\alpha \in \kappa^n$, and $c \in \kappa$.

a) Let $\kappa[x_1, \ldots, x_n]$ be a polynomial ring over κ and define
$$\Lambda : \kappa[x_1, \ldots, x_n] \to \mathrm{Map}(\kappa^n, \kappa)$$
by $\Lambda(f)(\alpha) = f(\alpha)$ for $f \in k[x_1, \ldots, x_n]$ and $\alpha \in \kappa^n$. Show that Λ is a κ-algebra homomorphism. The image, $\Lambda(\kappa[x_1, \ldots, x_n])$, is a subring of $\mathrm{Map}(\kappa^n, \kappa)$ which is called the *ring of polynomial functions on* κ^n.

b) Show that Λ is an isomorphism onto the polynomial functions of κ^n if and only if κ is an infinite field.

1.2. Field extensions

Suppose that K is a field and A is a K-algebra. Suppose that Λ is a subset of A. The set Λ is said to be algebraically independent over K if whenever we have a relation
$$f(t_1, \ldots, t_n) = 0$$
for some distinct $t_1, \ldots, t_n \in \Lambda$ and a polynomial f in the polynomial ring $K[x_1, \ldots, x_n]$, we have that $f = 0$ (all the coefficients of f are zero).

Suppose that K is a subfield of a field L and Λ is a subset of L. The subfield $K(\Lambda)$ of L is the smallest subfield of L which contains K and Λ.

A subset Λ of L which is algebraically independent over K and is maximal with respect to inclusions is called a transcendence basis of L over K. Transcendence bases always exist. Any set of algebraically independent elements in L over K can be extended to a transcendence basis of L over K. Any two transcendence bases of L over K have the same cardinality ([**95**, Theorem 1.1, page 356] or [**160**, Theorem 25, page 99]). This cardinality is called the transcendence degree of the field L over K and is written as $\mathrm{trdeg}_K L$.

Suppose that $L \subset M \subset N$ is a tower of fields. Then

(1.1) $$\mathrm{trdeg}_L N = \mathrm{trdeg}_M N + \mathrm{trdeg}_L M$$

by [**160**, Theorem 26, page 100].

1.2. Field extensions

An algebraic function field over a field K is a finitely generated field extension $L = K(y_1, \ldots, y_m)$ of K. After possibly permuting $y_1, \ldots, y_m$, there exists an integer r with $0 \leq r \leq m$ such that $y_1, \ldots, y_r$ is a transcendence basis of L over K. The field L is then said to be an r-dimensional algebraic function field. We have that L is finite over $K(y_1, \ldots, y_r)$. The field $K(y_1, \ldots, y_r)$ is isomorphic as a K-algebra to the quotient field of a polynomial ring over K in r variables, so $K(y_1, \ldots, y_r)$ is called a rational function field over K.

The field L is said to be separably generated over K if there exists a transcendence basis $z_1, \ldots, z_n$ of L over K such that L is separably algebraic over $K(z_1, \ldots, z_n)$. The set of elements $z_1, \ldots, z_n$ is then called a separating transcendence basis of L over K.

Theorem 1.14. *If K is a perfect field (K has characteristic 0, or K has characteristic $p > 0$ and all elements of K have a p-th root in K), then all finitely generated field extensions over K are separably generated over K.*

Proof. [160, Theorem 31, page 105]. □

In any algebraic extension of fields, there is a maximal separable extension.

Theorem 1.15. *Suppose that L is an algebraic extension of a field K. Then there exits a maximal subfield M of L which is separable algebraic over K and such that L is purely inseparable over M.*

Proof. [95, Theorem 4.5, page 241]. □

The M of the conclusions of Theorem 1.15 is called the separable closure of K in L. With the notation of the above theorem, we define

$$(1.2) \qquad [L:K]_s = [M:K] \quad \text{and} \quad [L:K]_i = [L:M].$$

The primitive element theorem gives a nice description of finite separable extensions.

Theorem 1.16 (Primitive element theorem). *Suppose that L is finite extension field of a field K. There exists an element $\alpha \in L$ such that $L = K(\alpha)$ if and only if there exist only a finite number of fields F such that $K \subset F \subset L$. If L is separable over K then there exists such an element α.*

Proof. [95, Theorem 4.6, page 243]. □

Suppose that L is a finite extension field of a field K. We will write $\text{Aut}(L/K)$ for the group of K-automorphisms of L. In the case that L is Galois over K, we will write $G(L/K)$ for the Galois group $\text{Aut}(L/K)$.

Exercise 1.17. Suppose that κ is a perfect field of characteristic $p > 0$ and $L = \kappa(s,t)$ is a rational function field over κ. Let $K = \{f^p \mid f \in L\}$. Show that $K = \kappa(s^p, t^p)$, L is finite algebraic over K, and L is not a primitive extension of K.

1.3. Modules

[13, Chapter 2] and [95, Chapters III and X] are good introductions to the theory of modules over a ring.

An R-module M is a finitely generated R-module if there exist $n \in \mathbb{Z}_+$ and $f_1, \ldots, f_n \in M$ such that $M = \{r_1 f_1 + \cdots + r_n f_n \mid r_1, \ldots, r_n \in R\}$.

The following is Nakayama's lemma.

Lemma 1.18. *Suppose that R is a ring, I is an ideal of R which is contained in all maximal ideals of R, M is a finitely generated R-module, and N is a submodule. If $M = N + IM$, then $M = N$.*

Proof. [95, Chapter X, Section 4] or [13, Proposition 2.6]. □

We will use the following lemma to determine the minimal number of generators of an ideal.

Lemma 1.19. *Suppose that R is a local ring with maximal ideal $\mathfrak{m}$ and M is a finitely generated R-module. Then the minimal number of elements $\mu(M)$ of M which generate M as an R-module is the $R/\mathfrak{m}$-vector space dimension*

$$\mu(M) = \dim_{R/\mathfrak{m}} M/\mathfrak{m}M.$$

Proof. Observe that $M/\mathfrak{m}M$ is an $R/\mathfrak{m}$-vector space by the well-defined map $R/\mathfrak{m} \times M/\mathfrak{m}M \to M/\mathfrak{m}M$ given by mapping the classes $[x]$ in $R/\mathfrak{m}$ of $x \in R$ and $[y] \in M/\mathfrak{m}M$ of $y \in M$ to the class $[xy]$ of xy in $M/\mathfrak{m}M$.

Suppose that $a_1, \ldots, a_r$ generate M as an R-module. Then the classes $[a_1], \ldots, [a_r] \in M/\mathfrak{m}M$ generate $M/\mathfrak{m}M$ as an $R/\mathfrak{m}$-vector space. Thus
$$\dim_{R/\mathfrak{m}} M/\mathfrak{m}M \leq \mu(M).$$
Suppose that $a_1, \ldots, a_r \in M$ are such that the classes $[a_1], \ldots, [a_r] \in M/\mathfrak{m}M$ generate $M/\mathfrak{m}M$ as an $R/\mathfrak{m}$-vector space. Let N be the submodule of M generated by $a_1, \ldots, a_r$. Then $N + \mathfrak{m}M = M$ so $N = M$ by Lemma 1.18. Thus
$$\mu(M) \leq \dim_{R/\mathfrak{m}} M/\mathfrak{m}M.$$
□

A chain of submodules of a module M is a sequence of submodules
$$(1.3) \qquad 0 = M_n \subset \cdots \subset M_1 \subset M_0 = M.$$

The length of (1.3) is n. If each module M_i/M_{i+1} has no submodules other than 0 and M_i/M_{i+1}, then (1.3) is called a composition series. If M has a composition series, then every composition series of M has length n, and every chain of submodules of M can be extended to a composition series [13, Proposition 6.7]. We define the length $\ell_R(M)$ of an R-module M to be the length of a composition series if a composition series exists, and we define $\ell_R(M) = \infty$ if a composition series does not exist. If R is a local ring with maximal ideal m_R containing a field κ such that $R/m_R \cong \kappa$, then any R-module M is naturally a κ-vector space, and

$$\ell_R(M) = \dim_\kappa M.$$

1.4. Localization

A multiplicatively closed (multiplicative) subset S of a ring R is a subset of R such that $1 \in S$ and S is closed under multiplication. Define an equivalence relation $\equiv$ on $R \times S$ by

$$(a, s) \equiv (b, t) \text{ if and only if } (at - bs)u = 0$$

for some $u \in S$. The localization of R with respect to S, denoted by $S^{-1}R$, is the set of equivalence classes $R \times S/\equiv$. The equivalence class of (a, s) is denoted by $\frac{a}{s}$. The localization $S^{-1}R$ is a ring with addition defined by

$$\frac{a}{s} + \frac{b}{t} = \frac{at + bs}{st}$$

and multiplication defined by

$$\left(\frac{a}{s}\right)\left(\frac{b}{t}\right) = \frac{ab}{st}.$$

This definition extends to localization $S^{-1}M$ of R-modules M, in particular for ideals in R [13, page 38].

There is a natural ring homomorphism $\phi : R \to S^{-1}R$ defined by $\phi(r) = \frac{r}{1}$ for $r \in R$.

We summarize a few facts from [13, Proposition 3.11]. The ideals in $S^{-1}R$ are the ideals $S^{-1}I = I(S^{-1}R)$ such that I is an ideal of R. We have that $S^{-1}I = S^{-1}R$ if and only if $S \cap I \neq \emptyset$. The prime ideals of $S^{-1}R$ are in 1-1 correspondence with the prime ideals of R which are disjoint from S.

Suppose that $f \in R$. Then $S = \{f^n \mid n \in \mathbb{N}\}$ is a multiplicatively closed set. The localization $S^{-1}R$ is denoted by R_f. Suppose that $\mathfrak{p}$ is a prime ideal in R. Then $S = R \setminus \mathfrak{p}$ is a multiplicatively closed set. The localization $S^{-1}R$ is denoted by $R_\mathfrak{p}$.

If $\mathfrak{p}$ is a prime ideal in a ring R, then $R_\mathfrak{p}$ is a local ring with maximal ideal $\mathfrak{p}_\mathfrak{p} = \mathfrak{p}R_\mathfrak{p}$.

If R is a domain, the quotient field of R is defined as $\operatorname{QF}(R) = R_{\mathfrak{p}}$ where $\mathfrak{p}$ is the zero ideal of R.

More basic properties of localization and localization of homomorphisms are established in [13, Chapter 3].

Exercise 1.20. Suppose that R is a domain and I is an ideal in R. Let R_I be the subset of the quotient field of R defined by

$$R_I = \left\{ \frac{f}{g} \mid f \in R, g \in R \setminus I \right\}.$$

Show that R_I is a ring if and only if I is a prime ideal in R.

Exercise 1.21. Suppose that S is a multiplicative set in a ring R. Show that the kernel of the natural homomorphism $\phi : R \to S^{-1}R$ is the ideal

$$\{g \in R \mid gs = 0 \text{ for some } s \in S\}.$$

Give an example of a ring R and $0 \neq f \in R$ such that the kernel of $R \to R_f$ is nonzero.

Exercise 1.22. Suppose that S and T are multiplicatively closed subsets of a ring R. Let U be the image of T in $S^{-1}R$. Show that $(ST)^{-1}R \cong U^{-1}(S^{-1}R)$, where $ST = \{st \mid s \in S \text{ and } t \in T\}$.

Exercise 1.23. Suppose that R is a ring and P is a prime ideal in R. Show that R_P is a local ring with maximal ideal $PR_P = P_P$. Let $\Lambda : R \to R_P$ be the natural homomorphism defined by $\Lambda(f) = \frac{f}{1}$ for $f \in R$. Show that $\Lambda^{-1}(PR_P) = P$.

1.5. Noetherian rings and factorization

Noetherian rings enjoy many good properties. They are ubiquitous throughout algebraic geometry.

Definition 1.24. A ring R is Noetherian if every ascending chain of ideals

$$I_1 \subset I_2 \subset \cdots \subset I_n \subset \cdots$$

is stationary (there exists n_0 such that $I_n = I_{n_0}$ for $n \geq n_0$).

Proposition 1.25. *A ring R is Noetherian if and only if every ideal I in R is finitely generated; that is, there exist $f_1, \ldots, f_n \in I$ for some $n \in \mathbb{Z}_+$ such that*

$$I = (f_1, \ldots, f_n) = f_1 R + \cdots + f_n R.$$

Proof. [13, Proposition 6.3]. □

1.5. Noetherian rings and factorization

We have the following fundamental theorem.

Theorem 1.26 (Hilbert's basis theorem). *If R is a Noetherian ring, then the polynomial ring $R[x]$ is Noetherian.*

Proof. [13, Theorem 7.5, page 81]. □

Corollary 1.27. *A polynomial ring over a field is Noetherian. A quotient of a Noetherian ring is Noetherian. A localization of a Noetherian ring is Noetherian.*

The following lemma will simplify some calculations.

Lemma 1.28. *Suppose that R is a Noetherian ring, $\mathfrak{m}$ is a maximal ideal of R, and N is an R-module such that $\mathfrak{m}^a N = 0$ for some positive integer a. Then $N_\mathfrak{m} \cong N$.*

Proof. Suppose that $f \in R \setminus \mathfrak{m}$. We will first prove that for any $r \in \mathbb{Z}_+$, there exists $e \in R$ such that $fe \equiv 1 \bmod \mathfrak{m}^r$.

The ring $R/\mathfrak{m}$ is a field, and the residue of f in $R/\mathfrak{m}$ is nonzero. Thus for any $h \in R$, there exists $g \in R$ such that $fg \equiv h \bmod \mathfrak{m}$. Taking $h = 1$, we obtain that there exists $e_0 \in R$ such that $fe_0 \equiv 1 \bmod \mathfrak{m}$.

Suppose that we have found $e \in R$ such that $fe \equiv 1 \bmod \mathfrak{m}^r$. Let $x_1, \ldots, x_n$ be a set of generators of $\mathfrak{m}$. There exists $h_{i_1,\ldots,i_n} \in R$ such that $fe - 1 = \sum_{i_1+\cdots+i_n=r} h_{i_1,\ldots,i_n} x_1^{i_1} \cdots x_n^{i_n}$. There exist $g_{i_1,\ldots,i_n} \in R$ such that $fg_{i_1,\ldots,i_n} \equiv h_{i_1,\ldots,i_n} \bmod \mathfrak{m}$. Thus

$$\sum_{i_1+\cdots+i_n=r} fg_{i_1,\ldots,i_n} x_1^{i_1} \cdots x_n^{i_n} \equiv \sum_{i_1+\cdots+i_n=r} h_{i_1,\ldots,i_n} x_1^{i_1} \cdots x_n^{i_n} \bmod \mathfrak{m}^{r+1}.$$

Set $e' = e - \sum g_{i_1,\ldots,i_n} x_1^{i_1} \cdots x_n^{i_n}$ to get $fe' \equiv 1 \bmod \mathfrak{m}^{r+1}$.

Consider the natural homomorphism $\Phi : N \to N_\mathfrak{m}$ defined by $\Phi(n) = \frac{n}{1}$ for $n \in N$. We will show that Φ is an isomorphism. Suppose that $\Phi(n) = 0$. Then there exists $f \in R \setminus \mathfrak{m}$ such that $fn = 0$. By the first part of the proof, there exists $e \in R$ such that $fe \equiv 1 \bmod \mathfrak{m}^a$. Thus $n = efn = 0$. Suppose $\frac{n}{f} \in N_\mathfrak{m}$. Then there exists $e \in R \setminus \mathfrak{m}$ such that $ef = 1 + h$ with $h \in \mathfrak{m}^a$, and $\Phi(ne) = \frac{n}{f}$. □

Suppose R is a domain. A nonzero element $f \in R$ is called irreducible if f is not a unit and whenever we have a factorization $f = gh$ with g and h in R, then g is a unit or h is a unit.

Since every ascending chain of principal ideals is stationary in a Noetherian ring, we have the following proposition.

Proposition 1.29. *Suppose R is a Noetherian domain. Then every nonzero nonunit $f \in R$ has a factorization $f = g_1 \cdots g_r$ for some positive integer r and irreducible elements $g_1, \ldots, g_r \in R$.*

Suppose R is a domain. A nonzero element $f \in R$ is called a prime if the ideal $(f) \subset R$ is a prime ideal.

Proposition 1.30. *Suppose R is a domain and $f \in R$. Then:*

1) *If f is prime, then f is irreducible.*

2) *If R is a unique factorization domain (UFD), then f is a prime if and only if f is irreducible.*

Proof. [84, Theorem 2.21]. □

Proposition 1.31. *Suppose that A is a UFD. Let K be the quotient field of A. Then the ring of polynomials in n variables $A[x_1, \ldots, x_n]$ is a UFD. Its units are precisely the units of A, and its prime elements are either primes of A or polynomials which are irreducible in $K[x_1, \ldots, x_n]$ and have content 1 (the greatest common divisor of the coefficients in A of the polynomial is 1).*

Proof. [95, Corollary 2.4, page 183]. □

Suppose that R is a ring and $R[x_1, \ldots, x_n]$ is a polynomial ring over R. If $f \in R[x_1, \ldots, x_n]$, then f has a unique expansion

$$f = \sum a_{i_1, \ldots, i_n} x_1^{i_1} \cdots x_n^{i_n}$$

with $a_{i_1, \ldots, i_n} \in R$. If f is nonzero, the (total) degree $\deg f$ of f is defined to be

$$\deg f = \max\{i_1 + \cdots + i_n \mid a_{i_1, \ldots, i_n} \neq 0\}.$$

The polynomial f is homogeneous of degree d if $a_{i_1, \ldots, i_n} = 0$ if $i_1 + \cdots + i_n \neq d$.

Suppose that $A = K[x, y, z, w]$ is a polynomial ring over a field K. The units in A are the nonzero elements of K. Let $f = xy - zw \in A$. Suppose that $f = gh$ with $g, h \in A$ nonunits. Since f is homogeneous of degree 2, we have that g and h are both homogeneous of degree 1, so $g = a_0 x + a_1 y + a_2 z + a_3 w$ and $h = b_0 x + b_1 y + b_2 z + b_3 w$ with $a_0, \ldots, a_3, b_0, \ldots, b_3 \in K$. We verify by expanding gh that there do not exist $a_0, \ldots, a_3, b_0, \ldots, b_3 \in K$ such that $gh = f$. Thus $xy - zw$ is irreducible in A. Since A is a UFD, we have that (f) is a prime ideal in A, and thus $R = A/(f)$ is a domain. For $u \in A$, let $\overline{u}$ denote the class of u in R. Then $R = K[\overline{x}, \overline{y}, \overline{z}, \overline{w}]$ where $\overline{x}, \overline{y}, \overline{z}, \overline{w}$ are the classes of x, y, z, w. Since f is homogeneous, the function $\deg \overline{g} = \deg g$ if $0 \neq g$ is well-defined on R (we will see that R is graded in Section 3.1). The units of R are the nonzero elements of K (they have degree 0) and since $\bar{x}, \bar{y}, \bar{z}, \bar{w}$ are all nonzero and they have degree 1, they must be irreducible in R. We have that the ideal $(f, x) = (zw, x)$ in A, so $R/(\bar{x}) \cong A/(zw, x) \cong K[y, z, w]/zw$ by Lemma 1.3, which is not a domain (the classes of z and w are zero divisors). In particular, $\bar{x}$ is an irreducible element of R which is not a prime. We see from Proposition 1.30 that R is not a UFD. We also have that

$$\overline{xy} = \overline{zw}$$

gives two factorizations in R by irreducible elements, none of which are associates, showing directly that R is not a UFD.

Exercise 1.32. Suppose that K is a field and $K[x_1, \ldots, x_n]$ is a polynomial ring over K. Let $f \in K[x_1, \ldots, x_n]$ be nonzero and homogeneous, and suppose that $g, h \in K[x_1, \ldots, x_n]$ are such that $f = gh$. Show that g and h are homogeneous and $\deg g + \deg h = \deg f$.

Exercise 1.33. Suppose that K is an algebraically closed field and $K[x_1, x_2]$ is a polynomial ring over K. Suppose that $f \in K[x_1, x_2]$ is homogeneous of positive degree. Show that f is a product of homogeneous polynomials of degree 1. Show that this is false if K is not algebraically closed.

Exercise 1.34. Let K be a field and $K[x, y, z]$ be a polynomial ring over K. Let $f = y^3 - x^3 + xz^2 \in K[x, y, z]$. Show that f is irreducible and that $R = K[x, y, z]/(f)$ is a domain. Show that R is not a UFD.

Exercise 1.35. Prove Euler's formula: Suppose that K is a field and F is a homogeneous polynomial of degree d in the polynomial ring $K[x_0, \ldots, x_n]$. Show that

$$\sum_{i=0}^{n} \frac{\partial F}{\partial x_i} x_i = dF.$$

1.6. Primary decomposition

Suppose that R is a ring. An ideal Q in R is primary if $Q \neq R$ and if for $x, y \in R$, $xy \in Q$ implies either $x \in Q$ or $y^n \in Q$ for some $n > 0$.

Proposition 1.36. *Let Q be a primary ideal in a ring R. Then $\sqrt{Q}$ is the smallest prime ideal of R containing Q.*

Proof. It suffices to show that $\sqrt{Q}$ is a prime ideal. Suppose $x, y \in R$ are such that $xy \in \sqrt{Q}$. Then $(xy)^n \in Q$ for some $n > 0$. Then either $x^n \in Q$ or $y^{mn} \in Q$ for some $m > 0$. Thus either $x \in \sqrt{Q}$ or $y \in \sqrt{Q}$. □

If $\mathfrak{p}$ is a prime ideal, an ideal Q is called $\mathfrak{p}$-primary if Q is primary and $\sqrt{Q} = \mathfrak{p}$.

Proposition 1.37. *If $\sqrt{I}$ is a maximal ideal $\mathfrak{m}$, then I is $\mathfrak{m}$-primary.*

Proof. [13, Proposition 4.2]. □

Lemma 1.38. *If the Q_i are $\mathfrak{p}$-primary, then $Q = \bigcap_{i=1}^{n} Q_i$ is $\mathfrak{p}$-primary.*

Proof. [13, Lemma 4.3]. □

A primary decomposition of an ideal I in R is an expression of I as a finite intersection of primary ideals,

$$(1.4) \qquad I = \bigcap_{i=1}^{n} Q_i.$$

The ideal I is called decomposable if it has a primary decomposition. If I is decomposable, then I has a minimal (or irredundant) primary decomposition, that is, an expression (1.4) where the $\sqrt{Q_i}$ are all distinct and $\bigcap_{j \neq i} Q_j \not\subset Q_i$ for all i. By Lemma 1.38, every decomposable ideal I has a minimal primary decomposition.

Theorem 1.39. *In a Noetherian ring R, every ideal has a primary decomposition (and hence has a minimal primary decomposition).*

Proof. [13, Theorem 7.13]. □

Let M be an R-module. A prime ideal $\mathfrak{p}$ is an associated prime of M if $\mathfrak{p}$ is the annihilator

$$\mathrm{Ann}(x) = \{r \in R \mid rx = 0\}$$

for some $x \in M$. The set of associated primes of M is denoted by $\mathrm{Ass}(M)$ or $\mathrm{Ass}_R(M)$. In the case of an ideal I of R, it is traditional to abuse notation and call the associated primes of R/I the associated primes of I.

An element a in a ring R is called a zero divisor for an R-module M if there exists a nonzero $x \in M$ such that $ax = 0$. Otherwise, a is M-regular.

Theorem 1.40. *Let A be a Noetherian ring and M a nonzero A-module.*

1) *Every maximal element of the family of ideals $F = \{\mathrm{Ann}(x) \mid 0 \neq x \in M\}$ is an associated prime of M.*

2) *The set of zero divisors for M is the union of all the associated primes of M.*

Proof. 1) We must show that if $\mathrm{Ann}(x)$ is a maximal element of F, then it is prime. If $a, b \in A$ are such that $abx = 0$ but $bx \neq 0$, then by maximality, $\mathrm{Ann}(bx) = \mathrm{Ann}(x)$. Hence $ax = 0$.

2) If $ax = 0$ for some $x \neq 0$, then $a \in \mathrm{Ann}(x) \in F$. By 1), there is an associated prime of M containing $\mathrm{Ann}(x)$. □

1.6. Primary decomposition

Another important set of prime ideals associated to a module M is the support of M, which is

$$\operatorname{Supp}(M) = \{\text{prime ideals } \mathfrak{p} \text{ of } R \mid M_\mathfrak{p} \neq 0\}.$$

Theorem 1.41. *Let R be a Noetherian ring and M a finitely generated R-module. Then:*

1) *$\operatorname{Ass}(M)$ is a finite set.*
2) *$\operatorname{Ass}(M) \subset \operatorname{Supp}(M)$.*
3) *Any minimal element of $\operatorname{Supp}(M)$ is in $\operatorname{Ass}(M)$.*

Proof. [**107**, (7.G) on page 52] and [**107**, Theorem 9], or [**106**, Theorem 6.5]. □

The minimal elements of the set $\operatorname{Ass}(M)$ are called minimal or isolated prime ideals belonging to M. The others are called embedded primes. We have that a prime ideal P of R is a minimal prime of an ideal I (a minimal prime of R/I) if $I \subset P$, and if Q is a prime ideal of R such that $I \subset Q \subset P$, then $Q = P$.

Theorem 1.42. *Let I be a decomposable ideal and let $I = \bigcap_{i=1}^n Q_i$ be a minimal primary decomposition of I. Then:*

1)
$$\operatorname{Ass}(R/I) = \{\sqrt{Q_i} \mid 1 \leq i \leq n\}.$$

2) *The isolated primary components (the primary components Q_i corresponding to minimal prime ideals $\mathfrak{p}_i$) are uniquely determined by I.*

Proof. [**13**, Theorem 4.5 and Corollary 4.11]. □

Proposition 1.43. *Let S be a multiplicatively closed subset of a ring R and let I be a decomposable ideal. Let $I = \bigcap_{i=1}^n Q_i$ be a minimal primary decomposition of I. Let $\mathfrak{p}_i = \sqrt{Q_i}$ and suppose that the Q_i are indexed so that $S \cap \mathfrak{p}_i \neq \emptyset$ for $m < i \leq n$ and $S \cap \mathfrak{p}_i = \emptyset$ for $1 \leq i \leq m$. Then*

$$S^{-1}I = \bigcap_{i=1}^m S^{-1}Q_i$$

is a minimal primary decomposition of $S^{-1}I$ in $S^{-1}R$, with $\sqrt{S^{-1}Q_i} = \mathfrak{p}_i S^{-1}R$.

Proof. [**13**, Proposition 4.9]. □

Exercise 1.44. Let K be a field and $R = K[x,y]$ be a polynomial ring. Let $I = (x^2y, xy^2)$. Compute a minimal primary decomposition of I. Compute the set $\operatorname{Ass}(R/I)$. Identify the minimal and embedded primes. Compute $\sqrt{I}$ and compute a minimal primary decomposition of $\sqrt{I}$. Identify the minimal and embedded primes. Compute the set $\operatorname{Ass}(R/\sqrt{I})$. Compute minimal primary decompositions of $I_\mathfrak{p}$ and the set $\operatorname{Ass}(I_\mathfrak{p})$ when $\mathfrak{p} = (x)$ and when $\mathfrak{p} = (x,y)$. Identify the minimal and embedded primes.

Exercise 1.45. Let K be a field and $R = K[x,y,z]/(z^2 - xy) = K[\bar{x}, \bar{y}, \bar{z}]$. Compute a minimal primary decomposition of the ideal $(\bar{z})$.

Exercise 1.46. Suppose that I is an ideal in a Noetherian ring R and $\sqrt{I} = I$. Show that all elements of $\operatorname{Ass}(R/I)$ are minimal and the minimal primary decomposition of I is

$$I = \bigcap_{\{\text{minimal primes } \mathfrak{p} \text{ of } I\}} \mathfrak{p}.$$

Exercise 1.47. Suppose that R is a Noetherian ring and $I \subset J$ are ideals of R. Show that $I = J$ if and only if $I_\mathfrak{m} = J_\mathfrak{m}$ for all maximal ideals $\mathfrak{m}$ of R.

1.7. Integral extensions

[**13**, Chapter 5], [**95**, Chapter VII, Section 1] and [**160**, Sections 1–4 of Chapter V] are good references for this section.

Definition 1.48. Suppose that R is a subring of a ring S. An element $u \in S$ is integral over R if u satisfies a relation

$$u^n + a_1 u^{n-1} + \cdots + a_{n-1} u + a_n = 0$$

with $a_1, \ldots, a_n \in R$.

Theorem 1.49. *Suppose that R is a subring of a ring S and $u \in S$. The following are equivalent:*

1) *u is integral over R.*

2) *$R[u]$ is a finitely generated R-module.*

3) *$R[u]$ is contained in a subring T of S such that T is a finitely generated R-module.*

Proof. [**13**, Proposition 5.1]. □

We have the following immediate corollaries.

Corollary 1.50. *Let $u_1, \ldots, u_n$ be elements of S which are each integral over R. Then the subring $R[u_1, \ldots, u_n]$ of S is a finitely generated R-module.*

1.7. Integral extensions

Corollary 1.51. *Suppose that R is a subring of a ring S. Let*
$$\overline{R} = \{u \in S \mid u \text{ is integral over } R\}.$$
Then $\overline{R}$ is a ring.

Proof. If $x, y \in \overline{R}$, then the subring $R[x, y]$ of S is a finitely generated R-module by Corollary 1.50. The elements $x+y$ and xy are in $R[x,y]$ so $x+y$ and xy are integral over R by Theorem 1.49. □

$\overline{R}$ is called the integral closure of R in S. This construction is particularly important when R is a domain and S is the quotient field of R. In this case, $\overline{R}$ is called the normalization of R. R is said to be normal if $\overline{R} = R$. If R is a domain and S is a field extension of the quotient field of R, then the integral closure of R in S is called the normalization of R in S.

We now state some theorems which will be useful.

Lemma 1.52. *Let $A \subset B$ be rings with B integral over A and let S be a multiplicatively closed subset of A. Then $S^{-1}B$ is integral over $S^{-1}A$.*

Proof. Suppose $b \in B$ satisfies a relation
$$b^n + a_1 b^{n-1} + \cdots + a_n = 0$$
with $a_1, a_2, \ldots, a_n \in A$ and $s \in S$. Then
$$\left(\frac{b}{s}\right)^n + \frac{a_1}{s}\left(\frac{b}{s}\right)^{n-1} + \cdots + \frac{a_n}{s} = 0.$$
□

Theorem 1.53 (Noether's normalization lemma). *Let R be a finitely generated L-algebra, where L is a field. Then there exist $y_1, \ldots, y_r \in R$ such that the subring $L[y_1, \ldots, y_r]$ of R is a polynomial ring over L and R is integral over $L[y_1, \ldots, y_r]$.*

Proof. We give a proof with the assumption that L is an infinite field. For a proof when L is a finite field, we refer to [**161**, Theorem 25 on page 200]. Write $R = L[x_1, \ldots, x_n]$. Suppose that R is not a polynomial ring over L, so there exists a nonzero f in the polynomial ring $L[z_1, \ldots, z_n]$ over L such that $f(x_1, \ldots, x_n) = 0$. Let $d = \deg f$ and let f_d be the homogeneous part of f of degree d, so that
$$f_d = z_n^d f_d\left(\frac{z_1}{z_n}, \ldots, \frac{z_{n-1}}{z_n}, 1\right).$$
By Theorem 1.4, there exist $c_1, \ldots, c_{n-1} \in L$ such that $f_d(c_1, \ldots, c_{n-1}, 1) \neq 0$. Set $y_i = x_i - c_i x_n$ for $1 \leq i \leq n-1$. Then
$$\begin{aligned} 0 = f(x_1, \ldots, x_n) &= f(y_1 + c_1 x_n, \ldots, y_{n-1} + c_{n-1} x_n, x_n) \\ &= f_d(c_1, \ldots, c_{n-1}, 1) x_n^d + g_1 x_n^{d-1} + \cdots + g_d \end{aligned}$$

with all $g_i \in L[y_1, \ldots, y_{n-1}]$, so that x_n is integral over $L[y_1, \ldots, y_{n-1}]$. The theorem then follows by induction on n. □

Theorem 1.54. *Let R be a domain which is a finitely generated algebra over a field K. Let Q be the quotient field of R and let L be a finite algebraic extension of Q. Then the integral closure R' of R in L is a finitely generated R-module and is also a finitely generated K-algebra.*

Lemma 1.55. *Suppose that R is a Noetherian ring, M is a finitely generated R-module, and N is a submodule of M. Then N is a finitely generated R-module.*

Proof. This follows from (1) of the "basic criteria" for a module to be Noetherian of [95, page 413] and [95, Proposition 1.4, page 415]. □

Let B be a ring and A be a subring. Let P be a prime ideal of A and let Q be a prime ideal of B. We say that Q lies over P if $Q \cap A = P$.

Proposition 1.56. *Let A be a subring of a ring B, let P be a prime ideal of A, and assume B is integral over A. Then $PB \neq B$ and there exists a prime ideal Q of B lying over P.*

Proof. We first show that $PB \neq B$. By Lemma 1.52, it suffices to show that $PS \neq S$ where $S = B_P$. Suppose that $PS = S$. Then there is a relation

$$1 = a_1 b_1 + \cdots + a_n b_n$$

with $a_i \in P$ and $b_i \in S$. Let $R = A_P$ and $S_0 = R[b_1, \ldots, b_n]$. Then $PS_0 = S_0$ and S_0 is a finitely generated R-module by Corollary 1.50, and so $S_0 = 0$ by Nakayama's lemma, Lemma 1.18, a contradiction.

Thus we have that PB_P is contained in a maximal ideal m of B_P. Then $PA_P \subset m \cap A_P$. But PA_P is the maximal ideal of A_P so $m \cap A_P = PA_P$. Let Q be the inverse image of m in B. We have that $P \subset Q \cap A$. Suppose $f \in Q \cap A$. Then $\frac{f}{1} \in m \cap A_P = PA_P$ and so $f \in P$. Thus $P = Q \cap A$. □

We will further develop the theory of integral extensions in Section 21.2.

Exercise 1.57. Suppose that R is a domain which is contained in a field K of characteristic $p > 0$. Suppose that $f \in K$ is such that $f^p \in R$. Let $S = R[f]$. Suppose that Q is a prime ideal in R. Show that $\sqrt{QS}$ is a prime ideal.

Exercise 1.58. Suppose that K is a field and A is a subring of K. Let B be the integral closure of A in K. Let S be a multiplicatively closed subset of A. Show that $S^{-1}B$ is the integral closure of $S^{-1}A$ in K.

Exercise 1.59. Let K be a field and R be a polynomial ring over K. Show that R is integrally closed in its quotient field.

1.8. Dimension

In this section we define the height of an ideal and the dimension (Krull dimension) of a ring.

Definition 1.60. The height, $\mathrm{ht}(P)$, of a prime ideal P in a ring R is the supremum of all natural numbers n such that there exists a chain

$$(1.5) \qquad P_0 \subset P_1 \subset \cdots \subset P_n = P$$

of distinct prime ideals. The dimension $\dim R$ of R is the supremum of the heights of all prime ideals in R.

If P is a prime ideal in a ring R, then $\dim R_P = \mathrm{ht}(P)$ by Proposition 1.43.

A chain (1.5) is maximal if the chain cannot be lengthened by adding an additional prime ideal somewhere in the chain.

Definition 1.61. The height of an ideal I in a ring R is

$$\mathrm{ht}(I) = \inf\{\mathrm{ht}(P) \mid P \text{ is a prime ideal of } R \text{ and } I \subset P\}.$$

Theorem 1.62. *Let B be a Noetherian ring and let A be a Noetherian subring over which B is integral. Then $\dim A = \dim B$.*

Proof. [**107**, Theorem 20, page 81]. □

Theorem 1.63. *Let K be a field and A be a finitely generated K-algebra which is a domain. Let L be the quotient field of A. Then $\dim A = \mathrm{trdeg}_K L$, the transcendence degree of L over K.*

Proof. The dimension of a polynomial ring over K in n variables is n by [**107**, Theorem 22]. The proof of the theorem now follows from Theorem 1.53 (Noether's normalization lemma) and Theorem 1.62. □

An example of a Noetherian ring which has infinite dimension is given in [**121**, Example 1 of Appendix A1, page 203]. Rings which are finitely generated K-algebras have the following nice property.

Theorem 1.64. *Let K be a field and A be a finitely generated K-algebra which is a domain. For any prime ideal $\mathfrak{p}$ in A we have that*

$$\mathrm{ht}(\mathfrak{p}) + \dim A/\mathfrak{p} = \dim A.$$

Proof. [**28**, Theorem A.16] or [**13**, Chapter 11]. □

There are Noetherian rings which do not satisfy the equality of Theorem 1.64 [**121**, Example 2, Appendix A1].

The following theorem is of fundamental importance.

Theorem 1.65 (Krull's principal ideal theorem). *Let A be a Noetherian ring, and let $f \in A$ be an element which is neither a zero divisor nor a unit. Then every minimal prime ideal $\mathfrak{p}$ containing f has height 1.*

Proof. [13, Corollary 11.17]. □

Proposition 1.66. *A Noetherian domain A is a UFD if and only if every prime ideal of height 1 in A is principal.*

Proof. [106, Theorem 20.1] or [23, Chapter 7, Section 3] or [50, Proposition 3.11]. □

1.9. Depth

Let R be a ring and M be an R-module. Elements $x_1, \ldots, x_r \in R$ are said to be an M-regular sequence if

1) for each $1 \leq i \leq r$, x_i is a nonzero divisor on $M/(x_1, \ldots, x_{i-1})M$ ($x_i y \neq 0$ for all nonzero $y \in M/(x_1, \ldots, x_{i-1})M$) and

2) $M \neq (x_1, \ldots, x_r)M$.

Definition 1.67. Let R be a Noetherian ring, I be an ideal in R, and M be a finitely generated R-module. We define $\text{depth}_I M$ to be the maximal length of an M-regular sequence $x_1, \ldots, x_r$ with all $x_i \in I$.

Definition 1.68. A Noetherian ring R is said to be Cohen-Macaulay if $\text{depth}_I R = \text{ht}(I)$ for every maximal ideal I of R.

We give some examples of Cohen-Macaulay rings in the following theorem and proposition.

Theorem 1.69. *Let A be a Cohen-Macaulay ring. Then the polynomial ring $A[x_1, \ldots, x_n]$ is a Cohen-Macaulay ring. In particular, a polynomial ring over a field is Cohen-Macaulay.*

Proof. [107, Theorem 33]. □

Proposition 1.70. *Let A be a Cohen-Macaulay ring and $J = (a_1, \ldots, a_r)$ be an ideal of height r. Then A/J^v is Cohen-Macaulay for every $v > 0$.*

Proof. [107, Proposition, page 112]. □

A proof of the following theorem is given in [50, Corollary 18.14] or [107, Theorem 32].

1.9. Depth

Theorem 1.71 (Unmixedness theorem). *Let R be a Cohen-Macaulay ring. If $I = (x_1, \ldots, x_n)$ is an ideal such that $\operatorname{ht}(I) = n$, then all associated primes of I are minimal primes of I and have height n.*

Lemma 1.72. *Suppose that R is a ring and $I, P_1, \ldots, P_r$ are ideals in R such that the P_i are prime ideals. Suppose that $I \not\subset P_i$ for each i. Then $I \not\subset \bigcup_i P_i$.*

Proof. We may omit the P_i which are contained in some other P_j and suppose that $P_i \not\subset P_j$ if $i \neq j$. We prove the lemma by induction on r. Suppose $r = 2$ and $I \subset P_1 \cup P_2$. Choose $x \in I \setminus P_2$ and $y \in I \setminus P_1$. Then $x \in P_1$ so $y + x \not\in P_1$. Thus y and $y + x \in P_2$ so $x \in P_2$, a contradiction.

Now suppose $r > 2$. Then $IP_1 \cdots P_{r-1} \not\subset P_r$ since P_r is a prime ideal. Choose $x \in IP_1 \cdots P_{r-1} \setminus P_r$. Let $S = I \setminus (P_1 \cup \cdots \cup P_{r-1})$. By induction, $S \neq \emptyset$. Suppose $I \subset P_1 \cup \cdots \cup P_r$. Then $S \subset P_r$. Suppose $s \in S$. Then $s + x \in S$ and thus both s and $s + x$ are in P_r, and so $x \in P_r$, a contradiction. □

Lemma 1.73. *Suppose that R is a Noetherian ring, $\mathfrak{m}$ is a maximal ideal of R, and M is a finite R-module. Then $\operatorname{depth}_{\mathfrak{m}} M = 0$ if and only if $\mathfrak{m} \in \operatorname{Ass}_R(M)$.*

Proof. If $\mathfrak{m}$ is an associated prime for M, then there exists $x \in M$ such that $\mathfrak{m} = \operatorname{Ann}(x)$. Thus $\operatorname{depth}_{\mathfrak{m}} M = 0$.

Suppose $\operatorname{depth}_{\mathfrak{m}} M = 0$. Then all elements of $\mathfrak{m}$ are zero divisors for M. Now the set of all zero divisors for M is the union of the finitely many associated primes of R by Theorems 1.40 and 1.41. Thus $\mathfrak{m}$ is an associated prime of M by Lemma 1.72. □

A proof of the following lemma is given in [**50**, Corollary 18.6].

Lemma 1.74. *Let R be a Noetherian ring, $\mathfrak{m}$ be a maximal ideal of R, and*
$$0 \to N' \to N \to N'' \to 0$$
be a short exact sequence of nonzero finitely generated R-modules. Then

1) $\operatorname{depth}_{\mathfrak{m}} N'' \geq \min\{\operatorname{depth}_{\mathfrak{m}} N, \operatorname{depth}_{\mathfrak{m}} N' - 1\}$,
2) $\operatorname{depth}_{\mathfrak{m}} N' \geq \min\{\operatorname{depth}_{\mathfrak{m}} N, \operatorname{depth}_{\mathfrak{m}} N'' + 1\}$.

Example 1.75. *There exists a domain A and a nonzero element $f \in A$ such that the ideal fA has an embedded prime.*

We now construct such an example. Let K be a field. We will first show that the two-dimensional domain
$$R = K[s^4, s^3t, st^3, t^4]$$

which is a subring of the polynomial ring $K[s,t]$ has $\text{depth}_m(R) = 1$ where $m = (s^4, s^3t, st^3, t^4)$ (so R is not Cohen-Macaulay). Let
$$S = K[s^4, s^3t, s^2t^2, st^3, t^4].$$
The domain S contains R as a subring, realizing $S = R + s^2t^2R$ as a finitely generated R-module. We have a short exact sequence of R-modules
$$0 \to R \to S \to M \to 0$$
where $M = S/R$. We have that $\text{depth}_m S \geq 1$ since S is a domain which is not a field. Let a be the class of s^2t^2 in M. Consider the surjective R-module homomorphism $\phi : R \to M$ defined by $\phi(f) = fa$ for $f \in R$. Since $ma = 0$, ϕ induces an isomorphism of R-modules $M \cong R/m$. By Lemma 1.74 we have that $\text{depth}_m R \leq 1$, so that $\text{depth}_m R = 1$ since R is a domain which is not a field.

Thus R has the following attribute: For every nonzero $f \in m$,
$$\text{depth}_m R/(f) = 0,$$
so by Lemma 1.73, m is an embedded prime for the ideal (f); that is, a minimal primary decomposition of fR_m is

(1.6) $$fR_m = Q_1 \cap \cdots \cap Q_t \cap Q_0$$

where the Q_i are P_i-primary for a height 1 prime P_i in R_m (a minimal prime of fR_m) and Q_0 is a nontrivial m_m-primary ideal.

The following theorem will be useful.

Theorem 1.76. *Suppose that R is a Cohen-Macaulay ring and $J = (g_1, \ldots, g_s)$ is an ideal in R such that $g_1, \ldots, g_s$ is an R-regular sequence. Then*
$$\text{gr}_J(R) = \bigoplus_{i \geq 0} J^i/J^{i+1} = R/J[\overline{g}_1, \ldots, \overline{g}_s]$$
is a polynomial ring over R/J in $\overline{g}_1, \ldots, \overline{g}_s$, where $\overline{g}_i$ is the class of g_i in J/J^2.

Proof. This follows from [**107**, Theorem 27, page 98] and the equivalence (***) on page 98 of [**107**]. □

1.10. Normal rings and regular rings

Normal and regular rings play an important role in algebraic geometry. In normal rings, the concepts of zeros and poles of a function are well-defined, and regular rings correspond to nonsingular spaces.

We begin this section with some properties of normal rings which we will use. A normal ring is defined in Section 1.7.

1.10. Normal rings and regular rings

Lemma 1.77. *Suppose that A is a domain with quotient field K. Then*
$$A = \bigcap_P A_P$$
where the intersection in K is over all maximal ideals P of A.

Proof. Suppose $x \in K$. Let $D = \{a \in A \mid ax \in A\}$. The element x is in A if and only if $D = A$, and x is in A_P if and only if $D \not\subset P$. Thus if $x \notin A$, there exists a maximal ideal P of A such that $D \subset P$, and so $x \notin A_P$. □

Corollary 1.78. *Suppose that A is a domain. Then A is normal if and only if A_P is normal for all maximal ideals P of A.*

Proof. If A is normal, then $S^{-1}A$ is normal for every multiplicatively closed subset of A not containing 0. Since $A = \bigcap A_P$ by Lemma 1.77, where the intersection is over all maximal ideals P of A, the domain A is normal if and only if A_P is normal for all maximal ideals P. □

A stronger intersection theorem holds for height 1 primes.

Theorem 1.79. *Let A be a Noetherian normal domain. Then:*

1) *All associated primes of a nonzero principal ideal have height 1.*
2)
$$A = \bigcap_{\mathfrak{p}} A_{\mathfrak{p}}$$
where the intersection in K is over all height 1 prime ideals $\mathfrak{p}$ of A.

Proof. [**107**, Theorem 38] or [**106**, Theorem 11.5]. □

We now develop some concepts to define a regular local ring.

Definition 1.80. Suppose that R is a local ring with maximal ideal m_R. The associated graded ring of R is
$$\operatorname{gr}_{m_R}(R) = \bigoplus_{i \geq 0} m_R^i / m_R^{i+1}.$$

Theorem 1.81. *Suppose that R is a Noetherian local ring with maximal ideal m_R. Then:*

1) $\dim \operatorname{gr}_{m_R}(R) = \dim R$.
2) $\dim_{R/m_R} m_R/m_R^2 \geq \dim R$.

Proof. Equation 1) is proven in [**107**, Theorem 17] or [**106**, Theorem 13.9], using the theory of Hilbert polynomials. Equation 2) follows from [**160**, Theorem 30, page 240, and Theorem 31, page 241] or [**107**, (12.J)]. □

If A is a local ring with maximal ideal m_A and residue field $\kappa = A/m_A$, then the tangent space of A is defined as

(1.7) $$T(A) = \operatorname{Hom}_\kappa(m_A/m_A^2, \kappa).$$

Definition 1.82. A Noetherian local ring R with maximal ideal m_R is a regular local ring if $\dim_{R/m_R} m_R/m_R^2 = \dim R$.

Since $\dim_\kappa T(R) = \dim_\kappa m_R/m_R^2$, we always have that $\dim_\kappa T(R) \geq \dim R$ and R is regular if and only if $\dim_\kappa T(R) = \dim R$.

We now state some useful properties of regular local rings and their relation to normal rings.

Theorem 1.83. *Let A be a ring such that for every prime ideal P of A the localization A_P is regular. Then the polynomial ring $A[x_1, \ldots, x_n]$ has the same property. In particular, every local ring of a polynomial ring over a field is a regular local ring.*

Proof. [**107**, Theorem 40] □

Theorem 1.84. *Suppose that R is a regular local ring. Then R is a Cohen-Macaulay normal domain.*

Proof. This follows from [**161**, Corollary 1 on page 302] and [**107**, Theorem 36]. □

The proofs of the following theorems are through homological algebra.

Theorem 1.85. *A Noetherian ring A is normal if and only if it satisfies the following two condtions:*

1) *For every prime ideal $\mathfrak{p} \subset A$ of height 1, $A_\mathfrak{p}$ is regular.*

2) *For every prime ideal $\mathfrak{p} \subset A$ of height ≥ 2, we have $\operatorname{depth} A_\mathfrak{p} \geq 2$.*

Proof. [**107**, Theorem 39, page 125]. □

Corollary 1.86. *Suppose that R is a regular local ring and $f \in R$ is nonzero and is not a unit. Then $R/(f)$ is normal if and only if $(R/(f))_\mathfrak{p}$ is regular for all prime ideals $\mathfrak{p}$ of $R/(f)$ of height 1.*

Proof. Let $A = R/(f)$. We must show that condition 2) of Theorem 1.85 holds. We have that R is a Cohen-Macaulay domain by Theorem 1.84. Since f is R-regular, we have that A_P is Cohen-Macaulay for all prime ideals P of A by [**107**, Theorem 30], and so

$$\operatorname{depth}(A_P) = \dim A_P = \operatorname{ht}(P).$$ □

Theorem 1.87. *Suppose that R is a normal Noetherian local ring of dimension 1. Then R is a regular local ring.*

1.10. Normal rings and regular rings

Proof. This follows from Theorem 1.85. □

Theorem 1.88. *Suppose that R is a regular local ring and P is a prime ideal in R. Then R_P is a regular local ring.*

Proof. [**107**, Corollary, page 139] or [**106**, Theorem 19.3]. □

Theorem 1.89 (Auslander and Buchsbaum)**.** *Suppose that R is a regular local ring. Then R is a UFD.*

Proof. [**15**] or [**107**, Theorem 48] or [**106**, Theorem 20.3]. □

Chapter 2

Affine Varieties

In this chapter we define affine and quasi-affine varieties and their regular functions and regular maps. We develop the basic properties of affine varieties. Recall that throughout this book, k will be a fixed algebraically closed field.

In Sections 2.1–2.4 we develop a correspondence between the commutative algebra of finitely generated k-algebras which are domains (or reduced) and the geometry of algebraic varieties (or algebraic sets) in an affine space. In Section 2.5, we study the open sets in the Zariski topology on an affine variety (which are the quasi-affine varieties) and the regular functions and regular maps on such open sets. We show in Lemma 2.83 and Proposition 2.93 that every affine variety X has the basis for the Zariski topology consisting of the open sets $D(f)$ for $f \in k[X]$ which are (isomorphic to) affine varieties.

In Section 2.6, we define rational maps on an affine variety X. A rational map on X is determined by a regular map on a dense open subset U of X.

2.1. Affine space and algebraic sets

Affine n-space over k is

$$\mathbb{A}^n = \mathbb{A}^n_k = \{(a_1, \ldots, a_n) \mid a_1, \ldots, a_n \in k\}.$$

An element $p = (a_1, \ldots, a_n) \in \mathbb{A}^n$ is called a point. The ring of regular functions on $\mathbb{A}^n$ is the k-algebra of polynomial mappings

$$k[\mathbb{A}^n] = \{f : \mathbb{A}^n \to \mathbb{A}^1 \mid f \in k[x_1, \ldots, x_n]\}.$$

Here $k[x_1, \ldots, x_n]$ is the polynomial ring over k in the variables $x_1, \ldots, x_n$.

Since an algebraically closed field is infinite, the natural surjective ring homomorphism $k[x_1,\ldots,x_n] \to k[\mathbb{A}^n]$ is an isomorphism by Theorem 1.4, as shown in Exercise 1.13. Thus we may identify the ring $k[\mathbb{A}^n]$ with the polynomial ring $k[x_1,\ldots,x_n]$.

The zeros of a regular function $f \in k[\mathbb{A}^n]$ are
$$Z(f) = \{p \in \mathbb{A}^n \mid f(p) = 0\}.$$

If $T \subset k[\mathbb{A}^n]$ is a subset, then the set of common zeros of the elements of T is
$$Z(T) = \{p \in \mathbb{A}^n \mid f(p) = 0 \text{ for all } f \in T\}.$$

A subset W of $\mathbb{A}^n$ is called an algebraic set if there exists a subset T of $k[\mathbb{A}^n]$ such that $W = Z(T)$.

If I is the ideal in $k[\mathbb{A}^n]$ generated by T, then $Z(T) = Z(I)$.

By Corollary 1.27 every algebraic set in $\mathbb{A}^n$ is the set of common zeros of a finite number of polynomials.

Proposition 2.1. *Suppose that I_1, I_2, $\{I_\alpha\}_{\alpha \in S}$ are ideals in*
$$k[\mathbb{A}^n] = k[x_1,\ldots,x_n].$$
Then:

1) $Z(I_1 I_2) = Z(I_1) \cup Z(I_2)$.
2) $Z(\sum_{\alpha \in S} I_\alpha) = \bigcap_{\alpha \in S} Z(I_\alpha)$.
3) $Z(k[\mathbb{A}^n]) = \emptyset$.
4) $\mathbb{A}^n = Z(0)$.

Proof of 1). Suppose that $p \in Z(I_1) \cup Z(I_2)$. Then $p \in Z(I_1)$ or $p \in Z(I_2)$. Thus for every $f \in I_1$ we have $f(p) = 0$ or for every $g \in I_2$ we have that $g(p) = 0$. If $f \in I_1 I_2$, then $f = \sum_{i=1}^r f_i g_i$ for some $f_1,\ldots,f_r \in I_1$ and $g_1,\ldots,g_r \in I_2$. Thus $f(p) = \sum f_i(p) g_i(p) = 0$, so that $p \in Z(I_1 I_2)$.

Now suppose that $p \in Z(I_1 I_2)$ and $p \notin Z(I_1)$. Then there exists $f \in I_1$ such that $f(p) \neq 0$. For any $g \in I_2$, we have $fg \in I_1 I_2$ so that $f(p)g(p) = 0$. Since $f(p) \neq 0$, we have that $g(p) = 0$. Thus $p \in Z(I_2)$. □

Proposition 2.1 tells us that:

1. The union of two algebraic sets is an algebraic set.
2. The intersection of any family of algebraic sets is an algebraic set.
3. $\emptyset$ and $\mathbb{A}^n$ are algebraic sets.

We thus have a topology on $\mathbb{A}^n$, defined by taking the closed sets to be the algebraic sets. The open sets are the complements of algebraic sets in $\mathbb{A}^n$ (any union of open sets is open, any finite intersection of open sets is open, the empty set is open, and $\mathbb{A}^n$ is open). This topology is called the Zariski topology. If Y is a subset of $\mathbb{A}^n$, we will denote the Zariski closure of Y in $\mathbb{A}^n$ by $\overline{Y}$.

Example 2.2. Suppose that I is a nontrivial ideal in $k[\mathbb{A}^1] = k[x]$; that is, $I \neq (0)$ and $I \neq k[x]$. Then $I = (f)$ where $f = (x - \alpha_1) \cdots (x - \alpha_r)$ for some $\alpha_1, \ldots, \alpha_r \in k$ since $k[x]$ is a PID and k is algebraically closed. Thus $Z(I) = \{\alpha_1, \ldots, \alpha_r\}$. The open sets in $\mathbb{A}^1$ are thus $\mathbb{A}^1$, the complement of finitely many points in $\mathbb{A}^1$, and $\emptyset$.

We see that the Zariski topology is not Hausdorff (to be Hausdorff, distinct points must have disjoint neighborhoods).

A nonempty subset Y of a topological space X is said to be irreducible if it cannot be expressed as a union $Y = Y_1 \cup Y_2$ of two proper subsets, each of which is closed in Y ($\emptyset$ is not irreducible).

Example 2.3. $\mathbb{A}^1$ is irreducible as all proper closed subsets are finite and $\mathbb{A}^1$ is infinite.

Definition 2.4. An affine algebraic variety is an irreducible closed subset of $\mathbb{A}^n$. An affine algebraic set is a closed subset of $\mathbb{A}^n$.

Given a subset Y of $\mathbb{A}^n$, the ideal of Y in $k[\mathbb{A}^n]$ is

$$I(Y) = \{f \in k[\mathbb{A}^n] \mid f(p) = 0 \text{ for all } p \in Y\}.$$

We now state Hilbert's Nullstellensatz.

Theorem 2.5. *Let F be an algebraically closed field, I be an ideal in the polynomial ring $R = F[x_1, \ldots, x_n]$, and $f \in R$ be a polynomial which vanishes at all points of $Z(I)$. Then $f^r \in I$ for some $r \in \mathbb{Z}_+$.*

Our proof is based on Lang's proof in [**95**, Chapter IX, Section 1]. To prove Theorem 2.5 we require some preliminary results.

Proposition 2.6. *Let A be a subring of a ring B and assume that B is integral over A. Let $\phi : A \to L$ be a homomorphism into a field L which is algebraically closed. Then ϕ has an extension to a homomorphism of B into L.*

Proof. Let P be the kernel of ϕ and $S = A \setminus P$. We have a natural commutative diagram

$$\begin{array}{ccc} B & \to & S^{-1}B \\ \uparrow & & \uparrow \\ A & \to & S^{-1}A = A_P \end{array}$$

and ϕ induces a natural homomorphism $\overline{\phi}$ of A_P into L which factors ϕ, by defining
$$\overline{\phi}\left(\frac{x}{y}\right) = \frac{\phi(x)}{\phi(y)}$$
for $x \in A$ and $y \in S$. Let $C = S^{-1}B$, which is integral over A_P. Let m be the maximal ideal of A_P. By Proposition 1.56, there exists a maximal ideal n of C which lies over m. Then C/n is a field which is an algebraic extension of A_P/m, and A_P/m is isomorphic to the subfield $\overline{\phi}(A_P)$ of L. Since the kernel of $\overline{\phi}$ is m, $\overline{\phi}$ induces a natural factorization
$$A_P \to A_P/m \to L$$
of $\overline{\phi}$. We can embed C/n into L since C/n is algebraic over A_P/m and L is algebraically closed [95, Theorem 2.8, page 233], to make a commutative diagram
$$\begin{array}{ccccc} C & \to & C/n & & \\ \uparrow & & \uparrow & \searrow & \\ A_P & \to & A_p/m & \to & L \end{array}$$
giving a homomorphism of B into L which extends ϕ. $\square$

Theorem 2.7. *Let F be a field and $F[y_1, \ldots, y_n]$ be a finitely generated F-algebra. If $F[y_1, \ldots, y_n]$ is a field, then $F[y_1, \ldots, y_n]$ is algebraic over F.*

Proof. Let L be an algebraic closure of F. Suppose that $F[y] = F[y_1, \ldots, y_n]$ is a field which is not algebraic over F. Let $t_1, \ldots, t_r$ (with $r \geq 1$) be a transcendence basis of $F[y]$ over F. The elements $y_1, \ldots, y_n$ are algebraic over $N = F(t_1, \ldots, t_r) = F(t)$. Let $f_i(x) \in N[x]$ be the minimal polynomial of y_i over N. If we multiply the f_i by a suitable nonzero element of $F[t] = F[t_1, \ldots, t_r]$, we get polynomials in $N[x]$, all of whose coefficients lie in $F[t]$. Let $a_1(t), \ldots, a_n(t)$ be the leading coefficients of these polynomials, and let
$$a(t) = a_1(t) \cdots a_n(t).$$
Since $a(t) \neq 0$, there exist $t'_1, \ldots, t'_r \in L$ such that $a(t') = a(t'_1, \ldots, t'_r) \neq 0$, by Theorem 1.4, so that $a_i(t') \neq 0$ for all i. Each y_i is integral over the ring
$$F\left[t_1, \ldots, t_r, \frac{1}{a_1(t)}, \ldots, \frac{1}{a_n(t)}\right].$$
Consider the F-algebra homomorphism $\Psi : F[t_1, \ldots, t_r] \to L$ such that Ψ is the identity on F and $\Psi(t_i) = t'_i$ for $1 \leq i \leq r$. Let P be the kernel of Ψ. We have an extension of Ψ to
$$\overline{\Psi} : F[t]_P \to L \text{ defined by } \overline{\Psi}\left(\frac{f}{g}\right) = \frac{\Psi(f)}{\Psi(g)}$$
for $f \in F[t]$ and $g \in F[t] \setminus P$. Since $a_i(t) \notin P$ for $1 \leq i \leq n$, we have that $y_1, \ldots, y_n$ are integral over $F[t]_P$. By Proposition 2.6, we have an extension

2.1. Affine space and algebraic sets

of $\overline{\Psi}$ to a homomorphism from $F[t]_P[y_1, \ldots, y_n]$ into L which restricts to Ψ, giving an F-algebra homomorphism $F[y] \to L$ which is an inclusion since $F[y]$ is a field. Thus $F[y]$ is algebraic over F, giving a contradiction. $\square$

Corollary 2.8. *Let F be an algebraically closed field and I be an ideal in the polynomial ring $R = F[x_1, \ldots, x_n]$. Then either $I = R$ or there exists $\alpha \in \mathbb{A}_F^n$ such that $f(\alpha) = 0$ for all $f \in I$.*

Proof. Suppose that $I \neq R$. Then I is contained in some maximal ideal m of R (as the ring R/I has a maximal ideal) and R/m is a field, which is a finitely generated F-algebra. By Theorem 2.7, this field is algebraic over F and so is equal to F as F is algebraically closed. Thus there exist $a_1, \ldots, a_n \in F$ such that $m = (x_1 - a_1, \ldots, x_n - a_n)$ and $f(a_1, \ldots, a_n) = 0$ for all $f \in I$ since $I \subset m$. $\square$

The above proof establishes the following corollary.

Corollary 2.9. *Suppose that F is an algebraically closed field and I is an ideal in the polynomial ring $F[x_1, \ldots, x_n]$. Then I is a maximal ideal if and only if there exist $a_1, \ldots, a_n \in F$ such that $I = (x_1 - a_1, x_2 - a_2, \ldots, x_n - a_n)$.*

We now prove Theorem 2.5. We may assume that $f \neq 0$. Let Y be a variable and let I' be the ideal in $R[Y]$ generated by I and $1 - fY$. By Corollary 2.8, the ideal $I' = R[Y]$, so there exist $g_i \in R[Y]$ and $h_i \in I$ such that
$$1 = g_0(1 - Yf) + g_1 h_1 + \cdots + g_r h_r.$$
Substitute $\frac{1}{f}$ for Y and multiply by an appropriate positive power f^m of f to clear denominators on the right-hand side, to conclude that $f^m \in I$.

The following proposition is proven in Exercise 2.14.

Proposition 2.10. *The following statements hold:*

1) *Suppose that Y is a subset of $\mathbb{A}^n$. Then $I(Y)$ is an ideal in $k[\mathbb{A}^n]$.*
2) *If $T_1 \subset T_2$ are subsets of $k[\mathbb{A}^n]$, then $Z(T_2) \subset Z(T_1)$.*
3) *If $Y_1 \subset Y_2$ are subsets of $\mathbb{A}^n$, then $I(Y_2) \subset I(Y_1)$.*
4) *For any two subsets Y_1, Y_2 of $\mathbb{A}^n$, we have $I(Y_1 \cup Y_2) = I(Y_1) \cap I(Y_2)$.*
5) *For any ideal $\mathfrak{a}$ of $k[\mathbb{A}^n]$, we have $I(Z(\mathfrak{a})) = \sqrt{\mathfrak{a}}$.*
6) *For any subset Y of $\mathbb{A}^n$, $Z(I(Y)) = \overline{Y}$, the Zariski closure of Y.*

Theorem 2.11. *A closed set $W \subset \mathbb{A}^n$ is irreducible if and only if $I(W)$ is a prime ideal.*

Proof. Suppose that W is irreducible and $f, g \in k[\mathbb{A}^n]$ are such that $fg \in I(W)$. Then $W \subset Z(fg) = Z(f) \cup Z(g)$. Thus $W = (Z(f) \cap W) \cup (Z(g) \cap W)$ expresses W as a union of closed sets. Since W is irreducible, we have $W \subset Z(f)$ or $W \subset Z(g)$. Thus $f \in I(W)$ or $g \in I(W)$. We have verified that $I(W)$ is a prime ideal.

Now suppose that W is not irreducible. Then $W = Z_1 \cup Z_2$ where Z_1 and Z_2 are proper closed subsets of W. The ideal $I(Z_1)$ is not a subset of $I(Z_2)$. If it were, then we would have

$$Z_2 = Z(I(Z_2)) \subset Z(I(Z_1)) = Z_1$$

by 2) and 6) of Proposition 2.10, which is impossible. Thus there exists $f_1 \in k[\mathbb{A}^n]$ which vanishes on Z_1 but not on Z_2. Similarly, there exists $f_2 \in k[\mathbb{A}^n]$ which vanishes on Z_2 and not on Z_1. We have $f_1 f_2 \in I(W)$, but $f_1, f_2 \notin I(W)$. Thus $I(W)$ is not a prime ideal. $\square$

Theorem 2.12. *Every closed set in $\mathbb{A}^n$ is the union of finitely many irreducible ones.*

Proof. Suppose that Z is an algebraic set in $\mathbb{A}^n$ which is not the union of finitely many irreducible ones. Then $Z = Z_1 \cup Z_2$ where Z_1 and Z_2 are proper closed subsets of Z and either Z_1 or Z_2 is not a finite union of irreducible closed sets. By induction, we can construct an infinite chain of proper inclusions

$$Z \supset W_1 \supset W_2 \supset \cdots$$

giving an infinite chain of proper inclusions

$$I(Z) \subset I(W_1) \subset I(W_2) \subset \cdots$$

of ideals in $k[\mathbb{A}^n]$ (by 3) and 6) of Proposition 2.10), a contradiction to Corollary 1.27. $\square$

Exercise 2.13. Prove 2), 3), and 4) of Proposition 2.1.

Exercise 2.14. Prove Proposition 2.10.

Exercise 2.15. Suppose that $X \subset \mathbb{A}^n$ is an affine algebraic set. Show that $\sqrt{I(X)} = I(X)$.

Exercise 2.16. Show that $\mathbb{A}^n$ is irreducible in the Zariski topology.

Exercise 2.17. Suppose that $k[x_1, \ldots, x_n]$ is a polynomial ring over k. Suppose that $I \subset k[x_1, \ldots, x_n]$ in an ideal. Let $R = k[x_1, \ldots, x_n]/I$. Let $\overline{x}_i$ be the class of x_i in R, so that $R = k[\overline{x}_1, \ldots, \overline{x}_n]$. Show that an ideal $\mathfrak{m}$ in R is a maximal ideal of R if and only if there exist $a_1, \ldots, a_n \in k$ such that $\mathfrak{m} = (\overline{x}_1 - a_1, \overline{x}_2 - a_2, \ldots, \overline{x}_n - a_n)$.

Exercise 2.18. Suppose that $F \in k[x_1, \ldots, x_n]$ is a nonzero nonunit ($\notin k$). Show that $Z(F) \subset \mathbb{A}^n$ is irreducible if and only if F is a positive power of an irreducible element of $k[x_1, \ldots, x_n]$. Warning: The conclusion of this exercise can be false if k is not algebraically closed.

Exercise 2.19. Let $Y = Z(x_1^2 - x_2 x_3, x_1 x_3 - x_1)$. Show that Y is a union of three irreducible components. Describe them and find their prime ideals.

Exercise 2.20. Identify $\mathbb{A}^2$ with $\mathbb{A}^1 \times \mathbb{A}^1$ in the natural way. Show that the Zariski topology on $\mathbb{A}^2$ is not the product topology of the Zariski topology on the two copies of $\mathbb{A}^1$.

Exercise 2.21. Suppose that X is an irreducible topological space.

a) Suppose that U is a nonempty open subset. Show that U is irreducible.

b) Suppose that U_1 and U_2 are nonempty open sets. Show that $U_1 \cap U_2 \neq \emptyset$.

2.2. Regular functions and regular maps of affine algebraic sets

Definition 2.22. Suppose $X \subset \mathbb{A}^n$ is a closed set. The regular functions on X are the polynomial maps on X,

$$k[X] = \{f : X \to \mathbb{A}^1 \mid f \in k[\mathbb{A}^n]\},$$

which is a subalgebra of the k-algebra $\text{Map}(X, \mathbb{A}^1)$ of maps from X to $\mathbb{A}^1$.

We have a natural surjective k-algebra homomorphism, given by restriction, $k[\mathbb{A}^n] \to k[X]$. An element $f \in k[\mathbb{A}^n]$ is in the kernel if and only if $f(q) = 0$ for all $q \in X$, which holds if and only if $f \in I(X)$. Thus

$$k[X] \cong k[\mathbb{A}^n]/I(X).$$

We have that $k[X]$ is a reduced ring by Exercise 2.15.

Definition 2.23. Suppose that X is an affine algebraic set. If $T \subset k[X]$, then

$$Z_X(T) = \{p \in X \mid f(p) = 0 \text{ for all } f \in T\},$$

a subset of X. Suppose that $Y \subset X$ is a subset. Then

$$I_X(Y) = \{f \in k[X] \mid f(p) = 0 \text{ for all } p \in Y\}.$$

We readily verify that $I_X(Y)$ is an ideal in $k[X]$.

When there is no ambiguity, we will usually write $Z(T)$ for $Z_X(T)$ and $I(Y)$ for $I_X(Y)$.

Lemma 2.24. *Suppose that X is a closed subset of $\mathbb{A}^n$. Let* res $: k[\mathbb{A}^n] \to k[X]$ *be the restriction map.*

1) *Suppose that $Y \subset X$. Then*
$$\text{res}^{-1}(I_X(Y)) = I_{\mathbb{A}^n}(Y).$$

2) *Suppose that I is an ideal in $k[X]$. Then*
$$Z_{\mathbb{A}^n}(\text{res}^{-1}(I)) = Z_X(I).$$

Proof. The map res $: k[\mathbb{A}^n] \to k[X]$ is surjective with kernel $I_{\mathbb{A}^n}(X)$. We first prove 1). Since $Y \subset X$, $f \in k[\mathbb{A}^n]$ vanishes on Y if and only if the restriction res(f) of f to X vanishes on Y. Thus formula 1) holds.

Now we prove 2). For $p \in Z_X(I)$ and $f \in k[\mathbb{A}^n]$, $f(p) = \text{res}(f)(p)$ since $Z_X(I) \subset X$. Thus $f \in \text{res}^{-1}(I)$ implies $f(p) = 0$ for all $p \in Z_X(I)$, so that $Z_X(I) \subset Z_{\mathbb{A}^n}(\text{res}^{-1}(I))$.

Now $0 \in I$ since I is an ideal, so $I_{\mathbb{A}^n}(X) \subset \text{res}^{-1}(I)$. Suppose that $p \in Z_{\mathbb{A}^n}(\text{res}^{-1}(I))$. Then $p \in Z_{\mathbb{A}^n}(I_{\mathbb{A}^n}(X)) = X$. Since res is surjective, $p \in X$, and $f(p) = 0$ for all $f \in \text{res}^{-1}(I)$, we have that $g(p) = 0$ for all $g \in I$. Thus $p \in Z_X(I)$. □

We see from Lemma 2.24 that the natural isomorphism of $k[X]$ with $k[\mathbb{A}^n]/I_{\mathbb{A}^n}(X)$ identifies the ideal $I_X(Y)$, for Y a subset of X, with the quotient $I_{\mathbb{A}^n}(Y)/I_{\mathbb{A}^n}(X)$.

Theorem 2.25. *Suppose that X is a closed subset of $\mathbb{A}^n$. Then the conclusions 1)–4) of Proposition 2.1 hold, with $\mathbb{A}^n$ replaced with X.*

We thus obtain a topology on a closed subset X of $\mathbb{A}^n$, where the closed sets are $Z_X(I)$ for ideals $I \subset k[X]$. This topology is the restriction topology of the Zariski topology on $\mathbb{A}^n$. We call this the Zariski topology on X.

If Y is a subset of X, $\overline{Y}$ will denote the Zariski closure of Y in X. A closed irreducible subset of an affine variety X is called a subvariety of X. An open subset of an affine variety is called a quasi-affine variety. An affine algebraic set is a closed subset of $\mathbb{A}^n$. A quasi-affine algebraic set is an open subset of a closed subset of $\mathbb{A}^n$.

Theorem 2.26. *Suppose that X is a closed subset of $\mathbb{A}^n$. Then the conclusions 1)–6) of Proposition 2.10 hold, with $\mathbb{A}^n$ replaced with X.*

Proof. We have already observed that the conclusion 1) holds. We will establish that 5) of Proposition 2.10 holds for algebraic sets. We first establish that

(2.1) $$\sqrt{\text{res}^{-1}(\mathfrak{a})} = \text{res}^{-1}(\sqrt{\mathfrak{a}}).$$

To prove this, observe that

$f \in \text{res}^{-1}(\sqrt{\mathfrak{a}})$
- if and only if $\text{res}(f^n) = \text{res}(f)^n \in \mathfrak{a}$ for some positive integer n
- if and only if $f^n \in \text{res}^{-1}(\mathfrak{a})$
- if and only if $f \in \sqrt{\text{res}^{-1}(\mathfrak{a})}$.

We have that
$$\text{res}^{-1}(I_X(Z_X(\mathfrak{a}))) = I_{\mathbb{A}^n}(Z_{\mathbb{A}^n}(\text{res}^{-1}(\mathfrak{a}))) \text{ by 1) and 2) of Lemma 2.24}$$
$$= \sqrt{\text{res}^{-1}(\mathfrak{a})} \text{ by 5) of Proposition 2.10.}$$

Thus
$$I_X(Z_X(\mathfrak{a})) = \text{res}(\text{res}^{-1}(I_X(Z_X(\mathfrak{a})))) = \text{res}(\sqrt{\text{res}^{-1}(\mathfrak{a})}) = \sqrt{\mathfrak{a}}$$
by (2.1). $\square$

Theorem 2.27. *Suppose that X is a closed subset of $\mathbb{A}^n$. A closed set $W \subset X$ is irreducible if and only if $I_X(W)$ is a prime ideal in $k[X]$.*

Definition 2.28. Suppose $X \subset \mathbb{A}^n$ is a closed set. A map $\phi : X \to \mathbb{A}^m$ is a regular map if there exist $f_1, \ldots, f_m \in k[X]$ such that $\phi = (f_1, f_2, \ldots, f_m)$.

A regular map $\phi = (f_1, \ldots, f_m) : X \to \mathbb{A}^m$ induces a k-algebra homomorphism $\phi^* : k[\mathbb{A}^m] \to k[X]$ by $\phi^*(g) = g \circ \phi$ for $g \in k[\mathbb{A}^m]$. Writing $k[\mathbb{A}^m] = k[y_1, \ldots, y_m]$, we see that ϕ^* is determined by $\phi^*(y_i) = f_i$ for $1 \leq i \leq m$. For $g = g(y_1, \ldots, y_m) \in k[\mathbb{A}^m]$, we have
$$\phi^*(g) = g(\phi^*(y_1), \ldots, \phi^*(y_m)) = g(f_1, \ldots, f_m).$$

Example 2.29. Let $C = Z(y^2 - x(x^2 - 1)) \subset \mathbb{A}^2$. Let $\phi : C \to \mathbb{A}^1$ be the projection on the first factor, so that $\phi(u, v) = u$ for $(u, v) \in C$.
$$\phi^* : k[\mathbb{A}^1] = k[t] \to k[C] = k[x, y]/(y^2 - x(x^2 - 1)) = k[\overline{x}, \overline{y}]$$
is the k-algebra homomorphism induced by $t \mapsto \overline{x}$. Here $\overline{x}$ is the class of x in $k[C]$ and $\overline{y}$ is the class of y in $k[C]$.

Example 2.30. Let $\psi : \mathbb{A}^1 \to \mathbb{A}^2$ be defined by $\psi(s) = (s^2, s^3)$ for $s \in \mathbb{A}^1$.
$$\psi^* : k[\mathbb{A}^2] = k[x, y] \to k[\mathbb{A}^1] = k[t]$$
is the k-algebra homomorphism induced by $x \mapsto t^2$ and $y \mapsto t^3$.

Proposition 2.31. *Suppose that X is a closed subset of $\mathbb{A}^n$, Y is a closed subset of $\mathbb{A}^m$, and $\phi : X \to \mathbb{A}^m$ is a regular map. Then $\phi(X) \subset Y$ if and only if*
$$I(Y) \subset \text{kernel } \phi^* : k[\mathbb{A}^m] \to k[X].$$

Proof. We have that $\phi(X) \subset Y$ holds if and only if $h(\phi(p)) = 0$ for all $h \in I(Y)$ and $p \in X$, which holds if and only if $\phi^*(h) = 0$ for all $h \in I(Y)$, which holds if and only if $I(Y) \subset \text{kernel } \phi^*$. □

Corollary 2.32. *Suppose that X is an affine algebraic set and $\phi : X \to \mathbb{A}^m$ is a regular map. Then $\sqrt{\text{kernel } \phi^*} = \text{kernel } \phi^*$, and $\overline{\phi(X)} = Z(\text{kernel } \phi^*)$, where $\overline{\phi(X)}$ is the Zariski closure of $\phi(X)$ in $\mathbb{A}^n$.*

Proof. The fact that $\sqrt{\text{kernel } \phi^*} = \text{kernel } \phi^*$ follows from the fact that $k[\mathbb{A}^m]/\text{kernel } \phi^*$ is isomorphic to a subring of the reduced ring $k[X]$.

Let
$$S = \{\text{closed subsets } Y \text{ of } \mathbb{A}^m \mid \phi(X) \subset Y\}.$$

A closed set Y is in S if and only if $I(Y) \subset \text{kernel } \phi^*$ by Proposition 2.31 and $W = Z(\text{kernel } \phi^*) \in S$ since $I(W) = \text{kernel } \phi^*$ by 5) of Proposition 2.10. Thus
$$\sum_{Y \in S} I(Y) = \text{kernel } \phi^*,$$
and
$$Z(\text{kernel } \phi^*) = Z\left(\sum_{Y \in S} I(Y)\right)$$
$$= \bigcap_{Y \in S} Z(I(Y)) \text{ by 2) of Proposition 2.1}$$
$$= \bigcap_{Y \in S} Y \text{ by 6) of Proposition 2.10}$$
$$= \overline{\phi(X)}.$$
□

Example 2.33. The image of a regular map may be neither closed nor open. Let $\phi : \mathbb{A}^2 \to \mathbb{A}^2$ be defined by $\phi(u,v) = (u, uv)$. Then
$$\phi(\mathbb{A}^2) = \mathbb{A}^2 \setminus \{(0,y) \mid y \neq 0\}.$$

Definition 2.34. Suppose that $X \subset \mathbb{A}^n$ and $Y \subset \mathbb{A}^m$ are closed sets. A map $\phi : X \to Y$ is a regular map if ϕ is the restriction of the range of a regular map $\tilde\phi : X \to \mathbb{A}^m$, such that $\tilde\phi(X) \subset Y$.

Suppose that $\phi : X \to Y$ is a regular map as in the definition. Let $\pi : k[\mathbb{A}^m] = k[y_1, \ldots, y_m] \to k[Y]$ be the restriction map, which has kernel $I(Y)$. We have that $\tilde\phi(X) \subset Y$, so $I(Y) \subset \text{kernel}(\tilde\phi^*)$ by Proposition 2.31. Thus $\tilde\phi^*$ induces a k-algebra homomorphism $\phi^* : k[Y] \cong k[\mathbb{A}^m]/I(Y) \to k[X]$.

Thus writing $\tilde\phi = (f_1, \ldots, f_m)$, where $f_1, \ldots, f_m \in k[X]$, and $k[Y] = k[\overline{y}_1, \ldots, \overline{y}_m]$, where $\overline{y}_i = \pi(y_i)$ for $1 \leq i \leq m$ are the restrictions of y_i to Y, we have that $f_i = \tilde\phi^*(\overline{y}_i) = \tilde\phi^*(y_i)$ for $1 \leq i \leq m$, and for $g(\overline{y}_1, \ldots, \overline{y}_m) \in k[Y]$, we have that $\phi^*(g) = g(\phi^*(\overline{y}_1), \ldots, \phi^*(\overline{y}_m)) = g(f_1, \ldots, f_m)$.

2.2. Regular functions and regular maps of affine algebraic sets

Definition 2.35. A regular map $\phi : X \to Y$ is dominant if $\phi(X)$ is dense in Y.

Proposition 2.36. *Suppose that $\phi : X \to Y$ is a regular map of affine algebraic sets and $Z \subset Y$ is a closed set. Then $\phi^{-1}(Z) = Z(\phi^*(I(Z)))$.*

Corollary 2.37. *Suppose that X and Y are affine algebraic sets and $\phi : X \to Y$ is a regular map. Then ϕ is continuous.*

Proposition 2.38. *Suppose that $\phi : X \to Y$ is a regular map of affine algebraic sets. Then $\phi^* : k[Y] \to k[X]$ is injective if and only if $\overline{\phi(X)} = Y$.*

Proof. We have that $\overline{\phi(X)} = Z(\text{kernel } \phi^*)$ by Corollary 2.32 and
$$Z(\text{kernel } \phi^*) = Y$$
if and only if kernel $\phi^* = I(Y) = (0)$. $\square$

Lemma 2.39. *Suppose that $\phi : X \to Y$ and $\psi : Y \to Z$ are regular maps of affine algebraic sets. Then $\psi \circ \phi : X \to Z$ is a regular map of affine algebraic sets. Further, $(\psi \circ \phi)^* = \phi^* \circ \psi^* : k[Z] \to k[X]$.*

Proposition 2.40. *Suppose that X and Y are affine algebraic sets and $\Lambda : k[Y] \to k[X]$ is a k-algebra homomorphism. Then there is a unique regular map $\phi : X \to Y$ such that $\phi^* = \Lambda$.*

Proof. We first prove existence. We have that Y is a closed subset of $\mathbb{A}^n$, giving a surjective k-algebra homomorphism $\pi : k[\mathbb{A}^n] = k[y_1, \ldots, y_n] \to k[Y]$. Let $\overline{y}_i = \pi(y_i)$ for $1 \leq i \leq n$, so that $k[Y] = k[\overline{y}_1, \ldots, \overline{y}_n]$. Define a regular map $\tilde{\phi} : X \to \mathbb{A}^n$ by $\tilde{\phi} = (\Lambda(\overline{y}_1), \ldots, \Lambda(\overline{y}_n))$.

Suppose $f(y_1, \ldots, y_n) \in k[\mathbb{A}^n]$. Then
$$\begin{aligned} \tilde{\phi}^*(f(y_1, \ldots, y_n)) &= f \circ \tilde{\phi} = f(\Lambda(\overline{y}_1), \ldots, \Lambda(\overline{y}_n)) \\ &= \Lambda(f(\overline{y}_1, \ldots, \overline{y}_n)) = \Lambda(\pi(f)) \end{aligned}$$
since Λ and π are k-algebra homomorphisms. Thus $\tilde{\phi}^* = \Lambda \circ \pi$, and so
$$I(Y) = \text{kernel } \pi \subset \text{kernel } \tilde{\phi}^*.$$
Thus $\tilde{\phi}(X) \subset Y$ by Proposition 2.31. Let $\phi : X \to Y$ be the induced regular map. The map ϕ^* is the homomorphism induced by $\tilde{\phi}^*$ on the quotient $k[y_1, \ldots, y_n]/I(Y) = k[Y]$. Thus $\phi^* = \Lambda$.

We now prove uniqueness. Suppose that $\phi : X \to Y$ and $\psi : X \to Y$ are regular maps such that $\phi^* = \psi^* = \Lambda$. Suppose that $\phi \neq \psi$. Then there exists $p \in X$ such that $\phi(p) \neq \psi(p)$. Let $q_1 = \phi(p)$ and $q_2 = \psi(p)$. There exists $f \in I(q_1) \setminus I(q_2)$ since $I(q_1)$ and $I(q_2)$ are distinct maximal ideals of $k[Y]$. Thus $f(q_1) = 0$ but $f(q_2) \neq 0$. We have
$$(\phi^* f)(p) = f(\phi(p)) = f(q_1) = 0$$

but
$$(\psi^* f)(p) = f(\psi(p)) = f(q_2) \neq 0.$$
Thus $\phi^* \neq \psi^*$, a contradiction, so we must have that $\phi = \psi$, and thus ϕ is unique. $\square$

Definition 2.41. Suppose that X and Y are affine algebraic sets. We say that X and Y are isomorphic if there are regular maps $\phi : X \to Y$ and $\psi : Y \to X$ such that $\psi \circ \phi = \text{id}_X$ and $\phi \circ \psi = \text{id}_Y$.

Proposition 2.42. *Suppose that $\phi : X \to Y$ is a regular map of affine algebraic sets. Then ϕ is an isomorphism if and only if $\phi^* : k[Y] \to k[X]$ is an isomorphism of k-algebras.*

Proof. First suppose that the regular map $\phi : X \to Y$ is an isomorphism. Then there exists a regular map $\psi : Y \to X$ such that $\psi \circ \phi = \text{id}_X$ and $\phi \circ \psi = \text{id}_Y$. Thus $(\psi \circ \phi)^* = \text{id}_{k[X]}$ and $(\phi \circ \psi)^* = \text{id}_{k[Y]}$. Now $(\psi \circ \phi)^* = \phi^* \circ \psi^*$ and $(\phi \circ \psi)^* = \psi^* \circ \phi^*$ by Lemma 2.39, so $\phi^* : k[Y] \to k[X]$ is a k-algebra isomorphism with inverse ψ^*.

Now assume that $\phi^* : k[Y] \to k[X]$ is a k-algebra isomorphism. Let $\Lambda : k[X] \to k[Y]$ be the k-algebra inverse of ϕ^*. By Proposition 2.40, there exists a unique regular map $\psi : Y \to X$ such that $\psi^* = \Lambda$. Now by Lemma 2.39,
$$(\psi \circ \phi)^* = \phi^* \circ \psi^* = \phi^* \circ \Lambda = \text{id}_{k[X]}$$
and
$$(\phi \circ \psi)^* = \psi^* \circ \phi^* = \Lambda \circ \phi^* = \text{id}_{k[Y]}.$$
Since $(\text{id}_X)^* = \text{id}_{k[X]}$, by uniqueness in Proposition 2.40, we have that $\psi \circ \phi = \text{id}_X$. Similarly, $\phi \circ \psi = \text{id}_Y$. Thus ϕ is an isomorphism. $\square$

Definition 2.43. Suppose that X is an affine algebraic set and $t_1, \ldots, t_r \in k[X]$ are such that $t_1, \ldots, t_r$ generate $k[X]$ as a k-algebra. Let $\phi : X \to \mathbb{A}^r$ be the regular map defined by $\phi = (t_1, \ldots, t_r)$. Then $t_1, \ldots, t_r$ are called coordinate functions on X and ϕ is called a closed embedding.

That the map ϕ of Definition 2.43 is called a closed embedding is justified by the following proposition.

Proposition 2.44. *Suppose that X is an affine algebraic set and $t_1, \ldots, t_r$ are coordinate functions on X. Let $\phi : X \to \mathbb{A}^r$ be the associated closed embedding $\phi = (t_1, \ldots, t_r)$, and let $Y = \phi(X)$. Then Y is a closed subset of $\mathbb{A}^r$ with ideal $I(Y) = \text{kernel } \phi^* : k[\mathbb{A}^r] \to k[X]$, and regarding ϕ as a regular map to Y, we have that $\phi : X \to Y$ is an isomorphism.*

Proof. Let $\overline{Y}$ be the Zariski closure of Y in $\mathbb{A}^r$. We have $I(\overline{Y}) = \text{kernel } \phi^*$ by Corollary 2.32. Thus $\phi^* : k[\mathbb{A}^r] \to k[X]$ is onto with kernel $I(\overline{Y})$, so that now regarding ϕ as a regular map from X to $\overline{Y}$, we have that $\phi^* : k[\overline{Y}] = k[\mathbb{A}^r]/I(\overline{Y}) \to k[X]$ is an isomorphism. Thus $\phi : X \to \overline{Y}$ is an isomorphism by Proposition 2.42, and so $Y = \overline{Y}$. □

Our definition of $\mathbb{A}^n = \{p = (a_1, \ldots, a_n) \mid a_1, \ldots, a_n \in k\}$ naturally gives us particular coordinate functions, namely the coordinate functions x_i for $1 \leq i \leq n$. If $B = (b_{ij})$ is an invertible $n \times n$ matrix with coefficients in k and $c = (c_1, \ldots, c_n)$ is a vector in k^n, then $y_i = \sum_{j=1}^n b_{ij} x_j + c_i$ for $1 \leq i \leq n$ defines another choice of coordinate functions $y_1, \ldots, y_n$ on $\mathbb{A}^n$.

We deduce the following from Proposition 2.36.

Lemma 2.45. *Suppose that $\phi : X \to Y$ is a regular map of affine algebraic sets and $t_1, \ldots, t_n$ are coordinate functions on Y (giving a closed embedding of Y in $\mathbb{A}^n$). Suppose that $p \in Y$. Then $I(p) = (t_1 - t_1(p), \ldots, t_n - t_n(p))$ and*

$$I(\phi^{-1}(p)) = \sqrt{(\phi^*(t_1) - t_1(p), \ldots, \phi^*(t_n) - t_n(p))}.$$

The results of this section show that there is an equivalence of categories between the category of reduced finitely generated k-algebras and their k-algebra homomorphisms, and the category of affine algebraic sets in $\mathbb{A}^n_k$ for some n and regular maps between affine algebraic sets. Further, this equivalence restricts to give an equivalence of categories between the category of finitely generated k-algebras which are domains and their k-algebra homomorphisms, and the category of affine varieties in $\mathbb{A}^n_k$ for some n and regular maps between affine varieties.

Exercise 2.46. Is formula 2) of Lemma 2.24 always true if I is replaced by a subset T of $k[X]$?

Exercise 2.47. Prove Theorem 2.25.

Exercise 2.48. Prove 4) and 6) of Theorem 2.26.

Exercise 2.49. Prove Theorem 2.27.

Exercise 2.50. Prove Proposition 2.36 and deduce Lemma 2.45.

Exercise 2.51. Prove Lemma 2.39.

Exercise 2.52. Let $k[\mathbb{A}^n] = k[x_1, \ldots, x_n]$.

  a) Suppose that $a, b, c, d, e, f \in k$ with $ae - bd \neq 0$. Show that the map $\phi : \mathbb{A}^2 \to \mathbb{A}^2$ defined by $\phi = (ax_1 + bx_2 + c, dx_1 + ex_2 + f)$ is an isomorphism. Give an explicit description of ϕ^{-1}.

b) Suppose that $\phi : \mathbb{A}^n \to \mathbb{A}^n$ is defined by $\phi = (f_1, \ldots, f_n)$ where

$$f_i = \sum_{j=1}^{n} a_{ij} x_j + b_i$$

with $a_{ij}, b_i \in k$ and $\mathrm{Det}(a_{ij}) \neq 0$. Show that ϕ is an isomorphism of $\mathbb{A}^n$. Give an explicit description of ϕ^{-1}.

Exercise 2.53. A quadratic polynomial in $k[x_1, x_2]$ is a polynomial all of whose terms are monomials of degree ≤ 2.

  a) Let $X = Z(x_2 - x_1^2) \subset \mathbb{A}^2$. Show that X is a variety and that $X \cong \mathbb{A}^1$.

  b) Let $Y = Z(x_1 x_2 - 1) \subset \mathbb{A}^2$. Show that Y is a variety and that $Y \not\cong \mathbb{A}^1$.

  c) Let f be any irreducible quadratic polynomial in $k[x_1, x_2]$, and let $W = Z(f)$. Assume that k has characteristic $\neq 2$. Show that W is isomorphic to X or to Y.

  d) Can you give a proof of c) which is valid when k has characteristic 2?

Exercise 2.54. Let $X = Z(x_2^2 - x_1^3) \subset \mathbb{A}^2$. Consider the regular map $\phi : \mathbb{A}^1 \to X$ defined by $\phi(t) = (t^2, t^3)$ for $t \in \mathbb{A}^1$.

  a) Show that ϕ is a bijection.

  b) Show that ϕ is not an isomorphism.

2.3. Finite maps

In this section we interpret the algebraic notion of a ring extension being finite geometrically.

Definition 2.55. Suppose that $f : X \to Y$ is a regular map of affine varieties. We say that f is a finite map if $k[X]$ is integral and thus a finitely generated module over the subring $f^*(k[Y])$.

In the case when $f : X \to Y$ is dominant, so that $f^* : k[Y] \to k[X]$ is injective, it may sometimes be convenient to abuse notation and identify $k[Y]$ with its isomorphic image $f^*(k[Y])$. In this way, we may sometimes write t for $f^*(t)$ if $t \in k[Y]$.

Theorem 2.56. *Suppose that $f : X \to Y$ is a finite map of affine varieties. Then $f^{-1}(p)$ is a finite set for all $p \in Y$.*

Proof. Let $t_1, \ldots, t_n$ be coordinate functions on X. It suffices to show that each t_i assumes only finitely many values on $f^{-1}(p)$. Since $k[X]$ is integral

2.3. Finite maps

over $f^*(k[Y])$, each t_i satisfies a dependence relation
$$t_i^m + f^*(b_{m-1})t_i^{m-1} + \cdots + f^*(b_0) = 0$$
with $m \in \mathbb{Z}_+$ and $b_0, \ldots, b_{m-1} \in k[Y]$. Suppose that $q \in f^{-1}(p)$. Then
$$\begin{aligned} 0 &= t_i(q)^m + f^*(b_{m-1})(q)t_i(q)^{m-1} + \cdots + f^*(b_0)(q) \\ &= t_i(q)^m + b_{m-1}(p)t_i(q)^{m-1} + \cdots + b_0(p). \end{aligned}$$
Thus $t_i(q)$ must be one of the $\leq m$ roots of this equation. $\square$

Theorem 2.57. *Suppose that $f : X \to Y$ is a dominant finite map of affine varieties. Then f is surjective.*

Proof. Let $q \in Y$. Let $\mathfrak{m}_q = I(q)$ be the ideal of q in $k[Y]$. Then $f^{-1}(q) = Z(f^*(\mathfrak{m}_q))$ by Proposition 2.36. Then $f^{-1}(q) = \emptyset$ if and only if $f^*(\mathfrak{m}_q)k[X] = k[X]$. By Proposition 1.56, $f^*(\mathfrak{m}_q)k[X]$ is a proper ideal of $k[X]$ since $f^*(k[Y]) \cong k[Y]$ and $\mathfrak{m}_q$ is a prime ideal of $k[Y]$. $\square$

Corollary 2.58. *A finite map $f : X \to Y$ of affine varieties is a closed map.*

Proof. It suffices to verify that if $Z \subset X$ is an irreducible closed subset, then $f(Z)$ is closed in Y. Let $W = \overline{f(Z)}$ be the closure of $f(Z)$ in Y. Let $\overline{f} = f|Z : Z \to W$. The homomorphism $f^* : k[Y] \to k[X]$ induces the homomorphism $\overline{f}^* : k[W] = k[Y]/I(W) \to k[X]/I(Z) = k[Z]$ by Proposition 2.31. The ring $k[Z]$ is integral over $\overline{f}^*(k[W])$ since $k[X]$ is integral over $f^*(k[Y])$. Thus $\overline{f} : Z \to W$ is a dominant finite map, which is surjective by Theorem 2.57. Thus $f(Z) = W$ is closed in Y. $\square$

Theorem 2.59. *Suppose that X is an affine algebraic set. Then there exists a dominant finite map $\phi : X \to \mathbb{A}^r$ for some r.*

Proof. There exist, by Theorem 1.53, $y_1, \ldots, y_r \in k[X]$ such that $k[y_1, \ldots, y_r]$ is a polynomial ring and $k[X]$ is integral over $k[y_1, \ldots, y_r]$. Define a regular map $\phi : X \to \mathbb{A}^r$ by $\phi(p) = (y_1(p), y_2(p), \ldots, y_r(p))$ for $p \in X$. Let $t_1, \ldots, t_r$ be the natural coordinate functions on $\mathbb{A}^r$. Then $\phi^* : k[\mathbb{A}^r] \to k[X]$ is the k-algebra homomorphism defined by $\phi^*(t_i) = y_i$ for $1 \leq i \leq r$. Thus ϕ^* is injective and $k[X]$ is integral over $k[\mathbb{A}^r]$, and so $\phi : X \to \mathbb{A}^r$ is dominant and finite. $\square$

Exercise 2.60. Suppose that $\phi : X \to Y$ is a regular map of affine varieties such that $\phi^{-1}(p)$ is a finite set for all $p \in Y$ and $\phi^* : k[Y] \to k[X]$ is injective. Is ϕ necessarily a finite map?

Exercise 2.61. Suppose that $\phi : X \to Y$ is a surjective regular map of affine varieties such that $\phi^{-1}(p)$ is a finite set for all $p \in Y$ and $\phi^* : k[Y] \to k[X]$ is injective. Is ϕ necessarily a finite map?

Exercise 2.62. Suppose that $\phi : X \to Y$ is a dominant regular map of varieties. Can there exist a prime ideal I in $k[Y]$ such that $Ik[X] = k[X]$? Compare this exercise with the conclusions of Proposition 1.56.

2.4. Dimension of algebraic sets

Theorem 2.59 gives us a geometric way to define the dimension of an affine variety: an affine variety X has dimension r if there is a dominant finite map from X to $\mathbb{A}^r$. We will give an algebraic definition of dimension and show that it agrees with the geometric definition. An introduction to dimension theory in rings can be found in Section 1.8.

An irreducible topological space is defined before Example 2.3.

Definition 2.63. Suppose that X is a topological space. The dimension of X, denoted $\dim X$, is the supremum of all numbers n such that there exists a chain

(2.2) $$Z_0 \subset Z_1 \subset \cdots \subset Z_n$$

of distinct irreducible closed subsets of X. The dimension of an affine algebraic set or quasi-affine algebraic set is its dimension as a topological space.

This definition of dimension works well for the Zariski topology but does not agree with the usual definition of dimension of $\mathbb{C}^n$ with the Euclidean topology since the only irreducible subsets in $\mathbb{C}^n$ (in the Euclidean topology) are the single points. We will see that the dimension of a complex variety is equal to its dimension in the Euclidean topology in Theorem 10.45.

Proposition 2.64. *Suppose that X is an affine algebraic set. Then the dimension of X is equal to the dimension of the ring $k[X]$ of regular functions on X.*

Proof. By Theorems 2.26, 2.25, and 2.27, chains

$$Z_0 \subset Z_1 \subset \cdots \subset Z_n$$

of distinct irreducible closed subsets of X correspond 1-1 to chains

$$I_X(Z_n) \subset I_X(Z_{n-1}) \subset \cdots \subset I_X(Z_0)$$

of distinct prime ideals in $k[X]$. □

From Theorem 1.63 we obtain the following two propositions.

Proposition 2.65. *The dimension of an affine variety is finite.*

This proposition follows since if X is an affine variety, then the domain $k[X]$ is a finitely generated k-algebra, so its quotient field is a finitely generated extension field of k, and thus it has a finite transcendence basis.

2.4. Dimension of algebraic sets

Proposition 2.66. *The dimension of $\mathbb{A}^n$ is $\dim \mathbb{A}^n = n$.*

Proposition 2.67. *Suppose that X is an affine algebraic set and $V_1, \ldots, V_r$ are the irreducible components of X (the distinct largest irreducible sets contained in X). Then*
$$\dim X = \max\{\dim V_i\}.$$

Proof. Suppose that (2.2) is a chain of irreducible closed subsets of X. Then Z_n is contained in V_i for some i since Z_n is irreducible and the V_i are the irreducible components of X. $\square$

A chain (2.2) is maximal if the chain cannot be lengthened by adding an additional irreducible closed set somewhere in the chain.

Corollary 2.68. *Suppose that X is an affine variety. Then every maximal chain of distinct prime ideals in $k[X]$ has the same finite length equal to the dimension of $k[X]$.*

Proof. The proof is by induction on the dimension of X. If $\dim X = 0$, then $k[X] = k$ is a field and the corollary is trivially true.

Suppose $\dim X = n > 0$ and the corollary is true for varieties of dimension $< n$. Suppose that
$$P_0 \subset P_1 \subset \cdots \subset P_m$$
is a maximal chain of distinct prime ideals in $k[X]$.
$$1 + \dim k[X]/P_1 = \mathrm{ht}(P_1) + \dim k[X]/P_1 = \dim k[X] = n$$
by Theorem 1.64, and so $\dim k[X]/P_1 = n - 1$. Now
$$P_1/P_1 \subset P_2/P_1 \subset \cdots \subset P_m/P_1$$
is a maximal chain of distinct prime ideals in $k[X]/P_1$, and $k[X]/P_1 = k[Z_X(P_1)]$ so by induction on n, the variety $Z_X(P_1)$ has dimension $m - 1$. Thus
$$n - 1 = \dim k[X]/P_1 = m - 1,$$
and so $m = n = \dim X$. $\square$

Some examples of noncatenary Noetherian rings (rings which do not satisfy the conclusions of Corollary 2.68) are given by Nagata in [**121**, Appendix A1].

Corollary 2.69. *Suppose that X is an affine variety. Then every maximal chain of distinct irreducible closed subsets of X has the same length (equal to $\dim X$).*

Proof. Suppose that (2.2) is a maximal chain of distinct irreducible closed subsets of X. Since X is irreducible, we must have that $Z_n = X$ and Z_0 is a point. Taking the sequence of ideals of (2.2), we have a maximal chain $(0) = I_X(Z_n) \subset \cdots \subset I_X(Z_0)$ of distinct prime ideals in $k[X]$. By Corollary 2.68 and Proposition 2.64, we have that $n = \dim k[X] = \dim X$. □

Proposition 2.70. *Suppose that X is an affine variety and Y is a nontrivial open subset of X. Then $\dim X = \dim Y$.*

Proof. Suppose that

(2.3) $$Z_0 \subset Z_1 \subset \cdots \subset Z_n$$

is a sequence of distinct closed irreducible subsets of Y. Let $\overline{Z}_i$ be the Zariski closure of Z_i in X for $0 \le i \le n$. Then

(2.4) $$\overline{Z}_0 \subset \overline{Z}_1 \subset \cdots \subset \overline{Z}_n$$

is a sequence of distinct closed irreducible subsets of X since $\overline{Z}_i \cap Y = Z_i$, as Z_i is closed in Y and Y is open in X. Thus $\dim Y \le \dim X$. In particular, $\dim Y$ is finite, so we can choose a maximal such chain (2.3). Since the chain is maximal, Z_0 is a point and $Z_n = Y$. Now if W is an irreducible closed subset of X such that the open subset $W \cap Y$ of W is nonempty, we then have that $W \cap Y$ is dense in W. In particular, if $A \subset B$ are irreducible closed subsets of X such that $A \cap Y \ne \emptyset$ and $A \cap Y = B \cap Y$, then we have that $A = \overline{A \cap Y} = \overline{B \cap Y} = B$. Thus we have that (2.4) is a maximal chain in X, and hence $\dim Y = \dim X$ by Corollary 2.69. □

Noether's normalization lemma, Theorem 1.53, can be used to compute the dimension of any affine variety (by Theorem 1.63). In fact, we see that if $\phi : X \to \mathbb{A}^r$ is a dominant finite map, then $r = \dim X$.

Suppose that R is a Noetherian ring and $I \subset R$ is an ideal. Since R is Noetherian, the ideal I has only a finite number of minimal primes $P_1, \ldots, P_r$ (Section 1.6). We have that $I \subset P_1 \cap \cdots \cap P_r$ and $I = P_1 \cap \cdots \cap P_r$ if and only if I is a radical ideal ($\sqrt{I} = I$).

Suppose that X is a closed subset of $\mathbb{A}^n$. Let $V_1, \ldots, V_r$ be the irreducible components of X; that is, $V_1, \ldots, V_r$ are the irreducible closed subsets of X such that $X = V_1 \cup \cdots \cup V_r$ and we have $V_i \not\subset V_j$ if $i \ne j$. Then the minimal primes of $I(X)$ are $P_i = I(V_i)$ for $1 \le i \le r$. The prime ideals $\overline{P}_i = I_X(V_i) = P_i/I(X)$ are the minimal primes of the ring $k[X]$, that is, the minimal primes of the zero ideal. We have that $\overline{P}_1 \cap \cdots \cap \overline{P}_r = (0)$ since $I(X)$ is a radical ideal.

The following theorem follows from Krull's principal ideal theorem (Theorem 1.65).

2.4. Dimension of algebraic sets

Theorem 2.71. *Suppose that X is an affine variety and $f \in k[X]$. Then:*

1) *If f is not 0 and is not a unit in $k[X]$, then $Z_X(f)$ is a nonempty algebraic set, all of whose irreducible components have dimension equal to $\dim X - 1$.*

2) *If f is a unit in $k[X]$, then $Z_X(f) = \emptyset$.*

3) *If $f = 0$, then $Z_X(f) = X$.*

Proposition 2.72. *Suppose that X is a variety in $\mathbb{A}^n$. Then X has dimension $n - 1$ if and only if $I(X) = (f)$ where $f \in k[\mathbb{A}^n] = k[x_1, \ldots, x_n]$ is an irreducible polynomial.*

Proof. Suppose that $I(X) = (f)$ where f is irreducible. Then (f) is a prime ideal by Proposition 1.30 (since $k[\mathbb{A}^n]$ is a unique factorization domain) which has height 1 by Theorem 1.65. Thus

$$\dim X = \dim k[X] = \dim k[\mathbb{A}^n] - \text{ht}(f) = \dim k[\mathbb{A}^n] - 1 = \dim \mathbb{A}^n - 1$$

by Theorem 1.64.

Now suppose that X has dimension $n - 1$. Then the prime ideal $I(X)$ has height 1 by Theorem 1.64. Since the polynomial ring $k[\mathbb{A}^n]$ is a unique factorization domain, $I(X)$ is a principal ideal generated by an irreducible element by Proposition 1.66. □

If Y is an affine or quasi-affine algebraic set contained in an affine variety X, then we define the codimension of Y in X to be $\text{codim}_X(Y) = \dim X - \dim Y$. If Y is a subvariety of an algebraic variety X, then we have that $\text{codim}_X(Y)$ is the height of the prime ideal $I_X(Y)$ in $k[X]$.

More generally, suppose that X is an n-dimensional affine variety and $Y \subset X$ is an algebraic set with irreducible components $Y_1, \ldots, Y_s$. We have

$$\begin{aligned}\text{codim}_X(Y) &= \dim(X) - \dim(Y) \\ &= \dim(X) - \max\{\dim(Y_i)\} \text{ (by Proposition 2.67)} \\ &= \min\{n - \dim(Y_i)\} \\ &= \min\{\text{height } I_X(Y_i)\}.\end{aligned}$$

We will call a one-dimensional affine variety a curve and a two-dimensional affine variety a surface. An n-dimensional affine variety is called an n-fold.

We see that if C is a curve, then the prime ideals in $k[C]$ are just the maximal ideals (corresponding to the points of C) and the zero ideal (corresponding to the curve C). If S is a surface and $k[S]$ is a UFD, then the prime ideals in $k[S]$ are the maximal ideals (corresponding to the points of S), principal ideals generated by an irreducible element (corresponding to the curves lying on S), and the zero ideal (corresponding to the surface S).

If X is an n-fold with $n \geq 3$, then the prime ideals in $k[X]$ are much more complicated. The prime ideals in $k[\mathbb{A}^3]$ are the maximal ideals (height 3), height 2 prime ideals, principal ideals generated by an irreducible element (height 1), and the zero ideal (height 0). The height 2 prime ideals $\mathfrak{p}$, which correspond to curves in $\mathbb{A}^3$, can be extremely complicated, although many times one has the nice case where $\mathfrak{p}$ is generated by two elements. A height 2 prime $\mathfrak{p}$ in $k[\mathbb{A}^3]$ requires at least two generators but there is no upper bound on the minimum number of generators required to generate such a prime $\mathfrak{p}$. Some examples showing this are given in [5]. A complete analysis of the generators of monomial space curves is given by Jürgen Herzog in [76]. The book [130] by Judith Sally gives an excellent introduction to the question of the number of generators of an ideal in a local ring.

Exercise 2.73. This exercise gives a criterion which can be used to determine the minimal number of generators of an ideal in a nonlocal ring. Suppose that $R = k[\overline{x}_1, \ldots, \overline{x}_n]$ is a finitely generated k-algebra, $I \subset R$ is an ideal, and $\mathfrak{m}$ is a maximal ideal of R. Let $\mu(I)$ be the minimal number of generators of I. Using Lemmas 1.19 and 1.28, show that

$$\mu(I) \geq \mu(I_\mathfrak{m}) = \dim_k I_\mathfrak{m}/\mathfrak{m} I_\mathfrak{m}.$$

Exercise 2.74. Define a regular map $\Phi : \mathbb{A}^1 \to \mathbb{A}^3$ by $\Phi(t) = (t, t^2, t^3)$ for $t \in \mathbb{A}^1$. Let X be the image of Φ.

a) Show that Φ is a finite map.

b) Show that X is a variety (the image of Φ is Zariski closed).

c) Show that Φ is injective.

d) Determine if $\Phi : \mathbb{A}^1 \to X$ is an isomorphism of varieties.

e) Find a minimal set of generators of the ideal $I(X)$; that is, find a set of generators of $I(X)$ with the smallest possible number of elements.

Exercise 2.75. Define a regular map $\Phi : \mathbb{A}^1 \to \mathbb{A}^3$ by $\Phi(t) = (t^2, t^3, t^4)$ for $t \in \mathbb{A}^1$. Let X be the image of Φ.

a) Show that Φ is a finite map.

b) Show that X is a variety (the image of Φ is Zariski closed).

c) Show that Φ is injective.

d) Determine if $\Phi : \mathbb{A}^1 \to X$ is an isomorphism of varieties.

e) Find a minimal set of generators of the ideal $I(X)$; that is, find a set of generators of $I(X)$ with the smallest possible number of elements. You may find that Exercise 2.73 and the method of the next problem (Exercise 2.76) will be useful in this problem.

2.4. Dimension of algebraic sets

Exercise 2.76. Define a regular map $\Phi : \mathbb{A}^1 \to \mathbb{A}^3$ by $\Phi(t) = (t^3, t^4, t^5)$ for $t \in \mathbb{A}^1$. Let X be the image of Φ.

  a) Show that Φ is a finite map.
  b) Show that X is a variety (the image of Φ is Zariski closed).
  c) Show that Φ is injective.
  d) Determine if $\Phi : \mathbb{A}^1 \to X$ is an isomorphism of varieties.
  e) Find a minimal set of generators of the ideal $I(X)$; that is, find a set of generators of $I(X)$ with the smallest possible number of elements. You may find the following outline of a solution helpful:

  i) Define a "weighting" on the variables x_1, x_2, x_3 by setting $\mathrm{wt}(x_1) = 3$, $\mathrm{wt}(x_2) = 4$, and $\mathrm{wt}(x_3) = 5$. Define the weight of the monomial $x_1^l x_2^m x_3^n$ to be $\mathrm{wt}(x_1^l x_2^m x_3^n) = 3l + 4m + 5n$. Say that an element $g = \sum a_{lmn} x_1^l x_2^m x_3^n$ is weighted homogeneous of degree d if $3l + 4m + 5n = d$ whenever $a_{lmn} \neq 0$. Every element $f \in k[x_1, x_2, x_3]$ has a unique expression $f = \sum_i F_i$ where F_i is weighted homogeneous of degree i.
  ii) Show that $f \in I(X)$ if and only if $F_i \in I(X)$ for all i. Conclude that $I(X)$ is generated by weighted homogeneous elements.
  iii) Show that $I(X)$ is generated by the set of "binomials" $A - B$ where A and B are monomials which have the same weighted degree.
  iv) Show that $I(X)$ is generated by the set of weighted homogeneous binomials which are of one of the following three types:
  $$x_1^l - x_2^m x_3^n, \quad x_2^m - x_1^l x_3^n, \quad x_3^n - x_1^l x_2^m.$$
  v) Make an (intelligent) guess of a set of minimal generators of $I(X)$, consisting of weighted homogeneous binomials. Let J be the ideal generated by this set. Show that J contains all weighted homogeneous binomials, by induction on the weighted degree. Conclude that $I(X) = J$ and your set generates $I(X)$.
  vi) Now use Exercise 2.73 to show that you have found a minimal generating set.

Exercise 2.77. Suppose that R is a ring. Recall (Section 1.5) that an element $f \in R$ is called irreducible if f is not a unit, and $f = ab$ with $a, b \in R$ implies a or b is a unit.

Suppose that X is an affine variety and $f \in k[X]$ is irreducible. Is $Z_X(f)$ necessarily irreducible? Either prove this or give a counterexample.

Exercise 2.78. Let X be the variety which is the image of $\mathbb{A}^1$ in $\mathbb{A}^3$ by the map $\phi(t) = (t^3, t^4, t^5)$ of Exercise 2.76. Compute the dimension of X and the height of $I(X)$ in $k[\mathbb{A}^3]$.

Exercise 2.79. This exercise shows that the assumption that A is a domain is necessary in the statement of Theorem 1.64.

Let $V_1 = Z(x)$ and $V_2 = Z(y,z)$ be algebraic sets in $\mathbb{A}^3$. The sets V_1 and V_2 are irreducible, with $I(V_1) = (x)$ and $I(V_2) = (y,z)$ (you do not need to show this). Compute the dimensions of V_1 and V_2 and the heights of the prime ideals $I(V_1)$ and $I(V_2)$ in $k[\mathbb{A}^3]$. Let $X = V_1 \cup V_2$. Compute $I(X)$. Compute the heights of the prime ideals $I_X(V_1)$ and $I_X(V_2)$ in $k[X]$.

Exercise 2.80. Let X be a variety, U an open subset of X, $0 \neq g \in k[X]$ a nonunit, and Z an irreducible component of $Z(g) \cap U$. Show that $\dim Z = \dim X - 1$.

Exercise 2.81. Suppose that $X \subset \mathbb{A}^n$ is a nonempty closed subset such that $I(X) = (f_1, \ldots, f_r)$ is generated by r elements. Show that $\text{codim}_{\mathbb{A}^n}(X) \leq r$.

Exercise 2.82. Suppose that C is a one-dimensional subvariety of $\mathbb{A}^3$. Let $k[\mathbb{A}^3] = k[x,y,z]$. Suppose that C is not a line parallel to the z-axis. Let $\pi : \mathbb{A}^3 \to \mathbb{A}^2$ be the projection $\pi(a,b,c) = (a,b)$ for $(a,b,c) \in \mathbb{A}^3$.

a) Show that the Zariski closure of $\pi(C)$ is a one-dimensional subvariety D of $\mathbb{A}^2$ and $I(D) = I(\pi(C))$ is a principal ideal (g) where g is an irreducible polynomial in $k[x,y]$.

b) Let $h = g_0(x,y)z^n + \cdots + g_n(x,y)$ be an element of $I(C)$ of smallest positive degree n in z. Prove that if $f \in I(C)$ has degree m as a polynomial in z, then we have an expression
$$fg_0^m = hq + v(x,y)$$
where $v(x,y)$ is divisible by $g(x,y)$.

c) Show that the algebraic set $Z(h,g)$ is the union of C and finitely many lines parallel to the z-axis.

d) Show that C can be defined by three equations by finding $t \in k[x,y,z]$ such that $C = Z(g,h,t)$.

2.5. Regular functions and regular maps of quasi-affine varieties

In this section we consider regular functions and regular maps on open subsets of an affine variety.

Lemma 2.83. *Suppose that X is an affine algebraic set. Then the open sets $D(f) = X \setminus Z(f)$ for $f \in k[X]$ form a basis of the Zariski topology on X.*

We will also denote the open set $D(f)$ by X_f.

2.5. Regular functions and regular maps of quasi-affine varieties

Proof. We must show that given an open subset U of X and a point $q \in U$, there exists $f \in k[X]$ such that $q \in X \setminus Z(f) \subset U$. Set $\mathfrak{m} = I(q)$. There exists an ideal I in $k[X]$ such that $U = X \setminus Z(I)$. The fact that $q \in U$ implies $q \notin Z(I)$ which implies $I \not\subset \mathfrak{m}$. Thus there exists $f \in I$ such that $f \notin \mathfrak{m}$. Then $Z(I) \subset Z(f)$ implies $X \setminus Z(f) \subset U$. Now $f \notin \mathfrak{m}$ implies $q = Z(\mathfrak{m}) \notin Z(f)$ so that $q \in X \setminus Z(f)$. $\square$

The process of localization is reviewed in Section 1.4. Suppose that R is a domain with quotient field K. If $0 \neq f \in R$, then R_f is the following subring of K:

$$R_f = R\left[\frac{1}{f}\right] = \left\{\frac{g}{f^n} \mid g \in R \text{ and } n \in \mathbb{N}\right\}.$$

If $\mathfrak{p}$ is a prime ideal in R, then $R_\mathfrak{p}$ is the following subring of K:

$$R_\mathfrak{p} = \left\{\frac{f}{g} \mid f \in R \text{ and } g \in R \setminus \mathfrak{p}\right\}.$$

$R_\mathfrak{p}$ is a local ring: its unique maximal ideal is $\mathfrak{p}R_\mathfrak{p} = \mathfrak{p}_\mathfrak{p}$.

Suppose that X is an affine variety. Let $k(X)$ be the quotient field of $k[X]$. The field $k(X)$ is called the field of rational functions on X, or the function field of X. For $p \in X$, we have that the localization

$$k[X]_{I(p)} = \left\{\frac{f}{g} \mid f, g \in k[X] \text{ and } g(p) \neq 0\right\}.$$

For $f \in k[X]_{I(p)}$, we have a value $f(p) \in k$, defined as follows. Write $f = \frac{g}{h}$ with $g, h \in k[X]$ and $h(p) \neq 0$ and define $f(p) = \frac{g(p)}{h(p)} \in k$. This value is independent of choice of g and h as above. We have natural isomorphisms $k[X]_{I(p)}/I(p)_{I(p)} \cong k[X]/I(p) \cong k$, identifying the value $f(p)$ with the residue of f in $k[X]_{I(p)}/I(p)_{I(p)}$.

Suppose that U is a nonempty open subset of X. Define the regular functions on U to be

$$\mathcal{O}_X(U) = \bigcap_{p \in U} k[X]_{I(p)},$$

where the intersection in $k(X)$ is over all $p \in U$. If $U = \emptyset$, we define $\mathcal{O}_X(U) = \mathcal{O}_X(\emptyset) = 0$.

Suppose that U is a nonempty open subset of X. Let $\text{Map}(U, \mathbb{A}^1)$ be the set of maps from U to $\mathbb{A}^1$. The set $\text{Map}(U, \mathbb{A}^1)$ is a k-algebra since $\mathbb{A}^1 = k$ is a k-algebra.

We have a natural k-algebra homomorphism $\phi : \mathcal{O}_X(U) \to \text{Map}(U, \mathbb{A}^1)$ defined by $\phi(f)(p) = f(p)$ for $f \in \mathcal{O}_X(U)$ and $p \in U$. We will show that ϕ is injective. Suppose $f \in \text{Kernel } \phi$ and $p \in U$. There exists an expression

$f = \frac{g}{h}$ where $g, h \in k[X]$ and $h(p) \neq 0$. For q in the nontrivial open set $U \setminus Z(h)$ we have that $\frac{g(q)}{h(q)} = f(q) = 0$. Thus $g(q) = 0$, and so

$$g \in I(U \setminus Z(h)) = I(X) = (0)$$

since $U \setminus Z(h)$ is a dense open subset of X. Thus $g = 0$, and so $f = \frac{g}{h} = 0$. Hence ϕ is injective. We may thus identify $\mathcal{O}_X(U)$ with the k-algebra $\phi(\mathcal{O}_X(U))$ of maps from U to $\mathbb{A}^1$.

An element f of the function field $k(X)$ of X induces a map $f : U \to \mathbb{A}^1$ on some open nonempty subset U of X.

For $p \in X$, we define

$$\mathcal{O}_{X,p} = \bigcup_{p \in U} \mathcal{O}_X(U),$$

where the union in $k(X)$ is over all open sets U in X containing p. An element $f \in \mathcal{O}_{X,p}$ thus induces a map $f : U \to \mathbb{A}^1$ on some open neighborhood U of p in X. We will see that $\mathcal{O}_{X,p}$ is a local ring (Proposition 2.86). We will denote the maximal ideal of $\mathcal{O}_{X,p}$ by $m_{X,p}$ or by m_p if there is no danger of confusion. We will denote $\mathcal{O}_{X,p}/m_p$ by $k(p)$. As a field, $k(p)$ is isomorphic to k. Also, $k(p)$ has a natural structure as an $\mathcal{O}_{X,p}$-module.

Suppose that $U \subset V$ are open subsets of X and $p \in V$. We then have injective restriction maps

(2.5) $$\mathcal{O}_X(V) \to \mathcal{O}_X(U)$$

and

(2.6) $$\mathcal{O}_X(V) \to \mathcal{O}_{X,p}.$$

Proposition 2.84. *Suppose that X is an affine variety and $0 \neq f \in k[X]$. Then $\mathcal{O}_X(D(f)) = k[X]_f$.*

Proof. We have $k[X]_f = \{\frac{g}{f^n} \mid g \in k[X]$ and $n \in \mathbb{N}\}$. If $\frac{g}{f^n} \in k[X]_f$, then $\frac{g}{f^n} \in k[X]_{I(p)}$ for all $p \in D(f)$ since then $f(p) \neq 0$. Thus $k[X]_f \subset \mathcal{O}_X(D(f))$.

Suppose that $h \in \mathcal{O}_X(D(f))$, which is a subset of $k(X)$. Let $B = \{g \in k[X] \mid gh \in k[X]\}$. If we can prove that $f^n \in B$ for some n, then we will have that $h \in k[X]_f$, and it follows that $k[X]_f = \mathcal{O}_X(D(f))$. By assumption, if $p \in D(f)$, then $h \in k[X]_{I(p)}$, so there exist functions $a, b \in k[X]$ such that $h = \frac{a}{b}$ with $b(p) \neq 0$. Then $bh = a \in k[X]$ so $b \in B$, and B contains an element not vanishing at p. Thus $Z(B) \subset Z(f)$. We have $f \in \sqrt{B}$ by the nullstellensatz 5) of Theorem 2.26. $\square$

In particular, we have that for any affine variety X,

(2.7) $$k[X] = \mathcal{O}_X(X).$$

2.5. Regular functions and regular maps of quasi-affine varieties

The above proposition shows that if $U = D(f)$ for some $f \in k[X]$, then every element of $\mathcal{O}_X(U)$ has the form $\frac{a}{b}$ where $a, b \in k[X]$ and $b(p) \neq 0$ for $p \in U$. The following example, from page 44 of [**116**], shows that the above desirable property fails in general for an open subset U of an affine variety X.

Example 2.85. There exists an open subset U of an affine variety X such that
$$\mathcal{O}_X(U) \neq \left\{ \frac{f}{g} \mid f, g \in k[X] \text{ and } g(p) \neq 0 \text{ for all } p \in U \right\}.$$

Proof. Write $k[\mathbb{A}^4] = k[x, y, z, w]$ and let $X = Z(xw - yz) \subset \mathbb{A}^4$ and $U = D(y) \cup D(w) = X \setminus Z(y, w)$. Write $k[X] = k[\mathbb{A}^4]/I(X) = k[\overline{x}, \overline{y}, \overline{z}, \overline{w}]$ where $\overline{x}, \overline{y}, \overline{z}, \overline{w}$ are the respective classes of x, y, z, w. Let $h \in \mathcal{O}_X(U)$ be defined by $h = \frac{\overline{x}}{\overline{y}}$ on $D(\overline{y})$ and $h = \frac{\overline{z}}{\overline{w}}$ on $D(\overline{w})$. We have that $\frac{\overline{x}}{\overline{y}} = \frac{\overline{z}}{\overline{w}}$ on $D(\overline{y}) \cap D(\overline{w}) = X - Z(\overline{yw})$, so that h is a well-defined function on U.

Now suppose that $h = \frac{f}{g}$ where $f, g \in k[X]$ and g does not vanish on U. We will derive a contradiction. Let $Z = Z_X(\overline{y}, \overline{w})$. Then Z is a plane in X ($k[Z] = k[x, y, z, w]/(xw - yz, y, w) \cong k[x, z]$). We have that $U = X \setminus Z$. Thus $Z_X(g) \subset Z$. Suppose that g does not vanish on X. Then $\frac{\overline{x}}{\overline{y}} = \frac{f}{g}$ so $\overline{x}g = f\overline{y}$. Now $p = (1, 0, 0, 0) \in X$ so $(\overline{x}g)(p) \neq 0$ but $(f\overline{y})(p) = 0$, which is a contradiction. Thus g is not a unit in $k[X]$. By Theorem 2.71 (X is irreducible) all irreducible components of $Z_X(g)$ have dimension $2 = \dim X - 1$. Since Z is irreducible of dimension 2, we have that $Z_X(g) = Z$. Let $Z' = Z_X(\overline{x}, \overline{z})$, which is another plane. We have that
$$\{(0, 0, 0, 0)\} = Z \cap Z' = Z_X(g) \cap Z'.$$
But (again by Theorem 2.71) a polynomial function vanishes on an algebraic set of dimension 1 on a plane, which is a contradiction since a point has dimension 0. $\square$

Proposition 2.86. *Suppose that X is an affine variety and $p \in X$. Then $\mathcal{O}_{X,p} = k[X]_{I(p)}$.*

Proof. We have that
$$k[X]_{I(p)} = \bigcup_{\{f \in k[X] \mid f(p) \neq 0\}} k[X]_f = \bigcup_{p \in D(f)} \mathcal{O}_X(D(f)) = \bigcup_{p \in U} \mathcal{O}_X(U).$$
The last equality is true since the open sets $D(f)$ are a basis for the topology of X. $\square$

Suppose that X is an affine variety. From the nullstellensatz, 5) of Theorem 2.26, we have that there is a 1-1 correspondence between the points in X and the maximal ideals in $k[X]$. Thus we have the following corollary.

Corollary 2.87. *Suppose that X is an affine variety. Then X is separated; that is, if $p, q \in X$ are distinct points, then $\mathcal{O}_{X,p} \neq \mathcal{O}_{X,q}$.*

We may express

$$(2.8) \quad k[X] = \mathcal{O}_X(X) = \bigcap_{p \in X} \mathcal{O}_{X,p} = \bigcap_{p \in X} k[X]_{I(p)} = \bigcap k[X]_{\mathfrak{m}}$$

where the last intersection is over the maximal ideals $\mathfrak{m}$ of $k[X]$.

Lemma 2.88. *Suppose that $\phi : X \to Y$ is a dominant regular map of affine varieties with induced k-algebra homomorphism $\phi^* : k[Y] \to k[X]$. Suppose that $p \in X$ and $q \in Y$. Then the following are equivalent:*

1) $\phi(p) = q$.
2) *The preimage* $(\phi^*)^{-1}(I(p)) = I(q)$.
3) $\phi^*(\mathcal{O}_{Y,q}) \subset \mathcal{O}_{X,p}$.

In 3), we consider ϕ^* to be its extension to an inclusion $\phi^* : k(Y) \to k(X)$.

Proof. For $p \in X$, we have that the preimage

$$\begin{aligned}(\phi^*)^{-1}(I(p)) &= \{f \in k[Y] \mid \phi^*(f) \in I(p)\} \\ &= \{f \in k[Y] \mid (f \circ \phi)(p) = 0\} = I(\phi(p)).\end{aligned}$$

Since Y is separated (or by the nullstellensatz 5) of Theorem 2.26), $\phi(p) = q$ if and only if $I(\phi(p)) = I(q)$, so we have established the equivalence of 1) and 2). The statement 2) is equivalent to 3) follows since

$$\begin{array}{ll}\phi^*(\mathcal{O}_{Y,q}) \subset \mathcal{O}_{X,q} & \text{if and only if} \quad (\phi^*)^{-1}(I(p)) \cap (k[Y] \setminus I(q)) = \emptyset \\ & \text{if and only if} \quad (\phi^*)^{-1}(I(p)) \subset I(q) \\ & \text{if and only if} \quad (\phi^*)^{-1}(I(p)) = I(q)\end{array}$$

since $(\phi^*)^{-1}(I(p))$ is a maximal ideal of $k[Y]$. $\square$

Lemma 2.89. *Suppose that U is a nontrivial open subset of an affine variety X. Then*

$$\bigcup_{p \in U} \mathcal{O}_{X,p} = k(X).$$

Proof. By Proposition 2.86,

$$\mathcal{O}_{X,p} = k[X]_{I(p)} \text{ for } p \in X.$$

So

$$\bigcup_{p \in U} \mathcal{O}_{X,p} \subset k(X).$$

2.5. Regular functions and regular maps of quasi-affine varieties

Suppose that $h \in k(X)$. Write $h = \frac{f}{g}$ with $f, g \in k[X]$ and $0 \neq g$. Then $Z(g) \cap U \neq U$ since U is Zariski dense in X. So there exists $p \in U \setminus Z(g)$. Thus $g(p) \neq 0$ and so $g \notin I(p)$,

$$\frac{f}{g} \in \mathcal{O}_{X,p} = k[X]_{I(p)},$$

and

$$h \in \mathcal{O}_{X,p} \subset \bigcup_{p \in U} \mathcal{O}_{X,p}.$$

Thus $k(X) \subset \bigcup_{p \in U} \mathcal{O}_{X,p}$. $\square$

We define the rational functions $k(U)$ on a quasi-affine variety U to be the quotient field of $\mathcal{O}_X(U)$, which is equal to $k(X)$, where X is the affine variety containing U as an open subset. We also say that $k(U)$ is the function field of U. We further define $\mathcal{O}_U(V) = \mathcal{O}_X(V)$ for an open subset V of U and

$$\mathcal{O}_{U,p} = \bigcup_{p \in V} \mathcal{O}_U(V)$$

where the union in $k(U)$ is over all open sets V in U containing p. We have that $\mathcal{O}_{U,p} = \mathcal{O}_{X,p}$ for $p \in U$. A quasi-affine variety is separated by Proposition 2.86.

Suppose that U is a quasi-affine variety with field $k(U)$ of rational functions on U. A function $f \in k(U)$ is said to be regular at a point $p \in U$ if $f \in \mathcal{O}_{U,p}$.

Lemma 2.90. *Suppose that U is a quasi-affine variety and $f \in k(U)$. Then*

$$V = \{p \in U \mid f \in \mathcal{O}_{U,p}\}$$

is a dense open subset of U.

Proof. The quasi-affine variety U is an open subset of an affine variety X. Suppose that $p \in V$. Since $f \in \mathcal{O}_{X,p}$, there exist $g, h \in k[X]$ with $h \notin I(p)$ such that $f = \frac{g}{h}$. For $q \in D(h) \cap U$ we have that $h \notin I_X(q)$, and thus $f = \frac{g}{h} \in \mathcal{O}_{X,q} = \mathcal{O}_{U,q}$ and $D(h) \cap U$ is an open neighborhood of p in V.

V is nonempty since we can always write $f = \frac{g}{h}$ for some $g, h \in k[X]$ with $h \neq 0$. We have that $U \cap D(h) \neq \emptyset$ since X is irreducible. Thus $\emptyset \neq D(h) \cap U \subset V$. $\square$

Lemma 2.91. *Suppose that U is a quasi-affine variety and $p \in U$. Let*

$$I_U(p) = \{f \in \mathcal{O}_U(U) \mid f(p) = 0\}.$$

Then $I_U(p)$ is a maximal ideal in $\mathcal{O}_U(U)$ and $\mathcal{O}_U(U)_{I_U(p)} = \mathcal{O}_{U,p}$.

Proof. The quasi-affine variety U is an open subset of an affine variety X, and $\mathcal{O}_U(U) = \mathcal{O}_X(U)$. We have injective restriction maps
$$k[X] \to \mathcal{O}_U(U) \to \mathcal{O}_{X,p} = k[X]_{I_X(p)}.$$
The ring $\mathcal{O}_{X,p}$ is a local ring with maximal ideal $\mathfrak{m} = I_X(p)\mathcal{O}_{X,p}$. We have that $\mathfrak{m} \cap \mathcal{O}_U(U) = I_U(p)$ and $\mathfrak{m} \cap k[X] = I_X(p)$, so we have inclusions
$$\mathcal{O}_{X,p} = k[X]_{I_X(p)} \subset \mathcal{O}_U(U)_{I_U(p)} \subset \mathcal{O}_{X,p}. \qquad \square$$

Definition 2.92. Suppose that Y is a quasi-affine variety. A regular map $\phi : Y \to \mathbb{A}^r$ is a map $\phi = (f_1, \ldots, f_r)$ where $f_1, \ldots, f_r \in \mathcal{O}_Y(Y)$. Suppose that $\phi(Y) \subset Z$ where Z is an open subset of an irreducible closed subset of $\mathbb{A}^r$ (a quasi-affine variety). Then ϕ induces a regular map $\phi : Y \to Z$. A regular map $\phi : Y \to Z$ of quasi-affine varieties is an isomorphism if there is a regular map $\psi : Z \to Y$ such that $\psi \circ \phi = \mathrm{id}_Y$ and $\phi \circ \psi = \mathrm{id}_Z$.

We point out that an affine variety is also a quasi-affine variety.

Proposition 2.93. *Suppose that X is an affine variety and $0 \neq f \in k[X]$. Then the quasi-affine variety $D(f)$ is isomorphic to an affine variety.*

Proof. A choice of coordinate functions on X gives us a closed embedding $X \subset \mathbb{A}^n$. We thus have a surjection $k[x_1, \ldots, x_n] \to k[X]$ with kernel $I(X)$. Let $g \in k[x_1, \ldots, x_n]$ be a function which restricts to f on X. Let J be the ideal in the polynomial ring $k[x_1, \ldots, x_n, x_{n+1}]$ generated by $I(X)$ and $1 - gx_{n+1}$. We will show that J is prime, and if Y is the affine variety $Y = Z(J) \subset \mathbb{A}^{n+1}$, then the projection of $\mathbb{A}^{n+1}$ onto its first n factors induces an isomorphism of Y with $D(f)$.

Using Exercise 1.7 of Section 1.1, we have
$$\begin{aligned} k[x_1, \ldots, x_n, x_{n+1}]/J &\cong (k[x_1, \ldots, x_n]/I(X))[x_{n+1}]/(x_{n+1}f - 1) \\ &\cong k[X][\tfrac{1}{f}] = \mathcal{O}_X(D(f)), \end{aligned}$$
which is an integral domain (it is a subring of the quotient field $k(X)$ of $k[X]$). Thus J is a prime ideal. Now projection onto the first n factors induces a regular map $\phi : Y \to \mathbb{A}^n$. We have that
$$Y = \{(a_1, \ldots, a_n, a_{n+1}) \mid (a_1, \ldots, a_n) \in X \text{ and } g(a_1, \ldots, a_n)a_{n+1} = 1\}.$$
Thus $\phi(Y) = D(f) \subset X$. In particular, we have a regular map $\phi : Y \to D(f)$. Now this map is injective and onto, but to show that it is an isomorphism we have to produce a regular inverse map. Let $\overline{x}_1, \ldots, \overline{x}_n$ be the restrictions of $x_1, \ldots, x_n$ to X. Then $\overline{x}_1, \ldots, \overline{x}_n, \tfrac{1}{f} \in \mathcal{O}_X(D(f))$. Thus $\psi : D(f) \to \mathbb{A}^{n+1}$ defined by $\psi = (\overline{x}_1, \ldots, \overline{x}_n, \tfrac{1}{f})$ is a regular map. The image of ψ is Y. We thus have an induced regular map $\psi : D(f) \to Y$. Composing the maps, we have that $\psi \circ \phi = \mathrm{id}_Y$ and $\phi \circ \psi = \mathrm{id}_{D(f)}$. Thus ϕ is an isomorphism. $\qquad \square$

2.5. Regular functions and regular maps of quasi-affine varieties

We will say that a quasi-affine variety is affine if it is isomorphic to an affine variety.

Definition 2.94. Suppose that $\phi : U \to V$ is a regular map of quasi-affine varieties.

1) The map ϕ is called a closed embedding if there exists a closed subvariety Z of V (an irreducible closed subset) such that $\phi(U) \subset Z$ and the induced regular map $\phi : U \to Z$ is an isomorphism.

2) The map ϕ is called an open embedding if there exists an open subset W of V such that $\phi(U) \subset W$ and the induced regular map $\phi : U \to W$ is an isomorphism.

The above definition of a closed embedding generalizes Definition 2.43.

In general, a map $\phi : X \to Y$ of affine varieties which is continuous (in the Zariski topology) is not regular. This can be seen most easily on $\mathbb{A}^1$. The closed subsets of $\mathbb{A}^1$ are the finite subsets and all of $\mathbb{A}^1$. Thus any bijection (or finite-to-one map) of sets from $\mathbb{A}^1$ to $\mathbb{A}^1$ is continuous.

Proposition 2.95. *Suppose that U, V are quasi-affine varieties and $\phi : U \to V$ is a continuous map. Let $\phi^* : \mathrm{Map}(V, \mathbb{A}^1) \to \mathrm{Map}(U, \mathbb{A}^1)$ be defined by $\phi^*(f) = f \circ \phi$ for $f : V \to \mathbb{A}^1$ a map. Then the following are equivalent:*

1) *ϕ^* maps $\mathcal{O}_V(V)$ into $\mathcal{O}_U(U)$, inducing a k-algebra homomorphism $\phi^* : \mathcal{O}_V(V) \to \mathcal{O}_U(U)$.*

2) *ϕ^* maps $\mathcal{O}_{V,\phi(p)}$ into $\mathcal{O}_{U,p}$ for all $p \in U$, inducing k-algebra homomorphisms $\phi^* : \mathcal{O}_{V,\phi(p)} \to \mathcal{O}_{U,p}$.*

3) *ϕ^* maps $\mathcal{O}_V(W)$ into $\mathcal{O}_U(\phi^{-1}(W))$ for all open subsets W of V, inducing k-algebra homomorphisms $\phi^* : \mathcal{O}_V(W) \to \mathcal{O}_U(\phi^{-1}(W))$.*

Proof. Suppose that 1) holds, $p \in U$, and $\phi(p) = q$. Then $(\phi^*)^{-1}(I_U(p)) = I_V(\phi(p))$. This follows since for $f \in k[V]$,

$$
\begin{array}{ll}
f \in I_V(\phi(p)) & \text{if and only if} \quad f(\phi(p)) = 0 \\
& \text{if and only if} \quad \phi^*(f)(p) = 0 \\
& \text{if and only if} \quad f \in (\phi^*)^{-1}(I_U(p)).
\end{array}
$$

Thus $\mathcal{O}_V(V)_{I_V(\phi(p))} \to \mathcal{O}_U(U)_{I_U(p)}$ and so $\phi^* : \mathcal{O}_{V,\phi(p)} \to \mathcal{O}_{U,p}$ by Lemma 2.91. Thus 2) holds.

Suppose that 2) holds. Then 3) follows from the definition of regular functions. Suppose that 3) holds. Then 1) follows by taking $W = V$. □

Proposition 2.96. *Suppose that U and V are quasi-affine varieties and $\phi : U \to V$ is a map. Then ϕ is a regular map if and only if ϕ is continuous and ϕ^* satisfies the equivalent conditions of Proposition 2.95.*

Proof. Suppose that ϕ is regular. Then V is an open subset of an affine variety Y which is a closed subset of $\mathbb{A}^n$, such that the extension $\tilde\phi : U \to \mathbb{A}^n$ of ϕ has the form $\tilde\phi = (f_1, \ldots, f_n)$ with $f_1, \ldots, f_n \in \mathcal{O}_U(U)$. We will first establish that $\tilde\phi$ is continuous. Let $\{U_i\}$ be a cover of U by affine open subsets U_i. Write $k[\mathbb{A}^n] = k[x_1, \ldots, x_n]$. It suffices to show that $\phi_i = \tilde\phi|U_i$ is continuous for all i. Now $\phi_i^* : k[\mathbb{A}^n] \to \mathcal{O}_U(U) \to k[U_i]$ is a k-algebra homomorphism. Since U_i and $\mathbb{A}^n$ are affine, there exists (by Proposition 2.40) a unique regular map $g_i : U_i \to \mathbb{A}^n$ such that $g_i^* = \phi_i^*$.

Suppose that $p \in U_i$ and $q \in \mathbb{A}^n$. We have that
(2.9)
$$\begin{aligned}
\phi_i^*(I_{\mathbb{A}^n}(q)) \subset I_{U_i}(p) \quad &\text{if and only if} \quad \phi_i^*(f)(p) = 0 \text{ for all } f \in I_{\mathbb{A}^n}(q) \\
&\text{if and only if} \quad f(\phi_i(p)) = 0 \text{ for all } f \in I_{\mathbb{A}^n}(q) \\
&\text{if and only if} \quad I_{\mathbb{A}^n}(q) \subset I_{\mathbb{A}^n}(\phi_i(p)) \\
&\text{if and only if} \quad I_{\mathbb{A}^n}(q) = I_{\mathbb{A}^n}(\phi_i(p)) \text{ since } I_{\mathbb{A}^n}(q) \\
&\qquad \text{and } I_{\mathbb{A}^n}(\phi_i(p)) \text{ are maximal ideals} \\
&\text{if and only if} \quad q = \phi_i(p) \text{ since } \mathbb{A}^n \text{ is separated} \\
&\qquad \text{(by Corollary 2.87).}
\end{aligned}$$

We have that $\phi_i(p) = q$ if and only if $\phi_i^*(I_{\mathbb{A}^n}(q)) \subset I_{U_i}(p)$ and similarly $g_i(p) = q$ if and only if $g_i^*(I_{\mathbb{A}^n}(q)) \subset I_{U_i}(p)$. Thus $\phi_i = g_i$. Since a regular map of affine varieties is continuous, we have that ϕ_i is continuous. Thus $\tilde\phi$ and ϕ are continuous.

Now let $\hat\phi$ be the extension $\hat\phi : U \to Y$ of ϕ. Then $\hat\phi^* : \mathcal{O}_Y(Y) \to \mathcal{O}_U(U)$ is a k-algebra homomorphism since $\mathcal{O}_Y(Y) = k[Y] = k[\overline{x}_1, \ldots, \overline{x}_n]$ and $\hat\phi^*(\overline{x}_i) = f_i$ for all i. By 1) implies 3) of Proposition 2.95, applied to $\hat\phi : U \to Y$, we have that
$$\phi^* = \hat\phi^* : \mathcal{O}_Y(V) = \mathcal{O}_V(V) \to \mathcal{O}_U(\phi^{-1}(V)) = \mathcal{O}_U(U)$$
is a k-algebra homomorphism. Thus $\phi : U \to V$ satisfies condition 1) of Proposition 2.95.

Now suppose that $\phi : U \to V$ is continuous and ϕ^* satisfies the equivalent conditions of Proposition 2.95. Then the extension $\tilde\phi : U \to \mathbb{A}^n$ satisfies $\tilde\phi^* : k[\mathbb{A}^n] \to \mathcal{O}_U(U)$ is a k-algebra homomorphism. Since $\tilde\phi = (\tilde\phi^*(x_1), \ldots, \tilde\phi^*(x_n))$, we have that ϕ is a regular map. $\square$

Suppose that R is an integral domain with quotient field L and I is an ideal in R. The ideal transform of I in R is
$$S(I; R) = \{f \in L \mid fI^n \subset R \text{ for some } n \in \mathbb{N}\} = \bigcup_{i=0}^{\infty} R :_L I^n = R :_L I^\infty.$$

We have the following algebraic interpretation of regular functions on a quasi-affine variety.

2.5. Regular functions and regular maps of quasi-affine varieties 57

Lemma 2.97. *Suppose that X is an affine variety and $I \subset k[X]$ is an ideal. Let $U = X \setminus Z(I)$. Then $\mathcal{O}_U(U) = S(I; k[X])$.*

Proof. Write $I = (g_1, \ldots, g_r)$ with $g_1, \ldots, g_r \in k[X]$. Suppose that f is in the quotient field of $k[X]$. We have that $f \in S(I; k[X])$ if and only if $I^m f \in k[X]$ for some $m > 0$, which holds if and only if $g_i^n f \in k[X]$ for some $n > 0$ and for all i with $1 \leq i \leq r$. This condition is equivalent to the statement that $f \in k[X]_{g_i}$ for $1 \leq i \leq r$, which is equivalent to the statement that $f \in \mathcal{O}_X(U)$ since $\mathcal{O}_X(U) = \bigcap_{1 \leq i \leq r} k[X]_{g_i}$. □

Example 2.98. There are examples of quasi-affine varieties U such that $\mathcal{O}_U(U)$ is not a finitely generated k-algebra.

Nagata gives examples in [122] showing this and discusses when $\mathcal{O}_U(U)$ is a finitely generated k-algebra. The simplest example of a quasi-affine variety U such that $\mathcal{O}_U(U)$ is not a finitely generated k-algebra that he presents is constructed from an example of Rees [125]. Nagata explains Rees's construction on page 48 of [122]. The existence of the example then follows from properties (1) and (2) of Rees's example, given on page 49 of [122], Proposition 4 on page 39 of [122], and the above Lemma 2.97.

Exercise 2.99. Suppose that U in an open subset an affine variety X and $f_1, \ldots, f_n \in k[X]$ are such that $U = D(f_1) \cup \cdots \cup D(f_n)$. Show that

$$\mathcal{O}_X(U) = k[X]_{f_1} \cap \cdots \cap k[X]_{f_n}$$

where the intersection is in the quotient field $k(X)$ of X.

Exercise 2.100. Let $U = \mathbb{A}^1 \setminus \{0\}$.

    a) Compute $\mathcal{O}_{\mathbb{A}^1}(U)$, the regular functions on the quasi-affine variety U.

    b) Is U (isomorphic) to an affine variety?

Exercise 2.101. Let $U = \mathbb{A}^2 \setminus \{(0,0)\}$.

    a) Compute $\mathcal{O}_{\mathbb{A}^2}(U)$, the regular functions on the quasi-affine variety U.

    b) Is U (isomorphic) to an affine variety?

Exercise 2.102. At what points of the subvariety $X = Z(x^2 + y^2 - 1)$ of $\mathbb{A}^2$ with regular functions $k[X] = k[x,y]/(x^2 + y^2 - 1) = k[\overline{x}, \overline{y}]$ is the rational function $\frac{1-\overline{y}}{\overline{x}}$ regular? Assume that the characteristic of k is $\neq 2$.

Exercise 2.103. Suppose that X is an affine variety and U is an open subset of X (so that U is a quasi-affine variety). Suppose that $p \in U$. Show that $\mathcal{O}_{U,p} = \mathcal{O}_{X,p}$.

2.6. Rational maps of affine varieties

In this section, we define a rational map of an affine variety X. We begin by noting some properties of restriction and extension of regular maps.

Lemma 2.104. *Suppose that X and Y are affine varieties and $U \subset V \subset X$ are nonempty open subsets. Suppose that $\phi, \psi : V \to Y$ are regular maps such that $\phi|U = \psi|U$. Then $\phi = \psi$.*

Proof. There exist a closed embedding $Y \subset \mathbb{A}^n$ and $f_1, \ldots, f_n \in \mathcal{O}_X(V)$, $g_1, \ldots, g_n \in \mathcal{O}_X(V)$ such that $\phi = (f_1, \ldots, f_n)$ and $\psi = (g_1, \ldots, g_n)$. Since $\phi|U = \psi|U$, we have that $f_i|U = g_i|U$ for $1 \le i \le n$. Thus $f_i = g_i$ for $1 \le i \le n$ since the restriction map $\mathcal{O}_X(V) \to \mathcal{O}_X(U)$ is injective. $\square$

Lemma 2.105. *Suppose that X, Y are affine varieties, $U \subset X$ is a nontrivial open subset, and $\phi : U \to Y$ is a regular map. Then there exists a largest open subset $W(\phi)$ of X such that there exists a regular map $\psi : W(\phi) \to X$ with the property that $\psi|U = \phi$. The map ψ is uniquely determined.*

Proof. There exist a closed embedding $Y \subset \mathbb{A}^n$ and $f_1, \ldots, f_n \in \mathcal{O}_X(U)$ such that $\phi = (f_1, \ldots, f_n) : U \to Y$. Let $V_i = \{p \in X \mid f_i \in \mathcal{O}_{X,p}\}$. The V_i are nontrivial open subsets of X by Lemma 2.90. Then $W(\phi) = \bigcap_{i=1}^n V_i$ is the largest open subset of X on which ϕ extends to a regular map ψ. The map ψ is uniquely determined by Lemma 2.104. $\square$

We can thus define a rational map ϕ between affine varieties X and Y to be a regular map on some nonempty open subset U of X to Y. By Lemma 2.105, ϕ has a unique extension as a regular map to a largest open subset $W(\phi)$ of X, and if ϕ and ψ are rational maps from X to Y which are regular on respective nonempty open subsets U and V such that ϕ and ψ agree on the intersection $U \cap V$, then $\phi = \psi$. It is usual to write $\phi : X \dashrightarrow Y$ for a rational map, to emphasize the fact that ϕ may not be regular everywhere. The rational maps $\phi : X \dashrightarrow \mathbb{A}^1$ can be identified with the rational functions $k(X)$.

We now formulate the concept of a rational map as a statement in algebra.

Definition 2.106. Suppose that X is an affine variety. A rational map $\phi : X \dashrightarrow \mathbb{A}^m$ is an m-tuple $\phi = (f_1, \ldots, f_m)$ with $f_1, \ldots, f_m \in k(X)$. Such a map ϕ induces a k-algebra homomorphism $\phi^* : k[\mathbb{A}^m] \to k(X)$ by $\phi^*(t_i) = f_i$ for $1 \le i \le m$ where $k[\mathbb{A}^m] = k[t_1, \ldots, t_m]$.

Suppose that Y is an affine variety which is a closed subvariety of an affine space $\mathbb{A}^m$. A rational map $\phi : X \dashrightarrow Y$ is a rational map $\phi : X \dashrightarrow \mathbb{A}^m$ such that $I(Y)$ is contained in the kernel of $\phi^* : k[\mathbb{A}^m] \to k(X)$. This induces a k-algebra homomorphism $\phi^* : k[Y] \to k(X)$.

2.6. Rational maps of affine varieties

In particular, a rational map $\phi : X \dashrightarrow Y$ of affine varieties can be understood completely by the corresponding k-algebra homomorphism $\phi^* : k[Y] \to k(X)$.

Suppose that $\phi = (f_1, \ldots, f_m) : X \dashrightarrow Y$ is a rational map of affine varieties. The set of points of X on which $f_1, \ldots, f_m$ are all regular is a nonempty open set $W(\phi)$ by Lemma 2.90. This is the set of points on which ϕ is regular (Lemma 2.105). If $U \subset W(\phi)$ is a nonempty open subset, we have that $\phi^* : k[Y] \to \mathcal{O}_X(U)$. A rational map ϕ is completely determined by its restriction to any open subset U of X. In particular, we have that if Y is a closed subvariety of $\mathbb{A}^m$ and $\phi : X \dashrightarrow \mathbb{A}^m$ is a rational map, then $\phi : X \dashrightarrow Y$ if and only if $\phi(W(\phi)) \subset Y$.

A rational map $\phi : X \dashrightarrow Y$ is called dominant if $\phi(U)$ is dense in Y when U is an open subset of X on which ϕ is a regular map.

Lemma 2.107. *Suppose that $\phi : X \dashrightarrow Y$ is a dominant rational map of affine varieties. Then ϕ induces an injective k-algebra homomorphism $\phi^* : k(Y) \to k(X)$ of function fields.*

Proof. There exists a nonempty open subset W of X on which ϕ is a regular map. W contains an open set $D(f)$ for some $f \in k[X]$ by Lemma 2.83. The open set $D(f)$ is affine with $k[D(f)] = k[X]_f$ by Propositions 2.93 and 2.84. Then we have an induced k-algebra homomorphism $\phi^* : k[Y] \to k[X]_f$ which is injective by Corollary 2.32. We thus have an induced k-algebra homomorphism of quotient fields. $\square$

If ϕ^* induces a well-defined injective homomorphism $\phi^* : k(Y) \to k(X)$, and if U is an affine open subset of X on which ϕ is regular, we have that $\phi^* : k[Y] \to k[U]$ is injective and thus $\phi(U)$ is dense in Y by Corollary 2.32. Thus the rational map $\phi : X \dashrightarrow Y$ is dominant.

Proposition 2.108. *Suppose that X and Y are affine varieties and $\Lambda : k(Y) \to k(X)$ is an injective k-algebra homomorphism. Then there is a unique (dominant) rational map $\phi : X \dashrightarrow Y$ such that $\phi^* = \Lambda$.*

Proof. Let $t_1, \ldots, t_m$ be coordinate functions on Y such that
$$k[Y] = k[t_1, \ldots, t_m].$$
Write $\Lambda(t_i) = \frac{f_i}{g_i}$ with $f_i, g_i \in k[X]$ (and $g_i \neq 0$) for $1 \leq i \leq m$. Let $g = g_1 g_2 \cdots g_m$. Then Λ induces a k-algebra homomorphism $\Lambda : k[Y] \to k[X]_g$. Now $k[X]_g = k[D(g)]$ where $D(g)$ is the affine open subset of X, by Proposition 2.84. By Proposition 2.40, there is a unique regular map $\phi : D(g) \to Y$ such that $\phi^* = \Lambda$. Since a rational map of varieties is uniquely determined by the induced regular map on a nontrivial open subset, there is a unique rational map $\phi : X \dashrightarrow Y$ inducing Λ. $\square$

Suppose that $\alpha : X \dashrightarrow Y$ is a dominant rational map and $\beta : Y \dashrightarrow Z$ is a rational map. Then $\beta \circ \alpha : X \dashrightarrow Z$ is a rational map. We see this as follows. Since α is dominant, α^* induces a homomorphism $k(Y) \to k(X)$ so the composition $\alpha^* \beta^* : k[Z] \to k(X)$ is well-defined.

There exist open sets U of X on which α is regular and V of Y on which β is regular. Since $\alpha(U)$ is dense in Y, we have that $\alpha(U) \cap V \neq \emptyset$. Thus the nonempty open set $(\alpha|U)^{-1}(V)$ is contained in the open set $W(\beta \circ \alpha)$ where $\beta \circ \alpha$ is regular.

Definition 2.109. A dominant rational map $\phi : X \dashrightarrow Y$ of affine varieties is birational if there is a dominant rational map $\psi : Y \dashrightarrow X$ such that $\psi \circ \phi = \mathrm{id}_X$ and $\phi \circ \psi = \mathrm{id}_Y$ (that is, $\phi^* \circ \psi^* : k[X] \to k(X)$ is equal to the inclusion id_X^* and $\psi^* \circ \phi^* : k[Y] \to k(Y)$ is equal to the inclusion id_Y^*).

Proposition 2.110. *A rational map $\phi : X \dashrightarrow Y$ is birational if and only if $\phi^* : k(Y) \to k(X)$ is a k-algebra isomorphism.*

Theorem 2.111. *A dominant rational map $\phi : X \dashrightarrow Y$ of affine varieties is birational if and only if there exist nonempty affine open subsets U of X and V of Y such that $\phi : U \to V$ is a regular map which is an isomorphism.*

Proof. Suppose that there exist open sets U of X and V of Y such that $\phi : U \to V$ is a regular map which is an isomorphism. Then there exists a regular map $\psi : V \to U$ such that $\psi \circ \phi = \mathrm{id}_U$ and $\phi \circ \psi = \mathrm{id}_V$. Since $\mathrm{id}_{k[V]} = (\phi \circ \psi)^* = \psi^* \circ \phi^*$ and $\mathrm{id}_{k[U]} = (\psi \circ \phi)^* = \phi^* \circ \psi^*$, we have that $\phi^* : k[V] \to k[U]$ is an isomorphism of k-algebras, so that ϕ^* induces an isomorphism of their quotient fields, which are respectively $k(Y)$ and $k(X)$.

Suppose that $\phi^* : k(Y) \to k(X)$ is an isomorphism. Then there exists a unique rational map $\psi : Y \dashrightarrow X$ such that ψ^* is the inverse of ϕ^* by Proposition 2.108. Let $t_1, \ldots, t_m$ be coordinate functions on Y and let $s_1, \ldots, s_n$ be coordinate functions on X. There are functions $a_1, \ldots, a_m, f \in k[X]$ and $b_1, \ldots, b_n, g \in k[Y]$, with $f, g \neq 0$, such that $\phi^*(t_i) = \frac{a_i}{f}$ for $1 \leq i \leq m$ and $\psi^*(s_j) = \frac{b_j}{g}$ for $1 \leq i \leq n$. Thus $\phi^*(k[Y]) \subset k[X]_f$ and $\psi^*(k[X]) \subset k[Y]_g$. Since ψ^* is the inverse of ϕ^*, we have that $\phi^*(k[Y]_{g\psi^*(f)}) \subset k[X]_{f\phi^*(g)}$ and $\psi^*(k[X]_{f\phi^*(g)}) \subset k[Y]_{g\psi^*(f)}$. Thus $\phi^* : k[Y]_{g\psi^*(f)} \to k[X]_{f\phi^*(g)}$ is an isomorphism, and $\phi : D(f\phi^*(g)) \to D(g\psi^*(f))$ is a regular map which is an isomorphism by Proposition 2.42. $\square$

Two affine varieties X and Y are said to be birationally equivalent if there exists a birational map $\phi : X \dashrightarrow Y$.

Proposition 2.112. *Every affine variety X is birationally equivalent to a hypersurface $Z(g) \subset \mathbb{A}^n$.*

2.6. Rational maps of affine varieties

Proof. Let $\dim(X) = r$. Then the quotient field $k(X)$ is a finite separable extension of a rational function field $L = k(x_1, \ldots, x_r)$ (by Theorem 1.14). By the theorem of the primitive element (Theorem 1.16), $k(X) \cong L[t]/(f(t))$ for some irreducible monic polynomial $f(t) \in L[t]$. Multiplying $f(t)$ by an appropriate element a of $k[x_1, \ldots, x_r]$, we obtain a primitive polynomial $g = af(t) \in k[x_1, \ldots, x_r, t]$, which is thus irreducible. The quotient field of $k[x_1, \ldots, x_r, t]/(g)$ is isomorphic to $k(X)$. Thus X is birationally equivalent to $Z(g) \subset \mathbb{A}^{r+1}$ by Proposition 2.110. $\square$

Exercise 2.113. Prove Proposition 2.110.

Exercise 2.114. Consider the regular map $\phi : \mathbb{A}^2 \to \mathbb{A}^2$ defined by

$$\phi(a_1, a_2) = (a_1, a_1 a_2) \quad \text{for } (a_1, a_2) \in \mathbb{A}^2.$$

a) Show that ϕ is dominant.
b) Show that ϕ is birational.
c) Show that ϕ is not an isomorphism.
d) Show that ϕ is not finite.

Exercise 2.115. Is a composition of rational maps always a rational map?

Chapter 3

Projective Varieties

In this chapter we define projective and quasi-projective varieties, their regular functions and regular maps. Recall that throughout this book, k will be a fixed algebraically closed field.

We develop a correspondence between the commutative algebra of standard graded k-algebras which are domains (or reduced) and the geometry of projective varieties (or projective algebraic sets) in Sections 3.1–3.2. In Section 3.4, we define the regular functions on a projective variety.

3.1. Standard graded algebras

In this section we discuss algebraic methods necessary for our study of projective space and projective varieties. We begin with some general definitions of graded rings and modules. Some references on graded algebras and modules are [**161**, Chapter VII] and [**28**].

A graded ring is a ring R with a decomposition $R = \bigoplus_{i \in \mathbb{Z}} R_i$ such that $R_i R_j \subset R_{i+j}$ for all $i, j \in \mathbb{Z}$.

A graded R-module is an R-module M together with a decomposition $M = \bigoplus_{i \in \mathbb{Z}} M_i$ such that $R_i M_j \subset M_{i+j}$ for all $i, j \in \mathbb{Z}$. The elements $x \in M_i$ are called homogeneous of degree i. The degree of x is denoted by $\deg x$. Every element $f \in M$ has a unique expression as a sum with finitely many nonzero terms $f = F_0 + F_1 + \cdots$ where the $F_i \in M_i$ are homogeneous of degree i, with $F_i = 0$ for all i sufficiently large.

If M is a graded R-module and $n \in \mathbb{Z}$, then the twisted R-module $M(n)$ is M, with the grading $M(n)_i = M_{n+i}$ for $i \in \mathbb{Z}$.

63

Suppose that R is a graded ring and M is a graded R-module. Suppose that $F \in R$ is homogeneous. Then the localization M_F is graded by

$$\deg\left(\frac{G}{F^n}\right) = \deg(G) - n\deg(F)$$

for $G \in M$ homogeneous. We define

$$M_{(F)} = \left\{ h = \frac{G}{F^n} \in M_F \mid G \text{ is homogeneous and } \deg(h) = 0 \right\}.$$

If $\mathfrak{p}$ is a homogeneous prime ideal in R, then we define

(3.1) $\qquad M_{(\mathfrak{p})} = \{\text{elements of degree 0 in } V^{-1}M\}$

where V is the multiplicatively closed subset of homogeneous elements of R not in $\mathfrak{p}$. The ring $R_{(\mathfrak{p})}$ is a local ring with maximal ideal $\mathfrak{p}_{(\mathfrak{p})}$. In particular, if R is a domain, so that the zero ideal $\mathfrak{q} = (0)$ is a prime ideal, then $R_{(\mathfrak{q})}$ is a field.

Suppose that R is a graded ring and d is a positive integer. The d-th Veronese ring of R is

(3.2) $\qquad\qquad R^{(d)} = \bigoplus_{i \in \mathbb{Z}} R_{id}.$

It is a graded ring with $R_i^{(d)} = R_{id}$ for $i \in \mathbb{Z}$.

Suppose that $A = \bigoplus_{i \in \mathbb{Z}} A_i$ and $B = \bigoplus_{j \in \mathbb{Z}} B_j$ are graded rings. A ring homomorphism $\phi : A \to B$ is said to be graded of degree s if $\phi(A_i) \subset B_{is}$ for all i. The inclusion of the Veronese ring $R^{(d)}$ into R is an example of a graded ring homomorphism of degree d.

Suppose that $M = \bigoplus_{i \in \mathbb{Z}} M_i$ and $N = \bigoplus_{j \in \mathbb{Z}} N_j$ are graded R-modules. An R-module homomorphism $\lambda : M \to N$ is graded of degree s if $\lambda(M_i) \subset N_{i+s}$ for all i.

The following lemma follows from [**153**, Theorem on page 151].

Lemma 3.1. *Suppose that R is a graded ring and $I \subset R$ is an ideal. Then the following are equivalent:*

1) *Suppose that $f \in I$ and $f = \sum F_i$ where F_i is homogeneous of degree i. Then $F_i \in I$ for all i.*
2) $I = \bigoplus_{i=-\infty}^{\infty}(I \cap R_i).$

An ideal satisfying the conditions of Lemma 3.1 is called a homogeneous ideal. An ideal I is homogeneous if and only if I has a homogeneous set of generators.

Lemma 3.2. *Suppose that P is a homogeneous ideal in a graded ring R. Then P is a prime ideal if and only if it has the property that whenever $F, G \in R$ are homogeneous and $FG \in P$, then F or G is in P.*

3.1. Standard graded algebras

Proof. Suppose that P is a homogeneous ideal in R and P has the property that whenever $F, G \in R$ are homogeneous and $FG \in P$, then F or G is in P. Let $f, g \in R$ and suppose that $f \notin P$ and $g \notin P$. We will show that $fg \notin P$. Let $f = f_r + f_{r+1} + \cdots$ and $g = g_s + g_{s+1} + \cdots$ be the decompositions of f and g into their homogeneous components. Let f_{r+a} and g_{s+b} be the first homogeneous components of f and g, respectively, which does not belong to P. Then $f_{r+a}, g_{s+b} \notin P$, and so

$$[f - (f_r + f_{r+1} + \cdots + f_{r+a-1})][g - (g_s + \cdots + g_{s+b-1})] \notin P$$

since P is homogeneous. Since $f_r + f_{r+1} + \cdots + f_{r+a-1}$ and $g_s + g_{s+1} + \cdots + g_{s+b-1}$ are in P, we have that $fg \notin P$. $\square$

The following lemma follows from [**161**, Theorem on page 152] and [**161**, Theorem 9 on page 153].

Lemma 3.3. *Suppose that R is a graded ring and I and J are homogeneous ideals in R. Then:*

1) *$I + J$ is a homogeneous ideal.*

2) *IJ is a homogeneous ideal.*

3) *$I \cap J$ is a homogeneous ideal.*

4) *$\sqrt{I}$ is a homogeneous ideal.*

5) *$I : J$ is homogeneous.*

6) *If I admits a primary decomposition, then I admits a homogeneous primary decomposition.*

We now consider the case of graded rings which are quotients of polynomial rings over a field. A thorough development of this material is in [**161**, Section 2, Chapter VII].

Let T be the polynomial ring $T = K[x_0, x_1, \ldots, x_n]$ over a field K. An element $f \in T$ is called homogeneous of (total) degree d if it is a K-linear combination of monomials of degree d. The degree of the monomial $x_0^{i_0} x_1^{i_1} \cdots x_n^{i_n}$ is $i_0 + i_1 + \cdots + i_n$. Let T_d be the K-vector space of all homogeneous polynomials of degree d (we include 0). Every polynomial $f \in T$ has a unique expression as a sum with finitely many nonzero terms $f = F_0 + F_1 + \cdots$ where the F_i are homogeneous of degree i, with $F_i = 0$ for all i sufficiently large. This is equivalent to the statement that $T = \bigoplus_{i=0}^{\infty} T_i$, where $T_0 = K$ and T_i is the K-vector space of homogeneous polynomials of degree i; that is, T is a graded ring. Since $T = K[T_1] = T_0[T_1]$ is generated by elements of degree 1 as a $T_0 = K$-algebra, we say that T is a standard graded K-algebra.

Suppose that $U = \bigoplus_{i=0}^{\infty} U_i$ is a homogeneous ideal in T. Then $S = T/U \cong \bigoplus_{i=0}^{\infty} S_i$ where $S_i = T_i/U_i$. The ring S is a standard graded K-algebra (elements of S_i have degree i and S is generated by S_1 as an $S_0 = K$-algebra). Every $f \in S$ has a unique expression as a sum with finitely many nonzero terms $f = F_0 + F_1 + \cdots$ where the $F_i \in S_i$ are homogeneous of degree i, with $F_i = 0$ for all i sufficiently large. The conclusions of Lemma 3.1 hold for ideals in S, so we can speak of homogeneous ideals in S.

Suppose that U is an ideal in the standard graded polynomial ring T and $S = T/U$. Then $S = K[\bar{x}_0, \ldots, \bar{x}_n]$ is the K-algebra generated by the classes $\bar{x}_i$ of the x_i in S. We can extend the grading of T to S if and only if U is a homogeneous ideal. In this case, we have that $S = \bigoplus_{i=0}^{\infty} S_i$ where S_i is the K-vector space generated by the monomials $\prod \bar{x}_j^{a_j}$ where $\sum a_j = i$. If U is not homogeneous, the concept of degree is not well-defined on S. As an example, if $U = (x - y^2) \subset K[x,y]$, then $\bar{x} = \bar{y}^2$ would have to have both degree 1 and degree 2 in $K[x,y]/U$.

Suppose that I is a homogeneous ideal in S and $I = \bigcap Q_j$ is a primary decomposition of I by homogeneous ideals. Let $\mathfrak{m}$ be the homogeneous maximal ideal $\mathfrak{m} = \bigoplus_{i>0} S_i$. The saturation of I is the homogeneous ideal I^{sat} which is the intersection of all the primary components of I which are not $\mathfrak{m}$-primary. Properties of the saturation are derived in [**161**, Section 2, Chapter VII], especially in Lemmas 4 and 5. We have that $I_n = (I^{\text{sat}})_n$ for $n \gg 0$ and $I^{\text{sat}} = I : \mathfrak{m}^{\infty} = \bigcup_{i=1}^{\infty} I : \mathfrak{m}^i$.

We have an expression of our standard graded K-algebra S as $S = K[\bar{x}_0, \ldots, \bar{x}_n]$ where $\bar{x}_0, \ldots, \bar{x}_n$ are homogeneous of degree 1. Suppose that $I \subset S$ is a homogeneous ideal. Then

$$S_{(\bar{x}_i)} = K\left[\frac{\bar{x}_0}{\bar{x}_i}, \ldots, \frac{\bar{x}_n}{\bar{x}_i}\right] \subset S_{\bar{x}_i}$$

for $0 \le i \le n$ and

$$I_{(\bar{x}_i)} = \left\{G\left(\frac{\bar{x}_0}{\bar{x}_i}, \ldots, \frac{\bar{x}_n}{\bar{x}_i}\right) \mid G \in I\right\}.$$

The following lemma will be useful.

Lemma 3.4. *Suppose that I and J are homogeneous ideals in S. Then $I_{(\bar{x}_i)} = J_{(\bar{x}_i)}$ for $0 \le i \le n$ if and only if*

$$I^{\text{sat}} = J^{\text{sat}}.$$

Proof. Suppose that $I_{(\bar{x}_i)} = J_{(\bar{x}_i)}$ for $0 \le i \le n$. We then have equality of localizations

$$I_{\bar{x}_i} = I_{(\bar{x}_i)} S = J_{(\bar{x}_i)} S = J_{\bar{x}_i}$$

for $0 \leq i \leq n$. In particular, there exists an integer t such that $\overline{x}_i^t J \subset I$ for all i, so that $J \subset I^{\text{sat}}$. Similarly, $I \subset J^{\text{sat}}$.

Suppose that $I^{\text{sat}} = J^{\text{sat}}$. Since $\overline{x}_i \in \mathfrak{m}$, we have by Proposition 1.43 that $I_{\overline{x}_i} = J_{\overline{x}_i}$ for all i. Thus $I_{(\overline{x}_i)} = J_{(\overline{x}_i)}$ for all i. $\square$

Exercise 3.5. Let K be a field and $S = K[\overline{x}_0, \ldots, \overline{x}_n]$ be a standard graded K-algebra. Show that if $\overline{x}_i$ is nilpotent in S, then $S_{(\overline{x}_i)} = 0$.

Exercise 3.6. Let K be a field and suppose that $A = \bigoplus_{i=0}^\infty A_i$ is a graded ring, which is a finitely generated $A_0 = K$-algebra. Show that there exists $d \in \mathbb{Z}_+$ such that $A^{(d)}$ is a standard graded K-algebra (generated by $A_1^{(d)}$).

Exercise 3.7. Let $A = \bigoplus_{i=0}^\infty A_i$ be a graded ring and X be the set of all homogeneous prime ideals in A which do not contain the ideal $A_+ = \bigoplus_{i>0} A_i$. For each subset E of homogeneous elements of A, let $V(E)$ be the set of all elements of X which contain E. Prove that:

a) If I is the homogeneous ideal generated by a set of homogeneous elements of E, then $V(E) = V(I) = V(\sqrt{I})$.

b) $V(0) = X$ and $V(1) = V(A_+) = \emptyset$.

c) If $\{E_s\}_{s \in S}$ is any family of subsets of homogeneous elements of A, then
$$V\left(\bigcup_{s \in S} E_s\right) = \bigcap_{s \in S} V(E_s).$$

d) $V(I \cap J) = V(IJ) = V(I) \cup V(J)$ for any homogeneous ideals I, J of A.

This exercise shows that the sets $V(E)$ satisfy the axioms for closed sets in a topological space. We call this topology on X the Zariski topology and write $\text{Proj}(A)$ for this topological space.

3.2. Projective varieties

In this section we define projective varieties and projective algebraic sets. [**161**, Sections 4 and 5 of Chapter VII] is a good reference to the algebra of this section. As usual, we assume throughout this chapter that k is an algebraically closed field.

Define an equivalence relation $\sim$ on $k^{n+1} \setminus \{(0, \ldots, 0)\}$ by
$$(a_0, a_1, \ldots, a_n) \sim (b_0, b_1, \ldots, b_n)$$
if there exists $0 \neq \lambda \in k$ such that $(b_0, b_1, \ldots, b_n) = (\lambda a_0, \lambda a_1, \ldots, \lambda a_n)$. Projective space $\mathbb{P}_k^n$ over k is defined as
$$\mathbb{P}_k^n = \mathbb{P}^n = \left(k^{n+1} \setminus \{(0, \ldots, 0)\}\right)/\sim.$$

The equivalence class of an element $(a_0, a_1, \ldots, a_n)$ in $k^{n+1} \setminus \{(0, \ldots, 0)\}$ is denoted by $(a_0 : a_1 : \ldots : a_n)$.

We define the homogeneous coordinate ring of $\mathbb{P}_k^n$ to be the standard graded polynomial ring $S(\mathbb{P}^n) = k[x_0, x_1, \ldots, x_n]$. We think of the x_i as "homogeneous coordinates" on $\mathbb{P}^n$.

Suppose that $F \in k[x_0, \ldots, x_n]$ is homogeneous of degree d, $\lambda \in k$, and $a_0, \ldots, a_n \in k$. Then

$$(3.3) \qquad F(\lambda a_0, \ldots, \lambda a_n) = \lambda^d F(a_0, \ldots, a_n).$$

Thus we see that if U is a set of homogeneous polynomials in $S(\mathbb{P}^n)$ (possibly of different degrees), then the set

$$Z(U) = \{(a_0 : \ldots : a_n) \in \mathbb{P}^n \mid F(a_0, \ldots, a_n) = 0 \text{ for all } F \in U\}$$

is a well-defined subset of $\mathbb{P}^n$.

If I is a homogeneous ideal in $S(\mathbb{P}^n)$, we define

$$Z(I) = \{(a_0 : \ldots : a_n) \in \mathbb{P}^n \mid F(a_0, \ldots, a_n) = 0 \text{ for all } F \in U\}$$

where U is a set of homogeneous generators of I. This is a well-defined set (independent of the choice of homogeneous generators U of I).

Definition 3.8. A subset Y of $\mathbb{P}^n$ is a projective algebraic set if there exists a set U of homogeneous elements of $S(\mathbb{P}^n)$ such that $Y = Z(U)$.

Proposition 3.9. *Suppose that I_1, I_2, $\{I_\alpha\}_{\alpha \in \Lambda}$ are homogeneous ideals in $S(\mathbb{P}^n)$. Then:*

1) $Z(I_1 I_2) = Z(I_1) \cup Z(I_2)$.
2) $Z(\sum_{\alpha \in \Lambda} I_\alpha) = \bigcap_{\alpha \in \Lambda} Z(I_\alpha)$.
3) $Z(S(\mathbb{P}^n)) = \emptyset$.
4) $\mathbb{P}^n = Z(0)$.

Proposition 3.9 tells us that:

1. The union of two algebraic sets is an algebraic set.
2. The intersection of any family of algebraic sets is an algebraic set.
3. $\emptyset$ and $\mathbb{P}^n$ are algebraic sets.

We thus have a topology on $\mathbb{P}^n$, defined by taking the closed sets to be the algebraic sets. The open sets are the complements of algebraic sets in $\mathbb{P}^n$. This topology is called the Zariski topology.

If X is an algebraic set in $\mathbb{P}^n$, then the Zariski topology on X is the subspace topology.

3.2. Projective varieties

Definition 3.10. A projective algebraic variety is an irreducible closed subset of $\mathbb{P}^n$. A quasi-projective variety is an open subset of a projective variety. A projective algebraic set is a closed subset of $\mathbb{P}^n$. A quasi-projective algebraic set is an open subset of a closed subset of $\mathbb{P}^n$. A subset X of a variety Y is called a subvariety of Y if X is a closed irreducible subset of Y.

Given a subset Y of $\mathbb{P}^n$, the ideal $I(Y)$ of Y in $S(\mathbb{P}^n)$ is the ideal in $S(\mathbb{P}^n)$ generated by the set

$$U = \{F \in S(\mathbb{P}^n) \mid F \text{ is homogeneous and } F(p) = 0 \text{ for all } p \in Y\}.$$

Proposition 3.11. *Suppose that $\mathfrak{a}$ is a homogeneous ideal in the standard graded polynomial ring $T = k[x_0, \ldots, x_n] = \bigoplus_{i=0}^{\infty} T_i$. Then the following are equivalent:*

1) $Z(\mathfrak{a}) = \emptyset$.
2) $\sqrt{\mathfrak{a}}$ *is either T or the ideal* $T_+ = \bigoplus_{d>0} T_d$.
3) $T_d \subset \mathfrak{a}$ *for some $d > 0$.*
4) $\mathfrak{a}^{\text{sat}} = T$.

Proof. We will prove the essential implication that $Z(\mathfrak{a}) = \emptyset$ implies $T_d \subset \mathfrak{a}$ for some $d > 0$. Suppose that $F_1, \ldots, F_r$ are homogeneous generators of $\mathfrak{a}$. Since $Z(\mathfrak{a}) = \emptyset$, we have that the polynomials $F_i(1, y_1, \ldots, y_n)$ have no common root. By the nullstellensatz in $k[y_1, \ldots, y_n]$ (Theorem 2.5), there exist polynomials $G_i(y_1, \ldots, y_n)$ such that

$$\sum_i G_i(y_1, \ldots, y_n) F_i(1, y_1, \ldots, y_n) = 1.$$

Substituting $y_i = \frac{x_i}{x_0}$ and multiplying by $x_0^{l_0}$ with l_0 sufficiently large, we obtain that $x_0^{l_0} \in \mathfrak{a}$. Similarly, we have $x_i^{l_i} \in \mathfrak{a}$ for $0 \leq i \leq n$. Let $l = \max\{l_0, \ldots, l_n\}$ and $d = (l-1)(n+1) + 1$. Then $T_d \subset \mathfrak{a}$. □

Theorem 3.12 (Homogeneous nullstellensatz). *Let $\mathfrak{a}$ be a homogeneous ideal in the polynomial ring $T = k[x_0, \ldots, x_n]$ such that $\sqrt{\mathfrak{a}} \neq (x_0, \ldots, x_n)$, and let $F \in T$ be a homogeneous polynomial which vanishes at all points of $Z(\mathfrak{a})$ in $\mathbb{P}^n$. Then $F^r \in \mathfrak{a}$ for some $r > 0$.*

Proof. We may suppose that $\mathfrak{a} \neq T$. Let $V = Z(\mathfrak{a}) \subset \mathbb{P}^n$ and $C(V) = Z_{\mathbb{A}^{n+1}}(\mathfrak{a}) \subset \mathbb{A}^{n+1}$ (regarding $k[x_0, \ldots, x_n]$ as the regular functions on $\mathbb{A}^{n+1}$). Since $\mathfrak{a}$ is a homogeneous ideal, a point $(a_0, \ldots, a_n)$ is in $C(V)$ if and only if $(ta_0, ta_1, \ldots, ta_n)$ is in $C(V)$ for all $t \in k$; further, the point $(a_0 : a_1 : \ldots : a_n) \in V$ if and only if $(ta_0, \ldots, ta_n) \in C(V)$ for all $t \in k$. Since $\sqrt{\mathfrak{a}} \neq (x_0, \ldots, x_n)$, we have that $C(V)$ contains a point other than $(0, \ldots, 0)$ by the affine nullstellensatz (Theorem 2.5) and thus $V \neq \emptyset$. Since $V \neq \emptyset$, $I(V) \subset I_{\mathbb{A}^{n+1}}(C(V))$.

If a polynomial $f \in I_{\mathbb{A}^{n+1}}(C(V))$, then $f(ta_0, \ldots, ta_n) = 0$ for all $(a_0 : \ldots : a_n) \in V$ and $t \in k$. Writing $f = \sum_{j=0}^{q} F_j$ where each F_j is a homogeneous form of degree j, we have

$$F_0(a_0, \ldots, a_n) + tF_1(a_0, \ldots, a_n) + \cdots + t^q F_q(a_1, \ldots, a_q) = 0$$

for all $t \in k$ which implies that $F_i(a_0, \ldots, a_n) = 0$ for all i since an algebraically closed field is infinite. Thus $f \in I(V)$ and so $I(V) = I_{\mathbb{A}^{n+1}}(C(V))$. By the affine nullstellensatz (Theorem 2.5), we have that $I(C(V)) = \sqrt{\mathfrak{a}}$, and the theorem follows. □

Proposition 3.13. *The following statements hold:*

1) *If $T_1 \subset T_2$ are subsets of $S(\mathbb{P}^n)$ consisting of homogeneous elements, then $Z(T_2) \subset Z(T_1)$.*
2) *If $Y_1 \subset Y_2$ are subsets of $\mathbb{P}^n$, then $I(Y_2) \subset I(Y_1)$.*
3) *For any two subsets Y_1, Y_2 of $\mathbb{P}^n$, we have $I(Y_1 \cup Y_2) = I(Y_1) \cap I(Y_2)$.*
4) *If $\mathfrak{a}$ is a homogeneous ideal in $S(\mathbb{P}^n)$ with $Z(\mathfrak{a}) \neq \emptyset$, then $I(Z(\mathfrak{a})) = \sqrt{\mathfrak{a}}$.*
5) *For any subset Y of $\mathbb{P}^n$, $Z(I(Y)) = \overline{Y}$, the Zariski closure of Y.*

The proofs of Theorem 3.14 and Proposition 3.15 are similar to those of Theorem 2.11 and Theorem 2.12 for affine space, using Proposition 3.13 instead of Proposition 2.10.

Theorem 3.14. *A closed set $W \subset \mathbb{P}^n$ is irreducible if and only if $I(W)$ is a prime ideal.*

Proposition 3.15. *Every closed set in $\mathbb{P}^n$ is the union of finitely many irreducible ones.*

Suppose that X is a closed subset of $\mathbb{P}^n$. We define the coordinate ring of X to be $S(X) = S(\mathbb{P}^n)/I(X)$, which is a standard graded ring. Suppose that U is a set of homogeneous elements of $S(X)$. Then we define

$$Z_X(U) = \{p \in X \mid F(p) = 0 \text{ for all } F \in U\}.$$

If J is a homogeneous ideal in $S(X)$, we define $Z_X(J) = Z_X(U)$ where U is a homogeneous set of generators of J. This set is independent of choice of homogeneous generating set U of J.

Given a subset Y of X, we define $I_X(Y)$ to be the ideal in $S(X)$ generated by the homogeneous elements of $S(X)$ which vanish at all points of Y. When there is no danger of confusion, we will sometimes write $Z(J)$ to denote $Z_X(J)$ and $I(Y)$ to denote $I_X(Y)$.

The above results in this section hold with $\mathbb{P}^n$ replaced by X. The Zariski topology on X is the topology whose closed sets are $Z_X(I)$ for $I \subset S(X)$ a homogeneous ideal (which is the subspace topology of X).

3.2. Projective varieties

Suppose that X is a projective algebraic set and $F \in S(X)$ is homogeneous. We define $D(F) = X \setminus Z(F)$, which is an open set in X. We will also denote the open set $D(F)$ by X_F.

The proof of the following lemma is like that of Lemma 2.83.

Lemma 3.16. *Suppose that $X \subset \mathbb{P}^n$ is a projective algebraic set. Then the open sets $D(F)$ for homogeneous $F \in S(X)$ form a basis for the topology of X.*

We now construct a correspondence between elements of $R = k[y_1, \ldots, y_n]$ (a polynomial ring in n variables) and homogeneous elements of the standard graded polynomial ring $T = k[x_0, \ldots, x_n]$. Fix i with $0 \le i \le n$. To $f(y_1, \ldots, y_n) \in R$ we associate the homogeneous polynomial

$$f^h = F(x_0, \ldots, x_n) = x_i^d f\left(\frac{x_0}{x_i}, \ldots, \frac{x_{i-1}}{x_i}, \frac{x_{i+1}}{x_i}, \ldots, \frac{x_n}{x_i}\right),$$

where d is the degree of f. To $F(x_0, \ldots, x_n) \in T$ we associate

$$F^a = F(y_1, \ldots, y_i, 1, y_{i+1}, \ldots, y_n).$$

This definition is valid for all $F \in T$. One has that

$$(f^h)^a = f$$

for all $f \in R$, and for all homogeneous $F \in T$,

$$(F^a)^h = x_i^{-m} F$$

where m is the highest power of x_i which divides F.

We extend h to a map from ideals in $k[y_1, \ldots, y_n]$ to homogeneous ideals in $k[x_0, \ldots, x_n]$ by taking an ideal I to the ideal I^h generated by the set of homogeneous elements $\{f^h \mid f \in I\}$. We also extend a to a map from homogeneous ideals in $k[x_0, \ldots, x_n]$ to ideals in $k[y_1, \ldots, y_n]$. A homogeneous ideal J is mapped to the ideal

$$J^a = \{f^a \mid f \in J\} = \{F^a \mid F \in J \text{ is homogeneous}\}.$$

The properties which are preserved by these correspondences of ideals are worked out in detail in [161, Section 5 of Chapter VII] (especially Theorem 17 on page 180 and Theorem 18 on page 182). The following formulas hold:

$$(I^h)^a = I \quad \text{for } I \text{ an ideal in } R$$

and

$$(J^a)^h = \bigcup_{j=0}^{\infty} J : x_i^j \quad \text{for } J \text{ a homogeneous ideal in } T$$

where $J : x_i^j = \{f \in T \mid f x_i^j \in J\}$. In particular, $(J^a)^h = J$ if J is a homogeneous prime ideal which does not contain x_i. We thus deduce from these formulas, [**161**, Theorem on page 180] and [**161**, Theorem on page 182], the following propositions.

Proposition 3.17. *The functions a and h give a 1-1 correspondence between prime ideals in $k[y_1, \ldots, y_n]$ and homogeneous prime ideals in $k[x_0, \ldots, x_n]$ which do not contain x_i.*

Recall that an ideal A is radical if $\sqrt{A} = A$.

Proposition 3.18. *The functions a and h give a 1-1 correspondence between radical ideals in $k[y_1, \ldots, y_n]$ and homogeneous radical ideals in $k[x_0, \ldots, x_n]$ all of whose associated prime ideals do not contain x_i.*

Theorem 3.19. *Suppose that i satisfies $0 \leq i \leq n$ and $D(x_i)$ is the open subset of $\mathbb{P}^n$. Then the maps $\phi : D(x_i) \to \mathbb{A}^n$ defined by*

$$\phi(a_0 : \ldots : a_n) = \left(\frac{a_0}{a_i}, \ldots, \frac{a_{i-1}}{a_i}, \frac{a_{i+1}}{a_i}, \ldots, \frac{a_n}{a_i}\right)$$

for $(a_0 : \ldots : a_n) \in D(x_i)$ and $\psi : \mathbb{A}^n \to D(x_i)$ defined by

$$\psi(a_1, a_2, \ldots, a_n) = (a_1 : \ldots : a_i : 1 : a_{i+1} : \ldots : a_n)$$

for $(a_1, \ldots, a_n) \in \mathbb{A}^n$ are inverse homeomorphisms.

Proof. The maps ϕ and ψ are inverse bijections of sets, so we need only show that ϕ and ψ are continuous.

Suppose that $Z \subset \mathbb{A}^n$ is closed. Then $Z = Z(I)$ for some ideal I in $k[y_1, \ldots, y_n]$. Then $\phi^{-1}(Z) = D(x_i) \cap Z(I^h)$ is closed in $D(x_i)$, and so ϕ is continuous. Suppose that $Z \subset D(x_i)$ is closed in $D(x_i)$. Then $Z = W \cap D(x_i)$ for some closed subset W of $\mathbb{P}^n$. We have that $W = Z(J)$ for some homogeneous ideal J of $k[x_0, \ldots, x_n]$ and so $\psi^{-1}(Z) = Z(J^a)$ is closed in $\mathbb{A}^n$. □

It follows from Theorem 3.19 and Proposition 3.17 that if X is an algebraic set (variety) in $\mathbb{A}^n$, then ψ induces a homeomorphism of X with $W \cap D(x_i)$, where W is the projective algebraic set (variety) $Z(I(X)^h)$. If W is a projective algebraic set (variety) in $\mathbb{P}^n$ which is not contained in $Z(x_i)$, then ψ induces a homeomorphism of the closed algebraic set (variety) $X = Z(I(W)^a)$ of $\mathbb{A}^n$ with $W \cap D(x_i)$.

Exercise 3.20. Prove Proposition 3.9.

Exercise 3.21. Prove Proposition 3.13.

Exercise 3.22. Suppose that $p = (a_0 : \ldots : a_n) \in \mathbb{P}^n$. Show that

$$I(p) = (a_i x_j - a_j x_i \mid 0 \leq i, j \leq n).$$

Exercise 3.23. In Theorem 3.19, we showed that the open set $D(x_i)$ of $\mathbb{P}^n$ is homeomorphic to $\mathbb{A}^n$. Suppose that Y is a closed subset of the open subset $D(x_i)$ of $\mathbb{P}^n$. The closure $\overline{Y}$ of Y in $\mathbb{P}^n$ is the intersection of all closed subsets of $\mathbb{P}^n$ containing Y. Show that
$$I_{\mathbb{A}^n}(\phi(Y))^h = I_{\mathbb{P}^n}(\overline{Y}).$$

Exercise 3.24. Suppose that $Y \subset \mathbb{P}^n$ is a closed subvariety of $D(x_0)$ such that $I(\phi(Y)) = (f)$ for some irreducible $f \in k[\mathbb{A}^n] = k[y_1, \ldots, y_n]$. Suppose that d is the degree of f, so that we have an expansion
$$f = \sum_{j_1+\cdots+j_n \leq d} a_{j_1,\ldots,j_n} y_1^{j_1} \cdots y_n^{j_n}$$
for some $a_{j_1,\ldots,j_n} \in k$ and so that some term $a_{j_1,\ldots,j_n} \neq 0$ with $j_1+\cdots+j_n = d$. Find generators of $I(\phi(Y))^h$.

Exercise 3.25. Let $\phi : k[x_0, x_1, x_2, x_3] \to k[s, t]$ be the k-algebra homomorphism of polynomial rings defined by
$$\phi(x_0) = s^3, \quad \phi(x_1) = ts^2, \quad \phi(x_2) = t^2 s, \quad \phi(x_3) = t^3.$$
We will say that a monomial $s^i t^j$ has bidegree (i, j) and a monomial $x_0^a x_1^b x_2^c x_3^d$ has bidegree $(3a + 2b + c, b + 2c + 3d)$. A k-linear combination of monomials of a common bidegree is called a bihomogeneous form. The map ϕ is bihomogeneous, as it takes forms of bidegree (i, j) to forms of bidegree (i, j). Let J be the kernel of ϕ. Use a variation of the method of Exercise 2.76 to show that
$$J = (x_1 x_2 - x_0 x_3, x_1^2 - x_0 x_2, x_2^2 - x_1 x_3).$$

Exercise 3.26. Let $I = (F_1 = y_1 y_2 - y_3, F_2 = y_1^2 - y_2)$. Show that I is a prime ideal in $R = k[y_1, y_2, y_3]$. Compute the ideal I^h in $T = k[x_0, x_1, x_2, x_3]$ and show that it is not equal to (F_1^h, F_2^h).

Hint: Consider the ideal of the "twisted cubic curve" $J = (x_1 x_2 - x_0 x_3, x_1^2 - x_0 x_2, x_2^2 - x_1 x_3)$ from Exercise 3.25 which is a prime ideal.

3.3. Grassmann varieties

The Grassmannian $\mathrm{Grass}(a, b)$ is the set of a-dimensional linear subspaces of k^b. We will show that $\mathrm{Grass}(a, b)$ naturally has a structure as a projective variety.

An a-dimensional linear subspace W of k^b is determined by a basis $v_1, \ldots, v_a$ of W and hence by the $a \times b$ matrix

(3.4)
$$A = \begin{pmatrix} v_1 \\ \vdots \\ v_a \end{pmatrix}$$

which has rank a. Two such matrices A and A' represent the same a-dimensional subspace W if and only if there exists an $a \times a$ matrix B with nontrivial determinant (an element of $\mathrm{GL}_a(k)$) such that $BA = A'$. This determines an equivalence relation $\sim$ on

$$S = \{a \times b \text{ matrices } A \text{ of rank } a\},$$

giving us a natural identification of the sets $\mathrm{Grass}(a,b)$ and $S/\sim$.

We now show, by means of the Plücker embedding, that $\mathrm{Grass}(a,b)$ has a natural structure as a projective variety. Define a map $\rho : S \to \mathbb{P}^{\binom{b}{a}-1}$ by sending the matrix A of (3.4) to the equivalence class (a_I) where the a_I are the minors of $a \times a$ submatrices of A (in some fixed order). At each point A of S, for at least one I, a_I is not zero. If $A \sim A'$, then $(a_I') = \mathrm{Det}(B)(a_I)$ as vectors in $k^{\binom{a}{b}}$. Thus $\rho(A) = \rho(A')$ so ρ induces a well-defined map

$$\overline{\rho} : \mathrm{Grass}(a,b) \to \mathbb{P}^{\binom{b}{a}-1}.$$

Now using linear algebra, we can show that $\overline{\rho}$ is injective and its image is a closed subvariety of $\mathbb{P}^{\binom{b}{a}-1}$, whose ideal is generated by quadrics. For more details about Grassmannians, see [**81**, Chapter VII of Volume 1], [**136**, Example 1 on page 42 of Volume 1], and [**62**, Section 5 of Chapter 1].

3.4. Regular functions and regular maps of quasi-projective varieties

Suppose that X is a projective variety, that is, a closed irreducible subset of $\mathbb{P}^n$. Then the coordinate ring $S(X) = S(\mathbb{P}^n)/I(X)$ of X is a standard graded k-algebra. We define

$$k(X) = \left\{ \frac{F}{G} \mid F, G \in S(X) \text{ are homogeneous of the same degree } d \text{ and } G \neq 0 \right\}.$$

It is readily verified that $k(X)$ is a field. The field $k(X)$ is called the field of rational functions on X or the function field of X. We have that $k(X) = S(X)_{(\mathfrak{q})}$, defined in (3.1), where $\mathfrak{q}$ is the prime ideal $\mathfrak{q} = (0)$ in $S(X)$, the elements of degree 0 in $T^{-1}S(X)$ where T is the multiplicative set of nonzero homogeneous elements of $S(X)$. By (3.3), if $f = \frac{F}{G} \in k(X)$ with F, G homogeneous of the same degree d and $p \in X$ is such that $G(p) \neq 0$, then the value $f(p) \in k$ is well-defined. Specifically, if $p = (a_0 : \ldots : a_n) = (b_0 : \ldots : b_n)$, then there exists $0 \neq \lambda \in k$ such that $a_i = \lambda b_i$ for $0 \leq i \leq n$ and

$$\begin{aligned} f(a_0, \ldots, a_n) &= \frac{F(a_0,\ldots,a_n)}{G(a_0,\ldots,a_n)} = \frac{F(\lambda b_0,\ldots,\lambda b_n)}{G(\lambda b_0,\ldots,\lambda b_n)} \\ &= \frac{\lambda^d F(b_0,\ldots,b_n)}{\lambda^d G(b_0,\ldots,b_n)} = \frac{F(b_0,\ldots,b_n)}{G(b_0,\ldots,b_n)} = f(b_0, \ldots, b_n). \end{aligned}$$

Suppose that $f \in k(X)$. We say that f is regular at $p \in X$ if there exists an expression $f = \frac{F}{G}$, where $F, G \in S(X)$ are homogeneous of the same degree d, such that $G(p) \neq 0$.

Lemma 3.27. *Suppose that $f \in k(X)$. Then the set*

$$U = \{p \in X \mid f \text{ is regular at } p\}$$

is an open subset of X.

Proof. Suppose that $p \in U$. Then there are $F, G \in S(X)$ which are homogeneous of the same degree d such that $f = \frac{F}{G}$ and $G(p) \neq 0$. Then $X \setminus Z_X(G)$ is an open neighborhood of p which is contained in U. Thus U is open. $\square$

For $p \in X$, we define

$$\mathcal{O}_{X,p} = \{f \in k(X) \mid f \text{ is regular at } p\}.$$

An element $f \in \mathcal{O}_{X,p}$ induces a map $f : U \to \mathbb{A}^1$ on some open neighborhood U of p in X. The ring $\mathcal{O}_{X,p}$ is a local ring. We will write its maximal ideal as $m_{X,p}$, or m_p if there is no danger of confusion. We have that $\mathcal{O}_{X,p} = S(X)_{(I(p))}$, defined in (3.1), the elements of degree 0 in the localization $T^{-1}S(X)$, where T is the multiplicative set of homogeneous elements of $S(X)$ which are not in $I(p)$. We will denote $\mathcal{O}_{X,p}/m_p$ by $k(p)$. As a field, $k(p)$ is isomorphic to k. Also, $k(p)$ has a natural structure as a $\mathcal{O}_{X,p}$-module.

Suppose that U is an open subset of X. Then we define

$$\mathcal{O}_X(U) = \bigcap_{p \in U} \mathcal{O}_{X,p}.$$

Here the intersection takes place in $k(X)$ and is over all $p \in U$. The ring $\mathcal{O}_X(U)$ is called the ring of regular functions on U.

Suppose that U is a nontrivial open subset of X. Let $\text{Map}(U, \mathbb{A}^1)$ be the set of maps from U to $\mathbb{A}^1$. The set $\text{Map}(U, \mathbb{A}^1)$ is a k-algebra since $\mathbb{A}^1 = k$ is a k-algebra. We have a natural k-algebra homomorphism $\phi : \mathcal{O}_X(U) \to \text{Map}(U, \mathbb{A}^1)$ defined by $\phi(f)(p) = f(p)$ for $f \in \mathcal{O}_X(U)$ and $p \in U$. We will show that ϕ is injective. Suppose $f \in \text{Kernel } \phi$ and $p \in U$. There exists an expression $f = \frac{F}{G}$ where $F, G \in S(X)$ are homogeneous of the same degree d and $G(p) \neq 0$. For q in the nontrivial open set $U \setminus Z(G)$ we have that $\frac{F(q)}{G(q)} = f(q) = 0$. Thus $F(q) = 0$, and so

$$F \in I(U \setminus Z(G)) = I(X) = \sqrt{(0)} = (0)$$

by Proposition 3.13. Thus $F = 0$, and so $f = \frac{F}{G} = 0$. Hence ϕ is injective. We may thus identify $\mathcal{O}_X(U)$ with the k-algebra $\phi(\mathcal{O}_X(U))$ of maps from U to $\mathbb{A}^1$.

An element f of the function field $k(X)$ of X induces a map $f : U \to \mathbb{A}^1$ on some open subset U of X. An element $f \in \mathcal{O}_{X,p}$ induces a map $f : U \to \mathbb{A}^1$ on some open neighborhood U of p in X.

There are examples of quasi-projective varieties U such that $\mathcal{O}_U(U)$ is not a finitely generated k-algebra, as discussed earlier for quasi-affine varieties in Example 2.98. A quasi-projective example is given by Zariski [159, page 456], based on his example which we will expound in Theorem 20.14.

We now define a regular map of a quasi-projective variety.

Definition 3.28. Suppose that X is a quasi-projective variety or a quasi-affine variety and Y is a quasi-projective variety or a quasi-affine variety. A regular map $\phi : X \to Y$ is a continuous map such that for every open subset U of Y, the map $\phi^* : \text{Map}(U, \mathbb{A}^1) \to \text{Map}(\phi^{-1}(U), \mathbb{A}^1)$ defined by $\phi^*(f) = f \circ \phi$ for $f \in \text{Map}(U, \mathbb{A}^1)$ gives a k-algebra homomorphism $\phi^* : \mathcal{O}_Y(U) \to \mathcal{O}_X(\phi^{-1}(U))$.

Definition 3.28 is consistent with our earlier definitions of regular maps of affine and quasi-affine varieties by Propositions 2.95 and 2.96.

We will show that $\mathcal{O}_X(X)$ can be identified with the ring of regular maps from X to $\mathbb{A}^1$ in Theorem 3.40.

Definition 3.29. Suppose that X is a quasi-projective variety or a quasi-affine variety and Y is a quasi-projective variety or a quasi-affine variety. A regular map $\phi : X \to Y$ is said to be an isomorphism if there exists a regular map $\psi : Y \to X$ such that $\psi \circ \phi = \text{id}_X$ and $\phi \circ \psi = \text{id}_Y$.

We will say that a quasi-projective variety is affine if it is isomorphic to an affine variety.

Theorem 3.30. *Suppose that W is a projective variety which is a closed subset of $\mathbb{P}^n$, with homogeneous coordinate ring $S(\mathbb{P}^n) = k[x_0, \ldots, x_n]$, and x_i is a homogeneous coordinate on $\mathbb{P}^n$ such that $W \cap D(x_i) \neq \emptyset$. Then $W \cap D(x_i)$ is an affine variety.*

Proof. Without loss of generality, we may suppose that $i = 0$. Write
$$S(W) = k[x_0, \ldots, x_n]/I(W) = k[\overline{x}_0, \ldots, \overline{x}_n]$$
where $\overline{x}_i$ is the class of x_i. The ring $S(W)$ is standard graded with the $\overline{x}_i$ having degree 1. By our assumption, $\overline{x}_0 \neq 0$, so $\frac{\overline{x}_j}{\overline{x}_0} \in k(W)$ for all j. We

3.4. Regular functions and regular maps of quasi-projective varieties

calculate
$$k(W) = S(W)_{(\mathfrak{q})} = \mathrm{QF}\left(k\left[\frac{\overline{x}_1}{\overline{x}_0}, \ldots, \frac{\overline{x}_n}{\overline{x}_0}\right]\right)$$
where $\mathfrak{q}$ is the zero ideal (0) of $S(W)$.

Let $\phi : W \cap D(x_0) \to X$ be the homeomorphism induced by the map of Theorem 3.19, as explained after the proof of Theorem 3.19, where X is the affine variety $X = Z(I(W)^a) \subset \mathbb{A}^n$. We have that
$$k[X] = k[\mathbb{A}^n]/I(X) = k[y_1, \ldots, y_n]/I(X) = k[\overline{y}_1, \ldots, \overline{y}_n],$$
where $\overline{y}_i$ is the class of y_i in $k[X]$. For a point
$$p = (a_0 : a_1 : \ldots : a_n) \in W \cap D(x_0),$$
we have that $\phi(p) = (\frac{a_1}{a_0}, \ldots, \frac{a_n}{a_0})$.

Define $\phi^* : k[\mathbb{A}^n] = k[y_1, \ldots, y_n] \to k(W)$ by

(3.5) $$\phi^*(f) = f \circ \phi = f\left(\frac{\overline{x}_1}{\overline{x}_0}, \ldots, \frac{\overline{x}_n}{\overline{x}_0}\right) = \frac{f^h(\overline{x}_0, \ldots, \overline{x}_n)}{(\overline{x}_0)^{\deg(f^h)}}$$

for $f \in k[y_1, \ldots, y_n]$.

The ideal $I(W)^a$ is a prime ideal by Proposition 3.17, so $I(X) = I(W)^a$ since $I(X) = I(Z(I(W)^a)) = \sqrt{I(W)^a}$. The kernel of ϕ^* is
$$\begin{aligned}\mathrm{Kernel}(\phi^*) &= \{f \in k[y_1, \ldots, y_n] \mid f^h(\overline{x}_0, \ldots, \overline{x}_n) = 0\} \\ &= \{f \in k[y_1, \ldots, y_n] \mid f^h \in I(W)\}.\end{aligned}$$

Suppose $f \in k[y_1, \ldots, y_n]$ and $f^h \in I(W)$. Then $f = f^{ha} \in I(W)^a = I(X)$. If $f \in I(X)$, then $f^h \in I(X)^h = I(W)$ by Proposition 3.17. Thus the kernel of ϕ^* is $I(X)$.

We thus have that ϕ^* induces an injective k-algebra homomorphism $\phi^* : k[X] = k[\overline{y}_1, \ldots, \overline{y}_n] \to k(W)$, giving an induced k-algebra homomorphism $\phi^* : k(X) \to k(W)$. By (3.5) we have that $\phi^*(f) \in \mathcal{O}_{W,p}$ for all $f \in k[X]$ and $p \in W \cap D(x_0)$.

Suppose that $p = (a_0 : \ldots : a_n) \in W \cap D(x_0)$ and $h \in \mathcal{O}_{X,\phi(p)}$. Then $h = \frac{f}{g}$ with $f, g \in k[X]$ and $g(\phi(p)) = g(\frac{a_1}{a_0}, \ldots, \frac{a_n}{a_0}) \neq 0$. Since $a_0 \neq 0$, we see from (3.5) that $\phi^*(g)(p) \neq 0$. Thus $\phi^*(h) \in \mathcal{O}_{W,p}$. We thus have that $\phi^*(\mathcal{O}_{X,\phi(p)}) \subset \mathcal{O}_{W,p}$. Suppose that $h \in \mathcal{O}_{W,p}$. Then $h = \frac{F(\overline{x}_0, \ldots, \overline{x}_n)}{G(\overline{x}_0, \ldots, \overline{x}_n)}$ where F, G are homogeneous of a common degree d and $G(p) \neq 0$. We have
$$\frac{F}{G} = \frac{\frac{F}{x_0^d}}{\frac{G}{x_0^d}} = \frac{F(1, \frac{\overline{x}_1}{\overline{x}_0}, \ldots, \frac{\overline{x}_n}{\overline{x}_0})}{G(1, \frac{\overline{x}_1}{\overline{x}_0}, \ldots, \frac{\overline{x}_n}{\overline{x}_0})} = \phi^*\left(\frac{F^a}{G^a}\right).$$

Now $G^a(\phi(p)) = \frac{G(p)}{a_0^d} \neq 0$ so $\frac{F^a}{G^a} \in \mathcal{O}_{X,\phi(p)}$. We thus have that

(3.6) $$\phi^* : \mathcal{O}_{X,\phi(p)} \to \mathcal{O}_{W,p}$$

is an isomorphism for all $p \in W \cap D(x_0)$.

Since ϕ is a homeomorphism and by the definitions of $\mathcal{O}_X$ and $\mathcal{O}_W$, we have for an open subset U of X an isomorphism
(3.7)
$$\mathcal{O}_X(U) = \bigcap_{q \in U} \mathcal{O}_{X,q} = \bigcap_{p \in \phi^{-1}(U)} \mathcal{O}_{X,\phi(p)} \xrightarrow{\phi^*} \bigcap_{p \in \phi^{-1}(U)} \mathcal{O}_{W,p} = \mathcal{O}_W(\phi^{-1}(U))$$

by (3.6). In particular, ϕ is a regular map.

Now we have that the inverse map ψ of ϕ is determined by the map defined in Theorem 3.19, as explained after the proof of Theorem 3.19, and ψ induces a map $\psi^* : k(W) \to k(X)$, defined by $\psi^*(f) = f \circ \psi$ for $f \in k(W)$. We have that

$$\psi^*\left(\frac{\overline{x_j}}{\overline{x_0}}\right) = y_j$$

for $1 \leq j \leq n$. Thus ψ^* is an isomorphism of function fields with inverse ϕ^*. By our above calculations, we then see that ψ is a regular map which is an inverse to ϕ. Thus ϕ is an isomorphism. $\square$

Corollary 3.31. *Suppose that X is a quasi-affine variety. Then X is isomorphic to a quasi-projective variety.*

Proof. The quasi-affine variety X is an open subset of an affine variety $Y \subset \mathbb{A}^n$. The affine variety $\mathbb{A}^n$ is isomorphic to the open subset $D(x_0)$ of $\mathbb{P}^n$ by Theorem 3.30. Let W be the Zariski closure of Y in $\mathbb{P}^n$. Then $W \cap D(x_0) \cong Y$ by Theorem 3.30. Since X is an open subset of Y and Y is isomorphic to an open subset of W, we have that X is isomorphic to an open subset of W and thus is isomorphic to a quasi-projective variety. $\square$

We will call an open subset U of a projective variety an affine variety if U is isomorphic to an affine variety. With this identification, we have that all quasi-affine varieties are quasi-projective.

Corollary 3.32. *Every point p in a quasi-projective variety has an open neighborhood which is isomorphic to an affine variety.*

Proof. Suppose that V is a quasi-projective variety and $p \in V$ is a point. Then V is an open subset of a projective variety W, which is itself a closed subset of a projective space $\mathbb{P}^n$. Writing $S(\mathbb{P}^n) = k[x_0, \ldots, x_n]$, there exists a homogeneous coordinate x_i on $\mathbb{P}^n$ such that $x_i(p) \neq 0$, so $D(x_i)$ contains p. By Theorem 3.30, $D(x_i) \cap W$ is isomorphic to an affine variety. We have that $V \cap (D(x_i) \cap W)$ is an open subset of $D(x_i) \cap W$ since V is open in

3.4. Regular functions and regular maps of quasi-projective varieties 79

W. Now by Lemma 2.83 and Proposition 2.93, there exists an affine open subset U of $D(x_i) \cap W$ which contains p and is contained in V. $\square$

The proof of Theorem 3.30 gives us the following useful formula. Suppose that $W \subset \mathbb{P}^n$ is a projective variety, and suppose that W is not contained in $Z(x_i)$. Letting $S(\mathbb{P}^n) = k[x_0, \ldots, x_n]$ and $S(W) = S(\mathbb{P}^n)/I(W) = k[\overline{x}_0, \ldots, \overline{x}_n]$, we have that

$$(3.8) \qquad \mathcal{O}_W(D(\overline{x}_i)) = k\left[\frac{\overline{x}_0}{\overline{x}_i}, \ldots, \frac{\overline{x}_n}{\overline{x}_i}\right] \cong k\left[\frac{x_0}{x_i}, \ldots, \frac{x_n}{x_i}\right]/J$$

where

$$\begin{aligned} J &= \{f(\tfrac{x_0}{x_i}, \ldots, \tfrac{x_n}{x_i}) \mid f \in I(W)\} \\ &= \{\tfrac{F}{x_i^d} \mid F \in I(W) \text{ is homogeneous of some degree } d\}. \end{aligned}$$

With the notation introduced before Lemma 3.4, we have that

$$\mathcal{O}_W(D(\overline{x}_i)) = S(W)_{(\overline{x}_i)} \cong S(\mathbb{P}^n)_{(x_i)}/I(W)_{(x_i)}.$$

We give a proof of this formula. Without loss of generality, we may assume that $i = 0$. By (3.7), the homomorphism $\phi^* : k(X) \to k(W)$ takes $\mathcal{O}_X(X) = k[X] = k[\overline{y}_1, \ldots, \overline{y}_n]$ to $\mathcal{O}_W(D(\overline{x}_0))$. In our construction of ϕ^*, we saw that $\phi^*(\overline{y}_j) = \frac{\overline{x}_j}{\overline{x}_0}$ for all j, so

$$\mathcal{O}_W(D(\overline{x}_0)) = k\left[\frac{\overline{x}_1}{\overline{x}_0}, \ldots, \frac{\overline{x}_n}{\overline{x}_0}\right].$$

Further, this last ring is the quotient of the polynomial ring

$$k\left[\frac{x_1}{x_0}, \ldots, \frac{x_n}{x_0}\right]$$

by $\phi^*(I(X)) = \phi^*(I(W)^a) = J$. Finally, we observe that by Euler's formula, if $F(x_0, \ldots, x_n)$ is homogeneous of degree d, then

$$F(x_0, \ldots, x_n) = F\left(x_i \frac{x_0}{x_i}, \ldots, x_i \frac{x_n}{x_i}\right) = x_i^d F\left(\frac{x_0}{x_i}, \ldots, \frac{x_n}{x_i}\right).$$

Example 3.33. Let $F = x_2^2 x_0 - x_1(x_1^2 - x_0^2) \in S(\mathbb{P}^2) = k[x_0, x_1, x_2]$. The form F is irreducible, so $C = Z(F) \subset \mathbb{P}^2$ is a projective variety (an elliptic curve). Let $y_1 = \frac{x_1}{x_0}$ and $y_2 = \frac{x_2}{x_0}$, which are coordinate functions on $D(x_0) \cong \mathbb{A}^2$. Then $I(C \cap D(x_0)) = (F)^a = (f)$ where $f = y_2^2 - y_1(y_1^2 - 1) \in k[D(x_0)] = k[y_1, y_2]$. We have that

$$C \cap Z(x_0) = Z(x_2^2 x_0 - x_1(x_1^2 - x_0^2), x_0) = Z(x_0, x_1) = \{Q\}$$

where $Q = (0 : 0 : 1)$. Thus C is the union of the affine curve $C \cap D(x_0)$ and the point $Q \in Z(x_0) \cong \mathbb{P}^1$. Figure 3.1 shows this curve over $\mathbb{C}$. We will see in Chapter 18 that C has genus 1, and thus (Section 18.8) C is topologically a sphere with one handle.

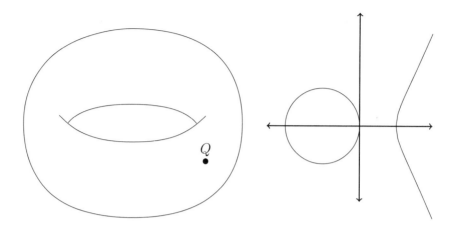

Figure 3.1. C over $\mathbb{C}$ (left) and the real part of $C \cap D(x_0)$ (right)

We can now calculate the regular functions on $\mathbb{P}^n$.

Proposition 3.34. *The regular functions on $\mathbb{P}^n$ are $\mathcal{O}_{\mathbb{P}^n}(\mathbb{P}^n) = k$.*

Proof. Since $\{D(x_i) \mid 0 \leq i \leq n\}$ is an open cover of $\mathbb{P}^n$,

$$\mathcal{O}_{\mathbb{P}^n}(\mathbb{P}^n) = \bigcap_{p \in \mathbb{P}^n} \mathcal{O}_{\mathbb{P}^n, p} = \bigcap_{i=0}^{n} \left(\bigcap_{p \in D(x_i)} \mathcal{O}_{\mathbb{P}^n, p} \right)$$

$$= \bigcap_{i=0}^{n} \mathcal{O}_{\mathbb{P}^n}(D(x_i)) = \bigcap_{i=0}^{n} k \left[\frac{x_0}{x_i}, \ldots, \frac{x_n}{x_i} \right].$$

Thus if $g \in \mathcal{O}_{\mathbb{P}^n}(\mathbb{P}^n)$, we have expressions

$$g = g_i \left(\frac{x_0}{x_i}, \ldots, \frac{x_n}{x_i} \right) \in k \left[\frac{x_0}{x_i}, \ldots, \frac{x_n}{x_i} \right].$$

For each i, there exists a smallest $d_i \in \mathbb{N}$ such that

$$x_i^{d_i} g_i \left(\frac{x_0}{x_i}, \ldots, \frac{x_n}{x_i} \right) = f_i(x_0, \ldots, x_n)$$

is a polynomial. Necessarily, we have that $x_i \nmid f_i$ in the UFD $k[x_0, \ldots, x_n]$. Further, d_i is the degree of f_i. If $d_0 = 0$, then we have that $g = g_0 \in k$, and we have established the proposition. Suppose that $d_0 > 0$. Since $g_i = g_j$ for all i, j, we have that $f_i x_j^{d_j} = f_j x_i^{d_i}$ for all i, j. Since the polynomial ring $k[x_0, \ldots, x_n]$ is a UFD and x_i, x_j are relatively prime for $i \neq j$, we have that $x_0 \mid f_0$, which is a contradiction. Thus $d_0 = 0$ and so $\mathcal{O}_{\mathbb{P}^n}(\mathbb{P}^n) = k$. $\square$

The statement of Proposition 3.34 is true for arbitrary projective varieties W (taking the intersection over the open sets $W \cap D(x_i)$ such that $D(x_i) \cap W \neq \emptyset$) but we need to be a little careful with the proof, as can

3.4. Regular functions and regular maps of quasi-projective varieties

be seen from the following example. Consider the standard graded domain $T = \mathbb{Q}[\overline{x}_0, \overline{x}_1] = \mathbb{Q}[x_0, x_1]/(x_0^2 + x_1^2)$. We compute $L = \mathbb{Q}[\frac{\overline{x}_0}{\overline{x}_1}] \cap \mathbb{Q}[\frac{\overline{x}_1}{\overline{x}_0}]$. We have that $\overline{x}_0^2 = -\overline{x}_1^2$ so $\frac{\overline{x}_1}{\overline{x}_0} = -\frac{\overline{x}_0}{\overline{x}_1}$.

$$L = \mathbb{Q}\left[\frac{\overline{x}_0}{\overline{x}_1}\right] \cong \mathbb{Q}[t]/(t^2 + 1) \cong \mathbb{Q}[\sqrt{-1}],$$

which is larger than $\mathbb{Q}$. This example shows that any proof that $\mathcal{O}_W(W) = k$ for a projective variety W must use the assumption that k is algebraically closed.

We will give a different proof of Theorem 3.35 in Corollary 5.15.

Theorem 3.35. *Suppose that W is a projective variety. Then the regular functions on W are $\mathcal{O}_W(W) = k$.*

Proof. W is a closed irreducible subset of $\mathbb{P}^n$ for some n. Let $S(\mathbb{P}^n) = k[x_0, \ldots, x_n]$ and $S(W) = k[\overline{x}_0, \ldots, \overline{x}_n]$ where $\overline{x}_i$ is the class of x_i in $S(W)$. We may suppose that $\overline{x}_i \neq 0$ for all i, for otherwise we have that $W \subset Z(x_i)$ so that W is a closed subset of $Z(x_i) \cong \mathbb{P}^{n-1} \subset \mathbb{P}^n$, and W is contained in a projective space of smaller dimension. Repeating this reduction at most a finite number of times, we eventually realize W as a closed subset of a projective space such that $W \not\subset Z(x_i)$ for all i. Suppose that

$$f \in \mathcal{O}_W(W) = \bigcap_{i=0}^n \mathcal{O}_W(D(x_i)) = \bigcap_{i=0}^n k\left[\frac{\overline{x}_0}{\overline{x}_i}, \ldots, \frac{\overline{x}_n}{\overline{x}_i}\right].$$

Then there exist $N_i \in \mathbb{N}$ and homogeneous elements $G_i \in S(W)$ of degree N_i such that

$$f = \frac{G_i}{\overline{x}_i^{N_i}} \quad \text{for } 0 \leq i \leq n.$$

Let S_i be the set of homogeneous forms of degree i in $S(W)$ (so that $S(W) \cong \bigoplus_{i=0}^\infty S_i$). We have that $\overline{x}_i^{N_i} f \in S_{N_i}$ for $0 \leq i \leq n$. Suppose that $N \geq \sum N_i$. Since S_N is spanned (as a k-vector space) by monomials of degree N in $\overline{x}_0, \ldots, \overline{x}_n$, for each such monomial at least one $\overline{x}_i$ has an exponent $\geq N_i$. Thus $S_N f \subset S_N$. Iterating, we have that $S_N f^q \subset S_N$ for all $q \in \mathbb{N}$. In particular, $\overline{x}_0^N f^q \in S(W)$ for all $q > 0$. Thus the subring $S(W)[f]$ of the quotient field of $S(W)$ is contained in $x_0^{-N} S(W)$, which is a finitely generated $S(W)$ module. Thus f is integral over $S(W)$ (by Theorem 1.49), and there exist m and $a_1, \ldots, a_m \in S(W)$ such that

$$f^m + a_1 f^{n-1} + \cdots + a_m = 0.$$

Since f has degree 0, we can replace the a_i with their homogeneous components of degree 0 and still have a dependence relation. But the elements of degree 0 in $S(W)$ consists of the field k. Now $k[f]$ is a domain since it is a subring of the quotient field of $S(W)$ and so $k[f]$ is a finite extension field of k. Thus $f \in k$ since k is algebraically closed. □

Proposition 3.36. *Suppose that W is a projective variety. Then W is separated (distinct points of W have distinct local rings).*

Proof. Suppose that W is a closed subset of a projective space $\mathbb{P}^n$. Write
$$S(W) = S(\mathbb{P}^n)/I(W) = k[x_0,\ldots,x_n]/I(W) = k[\overline{x}_0,\ldots,\overline{x}_n]$$
where $\overline{x}_i$ is the class of x_i in $S(W)$. Suppose that $p \in W$. Then $p \in D(x_i)$ for some i, and since $W \cap D(x_i)$ is affine with
$$k[W \cap D(x_i)] = k\left[\frac{\overline{x}_0}{\overline{x}_i},\ldots,\frac{\overline{x}_n}{\overline{x}_i}\right],$$
$\mathcal{O}_{W,p}$ is the localization of $k\left[\frac{\overline{x}_0}{\overline{x}_i},\ldots,\frac{\overline{x}_n}{\overline{x}_i}\right]$ at a maximal ideal $\mathfrak{m}$. Let $\pi : \mathcal{O}_{W,p} \to \mathcal{O}_{W,p}/\mathfrak{m}\mathcal{O}_{W,p} \cong k$ be the residue map. Let $\pi(\frac{\overline{x}_j}{\overline{x}_i}) = \alpha_j \in k$ for $0 \le j \le n$ (with $\alpha_i = 1$ since $\frac{\overline{x}_i}{\overline{x}_i} = 1$). We have that $\overline{x}_i(p) \ne 0$ and $\overline{x}_j(p) = \alpha_j \overline{x}_i(p)$. Thus
$$p = (\overline{x}_0(p):\ldots:\overline{x}_n(p)) = (\alpha_0:\ldots:\alpha_n) \in W$$
is uniquely determined by the ring $\mathcal{O}_{W,p}$. □

Proposition 3.37. *Suppose that X is a projective variety and $U \subset X$ is a nonempty quasi-affine open subset. Then the quotient field of $\mathcal{O}_X(U)$ is the function field $k(X)$ of X.*

Proof. Let $p \in U$. Then $\mathcal{O}_{X,p} = \mathcal{O}_{U,p}$ is a localization of $\mathcal{O}_U(U)$ at a maximal ideal (by Lemma 2.91). By the definition of $\mathcal{O}_{X,p}$, we have that $\mathcal{O}_{X,p} \subset k(X)$, so $\mathrm{QF}(\mathcal{O}_X(U)) \subset k(X)$. Suppose that $f \in k(X)$. The function $f = \frac{F}{G}$ where $F,G \in S(X)$ are homogeneous of a common degree d and $G \ne 0$. There exists a linear form $L \in S(X)$ such that $L(p) \ne 0$. Thus $\frac{F}{L^d}, \frac{G}{L^d} \in \mathcal{O}_{X,p}$ so $\frac{F}{G}$ is in the quotient field of $\mathcal{O}_{X,p}$ and hence is in the quotient field of $\mathcal{O}_U(U)$. □

It follows from Proposition 3.37 that if X is a projective variety and $p \in X$, then $k(X) = \mathrm{QF}(\mathcal{O}_{X,p})$.

Proposition 3.37 allows us to define the field of rational functions or the function field $k(Y)$ of a quasi-projective variety Y as $k(U)$, where U is a nontrivial affine open subset of Y.

Suppose that $\phi : X \to Y$ is a regular map of quasi-projective varieties. Extending our definition for quasi-affine varieties, we say that ϕ is dominant if the Zariski closure of $\phi(X)$ in Y is equal to Y.

3.4. Regular functions and regular maps of quasi-projective varieties

Proposition 3.38. *Suppose that X and Y are quasi-projective varieties and $\phi : X \to Y$ is a dominant regular map. Then the map $\phi^* : \mathrm{Map}(Y, \mathbb{A}^1) \to \mathrm{Map}(X, \mathbb{A}^1)$ defined by $\phi^*(f) = f \circ \phi$ for $f \in \mathrm{Map}(Y, \mathbb{A}^1)$ induces an injective k-algebra homomorphism $\phi^* : k(Y) \to k(X)$.*

Proof. Let V be an affine open subset of Y and U be an affine open subset of the open subset $f^{-1}(V)$ of X. Then the restriction of ϕ to a map of affine varieties $\phi : U \to V$ is dominant, so the k-algebra homomorphism $\phi^* : k[V] \to k[U]$ is injective. Taking the induced map on quotient fields, we obtain by Proposition 3.37 the desired homomorphism of function fields. □

The following proposition gives a useful criterion for a map of quasi-projective varieties to be regular.

Proposition 3.39. *Suppose that X and Y are quasi-projective varieties and $\phi : X \to Y$ is a map. Let $\{V_i\}$ be a collection of open affine subsets covering Y and $\{U_i\}$ be a collection of open subsets covering X, such that*

1. *$\phi(U_i) \subset V_i$ for all i and*
2. *the map $\phi^* : \mathrm{Map}(V_i, \mathbb{A}^1) \to \mathrm{Map}(U_i, \mathbb{A}^1)$ defined by $\phi^*(f) = f \circ \phi$ for $f \in \mathrm{Map}(V_i, \mathbb{A}^1)$ maps $\mathcal{O}_Y(V_i)$ into $\mathcal{O}_X(U_i)$ for all i.*

Then ϕ is a regular map.

Proof. Suppose that U is an affine subset of U_i. Then ϕ^* induces a k-algebra homomorphism $\phi^* : k[V_i] \to k[U]$ since the restriction map $\mathcal{O}_Y(U_i) \to \mathcal{O}_Y(U)$ is a k-algebra homomorphism. Thus we may refine our cover $\{U_i\}$ to assume that the U_i are affine for all i.

Let $\phi_i : U_i \to V_i$ be the restriction of ϕ. Consider the k-algebra homomorphism $\phi_i^* : k[V_i] \to k[U_i]$. Suppose that $p \in U_i$ and $q \in V_i$. We have that

(3.9)

$\phi_i^*(I_{V_i}(q)) \subset I_{U_i}(p)$ if and only if $\phi_i^*(f)(p) = 0$ for all $f \in I_{V_i}(q)$
                         if and only if $f(\phi_i(p)) = 0$ for all $f \in I_{V_i}(q)$
                         if and only if $I_{V_i}(q) \subset I_{V_i}(\phi_i(p))$
                         if and only if $I_{V_i}(q) = I_{V_i}(\phi_i(p))$ since $I_{V_i}(q)$
                                           and $I_{V_i}(\phi_i(p))$ are maximal ideals
                         if and only if $q = \phi_i(p)$ since the affine variety V_i
                                           is separated by Corollary 2.87.

Now there exists a regular map $g_i : U_i \to V_i$ such that $g_i^* = \phi_i^*$ (by Proposition 2.40). The calculation (3.9) shows that for $p \in U_i$ and $q \in V_i$ we have that $g_i(p) = q$ if and only if $g_i^*(I_{V_i}(q)) \subset I_{U_i}(p)$. Thus $\phi_i = g_i$ so that ϕ_i is a regular map. In particular the ϕ_i are all continuous so that ϕ is continuous.

Suppose that $q \in Y$ and $p \in \phi^{-1}(q)$. Then there exist U_i and V_i such that $p \in U_i$ and $q \in V_i$. For $f \in k[V_i]$,

$$\begin{aligned} f \in I_{V_i}(q) = I_{V_i}(\phi_i(p)) &\quad \text{if and only if} \quad f(\phi_i(p)) = 0 \\ &\quad \text{if and only if} \quad \phi_i^*(f)(p) = 0 \\ &\quad \text{if and only if} \quad f \in (\phi_i^*)^{-1}(I_{U_i}(p)). \end{aligned}$$

Thus $(\phi_i^*)^{-1}(I_{U_i}(p)) = I_{V_i}(q)$, and we have an induced k-algebra homomorphism

$$\phi_i^* : \mathcal{O}_{V_i,q} = k[V_i]_{I_{V_i}(q)} \to k[U_i]_{I_{U_i}(p)} = \mathcal{O}_{U_i,p}.$$

But this is just the statement that

$$\phi^* : \mathcal{O}_{Y,q} \to \mathcal{O}_{X,p}.$$

Thus

(3.10) $$\phi^* : \mathcal{O}_{Y,q} \to \bigcap_{p \in \phi^{-1}(q)} \mathcal{O}_{X,p}.$$

Suppose that U is an open subset of Y. Then

$$\mathcal{O}_Y(U) = \bigcap_{q \in U} \mathcal{O}_{Y,q}$$

and

$$\mathcal{O}_X(\phi^{-1}(U)) = \bigcap_{q \in U} \left(\bigcap_{p \in \phi^{-1}(q)} \mathcal{O}_{X,p} \right).$$

Thus by (3.10), we have that

$$\phi^* : \mathcal{O}_Y(U) \to \mathcal{O}_X(\phi^{-1}(U)).$$

We have established that ϕ satisfies Definition 3.28, and so ϕ is a regular map. $\square$

In the following theorem we show that the regular functions on a quasi-projective variety X are the regular maps from X to $\mathbb{A}^1$.

Theorem 3.40. *Suppose that X is a quasi-projective variety. Then $\mathcal{O}_X(X)$ is naturally isomorphic to the k-algebra of regular maps from X to $\mathbb{A}^1$.*

Proof. Let R be the k-algebra of regular maps from X to $\mathbb{A}^1$. Let t be the natural coordinate function on $\mathbb{A}^1$ defined by $t(q) = q$ for $q \in \mathbb{A}^1$. Suppose that $f \in \mathcal{O}_X(X)$. Then the association $p \mapsto f(p)$ is a well-defined map from X to $\mathbb{A}^1$, which we will denote by $f : X \to \mathbb{A}^1$. Since $f^*(t) = f$, the map $f^* : k[\mathbb{A}^1] = k[t] \to \mathcal{O}_X(X)$ is a well-defined k-algebra homomorphism. We have that $f : X \to \mathbb{A}^1$ is a regular map by Proposition 3.39, taking the trivial open cover $\{X\}$ of X and the trivial affine open cover $\{\mathbb{A}^1\}$ of $\mathbb{A}^1$. We thus have that the rule $\phi(f)(p) = f(p)$ for $f \in \mathcal{O}_X(X)$ and $p \in X$ determines a well-defined k-algebra homomorphism $\phi : \mathcal{O}_X(X) \to R$.

3.4. Regular functions and regular maps of quasi-projective varieties

The map ϕ is injective as shown before Definition 3.28. Now suppose that $g : X \to \mathbb{A}^1 \in R$. Then $g^* : k[\mathbb{A}^1] \to \mathcal{O}_X(X)$ is a k-algebra homomorphism. Let $f = g^*(t) \in \mathcal{O}_X(X)$. Then
$$\phi(f)(p) = [g^*(t)](p) = t(g(p)) = g(p)$$
for all $p \in X$. Thus ϕ is surjective. $\square$

Exercise 3.41. Let W be the projective variety (surface)
$$W = Z(x_0 x_1 - x_2 x_3) \subset \mathbb{P}^3.$$
We can write W as a union of an affine variety, $W \cap D(x_0)$, and the algebraic set $W \cap Z(x_0) \subset Z(x_0) \cong \mathbb{P}^2$. Let
$$y_1 = \frac{x_1}{x_0}, \qquad y_2 = \frac{x_2}{x_0}, \qquad y_3 = \frac{x_3}{x_0},$$
which are coordinate functions on $D(x_0) \cong \mathbb{A}^3$. Find the ideal $I(W \cap D(x_0)) \subset \mathcal{O}_{\mathbb{P}^3}(D(x_0)) = k[y_1, y_2, y_3]$. What is the algebraic set $W \cap Z(x_0)$ viewed as a subset of $\mathbb{P}^2$?

Exercise 3.42. Let U be the quasi-affine variety $U = \mathbb{A}^{n+1} \setminus \{(0, \ldots, 0)\}$. Consider the map
$$\pi : U \to \mathbb{P}^n$$
defined by $\pi(a_0, a_1, \ldots, a_n) = (a_0 : \ldots : a_n)$. Is π a regular map? Prove your answer.

Chapter 4

Regular and Rational Maps of Quasi-projective Varieties

In this chapter we define rational maps, give some useful ways to represent regular and rational maps, and give some examples. We show in Corollary 4.7 that every projective variety X has the basis of open sets $D(F)$ for $F \in S(X)$ which are (isomorphic to) affine varieties.

4.1. Criteria for regular maps

Lemma 4.1. *A composition of regular maps of quasi-projective varieties is a regular map.*

Proof. This follows from the definition of a regular map, Definition 3.28, since the composition of continuous functions is continuous and a composition of k-algebra homomorphisms is a k-algebra homomorphism. □

Proposition 4.2. *Suppose that U and V are quasi-projective varieties.*

1) *Suppose that U is an open subset of V. Then the inclusion*

$$i : U \to V$$

is a regular map.

2) *Suppose that U is a closed subset of V. Then the inclusion*

$$i : U \to V$$

is a regular map.

Proof. Let $\{V_i\}$ be a cover of V by affine open sets. Let $U_i = i^{-1}(V_i)$. Then $\{U_i\}$ is an open cover of U such that $i(U_i) \subset V_i$ for all i. In both cases 1) and 2), i^* is restriction of functions, so $i^* : \mathcal{O}_V(V_i) \to \mathcal{O}_U(U_i)$ for all i. By Proposition 3.39, i is a regular map. □

Generalizing Definition 2.94, we have the following definition.

Definition 4.3. *Suppose that $\phi : U \to V$ is a regular map of quasi-projective varieties.*

1) *The map ϕ is called a closed embedding if there exists a closed subvariety Z of V (an irreducible closed subset) such that $\phi(U) \subset Z$ and the induced regular map $\phi : U \to Z$ is an isomorphism.*

2) *The map ϕ is called an open embedding if there exists an open subset W of V such that $\phi(U) \subset W$ and the induced regular map $\phi : U \to W$ is an isomorphism.*

Proposition 4.4. *Suppose that X is a quasi-projective variety and*

$$\phi = (f_1, \ldots, f_n) : X \to \mathbb{A}^n$$

is a map. Then ϕ is a regular map if and only if f_i are regular functions on X for all i.

Proof. First suppose that $\phi : X \to \mathbb{A}^n$ is a regular map. Let $x_1, \ldots, x_n$ be the natural coordinate functions on $\mathbb{A}^n$, so that $k[\mathbb{A}^n] = k[x_1, \ldots, x_n]$. We have that $f_i = x_i \circ \phi : X \to \mathbb{A}^1$ is a regular map for $1 \leq i \leq n$ by Lemma 4.1, so $f_i \in \mathcal{O}_X(X)$ by Theorem 3.40.

Now suppose that $f_1, \ldots, f_n \in \mathcal{O}_X(X)$ and $\phi = (f_1, \ldots, f_n) : X \to \mathbb{A}^n$. Then the map $\phi^* : k[\mathbb{A}^n] \to \text{Map}(X, \mathbb{A}^1)$ has image in $O_X(X)$ and so ϕ is a regular map by Proposition 3.39 (taking the trivial open cover $\{X\}$ of X and the trivial affine open cover $\{\mathbb{A}^n\}$ of $\mathbb{A}^n$). □

Suppose that X is a quasi-projective variety and $\phi : X \to \mathbb{P}^n$ is a regular map. Let $S(\mathbb{P}^n) = k[y_0, \ldots, y_n]$. Suppose that $p \in X$. Then there exists a j such that $\phi(p) \in D(y_j)$. Let $V = D(y_j)$. Now $\phi^{-1}(V)$ is an open subset of X, so by Corollary 3.32, there exists an affine open neighborhood U of p in $\phi^{-1}(V)$. Consider the restriction $\phi : U \to V$. Then ϕ is a regular map of affine varieties, so on U, we can represent $\phi : U \to V \cong \mathbb{A}^n$ as $\phi = (f_1, \ldots, f_n)$ for some $f_1, \ldots, f_n \in k[U] = \mathcal{O}_X(U) \subset k(U) = k(X)$ (by

4.1. Criteria for regular maps

Proposition 3.37). Thus we have a representation $\phi = (f_1 : \ldots : f_j : 1 : f_{j+1} : \ldots : f_n)$ on the neighborhood U of p.

In summary, we have shown that there exists an open neighborhood U of p in X and regular functions $f_0, \ldots, f_n \in \mathcal{O}_X(U)$ such that

(4.1) $$\phi = (f_0 : \ldots : f_n)$$

on U and there are no points on U where all of the f_i vanish.

Suppose that $q \in X$ is another point and Y is an open neighborhood of q in X with regular functions $g_0, \ldots, g_n \in \mathcal{O}_X(Y)$ such that

(4.2) $$\phi = (g_0 : \ldots : g_n)$$

on Y and the g_i have no common zeros on Y. These two representations of ϕ must agree on $U \cap Y$, which happens if and only if

(4.3) $$f_i g_j - f_j g_i = 0 \quad \text{for } 0 \leq i, j \leq n$$

on $U \cap Y$ (which occurs if and only if $f_i g_j - f_j g_i = 0$ in $k(x)$).

We can also use this method to construct regular maps. Suppose that X is a quasi-projective variety, $\{U_s\}$ is an affine open cover of X, and for all U_s,

(4.4)
$f_{s,0}, \ldots, f_{s,n} \in \mathcal{O}_X(U_s)$ are functions that have no common zeros on U_s

and

(4.5) $$f_{s,i} f_{t,j} - f_{s,j} f_{t,i} = 0 \quad \text{for all } s, t \text{ and } 0 \leq i, j \leq n.$$

Then by Proposition 3.39 (and Proposition 2.96), the collection of maps (4.4) on an open cover $\{U_s\}$ of X satisfying (4.5) determines a regular map $\phi : X \to \mathbb{P}^n$.

We can thus think of a regular map $\phi : X \to \mathbb{P}^n$ as an equivalence class of expressions $(f_0 : f_1 : \ldots : f_n)$ with $f_0, \ldots, f_n \in k(X)$ and such that

$$(f_0 : f_1 : \ldots : f_n) \sim (g_0 : \ldots : g_n)$$

if and only if

$$f_i g_j - f_j g_i = 0 \quad \text{for } 0 \leq i, j \leq n.$$

We further have the condition that for each $p \in X$ there exists a representative $(f_0 : f_1 : \ldots : f_n)$ of ϕ such that $f_i \in \mathcal{O}_{X,p}$ for all i and some f_i does not vanish at p.

We can reinterpret this to give another useful way to represent a regular map $\phi : X \to \mathbb{P}^n$. The quasi-projective variety X is an open subset of a projective variety W which is a closed subvariety of a projective space $\mathbb{P}^m$. In the representation (4.1) of ϕ near $p \in X$, we can write the regular maps as ratios $f_i = \frac{F_i}{G_i}$ where F_i, G_i are homogeneous elements of a common degree d_i in the homogeneous coordinate ring $S(W)$ (which is a quotient of $S(\mathbb{P}^m)$)

and the G_i do not vanish near p. We can thus represent ϕ near p by the expression
$$(H_0 : H_1 : \ldots : H_n)$$
where the $H_i = (\prod_{j \neq i} G_j) F_i$ are homogeneous elements of $S(W)$ (or of $S(\mathbb{P}^m)$) of a common degree and at least one H_i does not vanish at p.

We can thus think of a regular map $\phi : X \to \mathbb{P}^n$ as an equivalence class of expressions $(F_0 : F_1 : \ldots : F_n)$ with $F_0, \ldots, F_n$ homogeneous elements of $S(W)$ (or of $S(\mathbb{P}^m)$) all having the same degree and such that
$$(F_0 : F_1 : \ldots : F_n) \sim (G_0 : \ldots : G_n)$$
if and only if
$$F_i G_j - F_j G_i = 0 \text{ for } 0 \leq i, j \leq n$$
in $S(W)$. We further have the condition that for each $p \in X$ there exists a representative $(F_0 : F_1 : \ldots : F_n)$ of ϕ such that some F_i does not vanish at p. It is not required that the common degree of the F_i be the same as the common degree of the G_j.

4.2. Linear isomorphisms of projective space

Suppose that $A = (a_{ij})$ is an invertible $(n+1) \times (n+1)$ matrix with coefficients in k (indexed as $0 \leq i, j \leq n$). Define homogeneous elements L_i of degree 1 in $S = k(\mathbb{P}^n) = k[x_0, \ldots, x_n] = \bigoplus S_i$ by $L_i = \sum_{j=0}^{n} a_{ij} x_j$ for $0 \leq i \leq n$. Then the L_i are a k-basis of S_1 so that $Z(L_0, \ldots, L_n) = Z(x_0, \ldots, x_n) = \emptyset$. Thus $\phi_A : \mathbb{P}^n \to \mathbb{P}^n$ defined by
$$\phi_A = (L_0 : \ldots : L_n)$$
is a regular map. If B is another invertible $(n+1) \times (n+1)$ matrix with coefficients in k, then we have that
$$\phi_A \circ \phi_B = \phi_{AB}.$$
Thus ϕ_A is an isomorphism of $\mathbb{P}^n$, with inverse map $\phi_{A^{-1}}$.

We will call $L_0, \ldots, L_n$ *homogeneous coordinates* on $\mathbb{P}^n$.

Proposition 4.5. *Suppose that W is a projective variety which is a closed subvariety of $\mathbb{P}^n$. Suppose that $L \in S(\mathbb{P}^n)$ is a linear homogeneous form, such that $D(L) \cap W \neq \emptyset$. Then $D(L) \cap W$ is an affine variety.*

Proof. Write $L = \sum_{j=0}^{n} a_{0j} x_j$ for some $a_{0j} \in k$ not all zero, and extend the vector $(a_{00}, \ldots, a_{0n})$ to a basis of k^{n+1}. Arrange this basis as the rows of the $(n+1) \times (n+1)$ matrix $A = (a_{ij})$. Here A is necessarily invertible. Now the isomorphism $\phi_A : \mathbb{P}^n \to \mathbb{P}^n$ maps $D(L)$ to $D(x_0)$ and W to a projective variety $\phi_A(W)$ which is not contained in $Z(x_0)$. We have that

4.3. The Veronese embedding

$\phi_A(W) \cap D(x_0)$ is an affine variety by Theorem 3.30. Thus $W \cap D(L)$ is affine since it is isomorphic to $\phi_A(W) \cap D(x_0)$. □

Composing the isomorphism $\phi_A^* : \mathcal{O}_{\phi_A(W)}(D(x_0)) \cong \mathcal{O}_W(D(L))$ of the above proof with the representation of $\mathcal{O}_{\phi_A(W)}(D(x_0))$ of (3.8), letting $S(\mathbb{P}^n) = k[x_0, \ldots, x_n]$ and $S(W) = S(\mathbb{P}^n)/I(W) = k[\overline{x}_0, \ldots, \overline{x}_n]$, we obtain that

$$(4.6) \qquad \mathcal{O}_W(D(L)) = k\left[\frac{\overline{x}_0}{\overline{L}}, \ldots, \frac{\overline{x}_n}{\overline{L}}\right] \cong k\left[\frac{x_0}{L}, \ldots, \frac{x_n}{L}\right]/J$$

where $J = \{f(\frac{x_0}{L}, \ldots, \frac{x_n}{L}) \mid f \in I(W)\}$. With the notation introduced before Lemma 3.4, we have that

$$\mathcal{O}_W(D(L)) = S(W)_{(\overline{L})} \cong S(\mathbb{P}^n)_{(L)}/I(W)_{(L)}.$$

4.3. The Veronese embedding

Suppose that d is a positive integer. Let $x_0^d, x_0^{d-1}x_1, \ldots, x_n^d$ be the set of all monomials of degree d in $S(\mathbb{P}^n) = k[x_0, \ldots, x_n]$. There are $\binom{n+d}{n}$ such monomials. Let $e = \binom{n+d}{d} - 1$. Since these monomials are a k-basis of S_d, we have that $Z(x_0^d, x_0^{d-1}x_1, \ldots, x_n^d) = \emptyset$. Thus we have a regular map

$$\Lambda : \mathbb{P}^n \to \mathbb{P}^e$$

defined by $\Lambda = (x_0^d : x_0^{d-1}x_1 : \ldots : x_n^d)$. Let W be the closure of $\Lambda(\mathbb{P}^n)$ in $\mathbb{P}^e$.

Let $\mathbb{P}^e$ have the homogeneous coordinates $y_{i_0 i_1 \ldots i_n}$ where $i_0, \ldots, i_n$ are nonnegative integers such that $i_0 + \cdots + i_n = d$. The map Λ is defined by the equations

$$(4.7) \qquad y_{i_0 i_1 \ldots i_n} = x_0^{i_0} x_1^{i_1} \cdots x_n^{i_n}.$$

We will establish that Λ is an isomorphism of $\mathbb{P}^n$ to W. Suppose that $q \in W$. We can verify that $q = \Lambda(p)$ for some $p \in \mathbb{P}^n$ (in Theorem 5.14 we will establish this generally), so that $x_j(p) \neq 0$ for some j. We have that $y_{0\ldots 0 d 0 \ldots 0}(q) = x_j^d(p) \neq 0$, where d is in the j-th place of $y_{0\ldots 0 d 0 \ldots 0}$. Thus the affine open sets $W_j = D(y_{0\ldots 0 d 0 \ldots 0})$ of $\mathbb{P}^e$ cover W. Let

$$S(W) = S(\mathbb{P}^e)/I(W) = k[\{\overline{y}_{i_0, \ldots, i_n}\}].$$

In $S(W)$, we have the identities

$$(4.8) \quad \overline{y}_{i_0 \ldots i_n} \overline{y}_{j_0 \ldots j_n} = \overline{y}_{k_0 \ldots k_n} \overline{y}_{l_0 \ldots l_n} \text{ if } i_0 + j_0 = k_0 + l_0, \ldots, i_n + j_n = k_n + l_n.$$

Now on each open subset $W_j = D(y_{0\ldots 0 d 0 \ldots 0}) \cap W$ of W we define a regular map

$$\Psi_j : W_j \to D(x_j) \subset \mathbb{P}^n$$

by

$$\Psi_j = (\overline{y}_{10\ldots 0(d-1)0\ldots 0} : \overline{y}_{010\ldots 0(d-1)0\ldots 0} : \cdots : \overline{y}_{0\ldots 0 d 0 \ldots 0} : \cdots : \overline{y}_{0\ldots 0(d-1)0\ldots 01}).$$

Now we have that
$$\Psi_j \circ \Lambda = (x_0 x_j^{d-1} : x_1 x_j^{d-1} : \ldots : x_n x_j^{d-1}) = (x_0 : x_1 : \ldots : x_n) = \mathrm{id}_{D(x_j)},$$
and the identities
$$\overline{y}_{i_0 \ldots i_n} \overline{y}_{0\ldots 0(d-1)0\ldots 0}^{d} = \overline{y}_{10\ldots 0(d-1)0\ldots 0}^{i_0} \overline{y}_{010\ldots 0(d-1)0\ldots 0}^{i_1} \cdots \overline{y}_{0\ldots 0(d-1)0\ldots 01}^{i_n}$$
whenever $i_0 + i_1 + \cdots + i_n = d$ (which are special cases of (4.8)) imply that
$$\Lambda \circ \Psi_j = \mathrm{id}_{W_j}.$$

Now we check, again using the identities (4.8), that $\Psi_j = \Psi_k$ on $W_j \cap W_k$. Thus the Ψ_j patch by Proposition 3.39 to give a regular map $\Psi : W \to \mathbb{P}^n$ which is an inverse to Λ.

The isomorphism $\Lambda : \mathbb{P}^n \to W$ is called the Veronese embedding.

Recalling the definition (3.2) of a Veronese ring of a graded ring, we have a graded, degree-preserving isomorphism
$$\Lambda^* : S(W) \to S(\mathbb{P}^n)^{(d)} = \bigoplus_{i \geq 0} S(\mathbb{P}^n)_{id}$$
where elements of $S(\mathbb{P}^n)^{(d)}_i = S(\mathbb{P}^n)_{id}$ have degree i. The map is the k-algebra homomorphism defined by
$$\Lambda^*(\overline{y}_{i_0,\ldots,i_n}) = x_0^{i_0} \cdots x_n^{i_n}.$$

We obtain the following result using the Veronese embedding, composed with a linear isomorphism, which generalizes Proposition 4.5.

Proposition 4.6. *Suppose that W is a projective variety which is a closed subvariety of $\mathbb{P}^n$. Suppose that $F \in S(\mathbb{P}^n)$ is a homogeneous form of degree $d > 0$ such that $D(F) \cap W \neq \emptyset$. Then $D(F) \cap W$ is an affine variety.*

Letting $S(\mathbb{P}^n) = k[x_0, \ldots, x_n]$ and $S(W) = S(\mathbb{P}^n)/I(W) = k[\overline{x}_0, \ldots, \overline{x}_n]$, we obtain that
(4.9)
$$\mathcal{O}_W(D(F))$$
$$= k[\tfrac{\overline{M}}{\overline{F}} \mid \overline{M} \text{ is a monomial in } \overline{x}_0, \ldots, \overline{x}_n \text{ of degree } d, \overline{F} = F(\overline{x}_0, \ldots, \overline{x}_n)]$$
$$\cong k[\tfrac{M}{F} \mid M \text{ is a monomial in } x_0, \ldots, x_n \text{ of degree } d]/J$$
where $J = \{\tfrac{G(x_0,\ldots,x_n)}{F^e} \mid G \in I(W) \text{ is a form of degree } de\}$. With the notation introduced before Lemma 3.4, we have that
$$\mathcal{O}_W(D(F)) = S(W)_{(\overline{F})} \cong S(\mathbb{P}^n)_{(F)}/I(W)_{(F)}.$$

Corollary 4.7. *Suppose that X is a projective variety. Then the set of affine open subsets $D(F)$ for $F \in S(X)$ is a basis of the Zariski topology on X.*

Proof. This follows from Proposition 4.6 and Lemma 3.16. □

Exercise 4.8. Suppose that k has characteristic $\neq 2$. Let W be the conic $W = Z(x_0^2 + x_1^2 + x_2^2) \subset \mathbb{P}^3$. Then W has the coordinate ring

$$S(W) = k[x_0, x_1, x_2]/(x_0^2 + x_1^2 + x_2^2).$$

Compute $\mathcal{O}_W(D(L))$ where L is the linear form $L = x_0 + x_1 + x_2$. Express your answer as a quotient of a polynomial ring by an ideal.

Exercise 4.9. A subset W of $\mathbb{P}^n$ is called a cone if there exists a closed subvariety Y of a linear subvariety $\mathbb{P}^r$ of $\mathbb{P}^n$ and a linear subvariety $Z \cong \mathbb{P}^{n-r-1}$ of $\mathbb{P}^n$ such that $\mathbb{P}^r \cap Z = \emptyset$ and W is the union of all lines joining a point of Y to a point of Z. Show that a cone W is a closed subvariety of $\mathbb{P}^n$ and that there exist homogeneous coordinates $x_0, \ldots, x_n$ on $\mathbb{P}^n$ such that $x_0, \ldots, x_r$ are homogeneous coordinates on $\mathbb{P}^r$, and if $I(Y)$ is the ideal of Y in $S(\mathbb{P}^r) = k[x_0, \ldots, x_r]$, then $I(W) = I(Y)k[x_0, \ldots, x_n]$ in $S(\mathbb{P}^n) = k[x_0, \ldots, x_n]$.

4.4. Rational maps of quasi-projective varieties

Suppose that X and Y are quasi-projective varieties. As in Section 2.6, we define a rational map $\phi : X \dashrightarrow Y$ to be a regular map on some nonempty open subset U of X to Y. As in the affine case, we have that ϕ has a unique extension as a regular map to a largest open subset $W(\phi)$ of X, and if ϕ and ψ are rational maps from X to Y which are regular on respective open subsets U and V such that ϕ and ψ agree on the intersection $U \cap V$, then $\phi = \psi$.

We now formulate the concept of a rational map of projective varieties in algebra.

The following definition extends Definition 2.106.

Definition 4.10. Suppose that X is a projective variety. A rational map $\phi : X \dashrightarrow \mathbb{P}^n$ is an equivalence class of $(n+1)$-tuples $\phi = (f_0 : \ldots : f_n)$ with $f_0, \ldots, f_n \in k(X)$ not all zero, where $(g_0 : \ldots : g_n)$ is equivalent to $(f_0 : \ldots : f_n)$ if $f_i g_j - f_j g_i = 0$ for $0 \leq i, j \leq n$.

A rational map $\phi : X \dashrightarrow \mathbb{P}^n$ is regular at a point $p \in X$ if and only if there exists a representation $(f_0 : \ldots : f_n)$ of ϕ such that all of the f_i are regular functions at p ($f_i \in \mathcal{O}_{X,p}$ for all i) and some $f_i(p) \neq 0$.

Let $W(\phi)$ be the open set of points of X on which ϕ is regular. Then $\phi : W(\phi) \to \mathbb{P}^n$ is a regular map.

Suppose that Y is a projective variety which is a closed subvariety of a projective space $\mathbb{P}^n$. A rational map $\phi : X \dashrightarrow Y$ is a rational map $\phi : X \dashrightarrow \mathbb{P}^n$ such that $\phi(W(\phi)) \subset Y$.

A rational map $\phi : X \dashrightarrow Y$ is called dominant if $\phi(U)$ is dense in Y when U is a (nontrivial) open subset of X on which ϕ is a regular map.

Definition 2.109 extends to define a birational map of quasi-projective varieties, as follows.

Definition 4.11. A dominant rational map $\phi : X \dashrightarrow Y$ of quasi-projective varieties is birational if there exists a dominant rational map $\psi : Y \dashrightarrow X$ such that $\psi \circ \phi = \mathrm{id}_X$ and $\phi \circ \psi = \mathrm{id}_Y$.

Two quasi-projective varieties X and Y are said to be birationally equivalent if there exists a birational map $\phi : X \dashrightarrow Y$.

Suppose that $\phi : X \dashrightarrow Y$ is a rational map. Let U be an open subset of X such that $\phi : U \to Y$ is a regular map. Let A be an open affine subset of Y such that $A \cap \phi(U) \neq \emptyset$, and let B be an open affine subset of $(\phi|U)^{-1}(A)$. Then ϕ induces a regular map of affine varieties $\phi : B \to A$ (by Definition 3.28 and Propositions 2.95 and 2.96). From this reduction, we obtain that the results of Section 2.6 on rational maps of affine varieties are also valid for projective varieties. We obtain the following generalization of Proposition 2.112.

Proposition 4.12. *Every projective variety X is birationally equivalent to a hypersurface $Z(G) \subset \mathbb{P}^n$.*

It is sometimes convenient to interpret rational maps in terms of equivalence classes of $(n+1)$-tuples of homogeneous forms $(H_0 : \ldots : H_n)$ of a common degree. Recalling our analysis of regular maps in the previous section, we see that a rational map $\phi : X \dashrightarrow \mathbb{P}^n$ can also be interpreted as an equivalence class of $(n+1)$-tuples $(F_0 : \ldots : F_n)$ with $F_0, \ldots, F_n$ forms of the same degree in $S(X)$ which are not all zero, where $(G_0 : \ldots : G_n)$ is equivalent to $(F_0 : \ldots : F_n)$ if $F_i G_j - F_j G_i = 0$ for $0 \leq i, j \leq n$. A rational map ϕ is regular at a point $p \in X$ if and only if there exists a representation $(F_0 : \ldots : F_n)$ of ϕ by forms in $S(X)$ of the same degree such that at least one of the F_i does not vanish at p.

The image of a rational map $\phi : X \dashrightarrow Y$ is the Zariski closure in Y of $\phi(U)$ where U is a nontrivial open subset of X on which ϕ is regular.

Lemma 4.13. *Suppose that X is a projective variety and $\phi : X \dashrightarrow \mathbb{P}^n$ is a rational map. Then the image of ϕ has the coordinate ring $k[H_0, \ldots, H_n]$ for any equivalence class of homogeneous forms $H_0, \ldots, H_n$ representing ϕ.*

Proof. Let W be the image of ϕ in $\mathbb{P}^n$. Let $V = X \setminus Z(H_0, \ldots, H_n)$. Then W is the Zariski closure of $\phi(V)$. Let $x_0, \ldots, x_n$ be our homogeneous coordinates on $\mathbb{P}^n$, so that $S(\mathbb{P}^n) = k[x_0, \ldots, x_n]$. We have a surjective

k-algebra homomorphism
$$\Phi : k[x_0, \ldots, x_n] \to k[H_0, \ldots, H_n]$$
defined by $\Phi(x_i) = H_i$ for $0 \le i \le n$. Let I be the kernel of this map.

For $0 \le i \le n$, we have regular maps
$$\phi_i = (\phi|X_{H_i}) : X_{H_i} \to \mathbb{P}^n_{x_i}$$
of affine varieties, with
$$\phi_i^* : k[\mathbb{P}^n_{x_i}] = k\left[\frac{x_0}{x_i}, \ldots, \frac{x_n}{x_i}\right] \to k[X_{H_i}]$$
defined by
$$\phi_i^*\left(\frac{x_j}{x_i}\right) = \frac{H_j}{H_i}$$
for $0 \le j \le n$. The kernel of ϕ_i^* is
$$I(W \cap \mathbb{P}^n_{x_i}) = \left\{ F\left(\frac{x_0}{x_i}, \ldots, \frac{x_n}{x_i}\right) \mid F \in I(W) \right\} = I(W)_{(x_i)}$$
and the image of ϕ_i^* is $k[\frac{H_0}{H_i}, \ldots, \frac{H_n}{H_i}]$. For all i we have short exact sequences
$$0 \to I_{(x_i)} \to S(\mathbb{P}^n)_{(x_i)} = k\left[\frac{x_0}{x_i}, \ldots, \frac{x_n}{x_i}\right] \xrightarrow{\phi_i^*} k\left[\frac{H_0}{H_i}, \ldots \frac{H_n}{H_i}\right] \to 0.$$

Thus $I_{(x_i)} = I(W)_{(x_i)}$ for $0 \le i \le n$, and thus $I^{\mathrm{sat}} = I(W)^{\mathrm{sat}}$ by Lemma 3.4. Since both $I(W)$ and I are prime ideals, we have that $I(W) = I$. $\square$

4.5. Projection from a linear subspace

A linear subspace E of a projective space $\mathbb{P}^n$ is the closed subset defined by the vanishing of a set of linear homogeneous forms. Such a subvariety is isomorphic to a projective space $\mathbb{P}^d$ for some $d \le n$. The ideal $I(E)$ is then minimally generated by a set of $n - d$ linear forms; in fact a set of linear forms $\{L_1, \ldots, L_{n-d}\}$ is a minimal set of generators of $I(E)$ if and only if $L_1, \ldots, L_{n-d}$ is a k-basis of the k-linear subspace S_1 of the homogeneous linear forms on $\mathbb{P}^n$ which vanish on E. We will say that E has dimension d.

Suppose that E is a d-dimensional linear subspace of $\mathbb{P}^n$. Let $L_1, \ldots, L_{n-d}$ be linear forms in $k[x_0, \ldots, x_n]$ which define E. The rational map $\phi : \mathbb{P}^n \dashrightarrow \mathbb{P}^{n-d-1}$ with $\phi = (L_1 : \ldots : L_{n-d})$ is called the projection from E. The map is regular on the open set $\mathbb{P}^n \setminus E$.

This map can be interpreted geometrically as follows: choose a linear subspace F of $\mathbb{P}^n$ of dimension $n-d-1$ which is disjoint from E (to find such an F, just extend $L_1, \ldots, L_{n-d}$ to a k-basis $L_1, \ldots, L_{n-d}, M_1, \ldots, M_{d+1}$ of S_1 (the vector space of linear forms in $S(\mathbb{P}^n)$), and let $F = Z(M_1, \ldots, M_{d+1})$. Suppose that $p \in \mathbb{P}^n \setminus E$. Let G_p be the unique linear subspace of $\mathbb{P}^n$ of dimension $d+1$ which contains p and E. We have that G_p intersects F in a unique point. This intersection point can be identified with $\phi(p)$.

This map depends on the choice of basis $L_1, \ldots, L_{n-d}$ of linear forms which define E. However, there is not a significant difference if a different basis $L'_1, \ldots, L'_{n-d}$ is chosen. In this case there is a linear isomorphism $\Lambda : \mathbb{P}^{n-d-1} \to \mathbb{P}^{n-d-1}$ such that $(L_1 : \ldots : L_{n-d}) = \Lambda \circ (L'_1 : \ldots : L'_{n-d})$.

Exercise 4.14. Suppose that $\phi : \mathbb{P}^m \to \mathbb{P}^n$ is a regular map. Show that there exist homogeneous forms $F_0, \ldots, F_n$ of $S(\mathbb{P}^m)$, all of a common degree, such that $Z(F_0, \ldots, F_n) = \emptyset$ and $\phi = (F_0 : \ldots : F_n)$. (The conclusions of this exercise are not true for $\mathbb{P}^m$ replaced with a projective variety W as is shown in the next exercise).

Exercise 4.15. Let $W = Z(xw - yz) \subset \mathbb{P}^3$, where
$$S(W) = k[x, y, z, w]/(xw - yz) = k[\overline{x}, \overline{y}, \overline{z}, \overline{w}].$$
Consider the rational map $\phi : W \dashrightarrow \mathbb{P}^1$ defined by $\phi = (\overline{x} : \overline{y})$.

a) Show that $(\overline{x} : \overline{y}) \sim (\overline{z} : \overline{w})$ as rational maps, so that ϕ is represented by both $(\overline{x} : \overline{y})$ and $(\overline{z} : \overline{w})$. Then show that ϕ is a regular map of W.

b) Show that there do not exist forms $F_0, F_1 \in S(W)$ of a common degree, such that ϕ is represented by $(F_0 : F_1)$ and $Z_W(F_0, F_1) = \emptyset$. To do this, assume that such F_0, F_1 do exist, and derive a contradiction. First show that $p \in W_{\overline{y}}$ implies $F_1(p) \neq 0$. Then show that $p \in W_{\overline{w}}$ implies $F_1(p) \neq 0$. Conclude that
$$Z_W(F_1) \cap (W_{\overline{y}} \cup W_{\overline{w}}) = \emptyset.$$
Consider the affine variety $X = Z(xw - yz) \subset \mathbb{A}^4$, which has regular functions $k[X] = k[\overline{x}, \overline{y}, \overline{z}, \overline{w}]$. Let $U = X_{\overline{y}} \cup X_{\overline{w}}$, an open subset of X. Show that $\frac{F_0}{F_1} \in \mathcal{O}_X(U)$. Now consider Example 2.85 and explain how the conclusions of this example give a contradiction.

Exercise 4.16. Let V be the vector space $V = k^{n+1}$, and let
$$\pi : V \setminus \{(0, \ldots, 0)\} \to \mathbb{P}^n$$
be the map $\pi(c_0, \ldots, c_n) = (c_0 : \ldots : c_n)$ for $(c_0, \ldots, c_n) \in V \setminus \{(0, \ldots, 0)\}$.

Writing $S(\mathbb{P}^n) = k[x_0, \ldots, x_n] = \bigoplus_{i \geq 0} S_i$, we have an identification of the linear forms S_1 on $\mathbb{P}^n$ with the linear forms on V; that is, S_1 is naturally isomorphic (as a k-vector space) to the dual space V^*.

a) Suppose that W is a linear subspace of V of dimension $d > 0$. Let $X = \pi(W \setminus \{(0, \ldots, 0)\})$. Show that X is a closed subset of $\mathbb{P}^n$

4.5. Projection from a linear subspace

which is isomorphic to $\mathbb{P}^{d-1}$. (We will say that X has dimension $d-1$). Define

$$W^\perp = \{L \in V^* \mid L(w) = 0 \text{ for all } w \in W\}.$$

Show that $X = Z(W^\perp)$.

b) Show that every linear subspace X of $\mathbb{P}^n$ is the image

$$\pi(W \setminus \{(0,\ldots,0)\})$$

for some linear subspace W of V.

c) Suppose that $w_0, \ldots, w_r \in V$ are nonzero. Define the linear span L of $\pi(w_0), \ldots, \pi(w_r)$ in $\mathbb{P}^n$ to be

$$L = \{\pi(c_0 w_0 + \cdots + c_n w_n) \mid c_0, \ldots, c_n \in k \text{ and } c_0 w_0 + \cdots + c_n w_n \neq (0,\ldots,0)\}.$$

Show that L is a linear subspace of $\mathbb{P}^n$.

Exercise 4.17. Suppose that

$$A = \begin{pmatrix} v_{0,0} & \cdots & v_{0,m} \\ \vdots & & \vdots \\ v_{n,0} & \cdots, & v_{n,m} \end{pmatrix}$$

is an $(n+1) \times (m+1)$ matrix with coefficients in our algebraically closed field k, such that $\operatorname{rank}(A) = m+1$. Let $v_i = (v_{0,i}, \ldots, v_{n,i}) \in V = k^{n+1}$ for $0 \leq i \leq m$, and let W be the span of $v_0, \ldots, v_m$ in V. Let F be the linear subspace $\pi(W \setminus \{(0,\ldots,0)\})$ of $\mathbb{P}^n$. Define

$$\phi_A : \mathbb{P}^m \to \mathbb{P}^n$$

by

$$\phi_A(c_0 : \ldots : c_m) = \pi(c_0 v_0 + \cdots + c_m v_m)$$
$$= (\textstyle\sum_{j=0}^{m} v_{0,j} c_j : \ldots : \sum_{j=0}^{m} v_{n,j} c_j)$$

for $(c_0, \ldots, c_m) \in \mathbb{P}^m$. Show that ϕ_A is a regular map, which is an isomorphism onto the linear subspace F of $\mathbb{P}^n$.

Exercise 4.18. The purpose of this exercise is to prove the geometric interpretation of projection from a linear subspace stated in this section. We begin by recalling notation.

Suppose that E is a d-dimensional linear subspace of $\mathbb{P}^n$. Let $L_1, \ldots, L_{n-d}$ be linear forms in $k[x_0, \ldots, x_n]$ which define E. The rational map $\phi : \mathbb{P}^n \to \mathbb{P}^{n-d-1}$ with $\phi = (L_1 : \ldots : L_n)$ is the projection from E. The map is regular on the open set $\mathbb{P}^n \setminus E$.

Choose a linear subspace F of $\mathbb{P}^n$ of dimension $n-d-1$ which is disjoint from E (to find such an F, just extend $L_1,\ldots,L_{n-d}$ to a k-basis $L_1,\ldots,L_{n-d},M_1,\ldots,M_{d+1}$ of S_1, the vector space of linear forms in $S(\mathbb{P}^n)$), and let $F = Z(M_1,\ldots,M_{d+1})$. Identifying $V = k^{n+1}$ with V^{**}. Let

$$\{v_1,\ldots,v_{n-d},w_1,\ldots,w_{d+1}\}$$

be the ordered dual basis to the ordered basis $\{L_1,\ldots,L_{n-d},M_1,\ldots,M_{d+1}\}$ of $S_1 = V^*$.

Suppose that $p \in \mathbb{P}^n \setminus E$. Let G_p be the unique linear subspace of $\mathbb{P}^n$ of dimension $d+1$ which contains p and E. Show that G_p intersects F in a unique point and

$$G_p = \phi_A(\phi(p)),$$

where

$$A = \begin{pmatrix} v_1 \\ \vdots \\ v_{n-d} \end{pmatrix}^t$$

(so that $\phi_A : \mathbb{P}^{n-d-1} \to \mathbb{P}^n$ is an isomorphism onto F).

Exercise 4.19. Suppose that $\phi : X \dashrightarrow Y$ is a dominant rational map of varieties, with induced inclusion of function fields $\phi^* : k(Y) \to k(X)$. Suppose that $p \in X$. Show that ϕ is regular at p if and only if there exists $q \in Y$ such that $\mathcal{O}_{Y,q} \subset \mathcal{O}_{X,p}$, and when this happens, $\phi(p) = q$.

Exercise 4.20. Suppose that $p = (a_0 : \ldots : a_n)$ and $q = (b_0 : \ldots : b_n)$ are points in $\mathbb{P}^n$. Show that there is a unique projective line L (a one-dimensional linear subvariety) in $\mathbb{P}^n$ containing p and q. Compute the ideal $I(L)$ in $S(\mathbb{P}^n)$.

Suppose that $p_i = (a_0(i) : \ldots : a_n(i))$ for $1 \leq i \leq m$ are points in $\mathbb{P}^n$ such that the matrix $(a_j(i))$ has rank m. Show that there is a unique linear subvariety M of $\mathbb{P}^n$ of dimension $n-m+1$ containing p_i for $1 \leq i \leq m$. Compute the ideal $I(M)$ in $S(\mathbb{P}^n)$.

Chapter 5

Products

In this chapter we define the product of two varieties and explore some of its basic properties. We define graphs of regular and rational maps.

5.1. Tensor products

The tensor product of two modules is defined as follows:

Definition 5.1. Let R be a ring and M, N be R-modules. A tensor product T of M and N is an R-module T and an R-bilinear mapping $g : M \times N \to T$ with the following universal property: given an R-module P and an R-bilinear map $f : M \times N \to P$, there exists a unique R-linear map $f' : T \to P$ such that $f = f' \circ g$.

Tensor products always exist and are uniquely determined by the universal property [**13**, Proposition 2.12]. The tensor product is denoted by $M \otimes_R N$.

In the section on "Tensor products of algebras" [**13**, pages 30–31], it is shown that the tensor product $A \otimes_R B$ of two R-algebras A and B naturally has the structure of an R-algebra.

Tensor products behave well with respect to localization. Let S be a multiplicative set in a ring A and let M be an A-module. Then $S^{-1}M \cong S^{-1}A \otimes_A M$ and by [**128**, Definition on page 100] and [**128**, Corollary 3.72] if M, N are two A-modules, then by [**128**, Lemma 3.77]

$$S^{-1}(M \otimes_A N) \cong S^{-1}M \otimes_{S^{-1}A} S^{-1}N.$$

In the case when $R = K$ is a field and $M = A$, $N = B$ are rings containing K, there is an alternate definition of the tensor product. This is developed by Zariski and Samuel [**160**, Section 14 of Chapter III].

Definition 5.2. Suppose that S is a ring containing a field K. Two K-subspaces L and L' of S are said to be linearly disjoint over K if the following condition is satisfied: whenever $x_1, \ldots, x_n$ are elements of L which are linearly independent over K and $x'_1, \ldots, x'_m$ are elements of L' which are linearly independent over K, then the mn products $x_i x'_j$ are also linearly independent over K.

Many important applications of this definition are given in [**160**, Section 15 of Chapter II]. A useful equivalent formulation is: the K-subspaces L and L' are linearly disjoint over K if and only if:

> Whenever $x_1, \ldots, x_n$ are elements of L which are linearly independent over K, these elements x_i are also linearly independent over L'.

Theorem 5.3. *Suppose that K is a field and A, B, C are K-algebras with K-algebra isomorphisms ϕ and ψ of A, respectively B, to K-subalgebras of C such that:*

1. *C is generated by $\phi(A)$ and $\psi(B)$ as a K-algebra.*
2. *$\phi(A)$ and $\phi(B)$ are linearly disjoint over K.*

Then C is a tensor product $A \otimes_K B$.

Theorem 5.3 follows from the observation that the construction of C in [**160**, Theorem 33 on page 179] is the same as the construction of the tensor product in [**13**, Proposition 2.12].

Lemma 5.4. *Suppose that K is an algebraically closed field and L, K' are any field extensions of K. Then $L \otimes_K K'$ is an integral domain.*

Lemma 5.4 is proven in [**160**, Corollary 1 to Theorem 40, page 198].

Proposition 5.5. *Suppose that R is a ring and G is an R-module. Let*

$$0 \to L \to M \to N \to 0$$

be a short exact sequence of R-modules. Then the sequence

(5.1) $$G \otimes_R L \to G \otimes_R M \to G \otimes_R N \to 0$$

is right exact.

Proof. [**95**, Proposition 2.6, page 610] or [**13**, Proposition 2.18]. □

An R-module G for which the sequence (5.1) is always short exact is called a flat R-module. Locally free modules are aways flat.

Theorem 5.6. *An R-module M is flat if and only if the following condition holds. Suppose that $a_i \in R$, $x_i \in M$, for $1 \le i \le r$ and $\sum_{i=1}^r a_i x_i = 0$. Then there exists an integer s and elements $b_{ij} \in R$ and $y_j \in M$, for $1 \le j \le s$, such that $\sum_i a_i b_{ij} = 0$ for all j and $x_i = \sum_j b_{ij} y_j$ for all i.*

Proof. [23, Corollary I.11.1, page 27]. □

5.2. Products of varieties

Now we define products of quasi-projective varieties. We continue to assume that k is a fixed algebraically closed field. Suppose that X and Y are varieties. We will put a structure of a variety on the set $X \times Y$ which has the property that the projections $\pi_1 : X \times Y \to X$ and $\pi_2 : X \times Y \to Y$ are regular maps.

We construct the product of $\mathbb{A}^m$ and $\mathbb{A}^n$ as the variety

$$\mathbb{A}^m \times \mathbb{A}^n = \mathbb{A}^{m+n}.$$

As sets, this gives a natural identification, and this identification makes $\mathbb{A}^m \times \mathbb{A}^n$ into an affine variety. The projections $\pi_1 : \mathbb{A}^m \times \mathbb{A}^n \to \mathbb{A}^m$ and $\pi_2 : \mathbb{A}^m \times \mathbb{A}^n \to \mathbb{A}^n$ are regular maps. In fact, we have that if $k[\mathbb{A}^m] = k[x_1, \ldots, x_m]$ and $k[\mathbb{A}^n] = k[y_1, \ldots, y_n]$, then

$$k[\mathbb{A}^m \times \mathbb{A}^n] = k[x_1, \ldots, x_m, y_1, \ldots, y_n].$$

This follows since the subrings $k[\mathbb{A}^m]$ and $k[\mathbb{A}^n]$ of $k[\mathbb{A}^m \times \mathbb{A}^n]$ are linearly disjoint and generate $k[\mathbb{A}^m \times \mathbb{A}^n]$ as a k-algebra.

Now suppose that X and Y are affine varieties, with X a closed subset of $\mathbb{A}^m$ and Y a closed subset of $\mathbb{A}^n$. The product $X \times Y$ can be naturally identified with a subset of $\mathbb{A}^m \times \mathbb{A}^n$, and we have that $X \times Y$ is a closed subset of $\mathbb{A}^m \times \mathbb{A}^n$, as we have that $X \times Y = Z(\pi_1^*(I(X)) \cup \pi_2^*(I(Y)))$. Let $R = k[\mathbb{A}^m \times \mathbb{A}^n]$. We identify $\pi_1^*(I(X))$ with $I(X)$ and $\pi_2^*(I(Y))$ with $I(Y)$ in the following proposition.

Proposition 5.7. *The ideal $I(X)R + I(Y)R$ is a prime ideal in R.*

Proof. We make use of properties of tensor products to prove this. We have that

$$R/(I(X)R + I(Y)R) \cong k[X] \otimes_k k[Y]$$

[160, Theorem 35, page 184]. We have that $k(X) \otimes_k k(Y)$ is a domain since $k(X)$ and $k(Y)$ are fields and k is algebraically closed, by Lemma 5.4. The subring of $k(X) \otimes_k k(Y)$ generated by $k[X]$ and $k[Y]$ is a tensor product of

$k[X]$ and $k[Y]$ over k by Theorem 5.3, so that $k[X] \otimes_k k[Y]$ is naturally a subring of the domain $k(X) \otimes_k k(Y)$. □

Thus $X \times Y$ is an affine variety, with prime ideal
$$I(X \times Y) = I(X)R + I(Y)R \text{ in } R = k[\mathbb{A}^m \times \mathbb{A}^n].$$

Products are much more subtle over nonalgebraically closed fields, as can be seen from the following example. Let $A = \mathbb{Q}[x]/(x^2 + 1)$ and $B = \mathbb{Q}[y]/(y^2 + 1)$, which are fields. We have that
$$A \otimes_\mathbb{Q} B \cong \mathbb{Q}[x,y]/(x^2 + 1, y^2 + 1) \cong \mathbb{Q}[i][y]/(y^2 + 1) = \mathbb{Q}[i][y]/(y-i)(y+i)$$
is not a domain.

We now construct a product $\mathbb{P}^m \times \mathbb{P}^n$. As a set, we can write
$$\mathbb{P}^m \times \mathbb{P}^n = \{(a_0 : \ldots : a_m; b_0 : \ldots : b_n) \mid (a_0 : \ldots : a_m) \in \mathbb{P}^m, (b_0 : \ldots : b_n) \in \mathbb{P}^n\}.$$

Let S be a polynomial ring in two sets of variables,
$$S = S(\mathbb{P}^m \times \mathbb{P}^n) = k[x_0, \ldots, x_m, y_0, \ldots, y_n].$$
We put a bigrading on S by $\mathrm{bideg}(x_i) = (1,0)$ for $0 \leq i \leq m$ and $\mathrm{bideg}(y_j) = (0,1)$ for $0 \leq j \leq n$. We have
$$S = \bigoplus_{k,l} S_{k,l}$$
where $S_{k,l}$ is the k-vector space generated by monomials $x_0^{i_0} \cdots x_m^{i_m} y_0^{j_0} \cdots y_n^{j_n}$ where $i_0 + \cdots + i_m = k$ and $j_0 + \cdots + j_n = l$. Elements of $S_{k,l}$ are called bihomogeneous of bidegree (k,l). The bigraded ring $S(\mathbb{P}^m \times \mathbb{P}^n)$ is called the bihomogeneous coordinate ring of $\mathbb{P}^m \times \mathbb{P}^n$. Suppose $F \in S$ is bihomogeneous of bidegree (k,l) and $(a_0 : \ldots : a_m; b_0 : \ldots : b_n) \in \mathbb{P}^m \times \mathbb{P}^n$. Suppose that $(c_0 : \ldots : c_m; d_0 : \ldots : d_n)$ is equal to $(a_0 : \ldots : a_m; b_0 : \ldots : b_n)$, so that there exist $0 \neq \alpha \in k$ and $0 \neq \beta \in k$ such that $c_i = \alpha a_i$ for $0 \leq i \leq m$ and $d_j = \beta b_j$ for $0 \leq j \leq n$. Then

(5.2) $\quad F(c_0, \ldots, c_m, d_0, \ldots, d_n) = \alpha^k \beta^l F(a_0, \ldots, a_m, b_0, \ldots, b_n).$

Thus the vanishing of such a form at a point is well-defined. We put a topology on the set $\mathbb{P}^m \times \mathbb{P}^n$ by taking the closed sets to be
$$Z(A) = \{(p,q) \in \mathbb{P}^m \times \mathbb{P}^n \mid F(p,q) = 0 \text{ for } F \in A\}$$
where A is a set of bihomogeneous forms. We can extend this definition to bihomogeneous ideals by considering the vanishing at a set of bihomogeneous generators.

5.2. Products of varieties

Given a subset Y of $\mathbb{P}^m \times \mathbb{P}^n$, the ideal $I(Y)$ of Y in S is the ideal in S generated by the set

$$U = \{F \in S \mid F \text{ is bihomogeneous and } F(p,q) = 0 \text{ for all } (p,q) \in Y\}.$$

The ideal $I(Y)$ is a bihomogeneous ideal (it is naturally bigraded as an S-module).

We define biprojective varieties, quasi-biprojective varieties, biprojective algebraic sets, quasi-biprojective algebraic sets, and subvarieties of biprojective varieties analogously to the projective case.

The ideal $I(W)$ of a biprojective subvariety W of $\mathbb{P}^m \times \mathbb{P}^n$ is a bigraded prime ideal in $S = S(\mathbb{P}^m \times \mathbb{P}^n)$. The bihomogeneous coordinate ring of W is $S(W) = S/I(W)$, which is a bigraded ring. The rational functions on W or the function field of W is

$$k(W) = \left\{ \frac{F}{G} \mid F, G \in S(W) \text{ are bihomogeneous of the same bidegree and } G \neq 0 \right\}.$$

If $f = \frac{F}{G} \in k(W)$ where F and G are bihomogeneous of the same bidegree and $(p,q) \in W$ is such that $G(p,q) \neq 0$, then the value of $f(p,q) \in k$ is well-defined by (5.2). The regular functions $\mathcal{O}_{W,(p,q)}$ at a point $(p,q) \in W$ are the quotients $\frac{F}{G}$ where $F, G \in S(W)$ are bihomogeneous of the same bidegree and $G(p,q) \neq 0$. We construct regular functions on an open subset U of W as

$$\mathcal{O}_W(U) = \bigcap_{(p,q) \in U} \mathcal{O}_{W,(p,q)}.$$

We expand our definition of regular maps (Definition 3.28) to include quasi-biprojective varieties (open subsets of biprojective varieties).

Suppose that $F \in S(W)$ is bihomogenous. We define $D(F) = W_F = W \setminus Z(F)$. The open sets $D(F) = W_F$ where F is bihomogenous of bidegree (a,b) with $a > 0$ and $b > 0$ are a basis for the topology of W.

The proof of Theorem 3.30 generalizes (working with a bigrading instead of a grading) to show that $W_{x_i y_j}$ is an affine variety, and we obtain the following explicit computation of $\mathcal{O}_W(W_{x_i y_j}) = k[W_{x_i y_j}]$. Write

$$S(W) = k[x_0, \ldots, x_m, y_0, \ldots, y_n]/I(W) = k[\overline{x}_0, \ldots, \overline{x}_m, \overline{y}_0, \ldots, \overline{y}_n]$$

where $I(W) = (G_1, \ldots, G_t)$ with $G_1, \ldots, G_t$ bihomogeneous generators of $I(W)$. Then

$$k[W_{x_i y_j}] = k[\tfrac{\overline{x}_0}{\overline{x}_i}, \ldots, \tfrac{\overline{x}_m}{\overline{x}_i}, \tfrac{\overline{y}_0}{\overline{y}_j}, \ldots, \tfrac{\overline{y}_n}{\overline{y}_j}]$$
$$\cong [\tfrac{x_0}{x_i}, \ldots, \tfrac{x_m}{x_i}, \tfrac{y_0}{y_j}, \ldots, \tfrac{y_n}{y_j}]/(F_1, \ldots, F_t),$$

where
$$F_a = G_a\left(\frac{x_0}{x_i}, \ldots, \frac{x_m}{x_i}, \frac{y_0}{y_j}, \ldots, \frac{y_n}{y_j}\right)$$
for $1 \leq a \leq t$.

If X is a projective variety which is a closed subset of $\mathbb{P}^m$ and Y is a projective variety which is a closed subset of $\mathbb{P}^n$, then $X \times Y = Z(I(X)S + I(Y)S)$ is a closed subset of $\mathbb{P}^m \times \mathbb{P}^n$. We have that $I(X)S + I(Y)S$ is a prime ideal in S by the proof of Proposition 5.7, so $X \times Y$ is a subvariety of $\mathbb{P}^m \times \mathbb{P}^n$, with $I(X \times Y) = I(X)S + I(Y)S$. The biprojective variety $X \times Y$ has a covering by open affine sets $(X \times Y) \cap (\mathbb{P}^m \times \mathbb{P}^n)_{x_i y_j} \cong (X \cap \mathbb{P}^m_{x_i}) \times (Y \cap \mathbb{P}^n_{y_j})$ for $0 \leq i \leq m$ and $0 \leq j \leq n$. We have that the open set W_F is (isomorphic to) an affine variety if and only if F is bihomogeneous of bidegree (a,b) with both $a > 0$ and $b > 0$ (Exercise 5.20). Proposition 3.39 is valid for maps between quasi-biprojective varieties; even the proof is valid in the larger setting of quasi-biprojective varieties. We deduce from this that if X, Y, Z, W are quasi-projective varieties and $\phi : Z \to X$ and $\psi : Z \to Y$ are regular maps, then $(\phi, \psi) : Z \to X \times Y$ is a regular map. If $\alpha : Z \to X$ and $\beta : W \to Y$ are regular maps, then $\alpha \times \beta : Z \times W \to X \times Y$ is a regular map.

Suppose that W is a closed subvariety of $\mathbb{P}^m \times \mathbb{P}^n$ with bihomogeneous coordinate ring
$$S(W) = S(\mathbb{P}^m \times \mathbb{P}^n)/I(W).$$

We can represent a rational map $\phi : W \dashrightarrow \mathbb{P}^l$ by equivalence classes $(f_0 : \ldots : f_l)$ with $f_0, \ldots, f_l \in k(W)$ not all zero. We have $(f_0 : \ldots : f_l) \sim (g_0 : \ldots : g_l)$ if $f_i g_j - f_j g_i = 0$ for $0 \leq i, j \leq l$. The rational map ϕ is regular at $p \in W$ if ϕ has a representative $(f_0 : \ldots : f_l)$ with $f_0, \ldots, f_l \in \mathcal{O}_{W,p}$ and $f_i(p) \neq 0$ for some i.

Alternatively, we can represent rational maps from W to a projective space $\mathbb{P}^l$ by equivalence classes $(F_0 : \ldots : F_l)$ where the $F_i \in S(W)$ are bihomogeneous of the same bidegree. The rational map is regular at a point $q \in W$ if there is a representative $(F_0 : \ldots : F_l)$ such that some $F_i(q) \neq 0$.

We have that the projections $\pi_1 : X \times Y \to X$ and $\pi_2 : X \times Y \to Y$ are regular maps.

If W is a closed subvariety of $\mathbb{P}^m \times \mathbb{P}^n$ and $U \subset W$ is a quasi-affine open subset, then $k(W)$ is the quotient field of $\mathcal{O}_W(U)$ (as in Proposition 3.37).

Using the fact that the affine open subsets $(\mathbb{P}^m \times \mathbb{P}^n)_{x_i y_j} \cong \mathbb{A}^m \times \mathbb{A}^n$ of $\mathbb{P}^m \times \mathbb{P}^n$ have regular functions
$$k[(\mathbb{P}^m \times \mathbb{P}^n)_{x_i y_j}] = k\left[\frac{x_0}{x_i}, \ldots, \frac{x_m}{x_i}, \frac{y_0}{y_j}, \ldots, \frac{y_n}{y_j}\right],$$

we see that $Y = (\mathbb{P}^m \times \mathbb{P}^n)_{x_i} \cong \mathbb{A}^m \times \mathbb{P}^n$ is covered by the affine open sets $(\mathbb{P}^m \times \mathbb{P}^n)_{x_i y_j}$ for $0 \leq j \leq n$. We can associate a coordinate ring

$$S(Y) = k\left[\frac{x_0}{x_i}, \ldots, \frac{x_m}{x_i}, y_0, \ldots, y_n\right]$$

to Y. An element of $k\left[\frac{x_0}{x_i}, \ldots, \frac{x_m}{x_i}\right]$ has bidegree $(0,0)$ and y_j has bidegree $(0,1)$ for $0 \leq j \leq n$. Thus the bidegree makes $S(Y)$ a graded ring (graded by the natural numbers $\mathbb{N}$). All of the theory that we have worked out above extends to this situation. We compute $k[Y_{y_j}] = S(Y)_{(y_j)}$, the elements of degree 0 in the localization $S(Y)_{y_j}$.

Given an affine variety X, we can realize X as a closed subvariety of $\mathbb{A}^m \cong \mathbb{P}^m_{x_i} \subset \mathbb{P}^m$, and then we have realized $X \times \mathbb{P}^n$ as a closed subvariety of $\mathbb{A}^m \times \mathbb{P}^n$. This gives us a graded coordinate ring for $X \times \mathbb{P}^n$ as

$$S(X \times \mathbb{P}^n) = k[X][y_0, \ldots, y_n].$$

Here, elements of $k[X]$ have degree 0, and the variables y_j have degree 1.

All of the theory we developed above goes through for this coordinate ring, with this grading.

Suppose that W is a closed subvariety of $X \times \mathbb{P}^n$ (where X is affine), with coordinate ring $S(W) = S(X \times \mathbb{P}^n)/I(W)$. We can represent rational maps from W to $\mathbb{P}^l$ by equivalence classes $(F_0 : \ldots : F_l)$ where the $F_i \in S(W)$ are homogeneous of the same degree (and not all zero). The rational map is regular at a point $q \in W$ if there is a representative $(F_0 : \ldots : F_l)$ such that some $F_i(q) \neq 0$.

If U is a quasi-affine open subset of W, then the quotient field of $\mathcal{O}_W(U)$ is $k(W)$.

If X is affine and W is a projective variety with coordinate ring $S(W) = k[\overline{y}_0, \ldots, \overline{y}_n]$, then $X \times W$ has the coordinate ring

$$S(X \times W) = k[X] \otimes_k S(W) = k[X][\overline{y}_0, \ldots, \overline{y}_n]$$

which is a domain as shown by the proof of Proposition 5.7. Also, $S(X)$ is graded by $\deg(f) = 0$ if $f \in k[X]$ and $\deg(\overline{y}_i) = 1$ for $1 \leq i \leq n$.

5.3. The Segre embedding

We define the Segre embedding

$$\phi : \mathbb{P}^m \times \mathbb{P}^n \to \mathbb{P}^N$$

where $N = (n+1)(m+1) - 1$ by

$$\phi(a_0 : \ldots : a_m; b_0 : \ldots : b_n) = (a_0 b_0 : a_0 b_1 : \ldots : a_i b_j \ldots : a_m b_n).$$

The map ϕ is a regular map (since $Z_{\mathbb{P}^m \times \mathbb{P}^n}(x_0 y_0, x_0 y_1, \ldots, x_m x_n) = \emptyset$), and it can be verified that its image is a closed subvariety of $\mathbb{P}^N$ and ϕ is an isomorphism onto this image (ϕ is a closed embedding). If we take w_{ij}, with $0 \leq i \leq m$ and $0 \leq j \leq n$, to be the natural homogeneous coordinates on $\mathbb{P}^N$, then the image of ϕ is the projective variety W whose ideal $I(W)$ is generated by $\{w_{ij} w_{kl} - w_{kj} w_{il} \mid 0 \leq i, k \leq m \text{ and } 0 \leq j, l \leq n\}$. This is proven in [81, Section 2 of Chapter XI of Volume 2].

Thus the product $X \times Y$ of two quasi-projective varieties is actually (isomorphic to) a quasi-projective variety, by the Segre embedding. In fact, any quasi-biprojective variety is (isomorphic to) a quasi-projective variety.

5.4. Graphs of regular and rational maps

Suppose that X and Y are quasi-projective varieties and $\phi : X \to Y$ is a regular map. Then we have a regular map $\psi : X \to X \times Y$ defined by $\psi(p) = (p, \phi(p))$ for $p \in X$. Let Γ_ϕ be the image $\psi(X)$ in $X \times Y$. We call Γ_ϕ the graph of ϕ.

Proposition 5.8. Γ_ϕ *is Zariski closed in* $X \times Y$.

Proof. We have an embedding of Y in a projective space $\mathbb{P}^n$, as an open subset of a projective subvariety. The map $\phi : X \to Y$ thus extends to a regular map $\tilde{\phi} : X \to \mathbb{P}^n$ and $\Gamma_\phi = \Gamma_{\tilde{\phi}} \cap (X \times Y)$. Thus it suffices to prove the proposition in the case when $Y = \mathbb{P}^n$. Let $i : \mathbb{P}^n \to \mathbb{P}^n$ be the identity map. Then $\phi \times i : X \times \mathbb{P}^n \to \mathbb{P}^n \times \mathbb{P}^n$ is a regular map. Let $\Delta_{\mathbb{P}^n} \subset \mathbb{P}^n \times \mathbb{P}^n$ be the "diagonal" $\{(q, q) \mid q \in \mathbb{P}^n\}$. We have that $S(\mathbb{P}^n \times \mathbb{P}^n) = k[u_0, \ldots, u_n, v_0, \ldots, v_n]$ where the u_i are homogeneous coordinates on the $\mathbb{P}^n$ of the first factor and the v_j are homogeneous coordinates on the $\mathbb{P}^n$ of the second factor. The diagonal $\Delta_{\mathbb{P}^n}$ is a closed subset of $\mathbb{P}^n \times \mathbb{P}^n$ with $\Delta_{\mathbb{P}^n} = Z(u_i v_j - u_j v_i \mid 0 \leq i, j \leq n)$. Since the preimage of a closed set by a regular map is closed, we have that $\Gamma_\phi = (\phi \times i)^{-1}(\Delta_{\mathbb{P}^n})$ is closed in $X \times \mathbb{P}^n$. $\square$

The graph $\Gamma_\phi = \psi(X)$ is irreducible since X is, so Γ_ϕ is a closed subvariety of $X \times Y$.

We can extend this construction to give a useful method of studying rational maps. Suppose that X and Y are quasi-projective varieties and $\phi : X \dashrightarrow Y$ is a rational map. Let U be a nontrivial open subset of X on which ϕ is regular. The graph Γ_ϕ of ϕ is defined to be the closure in $X \times Y$ of the image of the regular map $p \to (p, \phi(p))$ from U to $X \times Y$. The graph Γ_ϕ does not depend on the choice of open subset U on which ϕ is regular (since Γ_ϕ is irreducible and $(U \times Y) \cap \Gamma_\phi$ is a nontrivial open subset of Γ_ϕ).

5.4. Graphs of regular and rational maps

Further,

(5.3) $$\Gamma_{\phi|V} = \Gamma_\phi \cap (V \times Y) = \pi_1^{-1}(V)$$

for any open subset V of X, where $\pi_1 : \Gamma_\phi \to X$ is the projection.

Proposition 5.9. *Suppose that $\phi : X \dashrightarrow Y$ is a rational map of quasi-projective varieties. Then Γ_ϕ is a quasi-projective variety, and the projection $\pi_1 : \Gamma_\phi \to X$ is a birational map. If U is a nonempty open subset of X on which ϕ is regular, then the restriction*

$$\pi_1 : \Gamma_{\phi|U} = \Gamma_\phi \cap (U \times Y) \to U$$

is an isomorphism.

Proof. It suffices to prove this in the case when ϕ is itself a regular map, as Γ_ϕ is the Zariski closure in $X \times Y$ of the graph $\Gamma_{\phi|U}$ for any dense open subset U of X. Now Γ_ϕ is the image of the regular map $(i, \phi) : X \to X \times Y$ where $i : X \to X$ is the identity map. Since X is irreducible, its image Γ_ϕ is irreducible. The set Γ_ϕ is closed in $X \times Y$ by Proposition 5.8 so Γ_ϕ is a variety. The map (i, ϕ) is an inverse to π_1. Since both maps are regular maps, π_1 is an isomorphism, and π_1 is thus birational. $\square$

Proposition 5.10. *Suppose that $\phi : X \dashrightarrow Y$ is a rational map of quasi-projective varieties and Γ_ϕ is the graph of ϕ, with projection $\pi_1 : \Gamma_\phi \to X$. Suppose that $p \in X$. Then ϕ is a regular map at p if and only if the rational map π_1^{-1} is a regular map at p.*

Proof. Let U be an open subset of X on which ϕ is regular. Then $\pi_1^{-1} = i \times \phi : U \to \Gamma_\phi$ is a regular map where $i : X \to X$ is the identity map. Suppose that the rational map $\pi_1^{-1} : X \dashrightarrow \Gamma_\phi$ is regular at a point $p \in X$. Let V be an open neighborhood of p in which π_1^{-1} is regular. We have that $\pi_2 \pi_1^{-1} = \phi$ as a rational map where $\pi_2 : \Gamma_\phi \to Y$ is the projection and $\pi_2 \pi_1^{-1} : V \to Y$ is regular, so ϕ is regular on V. $\square$

Theorem 5.11 (Elimination theory). *For $1 \le i \le r$, let d_i be a positive integer, and let $F_i = \sum_{|J|=d_i} b^i_J x^J$ be a homogeneous polynomial in the variables $x_0, \ldots, x_n$ of degree d_i with indeterminate coefficients b^i_J, where the sum is over $J = (j_0, \ldots, j_n) \in \mathbb{N}^{n+1}$ with $|J| = j_0 + \cdots + j_n = d_i$ and $x^J = x_0^{j_0} \cdots x_n^{j_n}$. Then there exists a set of polynomials*

$$g_1, \ldots, g_t \in \mathbb{Z}[\{b^i_J \mid 1 \le i \le r, |J| = d_i\}]$$

which are homogeneous with respect to each set of variables b^i_J with fixed i, with the following property:

Let K be an algebraically closed field and suppose that $\overline{b}^i_J \in K$. Let

$$\overline{F}_i = \sum_{|J|=d_i} \overline{b}^i_J x^J \in K[x_0,\ldots,x_n] \quad \text{for } 1 \leq i \leq r.$$

Then a necessary and sufficient condition for the $\overline{F}_i$ to have a common zero in K^{n+1}, different from the trivial solution $(0,\ldots,0)$, is that the $\overline{b}^i_J$ are a common zero of the g_l.

Proof. [143, Section 80, page 8]. □

Theorem 5.12. *Suppose that Z is a closed subset of $\mathbb{P}^n \times \mathbb{A}^m$. Then the image of the second projection $\pi_2 : Z \to \mathbb{A}^m$ is Zariski closed.*

Proof. Let $x_0,\ldots,x_n$ be homogeneous coordinates on $\mathbb{P}^n$ and $y_1,\ldots,y_m$ be affine coordinates on $\mathbb{A}^m$. Then $I(Z) = (G_1,\ldots,G_r)$ for some $G_1,\ldots,G_r \in k[x_0,\ldots,x_n,y_1,\ldots,y_m]$ where the G_i are homogeneous in the x_i variables of some degrees d_i. Write $G_i = \sum_{|J|=d_i} a^i_J x^J$, where $a^i_J \in k[y_1,\ldots,y_m]$. Let $g_1,\ldots,g_t$ be the polynomials of the conclusions of Theorem 5.11, and let

$$h_l = g_l(a^i_J) \in k[y_1,\ldots,y_m] = k[\mathbb{A}^m].$$

For $P \in \mathbb{A}^m$, we have

$$\pi_2^{-1}(P) \cap Z = Z(G_1(x,P),\ldots,G_r(x,P)) \subset \mathbb{P}^n \times \{P\} \cong \mathbb{P}^n$$

where $G_i(x,P) = \sum a^i_J(P) x^J$. Thus $\pi_2^{-1}(P) \cap Z \neq \emptyset$ if and only if the polynomials $G_i(x,P)$ have a common nontrivial zero in k^{n+1}, which holds if and only if P is a common zero of the $h_l(y) = g_l(a^i_J)$ in k^m, and this holds if and only if $P \in Z(h_1,\ldots,h_t)$. Thus $\pi_2(Z) = Z(h_1,\ldots,h_t)$ is Zariski closed. □

Theorem 5.12 does not hold for closed subsets of $\mathbb{A}^n \times \mathbb{A}^m$. A simple example is to take Z to be $Z(xy-1) \subset \mathbb{A}^2 \cong \mathbb{A}^1 \times \mathbb{A}^1$. The projection of Z onto the y-axis is the nonclosed subset $\mathbb{A}^1 \setminus \{(0)\}$ of $\mathbb{A}^1$.

Corollary 5.13. *Suppose that X is a projective variety and Y is a quasi-projective variety. Then the second projection $\pi_2 : X \times Y \to Y$ takes closed sets to closed sets.*

Proof. Suppose that Z is a closed subset of $X \times Y$. After embedding X as a closed subset of a projective space $\mathbb{P}^n$, so that Z is a closed subset of $\mathbb{P}^n \times Y$, we may assume that $X = \mathbb{P}^n$. Let $Y = \bigcup_{i=1}^r V_i$ be an affine open cover of Y. We must show that $\pi_2(Z) \cap V_i$ is closed in V_i for all i. Let $Z_i = Z \cap (\mathbb{P}^n \times V_i)$, a closed subset of $\mathbb{P}^n \times V_i$. We have that $\pi_2(Z) \cap V_i = \pi_2(Z_i)$. Here V_i is isomorphic to a closed subset of $\mathbb{A}^{m_i}$ for some m_i. Thus Z_i is a closed subset of $\mathbb{P}^n \times \mathbb{A}^{m_i}$ under the natural inclusion. We have projections $\pi_2^i : \mathbb{P}^n \times \mathbb{A}^{m_i} \to \mathbb{A}^{m_i}$. By Theorem 5.12, $\pi_2^i(Z_i)$ is a closed subset of $\mathbb{A}^{m_i}$. Since $\pi_2^i(Z_i) \subset V_i$, we have that $\pi_2(Z_i) = \pi_2^i(Z_i)$ is closed in V_i for all i. $\square$

Theorem 5.14. *Suppose that $\phi : X \to Y$ is a regular map of projective varieties. Then the image of ϕ is a closed subset of Y.*

Proof. Apply the corollary to the closed subset Γ_ϕ of $X \times Y$. $\square$

We now give another proof of Theorem 3.35.

Corollary 5.15. *Suppose that X is a projective variety. Then $\mathcal{O}_X(X) = k$.*

Proof. Suppose that $f \in \mathcal{O}_X(X)$. Then f is a regular map $f : X \to \mathbb{A}^1$. After including $\mathbb{A}^1$ into $\mathbb{P}^1$, we obtain a regular map $f : X \to \mathbb{P}^1$. By Theorem 5.14, we have that the image $f(X)$ is closed in $\mathbb{P}^1$. Since f cannot be onto, $f(X)$ must be a finite union of points. Since X is irreducible, $f(X)$ is irreducible, so $f(X)$ is a single point. Thus $f \in k$. $\square$

Corollary 5.16. *Suppose that X is a projective variety and $\phi : X \to \mathbb{A}^n$ is a regular map. Then $\phi(X)$ is a point.*

Proof. Let $\pi_i : \mathbb{A}^n \to \mathbb{A}^1$ be projection onto the i-th factor. Then $\pi_i \circ \phi : X \to \mathbb{A}^1$ is a regular map, so $\pi_i \circ \phi$ is a constant map by the previous corollary for $1 \leq i \leq n$. Thus $\phi(X)$ is a point. $\square$

Exercise 5.17. Let t be an indeterminate, F be the field $\mathbb{Z}_p(t)$, and K, L be the fields

$$K = F[x]/(x^p - t), \qquad L = F[y]/(y^p - t).$$

Show that $K \otimes_F L$ is not reduced.

Exercise 5.18. Let $W \subset \mathbb{P}^3$ be the image of $\mathbb{P}^1 \times \mathbb{P}^1$ in $\mathbb{P}^3$ by the Segre map.

   i) Find the ideal $I(W) \subset k[\mathbb{P}^3]$.

   ii) Suppose that $p \in \mathbb{P}^1$. Find the ideal in $k[\mathbb{P}^3]$ of the image of $\{p\} \times \mathbb{P}^1$ in $\mathbb{P}^3$. Find the ideal in $k[\mathbb{P}^3]$ of the image of $\mathbb{P}^1 \times \{p\}$ in $\mathbb{P}^3$.

Exercise 5.19. Let a and b be positive integers. Consider the Veronese map ϕ_1 on $\mathbb{P}^m$ defined by the forms of degree a on $\mathbb{P}^m$ and the Veronese map ϕ_2 on $\mathbb{P}^n$ defined by the forms of degree b on $\mathbb{P}^n$. Let W_1 be the image of ϕ_1 and let W_2 be the image of ϕ_2.

  i) Show that $\phi_1 \times \phi_2 : \mathbb{P}^m \times \mathbb{P}^n \to W_1 \times W_2$ is an isomorphism onto the biprojective variety $W_1 \times W_2$.

  ii) Show that $S(W_1 \times W_2)_{(1,1)} = S(\mathbb{P}^m \times \mathbb{P}^n)_{(a,b)}$; that is, the forms of bidegree $(1,1)$ on $W_1 \times W_2$ are the forms of bidegree (a,b) on $\mathbb{P}^m \times \mathbb{P}^n$.

Exercise 5.20. Suppose that $F \in S(\mathbb{P}^m \times \mathbb{P}^n)$ with $m, n > 0$ is a bihomogeneous form of bidegreee (a, b).

  i) Show that $D(F) = (\mathbb{P}^m \times \mathbb{P}^n)_F$ is (isomorphic to) an affine variety if $a > 0$ and $b > 0$.

  ii) Show that $(\mathbb{P}^m \times \mathbb{P}^n)_F \cong \mathbb{P}^m_F \times \mathbb{P}^n$ if F has bidegree $(a, 0)$ and $(\mathbb{P}^m \times \mathbb{P}^n)_F \cong \mathbb{P}^m \times \mathbb{P}^n_F$ if F has bidegree $(0, b)$. Conclude that $D(F) = (\mathbb{P}^m \times \mathbb{P}^n)_F$ is not affine if $a = 0$ or $b = 0$.

Hint: Use the construction of Exercise 5.19, followed by a Segre embedding.

Exercise 5.21. Suppose that X is a quasi-projective variety and $U_1, U_2 \subset X$ are affine open subsets. Show that $U_1 \cap U_2$ is an affine open subset of X; that is, show that $U_1 \cap U_2$ is isomorphic to an affine variety. Hint: Consider the open subset $U_1 \times U_2 \subset X \times X$ and its intersection with the diagonal $\Delta_X = \{(p, p) \mid p \in X\} \subset X \times X$.

Chapter 6

The Blow-up of an Ideal

In this chapter we construct the blow-up of an ideal in an affine or projective variety. The more general construction of the blow-up of an ideal sheaf will be given in Chapter 12.

As usual, we will assume that k is a fixed algebraically closed field.

6.1. The blow-up of an ideal in an affine variety

Suppose that X is an affine variety and $f_0, \ldots, f_r \in k[X]$ are not all zero. We can define a rational map $\Lambda_{f_0,\ldots,f_r} : X \dashrightarrow \mathbb{P}^r$ by $\Lambda_{f_0,\ldots,f_r} = (f_0 : \ldots : f_r)$.

Proposition 6.1. *Suppose that $g_0, \ldots, g_s \in k[X]$ and*

$$(f_0, \ldots, f_r) = (g_0, \ldots, g_s)$$

are the same ideal $J \neq (0)$ in $k[X]$. Then there is a commutative diagram of regular maps

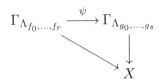

where the vertical arrows are the projections and ψ is an isomorphism.

Proof. This will follow from Theorem 6.4. In this theorem it is shown that there are graded $k[X]$-algebra isomorphisms

$$S(\Gamma_{\Lambda_{f_0,\ldots,f_r}}) = S(X \times \mathbb{P}^r)/I(\Gamma_{\Lambda_{f_0,\ldots,f_r}}) \cong \bigoplus_{i \geq 0} J^i$$

and

$$S(\Gamma_{\Lambda_{g_0,\ldots,g_s}}) = S(X \times \mathbb{P}^s)/I(\Gamma_{\Lambda_{g_0,\ldots,g_s}}) \cong \bigoplus_{i \geq 0} J^i,$$

which thus induce a graded $k[X]$-algebra isomorphism

$$\overline{\alpha} : S(\Gamma_{\Lambda_{g_0,\ldots,g_s}}) \to S(\Gamma_{\Lambda_{f_0,\ldots,f_r}}).$$

Write $S(\Gamma_{\Lambda_{g_0,\ldots,g_s}}) = k[X][\overline{y}_0,\ldots,\overline{y}_s]$ where the $\overline{y}_i$ are the restriction to $\Gamma_{\Lambda_{g_0,\ldots,g_s}}$ of the homogeneous coordinates on $\mathbb{P}^s$. Then define

$$\alpha : \Gamma_{\Lambda_{f_0,\ldots,f_r}} \to X \times \mathbb{P}^s$$

by

$$\alpha(p,q^*) = (p; \overline{\alpha}(\overline{y}_0)(p,q^*) : \ldots : \overline{\alpha}(\overline{y}_s)(p,q^*))$$

for $(p,q^*) \in \Gamma_{\Lambda_{f_0,\ldots,f_r}}$. We will now establish that α is an isomorphism onto $\Gamma_{\Lambda_{g_0,\ldots,g_s}}$.

Since $\overline{\alpha}$ is an isomorphism, we have that $\overline{\alpha}(\overline{y}_0),\ldots,\overline{\alpha}(\overline{y}_s)$ generate

$$[S(\Gamma_{\Lambda_{f_0,\ldots,f_r}})]_1$$

(the homogeneous forms of degree 1) as a $k[X]$-module.

Since $\overline{\alpha}(\overline{y}_0),\ldots,\overline{\alpha}(\overline{y}_s) \in S(\Gamma_{f_0,\ldots,f_r})$ are homogeneous of degree 1 and $Z(\overline{\alpha}(\overline{y}_0),\ldots,\overline{\alpha}(\overline{y}_s)) = \emptyset$, we have that

$$q \mapsto (\overline{\alpha}(\overline{y}_0)(q) : \ldots : \overline{\alpha}(\overline{y}_s)(q))$$

is a regular map from $\Gamma_{\Lambda_{f_0,\ldots,f_r}}$ to $\mathbb{P}^s$ (as explained before Section 5.3).

Since the first projection $\pi_1 : \Gamma_{\Lambda_{f_0,\ldots,f_r}} \to X$ is a regular map, the product α is a regular map $\alpha : \Gamma_{\Lambda_{f_0,\ldots,f_r}} \to X \times \mathbb{P}^s$.

Suppose that $F \in S(X \times \mathbb{P}^s)$ is a homogeneous form and that $q \in \Gamma_{\Lambda_{f_0,\ldots,f_r}}$. We have that

$$F(\alpha(q)) = (\overline{\alpha}(\overline{F}))(q)$$

where $\overline{F}$ is the residue of F in $S(\Gamma_{\Lambda_{g_0,\ldots,g_s}})$. Thus $F(\alpha(q)) = 0$ for all $F \in I(\Gamma_{\Lambda_{g_0,\ldots,g_s}})$, giving us that α is a regular map from $\Gamma_{\Lambda_{f_0,\ldots,f_r}}$ into $\Gamma_{\Lambda_{g_0,\ldots,g_s}}$.

6.1. The blow-up of an ideal in an affine variety

Since $\overline{\alpha}$ is an isomorphism, we can apply the above argument to $\overline{\beta} = (\overline{\alpha})^{-1}$ to construct a regular map $\beta : \Gamma_{\Lambda_{g_0,\ldots,g_s}} \to \Gamma_{\Lambda_{f_0,\ldots,f_r}}$ which has the property that

$$\alpha \circ \beta = \mathrm{id}_{\Gamma_{\Lambda_{g_0,\ldots,g_s}}} \quad \text{and} \quad \beta \circ \alpha = \mathrm{id}_{\Gamma_{\Lambda_{f_0,\ldots,f_r}}},$$

so that α is an isomorphism with inverse β. □

The following definition of the blow-up of an ideal is well-defined by Proposition 6.1.

Definition 6.2. Suppose that X is an affine variety and $I \subset k[X]$ is a nonzero ideal. Suppose that $I = (f_0, \ldots, f_r)$. Let $\Lambda : X \dashrightarrow \mathbb{P}^r$ be the rational map defined by $\Lambda = (f_0 : \ldots : f_r)$. The blow-up of I is $B(I) = \Gamma_\Lambda$, with projection $\pi : B(I) \to X$.

The blow-up $B(0)$ of the zero ideal is defined to be $B(0) = \emptyset$, with natural inclusion π of the empty set into X.

If $I = I(Y)$ is the ideal of a subvariety Y of X, then $B(I) \to X$ is called the blow-up of Y.

The blow-up $\pi : B(J) \to X$ has the property that π^{-1} is regular and is an isomorphism over the open set $X \setminus Z(J)$, as follows from Proposition 5.9. If $J \neq (0)$, then $X \setminus Z(J) \neq \emptyset$.

Lemma 6.3. Suppose that R is a Noetherian ring and $P \subset R$ is a prime ideal. Suppose that J, A are ideals in R such that $J \not\subset P$, $A \subset P$ and the localizations $A_Q = P_Q$ for Q a prime ideal in R such that $J \not\subset Q$. Then

$$P = A :_R J^\infty := \{f \in R \mid fJ^n \subset A \text{ for some } n \geq 0\}.$$

Proof. By Proposition 1.43, A has an irredundant primary decomposition

$$A = P \cap I_1 \cap \cdots \cap I_r$$

where for $1 \leq i \leq r$, I_i are Q_i-primary for prime ideals Q_i such that $J \subset Q_i$. Thus there exists $n > 0$ such that $J^n \subset I_i$ for all i. Thus $J^n P \subset A$, so $P \subset A : J^\infty$.

Now suppose that $f \in R$ is such that $fJ^n \subset P$ for some $n \geq 0$. Since P is a prime ideal and there exists an element of J^n which is not in P, we have that $f \in P$. Thus $A : J^\infty \subset P$. □

Theorem 6.4. Suppose that X is an affine variety and $J \subset k[X]$ is a nonzero ideal. Let $\pi : B(J) \to X$ be the blow-up of J. Suppose that $J = (f_0, \ldots, f_n)$, with the f_i all nonzero, so that $B(J) \subset X \times \mathbb{P}^n$. Then the coordinate ring of $B(J)$ (viewing $B(J)$ as a closed subvariety of $X \times \mathbb{P}^n$) is

$$S(B(J)) \cong \bigoplus_{i \geq 0} J^i$$

as a graded $k[X]$-algebra (with the degree 0 elements of $S(B(J))$ being $J^0 = k[X]$ and the degree i elements of $S(B(J))$ being J^i). Let $R = k[X]$ and $y_0, \ldots, y_n$ be homogeneous coordinates on $\mathbb{P}^n$. Then

$$(6.1) \qquad k[B(J)_{y_i}] = \mathcal{O}_{B(J)}(B(J)_{y_i}) = R\left[\frac{f_0}{f_i}, \ldots, \frac{f_n}{f_i}\right]$$

for $0 \leq i \leq n$. We thus have that

$$(6.2) \qquad Jk[B(J)_{y_i}] = f_i k[B(J)_{y_i}]$$

is a principal ideal for all i. Let $A = (y_i f_j - y_j f_i \mid 0 \leq i,j \leq n)$, a homogeneous ideal in $S(X \times \mathbb{P}^n) = k[X][y_0, \ldots, y_n]$. The ideal of $B(J)$ in $S(X \times \mathbb{P}^n)$ is

$$\begin{aligned} I_{X \times \mathbb{P}^n}(B(J)) &= A :_{S(X \times \mathbb{P}^n)} (JS(X \times \mathbb{P}^n))^\infty \\ &= \{f \in S(X \times \mathbb{P}^n) \mid fJ^n \subset A \text{ for some } n \geq 0\}. \end{aligned}$$

Proof. Let $\phi = (f_0 : \ldots : f_n) : X \dashrightarrow \mathbb{P}^n$. The variety $B(J)$ is defined to be the closure of $\Lambda(X \setminus Z_X(f_0, \ldots, f_n))$ in $X \times \mathbb{P}^n$, where $\Lambda(p) = (p; f_0(p) : \ldots : f_n(p))$ for $p \in X \setminus Z_X(f_0, \ldots, f_n)$. The coordinate ring of $X \times \mathbb{P}^n$ is $R[y_0, \ldots, y_n]$ which is graded by $\deg y_i = 1$ for $0 \leq i \leq n$. The ideal $A = (y_i f_j - y_j f_i \mid 0 \leq i,j \leq n)$ in $R[y_0, \ldots, y_n]$ is contained in $I(\Gamma_\phi)$.

Since f_i is a unit in R_{f_i}, ϕ is regular on X_{f_i} and we have that $\Gamma_{\phi|X_{f_i}} \subset X_{f_i} \times \mathbb{P}^n$ is isomorphic to the open subset X_{f_i} of X. Further, we calculate (using the fact that f_i is a unit in R_{f_i}) that A_{f_i} is a prime ideal in

$$R_{f_i}[y_0, \ldots, y_n] = S(X_{f_i} \times \mathbb{P}^n) = S(X \times \mathbb{P}^n)_{f_i}.$$

Now we will show that

$$I_{X \times \mathbb{P}^n}(\Gamma_\phi)_{f_i} = I_{X_{f_i} \times \mathbb{P}^n}(\Gamma_{\phi|X_{f_i}}) = A_{f_i}.$$

We have that $\Gamma_{\phi|X_{f_i}} = \{(p; f_0(p) : \cdots : f_n(p)) \mid p \in X_{f_i}\}$. Thus (as already observed) $A_{f_i} \subset I(\Gamma_{\phi|X_{f_i}})$. Now

$$(6.3) \qquad A_{f_i} = \left(y_j - \frac{f_j}{f_i}y_i \mid 0 \leq j \leq n\right).$$

Suppose $F \in S(X_{f_i} \times \mathbb{P}^n) = R_{f_i}[y_0, y_1, \ldots, y_n]$ is homogeneous of degree d. Then (6.3) implies that $F = gy_i^d + H$ with $H \in A_{f_i}$ homogeneous of degree d and $g \in R_{f_i}$. Suppose that $F \in I(\Gamma_{\phi|X_{f_i}})$. Then for all $p \in X_{f_i}$,

$$0 = F(p, f_0(p), \ldots, f_n(p)) = g(p)f_i(p)^d$$

which implies that $g(p) = 0$ for all $p \in X_{f_i}$, so that $g = 0$.

6.1. The blow-up of an ideal in an affine variety 115

If Q is a prime ideal in $S(X \times \mathbb{P}^n)$ such that $f_i \notin Q$, we have that
$$A_Q \cong (A_{f_i})_{Q_{f_i}} \cong (I(\Gamma_\phi)_{f_i})_{Q_{f_i}} \cong I(\Gamma_\phi)_Q$$
by Exercise 1.22. Since this is true for $0 \leq i \leq n$, we have that $A_Q = I(\Gamma_\phi)_Q$ for Q a prime ideal in $R[y_0, \ldots, y_n]$ such that $(f_0, \ldots, f_n) \not\subset Q$. We have that $J \not\subset I(\Gamma_\phi)$, since, otherwise, $\Gamma_\phi \subset \pi^{-1}(Z(J))$ implies $X \subset Z(J)$ which implies that $J \subset I(X) = (0)$, a contradiction to our assumption that $J \neq (0)$. By Lemma 6.3, we have that
$$A :_{S(X \times \mathbb{P}^n)} (JS(X \times \mathbb{P}^n))^\infty = I(\Gamma_\phi).$$
Let t be an indeterminate, which we give degree 1, and let P be the kernel of the graded k-algebra homomorphism
$$R[y_0, \ldots, y_n] \to R[tf_0, \ldots, tf_n] \subset R[t]$$
defined by mapping $y_j \mapsto tf_j$. Here P is a prime ideal since $R[t]$ is a domain. We have that $A \subset P$ and for a prime ideal Q in $R[y_0, \ldots, y_n]$, we have that $A_Q = P_Q$ if $J \not\subset Q$ (this follows since after localizing at such a Q, some f_i becomes invertible and so we can make a similar argument to the preceding paragraph). Assume that $J \subset P$. We will derive a contradiction. Then $P_{f_i} = R_{f_i}[y_0, \ldots, y_n]$ for all i, which implies $R_{f_i}[tf_0, \ldots, tf_n] = 0$, which is a contradiction since the f_i are assumed nonzero and R is a domain. Thus $J \not\subset P$. Thus by Lemma 6.3, $P = A :_{R[y_0, \ldots, y_n]} J^\infty = I(\Gamma_\phi)$, and the coordinate ring of $B(J)$ is
$$R[y_0, \ldots, y_n]/P \cong R[tf_0, \ldots, tf_n] \cong \bigoplus_{i=0}^\infty J^i.$$

We have
$$\begin{aligned}\mathcal{O}_{B(J)}(B(J) \cap (X \times \mathbb{P}^n_{y_j})) &= S(B(J))_{(y_j)} \cong R[tf_0, \ldots, tf_n]_{(tf_j)} \\ &\cong R[\tfrac{tf_0}{tf_j}, \ldots, \tfrac{tf_n}{tf_j}] = R[\tfrac{f_0}{f_j}, \ldots, \tfrac{f_n}{f_j}] \subset k(X).\end{aligned}$$
□

As a graded $k[X]$-algebra, we have
$$S(B(J)) \cong \bigoplus_{i \geq 0} J^i \cong k[X][tf_0, \ldots, tf_n] \subset k[X][t]$$
where $k[X][t]$ is the polynomial ring over $k[X]$, graded by $\deg(t) = 1$. The algebra $\bigoplus_{i \geq 0} J^i$ is called the Rees algebra of J.

We have the following useful proposition, which allows us to easily compute the coordinate ring of a blow-up in an important case.

Proposition 6.5. *With the notation of Theorem 6.4, suppose that the generators $f_0, \ldots, f_n$ of J are a $k[X]$-regular sequence. Then $I_{X \times \mathbb{P}^n}(B(J)) = A$ and $S(B(J)) \cong k[X][y_0, \ldots, y_n]/A$.*

Proof. This follows from Theorem 1 of [**108**]. □

There are sophisticated techniques to compute coordinate rings $S(B(J))$ of blow-ups of ideals in more general situations. Some important results on this topic are given in the papers [**77**] by Herzog, Simis, and Vasconcelos, [**82**] and [**83**] by Huneke, and [**138**] by Simis, Ulrich, and Vasconcelos.

Let W be a closed subset of the affine variety X. Then $\pi^{-1}(W \setminus Z(J)) \cong W \setminus Z(J)$ by Proposition 5.9. We will call the Zariski closure of $\pi^{-1}(W \setminus Z(J)) \cong W \setminus Z(J)$ in $B(J)$ the strict transform of W in $B(J)$.

Proposition 6.6. *Suppose that X is an affine variety and $J \subset k[X]$ is an ideal. Let W be a closed subvariety of X, and let $\overline{J} = Jk[W]$. Then the strict transform of W in $B(J)$ is isomorphic to $B(\overline{J})$.*

The ideal of $B(\overline{J})_{y_i}$ in $k[B(J)_{y_i}]$ is

$$(6.4) \quad \begin{aligned} I(B(\overline{J})_{y_i}) &= I(W)k[B(J)_{y_i}] :_{k[B(J)_{y_i}]} (Jk[B(J)_{y_i}])^\infty \\ &= I(W)k[B(J)_{y_i}] :_{k[B(J)_{y_i}]} (f_i k[B(J)_{y_i}])^\infty. \end{aligned}$$

Proof. We have natural projections $\pi : B(J) \to X$ and $\overline{\pi} : B(\overline{J}) \to W$.

First suppose $\overline{J} = (0)$, so that $B(\overline{J}) = \emptyset$. Then $J \subset I(W)$ so $W \subset Z(J)$. Thus $W \setminus Z(J) = \emptyset$ and the Zariski closure of $W \setminus Z(J)$ in $B(J)$ is $\emptyset$.

Now suppose that $\overline{J} \neq (0)$. Let $f_0, \ldots, f_n$ be a set of generators of J. Let $\overline{f}_i$ be the residues of the f_i in $k[W]$. Let $\phi : X \dashrightarrow X \times \mathbb{P}^n$ be the rational map $\phi = \mathrm{id} \times (f_0 : \ldots : f_n)$ and let $\overline{\phi} : W \dashrightarrow W \times \mathbb{P}^n$ be the rational map $\overline{\phi} = \mathrm{id} \times (\overline{f}_0 : \ldots : \overline{f}_n)$. We have a commutative diagram of regular maps, where the vertical maps are the natural inclusions:

$$\begin{array}{ccc} X \setminus Z(J) & \xrightarrow{\phi} & X \times \mathbb{P}^n \\ \uparrow & & \uparrow \\ W \setminus Z(\overline{J}) & \xrightarrow{\overline{\phi}} & W \times \mathbb{P}^n. \end{array}$$

Now $B(J)$ is the closure of $\phi(X \setminus Z(J))$ in $X \times \mathbb{P}^n$ and $B(\overline{J})$ is the closure of $\overline{\phi}(W \setminus Z(\overline{J}))$ in $W \times \mathbb{P}^n$, so the conclusions of the first paragraph of the proposition thus follow from the above diagram.

The final equation (6.4) follows from Lemma 6.3. □

We now discuss an important example.

Let p be the origin in $X = \mathbb{A}^2$. Let $\mathfrak{m} = (x_1, x_2)$ be the ideal of p in $k[\mathbb{A}^2] = k[x_1, x_2]$. Let $\pi : B = B(\mathfrak{m}) \to X$ be the blow-up of p (the blow-up of $\mathfrak{m}$). Let $E = \pi^{-1}(p)$.

Letting $\mathbb{P}^1$ have homogeneous coordinates y_0 and y_1, from formula (6.1), we know that $B(\mathfrak{m}) \subset X \times \mathbb{P}^1$ has the affine cover $\{B_1 = B_{y_0}, B_2 = B_{y_1}\}$

6.1. The blow-up of an ideal in an affine variety

where

(6.5) $$k[B_1] = k[\mathbb{A}^2]\left[\frac{x_2}{x_1}\right] = k\left[x_1, x_2, \frac{x_2}{x_1}\right] = k\left[x_1, \frac{x_2}{x_1}\right]$$

and

(6.6) $$k[B_2] = k[\mathbb{A}^2]\left[\frac{x_1}{x_2}\right] = k\left[x_2, \frac{x_1}{x_2}\right]$$

are polynomial rings (x_1, x_2 algebraically independent over k implies $x_1, \frac{x_2}{x_1}$ are algebraically independent over k and $x_2, \frac{x_1}{x_2}$ are algebraically independent over k), so $B_1 \cong \mathbb{A}^2$ and $B_2 \cong \mathbb{A}^2$.

We have that $B_1 \cong \mathbb{A}^2$ has coordinates $u_1 = x_1$, $u_2 = \frac{x_2}{x_1}$ and $B_2 \cong \mathbb{A}^2$ has coordinates $v_1 = x_2$, $v_2 = \frac{x_1}{x_2}$. On

$$B_1 \cap B_2 = B_1 \setminus Z(u_2) = B_2 \setminus Z(v_2),$$

we have $u_1 = v_1 v_2$ and $u_2 = \frac{1}{v_2}$. The map $\pi : B \to \mathbb{A}^2$ satisfies $\pi|B_1 = (u_1, u_1 u_2)$ and $\pi|B_2 = (v_1 v_2, v_1)$. Let $E = \pi^{-1}(p)$. We have that $E = Z(x_1, x_2)$, so $I(E \cap B_1) = (u_1)$ and $I(E \cap B_2) = (v_1)$.

The projection $\pi : B \setminus E \to X \setminus \{p\}$ is an isomorphism. Since $\mathfrak{m}k[B_1] = x_1 k[B_1]$ and $\mathfrak{m}k[B_2] = x_2 k[B_2]$ by (6.2), which are prime ideals, we have that $I(E \cap B_1) = x_1 k[B_1]$ and $I(E \cap B_2) = x_2 k[B_2]$ and

$$k[E \cap B_1] = k[B_1]/(x_1) \cong k\left[\frac{x_2}{x_1}\right] \quad \text{and} \quad k[E \cap B_2] = k[B_2]/(x_2) \cong k\left[\frac{x_1}{x_2}\right],$$

from which we see that $E \cap B_1 \cong \mathbb{A}^1$ and $E \cap B_2 \cong \mathbb{A}^1$. Since $k[E \cap B_1 \cap B_2] = k[\frac{x_1}{x_2}, \frac{x_2}{x_1}]$, the regular maps $(1 : \frac{x_2}{x_1}) : E \cap B_1 \to \mathbb{P}^1$ and $(\frac{x_1}{x_2} : 1) : E \cap B_2 \to \mathbb{P}^1$ agree on $E \cap B_1 \cap B_2$. Thus the maps patch to give a well-defined map $E \to \mathbb{P}^1$ which is a regular map by Proposition 3.39. By a similar argument, the inverse map $\mathbb{P}^1 \to E$ is a regular map, so $E \cong \mathbb{P}^1$. The blow-up $\pi : B \to E$ is illustrated in Figure 6.1. In the figure, $L_1 = Z(x_1)$, $L_2 = Z(x_2)$, $\tilde{L}_1 = Z(y_0) =$ strict transform of L_1, $\tilde{L}_2 = Z(y_1) =$ strict transform of L_2.

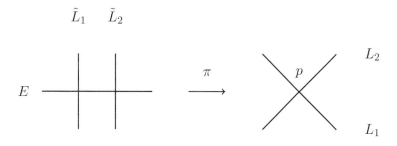

Figure 6.1. The blow-up $\pi : B \to X$

Suppose that $C \subset \mathbb{A}^2$ is a curve which contains p. Let $f(x_1, x_2) \in k[\mathbb{A}^2]$ be such that $I(C) = (f)$. Write

$$f = \sum_{i+j \geq r} a_{ij} x_1^i x_2^j$$

where $a_{ij} \in k$ and some $a_{ij} \neq 0$ with $i + j = r$ (r is the order of f). Let $\tilde{C}$ be the strict transform of C in B. In B_1, we have that

$$f = \sum a_{ij} x_1^{i+j} \left(\frac{x_2}{x_1}\right)^j = x_1^r f_1\left(x_1, \frac{x_2}{x_1}\right)$$

where

$$f_1 = \sum a_{ij} x_1^{i+j-r} \left(\frac{x_2}{x_1}\right)^j.$$

By (6.4), since x_1 does not divide f_1 in $k[B_1]$, we have that

$$I(\tilde{C} \cap B_1) = fk[B_1] :_{k[B_1]} (x_1 k[B_1])^\infty = (f_1).$$

Similarly,

$$f_2 = \sum a_{ij} \left(\frac{x_1}{x_2}\right)^i x_2^{i+j-r}$$

and $I(\tilde{C} \cap B_2) = (f_2)$.

By Proposition 6.6, we have that $B(\mathfrak{m}k[C]) \cong \tilde{C}$ (where $B(\mathfrak{m}k[C])$ is the blow-up of p in C). Thus $B(\mathfrak{m}k[C])$ is covered by the two affine open subsets $\tilde{C} \cap B_1$ and $\tilde{C} \cap B_2$. We have that

$$k[\tilde{C} \cap B_1] = k\left[x_1, \frac{x_2}{x_1}\right]/(f_1) \quad \text{and} \quad k[\tilde{C} \cap B_2] = k\left[x_2, \frac{x_1}{x_2}\right]/(f_2).$$

Write $k[C] = k[x_1, x_2]/(f) = k[\bar{x}_1, \bar{x}_2]$, From equation (6.1), we have that

$$k[\bar{x}_1, \bar{x}_2]\left[\frac{\bar{x}_2}{\bar{x}_1}\right] \cong k\left[x_1, \frac{x_2}{x_1}\right]/(f_1)$$

and

$$k[\bar{x}_1, \bar{x}_2]\left[\frac{\bar{x}_1}{\bar{x}_2}\right] \cong k\left[x_2, \frac{x_1}{x_2}\right]/(f_2),$$

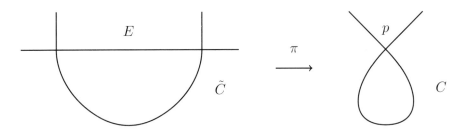

Figure 6.2. The curve C and its strict transform $\tilde{C}$

6.1. The blow-up of an ideal in an affine variety

from which we see that an understanding of the ring of (6.1) as isomorphic to a quotient of a polynomial ring is often not trivial (see Exercise 1.8). The curve C and its strict transform $\tilde{C}$ are illustrated in Figure 6.2.

From the above analysis, we are able to obtain all intrinsic properties of the blow-up $B(\mathfrak{m}k[C])$ of the point p in C. The methods demonstrated above are generally the best way to understand the geometry of blow-ups.

We will now compute the full coordinate ring $S(B(\mathfrak{m}))$ and indicate how we can compute $k[B_1]$ and $k[B_2]$ from the coordinate ring.

With the notation of Theorem 6.4, we have that
$$A = (y_1 x_1 - y_0 x_2) \subset S(X \times \mathbb{P}^1) = k[x_1, x_2, y_0, y_1]$$
is a prime ideal, so
$$I_{X \times \mathbb{P}^1}(B(\mathfrak{m})) = A : \mathfrak{m}^\infty = A.$$
This also follows from Proposition 6.5.

In this particular example, we have the desirable condition that $I(B) = (x_1 y_1 - x_2 y_0) = A$, so $I(B)$ is actually generated by the obvious relations A.

The coordinate ring of B (as a subvariety of $X \times \mathbb{P}^1$) is thus

(6.7) $$S(B) = k[x_1, x_2, y_0, y_1]/(y_1 x_1 - y_0 x_2).$$

We also have (by Theorem 6.4) that

(6.8) $$S(B) \cong \bigoplus_{i \geq 0} \mathfrak{m}^i \cong k[\mathbb{A}^2][tx_1, tx_2].$$

There is a natural isomorphism of graded $k[X]$-algebras from (6.7) to (6.8) by mapping $y_0 \mapsto tx_1$ and $y_1 \mapsto tx_2$.

We have that $\{X \times \mathbb{P}^1_{y_0}, X \times \mathbb{P}^1_{y_1}\}$ is an affine cover of $X \times \mathbb{P}^1$. Thus
$$\{B_1 = (X \times \mathbb{P}^1_{y_0}) \cap B, B_2 = (X \times \mathbb{P}^1_{y_1}) \cap B\}$$
is an affine cover of B. Now
$$k[B_1] = S(B)_{(tx_1)} = k[\mathbb{A}^2]\left[\frac{tx_2}{tx_1}\right] = k\left[x_1, x_2, \frac{x_2}{x_1}\right] = k\left[x_1, \frac{x_2}{x_1}\right]$$
since $x_2 = x_1 \frac{x_2}{x_1}$. Also, x_2 and $\frac{x_2}{x_1}$ are algebraically independent over k since x_1, x_2 are. Thus $k[B_1]$ is a polynomial ring in these two variables, so that B_1 is isomorphic to $\mathbb{A}^2$. Similarly, we have that
$$k[B_2] = S(B)_{(tx_2)} = k\left[x_2, \frac{x_1}{x_2}\right]$$
since $x_1 = x_2 \frac{x_1}{x_2}$ and B_2 is isomorphic to $\mathbb{A}^2$. We thus recover our calculations in (6.5) and (6.6).

$E = \pi^{-1}(p)$ is the algebraic set $Z_B(x_1, x_2) \subset B \subset X \times \mathbb{P}^1$. We compute
$$\begin{aligned} S(B)/(x_1, x_2)S(B) &= (k[\mathbb{A}^2][tx_1, tx_2])/(x_1, x_2) \\ &\cong (k[\mathbb{A}^2]/(x_1, x_2))[tx_1, tx_2] \\ &= k[tx_1, tx_2]. \end{aligned}$$

Since tx_1 and tx_2 are algebraically independent over k, this ring is isomorphic to a graded polynomial ring in two variables, which is the coordinate ring of $\mathbb{P}^1$. Since $E \subset \{p\} \times \mathbb{P}^1$, we have that $E = \{p\} \times \mathbb{P}^1$, with $I(E) = (x_1, x_2)$. We thus recover the calculations following (6.6).

6.2. The blow-up of an ideal in a projective variety

Suppose that $X \subset \mathbb{P}^n$ is a projective variety, with coordinate ring $S(X) = k[x_0, \ldots, x_n]$. Suppose that I is a homogeneous ideal in $S(X)$. Each open set $D(x_i)$ (for which $x_i \neq 0$) is affine with regular functions $k[D(x_i)] = k\left[\frac{x_0}{x_i}, \ldots, \frac{x_n}{x_i}\right]$. In $k[D(x_i)]$ we have an ideal

$$\tilde{I}(D(x_i)) = \left\{ f\left(\frac{x_0}{x_i}, \ldots, \frac{x_n}{x_i}\right) \mid f \in I \right\}.$$

This definition is such that for $p \in D(x_i) \cap D(x_j)$ we have equality of ideals

$$\tilde{I}(D(x_i))\mathcal{O}_{X,p} = \tilde{I}(D(x_j))\mathcal{O}_{X,p}.$$

We write $\tilde{I}_p$ for the ideal $\tilde{I}(D(x_i))\mathcal{O}_{X,p}$ in $\mathcal{O}_{X,p}$ if $p \in D(x_i)$. To every open subset U of X we can thus define an ideal

$$\tilde{I}(U) = \bigcap_{p \in U} \tilde{I}_p \subset \bigcap_{p \in U} \mathcal{O}_{X,p} = \mathcal{O}_X(U).$$

The method of the proof of Theorem 3.30 shows that we have

$$\bigcap_{p \in D(x_i)} \tilde{I}_p = \left\{ f\left(\frac{x_0}{x_i}, \ldots, \frac{x_n}{x_i}\right) \mid f \in I \right\},$$

which we initially defined to be $\tilde{I}(D(x_i))$, so our definition is consistent. If $Z(I) \neq \emptyset$, we have that $0 = \tilde{I}(X) \subset \mathcal{O}_X(X) = k$. We will see this construction again later as a special case of Theorem 11.25. We will call $\tilde{I}$ the ideal sheaf on X associated to I.

Lemma 6.7. *Suppose that $I \subset S(X)$ is a homogeneous ideal. Then there exists some $d \geq 1$ such that the ideal J which is generated by the elements I_d (the elements of I which are homogeneous of degree d) satisfies $\tilde{I} = \tilde{J}$.*

Proof. Let $F_0, \ldots, F_r$ be a homogeneous set of generators of I. Let $d_0, \ldots, d_r$ be their respective degrees. Suppose that $d \geq \max\{d_i\}$, and let J be the ideal generated by I_d. We will show that $\tilde{J} = \tilde{I}$. It suffices to show that

6.2. The blow-up of an ideal in a projective variety

$\tilde J(D(x_j)) = \tilde I(D(x_j))$ for all j. This follows since $x_j^{d-d_i} F_i \in J$ for all i, so that

$$F_i\left(\frac{x_0}{x_j}, \ldots, \frac{x_n}{x_j}\right) \in \tilde J(D(x_j)). \qquad \square$$

Definition 6.8. Suppose that X is a projective variety and $I \subset S(X) = k[x_0, \ldots, x_n]$ is a nonzero homogeneous ideal. Let d and J be as in the conclusions of Lemma 6.7. Suppose that $J = (F_0, \ldots, F_r)$, where $F_0, \ldots, F_r$ are homogeneous generators of degree d. Let $\Lambda : X \dashrightarrow \mathbb{P}^r$ be the rational map defined by $\Lambda = (F_0 : \ldots : F_r)$. The blow-up of I is $B(I) = \Gamma_\Lambda$, with projection $\pi : B(I) \to X$.

We have that $\pi^{-1}(D(x_i)) = \Gamma_{\Lambda | D(x_i)}$ by (5.3) and $\Lambda | D(x_i) = (\frac{F_0}{x_i^d} : \ldots : \frac{F_r}{x_i^d})$. Since

$$\tilde I(D(x_i)) = \tilde J(D(x_i)) = \left(\frac{F_0}{x_i^d}, \ldots, \frac{F_r}{x_i^d}\right),$$

we see that the restriction of π to $\pi^{-1}(D(x_i)) \to D(x_i)$ is the blow-up of the ideal $\tilde I(D(x_i))$ in $k[D(x_i)]$ for $0 \le i \le r$.

We thus have that Definition 6.8 is well-defined by Proposition 6.1 (applied on each affine open set $D(x_i)$), so that $B(I)$ is independent of choice of d and choice of generators of J (of the same degree d).

More generally, for any affine open subset U of X, the restriction of π to $\pi^{-1}(U) \to U$ is the blow-up of the ideal $\tilde I(U)$ in $k[U]$ (this will be proven in Lemma 12.3).

Theorem 6.9. *Suppose that $X \subset \mathbb{P}^m$ and $Y \subset \mathbb{P}^n$ are projective varieties and $\phi : Y \to X$ is a birational regular map. Then ϕ is the blow-up of a homogeneous ideal in $S(X)$.*

Proof. Let $\psi : X \dashrightarrow Y$ be the inverse rational map to ϕ. Define a regular map $\gamma : Y \to X \times Y$ by $\gamma(q) = (\phi(q), q)$ for $q \in Y$. The image of γ is closed in $X \times Y$ by Proposition 5.8 and the fact that the map $X \times Y \to Y \times X$ given by $(a, b) \mapsto (b, a)$ is an isomorphism. The map γ is an isomorphism onto its image since the inverse map is the projection onto Y. We have that $\pi_1(\gamma(q)) = \phi(q)$ for $q \in Y$.

By Theorem 2.111, there exist affine open subsets V of Y and U of X such that $\phi : V \to U$ is an isomorphism. For $q \in V$, $\gamma(q) = (p, \psi(p))$ where $p = \phi(q)$. Since $\phi : V \to U$ is an isomorphism, $\gamma : V \to \Gamma_{\psi|U}$ is an isomorphism. Thus $\gamma(Y) \subset \Gamma_\psi$ since Y is the closure of V, Y is irreducible, and Γ_ψ is closed. The image $\gamma(Y)$ contains $\Gamma_{\psi|U}$. Thus $\Gamma_\psi = \gamma(Y)$ since both Γ_ψ and $\gamma(Y)$ are closed and irreducible. We thus have a commutative

diagram

where γ is an isomorphism.

Choose forms $F_0, \ldots, F_r \in S(X)$ of a common degree, so that $(F_0 : \ldots : F_r)$ is a representative of the rational map ψ. Then $\pi : \Gamma_\psi \to X$ is the blow-up of the ideal $I = (F_0, \ldots, F_r)$. $\square$

Suppose that $\Lambda : X \dashrightarrow Y$ is a rational map of projective varieties, with $Y \subset \mathbb{P}^n$. Suppose that $(F_0 : \ldots : F_n)$ represents the rational map Λ; that is, $F_0, \ldots, F_n \in S(X)$ have the same degree, $U = X \setminus Z(F_0, \ldots, F_n) \neq \emptyset$, and $\Lambda|U = (F_0 : \ldots : F_n)$.

The graph Γ_Λ is the closure of $\Gamma_{\Lambda|V}$ in $X \times Y$ for any dense open subset V of X on which Λ is regular ($\Gamma_{\Lambda|V}$ is the image of V in $V \times Y$ of the map $p \mapsto (p, \Lambda(p))$). We thus have that

$$\Gamma_\Lambda = \Gamma_{(F_0:\ldots:F_n)} = B(I)$$

where I is the ideal $I = (F_0, \ldots, F_n) \subset S(X)$. We obtain the statement that if $(F_0 : \ldots : F_n)$ and $(G_0 : \ldots : G_n)$ are two representations of Λ, so that $F_0, \ldots, F_n$ are homogeneous of a common degree, $G_0, \ldots, G_n$ are homogeneous of a common degree, and

$$F_i G_j - F_j G_i = 0 \text{ for } 0 \leq i, j \leq n,$$

then $B(I)$ is isomorphic to $B(J)$, where

$$I = (F_0, \ldots, F_n) \quad \text{and} \quad J = (G_0, \ldots, G_n).$$

In general, $Z(I) \neq Z(J)$ for two such representations, and $\tilde{I} \neq \tilde{J}$.

The reason this works out is that two very different ideals can have the same blow-up. For instance, if X is affine, $I \subset k[X]$ is an ideal, and $0 \neq f \in k[X]$, then $B(I)$ is isomorphic to $B(fI)$. To prove this, suppose that $I = (g_0, \ldots, g_r) \subset k[X]$. The rational map $(g_0 : \ldots : g_r) : X \dashrightarrow \mathbb{P}^r$ is the same as the rational map $(fg_0 : \ldots, : fg_r) : X \dashrightarrow \mathbb{P}^r$ so the two ideals have the same blow-up. In particular, X is isomorphic to $B(fk[X])$ for any nonzero $f \in k[X]$.

As an example, consider the projective variety $A = Z(xy - zw) \subset \mathbb{P}^3$. Let $S(A) = k[x, y, z, w]/(xy - zw) = k[\overline{x}, \overline{y}, \overline{z}, \overline{w}]$. Consider the regular map $\phi : A \to \mathbb{P}^1$ which has the representations

$$\phi = (x : z) = (w : y).$$

Let $I = (x,z)$ and $J = (w,y)$. Since ϕ is regular, we have that $B(\tilde{I})$ and $B(\tilde{J})$ are isomorphic to A by Proposition 5.9. We can verify this by computing the blow-ups on the affine cover $\{D(x), D(y), D(z), D(w)\}$ of A. For instance, on $D(x)$, we have

$$k[A \cap D(x)] = k\left[\frac{y}{x}, \frac{z}{x}, \frac{w}{x}\right] = k\left[\frac{y}{x}, \frac{z}{x}, \frac{w}{x}\right] / \left(\frac{y}{x} - \frac{z}{x}\frac{w}{x}\right).$$

We have

$$\tilde{I}(D(x)) = \left(1, \frac{z}{x}\right) = k[A \cap D(x)]$$

and

$$\tilde{J}(D(x)) = \left(\frac{w}{x}, \frac{y}{x}\right) = \left(\frac{w}{x}, \frac{z}{x}\frac{w}{x}\right) = \left(\frac{w}{x}\right),$$

and we see that both ideals are principal ideals.

The blow-up of a subvariety Y of a projective variety X is the blow-up of a homogeneous ideal I, which has a set of generators of a common degree, such that $\tilde{I}(U) = \widetilde{I_X(Y)}(U)$ for all affine open sets $U \subset X$. We find such an ideal by applying Lemma 6.7 to $I_X(Y)$.

If $\pi : B(I) \to X$ is the blow-up of a homogeneous ideal in the coordinate ring $S(X)$ of a projective variety X and W is a closed subvariety of X, then as in the case when X is affine, the strict transform of W in $B(I)$ is defined to be the Zariski closure of $\pi^{-1}(W \setminus Z(I)) \cong W \setminus Z(I)$ in $B(I)$.

Definition 6.10. Suppose that $f : X \to Y$ is a birational regular map of quasi-projective varieties. Let U be the largest open subset of Y on which the inverse of f is a regular map. Suppose that Z is a subvariety of Y. The proper transform of Z by f is the closure of $f^{-1}(Z \cap U)$ in X.

In determining the difference between the strict transform and proper transform of a subvariety under a blow-up, the following lemma is useful.

Lemma 6.11. *Suppose that I is an ideal in the regular functions of an affine variety X and $\pi : B(I) \to X$ is the blow-up of I. Then the largest open subset U of X on which π^{-1} is regular (and thus an isomorphism) is the set*

$$U = \{p \in X \mid I\mathcal{O}_{X,p} \text{ is a principal ideal}\}.$$

Proof. Suppose that $I\mathcal{O}_{X,p}$ is a principal ideal. Then there exists an affine open neighborhood V of p such that $Ik[V]$ is principal. Suppose that $Ik[V] = (f)$. Then $\pi^{-1}(V) = B(Ik[V])$ is the graph of the rational map $\tilde{f} : V \dashrightarrow \mathbb{P}^0$ defined by f. But $\mathbb{P}^0$ is a single point, so $\tilde{f}$ is the regular map which contracts V to this point. The blow-up of $Ik[V]$ is the graph of the regular map $\tilde{f} : V \to \mathbb{P}^0$, so that $\pi^{-1}(V) = B(Ik[V]) \to V$ is an isomorphism.

Now suppose that V is an affine open subset of X such that $\pi^{-1}(V) = B(Ik[V]) \to V$ is an isomorphism. We have that $I\mathcal{O}_{B(I),q}$ is a principal ideal for all $q \in B(Ik[V])$ by (6.2), so $I\mathcal{O}_{X,p}$ is a principal ideal for all $p \in V$ since π is an isomorphism above V. $\square$

Use the methods of the example after Proposition 6.6 (before the computation of the coordinate ring of $B(\mathfrak{m})$) in doing these exercises. Describe the affine rings that come up in the computations as explicit quotients of polynomial rings. Lemma 6.11 should also be useful.

Exercise 6.12. Analyze the strict transform $\tilde{C}$ of the curve $C = Z(y^2 - x^3)$ under the blow-up $\pi : B(p) \to \mathbb{A}^2$ of the origin p, and compute the rings of regular functions on the natural affine cover of $\tilde{C}$. (See Exercise 1.9). Let $E = \pi^{-1}(p) \cong \mathbb{P}^1$. What is $E \cap \tilde{C}$?

Exercise 6.13. Analyze the strict transform $\tilde{C}$ of the curve $C = Z(x^2y - xy^2 + x^4 + y^4)$ under the blow-up $\pi : B(p) \to \mathbb{A}^2$ of the origin p, and compute the rings of regular functions on the natural affine cover of $\tilde{C}$. Let $E = \pi^{-1}(p) \cong \mathbb{P}^1$. What is $E \cap \tilde{C}$?

Exercise 6.14. Analyze the blow-up $\pi : B(W) \to \mathbb{A}^3$ of the following subvarieties of $\mathbb{A}^3$. Describe $\pi^{-1}(W)$.

a) $W = \{(0,0,0)\}$.

b) $W = Z(x_1, x_2)$.

c) Compute the strict transform of the surface $S = Z(x_1^3 + x_2 x_3^2)$ under each of these blow-ups.

d) Compute the strict transform of the surface $T = Z(x_1^3 + x_2^2)$ under each of these blow-ups.

e) Compute the strict transform of the curve $S = Z(x_3, x_1^2 - x_2^3)$ under each of these blow-ups.

Exercise 6.15. Consider the rational map $\phi : \mathbb{P}^2 \dashrightarrow \mathbb{P}^2$ which is represented by $(x_0 x_1 : x_0 x_2 : x_1 x_2)$. This rational map is called a quadratic transformation. Let $P_1 = (0:0:1), P_2 = (0:1:0), P_3 = (1:0:0) \in \mathbb{P}^2$. Let $L_{1,2}$ be the projective line in $\mathbb{P}^2$ containing P_1 and P_2 (that is, the one-dimensional linear subspace of $\mathbb{P}^3$ containing these two points), let $L_{1,3}$ be the projective line containing P_1 and P_3, and let $L_{2,3}$ be the projective line containing P_2 and P_3.

a) Show that $\phi^2 = \mathrm{id}$ as a rational map. Explain why this tells us that ϕ is birational.

b) Show that the ideal $I = (x_0x_1, x_0x_2, x_1x_2)$ in $S(\mathbb{P}^2) = k[x_0, x_1, x_2]$ is the intersection of prime ideals
$$I = I(P_1) \cap I(P_2) \cap I(P_3).$$

c) Let $\pi : B(I) \to \mathbb{P}^2$ be the blow-up of I. Show that $B(I)$ is the graph of ϕ.

d) Show that for $0 \leq i \leq 2$, $\pi^{-1}(D(x_i)) \to D(x_i)$ is isomorphic to the blow-up of a point in $\mathbb{A}^2$ which we analyzed at the end of this section; that is, show that there is a commutative diagram of regular maps

$$\begin{array}{ccc} \pi^{-1}(D(x_i)) & \xrightarrow{\alpha} & B \\ \pi \downarrow & & \downarrow \lambda \\ D(x_i) & \xrightarrow{\beta} & \mathbb{A}^2 \end{array}$$

where $\lambda : B \to \mathbb{A}^2$ is the blow-up of the origin and α, β are isomorphisms.

e) Determine the largest open subset of $\mathbb{P}^2$ on which ϕ is a regular map. Explain why part d) of this problem tells us the answer to this question.

f) Explain the geometry of the projections $\pi_1 : \Gamma_\phi \to \mathbb{P}^2$ and $\pi_2 : \Gamma_\phi \to \mathbb{P}^2$ in terms of the points P_1, P_2, P_3 and their preimages by π_1 and π_2, and the lines $L_{1,2}, L_{1,3}, L_{2,3}$ and their strict transforms by π_1 and π_2.

Exercise 6.16. Let $X = Z(xy - zw) \subset \mathbb{A}^4$, an affine 3-fold with
$$k[X] = k[x, y, z, w]/(xy - zw) = k[\overline{x}, \overline{y}, \overline{z}, \overline{w}].$$
Let S_1 be the affine surface $S_1 = Z(x, z)$ and S_2 be the affine surface $S_2 = Z(x, w)$. These surfaces are subvarieties of X.

a) Let $\pi : B(m) \to X$ be the blow-up of $m = (\overline{x}, \overline{y}, \overline{z}, \overline{w})$ (the blow-up of the point $p = (0,0,0,0)$). Show that $\pi : \pi^{-1}(X \setminus \{p\}) \to X \setminus \{p\}$ is an isomorphism and $E = \pi^{-1}(p)$ is a surface which is isomorphic to $\mathbb{P}^1 \times \mathbb{P}^1$.

b) Compute the strict and proper transforms of S_1 in $B(m)$. Are they the same? Compute the strict and proper transforms of S_2 in $B(m)$. Are they the same?

c) Let $\pi_1 : B(I_1) \to X$ be the blow-up of $I_1 = (\overline{x}, \overline{z})$ (the blow-up of S_1). Show that $\pi_1 : \pi_1^{-1}(X \setminus \{p\}) \to X \setminus \{p\}$ is an isomorphism and $F_1 = \pi_1^{-1}(p)$ is isomorphic to $\mathbb{P}^1$.

d) Compute the strict and proper transforms of S_1 in $B(I_1)$. Are they the same? Compute the strict and proper transforms of S_2 in $B(I_1)$. Are they the same?

e) Answer the questions of part c) for the blow-up $\pi_2 : B(I_2) \to X$ of the ideal $I_2 = (\overline{x}, \overline{w})$ (the blow-up of S_2), letting $F_2 = \pi_2^{-1}(p)$.

f) Answer the questions of part d) for the blow-up of I_2.

g) Show that there is a natural regular map $\phi_1 : B(m) \to B(I_1)$ such that $(\phi_1|E) : E \to F_1$ induces the projection on the first factor $\mathbb{P}^1 \times \mathbb{P}^1 \to \mathbb{P}^1$.

h) Show that there is a natural regular map $\phi_2 : B(m) \to B(I_2)$ such that $(\phi_2|E) : E \to F_2$ induces the projection on the second factor $\mathbb{P}^1 \times \mathbb{P}^1 \to \mathbb{P}^1$.

i) Show that the induced birational map $B(I_1) \dashrightarrow B(I_2)$ is not regular.

Chapter 7

Finite Maps of Quasi-projective Varieties

In this chapter we explore properties of finite maps.

7.1. Affine and finite maps

Definition 7.1. Suppose that X and Y are quasi-projective varieties and $\phi : X \to Y$ is a regular map.

1) The map ϕ is affine if for every $q \in Y$ there exists an affine neighborhood U of q in Y such that $\phi^{-1}(U)$ is an affine open subset of X.

2) The map ϕ is finite if for every $q \in Y$ there exists an affine neighborhood U of q in Y such that $\phi^{-1}(U)$ is an affine open subset of X and $\phi : \phi^{-1}(U) \to U$ is a finite map of affine varieties (as defined in Definition 2.55).

It follows from the more general statement of Theorem 7.5 below that if X and Y are affine varieties and $\phi : X \to Y$ is a finite map of quasi-projective varieties, then ϕ is a finite map of affine varieties (as defined in Definition 2.55).

Lemma 7.2. *Suppose that X is an affine variety. Then X is quasi-compact (every open cover has a finite subcover).*

Proof. Let $\{U_i\}_{i \in \Lambda}$ be an open cover of X. We may refine the cover by basic open sets $D(f)$ with $f \in k[X]$ and may so assume that each $U_i = D(f_i)$. Since $\bigcup_{i \in \Lambda} U_i = X$, we have that $Z(I) = \emptyset$, where I is the ideal $I = (f_i \mid i \in \Lambda)$. Thus $I = k[X]$ and so there exist a positive integer r and $i_1, \ldots, i_r \in \Lambda$ such that $(f_{i_1}, \ldots, f_{i_r}) = k[X]$. Thus $\{U_{i_1}, \ldots, U_{i_r}\}$ is a finite cover of X. □

Lemma 7.3. *Suppose that A is a ring and $f_1, \ldots, f_n \in A$ are such that the ideal $(f_1, \ldots, f_n) = A$. Suppose that N is a positive integer. Then $(f_1^N, \ldots, f_n^N) = A$.*

Proof. Since $(f_1, \ldots, f_n) = A$, there exist $g_1, \ldots, g_n \in A$ such that
$$f_1 g_1 + \cdots + f_n g_n = 1.$$
Thus
$$\begin{aligned} 1 &= (f_1 g_1 + \cdots + f_n g_n)^{nN} \\ &= \sum_{i_1 + \cdots + i_n = nN} \tfrac{(nN)!}{i_1! i_2! \cdots i_n!} (f_1 g_1)^{i_1} (f_2 g_2)^{i_2} \cdots (f_n g_n)^{i_n} \in (f_1^N, \ldots, f_n^N). \end{aligned}$$
□

Lemma 7.4. *Suppose that A is a domain which is a subring of a domain B and there exist $f_1, \ldots, f_n \in A$ such that*

1) *the ideal $(f_1, \ldots, f_n) = A$ and*
2) *the localization B_{f_i} is a finitely generated A_{f_i}-algebra for all i.*

Then B is a finitely generated A-algebra.

Further suppose that B_{f_i} is a finitely generated A_{f_i}-module for all i. Then B is a finitely generated A-module.

Proof. By assumption, there exist $r_i \in \mathbb{Z}_+$ for $1 \leq i \leq n$ and $z_{i1}, \ldots, z_{ir_i} \in B_{f_i}$ for $1 \leq i \leq n$ such that
$$B_{f_i} = A_{f_i}[z_{i1}, \ldots, z_{ir_i}].$$
After possibly multiplying the z_{ij} by a positive power of f_i, we may assume that $z_{ij} \in B$ for all i, j. Let
$$C = A[\{z_{ij}\}].$$
C is a finitely generated A-algebra which is a subring of B. We will show that $B = C$.

Suppose that $b \in B$. Then $b \in B_{f_i}$ implies there are polynomials $g_i \in A_{f_i}[x_1, \ldots, x_{r_i}]$ such that $b = g_i(z_{i1}, \ldots, z_{i,r_i})$ for $1 \leq i \leq n$. Since the polynomials g_i have only a finite number of nonzero coefficients, which are

7.1. Affine and finite maps

in A_{f_i}, there exists a positive integer N such that $f_i^N g_i \in A[x_1, \ldots, x_{r_i}]$ for $1 \le i \le n$. Thus
$$f_i^N b = f_i^N g_i(z_{i1}, \ldots, z_{ir_i}) \in A[z_{i1}, \ldots, z_{ir_i}] \subset C$$
for all i. By Lemma 7.3, there exist $c_i \in A$ such that $\sum c_i f_i^N = 1$. Thus
$$b = \left(\sum c_i f_i^N\right) b = \sum c_i f_i^N b \in C.$$

Now suppose that the B_{f_i} are finitely generated A_{f_i}-modules. Suppose that $b \in B$. Then b is integral over A_{f_i} for all i, so there exists $N > 0$ such that $f_i^N b$ is integral over A for all i. By Lemma 7.3, there exist $c_i \in A$ such that $\sum c_i f_i^N = 1$. Thus $b = \sum_i c_i f_i^N b$ is integral over A. $\square$

Theorem 7.5. *Suppose that $\phi : X \to Y$ is a regular map of quasi-projective varieties.*

1) *Suppose that ϕ is affine and U is an affine open subset of Y. Then $V = \phi^{-1}(U)$ is an affine open subset of X.*

2) *Suppose that ϕ is finite and U is an affine open subset of Y. Then $V = \phi^{-1}(U)$ is an affine open subset of X, and the restriction of ϕ to a regular map from V to U is a finite map of affine varieties.*

Proof. Let T be the Zariski closure of $\phi(X)$ in Y. Then for all affine open subsets U of Y, $T \cap U$ is affine and the inclusion of $T \cap U$ into U is a finite map of affine varieties. Thus we may assume that ϕ is dominant.

Suppose that ϕ is affine and dominant and U is an affine open subset of Y. We will first show that $\phi^{-1}(U) \to U$ is an affine map. Suppose that $q \in U$. Then there exists an affine neighborhood W of q in Y such that $\phi^{-1}(W)$ is affine. There exists $f \in k[W]$ such that $q \in W_f \subset U \cap W$ since such open sets are a basis of the topology on W. Thus $\phi^{-1}(W_f) = \phi^{-1}(W)_{\phi^*(f)}$ is affine by Proposition 2.93. We conclude that $\phi : \phi^{-1}(U) \to U$ is an affine map.

Let $A = k[U] = \mathcal{O}_Y(U)$. Suppose that $q \in U$ and let W be an affine neighborhood of q in U such that $\phi^{-1}(W)$ is affine. Then there exists $f \in A \subset k[W]$ such that $q \in U_f \subset W$ since open sets of this form are a basis for the topology on U. The inclusions $W_f \subset U_f \subset W$ imply that $W_f = U_f$ and so $\phi^{-1}(U_f) = \phi^{-1}(W_f) = \phi^{-1}(W)_{\phi^*(f)}$ is affine by Proposition 2.93. Since U is affine, it is quasi-compact (by Lemma 7.2) so there exist $f_1, \ldots, f_n \in A$ such that $\bigcup_{i=1}^n U_{f_i} = U$ and $\phi^{-1}(U_{f_i})$ is affine for all i. Thus $Z_U(f_1, \ldots, f_n) = \emptyset$, so $\sqrt{(f_1, \ldots, f_n)} = I(Z_U(f_1, \ldots, f_n)) = A$ by the nullstellensatz. Thus

(7.1) $$(f_1, \ldots, f_n) = A.$$

Let $V_i = \phi^{-1}(U_{f_i})$ for $1 \leq i \leq n$. The V_i are an affine cover of $V = \phi^{-1}(U)$. Let $B = \mathcal{O}_Y(V)$. The map ϕ gives us an injective k-algebra homomorphism $\phi^* : A \to B \subset k(X)$. Let $B_i = \mathcal{O}_Y(V_i) = k[V_i]$ for $1 \leq i \leq n$. The restriction of ϕ to V_i gives 1-1 k-algebra homomorphisms $\phi^* : k[U_{f_i}] = A_{f_i} \to B_i = k[V_i] \subset k(X)$, realizing the B_i as finitely generated A_{f_i}-algebras (B_i are finitely generated k-algebras since the V_i are affine). Now $V_i \cap V_j$ is precisely the open subset $(V_i)_{\phi^*(f_j)}$ of V_i, so for all i, j, $V_i \cap V_j$ is affine with regular functions

$$k[V_i \cap V_j] = (B_i)_{\phi^*(f_j)} = (B_j)_{\phi^*(f_i)} \subset k(X)$$

by Propositions 2.84 and 2.93.

Since $\phi^*(f_j)$ does not vanish on V_j, $\phi^*(f_j)$ is a unit in B_j, so $(B_j)_{\phi^*(f_j)} = B_j$, and

$$B_j \subset (B_j)_{\phi^*(f_i)} = (B_i)_{\phi^*(f_j)}$$

for all i, j. Now

$$B = \mathcal{O}_Y(V) = \bigcap_{i=1}^n \mathcal{O}_Y(V_i) = \bigcap_{i=1}^n B_i.$$

We compute

$$B_{\phi^*(f_j)} = \left(\bigcap_{i=1}^n B_i\right)_{\phi^*(f_j)} = \bigcap_{i=1}^n (B_i)_{\phi^*(f_j)} = B_j.$$

By Lemma 7.4, B is a finitely generated A-algebra, and since A is a finitely generated k-algebra, B is a finitely generated k-algebra. Thus there exists an affine variety Z such that $k[Z] = B$. Let $t_1, \ldots, t_m \in B$ generate B as a k-algebra (the t_i are coordinate functions on Z). Since $B = \mathcal{O}_X(V)$, $\alpha = (t_1, \ldots, t_m)$ induces a regular map $\alpha : V \to Z$. Now the $\alpha_i = \alpha \mid V_i$ induce isomorphisms $\alpha_i : V_i \to Z_{\phi^*(f_i)}$ of affine varieties for all i, since α_i^* induces an isomorphism of regular functions. We may thus define regular maps $\psi_i : Z_{\phi^*(f_i)} \to V_i$ which are isomorphisms by requiring that $(\psi_i)^* = (\alpha_i^*)^{-1}$. The ψ_i patch to give a continuous map $\psi : V \to Z$ which is a regular map by Proposition 3.39. Since ψ is an inverse to α, we have that $V \cong Z$ is an affine variety.

Now suppose that $X \to Y$ is a finite map of quasi-projective varieties. Then we may choose $f_1, \ldots, f_n \in A$ as above with the additional property that the B_{f_i} are finitely generated A_{f_i}-modules for all i. Thus B is a finite A-module by Lemma 7.4, and so $V = \phi^{-1}(U) \to U$ is a finite map of affine varieties. $\square$

Exercise 7.6. Suppose that X is a quasi-projective variety. Show that X is quasi-compact (every open cover has a finite subcover).

7.2. Finite maps

Theorem 7.7. *Suppose that X and Y are quasi-projective varieties and $\phi : X \to Y$ is a finite regular map. Then ϕ is a closed map, and if ϕ is dominant, then ϕ is surjective.*

Proof. There exists an affine cover $\{V_i\}$ of Y such that the maps $\phi : U_i \to V_i$ where $U_i = \phi^{-1}(V_i)$ are finite maps of affine varieties. We obtain this by either choosing the V_i so that $\phi^{-1}(V_i)$ are affine with $\phi : U_i \to V_i$ finite, which exist by the definition of a finite map, or we pick an arbitrary affine cover $\{V_i\}$ of Y and apply Theorem 7.5 to get this statement. We have that each $\phi : U_i \to V_i$ is a closed map by Corollary 2.58 and is surjective if ϕ is dominant by Theorem 2.57, so $\phi : X \to Y$ is a closed map, which is surjective if ϕ is dominant. $\square$

Theorem 7.8. *Suppose that X and Y are quasi-projective varieties and $\phi : X \to Y$ is a dominant regular map. Then $\phi(X)$ contains a nonempty open subset of Y.*

Proof. It suffices to prove the theorem for the map to an affine open subset V of Y from the restriction of ϕ to an affine open subset U contained in the preimage of V. Thus we may assume that X and Y are affine. Let $u_1, \ldots, u_r \in k[X]$ be a transcendence basis of $k(X)$ over $k(Y)$. Then

$$k[Y] \subset k[Y][u_1, \ldots, u_r] = k[Y \times \mathbb{A}^r] \subset k[X].$$

Thus ϕ factors as the composition $\phi = g \circ h$ of regular maps where $h : X \to Y \times \mathbb{A}^r$ and $g : Y \times \mathbb{A}^r \to Y$ is the projection onto the first factor.

Every element $v \in k[X]$ is algebraic over $k(Y \times \mathbb{A}^r)$. Thus there exists a polynomial $f(x) = x^s + b_1 x^{s-1} + \cdots + b_s$ with $b_i \in k(Y \times \mathbb{A}^r)$ such that $f(v) = 0$. Write $b_i = \frac{c_i}{a}$ with $a, c_i \in k[Y \times \mathbb{A}^r]$. Thus v is integral over $k[Y \times \mathbb{A}^r]_a$.

Let $v_1, \ldots, v_m$ be coordinate functions on X (so that $k[X] = k[v_1, \ldots, v_m]$). For each v_i choose $a_i \in k[Y \times \mathbb{A}^r]$ such that v_i is integral over $k[Y \times \mathbb{A}^r]_{a_i}$. Let $F = a_1 \cdots a_m$. Then $k[X]_{h^*(F)}$ is integral over $k[Y \times \mathbb{A}^r]_F$, so that $h : X_{h^*(F)} \to (Y \times \mathbb{A}^r)_F$ is finite and dominant. Thus $(Y \times \mathbb{A}^r)_F = h(X_{h^*(F)}) \subset h(X)$ by Theorem 7.7. It remains to show that $g((Y \times \mathbb{A}^r)_F)$ contains a set that is open in Y.

We have an expression

$$F = \sum f_{i_1, \ldots, i_r} u_1^{i_1} \ldots u_r^{i_r} \in k[Y \times \mathbb{A}^r] = k[Y][u_1, \ldots, u_r]$$

with $f_{i_1, \ldots, i_r} \in k[Y]$ not all zero. If $p \in Y$ and some $f_{i_1, \ldots, i_r}(p) \neq 0$, then there exists a point $q \in \mathbb{A}^r$ such that $F(p, q) \neq 0$ (by Theorem 1.4 or the affine nullstellensatz). Thus the nonempty open set

$$\bigcup Y_{f_{i_1,\ldots,i_r}} = Y \setminus Z(\{f_{i_1,\ldots,i_r}\}) \subset g((Y \times \mathbb{A}^r)_F) \subset \phi(X). \qquad \square$$

A short proof of Theorem 7.8 can be obtained by using some theorems from commutative algebra on flatness. We reduce to the case where X and Y are both affine. By the theorem of generic flatness, [**107**, Theorem 52 on page 158], there is a nontrivial open subset U of Y such that $\phi^{-1}(U) \to U$ is flat. By [**107**, Theorem 4 on page 33] and [**107**, Theorem 8 on page 48], $\phi = \phi^{-1}(U) \to U$ is an open map, so $\phi(X)$ contains a nontrivial open subset of X.

Theorem 7.9. *Suppose that X is a projective variety which is a closed subvariety of a projective space $\mathbb{P}^n$ and suppose that $X \subset \mathbb{P}^n \setminus E$ where E is a d-dimensional linear subspace. Then the projection $\pi : X \to \mathbb{P}^{n-d-1}$ from E determines a dominant finite map from X to the projective variety $\pi(X)$.*

Proof. Let $y_0, \ldots, y_{n-d-1}$ be homogeneous coordinates on $\mathbb{P}^{n-d-1}$ and let $L_0, \ldots, L_{n-d-1}$ be a basis of the vector space of linear forms vanishing on E. Define π by the formula $\pi = (L_0 : \ldots : L_{n-d-1})$. Here π is a regular map on X since $E \cap X = \emptyset$, so the forms $L_0, \ldots, L_{n-d-1}$ do not vanish simultaneously on X.

Let $U_i = \pi^{-1}(\mathbb{P}^{n-d-1}_{y_i}) \cap X = \mathbb{P}^n_{L_i} \cap X$. Then U_i is an affine open subset of X. We will show that for all i such that $U_i \neq \emptyset$, $U_i \to \pi(X) \cap \mathbb{P}^{n-d-1}_{y_i}$ is a finite map. The image $\pi(X)$ is a closed subset of $\mathbb{P}^{n-d-1}$ by Theorem 5.14. Hence $\pi(X)$ is a projective variety and $\pi(X)_{y_i} = \pi(X) \cap \mathbb{P}^{n-d-1}_{y_i}$ is an affine open subset of $\pi(X)$.

We will show that $k[U_i]$ is integral over the subring $k[\pi(X)_{y_i}]$ for all i, so that $\pi : U_i \to \pi(X)_{y_i}$ is finite for all i and thus $\pi : X \to \pi(X)$ is finite and dominant.

Suppose $g \in k[U_i]$. Then g is the restriction of a form $\frac{G}{L_i^m}$ where m is the degree of the homogeneous form $G \in S(\mathbb{P}^n)$ by formula (4.6). Let $z_0, \ldots, z_{n-d}$ be homogeneous coordinates on $\mathbb{P}^{n-d}$, and define a rational map $\pi_1 = (L_0^m : \ldots : L_{n-d-1}^m : G)$ from $\mathbb{P}^n$ to $\mathbb{P}^{n-d}$. The rational map π_1 induces a regular map of X and its image $\pi_1(X)$ is closed in $\mathbb{P}^{n-d}$ by Theorem 5.14. Let $F_1, \ldots, F_s$ be a set of generators of $I(\pi_1(X)) \subset S(\mathbb{P}^{n-d})$. As $X \cap E = \emptyset$, the forms $L_0, \ldots, L_{n-d-1}$ do not vanish simultaneously on X. Thus the point $(0 : \ldots : 0 : 1)$ is not contained in $\pi_1(X)$, so that

$$Z_{\mathbb{P}^{n-d}}(z_0, \ldots, z_{n-d-1}, F_1, \ldots, F_s) = \{(0 : \ldots : 0 : 1)\} \cap \pi_1(X) = \emptyset.$$

By Proposition 3.11, we have that $T_l \subset (z_0, \ldots, z_{n-d-1}, F_1, \ldots, F_s)$ for some $l > 0$, where T_l is the vector space of homogeneous forms of degree l on

7.2. Finite maps

$\mathbb{P}^{n-d}$. In particular, we have an expression

$$z_{n-d}^l = \sum_{j=0}^{n-d-1} z_j H_j + \sum_{j=1}^{s} F_j P_j$$

where $H_j, P_j \in S(\mathbb{P}^{n-d}) = k[z_0, \ldots, z_{n-d}]$ are polynomials. Denoting by $H^{(q)}$ the homogeneous component of H of degree q, let

$$\Phi(z_0, \ldots, z_{n-d}) = z_{n-d}^l - \sum_{j=0}^{n-d-1} z_j H_j^{(l-1)}.$$

We have that $\Phi \in I(\pi_1(X))$. The homogeneous polynomial Φ has degree l, and as a polynomial in z_{n-d} it has the leading coefficient 1, so that it has an expression

$$\Phi = z_{n-d}^l + \sum A_{l-j}(z_0, \ldots, z_{n-d-1}) z_{n-d}^j$$

where the A_{l-j} are homogeneous of degree $l-j$. Substitution of the defining formulas $\pi_1^*(z_i) = L_i^m$ for $0 \leq i \leq n-d-1$ and $\pi_1^*(z_{n-d}) = G$ induces a k-algebra homomorphism $\pi_1^* : S(\mathbb{P}^{n-d}) \to S(\mathbb{P}^n)$. Since the F_i vanish on $\pi_1(X)$, we have that $\pi_1^*(F_i) \in I(X)$. We thus have that

$$\pi_1^*(\Phi) = \Phi(L_0^m, \ldots, L_{n-d-1}^m, G) \in I(X)$$

is a homogeneous form of degree lm in $S(\mathbb{P}^n)$. Dividing this form by L_i^{ml}, we obtain a relation

$$\left(\frac{G}{L_i^m}\right)^l + \sum_{j=0}^{l-1} A_{l-j}\left(\left(\frac{L_0}{L_i}\right)^m, \ldots, 1, \ldots, \left(\frac{L_{n-d-1}}{L_i}\right)^m\right) \left(\frac{G}{L_i^m}\right)^j$$

$$\in I(X \cap \mathbb{P}_{L_i}^n) \subset k[\mathbb{P}_{L_i}^n].$$

The rational map π induces a regular map $\pi : \mathbb{P}_{L_i}^n \to \mathbb{P}_{y_i}^{n-d-1}$ of affine varieties, with $\pi^* : k[\mathbb{P}_{y_i}^{n-d-1}] \to k[\mathbb{P}_{L_i}^n]$ given by $\pi^*\left(\frac{y_j}{y_i}\right) = \frac{L_j}{L_i}$ for $0 \leq j \leq n-d-1$. We have that $(\pi^*)^{-1}(I(X \cap \mathbb{P}_{L_i}^n)) = I(\pi(X) \cap \mathbb{P}_{y_i}^{n-d-1})$. Since g is the residue of $\frac{G}{L_i^m}$ in $k[U_i] = k[X \cap \mathbb{P}_{L_i}^n] = k[\mathbb{P}_{L_i}^n]/I(X \cap \mathbb{P}_{L_i}^n)$, we obtain the desired dependence relation, showing that g is integral over $k[\pi(X)_{y_i}] = k[\mathbb{P}_{y_i}^{n-d-1}]/I(\pi(X)_{y_i})$. □

Remark 7.10. Looking back at the proof, we see that we have also proved the following theorem. Writing the coordinate ring of X as

$$S(X) = S(\mathbb{P}^n)/I(X)$$

and the coordinate ring of $\pi(X)$ as

$$S(\pi(X)) = k[\overline{y}_0, \ldots, \overline{y}_m] = S(\mathbb{P}^m)/I(\pi(X))$$

where $m = n-d-1$, we showed that the 1-1 graded k-algebra homomorphism

$$\pi^* : S(\pi(X)) \to S(X)$$

defined by $\pi^*(\overline{y}_i) = L_i$ for $0 \leq i \leq m$ makes $S(X)$ an integral extension of $S(\pi(X))$.

By applying this theorem to a Veronese embedding of X, or modifying the proof using formula (4.9) instead of (4.6), we obtain the following generalization.

Theorem 7.11. *Let $F_0, \ldots, F_s$ be linearly independent forms over k of degree $m > 0$ on $\mathbb{P}^n$ that do not vanish simultaneously on a closed subvariety $X \subset \mathbb{P}^n$. Then $\phi = (F_0 : \ldots : F_s)$ determines a dominant finite map $\phi : X \to \phi(X)$ of projective varieties.*

Corollary 7.12. *Suppose that X is a projective variety which is a closed subvariety of a projective space $\mathbb{P}^n$. Then there exists a surjective finite map $\phi : X \to \mathbb{P}^m$ for some m. The map ϕ is the restriction to X of a projection from a suitable linear subspace of $\mathbb{P}^n$.*

Proof. If $X \neq \mathbb{P}^n$, choose a point $p \in \mathbb{P}^n \setminus X$ and let $\pi : X \to \mathbb{P}^{n-1}$ be the projection from p. The induced regular map $\pi : X \to \pi(X)$ is a finite map and $\pi(X)$ is a projective variety which is a closed subset of $\mathbb{P}^{n-1}$ by Theorem 7.9. We continue until the image of X is the whole ambient projective space. A composition of finite maps is finite so the resulting map is finite. $\square$

Corollary 7.13 (Projective Noether normalization). *Suppose that R is the coordinate ring of a projective variety. Then there exist $m \geq 0$ and linear forms $L_0, \ldots, L_m$ in R such that the graded k-algebra homomorphism*

$$\phi^* : k[x_0, \ldots, x_m] \to R$$

is an integral extension, where $k[x_0, \ldots, x_m]$ is a polynomial ring and $\phi^*(x_i) = L_i$ for $0 \leq i \leq m$.

Proof. This statement follows from Corollary 7.12 and Remark 7.10. A purely algebraic proof is given in [28, Theorem I.5.17]. $\square$

Exercise 7.14. Suppose that $\phi : \mathbb{P}^n \to \mathbb{P}^m$ is a regular map. Show that either $\phi(\mathbb{P}^n)$ is a point or ϕ is a finite map onto a closed subvariety of $\mathbb{P}^m$. (Recall the conclusions of Exercise 4.14 and the assumption that $F_0, \ldots, F_s$ are linearly independent forms in Theorem 7.11.)

7.3. Construction of the normalization

In this section we construct the normalization of a quasi-projective variety in a finite extension of its function field.

Definition 7.15. Suppose that X is a quasi-projective variety and $p \in X$. The point p is called a normal point of X if $\mathcal{O}_{X,p}$ is integrally closed in its quotient field $k(X)$. The variety X is called normal if all points of X are normal points of X.

Proposition 7.16. *Suppose that X is a normal quasi-projective variety. Then $\mathcal{O}_X(X)$ is integrally closed in $k(X)$.*

Proof. Suppose that $f \in k(X)$ is integral over $\mathcal{O}_X(X)$. Then for all $p \in X$, f is integral over $\mathcal{O}_{X,p}$, so that $f \in \mathcal{O}_{X,p}$. Thus

$$f \in \bigcap_{p \in X} \mathcal{O}_{X,p} = \mathcal{O}_X(X). \qquad \square$$

The remainder of this section will be devoted to the proof of the following theorem (from the proof on page 177 of [**161**] and of [**116**, Theorem 4, Section III.8]).

Theorem 7.17 (Normalization, in a finite extension). *Suppose X is a quasi-projective variety and $\Lambda : k(X) \to L$ is a k-algebra homomorphism of fields, such that L is a finite extension of $k(X)$. Then there is a unique normal quasi-projective variety Y with function field $k(Y) = L$, and there is a dominant finite regular map $\pi : Y \to X$ such that $\pi^* : k(X) \to k(Y)$ is the homomorphism Λ.*

If X is affine, then Y is affine. If X is projective, then Y is projective.

The normal variety Y is called the normalization of X in L. If $L = k(X)$, then Y is called the normalization of X. We first prove uniqueness. Suppose that $\pi : Y \to X$ and $\pi' : Y' \to X$ each satisfy the conclusions of the theorem. Suppose that $p \in X$ and that U is an affine neighborhood of p in X. Then $V = \pi^{-1}(U)$ is an affine open subset of Y since π is finite. Further, $k[V]$ is integrally closed in L and is finite over $k[U]$ by Theorem 7.5. Thus $k[V]$ is the integral closure of $k[U]$ in L. We thus have that $k[V'] = k[V]$ where $V' = (\pi')^{-1}(U)$, so that the identity map is an isomorphism of the affine varieties V and V'. Since this holds for an affine cover of X, we have that $Y = Y'$ by Proposition 3.39.

We now prove existence for an affine variety X. Let R be the integral closure of $k[X]$ in L. Then R is a finitely generated k-algebra (by Theorem 1.54) which is a domain, so that $R = k[Y]$ for some affine variety Y. The inclusion $\Lambda : k[X] \to k[Y]$ induces a finite regular map $Y \to X$ by Proposition 2.40.

We now prove existence for a projective variety X, from which existence for a quasi-projective variety follows.

For a graded ring A and $d \in \mathbb{Z}$, recall that in (3.2), we define a graded ring $A^{(d)}$ by $A^{(d)} = \bigoplus_{i=-\infty}^{\infty} A_i^{(d)}$, where $A_i^{(d)} = A_{id}$.

Now suppose that $X \subset \mathbb{P}^n$ is a projective variety, with homogeneous coordinate ring

$$R = S(X) = k[x_0, \ldots, x_n]/P = k[\overline{x}_0, \ldots, \overline{x}_n].$$

Let $\alpha \in R_1$ be a nonzero element, and let $\Sigma \subset R$ be the multiplicatively closed set of nonzero homogeneous elements. Then the localization $\Sigma^{-1}R$ is graded.

Lemma 7.18. *There is an isomorphism of graded rings*

$$\Sigma^{-1}R \cong k(X)\left[\alpha, \frac{1}{\alpha}\right] \cong \bigoplus_{n \in \mathbb{Z}} k(X)\alpha^n,$$

which is the localization $k(X)[\alpha]_\alpha$ of the standard graded polynomial ring $K(X)[\alpha]$ in the variable α (which has degree 1) over $k(X)$.

Proof. We have that a homogeneous element $\beta \in \Sigma^{-1}R$ of degree d has an expression $\beta = \frac{a}{f}$ where $a \in R_i$ and $f \in R_m$ and $i - m = d$. Thus the elements of $\Sigma^{-1}R$ of degree 0 are exactly the elements of $k(X)$. If β has degree $d \neq 0$, then we have

$$\beta = \alpha^d \frac{a}{f\alpha^d},$$

where $\frac{a}{f\alpha^d}$ has degree 0. Thus $\frac{a}{f\alpha^d} \in k(X)$, and we have that $(\Sigma^{-1}R)_d = k(X)\alpha^d$. In particular, $\Sigma^{-1}R = k(X)[\alpha, \frac{1}{\alpha}]$. □

Lemma 7.19. *The integral closure S of R in $L[\alpha]$ is a graded ring and a finitely generated R-module.*

Proof. Since R is a finitely generated k-algebra and $L(\alpha)$ is a finite field extension of $k(X)(\alpha)$, we have that the integral closure of S in the field $L(\alpha)$ is a finite R module by Theorem 1.54. Thus the submodule S is a finitely generated R-module by Lemma 1.55. We observe that since $L[\alpha]$ is normal, S is the integral closure of R in the field $L(\alpha)$.

We first proof the lemma when $L = k(X)$. Suppose that $a \in S$. Write $a = a_s + a_{s+1} + \cdots + a_t$ where each $a_i \in L[\alpha]$ is homogeneous of degree i and $a_s \neq 0$. The homogeneous form a_s is called the initial component a. Since $S \subset k(X)[\alpha]$, every element of S can be written as a quotient of two elements of R such that the denominator is a homogeneous element. Since a is integral over R, $R[a]$ is a finitely generated R-module, by Theorem 1.49. Thus there exists a homogeneous element $0 \neq d \in R$ such that $dR[a] \subset R$.

7.3. Construction of the normalization

Hence for every $i \geq 0$, $da^i \in R$. We have that da_s^i is the initial component of da^i, so $da_s^i \in R$ since R is graded. Thus $a_s^i \in \frac{1}{d}R$ for all $i \in \mathbb{N}$, so $R[a_s] \subset \frac{1}{d}R$ is a finite R-module by Lemma 1.55. Thus a_s is integral over R, and so $a - a_s = a_{s+1} + \cdots + a_t$ is integral over R, and continuing this way, we establish that all a_i are integral over R. Thus S is a graded subring of $k(X)[\alpha]$.

Now assume that L is a finite extension of $k(X)$ and S is the integral closure of R in $L[\alpha]$. Let $S_q' = S \cap L[\alpha]_q$ for all $q \in \mathbb{N}$, and let $S' = \bigoplus S_q'$, a graded subring of $L[\alpha]$ which is contained in S. Suppose $\beta \in L$. There exist $n \in \mathbb{Z}_+$ and $c_i \in k(X)$ such that

$$\beta^n + c_1\beta^{n-1} + \cdots + c_n = 0.$$

Thus there exists a homogeneous element $h \in R$ such that $h\beta$ is integral over R. We have that $h\beta \in S'$ since $h\beta$ is homogeneous of degree $\deg h$. Thus the quotient field of S' is $L(\alpha)$. As S' contains R, the integral closure of S' in $L[\alpha]$ is S. But then $S = S'$ by the first part of the proof. □

Although S is graded, it may be that S is not generated in degree 1. By Exercise 3.6, there exists $d \in \mathbb{Z}_+$ such that $S^{(d)}$ is generated in degree 1.

We have that $R^{(d)} = R \cap k(X)[\alpha^d]$. We will show that $S^{(d)}$ is the integral closure of $R^{(d)}$ in $L[\alpha^d]$. If $x \in L[\alpha^d]$ is integral over $R^{(d)}$, then as an element of $L[\alpha]$, it is integral over R. Thus $x \in S \cap L[\alpha^d] = S^{(d)}$. If $x \in S^{(d)}$ is homogeneous, then x is integral over R and we then have a homogeneous equation of integral dependence

$$x^m + f_1 x^{m-1} + \cdots + f_m = 0$$

with $f_i \in R_{\deg(x)i}$. But d divides $\deg(x)$, so x is integral over $R^{(d)}$.

Choosing a basis $\overline{y}_0, \ldots, \overline{y}_m$ of $S_1^{(d)}$, we have an isomorphism

$$S^{(d)} \cong k[y_0, \ldots, y_m]/P^* = k[\overline{y}_0, \ldots, \overline{y}_m]$$

where P^* is a homogeneous prime ideal in the polynomial ring $k[y_0, \ldots, y_m]$ and the y_i all have degree 1. Let $Y \subset \mathbb{P}^m$ be the projective variety $Y = Z(P^*)$. We have $S(Y) \cong S^{(d)}$.

We will now show that Y is normal. The ring $S^{(d)}$ is integrally closed in $L[\alpha^d]$. Since $L[\alpha^d]$ is isomorphic to a polynomial ring over a field, it is integrally closed in its quotient field. Hence $S^{(d)}$ is integrally closed in its quotient field. Thus the localization $S_{y_i}^{(d)}$ is integrally closed in its quotient field (by Exercise 1.58). By (3.8),

$$\mathcal{O}_Y(D(y_i)) = \left\{ \frac{f}{y_i^m} \mid m \in \mathbb{Z}_+ \text{ and } f \in S_m^{(d)} \right\},$$

which is the set of elements of $S_{\bar{y}_i}^{(d)}$ of degree 0. Thus it is the intersection $S_{\bar{y}_i}^{(d)} \cap L$ taken within $L[\alpha^d]$, which is integrally closed in L since $S_{\bar{y}_i}^{(d)}$ is integrally closed. Since the local ring of every point of Y is a localization of one of the normal local rings $\mathcal{O}_Y(D(y_i))$, all of these local rings are integrally closed, so that Y is normal.

By the Veronese embedding $\phi : \mathbb{P}^n \to \mathbb{P}^e$ where $e = \binom{n+d}{n} - 1$, we have an isomorphism of X with a closed subset of $\mathbb{P}^e$, such that the coordinate ring of $\phi(X)$ satisfies $S(\phi(X)) \cong R^{(d)}$. Choosing a basis $\bar{z}_0, \ldots, \bar{z}_l$ of $R_1^{(d)}$, we have an isomorphism $R^{(d)} \cong k[z_0, \ldots, z_l]/P' = k[\bar{z}_0, \ldots, \bar{z}_l]$ where P' is a homogeneous prime ideal in the polynomial ring $k[z_0, \ldots, z_l]$ and the z_i all have degree 1.

Our graded inclusion $R^{(d)} \subset S^{(d)}$ gives us an expression
$$\bar{z}_i = \sum_j a_{ij} \bar{y}_i$$
with $a_{ij} \in k$ for all i, j.

Let $L_i = \sum_j a_{ij} y_i$ for $0 \le i \le l$. We will now show that $Z(L_1, \ldots, L_l) \cap Y = \emptyset$. Suppose that $p \in Z(L_1, \ldots, L_l) \cap Y$. Since $S^{(d)}$ is integral over $R^{(d)}$, for $0 \le i \le m$ we have that $\bar{y}_i$ is integral over $R^{(d)}$, by a homogeneous relation. Thus we have equations
$$y_i^{n_i} + b_{i1}(L_0, \ldots, L_l) y_i^{n_i - 1} + \cdots + b_{i, n_i}(L_0, \ldots, L_l) \in P^*$$
where the b_{ij} are homogeneous polynomials of degree j. Evaluating at p, we obtain that $y_i^{n_i}(p) = 0$, so that $y_i(p) = 0$ for all i, which is impossible. Thus $Z(L_1, \ldots, L_l) \cap Y = \emptyset$.

By Theorem 7.9, the rational map $\pi = (L_0 : \ldots : L_l)$ from Y to $\phi(X) \cong X$ is a finite regular map. By our construction, the induced map $\pi^* : k(X) \to k(Y)$ is Λ.

Exercise 7.20. Show that an affine variety X is normal if and only if $k[X]$ is integrally closed in $k(X)$.

Chapter 8

Dimension of Quasi-projective Algebraic Sets

8.1. Properties of dimension

Suppose that X is a quasi-projective algebraic set. We define the dimension of X to be its dimension as a topological space (Definition 2.63). The following proposition is proved in the same way as Proposition 2.67.

Proposition 8.1. *Suppose that X is a quasi-projective algebraic set and $V_1, \ldots, V_n$ are its irreducible components. Then $\dim X = \max\{\dim V_i\}$.*

Theorem 8.2. *Suppose that X is a projective variety. Then:*

1) $\dim X = \mathrm{trdeg}_k k(X)$.

2) *Any maximal chain of distinct irreducible closed subsets of X has length $n = \dim X$.*

3) *Suppose that U is a dense open subset of X. Then $\dim U = \dim X$.*

Proof. We have a closed embedding $X \subset \mathbb{P}^n$. Let $x_0, \ldots, x_n$ be homogeneous coordinates on $\mathbb{P}^n$. Suppose that

(8.1) $$W_0 \subset W_1 \subset \cdots \subset W_m$$

is a chain of distinct irreducible closed subsets of X. There exists an open set $D(x_i) = \mathbb{P}^n_{x_i}$ such that $W_0 \cap D(x_i) \neq \emptyset$. Then since the W_i are irreducible,

(8.2) $$W_0 \cap D(x_i) \subset \cdots \subset W_m \cap D(x_i)$$

is a chain of distinct irreducible closed subsets of the affine variety $U = X \cap D(x_i)$. Thus $m \leq \dim U = \operatorname{trdeg}_k k(X)$ by Proposition 2.64, Theorem 1.63, and Proposition 3.37.

Suppose that (8.1) is a maximal chain. Then (8.2) is a maximal chain. Thus $m = \operatorname{trdeg}_k k(X)$ by Proposition 2.64, Theorem 1.63, and Corollary 2.69.

Now the proof that all nontrivial open subsets of X have the same dimension as X follows from the proof of Proposition 2.70. □

Theorem 8.3. *Suppose that X is a projective variety, with homogeneous coordinate ring $S(X)$. Then $\dim X + 1 = \dim S(X)$.*

Proof. By Lemma 7.18, The localization $\Sigma^{-1} S(X) \cong k(X)[t, \frac{1}{t}]$ as graded rings, where Σ is the multiplicatively closed set of nonzero homogeneous elements of $S(X)$ and t is an indeterminate (with $\deg(t) = 1$ and the elements of $k(X)$ have degree 0). Thus the transcendence degree of the quotient field of $S(X)$ over k is equal to one plus the transcendence degree of $k(X)$ over k. □

Theorem 8.4. *Suppose that $W \subset \mathbb{P}^n$ is a projective variety of dimension ≥ 1 and $F \in S(\mathbb{P}^n)$ is a form which is not contained in $I(W)$. Then $W \cap Z(F) \neq \emptyset$ and all irreducible components of $Z(F) \cap W$ have dimension $\dim W - 1$.*

Proof. Suppose that X is an irreducible component of $W \cap Z(F)$. Then there exists an open subset $D(x_i)$ of $\mathbb{P}^n$ such that $X \cap D(x_i) \neq \emptyset$. Let d be the degree of F. Here $\frac{F}{x_i^d} \in \mathcal{O}_{\mathbb{P}^n}(D(x_i))$ and $X \cap D(x_i)$ is an irreducible component of $Z_{D(x_i)}(\frac{F}{x_i^d}) \cap (W \cap D(x_i))$. Since $\frac{F}{x_i^d}$ does not restrict to the zero element on $W \cap D(x_i)$, we have that $X \cap D(x_i)$ has dimension $\dim W - 1$ by Theorem 2.71.

Suppose that $Z(F) \cap W = \emptyset$. Then by Theorem 7.11, $\phi = (F)$ induces a finite regular map from W to $\mathbb{P}^0$, which is a point, so that $k(\mathbb{P}^0) = k$. Since ϕ is finite, $k(W)$ is a finite field extension of $k(\mathbb{P}^0)$, so that $k(W) = k$ and $\dim W = \operatorname{trdeg}_k k(W) = 0$. □

Corollary 8.5. *Suppose that $W \subset \mathbb{P}^n$ is a projective variety and $F_1, \ldots, F_r \in S(\mathbb{P}^n)$ are forms (of degree > 0). Then*

$$\dim G \geq \dim W - r$$

for all irreducible components G of $Z(F_1, \ldots, F_r) \cap W$. If $r \leq \dim W$, then $Z(F_1, \ldots, F_r) \cap W \neq \emptyset$. (The dimension of $\emptyset$ is -1.)

Proof. If $r > \dim W$, then the corollary is certainly true, so suppose $r \leq \dim W$. Inductively define subvarieties $W_{i_1, \ldots, i_s}$ of $\mathbb{P}^n$ for $1 \leq s \leq r$ and

natural numbers $\sigma(i_1, \ldots, i_s)$ by
$$W \cap Z(F_1) = W_1 \cup \cdots \cup W_{\sigma(0)}$$
and
$$W_{i_1,\ldots,i_s} \cap Z(F_{s+1}) = W_{i_1,\ldots,i_s,1} \cup \cdots \cup W_{i_1,\ldots,i_s,i_{\sigma(i_1,\ldots,i_s)}}$$
for $s \geq 1$, where
$$W_{i_1,\ldots,i_s,1}, \ldots, W_{i_1,\ldots,i_s,i_{\sigma(i_1,\ldots,i_s)}}$$
are the irreducible components of $W_{i_1,\ldots,i_s} \cap Z(F_{s+1})$.

If $F_{s+1} \in I(W_{i_1,\ldots,i_s})$, then $W_{i_1,\ldots,i_s} \cap Z(F_{s+1}) = W_{i_1,\ldots,i_s}$, so that
$$\sigma(i_1, \ldots, i_s) = 1,$$
$W_{i_1,\ldots,i_s,1} = W_{i_1,\ldots,i_s}$ and $\dim(W_{i_1,\ldots,i_s} \cap Z(F_{s+1})) = \dim(W_{i_1,\ldots,i_s})$.

If $F_{s+1} \notin I(W_{i_1,\ldots,i_s})$, then $\dim W_{i_1,\ldots,i_s,j} = \dim W_{i_1,\ldots,i_s} - 1$ for all j by Theorem 8.4 since $\dim W_{i_1,\ldots,i_s} \geq 1$ for $s < r = \dim W$ by induction on s.

Since $Z(F_1, \ldots, F_r) \cap W = \bigcup W_{i_1,\ldots,i_r}$, we have that $G = W_{i_1,\ldots,i_r}$ for some $i_1, \ldots, i_r$, and thus $\dim G \geq \dim W - r$. $\square$

Corollary 8.6. *Suppose that W is a quasi-projective variety and $f_1, \ldots, f_r \in \mathcal{O}_W(W)$. Suppose that $Z_W(f_1, \ldots, f_r) \neq \emptyset$. Then*
$$\dim G \geq \dim W - r$$
for all irreducible components G of $Z_W(f_1, \ldots, f_r)$.

If Y is a quasi-projective algebraic set, contained in a quasi-projective variety X, then we define the codimension of Y in X to be
$$\operatorname{codim}_X(Y) = \dim X - \dim Y.$$

Exercise 8.7. Show that the m in Corollary 7.12 is $m = \dim X$.

Exercise 8.8. Show that the definition of the dimension of a linear subspace of a projective variety, defined in Section 4.5, agrees with Definition 2.63.

Exercise 8.9. Suppose that L is an $(n-1)$-dimensional linear subspace of $\mathbb{P}^n$, $X \subset L$ is a closed subvariety, and $y \in \mathbb{P}^n \setminus L$. Let Y be the union of all lines containing y and a point of X. Recall from Exercise 4.9 that Y is a projective subvariety of $\mathbb{P}^n$. Show that $\dim Y = \dim X + 1$.

8.2. The theorem on dimension of fibers

Lemma 8.10. *Suppose that $p_1, \ldots, p_s, q_1, \ldots, q_r \in \mathbb{P}^n$ are distinct points for some s, r, and n. Then there exists a homogeneous form $F \in S(\mathbb{P}^n)$ such that $F(p_1) = \cdots = F(p_s) = 0$ and $F(q_i) \neq 0$ for $1 \leq i \leq r$.*

Proof. By the homogeneous nullstellensatz (Theorem 3.12), there exist homogeneous forms
$$F_i \in I(\{p_1, \ldots, p_s, q_1, \ldots, q_{i-1}, q_{i+1}, \ldots, q_r\})$$
for $1 \leq i \leq r$ such that $F_i(q_i) \neq 0$ for $1 \leq i \leq r$. Let d_i be the degree of F_i for $1 \leq i \leq r$, and set $d = d_1 d_2 \cdots d_r$. The homogeneous form
$$F = \sum_{i=1}^{r} F_i^{\frac{d}{d_i}}$$
satisfies the conclusions of the lemma. □

Proposition 8.11. *Suppose that X is a quasi-projective variety of dimension $m \geq 1$ and $p \in X$. Then there exist an affine neighborhood U of p in X and $f_1, \ldots, f_m \in \mathcal{O}_X(U)$ such that $Z_U(f_1, \ldots, f_m) = \{p\}$.*

Proof. X is an open subset of a projective variety $W \subset \mathbb{P}^n$. Choose a point $q_1 \in W \setminus \{p\}$. By Lemma 8.10, there exists a form $F_1 \in S(\mathbb{P}^n)$ such that $F_1(q_1) \neq 0$ and $F_1(p) = 0$. Let $X_1 = Z(F_1) \cap W$. By Theorem 8.4, $X_1 = X_{1,1} \cup \cdots \cup X_{1,r}$ is a union of irreducible components each of dimension $m - 1$. At least one of the components necessarily contains p. If $m > 1$, we continue, choosing points $q_{1,i} \in X_{1,i}$ for $1 \leq i \leq r$, none of which are equal to p. By Lemma 8.10, there exists a form $F_2 \in S(\mathbb{P}^n)$ such that $F_2(q_{1,i}) \neq 0$ for $1 \leq i \leq r$ and $F_2(p) = 0$. By Theorem 8.4, for each i, $Z(F_2) \cap X_{1,i} = X_{2,i,1} \cup \cdots \cup X_{2,i,s_i}$ is a union of irreducible components each of dimension $m - 2$. Thus $Z(F_1, F_2) \cap W = \bigcup X_{2,i,j}$ is a union of irreducible components of dimension $m - 2$, at least one of which contains p. Continuing by induction, we find homogeneous forms $F_1, \ldots, F_m \in S(\mathbb{P}^n)$ such that $Z(F_1, \ldots, F_m) \cap W$ is a zero-dimensional algebraic set which contains p. Thus $Z(F_1, \ldots, F_m) \cap W = \{a_0, a_1, \ldots, a_t\}$ for some points $a_0 = p, a_1, \ldots, a_t \in W$. Now by Lemma 8.10, there exists a form $G \in S(\mathbb{P}^n)$ such that $G(a_i) = 0$ for $1 \leq i \leq t$ and $G(p) \neq 0$. Let L be a linear form on $\mathbb{P}^n$ such that $L(p) \neq 0$. Let d_i be the degree of F_i and e be the degree of G. Then
$$f_1 = \frac{F_1}{L^{d_1}}, \ldots, f_m = \frac{F_m}{L^{d_m}}, g = \frac{G}{L^e} \in \mathcal{O}_{\mathbb{P}^n}(D(L)).$$
Let V be an affine neighborhood of p in X such that $V \subset X \cap D(L)$. Then $Z(f_1, \ldots, f_m) \cap V \subset \{a_0, \ldots, a_t\}$. Let U be an affine neighborhood of p in $(V \setminus Z(g)) \cap X$. Then $Z_U(f_1, \ldots, f_m) = \{p\}$. □

A set of elements $\{f_1, \ldots, f_d\}$ in a local ring R with maximal ideal m_R of dimension d such that the ideal $(f_1, \ldots, f_d)$ is m_R-primary is called a system of parameters in R.

Theorem 8.12. *Suppose that A is a Noetherian local ring. Then A has a system of parameters.*

Proof. [**161**, Theorem 20, page 288] or [**50**, Corollary 10.7]. □

Proposition 8.11 and the nullstellensatz give a direct proof that a system of parameters exists in a local ring $\mathcal{O}_{X,p}$ of a point p on an algebraic variety X.

Theorem 8.13. *Let $\phi : X \to Y$ be a dominant regular map between quasi-projective varieties. Let $\dim X = n$ and $\dim Y = m$. Then $m \leq n$ and:*

1) *Suppose that $p \in Y$. Then $\dim F \geq n - m$ for all irreducible components F of $\phi^{-1}(p)$.*

2) *There exists a nonempty open subset U of Y such that all irreducible components of $\phi^{-1}(p)$ have dimension $n - m$ for $p \in U$.*

Proof. From the inclusion $k(Y) \to k(X)$ we see that

$$m = \dim Y = \mathrm{trdeg}_k k(Y) \leq \mathrm{trdeg}_k k(X) = \dim X = n.$$

We first prove 1). The conclusion of 1) is local in Y, so we can replace Y with an affine open neighborhood U of p in Y and X with $\phi^{-1}(U)$. By Proposition 8.11, we may assume that there exist $f_1, \ldots, f_m \in k[Y]$ such that $Z_Y(f_1, \ldots, f_m) = \{p\}$. Thus the equations $\phi^*(f_1) = \cdots = \phi^*(f_m) = 0$ define $\phi^{-1}(p)$ in X. By Corollary 8.6, all irreducible components F of $\phi^{-1}(p)$ satisfy $\dim F \geq n - m$.

Now we prove 2). We may replace Y with an affine open subset W, X by an affine open subset $V \subset \phi^{-1}(W)$, and ϕ with the restriction of ϕ to V. (The theorem follows for X if it holds for each member of a finite affine open cover of $\phi^{-1}(W)$.) Since ϕ is dominant, ϕ determines an inclusion $\phi^* : k[W] \to k[V]$, hence an inclusion $k(W) = k(Y) \subset k(X) = k(V)$. Let $S = k[W]$. Consider the subring R of $k(V)$ generated by $k(W)$ and $k[V]$. This is a domain which is a finitely generated $k(W)$-algebra. Further, the quotient ring of R is $k(V)$. Now $k(W)$ is not algebraically closed, but Noether's normalization lemma does not need this assumption. By Noether's normalization lemma (Theorem 1.53) we have that there exist $t_1, \ldots, t_r$ in R such that $t_1, \ldots, t_r$ are algebraically independent over $k(W)$ and R is integral over the polynomial ring $k(W)[t_1, \ldots, t_r]$. We may assume, after multiplying by an element of $k[W]$, that $t_1, \ldots, t_r \in k[V]$. Since the quotient field of R is $k(V)$, we have by (1.1) that

$$r = \mathrm{trdeg}_{k(W)} k(V) = \mathrm{trdeg}_k k(V) - \mathrm{trdeg}_k k(W) = \dim X - \dim Y = n - m.$$

Now consider the subring $S[t_1, \ldots, t_r]$ of $k[V]$. Here $S[t_1, \ldots, t_r]$ is a polynomial ring over S, so $S[t_1, \ldots, t_r] = k[W \times \mathbb{A}^r]$, and we have a factorization of ϕ by

$$V \xrightarrow{\pi} W \times \mathbb{A}^r \xrightarrow{\psi} W.$$

We have that $k[V]$ is a finitely generated k-algebra, so it is a finitely generated $S[t_1, \ldots, t_r]$-algebra, say generated by $v_1, \ldots, v_l$ as an $S[t_1, \ldots, t_r]$-algebra. Since R is integral over $k(W)[t_1, \ldots, t_r]$, there exist polynomials $F_i(x)$ in the indeterminate x,

$$F_i(x) = x^{d_i} + P_{i,1}(t_1, \ldots, t_r)x^{d_i-1} + \cdots + P_{i,d_i}(t_1, \ldots, t_r)$$

where the $P_{i,j}$ are polynomials with coefficients in $k(W)$, such that

$$F_i(v_i) = 0 \text{ for } 1 \leq i \leq l.$$

Let $g \in S = k[W]$ be a common denominator of the polynomials $P_{i,j}$. Then $P_{i,j} \in (S_g)[t_1, \ldots, t_r]$ for all i, j. Thus $k[V]_g$ is finite over $(S_g)[t_1, \ldots, t_r]$.

Let $U = D(g) \subset W$. We then have a factorization of ϕ restricted to $\phi^{-1}(U)$ as

$$\phi^{-1}(U) \xrightarrow{\pi} U \times \mathbb{A}^r \xrightarrow{\psi} U$$

where U is affine with regular functions $k[U] = k[W]_g = S_g$ and $\phi^{-1}(U)$ is affine with regular functions $k[\phi^{-1}(U)] = k[V]_g$ and π is a finite map. For $y \in U$, we have that $\psi^{-1}(y) = \{y\} \times \mathbb{A}^r$ has dimension r. Suppose that A is an irreducible closed subset of $\phi^{-1}(U)$ which maps into $\{y\} \times \mathbb{A}^r$. Then $\pi(A)$ is closed in $U \times \mathbb{A}^r$ and the restriction of ϕ from A to $\pi(A)$ is finite, so that the extension $k(A)$ of $k(\pi(A))$ is an algebraic extension. Thus $\dim A = \dim \pi(A)$. Since $\pi(A)$ is a subvariety of $\{y\} \times \mathbb{A}^r \cong \mathbb{A}^r$, we have that $\dim A \leq r = n - m$.

By part 1) of this theorem, $\dim A = n - m$ for all irreducible components A of $\phi^{-1}(y)$. $\square$

A short proof of Theorem 8.13 can be obtained by using some theorems from commutative algebra, particularly on flatness, as we now indicate. We reduce to the case where X and Y are both affine. Conclusion 1) of the theorem follows from [**107**, (1) of Theorem 19 on page 79]. By the theorem of generic flatness, [**107**, Theorem 52 on page 158], there is a nontrivial open subset U of Y such that $\phi^{-1}(U) \to U$ is flat. By consideration of an open affine cover of $\phi^{-1}(U)$, we obtain conclusion 2) of Theorem 8.13 from [**107**, Theorem 4 on page 33] and [**107**, (2) of Theorem 19 on page 79].

Corollary 8.14. *Suppose that $\phi : X \to Y$ is a dominant regular map between quasi-projective varieties. Then the sets*

$$Y_k = \{p \in Y \mid \dim \phi^{-1}(p) \geq k\}$$

are closed in Y.

Proof. Let $\dim X = n$ and $\dim Y = m$. By Theorem 8.13, $Y_{n-m} = Y$, and there exists a proper closed subset Y' of Y such that $Y_k \subset Y'$ if $k > n-m$. If Z_i are the irreducible components of Y' and $\phi_i : \phi^{-1}(Z_i) \to Z_i$ the restrictions of ϕ, then $\dim Z_i < \dim Y$, and the corollary follows by induction on $\dim Y$, applied to the restriction of ϕ_i to the irreducible components of $\phi^{-1}(Z_i)$. □

Chapter 9

Zariski's Main Theorem

In this chapter we prove Zariski's main theorem and give some applications.

Proposition 9.1. *Suppose that X and Y are quasi-projective varieties.*

1) *Suppose that $\phi : X \to Y$ is a dominant regular map. Then*

$$\Lambda = \phi^* : k(Y) \to k(X)$$

has the property that for all $p \in X$,

$$\Lambda : \mathcal{O}_{Y,\phi(p)} \to \mathcal{O}_{X,p}$$

is a local homomorphism ($\Lambda(\mathcal{O}_{Y,\phi(p)}) \subset \mathcal{O}_{X,p}$ and $\Lambda^{-1}(m_p) = m_{\phi(p)}$). Further, $q = \phi(p)$ is the unique point q' in Y such that

$$\Lambda : \mathcal{O}_{Y,q'} \to \mathcal{O}_{X,p}$$

is a local homomorphism.

2) *Conversely, suppose that $\Lambda : k(Y) \to k(X)$ is a k-algebra homomorphism. Further suppose that for all $p \in X$, there exists a unique point $q \in Y$ such that $\Lambda : \mathcal{O}_{Y,q} \to \mathcal{O}_{X,p}$ and $\Lambda^{-1}(m_p) = m_q$. Then there exists a unique dominant regular map $\phi : X \to Y$ such that $\phi^* : k(Y) \to k(X)$ is the map Λ.*

Proof. We first prove statement 1). The fact that for all $p \in X$, Λ induces a local homomorphism

$$\Lambda : \mathcal{O}_{Y,\phi(p)} \to \mathcal{O}_{X,p}$$

follows from Definition 3.28 of a regular map. Suppose that $q, q' \in Y$ have the properties that $\Lambda : \mathcal{O}_{Y,q} \to \mathcal{O}_{X,p}$ with $\Lambda^{-1}(m_p) = m_q$ and

$$\Lambda : \mathcal{O}_{Y,q'} \to \mathcal{O}_{X,p}$$

with $\Lambda^{-1}(m_p) = m_{q'}$. We will show that $q = q'$. We have that Y is an open subset of a projective variety $\overline{Y}$. Thus there exists an affine open subset U of $\overline{Y}$ such that $q, q' \in U$. There exists a closed subset Z of U such that $Y \cap U = U \setminus Z$. By the nullstellensatz, there exists $f \in k[U]$ such that $f(Z) = 0$ but $f(q) \neq 0$ and $f(q') \neq 0$. Then $V = U_f$ is an affine open subset of Y containing q and q'. From the homomorphism $\Lambda : k[V] \to \mathcal{O}_{X,p}$, we see that $I_V(q) = \Lambda^{-1}(m_p) = I_V(q')$. Thus $q = q'$ since an affine variety is separated by Corollary 2.87.

We now prove 2). Define a map $\Sigma : X \to Y$ by $\Sigma(p) = q$ for $p \in X$ if $\Lambda^{-1}(m_p) = m_q$. We will show that Σ is a regular map. Fix $p \in X$. Let $q = \Sigma(p)$. Let U_q be an affine neighborhood of q in Y and V_p be an affine neighborhood of p in X. Since $\mathcal{O}_{Y,q}$ is a localization of $k[U_q]$, we have that Λ induces a homomorphism $\Lambda : k[U_q] \to \mathcal{O}_{X,p}$. Let $k[U_q] = k[\overline{x}_1, \ldots, \overline{x}_n]$. Since $\mathcal{O}_{X,p}$ is a localization of $k[V_p]$, there exists $f \in k[V_p] \setminus I_{V_p}(p)$ such that $\Lambda(\overline{x}_i) \in k[V_p]_f$ for $1 \leq i \leq n$. Thus Λ induces a homomorphism $\Lambda : k[U_q] \to k[W_p]$ where $W_p = D_{V_p}(f)$ is an affine neighborhood of p in X. Thus there exists a unique regular map $\psi_p : W_p \to U_q$ such that $\psi_p^* = \Lambda : k[U_q] \to k[W_p]$ by Proposition 2.40. By Lemma 2.88, for $a \in W_p$, $\psi_p(a) = b$ if and only if $\Lambda^{-1}(m_a) = m_b$, so that $\Sigma|W_p = \psi_p$. Now by Proposition 3.39, Σ is a regular map. □

We will need the following local version of Zariski's main theorem, which was first proven by Zariski in [**152**, Theorem 14].

Theorem 9.2. *Let K be an algebraic function field over a field κ. Suppose that R and S are local domains which have K as their common quotient field such that S dominates R ($R \subset S$ and $m_S \cap R = m_R$ where m_R and m_S are the respective maximal ideals of R and S) and R is normal. Suppose that R and S are localizations of finitely generated κ-algebras, $\dim R = \dim S$, S/m_S is a finite extension of R/m_R, and $m_R S$ is m_S-primary. Then $R = S$.*

Theorem 9.2 is an immediate consequence of Proposition 21.54, which we will prove in Section 21.6. The following theorem was proven by Zariski in [**152**] and [**156**].

Theorem 9.3 (Zariski's main theorem). *Let $\phi : X \to Y$ be a birational regular map of projective varieties. Suppose that Y is normal. Let $U = \{q \in Y \mid \phi^{-1}(q) \text{ is a finite set}\}$. Then U is a dense open subset of Y and $\phi : \phi^{-1}(U) \to U$ is an isomorphism.*

Proof. The set U is a dense open subset of Y by Theorem 8.13 and Corollary 8.14. The regular map ϕ induces an isomorphism $\phi^* : k(Y) \to k(X)$ which allows us to identify $k(Y)$ with $k(X)$. Suppose $q \in U$. Since X is projective, $q = \phi(p)$ for some $p \in \phi^{-1}(U)$ by Theorem 5.14. Thus $\mathcal{O}_{X,p}$ dominates $\mathcal{O}_{Y,q}$. We have that $m_{Y,q}\mathcal{O}_{X,p}$ is $m_{X,p}$-primary since $\phi^{-1}(q)$ is a finite set. By Theorem 9.2, $\mathcal{O}_{Y,q} = \mathcal{O}_{X,p}$.

Suppose that $p' \in X$ is such that $\mathcal{O}_{X,p'}$ dominates $\mathcal{O}_{Y,q}$. Then $\mathcal{O}_{X,p} \subset \mathcal{O}_{X,p'}$ and so $\mathcal{O}_{X,p'} = \mathcal{O}_{X,p}$ since the dimension of both rings is equal to $\dim X$. Thus $p = p'$ since X is separated by Proposition 3.36. By Proposition 9.1, there exists a unique dominant regular map $\psi : U \to \phi^{-1}(U)$ such that $\psi(q) = p$ if and only if $\phi(p) = q$. Thus $\phi : \phi^{-1}(U) \to U$ is an isomorphism with inverse $\psi : U \to \phi^{-1}(U)$. □

Theorem 9.4. *Suppose that $\phi : X \to Y$ is a dominant regular map of projective varieties and X is normal. Suppose that $k(X)$ is a finite field extension of $k(Y)$. Let $\lambda : \overline{Y} \to Y$ be the normalization of Y in $k(X)$ (Theorem 7.17). Then there exists a unique regular map $\psi : X \to \overline{Y}$ making a commutative diagram*

Proof. Let $\Gamma = \Gamma_\psi \subset X \times \overline{Y}$ be the graph of the natural rational map $\psi : X \dashrightarrow \overline{Y}$ (since $k(\overline{Y}) = k(X)$). Let $\pi_1 : \Gamma \to X$ and $\pi_2 \to \overline{Y}$ be the projections. The map π_1 is birational.

We will show that we have a commutative diagram of regular maps

(9.1)

By the construction of the normalization $\overline{Y}$, we have a commutative diagram

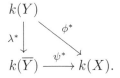

There exists a nontrivial open subset U of X such that $\psi : U \to \overline{Y}$ is a regular map. The regular maps $\lambda\psi : U \to Y$ and $\phi : U \to Y$ are equal as rational maps by Proposition 2.108 (which is valid for arbitrary varieties) so $\lambda\psi : U \to Y$ and $\phi : U \to Y$ are the same regular map by Lemma 2.104. Thus when restricting to the open subset $\Gamma_{\psi|U}$ of Γ, we have a commutative diagram of regular maps

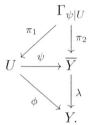

Thus $\phi\pi_1 = \lambda\pi_2$ as rational maps from Γ to Y, so that they are the same regular maps from Γ to Y by Proposition 2.108 and Lemma 2.104, establishing the commutativity of the diagram (9.1).

Suppose that $p \in X$. Then $\pi_1^{-1}(p) \subset \{p\} \times \lambda^{-1}(\phi(p))$, which is a finite set for all $p \in X$ since λ is a finite map. Thus π_1 is a birational regular map of projective varieties such that all fibers are finite. Since X is normal, we have that π_1 is an isomorphism by Theorem 9.3, and thus the rational map $\psi : X \dashrightarrow \overline{Y}$ is a regular map. $\square$

Proposition 9.5. *Suppose that $\phi : X \to Y$ is a dominant regular map of projective varieties such that $\phi^{-1}(q)$ is a finite set for all $q \in Y$. Then ϕ is an affine map.*

Proof. We have a closed embedding $X \subset \mathbb{P}^m$, giving closed embeddings
$$X \cong \Gamma_\phi \subset X \times Y \subset \mathbb{P}^m \times Y.$$
Let $\pi : \mathbb{P}^m \times Y \to Y$ be the projection. Suppose that $q \in Y$. Then $\phi^{-1}(q) \cong X \cap (\mathbb{P}^m \times \{q\})$ is a finite set of points. Let H be a hyperplane section of Y such that $q \notin H$ and let L be a hyperplane section of $\mathbb{P}^m$ such that
$$(L \times \{q\}) \cap X \cap (\mathbb{P}^m \times \{q\}) = \emptyset.$$
Then $U = X \cap (\mathbb{P}^m_L \times Y_H)$ is a closed subset of an affine space so it is affine.

Let $Z = \pi(X \cap (L \times Y_H))$. We have that Z is a closed subset of Y_H, by Corollary 5.13, which does not contain q. We have that
$$\phi^{-1}(Y_H \setminus Z) \cong X \cap (\mathbb{P}^m \times (Y_H \setminus Z)) = X \cap (\mathbb{P}^m_L \times (Y_H \setminus Z)) = U \cap (\pi^{-1}(Y_H \setminus Z)).$$
Let $f \in k[Y_H]$ be such that $q \in D_{Y_H}(f) \subset Y_H \setminus Z$. Let $V = D_{Y_H}(f)$. Then V is an affine neighborhood of q and $\phi^{-1}(V) \cong U \cap \pi^{-1}(V) = D_U(\phi^*(f))$ is affine with $k[\phi^{-1}(V)] = k[U]_{\phi^*(f)}$. Applying this construction to all $q \in Y$, we see that ϕ is an affine map. $\square$

Theorem 9.6. *Suppose that $\phi : X \to Y$ is a dominant regular map of projective varieties such that $\phi^{-1}(q)$ is a finite set for all $q \in Y$. Then ϕ is a finite map.*

Proof. We have that $\dim X = \dim Y$ by Theorem 8.13. Now Theorem 8.2 implies $k(X)$ is a finite extension of $k(Y)$. Let $\alpha : \overline{X} \to X$ be the normalization of X in $k(X)$ and $\beta : \overline{Y} \to Y$ be the normalization of Y in $k(X)$ (Theorem 7.17). Then we have a commutative diagram of regular maps of projective varieties

$$\begin{array}{ccc} \overline{X} & \xrightarrow{\psi} & \overline{Y} \\ \alpha \downarrow & & \downarrow \beta \\ X & \xrightarrow{\phi} & Y \end{array}$$

by Theorem 9.4. Now α and β are finite maps so they have finite fibers by Theorem 2.56. Thus ψ has finite fibers and is birational so ψ is an isomorphism by Theorem 9.3. Let $V \subset Y$ be an affine open subset. Then $(\phi\alpha)^{-1}(V) \cong \beta^{-1}(V)$ is affine and $k[(\phi\alpha)^{-1}(V)]$ is a finitely generated $k[V]$-algebra since β is finite. Now $\phi^{-1}(V)$ is affine by Proposition 9.5, and $k[\phi^{-1}(V)]$ is a finitely generated $k[V]$-module by Lemma 1.55, since it is a submodule of $k[(\phi\alpha)^{-1}(V)]$. Thus ϕ is finite. $\square$

We also have the related theorem, which was first proven by Zariski in [**156**].

Theorem 9.7 (Zariski's connectedness theorem). *Let X and Y be projective varieties and $\phi : X \to Y$ be a birational regular map. Suppose that Y is normal at a point $q \in Y$. Then $\phi^{-1}(q)$ is connected.*

We mention a nice generalization of Zariski's main theorem, which is proven in [**67**] or [**124**, Chapter IV] and [**109**, Theorem 1.8].

Theorem 9.8 (Grothendieck). *Let $\phi : X' \to X$ be a regular map of varieties with finite fibers. Then ϕ is a composition of an open embedding of X' into a variety Y and a finite map from Y to X.*

Exercise 9.9. Suppose that X and Y are projective varieties such that X is normal and $\phi : X \dashrightarrow Y$ is a rational map. Let Γ_ϕ be the graph of ϕ, with projections $\pi_1 : \Gamma_\phi \to X$ and $\pi_2 : \Gamma_\phi \to Y$. Suppose that $p \in X$ is such that ϕ is not regular at p. Show that there exists a curve $C \subset \pi_1^{-1}(p)$ such that $\pi_2(C)$ is a curve of Y.

Chapter 10

Nonsingularity

In this chapter, we explore the concept of nonsingularity of a variety. The tangent space $T_p(X)$ to a point p on a variety X is first defined in Definition 10.7 extrinsically for an affine variety embedded in $\mathbb{A}^n$ and then given an equivalent intrinsic definition, $T_p(X) = \operatorname{Hom}_k(m_p/m_p^2, k)$, in Definition 10.8. We show that $\dim X \le \dim T_p(X)$ in Theorem 10.11, and in Definition 10.12 we define X to be nonsingular at p if $\dim X = \dim_k T_p(X)$. In Proposition 10.13, we show that X is nonsingular at p if and only if $\mathcal{O}_{X,p}$ is a regular local ring. Our definition of nonsingularity extends to varieties over arbitrary fields and general schemes. We prove the Jacobian criterion of nonsingularity, Proposition 10.14, in our situation of varieties over an algebraically closed field. The Jacobian criterion of nonsingularity is valid for varieties over a perfect field but can fail for varieties over nonperfect fields (Exercise 10.21). In Theorem 10.16, we show that the set of nonsingular points of X is a dense open subset of X.

In the remainder of the chapter, we give applications of the above concepts, including the factorization of birational regular maps of nonsingular projective surfaces, Theorem 10.32, and the fact (Theorem 10.38) that a nonsingular projective variety of dimension n can be embedded in $\mathbb{P}^{2n+1}$. We end the chapter with a proof (Theorem 10.45) that a nonsingular complex variety is a complex analytic manifold.

10.1. Regular parameters

Suppose that R is a regular local ring, with maximal ideal $\mathfrak{m}$ and residue field $\kappa = R/\mathfrak{m}$. Let $d = \dim R$, so that $\dim_\kappa \mathfrak{m}/\mathfrak{m}^2 = d$. Elements $u_1, \ldots, u_d \in \mathfrak{m}$

such that the classes of $u_1, \ldots, u_d$ are a κ-basis of $\mathfrak{m}/\mathfrak{m}^2$ are called regular parameters (or a regular system of parameters) in R.

Lemma 10.1. *Suppose that R is a regular local ring of dimension d with maximal ideal $\mathfrak{m}$ and that $\mathfrak{p} \subset R$ is a prime ideal such that $\dim R/\mathfrak{p} = a$. Then*

$$\dim_{R/\mathfrak{m}}(\mathfrak{p} + \mathfrak{m}^2/\mathfrak{m}^2) \leq d - a,$$

with equality if and only if $R/\mathfrak{p}$ is a regular local ring.

Proof. Let $R' = R/\mathfrak{p}$, $\mathfrak{m}' = \mathfrak{m} R'$, and $\kappa = R/\mathfrak{m} \cong R'/\mathfrak{m}'$. There is a short exact sequence of κ-vector spaces

(10.1) $\quad 0 \to \mathfrak{p}/\mathfrak{p} \cap \mathfrak{m}^2 \cong \mathfrak{p} + \mathfrak{m}^2/\mathfrak{m}^2 \to \mathfrak{m}/\mathfrak{m}^2 \to \mathfrak{m}/(\mathfrak{p}+\mathfrak{m}^2) = \mathfrak{m}'/(\mathfrak{m}')^2 \to 0.$

The lemma now follows since $\dim_\kappa \mathfrak{m}'/(\mathfrak{m}')^2 \geq a$ by Theorem 1.81 and R' is regular if and only if $\dim_\kappa \mathfrak{m}'/(\mathfrak{m}')^2 = a$. $\quad\square$

Lemma 10.2. *Suppose that R is a regular local ring, with maximal ideal $\mathfrak{m}$ and residue field $\kappa = R/\mathfrak{m}$. Let $d = \dim R$, and suppose that $u_1, \ldots, u_d$ are regular parameters in R. Then $I_i = (u_1, \ldots, u_i)$ is a prime ideal in R of height i for all i, R/I_i is a regular local ring of dimension $d - i$, and the residues of $u_{i+1}, \ldots, u_d$ in R/I_i are regular parameters in R/I_i.*

Proof. Let $\mathfrak{m}_i = \mathfrak{m}(R/I_i)$ be the maximal ideal of R/I_i. Then

$$d - i = \dim_\kappa \mathfrak{m}_i / \mathfrak{m}_i^2 \geq \dim R/I_i$$

by Theorem 1.81. Since none of the u_i are units in R, we have by Krull's principal ideal theorem, Theorem 1.65, that

$$\dim R/I_i = \operatorname{ht} \mathfrak{m}_i \geq d - i,$$

so that R/I_i is a regular local ring of dimension $d - i$. $\quad\square$

Lemma 10.3. *Suppose that R is a regular local ring of dimension n with maximal ideal $\mathfrak{m}$, $\mathfrak{p} \subset R$ is a prime ideal such that $R/\mathfrak{p}$ is regular, and $\dim R/\mathfrak{p} = n - r$. Then there exist regular parameters $u_1, \ldots, u_n$ in R such that $\mathfrak{p} = (u_1, \ldots, u_r)$ and $u_{r+1}, \ldots, u_n$ map to regular parameters in $R/\mathfrak{p}$.*

Proof. Consider the exact sequence (10.1). Since R and R' are regular, there exist regular parameters $u_1, \ldots, u_n$ in R such that $u_{r+1}, \ldots, u_n$ map to regular parameters in R' and $u_1, \ldots, u_r$ are in $\mathfrak{p}$. By Lemma 10.2, $(u_1, \ldots, u_r)$ is a prime ideal of height r contained in $\mathfrak{p}$. Thus $\mathfrak{p} = (u_1, \ldots, u_r)$. $\quad\square$

Proposition 10.4. *Suppose that $x_1, \ldots, x_d$ is a regular system of parameters in a regular local ring R. For $i \leq d$, let $P_i = (x_1, \ldots, x_i)$. Then P_i is a prime ideal of height i and we have equality of ordinary and symbolic powers*

$$P_i^{(n)} = P_i^n$$

for all $n > 0$.

Proof. The ideals P_i are prime ideals of height i by Lemma 10.2. The regular system of parameters $x_1, \ldots, x_d$ is an R-regular sequence by Lemma 10.2, and hence R is a Cohen-Macaulay local ring. Thus we have equality of ordinary and symbolic powers by Proposition 1.70 and Theorem 1.71. □

10.2. Local equations

Suppose that X is a quasi-projective variety and Y is a closed subvariety. Define $\mathcal{I}_Y$ by

$$\mathcal{I}_Y(U) = I_U(Y \cap U) \subset k[U]$$

if U is an affine open subset of X, and for $p \in X$, define

$$\mathcal{I}_{Y,p} = I_U(Y \cap U) k[U]_{I(p)}$$

if U is an affine open subset of X containing p. We have that $\mathcal{I}_{Y,p}$ is independent of the choice of affine open subset U containing p.

Letting $\overline{X}$ be the closure of X in a projective space $\mathbb{P}^n$ and $\overline{Y}$ be the closure of Y in $\mathbb{P}^n$, we can construct $\mathcal{I}_Y$ as in Section 6.2, by taking $\mathcal{I}_Y$ to be the restriction of $\widetilde{I_{\overline{X}}(\overline{Y})}$ to X, where $I_{\overline{X}}(\overline{Y})$ is the homogeneous ideal of $\overline{Y}$ in the coordinate ring $S(\overline{X})$. This construction will be examined in more generality in Chapter 11.

Definition 10.5. Suppose that X is a quasi-projective variety, Y is a closed subvariety of X, and $p \in X$. Functions $f_1, \ldots, f_n \in \mathcal{O}_{X,p}$ are called local equations of Y in X at p (we also say $f_1 = \cdots = f_n = 0$ are local equations of Y in X at p) if there exists an affine neighborhood U of p in X such that $f_1, \ldots, f_n$ generate the ideal $\mathcal{I}_Y(U)$ of $Y \cap U$ in U.

Lemma 10.6. *Functions $f_1, \ldots, f_n \in \mathcal{O}_{X,p}$ are local equations of Y in some affine neighborhood of $p \in X$ if and only if $\mathcal{I}_{Y,p} = (f_1, \ldots, f_n)$, the ideal generated by $f_1, \ldots, f_n$ in $\mathcal{O}_{X,p}$.*

Observe that $\mathcal{I}_{Y,p} = \mathcal{O}_{X,p}$ if and only if $p \notin Y$, and this holds if and only if $f_i(p) \neq 0$ for some i.

Proof. Suppose that $f_1, \ldots, f_n$ are local equations of Y at p. Then there exists an affine neighborhood U of p such that $\mathcal{I}_Y(U) = (f_1, \ldots, f_n)$, so

$$\mathcal{I}_{Y,p} = (f_1, \ldots, f_n) k[U]_{I(p)} = (f_1, \ldots, f_n) \mathcal{O}_{X,p},$$

where $I(p)$ is the ideal of p in U.

Now suppose that $\mathcal{I}_{Y,p} = (f_1, \ldots, f_n)$. Let U be an affine neighborhood of p, and suppose that $\mathcal{I}_Y(U) = (g_1, \ldots, g_m)$. Then we have that $(g_1, \ldots, g_m)\mathcal{O}_{X,p} = (f_1, \ldots, f_n)$, so there exist expressions

$$(10.2) \qquad g_i = \sum_{j=1}^n a_{i,j} f_j$$

for $1 \leq i \leq m$, where $a_{i,j} \in \mathcal{O}_{X,p}$ for all i, j, and we have expressions

$$(10.3) \qquad f_k = \sum_{l=1}^m b_{k,l} g_l$$

for $1 \leq k \leq n$, where $b_{k,l} \in \mathcal{O}_{X,p}$ for all k, l. After possibly replacing U with a smaller affine neighborhood of p, we may assume that all $f_i, a_{i,j}, b_{k,l}$ are in $k[U]$. Then

$$(f_1, \ldots, f_n) k[U] = (g_1, \ldots, g_m) k[U] = \mathcal{I}_Y(U). \qquad \square$$

10.3. The tangent space

Suppose that $p = (b_1, \ldots, b_n) \in \mathbb{A}^n$. Let $\overline{x}_i = x_i - b_i$ for $1 \leq i \leq n$. Translation by p is an isomorphism of $\mathbb{A}^n$, and we have that $k[\mathbb{A}^n] = k[x_1, \ldots, x_n] = k[\overline{x}_1, \ldots, \overline{x}_n]$ is a polynomial ring. Suppose that $f \in k[\mathbb{A}^n]$. Then f has a unique expansion

$$f = \sum a_{i_1, \ldots, i_n} (x_1 - b_1)^{i_1} \cdots (x_n - b_n)^{i_n}$$

with $a_{i_1, \ldots, i_n} \in k$. If $f(p) = 0$, we have that $a_{0, \ldots, 0} = 0$, and

$$f \equiv L_p(f) \bmod I(p)^2,$$

where

$$(10.4) \qquad \begin{aligned} L_p(f) &= a_{1,0,\ldots,0}(x_1 - b_1) + \cdots + a_{0,\ldots,0,1}(x_n - b_n) \\ &= \tfrac{\partial f}{\partial x_1}(p)(x_1 - b_1) + \cdots + \tfrac{\partial f}{\partial x_n}(p)(x_n - b_n). \end{aligned}$$

Definition 10.7 (Extrinsic definition of the tangent space). Suppose that X is an affine variety, which is a closed subvariety of $\mathbb{A}^n$, and that $p \in X$. The tangent space to X at p is the linear subvariety $T_p(X)$ of $\mathbb{A}^n$, defined by

$$T_p(X) = Z(L_p(f) \mid f \in I(X)).$$

10.3. The tangent space

We have that if $I = (f_1, \ldots, f_r)$, then $I(T_p(X)) = (L_p(f_1), \ldots, L_p(f_r))$.

The ideal $I(p)$ of the point p is $I(p) = (\overline{x}_1, \ldots, \overline{x}_n) \subset k[\mathbb{A}^n]$. Let V be the n-dimensional k-vector space which is spanned by $\overline{x}_1, \ldots, \overline{x}_n$ in $k[\mathbb{A}^n]$. For $f \in I(X)$, we have that $L_p(f) \in V$. Let W be the subspace of V defined by $W = \{L_p(f) \mid f \in I(X)\}$.

Let m_p be the maximal ideal of the local ring $\mathcal{O}_{X,p}$. By Lemma 1.28,

$$m_p/m_p^2 \cong I(p)/(I(p)^2 + I(X)) \cong I(p)/(I(p)^2 + I(T_p(X))) \cong V/W.$$

We can naturally identify the set of points of $\mathbb{A}^n$ with the dual vector space $\operatorname{Hom}_k(V, k)$ by associating to $q \in \mathbb{A}^n$ the linear map $L \mapsto L(q)$ for $L \in V$. Now

$$\begin{aligned}
\operatorname{Hom}_k(V/W, k) &= \{\phi \in \operatorname{Hom}_k(V, k) \mid \phi(W) = 0\} \\
&= \{q \in \mathbb{A}^n \mid L_p(f)(q) = 0 \text{ for all } f \in I(X)\} \\
&= T_p(X).
\end{aligned}$$

This gives us the following alternate definition of the tangent space.

Definition 10.8 (Intrinsic definition of the tangent space). Suppose that X is a quasi-projective variety and $p \in X$. The tangent space to X at p is the k-vector space $T_p(X)$ defined by

$$T_p(X) = \operatorname{Hom}_k(m_p/m_p^2, k),$$

where m_p is the maximal ideal of $\mathcal{O}_{X,p}$.

The vector space $T_p(X)$ is the tangent space $T(\mathcal{O}_{X,p})$ of the local ring $\mathcal{O}_{X,p}$, as defined in (1.7).

Suppose that $\phi : X \to Y$ is a regular map of quasi-projective varieties and $p \in X$. Let $q = \phi(p)$. Then $\phi^* : \mathcal{O}_{Y,q} \to \mathcal{O}_{X,p}$ induces a k-vector space homomorphism $m_q/m_q^2 \to m_p/m_p^2$, where m_q and m_p are the maximal ideals of $\mathcal{O}_{Y,q}$ and $\mathcal{O}_{X,p}$ respectively, and thus we have a k-vector space homomorphism $d\phi_p : T_p(X) \to T_q(Y)$.

Suppose that $\phi : X \to Y$ is a regular map of affine varieties, $Z \subset X$ is a subvariety, $W \subset Y$ is a subvariety of Y such that $\phi(Z) \subset W$, and $p \in Z$. Let $q = \phi(p)$. Let $\overline{\phi} : Z \to W$ be the restricted map. We have prime ideals $I(W) \subset I(q) \subset k[Y]$ and $I(Z) \subset I(p) \subset k[X]$. Here $\phi^* : k[Y] \to k[X]$ induces $\overline{\phi}^* : k[W] = k[Y]/I(W) \to k[X]/I(Z) = k[Z]$.

We have a commutative diagram of k-vector spaces

$$\begin{array}{ccc}
I(q)/I(q)^2 & \to & I(p)/I(p)^2 \\
\downarrow & & \downarrow \\
I(q)/I(q)^2 + I(W) & \to & I(p)/I(p)^2 + I(Z),
\end{array}$$

where the vertical arrows are surjections. Taking the associated diagram of dual k-vector spaces (applying $\mathrm{Hom}_k(*, k)$), we get a commutative diagram

(10.5)
$$\begin{array}{ccc} T_q(Y) & \xleftarrow{d\phi_p} & T_p(X) \\ \uparrow & & \uparrow \\ T_q(W) & \xleftarrow{d\overline{\phi}_p} & T_p(Z) \end{array}$$

where the vertical arrows are the natural inclusions.

Example 10.9. Suppose that $\phi : \mathbb{A}^m \to \mathbb{A}^n$ is a regular map, defined by $\phi = (f_1, \ldots, f_n)$ for some $f_i \in k[\mathbb{A}^m] = k[x_1, \ldots, x_m]$. Suppose that $\alpha \in \mathbb{A}^m$ and $\beta = \phi(\alpha)$. Then

$$d\phi_\alpha : T_\alpha(\mathbb{A}^m) \to T_\beta(\mathbb{A}^n)$$

is the linear map $k^m \to k^n$ given by multiplication by the $n \times m$ matrix $(\frac{\partial f_i}{\partial x_j}(\alpha))$.

Proof. $\phi^* : k[y_1, \ldots, y_n] = k[\mathbb{A}^n] \to k[x_1, \ldots, x_m] = k[\mathbb{A}^m]$ is defined by $\phi^*(y_i) = f_i$ for $1 \leq i \leq n$. Let $\alpha = (\alpha_1, \ldots, \alpha_m)$ and $\beta = (\beta_1, \ldots, \beta_n)$. We have expressions for $1 \leq i \leq n$,

(10.6)
$$f_i = f_i(\alpha) + \sum_{j=1}^m \frac{\partial f_i}{\partial x_j}(\alpha)(x_j - \alpha_j) + h_i$$

where $h_i \in I(\alpha)^2$. We have that $f_i(\alpha) = \beta_i$. Here $\{y_1 - \beta_1, \ldots, y_n - \beta_n\}$ is a k-basis of $I(\beta)/I(\beta)^2$, and $\{x_1 - \alpha_1, \ldots, x_m - \alpha_m\}$ is a k-basis of $I(\alpha)/I(\alpha)^2$. By (10.6) and since $\phi^*(y_i - \beta_i) = f_i - f_i(\alpha)$, we have that the induced map $\phi^* : I(\beta)/I(\beta)^2 \to I(\alpha)/I(\alpha)^2$ is given by

(10.7)
$$\phi^*(y_i - \beta_i) = \sum_{j=1}^m \frac{\partial f_i}{\partial x_j}(\alpha)(x_j - \alpha_j),$$

for $1 \leq i \leq n$. Let $\{\delta_1, \ldots, \delta_m\}$ be the dual basis to $\{x_1 - \alpha_1, \ldots, x_m - \alpha_m\}$, and let $\{\epsilon_1, \ldots, \epsilon_n\}$ be the dual basis to $\{y_1 - \beta_1, \ldots, y_n - \beta_n\}$. That is,

$$\delta_s(x_t - \alpha_t) = \begin{cases} 1 & \text{if } s = t, \\ 0 & \text{if } s \neq t \end{cases}$$

and

$$\epsilon_s(y_t - \beta_t) = \begin{cases} 1 & \text{if } s = t, \\ 0 & \text{if } s \neq t. \end{cases}$$

Now we compute the dual map

$$\phi^* : \mathrm{Hom}_k(I(\alpha)/I(\alpha)^2, k) \to \mathrm{Hom}_k(I(\beta)/I(\beta)^2, k).$$

For $1 \leq s \leq m$, we have commutative diagrams

$$\begin{array}{c} I(\beta)/I(\beta)^2 \\ \phi^* \downarrow \qquad \searrow \phi^*(\delta_s) \\ I(\alpha)/I(\alpha)^2 \xrightarrow[\delta_s]{} k. \end{array}$$

For $1 \leq t \leq n$ and $1 \leq s \leq m$, we have

$$\delta_s(\phi^*(y_t - \beta_t)) = \delta_s\left(\sum_{j=1}^m \frac{\partial f_t}{\partial x_j}(\alpha)(x_j - \alpha_j)\right) = \frac{\partial f_t}{\partial x_s}(\alpha).$$

Thus

$$\phi^*(\delta_s) = \sum_{t=1}^n \frac{\partial f_t}{\partial x_s}(\alpha)\epsilon_t$$

for $1 \leq s \leq m$. □

Exercise 10.10. Suppose that k is an algebraically closed field of characteristic $p > 0$. Show that the regular map $\phi : \mathbb{A}^1 \to \mathbb{A}^1$ defined by $\phi(\alpha) = \alpha^p$ is a homeomorphism. Show that $d\phi_q : T_q(\mathbb{A}^1) \to T_{\phi(q)}(\mathbb{A}^1)$ is the zero map for all $q \in \mathbb{A}^1$.

10.4. Nonsingularity and the singular locus

Theorem 10.11. *Suppose that X is a quasi-projective variety and $p \in X$. Then*

$$\dim_k T_p(X) \geq \dim X.$$

Proof. Let $d = \dim X$ and $n = \dim_k T_p(X) = \dim_k m_p/m_p^2$, where m_p is the maximal ideal of $\mathcal{O}_{X,p}$. The ideal m_p is generated by n elements as an $\mathcal{O}_{X,p}$-module by Nakayama's lemma, Lemma 1.18. Let $m_p = (f_1, \ldots, f_n)$. There exists an affine neighborhood U of p in X such that $I_U(p) = (f_1, \ldots, f_n)$ by Lemma 10.6. Without loss of generality, we may assume that X is affine with this property. Let $\pi : B(p) \to X$ be the blow-up of p. Since $B(p)$ is the graph of the rational map $(f_1 : \ldots : f_n) : X \dashrightarrow \mathbb{P}^{n-1}$, $B(p)$ is a closed subvariety of $X \times \mathbb{P}^{n-1}$, and so $\pi^{-1}(p) \subset \{p\} \times \mathbb{P}^{n-1}$ has dimension $\leq n-1$. But $\dim B(p) = d$ and so $\dim \pi^{-1}(p) = d - 1$, by Krull's principal ideal theorem (Theorem 1.65) since $I(p)\mathcal{O}_{B(p),q}$ is a principal nonzero ideal in the domain $\mathcal{O}_{B(p),q}$ for all $q \in \pi^{-1}(p)$ (by (6.2)). Thus $d \leq n$. □

Suppose that X is affine and let $S = S(B(p))$ be the coordinate ring of the blow-up of p in the above proof. We will show that

$$\dim S/I(p)S = \dim \pi^{-1}(p) + 1 = d.$$

Let $V_1, \ldots, V_n$ be the irreducible components of $\pi^{-1}(p)$. Then
$$\sqrt{I(p)S} = \bigcap_i I(V_i)$$
where $I(V_i)$ is the homogeneous ideal of V_i in S. We have that
$$\begin{aligned}
\dim S/I(p)S &= \max_i\{\dim(S/I(V_i))\} \\
&\quad \text{since } I(V_i) \text{ are the minimal primes of } I(p)S \\
&= \max_i\{\dim(V_i) + 1\} \text{ by Theorem 8.3} \\
&= \max_i\{\dim(V_i)\} + 1 = \dim \pi^{-1}(p) + 1 \\
&\quad \text{by Proposition 8.1} \\
&= d \text{ by the proof of Theorem 10.11.}
\end{aligned}$$

By Theorem 6.4, $S \cong \bigoplus_{i \geq 0} I(p)^i$, and thus
$$S/I(p)S \cong \bigoplus_{i \geq 0} I(p)^i/I(p)^{i+1} \cong \bigoplus_{i \geq 0} m_p^i/m_p^{i+1}$$
by Lemma 1.28, where m_p is the maximal ideal of $\mathcal{O}_{X,p}$. This last ring is the associated graded ring of m_p,
$$\operatorname{gr}_{m_p}(\mathcal{O}_{X,p}) = \bigoplus_{i \geq 0} m_p^i/m_p^{i+1}.$$

We obtain that
$$\dim \operatorname{gr}_{m_p}(\mathcal{O}_{X,p}) = \dim \mathcal{O}_{X,p}.$$
From Theorem 10.11, we have that $\dim_{\mathcal{O}_{X,p}/m_p} m_p/m_p^2 \geq \dim \mathcal{O}_{X,p}$, so we recover the statements of Theorem 1.81 in our geometric setting.

Definition 10.12. A point p of a quasi-projective variety X is a nonsingular point of X if $\dim_k T_p(X) = \dim X$. A quasi-projective variety X is said to be nonsingular if all points of X are nonsingular points of X.

We have the following proposition, which is immediate from the definition of a regular local ring, giving us an alternate algebraic definition of a nonsingular point.

Proposition 10.13. *A point p of a quasi-projective variety X is a nonsingular point of X if and only if $\mathcal{O}_{X,p}$ is a regular local ring.*

Proposition 10.14 (Jacobian criterion)**.** *Suppose that X is an affine variety of dimension r, which is a closed subvariety of $\mathbb{A}^n$, and $f_1, \ldots, f_t \in k[\mathbb{A}^n] = k[x_1, \ldots, x_n]$ are a set of generators of $I(X)$. Suppose that $p \in X$. Then*
$$\dim_k T_p(X) = n - s$$

10.4. Nonsingularity and the singular locus

where s is the rank of the $t \times n$ matrix

$$A = \left(\frac{\partial f_i}{\partial x_j}(p)\right).$$

In particular, $s \leq n - r$, and p is a nonsingular point of X if and only if $s = n - r$.

Proof. Going back to our analysis of $T_p(X)$, we have that $\bar{x}_1, \ldots, \bar{x}_n$ is a k-basis of V and W is the subspace of V spanned by $\{L_p(f_1), \ldots, L_p(f_t)\}$. This subspace has dimension equal to the rank of A by (10.4). Since $T_p(X)$ and V/W are k-vector spaces of the same dimension, we have that $\dim_k T_p(X) = n - s$. □

Exercise 10.21 shows that the Jacobian criterion for nonsingularity of Proposition 10.14 does not always hold for a nonsingular point on a variety over a nonalgebraically closed field in positive characteristic. An extensive study of nonsingularity of varieties over arbitrary fields is made by Zariski in his paper [**154**].

Corollary 10.15. *Suppose that X is a quasi-projective variety. Then the set of nonsingular points of X is an open subset of X.*

Proof. Since X has an open cover by affine open subsets, we may assume that X is affine, so that X is a subvariety of $\mathbb{A}^n$ for some n. Let $I(X) = (f_1, \ldots, f_t)$ and let B be the $t \times n$ matrix $B = (\frac{\partial f_i}{\partial x_j})$. Let $r = \dim X$ and let $I_{n-r}(B)$ be the ideal generated by the determinants of $(n-r) \times (n-r)$ submatrices of B in $k[\mathbb{A}^n]$. By Proposition 10.14, $q \in X$ is a nonsingular point if and only if $q \notin Z(I_{n-r}(B))$. Thus the set of nonsingular points of X is open in X. □

Theorem 10.16. *Suppose that X is a quasi-projective variety. Then the set of nonsingular points of X is a dense open subset of X.*

Proof. By Proposition 2.112, there is a birational map from X to a hypersurface $Z(f) \subset \mathbb{A}^n$, where f is irreducible in $k[\mathbb{A}^n] = k[x_1, \ldots, x_n]$. We will show that the nonsingular locus of $Z(f)$ is nontrivial. Then the conclusions of the theorem follow since the nonsingular locus is open, any nontrivial open subset of a variety is dense, and birational varieties have isomorphic open subsets (by Theorem 2.111). Thus we may assume that $X = Z(f)$. Suppose that every point of X is singular. Then $Z(f, \frac{\partial f}{\partial x_1}, \ldots, \frac{\partial f}{\partial x_n}) = Z(f)$, so $\frac{\partial f}{\partial x_i} \in I(X) = (f)$ for all i. Since $\deg(\frac{\partial f}{\partial x_i}) < \deg(f)$ for all i (here the degree of a polynomial is the largest total degree of a monomial appearing in the polynomial), the only way this is possible is if $\frac{\partial f}{\partial x_i} = 0$ for all i. If the characteristic of k is zero, this implies that $f \in k$, which is impossible.

If k has positive characteristic, then f must be a polynomial in $x_1^p, \ldots, x_n^p$ with coefficients in k. Since the p-th roots of these coefficients are in k (as k is algebraically closed), we have that $f = g^p$ for some $g \in k[x_1, \ldots, x_n]$, contradicting the fact that f is irreducible. □

Suppose X is a quasi-projective variety. The closed algebraic set of singular points of X is called the singular locus of X.

Theorem 10.17. *Suppose that X is a normal quasi-projective variety. Let Z be the singular locus of X. Then $\operatorname{codim}_X Z \geq 2$.*

Proof. We may assume that X is affine. Let Y be a codimension 1 subvariety of X with ideal $I(Y) \subset k[X]$. The ideal $I(Y)$ is a height 1 prime ideal and the property of being normal localizes by Exercise 1.58, so the local ring $R = k[X]_{I(Y)}$ is normal of dimension 1. Thus R is a regular local ring by Theorem 1.87. Let f be a generator of the maximal ideal of R. After possibly multiplying f by an element of $k[X]$ which is in $k[X] \setminus I(Y)$, we may assume that $f \in k[X]$. The ideal (f) in $k[X]$ thus has a minimal primary decomposition $(f) = I(Y) \cap Q_1 \cap \cdots \cap Q_r$ where $Q_1, \ldots, Q_r$ are primary ideals in $k[X]$ which are A_i-primary for respective prime ideals $A_1, \ldots, A_r$ which are not contained in $I(Y)$. Thus there exists $q \in Y$ such that $q \notin Z(A_i)$ for $1 \leq i \leq r$ and so $I(Y)_{m_q} = f\mathcal{O}_{Y,q}$. We then have that there exists an affine open subset U of X such that $U \cap Y \neq \emptyset$ and $I_U(Y) = (f)$ by Lemma 10.6. The variety $Y \cap U$ has a nonsingular point $p \in Y \cap U$, so that $\mathcal{O}_{Y,p} \cong \mathcal{O}_{X,p}/(f)$ is a regular local ring. Let m be the maximal ideal of $\mathcal{O}_{X,p}$ and let n be the maximal ideal of $\mathcal{O}_{Y,p}$. There exist $\overline{v}_1, \ldots, \overline{v}_{d-1} \in \mathcal{O}_{Y,p}$ such that $n = (\overline{v}_1, \ldots, \overline{v}_{d-1})$ where $d = \dim X$. Let $v_1, \ldots, v_{d-1}$ be lifts of $\overline{v}_1, \ldots, \overline{v}_{d-1}$ to $\mathcal{O}_{X,p}$. The ideal $m = (v_1, \ldots, v_{d-1}, f)$ so that $\dim_k m/m^2 \leq d$. Thus $\mathcal{O}_{X,p}$ is a regular local ring by Theorem 10.11 or Theorem 1.81 and so p is a nonsingular point of X. Thus Y is not contained in the singular locus Z of X, and so $\operatorname{codim}_X Z \geq 2$. □

Lemma 10.18. *Suppose that Y is an irreducible codimension 1 subvariety of a quasi-projective variety X. Suppose that $p \in Y$ is a nonsingular point of X. Then there exists an irreducible element $f \in \mathcal{O}_{X,p}$ which is a local equation of Y at p.*

Proof. This follows from Lemma 10.6, Theorem 1.89, and Proposition 1.66. □

Theorem 10.19. *Suppose that X is an n-dimensional nonsingular quasi-projective variety and Y is a nonsingular subvariety of X. Let $\pi : B \to X$ be the blow-up of Y. Then B is nonsingular, and $E = \pi^{-1}(Y)$ is an irreducible, nonsingular codimension 1 subvariety of B. If $p \in Y$, then $\pi^{-1}(p) \cong \mathbb{P}^{r-1}$, where $r = \operatorname{codim}_X Y$. Suppose that $x_1, \ldots, x_n$ are regular parameters in*

10.4. Nonsingularity and the singular locus

$\mathcal{O}_{X,p}$, such that $x_1 = \cdots = x_r = 0$ are local equations of Y at p and $q \in \pi^{-1}(p)$. Then there exists a j with $1 \leq j \leq r$ and there exist $\alpha_1, \ldots, \alpha_r \in k$ such that $\mathcal{O}_{B,q}$ has regular parameters $y_1, \ldots, y_n$ such that

$$x_i = \begin{cases} y_j(y_i + \alpha_i) & \text{for } 1 \leq i \leq r \text{ and } i \neq j, \\ y_j & \text{for } i = j, \\ y_i & \text{for } i > r. \end{cases}$$

A local equation of E in B at q is $y_j = 0$.

With the hypotheses of Theorem 10.19, we have by Proposition 6.5 that

$$S(B) \cong k[X][y_1, \ldots, y_r]/J$$

where $J = (y_i x_j - x_i y_j \mid 1 \leq i < j \leq r)$.

Proof. Suppose that $p \in Y$. By Lemmas 10.3 and 10.6, there exist regular parameters $x_1, \ldots, x_n$ in $\mathcal{O}_{X,p}$ and an affine neighborhood U of p in X such that $x_1 = \cdots = x_r = 0$ are local equations of Y in U, $x_1 = \cdots = x_n = 0$ are local equations of p in U, and $\mathcal{I}_Y(U) = (x_1, \ldots, x_r)$. We have by Theorem 6.4 that $\pi^{-1}(U)$ has the affine cover $\{V_1, \ldots, V_r\}$ where

$$k[V_j] = k[U]\left[\frac{x_1}{x_j}, \ldots, \frac{x_r}{x_j}\right]$$

for $1 \leq j \leq r$. Suppose that $q \in \pi^{-1}(p)$. Then $q \in V_j$ for some j. Let $\mathfrak{n} = I_{V_j}(q)$ and $\mathfrak{m} = I_U(p)$. Without loss of generality, we may assume that $j = r$. We have $\mathfrak{n} \cap k[U] = \mathfrak{m}$ since $\pi(q) = p$. Now

$$k \cong k[V_r]/\mathfrak{n} \cong (k[U]/\mathfrak{m})[t_1, \ldots, t_{r-1}] \cong k[t_1, \ldots, t_{r-1}]$$

where $t_1, \ldots, t_{r-1}$ are the residues of $\frac{x_1}{x_r}, \ldots, \frac{x_{r-1}}{x_r}$ in $k[V_r]/\mathfrak{n}$. Since this k-algebra is isomorphic to k, we must have that $t_i = \alpha_i$ for some $\alpha_i \in k$, so that

$$\mathfrak{n} = \mathfrak{m}k[V_r] + \left(\frac{x_1}{x_r} - \alpha_1, \ldots, \frac{x_{r-1}}{x_r} - \alpha_{r-1}\right).$$

Since $\mathfrak{m}k[V_r] = (x_r, x_{r+1}, \ldots, x_n)$, we have that the n functions

(10.8) $\quad x_r, x_{r+1}, \ldots, x_n, \frac{x_1}{x_r} - \alpha_1, \ldots, \frac{x_{r-1}}{x_r} - \alpha_{r-1}$

generate the maximal ideal of $\mathcal{O}_{B,q}$, so that letting $\mathfrak{a} = \mathfrak{n}\mathcal{O}_{B,q}$ be the maximal ideal of $\mathcal{O}_{B,q}$, we have that $\dim_k \mathfrak{a}/\mathfrak{a}^2 \leq n$. Since $\pi : B \to X$ is a birational map, and by Theorem 10.11, we have that

$$n = \dim \mathcal{O}_{B,q} \leq \dim_k \mathfrak{a}/\mathfrak{a}^2 \leq n.$$

Thus $\mathcal{O}_{B,q}$ is a regular local ring, with regular parameters (10.8).

We have that $\pi^{-1}(U)$ is a closed subvariety of $U \times \mathbb{P}^{r-1}$, and $x_r = 0$ is a local equation in V_r of $\pi^{-1}(Y \cap U) \subset (Y \cap U) \times \mathbb{P}^{r-1}$. By Krull's principal ideal theorem, Theorem 1.65, all irreducible components of $\pi^{-1}(Y \cap U)$ have codimension 1 in $\pi^{-1}(U)$. Thus $\pi^{-1}(Y \cap U) = (Y \cap U) \times \mathbb{P}^{r-1}$.

Since $B \setminus E \to X \setminus Y$ is an isomorphism and X is nonsingular, we have that B is nonsingular. □

Exercise 10.20. Consider the curve $X = Z(y^2 - x^3) \subset \mathbb{A}^2$, where k is an algebraically closed field of characteristic 0.

- a) Show that at the point $p_1 = (1, 1)$, the tangent space $T_{p_1}(X)$ is the line $Z(-3(x-1) + 2(y-1))$.
- b) Show that at the point $p_2 = (0, 0)$, the tangent space $T_{p_2}(X)$ is the entire plane $\mathbb{A}_k^2$.
- c) Show that the curve is singular only at the origin p_2.
- d) Show that the blow-up of p_2 in C is nonsingular everywhere. (This blow-up was constructed in Exercise 6.12).

Exercise 10.21. Let p be a prime number, and let $K = F_p(t)$ where F_p is the finite field with p elements and t is transcendental over F_p. Show that $R = K[x, y]/(x^p + y^p - t)$ is a regular ring (all localizations at prime ideals are regular). Hint: You can use the fact that a polynomial ring over a field is a regular ring.

Let $f = x^p + y^p - t$. Show that the matrix

$$\left(\frac{\partial f}{\partial x}, \frac{\partial f}{\partial y}\right) = (0, 0).$$

Conclude that the Jacobian criterion for nonsingularity of Proposition 10.14 is not valid over fields of characteristic $p > 0$ which are not algebraically closed (not perfect).

Exercise 10.22. Let $X \subset \mathbb{P}^n$ be a projective variety of dimension r. Let $I(X) = (F_1, \ldots, F_t) \subset S(\mathbb{P}^n) = k[x_0, \ldots, x_n]$ where $F_1, \ldots, F_t$ are homogeneous forms. Show that the singular locus of X is the algebraic set $Z(I(X) + I_{n-r}(M)) \subset \mathbb{P}^n$ where M is the $t \times (n+1)$ matrix

$$M = \left(\frac{\partial F_i}{\partial x_j}\right)$$

and $I_{n-r}(M)$ is the ideal generated by the $n - r$ minors of M. Hint: Use Euler's formula, Exercise 1.35.

10.5. Applications to rational maps

Theorem 10.23. *Suppose that $\phi : X \dashrightarrow Y$ is a rational map of projective varieties and X is normal. Let Z be the closed subset of X consisting of the points where ϕ is not a regular map. Then Z has codimension ≥ 2 in X.*

Proof. The variety Y is a closed subset of a projective space $\mathbb{P}^n$, so after composing ϕ with a closed embedding, we may suppose that $Y = \mathbb{P}^n$. Let W be the set of singular points of X. The closed set W has codim ≥ 2 in X by Theorem 10.17 since X is normal. Suppose that $p \in Z \cap (X \setminus W)$. We will find an affine open neighborhood U_p of p in X and a representative $(f_0 : \ldots : f_n)$ of ϕ such that $f_0, \ldots, f_n \in k[U_p]$ and $Z_{U_p}(f_0, \ldots, f_n)$ has codimension ≥ 2 in U_p. Since $Z \cap U_p \subset Z_{U_p}(f_0, \ldots, f_n)$ and there is an open cover of $X \setminus W$ by sets of this form, we will have that Z has codimension ≥ 2 in X.

We will now prove the assertion. Suppose that $p \in X \setminus W$. Let $f_0, \ldots, f_n \in k(X)$ be such that $(f_0 : \ldots : f_n)$ is a representative of the rational map ϕ. After multiplying all of the f_i by a suitable element of $\mathcal{O}_{X,p}$, we may assume that $f_0, \ldots, f_n \in \mathcal{O}_{X,p}$. Since $\mathcal{O}_{X,p}$ is a UFD (Theorem 1.89), the greatest common divisor of a set of elements in $\mathcal{O}_{X,p}$ is defined. Let g be the greatest common divisor of the elements $f_0, \ldots, f_n$ in $\mathcal{O}_{X,p}$. Let $\overline{f}_i = \frac{f_i}{g} \in \mathcal{O}_{X,p}$.

We will show that the ideal $(\overline{f}_0, \ldots, \overline{f}_n)$ in $\mathcal{O}_{X,p}$ has height ≥ 2. Lemma 10.18 shows that for every codimension 1 subvariety S of X which contains p, there exists an irreducible $h \in \mathcal{O}_{X,p}$ which is a local equation of S at p. Let U be an affine neighborhood of p in X such that $\overline{f}_0, \ldots, \overline{f}_n \in k[U]$ and h is a local equation of S in U. If $S \cap U \subset Z_U(\overline{f}_0, \ldots, \overline{f}_n)$, then by the nullstellensatz, some power of $\overline{f}_j$ is in $I_U(S \cap U) = hk[U]$ for all j, so h divides $\overline{f}_j$ in $\mathcal{O}_{X,p}$ since h is irreducible in $\mathcal{O}_{X,p}$, giving a contradiction. Since $(\overline{f}_0 : \ldots : \overline{f}_n)$ is a representative of ϕ, we have the desired conclusion. $\square$

Corollary 10.24. *Every rational map $\phi : X \dashrightarrow Y$ of projective varieties such that X is a nonsingular projective curve is regular.*

Corollary 10.25. *Every birational map of nonsingular projective curves is an isomorphism.*

Corollary 10.26. *Every dominant rational map $\phi : X \dashrightarrow Y$ of projective curves such that X is nonsingular is regular and finite.*

Proof. The map $\phi : X \to Y$ is regular by Corollary 10.24. Consider the injective k-algebra homomorphism $\phi^* : k(Y) \to k(X)$. Let Z be the normalization of Y in $k(X)$ constructed in Theorem 7.17, with finite regular map $\pi : Z \to Y$. The variety Z is a normal projective curve by Theorem 7.17 and by 1) of Theorem 8.2. Since Z has dimension 1 and is normal, Z is nonsingular by Theorem 1.87. By Proposition 2.108 (as commented in Section 4.4, this result is valid for rational maps of projective varieties), there is a rational map $\psi : Z \dashrightarrow X$ such that $\psi^* : k(Z) \to k(X)$ is the identity map. By Proposition 2.110 (as commented in Section 4.4, this result is valid for rational maps of projective varieties), ψ is birational. Since ψ is a birational map of nonsingular projective curves, ψ is an isomorphism by Corollary 10.25. □

Proposition 10.27. *Suppose that $\phi : X \to Y$ is a birational regular map of quasi-projective varieties which is not an isomorphism. Suppose that $p \in X$ with $q = \phi(p)$, $\psi = \phi^{-1}$ is not regular at q and $\mathcal{O}_{Y,q}$ is a UFD (for instance if q is a nonsingular point of Y). Then there exists a subvariety Z of X with $p \in Z$ such that $\mathrm{codim}_X Z = 1$ and $\mathrm{codim}_Y \phi(Z) \geq 2$.*

Proof. The homomorphism $\phi^* : k(Y) \to k(X)$ is an isomorphism, with inverse ψ^*. Replacing Y with an affine open neighborhood U of q in Y and X with an affine open neighborhood of p in the preimage of U, we may assume that X and Y are affine.

We have that X is a closed subset of an affine space $\mathbb{A}^n$. Let $t_1, \ldots, t_n$ be the coordinate functions on X (so that $k[X] = k[t_1, \ldots, t_n]$). We may represent the rational map $\psi : Y \dashrightarrow X$ by $\psi = (g_1, \ldots, g_n)$ where the $g_i \in k(Y)$ are rational functions on Y, so that $\phi^*(g_i) = t_i$ for $1 \leq i \leq n$ (since $\phi^* : k(Y) \to k(X)$ is the inverse of $\psi^* : k(X) \to k(Y)$). Since ψ is not regular at q, at least one of the g_i is not regular at q, say $g_1 \notin \mathcal{O}_{Y,q}$. We have that $\mathcal{O}_{Y,q}$ is a UFD by assumption. Thus we have an expression $g_1 = \frac{u}{v}$ where $u, v \in \mathcal{O}_{Y,q}$ and u, v are relatively prime. We necessarily have that $v(q) = 0$. We may if necessary replace Y and X with smaller affine neighborhoods of q and p to obtain that $u, v \in k[Y]$. Now $Z_Y(u, v)$ cannot contain an irreducible component D which has codimension 1 in Y and contains q, since if it did, a local equation $f = 0$ of D at q would satisfy $f \mid u$ and $f \mid v$ in $\mathcal{O}_{Y,q}$, which is impossible since u and v are relatively prime. Thus replacing Y and X with possibly smaller affine neighborhoods of q and p, we may assume that $Z_Y(u, v)$ has codimension ≥ 2 in Y.

We have that $t_1 = \phi^*(g_1) = \phi^*(u)/\phi^*(v)$, so that

(10.9) $$\phi^*(v) t_1 = \phi^*(u)$$

in $k(X)$. We have that $t_1, \phi^*(u), \phi^*(v) \in k[X]$ and $\phi^*(v)(p) = 0$, so that $p \in Z_X(\phi^*(v))$. Set $Z = Z_X(\phi^*(v))$. Then $\mathrm{codim}_X Z = 1$ since $p \in Z$. By (10.9) it follows that $\phi^*(u) \in I_X(Z)$. Thus $u, v \in I_Y(\phi(Z))$, so that $\phi(Z) \subset Z_Y(u, v)$, which we have shown has codimension ≥ 2 in Y. □

10.5. Applications to rational maps

The conclusions of Proposition 10.27 may be false if $\mathcal{O}_{Y,q}$ is normal but is not a UFD, as is shown by the example computed in Exercise 6.16 c). In this example we construct the blow-up $\pi_1 : B(I_1) \to X$ where $X = Z(xy - zw) \subset \mathbb{A}^4$ and $I_1 = (x, z)k[X]$ and show that $B(I_1) \setminus F_1 \cong X \setminus \{p\}$ where p is the origin and $F_1 = \pi_1^{-1}(p) \cong \mathbb{P}^1$.

Theorem 10.28. *Suppose that $\phi : X \to Y$ is a birational regular map of projective varieties and Y is nonsingular. Let $C = \{q \in Y \mid \dim \phi^{-1}(q) > 0\}$. The set C is a closed subset of Y by Corollary 8.14. Let $G = \phi^{-1}(C)$, a closed subset of X. Then:*

1. *$\phi : X \setminus G \to Y \setminus C$ is an isomorphism.*
2. *$\operatorname{codim}_Y C \geq 2$.*
3. *The closed set G is a union of codimension 1 subvarieties of X. If E is one of these components, then $\operatorname{codim}_Y \phi(E) \geq 2$.*

The set G is called the exceptional locus of ϕ.

Proof. We have that $\phi(X \setminus G) = Y \setminus C$ by Theorem 5.14 since X and Y are projective. Proposition 10.27 now implies $\phi : X \setminus G \to Y \setminus C$ is an isomorphism. The second statement of the theorem follows from Theorem 10.23 applied to the rational map ϕ^{-1}. Now we prove the third statement. Suppose that E is an irreducible component of G and $p \in E$ is a point which is not contained in any other irreducible component of G. Let $q = \phi(p)$. Let B be an affine neighborhood of q in Y, and let A be an affine neighborhood of p in X such that A does not contain points of any irreducible component of G except for E and $\phi(A) \subset B$. Proposition 10.27 applied to $\phi : A \to B$ tells us that there exists a codimension 1 subvariety F of A such that $\operatorname{codim}_Y \phi(F) \geq 2$. We must have that $F \subset G \cap A = E \cap A$. Since E is irreducible, we have that $F = E \cap A$. $\square$

Exercise 10.29. Let $U = \mathbb{A}^2$ with regular functions $k[U] = k[x, y]$. Let $\pi : X \to U$ be the blow-up of the origin p. Let $E = \pi^{-1}(p) \cong \mathbb{P}^1$, and let S be the set of lines through the origin in U. In this problem, the analysis of $\pi : X \to U$ given at the end of Section 6.1 and a careful reading of the proof of Lemma 10.30 will be helpful.

  a) Show that the map (of sets) $\Lambda : \mathbb{P}^1 \to S$ defined by $\Lambda(\alpha : \beta) = L_{(\alpha:\beta)} = Z(\alpha x + \beta y)$ for $(\alpha : \beta) \in \mathbb{P}^1$ is injective and onto.

  b) Suppose that L is a line through the origin in U. Let L' be the strict transform of L in X. Show that the map $\mathbb{P}^1 \to E$ defined by
  $$(\alpha : \beta) \mapsto (L_{(\alpha:\beta)})' \cap E$$
  is an injective and onto map of sets.

c) Show that the map defined in part b) is an isomorphism of projective varieties.

d) Using the facts that X is covered by two affine open sets isomorphic to $\mathbb{A}^2$ and that the strict transform L' of a line L through p is a line in one of these charts, show that if p' is the point $E \cap L'$, then the map
$$d\pi_{p'} : T_{p'}(L') \to T_p(U)$$
is injective with image equal to $T_p(L)$.

e) Show that for $p' = L' \cap E$ as in part d), the image $d\pi_{p'}(T_{p'}(X)) = T_p(L)$.

10.6. Factorization of birational regular maps of nonsingular surfaces

The proof in this section is based on the proof by Shafarevich in [**136**]. The first proof of Theorem 10.32 was by Zariski [**151**].

Lemma 10.30. *Suppose that X is a nonsingular projective surface and $\pi : Y \to X$ is the blow-up of a point $p \in X$. Let $E = \pi^{-1}(p) \cong \mathbb{P}^1$. Then there is a 1-1 correspondence between points q of E and one-dimensional subspaces L of $T_p(X)$, given by $q \mapsto d\pi_q(T_q(Y))$.*

Proof. Let u, v be regular parameters in $\mathcal{O}_{X,p}$. Let m_p be the maximal ideal of $\mathcal{O}_{X,p}$. Then the k-vector space $m_p/m_p^2 \cong ku \oplus kv$. By Theorem 10.19, the distinct points q of E have regular parameters u_1, v_1 in $\mathcal{O}_{Y,q}$ which have the forms

(10.10) $\qquad u = u_1, \qquad v = u_1(v_1 + \alpha) \quad \text{for } \alpha \in k,$

or

(10.11) $\qquad u = u_1 v_1, \qquad v = v_1.$

Let m_q be the maximal ideal of $\mathcal{O}_{Y,q}$. We have a k-vector space isomorphism $m_q/m_q^2 \cong ku_1 \oplus kv_1$. The k-linear map $\pi^* : m_p/m_p^2 \to m_q/m_q^2$ is defined by
$$\pi^*(u) = u_1 \text{ and } \pi^*(v) = \alpha u_1 \text{ if (10.10) holds}$$
and
$$\pi^*(u) = 0 \text{ and } \pi^*(v) = v_1 \text{ if (10.11) holds.}$$
Let $\delta_{u_1}, \delta_{v_1}$ be the dual basis of $T_q(Y)$ to u_1, v_1, and let δ_u, δ_v be the dual basis of $T_p(X)$ to u, v.

Suppose that (10.10) holds. Then
$$d\pi_q(\delta_{u_1}) = \delta_{u_1} \pi^* = \delta_u + \alpha \delta_v \quad \text{and} \quad d\pi_q(\delta_{v_1}) = \delta_{v_1} \pi^* = 0.$$
Thus $d\pi_q(T_q(Y)) = (\delta_u + \alpha \delta_v)k$.

10.6. Factorization of birational regular maps of nonsingular surfaces

Suppose that (10.11) holds. Then
$$d\pi_q(\delta_{u_1}) = \delta_{u_1}\pi^* = 0 \quad \text{and} \quad d\pi_q(\delta_{v_1}) = \delta_{v_1}\pi^* = \delta_v.$$
Thus $d\pi_q(T_q(Y)) = \delta_v k$. □

Lemma 10.31. *Suppose that $\phi : X \dashrightarrow Y$ is a birational map of nonsingular projective surfaces. Let $\Gamma \subset X \times Y$ be the graph of ϕ, with projections $\pi_1 : \Gamma \to X$ and $\pi_2 : \Gamma \to Y$. Suppose that ϕ^{-1} is not regular at a point $q \in Y$. Then there exists a curve $D \subset \Gamma$ such that $\pi_1(D) = C$ is a curve of X and $\pi_2(D) = q$.*

Proof. By Theorem 5.14, we have that $\pi_2(\Gamma) = Y$. We have that π_2^{-1} is not regular at q since ϕ^{-1} is not regular at q. By Proposition 10.27, there exists a curve $D \subset \Gamma$ such that $\pi_2(D) = q$. If $\pi_1(D)$ is not a curve, we must have that $\pi_1(D)$ is a point p. But then $D \subset \pi_1^{-1}(p) \cap \pi_2^{-1}(q) = \{(p,q)\}$, which is a point, giving a contradiction. □

Theorem 10.32. *Suppose that $\phi : X \to Y$ is a birational regular map of nonsingular projective surfaces. Then ϕ has a factorization*
$$X = Y_n \to Y_{n-1} \to \cdots \to Y_1 \to Y_0 = Y,$$
where each $Y_{i+1} \to Y_i$ is the blow-up of a point.

Proof. Suppose that ϕ is not an isomorphism, so that ϕ^{-1} is not regular at some point $q \in Y$. Let $\sigma : Y' \to Y$ be the blow-up of q, and let $E = \sigma^{-1}(q) \cong \mathbb{P}^1$ be the exceptional locus of σ. Let $\phi' : X \dashrightarrow Y'$ be the rational map $\phi' = \sigma^{-1}\phi$. We will show that ϕ' is a regular map. Suppose that ϕ' is not a regular map. Let $\Gamma \subset X \times Y'$ be the graph of ϕ', and suppose that $p \in X$ is a point where ϕ' is not regular. Then by Lemma 10.31, there exists a curve $D \subset \pi_1^{-1}(p)$ such that $\pi_2(D)$ is a curve C in Y'. Since $\sigma \mid (Y' \setminus E)$ is an isomorphism onto $Y \setminus \{q\}$, we have that $C \subset E$, so $C = E$ since both C and E are irreducible curves. Let $\psi : Y' \dashrightarrow X$ be the rational map $\psi = (\phi')^{-1}$. By Theorem 10.23, there exists a finite set $T \subset Y'$ such that $\psi \mid (Y' \setminus T)$ is a regular map. We have that $\psi(E \setminus T) = \pi_1(D \setminus \pi_2^{-1}(T)) = p$ and $\phi(p) = q$.

We will show that

(10.12) $$d\phi_p : T_p(X) \to T_q(Y)$$

is an isomorphism. Suppose not. Then there exists a one-dimensional subspace $L \subset T_q(Y)$ such that $d\phi_p(T_p(X)) \subset L$. Since $\psi(E \setminus T) = p$, we have that $d\sigma_{q'}(T_{q'}(Y')) \subset L$ for $q' \in E \setminus T$. But this is a contradiction to the conclusions of Lemma 10.30. Thus (10.12) is an isomorphism. Now ϕ^{-1} is not regular at q, so by Proposition 10.27, there exists a curve $F \subset X$ such that $p \in F$ and $\phi(F) = q$. But then $T_p(F) \subset T_p(X)$ has dimension ≥ 1, and

$d\phi_p(T_p(F)) = 0$, a contradiction to the fact that (10.12) is an isomorphism. This contradiction shows that ϕ' is a regular map.

The exceptional locus of ϕ is a union of irreducible curves by Theorem 10.28. Let r be the number of irreducible components of the exceptional locus of ϕ. The regular map ϕ' maps $\phi^{-1}(q)$ onto $\sigma^{-1}(q) = E$. Thus there exists a curve $G \subset \phi^{-1}(q)$ which maps onto E, so the number of irreducible components of the exceptional locus of ϕ' is $\leq r - 1$. By induction, after enough blow-ups of points, we obtain the desired factorization of ϕ. □

There is a very general local form of this theorem by Abhyankar [3], from which we can also deduce Theorem 10.32.

Theorem 10.33 (Abhyankar). *Suppose that R and S are two-dimensional regular local rings with a common quotient field K such that S dominates R ($R \subset S$ and $m_S \cap R = m_R$, where m_R and m_S are the respective maximal ideals of R and S). Then $R \to S$ factors by a finite sequence of quadratic transforms*

$$R = R_0 \to R_1 \to \cdots \to R_n = S.$$

A quadratic transform of a regular local ring R is a local ring of the blow-up of the maximal ideal; so if x, y are regular parameters in a two-dimensional regular local ring R, then a quadratic transform of $R \to R_1$ is a local ring (a localization at a maximal ideal) of $R[\frac{x}{y}]$ or $R[\frac{y}{x}]$.

10.7. Projective embedding of nonsingular varieties

Lemma 10.34. *Suppose that R and S are local Noetherian rings with respective maximal ideals m_R and m_S. Let $f : R \to S$ be a local homomorphism of local Noetherian rings such that*

1) $R/m_R \to S/m_S$ *is an isomorphism,*
2) $m_R \to m_S/m_S^2$ *is surjective, and*
3) S *is a finitely generated R-module.*

Then f is surjective.

Proof. By 2), we have that $m_S = m_R S + m_S^2$, so by Nakayama's lemma (Lemma 1.18), we have that $m_R S = m_S$. By 1),

$$S = f(R) + m_S = f(R) + m_R S,$$

and by 3), S is a finitely generated R-module, so again by Nakayama's lemma, we have that $S = f(R)$. □

10.7. Projective embedding of nonsingular varieties

Lemma 10.35. *Suppose that $\phi : X \to Y$ is a regular map of quasi-projective varieties such that*

1) *ϕ is bijective and*
2) *$\phi^* : \mathcal{O}_{Y,\phi(p)} \to \mathcal{O}_{X,p}$ is an isomorphism for all $p \in X$.*

Then ϕ is an isomorphism.

Proof. We will find an affine cover $\{V_i\}$ of Y such that for all i, $U_i = \phi^{-1}(V_i)$ is an affine open subset of X and $\phi^* : k[V_i] \to k[U_i]$ is an isomorphism. This is enough to conclude that ϕ is an isomorphism, since each map $\phi : U_i \to V_i$ is then an isomorphism, with regular inverse $\psi_i : V_i \to U_i$ by Proposition 2.40 and Proposition 2.42. Now the ψ_i patch to give a continuous map $\psi : Y \to X$, which is an inverse to ϕ by Proposition 3.39.

Suppose that $p \in X$. Let $q = \phi(p) \in Y$ and let V be an affine neighborhood of q in Y. Let $U \subset \phi^{-1}(V)$ be an affine neighborhood of p in X. We then have (since ϕ is dominant) that $\phi^* : k[V] \to k[U]$ is 1-1. Let $I(q)$ be the ideal of q in V, and let $I(p)$ be the ideal of p in U. By assumption, ϕ^* induces an isomorphism $k[V]_{I(q)} \to K[U]_{I(p)}$. Suppose that $t_1, \ldots, t_r$ generate $k[U]$ as a k-algebra. Since $t_i \in k[U]_{I(p)} = k[V]_{I(q)}$, there exists $h \in k[V] \setminus I(q)$ and $f_1, \ldots, f_r \in k[U]$ such that $t_i = \frac{f_i}{h}$. Thus $k[V]_h \to k[U]_h$ is an isomorphism. Now V_h is an affine neighborhood of q in Y, U_h is an affine neighborhood of p in X, and $\phi^* : k[V_h] \to k[U_h]$ is an isomorphism. $\square$

Theorem 10.36. *Suppose that X and Y are projective varieties and $\phi : X \to Y$ is a regular map which is injective and such that $d\phi_p : T_p(X) \to T_{\phi(p)}(Y)$ is injective for all $p \in X$. Then ϕ is a closed embedding.*

Proof. We have that $\phi(X)$ is a closed subvariety of Y by Theorem 5.14 since X is projective. Further, the map $d\phi_p$ factors through the inclusion $T_q(\phi(X)) \subset T_q(Y)$, so without loss of generality, we may assume that $Y = \phi(X)$. Here X is a closed subvariety of $\mathbb{P}^n$ for some n. We have a sequence of maps

$$X \xrightarrow{\lambda} \Gamma_\phi \subset X \times Y \subset \mathbb{P}^n \times Y$$

and a commutative diagram

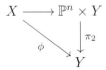

where $\lambda : X \to \Gamma_\phi$, defined by $\lambda(p) = (p, \phi(p))$ for $p \in X$, is an isomorphism, π_2 is the projection on the second factor, and the inclusions in the top row are all closed embeddings. Suppose that $p \in X$, with $q = \phi(p) \in Y$, and U is an affine open subset of Y which contains q. Let $Z = \Gamma_{\phi|\phi^{-1}(U)} = \Gamma_\phi \cap \pi_2^{-1}(U)$

which is a closed subset of $\pi_2^{-1}(U) \cong \mathbb{P}^n \times U$. Let $x_0, \ldots, x_n$ be homogeneous coordinates on $\mathbb{P}^n$. Then $S = k[U][x_0, \ldots, x_n]$ is the homogeneous coordinate ring of $\mathbb{P}^n \times U$, which is a graded ring, where elements of $k[U]$ have degree 0 and the x_i have degree 1 for $0 \le i \le n$. Let $I(Z)$ be the homogeneous ideal of Z in S. Then

$$\mathcal{O}_X(\phi^{-1}(U)) \cong \mathcal{O}_Z(Z) = \bigcap_{i=0}^{n} (S/I(Z))_{(x_i)},$$

where $(S/I(Z))_{(x_i)}$ denotes the elements of degree 0 in the localization of $S/I(Z)$ with respect to x_i. Now the proof of Theorem 3.35 (with k replaced with the ring $k[U]$) or Theorem 11.47 shows that $\mathcal{O}_X(\phi^{-1}(U))$ is finite over $k[U]$.

Further, for $q \in Y$, we can compute

$$\bigcap_{q \in V} \mathcal{O}_X(\phi^{-1}(V)) = \bigcap_{i=0}^{n} T_{(x_i)},$$

where $T = \mathcal{O}_{Y,q}[x_0, \ldots, x_n]/I(Z)\mathcal{O}_{Y,q}[x_0, \ldots, x_n]$ and the intersection is over all affine open subsets $V \subset U$ which contain q (since $\bigcap_{q \in V} \mathcal{O}_Y(V) = \mathcal{O}_{Y,q}$). Again, the proof of Theorem 3.35, with k replaced with the ring $\mathcal{O}_{Y,q}$, shows that $\bigcap_{q \in V} \mathcal{O}_X(\phi^{-1}(V))$ is finite over $\mathcal{O}_{Y,q}$, where the intersection is over the affine open subsets V of U containing q. Since ϕ is bijective and $\bigcap_{q \in V} V = \{q\}$, so that $\bigcap_{q \in V} \phi^{-1}(V) = \{p\}$, we have that $\bigcap_{q \in V} \mathcal{O}_X(\phi^{-1}(V)) = \mathcal{O}_{X,p}$. Thus $\mathcal{O}_{X,p}$ is finite over $\mathcal{O}_{Y,\phi(p)}$ for $p \in X$. Since $T_p(X) \to T_q(Y)$ is an injective homomorphism of finite-dimensional k-vector spaces, we have that the natural homomorphism $m_q/m_q^2 \to m_p/m_p^2$ is a surjection, where m_p is the maximal ideal of $\mathcal{O}_{X,p}$ and m_q is the maximal ideal of $\mathcal{O}_{Y,q}$. Thus by Lemma 10.34, we have that $\mathcal{O}_{X,p} = \mathcal{O}_{Y,q}$ for all $p \in X$. Finally, we have by Lemma 10.35 that ϕ is an isomorphism. □

Corollary 10.37. *Suppose that $X \subset \mathbb{P}^n$ is a projective variety and $p \in \mathbb{P}^n \setminus X$. Suppose that every line through p intersects X in at most one point and p is not contained in the (Zariski closure in $\mathbb{P}^n$ of the) tangent space to X at any point. Then the projection π from p is a closed embedding of X into $\mathbb{P}^{n-1}$.*

Proof. Suppose that $a \in X$. Let $b = \pi(a)$. Since $a \ne p$, we can make a linear change of coordinates in $\mathbb{P}^n$ so that $p = (0 : \ldots : 0 : 1)$ and $a = (1 : 0 : \ldots : 0)$. Now the projection π from p is defined by the rational map $(x_0 : \ldots : x_{n-1})$. Thus the restriction of π to $D(x_0)$ is just the map $\pi : \mathbb{A}^n \to \mathbb{A}^{n-1}$ which is the projection onto the first $n-1$ factors. Since a is the origin in $\mathbb{A}^n$, $d\pi_a : T_a(\mathbb{P}^n) \to T_b(\mathbb{P}^{n-1})$, which we can identify with $\pi : \mathbb{A}^n \to \mathbb{A}^{n-1}$, is just the projection onto the first $n-1$ factors. The kernel of $d\pi_a$ is the line $Z_{D(x_0)}(\frac{x_1}{x_0},\ldots,\frac{x_{n-1}}{x_0})$, which is the intersection of the line containing p and a with $D(x_0)$. From the commutative diagram (10.5) and the assumptions of the theorem, we have that $d\pi_a : T_a(X) \to T_b(\mathbb{P}^{n-1})$ is injective. Now we apply Theorem 10.36 to obtain the conclusions of the corollary. $\square$

Theorem 10.38. *Suppose that X is a nonsingular projective variety of dimension n. Then X is isomorphic to a subvariety of $\mathbb{P}^{2n+1}$.*

Proof. It suffices to prove that if $X \subset \mathbb{P}^N$ with $N > 2n+1$, then there exists $p \in \mathbb{P}^N \setminus X$ satisfying the hypotheses of Corollary 10.37. Let $U_1, U_2 \subset \mathbb{P}^N$ be the respective sets of points not satisfying the respective assumptions of Corollary 10.37.

Let $a = (a_0 : \ldots : a_N), b = (b_0 : \ldots : b_N), c = (c_0 : \ldots : c_N) \in \mathbb{P}^N$. Let

$$A = \begin{pmatrix} a_0 & \cdots & a_N \\ b_0 & \cdots & b_N \\ c_0 & \cdots & c_N \end{pmatrix}.$$

The coefficients of the linear forms which vanish on the three points a, b, c are the elements of the kernel of the linear map $A : k^{N+1} \to k^3$. The condition that a, b, c be collinear is that there are at least $N-1$ independent forms in the kernel; that is, A has rank ≤ 2.

Let $x_0, \ldots, x_N$ be our homogeneous coordinates on $\mathbb{P}^N$ and let $y_0, \ldots, y_N$ be the induced homogeneous coordinates on X. Let $z_0, \ldots, z_N$ be the corresponding homogeneous coordinates on a copy of X. Then

$$x_0, \ldots, x_N, y_0, \ldots, y_N, z_0, \ldots, z_N$$

are trihomogeneous coordinates on $\mathbb{P}^n \times X \times X$ with trigraded coordinate ring

$$S(\mathbb{P}^N \times X \times X) = k[x_0, \ldots, x_N, y_0, \ldots, y_N, z_0, \ldots, z_N].$$

Let

$$W = \{(a, b, c) \mid a \in \mathbb{P}^N, b, c \in X, \text{ and } a, b, c \text{ are collinear in } \mathbb{P}^N\}.$$

Then W is the closed set $W = Z(I)$ of $\mathbb{P}^n \times X \times X$ where

$$I = I_3 \begin{pmatrix} x_0 & \cdots & x_N \\ y_0 & \cdots & y_N \\ z_0 & \cdots & z_N \end{pmatrix},$$

the ideal generated by the determinants of 3×3 submatrices.

In $\mathbb{P}^N \times X \times X$, consider the set Γ, which is the Zariski closure in $\mathbb{P}^N \times X \times X$ of triples (a, b, c) with $a \in \mathbb{P}^N$, $b, c \in X$ such that $b \neq c$

and a, b, c are collinear. Let Γ' be an irreducible component of Γ. Then Γ' necessarily contains a point (a, b, c) such that a, b, c are collinear and $b \neq c$. Let $y = (b, c)$ for these values of b and c.

The projections $\mathbb{P}^N \times X \times X$ to $\mathbb{P}^N$ and $X \times X$ define regular maps $\phi : \Gamma' \to \mathbb{P}^N$ and $\psi : \Gamma' \to X \times X$. Now $\psi^{-1}(y)$ is a subset of $(\mathbb{P}^N \times \{y\}) \cap W$ which is the set of points (a, b, c) where a is any point of the line through b and c. Hence $\dim \psi^{-1}(y) \leq 1$, and it follows from Theorem 8.13 that $\dim \Gamma' \leq 2n+1$. In particular, $\dim \Gamma \leq 2n+1$. By the definitions of U_1 and Γ, $U_1 \subset \phi(\Gamma)$, and thus $\dim U_1 \leq \dim \Gamma \leq 2n+1$.

Let $x_0, \ldots, x_N$ be homogeneous coordinates on $\mathbb{P}^N$ and suppose that $p = (b_0 : b_1 : \ldots : b_N) \in X$ is such that $b_i \neq 0$. Suppose that $F \in S(\mathbb{P}^N) = k[x_0, \ldots, x_N]$ is a homogeneous form of degree d. We will show that the Zariski closure of the tangent space to $Z(F) \cap D(x_i)$ in $D(x_i) \cong \mathbb{A}^N$ at p is the projective hyperplane in $\mathbb{P}^N$ with equation

$$\sum_{j=0}^{N} \frac{\partial F}{\partial x_j}(b) x_j = 0.$$

Let $u_0, \ldots, u_{i-1}, u_{i+1}, \ldots, u_N$ be coordinates on $D(x_i)$, defined by $u_j = \frac{x_j}{x_i}$ if $j \neq i$ and $f = F(u_0, \ldots, u_{i-1}, 1, u_{i+1}, \ldots, u_N)$. For $j \neq i$, we have

$$\frac{\partial f}{\partial u_j} = \frac{\partial F}{\partial x_j}(u_0, \ldots, u_{i-1}, 1, u_i, \ldots, u_N)$$

so

$$\frac{\partial F}{\partial x_j}(b_0, \ldots, b_N) = b_i^{d-1} \frac{\partial f}{\partial u_j}\left(\frac{b_0}{b_i}, \ldots, \frac{b_N}{b_i}\right).$$

By Euler's formula, Exercise 1.35,

$$x_i \frac{\partial F}{\partial x_i} = dF - \sum_{j \neq i} x_j \frac{\partial F}{\partial x_j}.$$

Thus,

$$\begin{array}{rcl}
\sum_{j=0}^{N} \frac{\partial F}{\partial x_j}(b_0, \ldots, b_N) x_j & = & \sum_{j \neq i} \frac{\partial F}{\partial x_j}(b_0, \ldots, b_N)(x_j - \frac{b_j}{b_i} x_i) \\
& = & b_i^{d-1} \left(\sum_{j \neq i} \frac{\partial f}{\partial u_j}\left(\frac{b_0}{b_i}, \ldots, \frac{b_N}{b_i}\right) \right) \left(x_i u_j - \frac{b_j}{b_i} x_i \right) \\
& = & x_i b_i^{d-1} \left(\sum_{j \neq i} \frac{\partial f}{\partial u_j}\left(\frac{b_0}{b_i}, \ldots, \frac{b_N}{b_i}\right) \right) \left(u_j - \frac{b_j}{b_i} \right),
\end{array}$$

proving the assertion.

Now let Γ_1 be the subset of $\mathbb{P}^N \times X$ consisting of points (a, b) such that a is in the Zariski closure of $T_b(X)$.

Let $a = (a_0 : \ldots : a_N) \in \mathbb{P}^N$ and $b = (b_0 : \ldots : b_N) \in X$. Let $I(X) = (F_1, \ldots, F_t)$ where $F_1, \ldots, F_t$ are homogeneous. Then the condition that a is in the Zariski closure of $T_b(X)$ in $\mathbb{P}^N$ is that

$$\sum_{i=0}^{N} \frac{\partial F_j}{\partial x_i}(b)a_i = 0 \quad \text{for } 1 \leq j \leq t.$$

Thus

$$\Gamma_1 = Z\left(\sum_{i=0}^{N} \frac{\partial F_j(y_0,\ldots,y_N)}{\partial x_i}(y_0,\ldots,y_N)x_i \mid 1 \leq j \leq t\right) \subset \mathbb{P}^n \times X$$

is a Zariski closed subset.

Let Γ_1' be an irreducible component of Γ_1. We necessarily have $(a,b) \in \Gamma_1'$ for some $b \in X$. We have projections $\psi : \Gamma_1' \to X$ and $\phi : \Gamma_1' \to \mathbb{P}^N$. For our $b \in X$, we have $\dim \psi^{-1}(b) \leq n$ since X is nonsingular, and hence $\dim \Gamma_1' \leq 2n$, and since $U_2 = \phi(\Gamma_1)$, we have $\dim U_2 \leq 2n$.

We have shown that $\dim U_1 \leq 2n+1$ and $\dim U_2 \leq 2n$. Thus if $N > 2n+1$ we have that $U_1 \cup U_2 \neq \mathbb{P}^N$. □

10.8. Complex manifolds

We have defined $\mathbb{A}_k^n$ to be k^n (as a set) with the Zariski topology. If we take $k = \mathbb{C}$, the complex numbers, we have a finer topology, the Euclidean topology on $\mathbb{C}^n$. We also have the theory of analytic functions on (Euclidean) open subsets of $\mathbb{C}^n$. If $U \subset \mathbb{C}^n$ is a Zariski open subset and f is a regular function (in the Zariski topology) on U, then f is also an analytic function on U.

Definition 10.39. A complex manifold of dimension n is a Hausdorff topological space M such that M has a covering $\{U_i\}$ by open subsets with homeomorphisms $\phi_i : U_i \to V_i$ between U_i and open subsets V_i of $\mathbb{C}^n$ such that $\phi_j \circ \phi_i^{-1} : \phi_i(U_i \cap U_j) \to \phi_j(U_i \cap U_j)$ are analytic (and hence bianalytic).

Complex projective space $\mathbb{P}_\mathbb{C}^n$ has a covering by Zariski open subsets $U_i = \mathbb{A}^n$, where $\mathbb{A}^n$ is just $\mathbb{C}^n$ with the Zariski topology. The Euclidean topology on $\mathbb{C}^n$ thus gives us the Euclidean topology on each U_i. These topologies agree on $U_i \cap U_j$ for $i \neq j$. This defines the Euclidean topology on $\mathbb{P}_\mathbb{C}^n$. We have that the Euclidean topology is finer than the Zariski topology (a Zariski open subset is open in the Euclidean topology). Suppose that X is a quasi-projective algebraic set which is contained in $\mathbb{P}^n$. Then the Euclidean topology on X is the subspace topology of the Euclidean topology of $\mathbb{P}_\mathbb{C}^n$. In particular, any quasi-projective complex variety X (the base field k is $\mathbb{C}$) has the Euclidean topology, as we can transcribe the Euclidean topology to X by an embedding $\phi : X \to \mathbb{P}^n$ by prescribing that $\phi^{-1}(W)$ is open in the Euclidean topology on X whenever W is a Euclidean open subset of $\mathbb{P}_\mathbb{C}^n$.

Suppose that X is a quasi-projective variety. As commented above, we can consider X to be an open subset of a closed subvariety of a projective space $\mathbb{P}^n$ and give X the induced Euclidean topology. Now $\mathbb{P}^n$ has an open covering by open subsets $U_i \cong \mathbb{A}^n$ as above. Suppose that V is a Zariski open subset of X. Then V has an affine covering by Zariski open subsets $W_{i,j} = (U_i)_{g_j} \cap \overline{X}$ where $\overline{X}$ is the Zariski closure of X in $\mathbb{P}^n$ and $g_j \in \mathbb{C}[U_i]$. Suppose that $f \in \mathcal{O}_X(V)$ is a regular function. Then each restriction of f to each W_{ij} is a continuous map from W_{ij} to $\mathbb{C}$ in the Euclidean topology, so $f: V \to \mathbb{C}$ is continuous in the Euclidean topology.

The following lemma is [**120**, Exercise 13, page 100].

Lemma 10.40. *Suppose that X is a topological space. Then X is Hausdorff if and only if the diagonal $\Delta = \{(p,p) \mid p \in X\}$ is closed in $X \times X$ (where $X \times X$ has the product topology).*

Theorem 10.41. *Suppose that X is a quasi-projective variety. Then X is Hausdorff in the Euclidean topology.*

Proof. The diagonal Δ of $X \times X$ is closed in the Zariski topology by Proposition 5.8. The product topology of the Euclidean topologies on X is the Euclidean topology on $X \times X$, which is finer than the Zariski topology on $X \times X$. Thus Δ is closed in $X \times X$ in the product topology of the Euclidean topology on X, so that X is Hausdorff in the Euclidean topology by Lemma 10.40. $\square$

We have inclusions of rings

$$\mathbb{C}[x_1,\ldots,x_n] \subset \mathbb{C}[x_1,\ldots,x_n]_{(x_1,\ldots,x_n)} \subset \mathbb{C}\{x_1,\ldots,x_n\} \subset \mathbb{C}[[x_1,\ldots,x_n]]$$

where $\mathbb{C}[x_1,\ldots,x_n]$ is the polynomial ring in n variables, $\mathbb{C}[[x_1,\ldots,x_n]]$ is the ring of formal power series, and $\mathbb{C}\{x_1,\ldots,x_n\}$ is the ring of formal power series which have a positive radius of convergence (the germs of analytic functions at the origin in $\mathbb{C}^n$).

Theorem 10.42 (Analytic implicit function theorem). *Suppose that $f \in \mathbb{C}\{x_1,\ldots,x_n\}$ is such that $f = \sum_{i=1}^n a_i x_i + (\text{higher-order terms})$ with $a_i \in \mathbb{C}$ and $a_1 \neq 0$. Then there exist $g \in \mathbb{C}\{x_2,\ldots,x_n\}$ and a unit series $u \in \mathbb{C}\{x_1,\ldots,x_n\}$ such that $f = u(x_1 - g(x_2,\ldots,x_n))$.*

This is a special case of the Weierstrass preparation theorem ([**102**, Section C.2.4], [**62**, page 8], or [**161**, pages 142–145]).

10.8. Complex manifolds

Corollary 10.43. *Suppose that $f \in \mathbb{C}\{x_1, \ldots, x_n\}$ is such that*

$$f = \sum_{i=1}^{n} a_i x_i + (\text{higher-order terms})$$

with $a_i \in \mathbb{C}$ and $a_1 \neq 0$. Then $\mathbb{C}\{x_2, \ldots, x_n\} \to \mathbb{C}\{x_1, \ldots, x_n\}/(f)$ is an isomorphism.

Proof. With the notation of Theorem 10.42, we have that the ideal $(f) = (x_1 - g(x_2, \ldots, x_n))$, so we can eliminate x_1. □

Corollary 10.44. *Suppose that $f_1, \ldots, f_r \in \mathbb{C}\{x_1, \ldots, x_n\}$ are such that*

$$f_i = \sum_{j=1}^{r} a_{ij} x_j + (\text{higher-order terms})$$

with $\operatorname{Det}(a_{ij}) \neq 0$. Then the map

$$\mathbb{C}\{x_{r+1}, \ldots, x_n\} \to \mathbb{C}\{x_1, \ldots, x_n\}/(f_1, \ldots, f_r)$$

is an isomorphism.

Proof. Let B be the inverse of the matrix (a_{ij}). Define $f'_1, \ldots, f'_r \in \mathbb{C}\{x_1, \ldots, x_n\}$ by

$$\begin{pmatrix} f'_1 \\ f'_2 \\ \vdots \\ f'_r \end{pmatrix} = B \begin{pmatrix} f_1 \\ f_2 \\ \vdots \\ f_r \end{pmatrix}.$$

Then we have an equality of ideals $(f_1, \ldots, f_r) = (f'_1, \ldots, f'_r)$, so we may replace the f_i with the f'_i and assume that (a_{ij}) is the identity matrix. By Theorem 10.42,

$$f_1 = u_1(x_1 - g_1(x_2, x_3, \ldots, x_n))$$

where $u_1 \in \mathbb{C}\{x_1, \ldots, x_n\}$ is a unit and by considering the linear term in f_1, we see that $g_1 \in (x_2, \ldots, x_n)^2 \mathbb{C}\{x_2, \ldots, x_n\}$. Now the ideal $(f_1, \ldots, f_r) = (x_1 - g_1, f_2, \ldots, f_r)$, so

$$\mathbb{C}\{x_1, \ldots, x_n\}/(f_1, \ldots, f_r)$$
$$\cong \mathbb{C}\{x_2, \ldots, x_n\}/(f_2(g_1, x_2, \ldots, x_n), \ldots, f_r(g_1, x_2, \ldots, x_n))$$

and for $2 \leq i \leq n$,

$$f_i(g_1, x_2, \ldots, x_n) = x_i + (\text{higher-order terms in } x_2, \ldots, x_n).$$

Thus we have the assumptions of the corollary with a reduction of r to $r-1$ and n to $n-1$ and further have that the new matrix A giving the linear part of the expression of $f_2, \ldots, f_r$ in terms of the variables $x_2, \ldots, x_n$ is the identity matrix. By induction on r, repeating the above argument, we obtain the conclusions of the corollary. □

Theorem 10.45. *Suppose that X is a nonsingular quasi-projective complex variety of dimension r. Then X is a complex manifold of dimension r in the Euclidean topology.*

Proof. X is Hausdorff in the Euclidean topology by Theorem 10.41. Suppose that $p \in X$. Then there exists an affine neighborhood W of p such that $\mathbb{C}[W] = \mathbb{C}[x_1, \ldots, x_n]/\mathfrak{p}$ for some prime ideal $\mathfrak{p}$ in $\mathbb{C}[x_1, \ldots, x_n]$. We have a closed embedding $\lambda = (x_1, \ldots, x_n) : W \to \mathbb{C}^n$ which is a homeomorphism in the Euclidean topology onto its image since the restrictions of the x_i to W are regular functions on W and we have seen that they are thus continuous in the Euclidean topology. We can translate p so that $\lambda(p)$ is the origin in $\mathbb{C}^n$, and thus $I(\lambda(p)) = (x_1, \ldots, x_n)$. Let $\mathfrak{p} = (f_1, \ldots, f_m)$. Since X is nonsingular at p, the matrix

$$A = \left(\frac{\partial f_i}{\partial x_j}(\lambda(p)) \right)$$

has rank $n - r$. After possibly permuting the variables $x_1, \ldots, x_n$ and the functions $f_1, \ldots, f_m$, we may assume that if B is the $(n-r) \times (n-r)$ submatrix of A consisting of the first $n - r$ rows and columns of A, then $\text{Det}(B) \neq 0$, so that the classes of $f_1, \ldots, f_{n-r}$ are linearly independent over $\mathbb{C}$ in $I(p)/I(p)^2$. Let $R = \mathbb{C}[x_1, \ldots, x_n]_{(x_1, \ldots, x_n)}$. We can extend $f_1, \ldots, f_{n-r}$ to regular parameters $f_1, \ldots, f_{n-r}, g_1, \ldots, g_r$ in the regular local ring R. By Lemma 10.2, the ideal $I = (f_1, \ldots, f_{n-r})R$ is a prime ideal in R of height $n - r$. But $I \subset \mathfrak{p}R$ and I and $\mathfrak{p}R$ are prime ideals of the same height so

$$\mathfrak{p}\mathbb{C}[x_1, \ldots, x_n]_{(x_1, \ldots, x_n)} = (f_1, \ldots, f_{n-r})\mathbb{C}[x_1, \ldots, x_n]_{(x_1, \ldots, x_n)},$$

and thus

$$\mathfrak{p}\mathbb{C}\{x_1, \ldots, x_n\} = (f_1, \ldots, f_{n-r})\mathbb{C}\{x_1, \ldots, x_n\}.$$

Let

$$L_{r+i} = \sum_{j=1}^{n} \frac{\partial f_i}{\partial x_j}(p) x_j \text{ for } 1 \leq i \leq n - r$$

and let $L_1, \ldots, L_r$ be linear forms in $x_1, \ldots, x_n$ such that $\{L_1, \ldots, L_n\}$ is a basis of $\mathbb{C}x_1 + \cdots + \mathbb{C}x_n$. We have that $\mathbb{C}[x_1, \ldots, x_n] = \mathbb{C}[L_1, \ldots, L_n]$, so we may replace the x_i with L_i for $1 \leq i \leq n$. By Corollary 10.44, we have that the map

$$\mathbb{C}\{x_1, \ldots, x_r\} \to \mathbb{C}\{x_1, \ldots, x_n\}/\mathfrak{p}\mathbb{C}\{x_1, \ldots, x_n\}$$

is an isomorphism, so there exist functions $h_{r+1}, \ldots, h_n \in \mathbb{C}\{x_1, \ldots, x_r\}$ such that $x_i - h_i \in \mathfrak{p}\mathbb{C}\{x_1, \ldots, x_n\}$ for $r+1 \leq i \leq n$. All of the h_i converge within a polydisc

$$V = \{(a_1, \ldots, a_r) \in \mathbb{C}^r \mid ||a_i|| < \epsilon \text{ for all } i\}$$

10.8. Complex manifolds

for some small ϵ. Let $\pi : \mathbb{C}^n \to \mathbb{C}^r$ be the projection onto the first r factors (which is analytic), and let $\overline{U} = \pi^{-1}(V) \cap \lambda(W)$. Then $\pi : \overline{U} \to V$ has the analytic inverse

$$(a_1, \ldots, a_r) \to (a_1, \ldots, a_r, h_{r+1}(a_1, \ldots, a_r), \ldots, h_n(a_1, \ldots, a_r)),$$

for $(a_1, \ldots, a_r) \in V$. Let $U = \lambda^{-1}(\overline{U})$ and $\phi = \pi\lambda : U \to V$. The map ϕ is a homeomorphism in the Euclidean topology since it is a composition of homeomorphisms in the Euclidean topology.

Repeating this for every point $p \in X$, we obtain an open covering of X by open sets $\{U_i\}$ (in the Euclidean topology) with homeomorphisms $\phi_i : U_i \to V_i$, with V_i an open subset (in the Euclidean topology) of $\mathbb{C}^r$ as above. We will show that they satisfy the condition of the definition of a complex manifold. Now we introduce some notation on our construction of $\phi_i : U_i \to V_i$. There exists an affine open subset W_i of X such that U_i is an open subset of W_i (in the Euclidean topology). We have a representation

$$\mathbb{C}[W_i] = \mathbb{C}[x_{i,1}, \ldots, x_{i,n_i}]/\mathfrak{p}_i = \mathbb{C}[\overline{x}_{i,1}, \ldots, \overline{x}_{i,n_i}]$$

such that $\phi_i(p) = (\overline{x}_{i,1}(p), \ldots, \overline{x}_{i,r}(p))$ for $p \in U_i$ and there exist analytic functions h^i_j on V_i for $r+1 \le j \le n_i$ such that $\overline{x}_{i,j}(p) = h^i_j(\overline{x}_{i,1}(p), \ldots, \overline{x}_{i,r}(p))$.

Given $i \ne j$, there exist $f_k \in \mathbb{C}[W_i]$ such that $W_i \cap W_j = \bigcup_k (W_i)_{f_k}$. The restriction map

$$\mathbb{C}[W_j] \to \mathbb{C}[(W_i)_{f_k}] = \mathbb{C}[W_i]_{f_k} \text{ takes } \overline{x}_{j,l} \text{ to } \frac{g_l(\overline{x}_{i,1}, \ldots, \overline{x}_{i,n_i})}{f_k(\overline{x}_{i,1}, \ldots, \overline{x}_{i,n_i})^{t_l}},$$

for $1 \le l \le n_j$, where $t_l \in \mathbb{N}$ and g_l and f_k are polynomials in $\overline{x}_{i,1}, \ldots, \overline{x}_{i,n_i}$.

We will show that $\phi_j \phi_i^{-1} : \phi_i(U_i \cap U_j) \to \phi_j(U_i \cap U_j)$ is an analytic map, showing that X is a complex manifold. To show this, it suffices to show that

$$\phi_j \phi_i^{-1} : \phi_i((W_i)_{f_k} \cap U_i \cap U_j) \to \phi_j((W_i)_{f_k} \cap U_i \cap U_j)$$

is an analytic map for all k.

Suppose that $(a_1, \ldots, a_r) \in \phi_i((W_i)_{f_k} \cap U_i \cap U_j)$. Let $p = \phi_i^{-1}(a_1, \ldots, a_r)$. Then

$$\overline{x}_{i,1}(p) = a_1, \ldots, \overline{x}_{i,r}(p) = a_r$$

and

$$\overline{x}_{i,r+1}(p) = h^i_{r+1}(a_1, \ldots, a_r), \ldots, \overline{x}_{i,n_i}(p) = h^i_{n_i}(a_1, \ldots, a_r).$$

Thus

$$\phi_j \phi_i^{-1}(a_1, \ldots, a_r)$$
$$= \phi_j(p)$$
$$= (\overline{x}_{j,1}(p), \ldots, \overline{x}_{j,r}(p))$$
$$= \left(\frac{g_1(\overline{x}_{i,1}(p), \ldots, \overline{x}_{i,n_i}(p))}{f_k(\overline{x}_{i,1}(p), \ldots, \overline{x}_{i,n_i}(p))^{t_1}}, \ldots, \frac{g_r(\overline{x}_{i,1}(p), \ldots, \overline{x}_{i,n_i}(p))}{f_k(\overline{x}_{i,1}(p), \ldots, \overline{x}_{i,n_i}(p))^{t_r}} \right)$$
$$= (\sigma_1, \ldots, \sigma_r)$$

where

$$\sigma_b = \frac{g_b(a_1, \ldots, a_r, h_{r+1}^i(a_1, \ldots, a_r), \ldots, h_{n_i}^i(a_1, \ldots, a_r))}{f_k(a_1, \ldots, a_r, h_{r+1}^i(a_1, \ldots, a_r), \ldots, h_{n_i}^i(a_1, \ldots, a_r))^{t_1}}$$

for $1 \leq b \leq r$, showing that the homeomorphism $\phi_j \phi^{-1}$ is analytic on $(W_i)_k \cap U_i \cap U_j$. □

Exercise 10.46. We know that if X is a variety, then the diagonal $\Delta_X = \{(p, p) \mid p \in X\}$ is closed in $X \times X$ in the Zariski topology (proof of Proposition 5.8) and that X is not Hausdorff in the Zariski topology if X has positive dimension (proved for $\mathbb{A}^1$ in Example 2.3). Why does this not contradict Lemma 10.40?

Chapter 11

Sheaves

In this chapter we introduce the formalism of sheaves on a topological space. The most important concepts from this section are the invertible sheaves and coherent sheaves. Further discussion of sheaves can be found in Godement [59] and Hartshorne [73].

11.1. Limits

In this section we define direct and inverse limits of systems of algebraic structures.

A directed set I is a set with a partial order $\leq$ such that for any $i, j \in I$, there exists $k \in I$ such that $i \leq k$ and $j \leq k$.

A directed system of Abelian groups is a set of Abelian groups $\{A_i\}$, indexed by a directed set I, such that if $i \leq j$, then there is a homomorphism $\phi_{ij} : A_i \to A_j$ which satisfies $\phi_{ii} = \mathrm{id}_{A_i}$ and $\phi_{ik} = \phi_{jk}\phi_{ij}$ if $i \leq j \leq k$.

Proposition 11.1. *Suppose that $\{A_i\}_{i \in I}$ is a directed system of Abelian groups. Then there exists a group $\lim_{\to} A_i$ with homomorphisms $\phi_i : A_i \to \lim_{\to} A_i$ such that $\phi_i = \phi_j \phi_{ij}$ for all $i \leq j \in I$, which satisfies the following universal property: suppose that B is an Abelian group with homomorphisms $\tau_i : A_i \to B$ such that $\tau_i = \tau_j \phi_{ij}$ for $i \leq j \in I$. Then there exists a unique homomorphism $\tau : \lim_{\to} A_i \to B$ such that $\tau \phi_i = \tau_i$ for $i \in I$. The group $\lim_{\to} A_i$ with homomorphisms $\phi_i : A_i \to \lim_{\to} A_i$ is uniquely determined up to isomorphism. It is called the direct limit of $\{A_i\}$.*

The direct limit can be constructed as follows: let $M = \bigoplus_{i \in I} A_i$, and let N be the subgroup generated by elements of the form $a - \phi_{ij}(a)$ such that $a \in A_i$ and $i \leq j$. Let $\phi_i : A_i \to M/N$ be the natural map. Then M/N is the direct limit of the $\{A_i\}$.

We give an alternate construction of the direct limit. Let $\{A_i, \phi_{ij}\}$ be a directed system of Abelian groups. Let B be the disjoint union of pairs (A_i, a_i) such that $a_i \in A_i$. Define a relation $\sim$ on B by $(A_i, a_i) \sim (A_j, a_j)$ if there is a $k \geq i, j$ with $\phi_{ik}(a_i) = \phi_{jk}(a_j)$. The relation $\sim$ is an equivalence relation on B. Let $C = B/\sim$ be the set of equivalence classes. Let $[A_i, a_i]$ denote the equivalence class of (A_i, a_i). The set C is a group under the following operation:
$$[A_i, a_i] + [A_j, a_j] = [A_k, \phi_{ik}(a_i) + \phi_{jk}(a_j)]$$
where k is any index with $k \geq i, j$. Let $\phi_i : A_i \to C$ be the map $\phi_i(a) = [A_i, a]$ for $a \in A_i$. The map ϕ_i is a group homomorphism, which satisfies $\phi_i = \phi_j \phi_{ij}$ if $i \leq j$.

Lemma 11.2. *The group C with the homomorphisms ϕ_i constructed above is the direct limit* $\lim_\to A_i$.

Proof. We will show that C with the homomorphisms ϕ_i constructed above satisfies the universal property of Proposition 11.1. Suppose that G is an Abelian group with homomorphisms $\tau_i : A_i \to G$ such that $\tau_i = \tau_j \phi_{ij}$ for $i \leq j \in I$. Define $\tau : C \to G$ by the rule $\tau([A_i, a_i]) = \tau_i(a_i)$. This is well-defined since if $[A_i, a_i] = [A_j, a_j]$, then there is a k with $i, j \leq k$ and $\phi_{ik}(a_i) = \phi_{jk}(a_j)$. Thus
$$\tau_i(a_i) = \tau_k(\phi_{ik}(a_i)) = \tau_k(\phi_{jk}(a_j)) = \tau_j(a_j).$$
The map τ is a group homomorphism satisfying $\tau_i = \tau \phi_i$. If $\tau' : C \to G$ is a group homomorphism satisfying $\tau_i = \tau' \phi_i$ for each i, then $\tau'([A_i, a_i]) = \tau'(\phi_i(a_i)) = \tau_i(a_i) = \tau([A_i, a_i])$. Thus $\tau' = \tau$. Thus C satisfies the universal property of the direct limit, so C with the maps ϕ_i is the direct limit of the $\{A_i\}$. □

Proposition 11.3. *Let $\lim_\to A_i$ be the direct limit of a directed system of Abelian groups $\{A_i, \phi_{ij}\}$. Then:*

1) *Suppose that $x \in \lim_\to A_i$. Then $x = \phi_i(a)$ for some i and $a \in A_i$.*
2) *Suppose that $a \in A_i$ satisfies $\phi_i(a) = 0$. Then there is a $j \geq i$ such that $\phi_{ij}(a) = 0$.*

Proof. This follows easily from the construction C of $\lim_\to A_i$ given above. The first statement follows since every element of C has the form $[A_i, a_i] = \phi_i(a_i)$ for some i and $a_i \in A_i$. The second statement follows since if $[A_i, a_i] = 0$, then $(A_i, a_i) \sim (A_i, 0)$, so by the definition of the relation, there is a $j \geq i$ such that $\phi_{ij}(a_i) = \phi_{ij}(0) = 0$. □

11.1. Limits

The above construction and proofs carry through for commutative rings $\{A_i\}$. If M_i are modules over the A_i, then we can construct a direct limit of the modules $\{M_i\}$ which is a $\lim_\to A_i$ module. All rings and modules are Abelian groups, so we just apply the above construction and keep track of the fact that all of the groups and homomorphisms constructed have appropriate extra structure.

An important special case is when we have a direct system of groups, rings, or modules which are all subgroups, subrings, or submodules M_i of a larger group, ring, or module M and the maps in the system are inclusion maps. Then the direct limit is just the union $\bigcup M_i$ inside M.

Suppose that R, R' are rings and $\lambda : R \to R'$ is a homomorphism. Suppose M, N are R-modules and M', N' are R'-modules. Then λ makes M', N' and $M' \otimes_{R'} N'$ into R-modules. Suppose that $\phi : M \to M'$ and $\psi : N \to N'$ are R-module homomorphisms. We then have an R-bilinear map $M \times N \to M' \otimes_{R'} N'$ defined by $(x, y) \mapsto \phi(x) \otimes \psi(y)$ for $x \in M$ and $y \in N$. By the universal property of the tensor product (Definition 5.1), there exists a unique homomorphism of R-modules

$$(11.1) \qquad \phi \otimes \psi : M \otimes_R N \to M' \otimes_{R'} N'$$

satisfying $(\phi \otimes \psi)(x \otimes y) = \phi(x) \otimes \psi(y)$ for $x \in M$, $y \in N$.

Lemma 11.4. *Suppose that $\{R_\alpha\}_{\alpha \in I}$ is a directed system of rings and*

$$\{M_\alpha\}_{\alpha \in I}, \quad \{N_\alpha\}_{\alpha \in I}$$

are directed systems of R_α-modules. Then there is a natural isomorphism of $\lim_\to R_\alpha$-modules

$$\lim_\to (M_\alpha \otimes_{R_\alpha} N_\alpha) \cong \lim_\to M_\alpha \otimes_{\lim_\to R_\alpha} \lim_\to N_\alpha.$$

Proof. Let $\lambda_{\alpha'\alpha} : R_\alpha \to R_{\alpha'}$, $\phi_{\alpha'\alpha} : M_\alpha \to M_{\alpha'}$, $\psi_{\alpha'\alpha} : N_\alpha \to N_{\alpha'}$ for $\alpha < \alpha'$ be the homomorphisms in the directed systems. Let $R = \lim_\to R_\alpha$, $M = \lim_\to M_\alpha$, $N = \lim_\to N_\alpha$, $D = \lim_\to (M_\alpha \otimes_{R_\alpha} N_\alpha)$, with homomorphisms $\lambda_\alpha : R_\alpha \to R$, $\phi_\alpha : M_\alpha \to M$, $\psi_\alpha : N_\alpha \to N$, and $\chi_\alpha : M_\alpha \otimes_{R_\alpha} N_\alpha \to D$ for $\alpha \in I$.

The homomorphisms of (11.1), $\phi_\alpha \otimes \psi_\alpha : M_\alpha \otimes_{R_\alpha} N_\alpha \to M \otimes_R N$, induce homomorphisms $\mu : D \to M \otimes_R N$ by the universal property of limits (Proposition 11.1).

We will now define an R-bilinear homomorphism $u : M \times N \to D$. Suppose $x \in M$, $y \in N$. There exists an index $\alpha \in I$ and $x_\alpha \in M_\alpha$, $y_\alpha \in N_\alpha$ such that $x = \phi_\alpha(x_\alpha)$, $y = \psi_\alpha(y_\alpha)$ by Proposition 11.3. Define $u(x, y) = \chi_\alpha(x_\alpha \otimes y_\alpha) \in D$. Since the R_α, M_α, N_α, and $M_\alpha \otimes_{R_\alpha} N_\alpha$ are directed systems, $u(x, y)$ is independent of choices of α, x_α, y_α, so u is well-defined. Further, u is R-bilinear. Thus, by the universal property of tensor products, Definition 5.1, there exists a unique R-module homomorphism $\xi : M \otimes_R N \to D$ satisfying $\xi(x \times y) = u(x,y)$ for $x \in M$, $y \in N$. Here $\xi\mu$ and $\mu\xi$ are identity maps by our construction, so μ is an isomorphism with inverse ξ. $\square$

An inverse system of Abelian groups is a set of Abelian groups $\{A_i\}$, indexed by a directed set I, such that if $i \leq j$, then there is a homomorphism $\psi_{ij} : A_j \to A_i$ which satisfies $\psi_{ii} = \mathrm{id}_{A_i}$ and $\psi_{ik} = \psi_{ij}\psi_{jk}$ if $i \leq j \leq k$.

Proposition 11.5. *Suppose that $\{A_i\}_{i \in I}$ is an inverse system of Abelian groups. Then there exists a group $\varprojlim A_i$ with homomorphisms*

$$\psi_i : \varprojlim A_i \to A_i$$

such that $\psi_i = \psi_{ij}\psi_j$ for $i \leq j$ in I, which satisfies the following universal property: suppose that B is an Abelian group with homomorphisms $\sigma_i : B \to A_i$ such that $\sigma_i = \psi_{ij}\sigma_j$ for $i \leq j$ in I. Then there exists a unique homomorphism $\sigma : B \to \varprojlim A_i$ such that $\sigma_i = \sigma\psi_i$ for $i \in I$. The group $\varprojlim A_i$ with homomorphisms $\psi_i : \varprojlim A_i \to A_i$ is uniquely determined up to isomorphism. It is called the inverse limit of the system $\{A_i\}$.

The inverse limit can be defined as

$$(11.2) \qquad \varprojlim A_i = \left\{ (a_i) \in \prod_{i \in I} A_i \mid \psi_{ij}(a_j) = a_i \text{ for } i \leq j \right\}.$$

The construction of inverse limits carries through for inverse systems of commutative rings and for inverse systems of modules over a fixed commutative ring.

In the case when we have an inverse system of groups, rings, or modules which are all subgroups, subrings, or submodules M_i of a larger group, ring, or module M and the maps in the system are all inclusions maps, the inverse limit is the intersection $\bigcap M_i$ inside of M.

Exercise 11.6. If I is a directed set and J is a subset of I, then J is called cofinal in I if for every $i \in I$ there exists $j \in J$ such that $i \leq j$. Suppose $\{A_i\}_{i \in I}$ is a directed system of Abelian groups and J is a cofinal subset of I. Show that there is a natural isomorphism

$$\varinjlim A_j \to \varinjlim A_i$$

where the first limit is over the directed set J and the second limit is over the directed set I.

11.2. Presheaves and sheaves

In this section, we define presheaves and sheaves on a topological space.

Definition 11.7. Suppose that X is a topological space. A presheaf PF of Abelian groups on X associates to every open subset U of X an Abelian group $PF(U)$, with $PF(\emptyset) = 0$, and to every pair of open sets $U_1 \subset U_2$ a restriction map $\operatorname{res}_{U_2 U_1} : PF(U_2) \to PF(U_1)$ which is a group homomorphism and such that $\operatorname{res}_{U,U} = \operatorname{id}_{PF(U)}$ for all U and if $U_1 \subset U_2 \subset U_3$, then the diagram

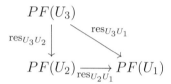

is commutative.

We will often write $f|U_1$ for $\operatorname{res}_{U_2 U_1}(f)$ and say that $f|U_1$ is the restriction of f to U_1. We will write $PF|U_1$ for the restriction of PF to U_1.

Definition 11.8. Suppose that PF_1 and PF_2 are presheaves of Abelian groups on a topological space X. A homomorphism of presheaves of Abelian groups $\phi : PF_1 \to PF_2$ is a collection of homorphisms $\phi(U) : PF_1(U) \to PF_2(U)$ for each open subset U of X, such that if $U \subset V$, then the diagram

$$\begin{array}{ccc} PF_1(V) & \xrightarrow{\phi(V)} & PF_2(V) \\ {\scriptstyle \operatorname{res}_{V,U}}\downarrow & & \downarrow{\scriptstyle \operatorname{res}_{V,U}} \\ PF_1(U) & \xrightarrow{\phi(U)} & PF_2(U) \end{array}$$

is commutative.

A homomorphism $\phi : PF_1 \to PF_2$ is an isomorphism if there exists a homomorphism $\psi : PF_2 \to PF_1$ such that $\psi(U) \circ \phi(U) = \operatorname{id}_{PF_1(U)}$ and $\phi(U) \circ \psi(U) = \operatorname{id}_{PF_2(U)}$ for all open subsets U of X.

The stalk of a presheaf PF at a point p is $\lim_{\to} PF(U)$, where we take the direct limit over the open sets U of X which contain p. Elements of PF_p are called germs. If $U \subset X$ is an open subset, $t \in PF(U)$, and $p \in U$, then t_p will denote the germ which is the image of t in PF_p.

If $\phi : PF_1 \to PF_2$ is a homomorphism of presheaves and $p \in X$, then for open neighborhoods $V \subset U$ of p, we have a natural commutative diagram

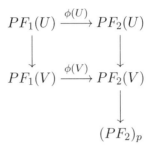

so we have a unique induced homomorphism of stalks $\phi_p : (PF_1)_p \to (PF_2)_p$ by the universal property of direct limits.

Lemma 11.9. *Suppose that PF is a presheaf on a topological space X, U is an open subset of X, and $p \in U$. Then $(PF|U)_p = PF_p$.*

Proof. By the universal property of direct limits of Proposition 11.1, we have a homomorphism $\pi : (PF|U)_p \to PF_p$ such that for V an open subset of U, we have a commutative diagram

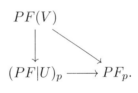

Now the injectivity and surjectivity of π follow from 2) and 1) of Proposition 11.3, respectively. Alternatively, the proof follows from Exercise 11.6. $\square$

Definition 11.10. A presheaf of Abelian groups F on a topological space X is a sheaf of Abelian groups on X if for every open subset U of X and collection $\{U_i\}_{i \in I}$ of open sets in X with $U = \bigcup U_i$:

1) If $x_1, x_2 \in F(U)$ and $\mathrm{res}_{U,U_i} x_1 = \mathrm{res}_{U,U_i} x_2$ for all i, then $x_1 = x_2$.

2) If $x_i \in F(U_i)$ for $i \in I$ are such that $\mathrm{res}_{U_i, U_i \cap U_j} x_i = \mathrm{res}_{U_j, U_i \cap U_j} x_j$ for all i and j, then there is an $x \in F(U)$ such that $\mathrm{res}_{U, U_i} x = x_i$ for all i.

We will often call condition 1) the "first sheaf axiom" and condition 2) the "second sheaf axiom".

It is common to denote $F(U)$ by $\Gamma(U, F)$. An element $\sigma \in \Gamma(X, F)$ is called a global section of F. If $\mathcal{F}$ and $\mathcal{G}$ are sheaves on a topological space X, then a homomorphism of sheaves of Abelian groups $\phi : \mathcal{F} \to \mathcal{G}$ is just

11.2. Presheaves and sheaves

a homomorphism of presheaves of Abelian groups, and an isomorphism of sheaves is an isomorphism of presheaves.

Example 11.11. Suppose that X is a quasi-projective variety. Then the association $U \mapsto \mathcal{O}_X(U)$ for U an open subset of X is a sheaf on X.

Proof. We have that $\mathcal{O}_X$ is a presheaf with restriction of functions. We will show that $\mathcal{O}_X$ is a sheaf. Suppose that U is an open subset of X and $\{U_i\}$ is an open cover of X. The restriction map $\mathcal{O}_X(U) \to \mathcal{O}_X(U_i)$ is injective, so we immediately get condition 1) of the definition of a sheaf. Now if $x_i \in \mathcal{O}_X(U_i)$ and $x_j \in \mathcal{O}_X(U_j)$ are such that they have the same restriction in $\mathcal{O}_X(U_i \cap U_j)$, then $x_i = x_j$ (as elements of $k(X)$). Thus if $\text{res}_{U_i, U_i \cap U_j} x_i = \text{res}_{U_j, U_i \cap U_j} x_j$ for all i, j, then the elements x_i are a common element $x \in \bigcap_i \mathcal{O}_X(U_i) = \mathcal{O}_X(U)$, so condition 2) of the definition of a sheaf holds. $\square$

In the above example, the stalk of $\mathcal{O}_X$ at a point $p \in X$ is the direct limit over open sets U containing p,

$$\lim_{p \in U} \mathcal{O}_X(U) = \bigcup_{p \in U} \mathcal{O}_X(U) = \mathcal{O}_{X,p},$$

as defined earlier.

Proposition 11.12. *Suppose that PF is a presheaf of Abelian groups on a topological space X. Then there is a sheaf of Abelian groups F on X and a homomorphism $f : PF \to F$ of presheaves such that if F' is a sheaf of Abelian groups on X and $g : PF \to F'$ is a homomorphism of presheaves on X, then there is a unique homomorphism of sheaves $h : F \to F'$ such that $g = hf$.*

The sheaf F of Proposition 11.12 is uniquely determined up to isomorphism (since it satisfies the stated universal property). It is called the sheafification of PF. The sheafification F of PF has the property that the stalks $PF_p = F_p$ for all $p \in X$.

Proof. For an open subset U of X, define $F(U)$ to be the set of maps

$$s : U \to \coprod PF_p$$

where $\coprod PF_p$ is the disjoint union of the stalks PF_p for $p \in X$ such that:

1) For each $p \in U$, $s(p) \in PF_p$.
2) For each $p \in U$, there exists a neighborhood V of p contained in U and an element $t \in PF(V)$ such that for all $Q \in V$, the germ $t_Q = s(Q)$.

We have that F is a presheaf of Abelian groups with the natural restriction maps (since each stalk PF_p is a group). Suppose that U is an open subset of X and U_i are open subsets of X with $\bigcup_i U_i = U$. The first sheaf axiom holds since $F(U)$ is completely determined by its stalks (condition 1) above). Since $\bigcup U_i = U$, elements $x_i \in F(U_i)$ satisfying the assumptions of the second sheaf axiom induce a well-defined map $x : U \to \coprod PF_p$ by prescribing $x(p) = (x_i)_p \in PF_p$ if $p \in U_i$. Since the $x_i \in F(U_i)$ satisfy condition 2) above on U_i for all i, we have that x also satisfies condition 2), and thus $x \in F(U)$ and the second sheaf axiom is satisfied. Thus $F(U)$ is a sheaf. We have a natural homomorphism $f : PF \to F$ of presheaves defined by mapping $t \in PF(U)$ to the map $p \mapsto t_p$ for $p \in U$.

Now suppose that F' is a sheaf of Abelian groups on X and $g : PF \to F'$ is a homomorphism of presheaves. The extension $h : F \to F'$ is defined as follows. Suppose that U is an open subset of X and $s \in F(U)$. By condition 2), there exists an open cover $\{U_i\}$ of U and $t_i \in PF(U_i)$ such that $s|U_i = f(t_i)$. We necessarily have that $s|U_i \cap U_j = f(t_i)|U_i \cap U_j = f(t_j)|U_i \cap U_j$ for all i, j. Thus the germs $(t_i)_p = (t_j)_p$ in $PF_p = F_p$ for all $p \in U_i \cap U_j$.

Let $u_i = g(U_i)(t_i) \in F'(U_i)$. Suppose that $p \in U_i \cap U_j$. Then the germ $(u_i)_p = g_p((t_i)_p) = g_p((t_j)_p) = (u_j)_p$. Using the second statement of Proposition 11.3, we find that there exists an open neighborhood V_p of p in $U_i \cap U_j$ such that $u_i \mid V_p = u_j \mid V_p$. Thus $u_i|U_i \cap U_j = u_j|U_i \cap U_j$ since F' is a sheaf (the first sheaf axiom) and thus there exists $a \in F'(U)$ such that $a|U_i = u_i = g(U_i)(t_i)$ for all i, again since F' is a sheaf (the second sheaf axiom).

We define $h(U)(s) = a$. The element a is uniquely determined by the first sheaf axiom since we must have that $h(U_i)(s|U_i) = h(U_i)(f(t_i)) = g(U_i)(t_i) = u_i$ for all i. $\square$

Proposition 11.13. *Let $\phi : F \to G$ be a homomorphism of sheaves of Abelian groups on a topological space X. Then ϕ is an isomorphism if and only if the induced map on stalks $\phi_p : F_p \to G_p$ is an isomorphism for every $p \in X$.*

Proof. If ϕ is an isomorphism, then $\phi(U)$ is a group isomorphism for all open subsets U of X so ϕ_p is an isomorphism for all $p \in X$.

Suppose that ϕ_p is an isomorphism for all $p \in X$. To show that ϕ is an isomorphism, we will show that $\phi(U)$ is an isomorphism for all open subsets U of X. We can then define the inverse map ψ to ϕ by defining $\psi(U)$ to be the inverse to $\phi(U)$ for all open subsets U of X. To show that $\phi(U)$ is an isomorphism, we must show that $\phi(U)$ is injective and surjective.

We will first show that $\phi(U)$ is injective. Suppose that $s \in F(U)$ and $\phi(U)(s) = 0$. Then for all $p \in U$, $0 = \phi(U)(s)_p = \phi_p(s_p) = 0$ in G_p. Thus

11.2. Presheaves and sheaves

for all $p \in U$, $s_p = 0$ in F_p since ϕ_p is injective. Thus there exists an open neighborhood W_p of p in U such that $s|W_p = 0$, by 2) of Proposition 11.3. The open set U is covered by the open sets W_p for $p \in U$, so by the first sheaf axiom, $s = 0$ in $F(U)$. Thus $\phi(U)$ is injective.

Now we will show that $\phi(U)$ is surjective. Suppose $t \in G(U)$. For $p \in U$, let $t_p \in G_p$ be the germ of t at p. Since ϕ_p is surjective, there exists $s_p \in F_p$ such that $\phi_p(s_p) = t_p$. By 1) of Proposition 11.3, there exists an open neighborhood V_p of p in U and $h(p) \in F(V_p)$ such that the germ of $h(p)$ at p is s_p. By 2) of Proposition 11.3, $\phi_p(h(p)_p) = t_p$ implies that there exists a neighborhood W_p of p in V_p such that $\phi(W_p)(h(p)|W_p) = t|W_p$. So replacing V_p with W_p, we may assume that $\phi(V_p)(h(p)) = t|V_p$. If $p, q \in U$ are two points, then

$$\phi(V_p \cap V_q)(h(p)) = t|V_p \cap V_q = \phi(V_p \cap V_q)(h(q)).$$

Since $\phi(V_p \cap V_q)$ was shown to be injective, we have that $h(p)|V_p \cap V_q = h(q)|V_p \cap V_q$. Thus by the second sheaf axiom, there exists $s \in F(U)$ such that $s|V_p = h(p)$ for all $p \in U$. Now

$$\phi(U)(s)|V_p = \phi(V_p)(h(p)) = t|V_p$$

for all $p \in X$ and $\{V_p\}$ is an open cover of U so $\phi(U)(s) = t$ by the first sheaf axiom. $\square$

Proposition 11.14 (The constant sheaf). *Suppose that G is an Abelian group and X is a topological space which has the property that if U is an open subset of X, then all connected components of U are open. Define a presheaf PG_X on X by*

$$PG_X(U) = G$$

whenever U is a nonempty subset of X. Define the restriction maps to be the identity. Let G_X be the sheaf on X associated to PG_X.

Suppose that U is an open subset of X and $\{U_i\}_{i \in I}$ are the connected components of U. Then

$$G_X(U) \cong \prod_{i \in I} G.$$

Proof. Let $PF = PG_X$ and $F = G_X$. By Proposition 11.12, there is a homomorphism of presheaves $\lambda : PF \to F$ such that for all open subsets U of X and $p \in U$, there is a commutative diagram

$$\begin{array}{ccc} PF(U) & \to & F(U) \\ \downarrow & & \downarrow \\ PF_p & \stackrel{\cong}{\to} & F_p. \end{array}$$

Since $PF(U) = PF_p = F_p = G$, we have that $PF(U) \to F(U)$ is an injective homomorphism for all open subsets of X. In particular, this gives us natural inclusions $\lambda_U : G \to F(U)$ for all U, which is compatible with restriction for $V \subset U$,

$$\begin{array}{ccc} G & \xrightarrow{\lambda_U} & F(U) \\ & \lambda_V \searrow & \downarrow \\ & & F(V). \end{array}$$

We thus have that the restriction map $F(U) \to F_p = G$ defined by $\phi \mapsto \phi_p$ is surjective for all open subsets U of X which contain p.

Suppose that $p \in X$ and U is an open subset of X containing p. Suppose that $\phi \in F(U)$. Let $g = \phi_p$. Then $(\phi - \lambda_U(g))_p = 0$. We have that

$$F_p = \varinjlim F(V),$$

where the limit is over open subsets V of U containing p. Thus by the second statement of Proposition 11.3, there exists an open subset V of U containing p such that the restriction $(\phi - \lambda_V(g))|V = 0$. Thus $\phi|V = \lambda_V(g)$, so that $\phi_q = (\lambda_V(g))_q = g$ for all $q \in V$.

We have shown that given an open subset U of X and $\phi \in F(U)$, the map $U \to G$ defined by $p \mapsto \phi_p$ is continuous if we give G the discrete topology (a point of G is an open set).

Suppose that U is a connected open subset of X. For $g \in G$ and $\phi \in F(U)$,

$$W_g^\phi = \{q \in U \mid \phi_q = g\}$$

is an open subset of U. Further, $\{W_g^\phi \mid g \in G\}$ is an open cover of U by disjoint open sets. Thus there exists $g \in G$ such that $W_g^\phi = U$. In particular if we take any $p \in U$, $\phi_q = \phi_p$ for all $q \in U$.

Consider the restriction homomorphism

$$\Psi : F(U) \to G$$

defined by $\Psi(\phi) = \phi_p$.

Suppose that $\phi_p = 0$. Then $\phi_q = 0$ for all $q \in U$, and so there exists an open cover $\{V_j\}$ of U such that $\phi|V_j = 0$ for all j (by the second statement of Proposition 11.3). Thus $\phi = 0$ by the first sheaf axiom. We have already established that Ψ is onto. We have shown that if U is a connected open set and $p \in U$, then the restriction map $F(U) \to G$ is an isomorphism.

Suppose that U is an open subset of X, and let $\{U_i\}_{i \in I}$ be the connected components of X, which by assumption are open. Consider the group homomorphism

$$\Lambda : F(U) \to \prod_{i \in I} F(U_i)$$

defined by $\phi \mapsto \{\phi|U_i\}$. By the first sheaf axiom Λ is injective. Since $U_i \cap U_j = \emptyset$ if $i \neq j$, so that $F(U_i \cap U_j) = (0)$, we have by the second sheaf axiom that Λ is onto. Thus Λ is an isomorphism. Thus
$$F(U) \cong \prod_{i \in I} G. \qquad \square$$

Suppose that $f : X \to Y$ is a continuous map of topological spaces and F is a sheaf of groups on X. We define a presheaf f_*F on Y by $f_*F(U) = F(f^{-1}(U))$ for U an open subset of Y. Then f_*F is actually a sheaf.

We can now recognize that the analysis in the proof of Theorem 10.36 is actually of the sheaf $\phi_*\mathcal{O}_x$.

We extend our definitions of presheaves and sheaves of Abelian groups to presheaves and sheaves of rings.

Definition 11.15. A locally ringed space is a pair $(X, \mathcal{O}_X)$ where X is a topological space and $\mathcal{O}_X$ is a sheaf of rings on X such that for each $p \in X$ the stalk $\mathcal{O}_{X,p}$ is a local ring. The space X is called the underlying topological space of $(X, \mathcal{O}_X)$ and $\mathcal{O}_X$ is called the structure sheaf of X. A morphism of locally ringed spaces from $(X, \mathcal{O}_X)$ to $(Y, \mathcal{O}_Y)$ is a pair $(f, f^\#)$ such that $f : X \to Y$ is a continuous map and $f^\# : \mathcal{O}_Y \to f_*\mathcal{O}_X$ is a map of sheaves of rings on Y such that for all $p \in X$, the induced map $f_p^\# : \mathcal{O}_{Y, f(p)} \to (f_*\mathcal{O}_X)_{f(p)} \to \mathcal{O}_{X,p}$ is a local homomorphism of local rings. A morphism of locally ringed spaces $X \to Y$ is an isomorphism if there exists a morphism of locally ringed spaces $Y \to X$ which is a two-sided inverse.

If X is a quasi-projective variety, then X with its sheaf of regular functions $\mathcal{O}_X$ is a locally ringed space. If X and Y are two quasi-projective varieties, then the regular maps $\varphi : X \to Y$ are morphisms, with $\varphi^\#$ induced by φ^*.

Definition 11.16. Suppose that X is a locally ringed space. A sheaf of $\mathcal{O}_X$-modules (an $\mathcal{O}_X$-module) is a sheaf $\mathcal{F}$ on X such that for each open subset U of X the group $\mathcal{F}(U)$ is an $\mathcal{O}_X(U)$-module, and for each inclusion of open subsets $V \subset U$ of X, the restriction homomorphism $\mathcal{F}(U) \to \mathcal{F}(V)$ is compatible with the module structures by the restriction ring homomorphism $\mathcal{O}_X(U) \to \mathcal{O}_X(V)$. A homomorphism $\mathcal{F} \to \mathcal{G}$ of sheaves of $\mathcal{O}_X$-modules is a homomorphism of sheaves such that for all open subsets U of X, $\mathcal{F}(U) \to \mathcal{G}(U)$ is a homomorphism of $\mathcal{O}_X(U)$-modules.

If $f : X \to Y$ is a morphism of locally ringed spaces and $\mathcal{F}$ is a sheaf of $\mathcal{O}_X$-modules, then $f_*\mathcal{F}$ is a sheaf of $\mathcal{O}_Y$-modules.

A subsheaf of a sheaf $\mathcal{F}$ is a sheaf $\mathcal{F}'$ such that for every open subset U of X, $\mathcal{F}'(U)$ is a subgroup of $\mathcal{F}(U)$ and the restriction maps of the sheaf

$\mathcal{F}'$ are induced by those of $\mathcal{F}$. If $\mathcal{F}$ and $\mathcal{F}'$ are $\mathcal{O}_X$-modules, then $\mathcal{F}'$ is a sub-$\mathcal{O}_X$-module, or an $\mathcal{O}_X$-submodule of $\mathcal{F}$ if $\mathcal{F}'$ is a subsheaf of $\mathcal{F}$ such that $\mathcal{F}'(U)$ is an $\mathcal{O}_X(U)$-submodule of $\mathcal{F}(U)$ for all open subsets U of X.

Example 11.17 (Ideal sheaf, on an affine variety). Suppose that X is an affine variety and $J \subset k[X]$ is an ideal. We associate to J a presheaf $\tilde{J}$ on X defined by
$$\tilde{J}(U) = \bigcap_{p \in U} J_{I(p)}$$
for an open subset U of X, where $I(p)$ is the ideal of p in $k[X]$ and
$$J_{I(p)} = Jk[X]_{I(p)} = J\mathcal{O}_{X,p}$$
is the localization of J at the maximal ideal $I(p)$.

Using the definition of $\tilde{J}$ in Example 11.17, we verify that $\tilde{J}$ is a sheaf of $\mathcal{O}_X$-modules. We calculate that the stalk
$$\tilde{J}_p = J_{I(p)}$$
for $p \in X$. We also observe that, by the definition, if $U \subset X$ is an open set, then
$$\tilde{J}(U) = J\mathcal{O}_X(U),$$
and if U is an affine open subset of X, then the restriction $\tilde{J}|U$ is in fact the tilde on U of the ideal $Jk[U]$ in $k[U]$; that is, $\tilde{J}|U = \widetilde{Jk[U]}$.

Example 11.18 (Ideal sheaf, on a projective variety). Suppose that $X \subset \mathbb{P}^n$ is a projective variety, with homogeneous coordinates $x_0, \ldots, x_n$ and homogeneous coordinate ring $S(X)$. We may suppose that none of the x_i vanish everywhere on X. Suppose that J is a homogeneous ideal in $S(X)$. We associate to J a presheaf $\tilde{J}$ on X defined by
$$\tilde{J}(U) = \bigcap_{p \in U} J_{(I(p))}$$
for an open subset U of X, where the ideal $J_{(I(p))}$ in $\mathcal{O}_{X,p} \subset k(X)$ is defined to be the elements of degree 0 in the localization $T^{-1}J$ where T is the multiplicative set of homogeneous elements of $S(X)$ which are not in $I(p)$.

Using the definition of $\tilde{J}$ in Example 11.18, we verify that $\tilde{J}$ is a sheaf of $\mathcal{O}_X$-modules. We calculate that the stalk
$$\tilde{J}_p = J_{(I(p))}$$
for $p \in X$.

Looking back over the analysis we made of $\mathcal{O}_X$ in Section 3.2, we find that

(11.3) $\tilde{J}(X_{x_i}) = J_{(x_i)}$ is the dehomogenization of J for $0 \leq i \leq n$.

Here $J_{(x_i)}$ denotes the elements of degree 0 in the localization J_{x_i}. In particular, if the ideal J has the homogeneous generators $F_1, \ldots, F_m$, then $\tilde{J}(X_{x_i})$ is the ideal in $\mathcal{O}_X(X_{x_i}) = k[\frac{x_0}{x_i}, \ldots, \frac{x_n}{x_i}]$ generated by

$$F_1\left(\frac{x_0}{x_i}, \ldots, \frac{x_n}{x_i}\right), \ldots, F_m\left(\frac{x_0}{x_i}, \ldots, \frac{x_n}{x_i}\right).$$

We deduce that the restriction of $\tilde{J}$ to X_{x_i} is the tilde on the affine variety X_{x_i} of the ideal $\Gamma(X_{x_i}, \tilde{J})$ in $k[X_{x_i}]$; that is,

$$\tilde{J}|X_{x_i} = \widetilde{\Gamma(X_{x_i}, \tilde{J})}.$$

In fact, if $U \subset X$ is any affine open subset, then the restriction of $\tilde{J}$ to U is the tilde on U of the ideal $\Gamma(J, U) \subset k[U]$; that is,

$$\tilde{J}|U = \widetilde{\Gamma(U, \tilde{J})}.$$

The ideal sheaf $\tilde{I}$ which we have just defined has been previously encountered in Chapter 6 (before Lemma 6.7).

Suppose that $\mathcal{A}$ is a subsheaf of a sheaf $\mathcal{B}$ on a topological space X. Then $\mathcal{B}/\mathcal{A}$ will denote the sheaf associated to the presheaf $U \mapsto \mathcal{B}(U)/\mathcal{A}(U)$ for U an open subset of X.

Lemma 11.19. *Suppose that $\mathcal{A}$ is a subsheaf of a sheaf $\mathcal{B}$ on a topological space X and $p \in X$. Then $(\mathcal{B}/\mathcal{A})_p \cong \mathcal{B}_p/\mathcal{A}_p$.*

Proof. We have that the stalk of the sheaf $\mathcal{B}/\mathcal{A}$ at p is the stalk at p of the presheaf $U \mapsto \mathcal{B}(U)/\mathcal{A}(U)$ for U an open subset X by the comment after Proposition 11.12.

By the universal property of direct limits, Proposition 11.1, there is a unique homomorphism $\mathcal{B}_p \to (\mathcal{B}/\mathcal{A})_p$ such that for all open subsets U of X containing p, we have a commutative diagram

$$\begin{array}{ccc} \mathcal{B}(U) & \to & \mathcal{B}(U)/\mathcal{A}(U) \\ \downarrow & & \downarrow \\ \mathcal{B}_p & \to & (\mathcal{B}/\mathcal{A})_p \end{array}$$

which is compatible with restriction. Now

$$0 \to \mathcal{A}(U) \to \mathcal{B}(U) \to \mathcal{B}(U)/\mathcal{A}(U) \to 0$$

is exact for all open subsets U of X so a diagram chase using Proposition 11.3 shows that

$$0 \to \mathcal{A}_p \to \mathcal{B}_p \to (\mathcal{B}/\mathcal{A})_p \to 0$$

is exact. □

If $\alpha : \mathcal{F} \to \mathcal{G}$ is a homomorphism of sheaves on a topological space X, then we have sheaves Kernel(α), Image(α) and Cokernel(α) associated to the respective presheaves defined by $U \mapsto \text{Kernel}(\alpha(U))$, $U \mapsto \text{Image}(\alpha(U))$, and $U \mapsto \text{Cokernel}(\alpha(U))$ for U an open subset of X.

A homomorphism $\alpha : \mathcal{F} \to \mathcal{G}$ of sheaves on a topological space X is called injective if $\alpha_p : \mathcal{F}_p \to \mathcal{G}_p$ is injective for all $p \in X$. The homomorphism α is called surjective if $\alpha_p : \mathcal{F}_p \to \mathcal{G}_p$ is surjective for all $p \in X$. A sequence

(11.4) $$0 \to \mathcal{A} \to \mathcal{B} \to \mathcal{C} \to 0$$

of homomorphisms of sheaves is called short exact if for all $p \in X$, the sequence of homomorphisms of groups

$$0 \to \mathcal{A}_p \to \mathcal{B}_p \to \mathcal{C}_p \to 0$$

is short exact. Proving the following proposition is Exercise 11.22.

Proposition 11.20. *Suppose that (11.4) is a short exact sequence of sheaves and U is an open subset of X. Then*

(11.5) $$0 \to \mathcal{A}(U) \to \mathcal{B}(U) \to \mathcal{C}(U)$$

is exact. In particular, the presheaf $U \mapsto \text{Kernel}(\alpha(U))$ is a sheaf for any homomorphism of sheaves $\alpha : \mathcal{F} \to \mathcal{G}$.

Example 11.21. Suppose that $\mathcal{A}$ is a subsheaf of a sheaf $\mathcal{B}$. Then the presheaf $U \mapsto \mathcal{B}(U)/\mathcal{A}(U)$ may not be a sheaf. Further, if $\alpha : \mathcal{A} \to \mathcal{B}$ is a homomorphism of sheaves, then the presheaf $U \mapsto \text{Image}(\alpha(U))$ may not be a sheaf, and if (11.4) is a short exact sequence of sheaves, then $\mathcal{B}(U) \to \mathcal{C}(U)$ may not be surjective for some open subset U of X.

Proof. Let p be a nonsingular point on a projective curve X. Let $\mathcal{I}_p$ be the ideal sheaf of p in X. For $q \in X$, we have that

$$\mathcal{I}_{p,q} = \begin{cases} \mathcal{O}_{X,q} & \text{if } q \neq p, \\ \text{the maximal ideal } m_p \subset \mathcal{O}_{X,p} & \text{if } q = p. \end{cases}$$

Let PF be the presheaf defined by

$$PF(U) = \mathcal{I}_p(U)/\mathcal{I}_p(U)^2$$

for U an open set in X. Let F be the sheaf associated to PF. We calculate, from Lemma 11.19, that the stalk

$$F_q = \begin{cases} 0 & \text{if } q \neq p, \\ m_p/m_p^2 \cong k & \text{if } q = p. \end{cases}$$

We will now establish that for an open subset U of X,

(11.6) $$F(U) = \begin{cases} k & \text{if } p \in U, \\ 0 & \text{if } p \notin U. \end{cases}$$

11.2. Presheaves and sheaves

Suppose that $p \notin U$ and $x \in F(U)$. Then for each $q \in U$, the stalk $F_q = 0$. Thus the image of x in F_q is zero, so there exists an open neighborhood U_q of q, which is contained in U, such that the restriction of x in $F(U_q)$ is zero (by part 2) of Proposition 11.3). Thus we have a cover $\{U_q\}$ of U by open subsets of U such that the restriction of x in each $F(U_q)$ is zero. Now $0 \in F(U)$ also has this property, so $x = 0$ by the first sheaf axiom. We have established that $F(U) = (0)$ if $p \notin U$.

Suppose that $p \in U$. We will show that the restriction homomorphism $\Lambda : F(U) \to F_p \cong k$ is an isomorphism. Let $r \in k$. Then there exist an open neighborhood U_p of p in U and $\phi \in F(U_p)$ such that the restriction of ϕ to F_p is r (by part 1) of Proposition 11.3). Let $V = U \setminus \{p\}$, which is an open subset of U. The point $p \notin V$, so $F(V) = F(V \cap U_p) = (0)$. Since the restriction of ϕ to $F(V \cap U_p)$ is zero, which is the restriction of $0 \in F(V)$ to $F(V \cap U_p)$, we have by the second sheaf axiom that there exists $x \in F(U)$ which restricts to 0 in $F(V)$ and restricts to ϕ in $F(U_p)$ so necessarily restricts to $r \in F_p$. Thus Λ is surjective.

Suppose that $x' \in F(U)$ and $\Lambda(x') = 0$. Then there exists an open neighborhood U'_p of p in U such that the restriction of x' to $F(U'_p)$ is zero, by 2) of Proposition 11.3. Since the restriction of x' to $V = U \setminus \{p\}$ is necessarily zero, we have that $x' = 0$ in $F(U)$ by the first sheaf axiom. Thus Λ is injective and is necessarily an isomorphism.

From the natural inclusions of sheaves of $\mathcal{O}_X$-modules, $\mathcal{I}_p^2 \subset \mathcal{I}_p \subset \mathcal{O}_X$, we have (by applying Proposition 11.20 to each inclusion) inclusions of modules

$$\mathcal{I}_p^2(X) \subset \mathcal{I}_p(X) \subset \mathcal{O}_X(X).$$

Now $\mathcal{O}_X(X) = k$ (by Theorem 3.35). Further, $\mathcal{I}_p(X)$ cannot contain a nonzero element of k since every element of $\mathcal{I}_p(X)$ must restrict to an element of $\mathcal{I}_{p,p} = m_p$ and hence must vanish at p. Thus $\mathcal{I}_p^2(X) = \mathcal{I}_p(X) = 0$, and

$$PF(X) = \mathcal{I}_p(X)/\mathcal{I}_p^2(X) = 0.$$

In contrast, by (11.6), $F(X) \cong k$. Thus $F \neq PF$.

We have (by Proposition 11.12) a natural exact sequence of sheaves

$$0 \to \mathcal{I}_p^2 \to \mathcal{I}_p \xrightarrow{\alpha} F \to 0.$$

The image of α is the presheaf PF, which we have already established is not a sheaf. Further, the evaluation of the above short exact sequence at X is

$$0 \to 0 \to 0 \to k \to 0$$

which is not short exact. $\square$

Exercise 11.22. Prove Proposition 11.20.

Exercise 11.23. Let X be a topological space, let $\{U_i\}$ be an open cover of X, and suppose that for each i we have a sheaf $\mathcal{F}_i$ on U_i, and for each i, j an isomorphism $\phi_{ij} : \mathcal{F}_i|U_i \cap U_j \to \mathcal{F}_j|U_i \cap U_j$ such that for each i, $\phi_{ii} = \text{id}$ and for each i, j, k, $\phi_{ik} = \phi_{jk} \circ \phi_{ij}$ on $U_i \cap U_j \cap U_k$. Show that there exists a unique sheaf $\mathcal{F}$ on X, with isomorphisms $\psi_i : \mathcal{F}|U_i \to \mathcal{F}_i$ such that for each i, j, $\psi_j = \phi_{ij} \circ \psi_i$ on $U_i \cap U_j$.

11.3. Some sheaves associated to modules

Theorem 11.24 (Sheafification of a module on an affine variety)). *Suppose that X is an affine variety and M is a $k[X]$-module. Then there is a unique sheaf $\tilde{M}$ of $\mathcal{O}_X$-modules on X which has the property that*

$$\text{(11.7)} \qquad \tilde{M}(X_f) = M_f \text{ for } f \in k[X],$$

and the restriction map $\tilde{M}(X) = \tilde{M}(X_1) = M \to \tilde{M}(X_f) = M_f$ is the map $a \mapsto \frac{a}{1}$.

For $p \in X$, the stalk

$$\tilde{M}_p = \lim_{\substack{\longrightarrow \\ p \in U}} \tilde{M}(U) = \lim_{\substack{\longrightarrow \\ f \in k[X] \setminus I(p)}} M_f = M_{I(p)},$$

where $I(p)$ is the maximal ideal in $k[X]$ of the point p and $M_{I(p)}$ is the localization of M at this prime ideal.

Proof. The property (11.7) and the sheaf axioms uniquely determine $\tilde{M}(U)$ for U an arbitrary open subset of X. In fact, U is a finite union of open sets $U = X_{f_1} \cup X_{f_2} \cup \cdots \cup X_{f_n}$ for some $f_i \in k[X]$. By the sheaf axioms, we have an exact sequence of $k[X]$-module homomorphisms

$$\text{(11.8)} \qquad 0 \to \tilde{M}(U) \xrightarrow{\alpha} \bigoplus_{i=1}^{n} \tilde{M}(X_{f_i}) \xrightarrow{\beta} \bigoplus_{1 \le i < j \le n} \tilde{M}(X_{f_i} \cap X_{f_j}),$$

where

$$\alpha(g) = \{\text{res}_{U, X_{f_i}}(g)\}$$

for $g \in \tilde{M}(U)$ and

$$\beta(\{h_l\}) = \{\text{res}_{X_{f_j}, X_{f_i} \cap X_{f_j}}(h_j) - \text{res}_{X_{f_i}, X_{f_i} \cap X_{f_j}}(h_i)\}$$

for $\{h_l\} \in \bigoplus_{i=1}^{n} \tilde{M}(X_{f_i})$.

Since $X_{f_i} \cap X_{f_j} = X_{f_i f_j}$, (11.8) tells us that $\tilde{M}(U)$ can be identified with the kernel of β, which has the explicit form

$$\bigoplus_{i=1}^{n} M_{f_i} \xrightarrow{\beta} \bigoplus_{1 \le i < j \le n} M_{f_i f_j}.$$

11.3. Some sheaves associated to modules

We now construct the sheaf $\tilde{M}$ satisfying (11.7). Let U be an open subset of the affine variety X, with regular functions $A = k[X]$. Define $\tilde{M}(U)$ to be the set of functions $s : U \to \coprod_{p \in U} M_{I(p)}$ such that $s(p) \in M_{I(p)}$ for each p in U and such that for each $p \in U$ there is an open neighborhood V of p in U, $a \in M$, and $f \in A$ such that $Z(f) \cap V = \emptyset$ and for each $q \in V$, $s(q) = \frac{a}{f} \in M_{I(q)}$. We have that $\tilde{M}$ is a sheaf on X as it satisfies the sheaf axioms.

For U an open subset of X, $\mathcal{O}_X(U) = \bigcap_{p \in U} A_{I(p)} \subset k(X)$. It follows that the map of sheaves $\mathcal{O}_X \to \tilde{A}$ defined by associating to $t \in \mathcal{O}_X(U)$ the map $s : U \to \tilde{A}(U)$ defined by $s(p) = t$ for $p \in U$ is an isomorphism of sheaves of k-algebras. We thus have that $\tilde{A} \cong \mathcal{O}_X$.

We have that property (11.7) holds for $\tilde{A}$ by Proposition 2.84. For A-modules M, the sheaf $\tilde{M}$ is naturally a sheaf of $\tilde{A} \cong \mathcal{O}_X$-modules.

The fact that property (11.7) holds for $\tilde{M}$ follows from a careful analysis of the natural map $M_f \to \tilde{M}(D(f))$ for $f \in A$, as we now verify. Define an A_f-module homomorphism

$$\psi : M_f \to \tilde{M}(D(f))$$

by $\psi(\frac{a}{f^n}) = s$ where $s(p) = \frac{a}{f^n} \in M_{I(p)}$ for $p \in D(f)$.

We first will show that ψ is injective. Suppose that $\psi(\frac{a}{f^n}) = \psi(\frac{b}{f^m})$ for some $\frac{a}{f^n}, \frac{b}{f^m} \in M_f$. Then $\frac{a}{f^n} = \frac{b}{f^m}$ in $M_{I(p)}$ for all $p \in D(f)$, so for each p, there exists $h_p \in A$ with $h_p \notin I(p)$ such that $h_p(f^m a - f^n b) = 0$ in M. Let $J = \mathrm{Ann}(f^m a - f^n b)$ be the annihilator of $f^m a - f^n b$ in A. Then $h_p \in J$ and $h_p \notin I(p)$ so J is not contained in $I(p)$. Since this is true for all $p \in D(f)$, we have $Z(J) \cap D(f) = \emptyset$. Thus $f \in \sqrt{J}$ so $f^l \in J$ for some positive power l, and so $f^l(f^m a - f^n b) = 0$, implying $\frac{a}{f^n} = \frac{b}{f^m}$ in M_f, and so ψ is injective.

Now we will show that ψ is surjective. Let $s \in \tilde{M}(D(f))$. We can cover $D(f)$ with open sets V_i so that there are $a_i \in M$ and $g_i \in A$ such that $Z(g_i) \cap V_i = \emptyset$ and $s(p) = \frac{a_i}{g_i} \in M_{I(p)}$ for all $p \in V_i$. The open sets $D(h)$ with $h \in A$ are a basis of the topology of X (Lemma 2.83) so we may assume that each $V_i = D(h_i)$ for some $h_i \in A$. We have $D(h_i) \subset D(g_i)$ for all i so $\sqrt{(h_i)} \subset \sqrt{(g_i)}$ by the nullsetellensatz. Thus $h_i^{n_i} = c_i g_i$ for some positive power n_i and $c_i \in A$, so that $\frac{a_i}{g_i} = \frac{c_i a_i}{h_i^{n_i}}$ in M_{h_i}. Replacing h_i with $h_i^{n_i}$ (we have $D(h_i) = D(h_i^{n_i})$) and a_i by $c_i a_i$, we may assume that $D(f)$ is covered by open sets $D(h_i)$ such that $s(p) = \frac{a_i}{h_i}$ for $p \in D(h_i)$.

Since $D(f) \subset \bigcup D(h_i)$, we have that $Z(\{h_i\}) = \bigcap Z(h_i) \subset Z(f)$. Since A is Noetherian, the ideal $(\{h_i\})$ is generated by a finite number of the h_i, say $h_1, \ldots, h_r$, so $D(f) \subset D(h_1) \cup \cdots \cup D(h_r)$ and we have that $f^n \in (h_1, \ldots, h_r)$

for some positive power n by the nullstellensatz. Thus we have an expression

(11.9) $$f^n = b_1 h_1 + \cdots + b_r h_r$$

for some $b_1, \ldots, b_r \in A$.

For $p \in D(h_i) \cap D(h_j) = D(h_i h_j)$, we have that $s(p) = \frac{a_i}{h_i} = \frac{a_j}{h_j} \in M_{I(p)}$. By our proof of injectivity, ψ is injective when restricted to $D(h_i h_j)$, so we have that $\frac{a_i}{h_i} = \frac{a_j}{h_j}$ in $M_{h_i h_j}$. Hence,

$$(h_i h_j)^n (h_j a_i - h_i a_j) = 0$$

in M for some n. We may pick n sufficiently large so that this equation and (11.9) are valid for all i, j. Rewrite the equation as

$$h_j^{n+1}(h_i^n a_i) - h_i^{n+1}(h_j^n a_j) = 0.$$

Then replace h_i by h_i^{n+1} and a_i by $h_i^n a_i$ (we have that $D(h_i) = D(h_i^{n+1})$) so that we still have that $s(p) = \frac{a_i}{h_i}$ for $p \in D(h_i)$, and we now have that $h_j a_i = h_i a_j$ for all i, j.

Let $a = \sum_i b_i a_i$ where the b_i are from (11.9). Then for each j, we have

$$h_j a = \sum_i b_i h_j a_i = \sum_i b_i h_i a_j = f^n a_j$$

so that $\frac{a}{f^n} = \frac{a_j}{h_j}$ on $D(h_j)$, and thus $\psi(\frac{a}{f^n}) = s$, so that ψ is surjective and hence is an isomorphism. $\square$

Important special cases of the construction of Theorem 11.24 are that $\widetilde{k[X]} = \mathcal{O}_X$ and if Y is a subvariety of X, then the ideal sheaf $\mathcal{I}_Y = \widetilde{I(Y)}$ where $I(Y)$ is the prime ideal in $k[X]$ of Y. More generally, we have a sheaf of $\mathcal{O}_X$-modules $\tilde{I}$ (an ideal sheaf) associated to any ideal $I \subset k[X]$ (Example 11.17).

In the case when M is a $k[X]$-submodule of $k(X)$ and $f \in k[X]$, we have that

(11.10) $$\tilde{M}(X_f) = M_f = \bigcap_{p \in D(f)} M_p$$

where the intersection takes place in $k(X)$. This follows from Lemma 1.77.

We have a corresponding construction for projective varieties. Suppose that Y is an affine variety and X is a closed subvariety of $Y \times \mathbb{P}^r$. The variety $Y \times \mathbb{P}^r$ has the coordinate ring $S(Y \times \mathbb{P}^r) = A[x_0, \ldots, x_r]$ where $A = k[Y]$ and the polynomial ring $A[x_0, \ldots, x_r]$ over A is graded by $\deg(x_i) = 1$ for all i. Let $\mathfrak{p} \subset A[x_0, \ldots, x_r]$ be the graded prime ideal $\mathfrak{p} = I(X)$. Then the coordinate ring of X is $S(X) = A[x_0, \ldots, x_r]/\mathfrak{p} = \bigoplus_{i=0}^{\infty} S_i$.

11.3. Some sheaves associated to modules

Theorem 11.25 (Sheafification of a graded module on a projective variety). *Suppose that X is a projective variety, or, more generally, Y is an affine variety and X is a closed subvariety of $Y \times \mathbb{P}^r$, with coordinate ring $S(X)$, and N is a graded $S(X)$-module. Then there is a unique sheaf $\tilde{N}$ of $\mathcal{O}_X$-modules on X which has the property that*

(11.11) $\qquad \tilde{N}(X_F) = N_{(F)} \quad \text{for homogeneous } F \in S(X)$

and for forms $F, G \in S(X)$ with $X_F \subset X_G$, the restriction map $\tilde{N}(X_G) \to \tilde{N}(X_F)$ is the natural map $N_{(G)} \to N_{(F)}$ induced by localization.

Recall that $N_{(F)}$ denotes the set of elements of degree 0 in the localization N_F.

Using the sheaf axioms, we can give an explicit formula similar to (11.8) for the calculation of $\tilde{N}(U)$, when $U = X_{F_1} \cup \cdots \cup X_{F_n}$ for some homogeneous forms $F_i \in S(X)$

We calculate that for $p \in X$, the stalk

(11.12) $\qquad \tilde{N}_p = N_{(I(p))},$

where $N_{I(p)}$ denotes the elements of degree 0 in the localization $T^{-1}N$ where T is the multiplicative system of homogeneous elements of $S(X) \setminus I(p)$.

In fact, we have that the restriction of the sheaf $\tilde{N}$ to the affine open subset X_F (where F is homogeneous of positive degree) is just the sheaf

(11.13) $\qquad \tilde{N}|X_F = \widetilde{N_{(F)}}$

on the affine variety X_F (which has regular functions $k[X_F] = S(X)_{(F)}$), as follows from a comparison of the definition of $\tilde{N}$ below, the definition of $\widetilde{N_{(F)}}$ from Theorem 11.24, and (11.12).

The sheaf $\tilde{N}$ of Theorem 11.25 is constructed for general graded modules N as follows. Let U be an open subset of the projective variety X, with coordinate ring $S = S(X)$, and suppose that N is a graded S-module. Suppose that U is an open subset of X. Define $\tilde{N}(U)$ to be the set of functions $s : U \to \coprod_{p \in U} N_{(I(p))}$ such that $s(p) \in N_{I(p)}$ for each $p \in U$ and such that for each $p \in U$ there is an open neighborhood V of p in U and homogeneous elements $a \in N$ and $f \in S$ of the same degree such that $Z(f) \cap V = \emptyset$ and $s(q) = \frac{a}{f} \in N_{(I(q))}$. Here $I(p)$ is the homogeneous ideal of the point p in S and $N_{(I(p))}$ is the set of elements of degree 0 in the localization $T^{-1}N$ where T is the multiplicative set of homogeneous elements of S which are not in $I(p)$. We have that $\tilde{N}$ is a sheaf on X as it satisfies the sheaf axioms.

For U an open subset of X, $\mathcal{O}_X(U) = \bigcap_{p \in U} S_{(I(p))} \subset k(X)$. It follows that the map of sheaves $\mathcal{O}_X \to \tilde{S}$ defined by associating to $t \in \mathcal{O}_X(U)$ the map $s: U \to \tilde{S}(U)$ defined by $s(p) = t$ for $p \in U$ is an isomorphism of sheaves of k-algebras. We thus have that $\tilde{S} \cong \mathcal{O}_X$. We have that property (11.11) holds for $\tilde{S}$ by Proposition 4.6.

For general modules N, the sheaf $\tilde{N}$ is naturally a sheaf of $\tilde{S} \cong \mathcal{O}_X$-modules. The fact that property (11.11) holds for N follows from a careful analysis of the natural map $N_{(F)} \to \tilde{N}(D(F))$ for $F \in S$ homogeneous, generalizing the proof of Theorem 11.24.

It follows from the above analysis that the definition in Theorem 11.25 is consistent with Example 11.18.

An important example which we have encountered before is the $\mathcal{O}_X$-module $\tilde{I}$ (ideal sheaf) associated to any homogeneous ideal I of $S(X)$. In particular, if Y is a closed algebraic set in X, with homogeneous reduced ideal $I(Y)$ in $S(X)$, the ideal sheaf of Y is $\mathcal{I}_Y = \widetilde{I(Y)}$. If X is a quasi-projective variety and Y is a closed algebraic set in X, then X is an open subset of projective variety $\overline{X}$. Letting $\overline{Y}$ be the closed algebraic set in $\overline{X}$ which is the closure of Y in $\overline{X}$, we have that $\overline{Y} \cap X = Y$. We define the ideal sheaf $\mathcal{I}_Y$ on X to be the restriction $\mathcal{I}_{\overline{Y}}|X$.

Suppose that X is a projective variety or a closed subvariety of $Y \times \mathbb{P}^r$ where Y is an affine variety, with a closed embedding $i: X \to Y \times \mathbb{P}^r$. Let $S(Y \times \mathbb{P}^r)$ and $S(X) = S(Y \times \mathbb{P}^r)/I(X)$ be the respective coordinate rings. Suppose that M is a graded $S(X)$-module. Let $\mathcal{M}$ be the sheafification of M as a graded $S(X)$-module. Then
(11.14)
$i_*\mathcal{M}$ is the sheafification of M regarded as a graded $S(Y \times \mathbb{P}^r)$-module.

Exercise 11.26. Suppose that $\phi: X \to Y$ is regular map of affine varieties, $M_{k[X]}$ is a $k[X]$-module, and $\mathcal{F} = \widetilde{M_{k[X]}}$ is the induced sheaf on X. Let $M_{k[Y]}$ be the $k[Y]$-module which is $M_{k[X]}$ with the $k[Y]$-module structure induced by the homomorphism $\phi^*: k[Y] \to k[X]$. Let $\mathcal{G} = \widetilde{M_{k[Y]}}$ be the induced sheaf on Y. Show that $\phi_*\mathcal{F} = \mathcal{G}$.

Exercise 11.27. Prove formula (11.14).

11.4. Quasi-coherent and coherent sheaves

Definition 11.28. Let X be a quasi-projective variety. A sheaf of $\mathcal{O}_X$-modules $\mathcal{F}$ is quasi-coherent if X can be covered by affine open sets U_i such that for all i, the restriction $\mathcal{F}|U_i$ of $\mathcal{F}$ to U_i is isomorphic as a sheaf of $\mathcal{O}_{U_i}$-modules to a sheaf $\tilde{M}_i$ for some $k[U_i]$-module M_i. The sheaf $\mathcal{F}$ is coherent if the M_i are all finitely generated $k[U_i]$-modules.

11.4. Quasi-coherent and coherent sheaves

The conclusions of the following example follow from the remark that $\tilde{N}|X_F \cong \widetilde{N_{(F)}}$ for homogeneous $F \in S(X)$ of positive degree in equation (11.13).

Example 11.29. Suppose that Y is an affine variety and X is a closed subvariety of $Y \times \mathbb{P}^r$. Let $S(X)$ be the graded coordinate ring of X, and let N be a graded $S(X)$-module. Then $\tilde{N}$ is a quasi-coherent $\mathcal{O}_X$-module. If N is a finitely generated $S(X)$-module, then $\tilde{N}$ is coherent.

Suppose that X is a locally ringed space and $\mathcal{F}$, $\mathcal{G}$ are sheaves of $\mathcal{O}_X$-modules on X. Then $\text{Hom}_{\mathcal{O}_X}(\mathcal{F}, \mathcal{G})$ denotes the $\mathcal{O}_X(X)$-module of $\mathcal{O}_X$-module sheaf homomorphisms from $\mathcal{F}$ to $\mathcal{G}$.

Lemma 11.30. *Suppose that X is an affine variety, $R = \Gamma(X, \mathcal{O}_X)$, and M, N are R-modules. Then the natural map*

$$\text{Hom}_{\mathcal{O}_X}(\tilde{M}, \tilde{N}) \to \text{Hom}_R(M, N)$$

defined by $\psi \mapsto \psi(X)$ is an isomorphism of R-modules.

Proof. Suppose that $\phi \in \text{Hom}_R(M, N)$. We will construct $\tilde{\phi} \in \text{Hom}_{\mathcal{O}_X}(\tilde{M}, \tilde{N})$ giving an inverse to the natural map $\text{Hom}_{\mathcal{O}_X}(\tilde{M}, \tilde{N}) \to \text{Hom}_R(M, N)$.

Let U be an open subset of X. Then $U = X_{f_1} \cup \cdots \cup X_{f_n}$ for some $f_1, \ldots, f_n \in k[X]$. The homomorphism ϕ induces a commutative diagram

$$\begin{array}{ccc} \bigoplus_i M_{f_i} & \xrightarrow{\beta_1} & \bigoplus_{i<j} M_{f_i f_j} \\ \oplus \phi_{f_i} \downarrow & & \downarrow \oplus \phi_{f_i f_j} \\ \bigoplus_i N_{f_i} & \xrightarrow{\beta_2} & \bigoplus_{i<j} N_{f_i f_j} \end{array}$$

where the horizontal maps are defined as in equation (11.8). Hence we have a natural homomorphism

$$\tilde{\phi}(U) : \tilde{M}(U) = \text{Kernel } \beta_1 \to \text{Kernel } \beta_2 = \tilde{N}(U),$$

which is compatible with restrictions. $\square$

Lemma 11.31. *Suppose that X is an affine variety and $R = \Gamma(X, \mathcal{O}_X) = k[X]$. The sequence of R-modules*

$$M \to N \to P$$

is exact if and only if the sequence of sheaves of $\mathcal{O}_X$-modules

$$\tilde{M} \to \tilde{N} \to \tilde{P}$$

is exact.

Thus if $\tilde{M} \to \tilde{N}$ is a homomorphism of $\mathcal{O}_X$-modules, then its kernel, cokernel, and image are of the form $\tilde{K}$ for some R-module K.

Proof. The sequence of sheaves is exact if and only if it is exact at all stalks; that is,
$$M_\mathfrak{m} \to N_\mathfrak{m} \to P_\mathfrak{m}$$
is exact for all maximal ideals $\mathfrak{m}$ of R. This is equivalent to the exactness of $M \to N \to P$ by [**13**, Proposition 3.9]. $\square$

Theorem 11.32. *Suppose that $\mathcal{F}$ is a sheaf of $\mathcal{O}_X$-modules on a quasi-projective variety X. Then $\mathcal{F}$ is quasi-coherent (coherent) if and only if for every affine open subset U of X, there exists a (finitely generated) $k[U]$-module M such that $\mathcal{F}|U \cong \tilde{M}$.*

If $\mathcal{F}$ is a quasi-coherent sheaf on X and U is an affine open subset of X, then the above theorem tells us that
$$\mathcal{F}|U = \widetilde{\Gamma(U, \mathcal{F})},$$
by Theorem 11.24.

Proof. We will first assume that $\mathcal{F}$ is quasi-coherent and that U is an affine open subset of X.

We observe that if V is an affine open subset of X such that $\mathcal{F}|V \cong \tilde{M}$ for some $k[V]$-module M and $h \in k[V]$, then $\mathcal{F}|V_h \cong \tilde{M}_h$. Since $U \cap V$ is covered by affine open subsets V_h, we have that $\mathcal{F}|U$ is quasi-coherent since $\mathcal{F}$ is quasi-coherent.

Let $R = k[U]$. If $V \subset U$ is an affine open subset and $p \in V$, then there exists $g \in R$ such that $p \in U_g \subset V$, and $U_g = V_g$. Thus U can be covered by a finite number of open sets $U_i = U_{g_i}$ such that there are $k[U_i] = R_{g_i}$-modules M_i with $\mathcal{F}|U_i \cong \tilde{M}_i$.

We have
$$\mathcal{F}|U_{g_i} \cap U_{g_j} = \mathcal{F}|U_{g_i g_j} \cong \widetilde{(M_i)}_{g_j}.$$
For every open subset W of U, the sequence
$$0 \to \Gamma(W, \mathcal{F}) \to \prod_i \Gamma(W \cap U_{g_i}, \mathcal{F}) \to \prod_{i<j} \Gamma(W \cap U_{g_i} \cap U_{g_j}, \mathcal{F})$$
is exact by the sheaf axioms. Define new sheaves $\mathcal{F}_i^*$ and $\mathcal{F}_{i,j}^*$ on U by
$$\Gamma(W, \mathcal{F}_i^*) = \Gamma(W \cap U_{g_i}, \mathcal{F})$$
and
$$\Gamma(W, \mathcal{F}_{i,j}^*) = \Gamma(W \cap U_{g_i} \cap U_{g_j}, \mathcal{F})$$
for an open subset W of U, so the sequence of sheaves
$$0 \to \mathcal{F} \to \prod_i \mathcal{F}_i^* \to \prod_{i<j} \mathcal{F}_{i,j}^*$$

11.4. Quasi-coherent and coherent sheaves

is exact. So to prove that $\mathcal{F}$ is of the form $\tilde{M}$ for some R-module, it suffices to prove this for $\mathcal{F}_i^*$ and $\mathcal{F}_{i,j}^*$, by Lemma 11.31, as a direct sum of tildes of modules on an affine variety is the tilde of the sum. Viewing M_i as an R-module, for all $g \in R$ we have

$$\Gamma(U_g, \mathcal{F}_i^*) = \Gamma(U_g \cap U_{g_i}, \mathcal{F}) = \Gamma((U_{g_i})_g, \mathcal{F}|U_{g_i}) = (M_i)_g = \Gamma(U_g, \tilde{M}_i)$$

so that $\mathcal{F}_i^* = \tilde{M}_i$. The same argument shows that $\mathcal{F}_{i,j}^* = \widetilde{(M_i)_{g_j}}$. Thus $\mathcal{F}$ is the tilde of an R-module M.

The conclusions of the theorem for coherent sheaves reduces by the above arguments to the statement that if U is affine, with $R = k[U]$, $f_1, \ldots, f_n \in R$ are such that $U_{f_1}, \ldots, U_{f_n}$ is an affine cover of U, and M is an R-module such that M_{f_i} is a finitely generated R_{f_i}-module for $1 \leq i \leq n$, then M is a finitely generated R-module. We will now establish this statement, which is a minor extension of Lemma 7.4.

For $1 \leq i \leq n$, there exist elements $\sigma_{i1}, \ldots, \sigma_{it_i}$ in M which generate M_{f_i} as an R_{f_i}-module. Let N be the R-submodule of M generated by σ_{ij} for $1 \leq i \leq n$ and $1 \leq j \leq t_i$. Since $U = \bigcup_i U_{f_i}$, we have that $Z_U(f_1, \ldots, f_n) = \emptyset$, so that the ideal $(f_1, \ldots, f_n) = R$. Suppose $m \in M$. Then for $1 \leq i \leq n$, there exist $r_{ij} \in R$ and $\lambda \in \mathbb{N}$ such that

$$f_i^\lambda \left(m - \sum_j r_{ij} \sigma_{ij} \right) = 0$$

in M, so that $(f_1^\lambda, \ldots, f_n^\lambda)m \subset N$. But $(f_1^\lambda, \ldots, f_n^\lambda) = R$ by Lemma 7.3 so $m \in N$. Thus $M = N$ is a finitely generated R-module. □

Definition 11.33. Suppose that X is a quasi-projective variety and $\mathcal{F}$ is a coherent sheaf of $\mathcal{O}_X$-modules. The sheaf $\mathcal{F}$ is said to be invertible if there exists an open cover $\{U_i\}$ of X and $\mathcal{O}_{U_i}$-module isomomorphisms $\phi_i : \mathcal{O}_{U_i} \to \mathcal{F}|U_i$ for all U_i in the cover.

Exercise 11.34. Suppose that X is a locally ringed space and $\mathcal{F}$ is a sheaf of $\mathcal{O}_X$-modules on X. Show that there is a natural isomorphism of $\mathcal{O}_X(X)$-modules $\text{Hom}_{\mathcal{O}_X}(\mathcal{O}_X, \mathcal{F}) \to \mathcal{F}(X)$.

Exercise 11.35. Suppose that X is a quasi-projective variety, $\mathcal{F}$, $\mathcal{G}$ are $\mathcal{O}_X$-modules, and $\phi : \mathcal{F} \to \mathcal{G}$ is a homomorphism of $\mathcal{O}_X$-modules.

    a) Suppose that $\mathcal{F}$ and $\mathcal{G}$ are quasi-coherent. Show that Kernel(ϕ), Image(ϕ), and Cokernel(ϕ) are quasi-coherent $\mathcal{O}_X$-modules.

    b) Suppose that $\mathcal{F}$ and $\mathcal{G}$ are coherent. Show that Kernel(ϕ), Image(ϕ), and Cokernel(ϕ) are coherent $\mathcal{O}_X$-modules.

Exercise 11.36. Suppose that X is a quasi-projective variety. Suppose that $\mathcal{F}$ is a sheaf on X. Define the support of $\mathcal{F}$ by
$$\mathrm{Supp}(\mathcal{F}) = \{p \in X \mid \mathcal{F}_p \neq 0\}.$$
Suppose that $\mathcal{F}$ is a coherent sheaf on X. Show that $\mathrm{Supp}(\mathcal{F})$ is a closed set. Hint: You may use the following lemma from commutative algebra [**13**, Exercise 19, page 46]: suppose that Y is an affine variety. Let $A = k[Y]$, and suppose that M is a finitely generated A-module. Then
$$\{p \in Y \mid M_{I_Y(p)} \neq 0\}$$
is a closed subset of Y.

11.5. Constructions of sheaves from sheaves of modules

In this section we give some constructions of sheaves from sheaves of modules. Our primary interest in these sheaves is in the case of coherent or quasi-coherent $\mathcal{O}_X$-modules. In this case, the equations (11.15), (11.16), (11.17), and (11.18) should be taken as the definitions of these sheaves. These formulas, which give local realizations of the sheaves in terms of commutative algebra, are all that is needed to work effectively with these sheaves.

Assume that X is a quasi-projective variety and $\mathcal{F}$ and $\mathcal{G}$ are quasi-coherent sheaves on X. The tensor product $\mathcal{F} \otimes_{\mathcal{O}_X} \mathcal{G}$ is a quasi-coherent sheaf of $\mathcal{O}_X$-modules which is uniquely determined by the following property: if U is an affine open subset of X and $L = \Gamma(U, \mathcal{F})$, $M = \Gamma(U, \mathcal{G})$, then

(11.15) $$(\mathcal{F} \otimes_{\mathcal{O}_X} \mathcal{G})|U = \tilde{N}$$

where $N = L \otimes_{k[U]} M$. We can use the sheaf axioms to show that the condition (11.15) determines a unique sheaf on X. If $\mathcal{F}$ and $\mathcal{G}$ are coherent, then $\mathcal{F} \otimes_{\mathcal{O}_X} \mathcal{G}$ is coherent. The stalk $(\mathcal{F} \otimes_{\mathcal{O}_X} \mathcal{G})_p \cong \mathcal{F}_p \otimes_{\mathcal{O}_{X,p}} \mathcal{G}_p$ for $p \in X$. We sometimes denote $\mathcal{F} \otimes_{\mathcal{O}_X} \mathcal{G}$ by $\mathcal{F} \otimes \mathcal{G}$ when there is no danger of confusion.

Assume that X is a quasi-projective variety and $\mathcal{F}$ is a quasi-coherent sheaf on X and $\mathcal{G}$ is a coherent sheaf on X. The sheaf $\mathrm{Hom}_{\mathcal{O}_X}(\mathcal{F}, \mathcal{G})$ is a quasi-coherent sheaf of $\mathcal{O}_X$-modules which is uniquely determined by the property that if U an affine open subset of X, then

(11.16) $$\mathrm{Hom}_{\mathcal{O}_X}(\mathcal{F}, \mathcal{G})|U = \tilde{G}$$

where
$$G = \mathrm{Hom}_{k[U]}(\Gamma(U, \mathcal{F}), \Gamma(U, \mathcal{G}))$$
is the $k[U]$-module of $k[U]$-module homomorphisms from $\Gamma(U, \mathcal{F})$ to $\Gamma(U, \mathcal{G})$. We can use the sheaf axioms to show that the condition (11.16) determines a unique sheaf on X. If $\mathcal{F}$ is coherent, then $\mathrm{Hom}_{\mathcal{O}_X}(\mathcal{F}, \mathcal{G})$ is coherent. For $p \in X$, the stalk $\mathrm{Hom}_{\mathcal{O}_X}(\mathcal{F}, \mathcal{G})_p = \mathrm{Hom}_{\mathcal{O}_{X,p}}(\mathcal{F}_p, \mathcal{G}_p)$, the $\mathcal{O}_{X,p}$-module homomorphisms from $\mathcal{F}_p$ to $\mathcal{G}_p$, as follows from [**50**, Proposition 1.10].

11.5. Constructions of sheaves from sheaves of modules

Suppose that $\phi : X \to Y$ is a regular map of quasi-projective varieties and $\mathcal{M}$ is a quasi-coherent sheaf of $\mathcal{O}_Y$-modules. The inverse image $\phi^*\mathcal{M}$ of $\mathcal{M}$ by ϕ is a quasi-coherent sheaf of $\mathcal{O}_X$-modules, which has the following defining property: if U is an affine open subset of X and V is an affine open subset of Y such that $\phi(U) \subset V$, then

$$\phi^*\mathcal{M}|U = \widetilde{(M \otimes_{k[V]} k[U])} \tag{11.17}$$

where $M = \Gamma(V, \mathcal{M})$. We can use the sheaf axioms and basic properties of tensor products to show that the condition (11.17) determines a unique sheaf $\phi^*(\mathcal{M})$ on X. It follows from (11.17) that $\phi^*\mathcal{M}$ is coherent if $\mathcal{M}$ is coherent. If $p \in X$ and $q = \phi(p)$, then $(\phi^*\mathcal{M})_p \cong M_q \otimes_{\mathcal{O}_{Y,q}} \mathcal{O}_{X,p}$ (by Lemma 11.4).

Definition 11.37. Suppose that X is a variety. A coherent $\mathcal{O}_X$-module $\mathcal{I}$ is called an ideal sheaf if $\mathcal{I}$ is an $\mathcal{O}_X$-submodule of $\mathcal{O}_X$.

If $\mathcal{I}$ is an ideal sheaf on an affine variety X, we see from Theorem 11.32 that $I = \Gamma(X, \mathcal{I}) \subset k[X]$ is such that $\mathcal{I} = \tilde{I}$, and we will see from Theorem 11.48 that if X is projective with coordinate ring $S(X)$ and $\mathcal{I}$ is an ideal sheaf on X, then there exists a homogeneous ideal I of $S(X)$ such that $\mathcal{I} = \tilde{I}$.

We will see, by Proposition 11.53, that Definition 11.37 is consistent with Examples 11.17 and 11.18.

Suppose that $\phi : X \to Y$ is a regular map of quasi-projective varieties and $\mathcal{I}$ is an ideal sheaf on Y. We define the inverse image ideal sheaf $\mathcal{IO}_X$ to be the natural image of $\phi^*\mathcal{I}$ in $\mathcal{O}_X$ induced by the inclusion $\mathcal{I} \subset \mathcal{O}_Y$, giving a map $\phi^*\mathcal{I} \to \phi^*\mathcal{O}_Y \cong \mathcal{O}_X$. The sheaf $\mathcal{IO}_X$ is coherent since it is the image of a homomorphism of coherent sheaves (Exercise 11.35). This ideal sheaf has the defining property that when U an affine open subset of X and V an affine open subset of Y such that $\phi(U) \subset V$,

$$\mathcal{IO}_X|U = \widetilde{Ik[U]}, \tag{11.18}$$

where $I = \Gamma(V, \mathcal{I})$. If $p \in X$ and $q = \phi(p)$, then $(\mathcal{IO}_X)_p = I_q \mathcal{O}_{X,p}$.

If $\mathcal{I}$ is an ideal sheaf on Y, then there exists a natural surjection of $\mathcal{O}_X$-modules

$$\phi^*\mathcal{I} \to \mathcal{IO}_X,$$

but in general this map is not injective. We do have that

$$\phi^*\mathcal{I}/\mathcal{T}(\phi^*(\mathcal{I})) \cong \mathcal{IO}_X,$$

where $\mathcal{T}(\phi^*(\mathcal{I}))$ is the $\mathcal{O}_X$-torsion of $\phi^*(\mathcal{I})$ (Exercise 11.43). The fact that $\phi^*\mathcal{I} \to \mathcal{IO}_X$ is in general not injective can be seen in the following simple example. Let $A = k[x, y]$ and $B = k[x_1, y_1]$ be polynomial rings, and

consider the injective k-algebra homomorphism $A \to B$ defined by $x = x_1$, $y = x_1 y_1$. Let $I = (x, y) \subset A$. We have a short exact sequence of A-modules
$$0 \to A \to A^2 \to I \to 0,$$
where the first map is defined by $1 \mapsto (y, -x)$ and the second map is defined by $(1, 0) \mapsto x$ and $(0, 1) \mapsto y$. Tensoring this sequence with B, we have the right exact sequence (tensoring with a module is right exact by Proposition 5.5)
$$B \to B^2 \to I \otimes_A B \to 0,$$
so that
$$I \otimes_A B \cong B^2/(x_1 y_1, -x_1) B.$$
The class of $(y_1, -1)$ is nonzero in $I \otimes_A B$, but $x_1(y_1, -1) = 0$. Thus $I \otimes_A B$ has B-torsion, which must be in the kernel of the surjection onto the ideal IB.

The above sheaves can be defined more generally for sheaves of $\mathcal{O}_X$-modules on a locally ringed space X. We will give an outline of these constructions here for the interested reader.

Suppose that $\mathcal{F}$ and $\mathcal{G}$ are $\mathcal{O}_X$-modules on a locally ringed space X. Then we have a presheaf
$$U \mapsto \mathcal{F}(U) \otimes_{\mathcal{O}_X(U)} \mathcal{G}(U).$$
We denote by $\mathcal{F} \otimes_{\mathcal{O}_X} \mathcal{G}$ (or sometimes $\mathcal{F} \otimes \mathcal{G}$ when there is no danger of confusion) the sheaf associated to this presheaf. The stalk $(\mathcal{F} \otimes \mathcal{G})_p \cong \mathcal{F}_p \otimes_{\mathcal{O}_{X,p}} \mathcal{G}_p$ for $p \in X$ (by Lemma 11.4).

To prove that this definition gives formula (11.15) in the case when X is quasi-projective and $\mathcal{F}$ and $\mathcal{G}$ are quasi-coherent, observe that we have a natural restriction homomorphism from $\tilde{N}$ to the presheaf $V \mapsto \mathcal{F}(V) \otimes_{\mathcal{O}_X(V)} \mathcal{G}(V)$ for V an open subset of U. Then by Proposition 11.12, there is a natural homomorphism of sheaves $\tilde{N} \to (\mathcal{F} \otimes_{\mathcal{O}_X} \mathcal{G})|U$. This map is an isomorphism on stalks, so it is an isomorphism of sheaves by Proposition 11.13. The quasi-coherence (or coherence) of $\mathcal{F} \otimes_{\mathcal{O}_X} \mathcal{G}$ then follows from Theorem 11.32.

Suppose that $\mathcal{F}$ and $\mathcal{G}$ are $\mathcal{O}_X$-modules on a locally ringed space X. We have a presheaf
$$U \mapsto \mathrm{Hom}_{\mathcal{O}_X|U}(\mathcal{F}|U, \mathcal{G}|U),$$
for U an open subset of X, the $\mathcal{O}_X(U)$-module of $\mathcal{O}_X|U$-module sheaf homomorphisms $\mathcal{F}|U \to \mathcal{G}|U$. This presheaf is a sheaf (since $\mathcal{F}$ and $\mathcal{G}$ are sheaves), which we write as $\mathcal{H}om_{\mathcal{O}_X}(\mathcal{F}, \mathcal{G})$.

Lemma 11.38. *Suppose that X is a quasi-projective variety, $\mathcal{F}$ is quasi-coherent and $\mathcal{G}$ is coherent on X. Then $\mathcal{H}om_{\mathcal{O}_X}(\mathcal{F}, \mathcal{G})$ is quasi-coherent. Further, if $\mathcal{F}$ and $\mathcal{G}$ are both coherent, then $\mathcal{H}om_{\mathcal{O}_X}(\mathcal{F}, \mathcal{G})$ is coherent.*

Proof. Let $\mathcal{H} = Hom_{\mathcal{O}_X}(\mathcal{F}, \mathcal{G})$. It suffices to prove the lemma when X is affine. Let $R = k[X]$, $M = \mathcal{F}(X)$, and $N = \mathcal{G}(X)$. Then N is a finitely generated R-module, $\mathcal{F} = \tilde{M}$, and $\mathcal{G} = \tilde{N}$. Let $A = Hom_R(M, N)$. We have a natural homomorphism of sheaves $\tilde{A} \to \mathcal{H}$. For $f \in R$, we have

$$\tilde{A}(D(f)) = Hom_R(M, N)_f \cong Hom_{R_f}(M_f, N_f) \text{ (by [50, Proposition 1.10])}$$
$$\cong Hom_{\tilde{R}_f}(\tilde{M}_f, \tilde{N}_f) \text{ by Lemma 11.30}$$
$$\cong \mathcal{H}(D(f)).$$

Since any affine open subset of X is a finite union of basic open sets $D(f_i)$, we thus have that $\tilde{A}(U) \to \mathcal{H}(U)$ is an isomorphism for all open subsets U of X by the sheaf axioms. Thus $\mathcal{H} \cong \tilde{A}$ is quasi-coherent. Further, if $\mathcal{F}$ and $\mathcal{G}$ are both coherent, so that M and N are both finitely generated R-modules, then A is a finitely generated R-module, so $\mathcal{H} \cong \tilde{A}$ is coherent. $\square$

From Lemmas 11.30, 11.32, and 11.38 we see that the formula 11.16 holds.

We now give an outline of the general construction of the pull-back $f^*\mathcal{G}$ of a sheaf of $\mathcal{O}_Y$-modules by a morphism of ringed spaces $f : X \to Y$.

Suppose that $f : X \to Y$ is a continuous map of topological spaces and $\mathcal{G}$ is a sheaf on Y. The inverse image sheaf $f^{-1}\mathcal{G}$ on X is the sheaf associated to the presheaf

$$U \mapsto \lim_{f(U) \subset V} \mathcal{G}(V)$$

where U is an open subset of X and the limit is over the open sets V of Y which contain $f(U)$.

Let $f : X \to Y$ be a morphism of ringed spaces and $\mathcal{G}$ be an $\mathcal{O}_Y$-module. Then $f^{-1}\mathcal{G}$ is an $f^{-1}\mathcal{O}_Y$-module and we have a natural homomorphism $f^{-1}\mathcal{O}_Y \to \mathcal{O}_X$ of sheaves of rings on X. We define the inverse image of $\mathcal{G}$ by the morphism f to be

$$f^*\mathcal{G} = f^{-1}\mathcal{G} \otimes_{f^{-1}\mathcal{O}_Y} \mathcal{O}_X.$$

The sheaf $f^*\mathcal{G}$ is a sheaf of $\mathcal{O}_X$-modules. In the case when $f : X \to Y$ is a regular maps of quasi-projective varieties and $\mathcal{G}$ is quasi-coherent, we have the formula (11.17).

Exercise 11.39. Suppose that X is a closed subvariety of a quasi-projective variety Y. Let $i : X \to Y$ be the inclusion, and let $\mathcal{I}_X$ be the ideal sheaf of X in Y. Show that there is a short exact sequence of sheaves of $\mathcal{O}_Y$-modules

$$0 \to \mathcal{I}_X \to \mathcal{O}_Y \to i_*\mathcal{O}_X \to 0.$$

Hint: First solve this when Y is affine.

Exercise 11.40. Suppose that X is a quasi-projective variety of dimension > 0 and $p \in X$. Define a presheaf $\mathcal{F}$ on X by

$$\mathcal{F}(U) = \begin{cases} \mathcal{O}_X(U) & \text{if } p \notin U, \\ 0 & \text{if } p \in U \end{cases}$$

for U an open subset of X.

   a) Show that $\mathcal{F}$ is a sheaf of $\mathcal{O}_X$-modules.

   b) Show that $\mathcal{F}$ is not coherent.

   c) Observe that $\mathcal{F}$ is a sub-$\mathcal{O}_X$-module of $\mathcal{O}_X$ which is not an ideal sheaf.

Exercise 11.41. Suppose that X and Y are varieties, $\phi : X \to Y$ is a regular map, and $\mathcal{F}, \mathcal{G}$ are $\mathcal{O}_X$-modules.

   a) Show that there is a natural homomorphism of $\mathcal{O}_Y$-modules

$$\phi_*\mathcal{F} \otimes_{\mathcal{O}_Y} \phi_*\mathcal{G} \to \phi_*(\mathcal{F} \otimes_{\mathcal{O}_X} \mathcal{G}).$$

   b) Suppose that $\mathcal{V}$ is a locally free sheaf on Y (every point $q \in Y$ has a neighborhood U such that $\mathcal{V}|U \cong \mathcal{O}_U^n$ for some n). Show that there is a natural isomorphism of $\mathcal{O}_Y$-modules

$$\phi_*(\mathcal{F} \otimes_{\mathcal{O}_X} \phi^*\mathcal{V}) \cong \phi_*\mathcal{F} \otimes_{\mathcal{O}_Y} \mathcal{V}.$$

Exercise 11.42. Suppose that X is a quasi-projective variety and $\mathcal{I}$ and $\mathcal{J}$ are ideal sheaves on X. Suppose that $p \in X$ and $\mathcal{I}_p = \mathcal{J}_p$. Show that there exists a neighborhood U of p in X such that $\mathcal{I}|U = \mathcal{J}|U$.

Exercise 11.43. If A is an integral domain and M is an A-module, then the A-torsion submodule of M is

$$T_A(M) = \{x \in M \mid \text{Ann}_A(x) \neq 0\}.$$

The fact that $T_A(M)$ is a submodule of M is shown in [13, Exercise 12, Chapter 3]. It has the property that $M/T_A(M)$ is A-torsion free (the torsion submodule is 0).

Suppose that X is a quasi-projective variety and $\mathcal{F}$ is a quasi-coherent $\mathcal{O}_X$-module. Suppose that $\sigma \in \Gamma(X, \mathcal{F})$. Then we have a coherent ideal sheaf $\text{Ann}_X(\sigma)$ on X defined by

$$\Gamma(U, \text{Ann}_X(\sigma)) = \{\tau \in \Gamma(U, \mathcal{O}_X) \mid \tau(\sigma|U) = 0\}$$

for U an open subset of X.

Define a presheaf $\mathcal{T}(\mathcal{F})$ on X by
$$\mathcal{T}(\mathcal{F})(U) = \{\sigma \in \mathcal{F}(U) \mid \text{the sheaf } \text{Ann}_U(\sigma) \neq 0\}$$
for U an open subset of X. Show that $\mathcal{T}(\mathcal{F})$ is a quasi-coherent $\mathcal{O}_X$-module which has the property that
$$\mathcal{T}(\mathcal{F})(U) = T_{\Gamma(U,\mathcal{O}_X)}(\Gamma(U, \mathcal{F}))$$
if U is affine. The sheaf $\mathcal{T}(\mathcal{F})$ is called the sheaf of $\mathcal{O}_X$-torsion of $\mathcal{F}$.

11.6. Some theorems about coherent sheaves

The principle results of this section (through Theorem 11.51) were originally proven by Serre in [**132**]. In this section, we first define the twisted sheaf $\mathcal{O}_X(n)$. We show that, on a projective variety X with coordinate ring $S(X)$, a quasi-coherent sheaf is isomorphic to $\tilde{N}$ where N is a graded $S(X)$-module and a coherent sheaf is isomorphic to $\tilde{N}$ where N is a finitely generated graded $S(X)$-module (Theorem 11.46). We show that if $\phi : X \to Y$ is a regular map of projective varieties and $\mathcal{F}$ is a coherent sheaf on X, then $\phi_*\mathcal{F}$ is a coherent sheaf on Y (Theorem 11.51).

Suppose that X is a projective variety or more generally a closed subvariety of $Y \times \mathbb{P}^r$ where Y is an affine variety. We have a natural closed embedding $i : X \to \mathbb{P}^r$ or more generally a closed embedding $i : X \to Y \times \mathbb{P}^r$ so that X has the coordinate ring $S(X) = S(Y \times \mathbb{P}^r)/I(X)$. Here $S(Y \times \mathbb{P}^r) = k[Y][y_0, \ldots, y_r]$ is a polynomial ring in the variables $y_0, \ldots, y_r$, with the grading that the elements of $k[Y]$ have degree 0 and the y_i have degree 1. The graded ideal $I(X) = \bigoplus_{j \geq 0} I(X)_j$, so that

(11.19) $$S(X) = S(Y \times \mathbb{P}^r)/I(X) = R[x_0, \ldots, x_r]$$

where $R = k[Y]/I(X)_0$ and x_i is the class of y_i in $S(X)$ (which has degree 1).

In the case that X is a projective variety, we have a closed embedding of X in $\mathbb{P}^r$ for some r. Let p be a point. Then p is an affine variety with $k[p] = k$. Thus we have a natural closed embedding of X in $p \times \mathbb{P}^r \cong \mathbb{P}^r$, which has the coordinate ring
$$S(p \times \mathbb{P}^r) = k[p] \otimes_k k[y_0, \ldots, y_r] \cong k[y_0, \ldots, y_r] = S(\mathbb{P}^r).$$

The following theorems, which are stated for subvarieties of $Y \times \mathbb{P}^r$, are thus valid for the case of a subvariety of $\mathbb{P}^r$, taking $Y = p$ and $k[Y] = k[p] = k$ in the proofs. We require the more general statements about subvarieties of $Y \times \mathbb{P}^r$ with Y an affine variety in the proof of Theorem 11.51 and other later applications.

Suppose that N is a graded $S(X)$-module. Recall (Section 3.1) that for $n \in \mathbb{Z}$, we define $N(n)$ to be the module N, but with a different grading,
$$N(n)_i = N_{n+i} \quad \text{for } i \in \mathbb{Z}.$$

We have a quasi-coherent sheaf of $\mathcal{O}_X$-modules $\widetilde{N(n)}$ on X, which is coherent if N is a finitely generated $S(X)$-module (Example 11.29). We define a coherent sheaf $\mathcal{O}_X(n) = \widetilde{S(X)(n)}$.

Lemma 11.44. *Suppose that R is a ring and $S = R[x_0, \ldots, x_n]$ is a graded R-algebra with $\deg x_i = 1$ for $0 \leq i \leq n$. Let N be a graded S-module. Then*
$$N_{x_i} = \bigoplus_{j \in \mathbb{Z}} N_{(x_i)} x_i^j.$$

Proof. Write the graded S_{x_i}-module N_{x_i} as $N_{x_i} = \bigoplus_{j \in \mathbb{Z}} \overline{N}_j$, so that $\overline{N}_0 = N_{(x_i)}$. Suppose that $j \in \mathbb{Z}$. We have that $N_{(x_i)} x_i^j \subset \overline{N}_j$. Suppose that $f \in \overline{N}_j$. Then $f = \frac{g}{x_i^l}$ with $l \in \mathbb{N}$ and $g \in N_a$ with $a - l = j$. Now
$$f = \left(\frac{g}{x_i^{l+j}}\right) x_i^j \in N_{(x_i)} x_i^j. \qquad \square$$

Suppose that N is a graded $S(X)$-module. We have that

(11.20) $\qquad \widetilde{N(n)} \cong \mathcal{F} \otimes \mathcal{O}_X(n) \text{ if } \mathcal{F} = \tilde{N}.$

Letting $S = S(X)$, this formula follows from the identities $N(n) \cong N \otimes_S S(n)$ and $\widetilde{N(n)} \cong \widetilde{N \otimes_S S(n)} \cong \tilde{N} \otimes_{\tilde{S}} \widetilde{S(n)}$. This last equality of sheaves follows from the sheaf axioms, since we have, by Lemma 11.44, natural isomorphisms $N_{(\overline{x}_i)} \otimes_{S_{(\overline{x}_i)}} S(n)_{(\overline{x}_i)} \cong N(n)_{(\overline{x}_i)}$ which are compatible with localization.

More generally, we define $\mathcal{F}(n) = \mathcal{F} \otimes \mathcal{O}_X(n)$ if $\mathcal{F}$ is a sheaf of $\mathcal{O}_X$-modules on X.

We have the following useful formula, which shows that $\mathcal{O}_X(n)$ is an invertible sheaf (Definition 11.33):

(11.21) $\qquad \Gamma(X_{x_i}, \mathcal{O}_X(n)) = \Gamma(X_{x_i}, \mathcal{O}_X) x_i^n$

for $0 \leq i \leq n$ and $n \in \mathbb{Z}$.

We now prove formula (11.21). By Lemma 11.44, the localization
$$S(X)_{x_i} = \bigoplus_{n \in \mathbb{Z}} S(X)_{(x_i)} x_i^n$$
as graded rings. Thus
$$\begin{aligned} \Gamma(X_{x_i}, \mathcal{O}_X(n)) &= \Gamma(X_{x_i}, \widetilde{S(X)(n)}) = S(X)_{(x_i)} x_i^n \\ &= \Gamma(X_{x_i}, \widetilde{S(X)}) x_i^n. \end{aligned}$$

We have the following formula:

(11.22) $\qquad i^*(\mathcal{O}_{Y \times \mathbb{P}^r}(n)) \cong \mathcal{O}_X(n).$

11.6. Some theorems about coherent sheaves

We now establish formula (11.22). By Lemma 11.44, we have that

$$I(X)_{y_i} = \bigoplus_{n \in \mathbb{Z}} I(X)_{(y_i)} y_i^n$$

and

$$\Gamma((Y \times \mathbb{P}^r)_{y_i}, \widetilde{I(X)(n)}) = I(X)_{(y_i)} y_i^n.$$

We have short exact sequences of $S(Y \times \mathbb{P}^r)_{(y_i)}$-modules

$$0 \to I(X)_{(y_i)} y_i^n \to S(Y \times \mathbb{P}^r)_{(y_i)} y_i^n \to S(X)_{(x_i)} x_i^n \to 0$$

so we have natural isomorphisms

$$\begin{aligned}
\Gamma(X_{x_i}, \mathcal{O}_X(n)) &= S(X)_{(x_i)} x_i^n = S(Y \times \mathbb{P}^r)_{(y_i)} y_i^n / I(X)_{(y_i)} y_i^n \\
&\cong S(Y \times \mathbb{P}^r)_{(y_i)} y_i^n \otimes_{S(Y \times \mathbb{P}^r)_{(y_i)}} (S(Y \times \mathbb{P}^r)_{(y_i)} / I(X)_{(y_i)}) \\
&\cong S(Y \times \mathbb{P}^r)_{(y_i)} y_i^n \otimes_{S(Y \times \mathbb{P}^r)_{(y_i)}} S(X)_{(x_i)} \\
&\cong \Gamma((Y \times \mathbb{P}^r)_{y_i}, \mathcal{O}_{Y \times \mathbb{P}^r}(n)) \otimes_{\Gamma((Y \times \mathbb{P}^r)_{y_i}, \mathcal{O}_{Y \times \mathbb{P}^r})} \Gamma(X_{x_i}, \mathcal{O}_X) \\
&\cong \Gamma(X_{x_i}, i^*(\mathcal{O}_{Y \times \mathbb{P}^r}(n))).
\end{aligned}$$

These isomorphisms are compatible with localization, so by the sheaf axioms, we have a natural isomorphism $\mathcal{O}_X(n) \cong i^*(\mathcal{O}_{Y \times \mathbb{P}^r}(n))$.

Suppose that $\mathcal{G}$ is a sheaf of $\mathcal{O}_{Y \times \mathbb{P}^r}$-modules and $n \in \mathbb{Z}$. Then

$$(11.23) \qquad i^*(\mathcal{G}(n)) \cong (i^*\mathcal{G})(n).$$

We now establish (11.23). We have that $\mathcal{G}(n) = \mathcal{G} \otimes_{\mathcal{O}_{Y \times \mathbb{P}^r}} (\mathcal{O}_{Y \times \mathbb{P}^r}(n))$ and

$$(11.24) \quad i^*(\mathcal{G}(n)) = \left(\mathcal{G} \otimes_{\mathcal{O}_{Y \times \mathbb{P}^r}} \mathcal{O}_{Y \times \mathbb{P}^r}(n) \right) \otimes_{\mathcal{O}_{Y \times \mathbb{P}^r}} \mathcal{O}_X \cong \mathcal{G} \otimes_{\mathcal{O}_{Y \times \mathbb{P}^r}} \mathcal{O}_X(n)$$

by (11.22). We have $i^*\mathcal{G} = \mathcal{G} \otimes_{\mathcal{O}_{Y \times \mathbb{P}^r}} \mathcal{O}_X$ and

$$(i^*\mathcal{G})(n) = \left(\mathcal{G} \otimes_{\mathcal{O}_{Y \times \mathbb{P}^r}} \mathcal{O}_X \right) \otimes_{\mathcal{O}_X} \mathcal{O}_X(n) \cong \mathcal{G} \otimes_{\mathcal{O}_{Y \times \mathbb{P}^r}} \mathcal{O}_X(n) \cong i^*(\mathcal{G}(n))$$

by equation (11.24).

Suppose $\mathcal{F}$ is a sheaf of $\mathcal{O}_X$-modules. For all $n \in \mathbb{Z}$, we have natural isomorphisms

$$(11.25) \qquad (i_*\mathcal{F})(n) \cong i_*(\mathcal{F}(n)).$$

We now establish formula (11.25). We have that $(i_*\mathcal{F})(n)$ is the sheaf associated to the presheaf

$$U \mapsto \Gamma(U \cap X, \mathcal{F}) \otimes_{\Gamma(U, \mathcal{O}_{Y \times \mathbb{P}^r})} \Gamma(U, \mathcal{O}_{Y \times \mathbb{P}^r}(n))$$

for U an open subset of $Y \times \mathbb{P}^r$, and we have that $i_*(\mathcal{F}(n))$ is the sheaf associated to the presheaf

$$U \mapsto \Gamma(U \cap X, \mathcal{F}) \otimes_{\Gamma(U \cap X, \mathcal{O}_X)} \left(\Gamma(U \cap X, \mathcal{O}_X) \otimes_{\Gamma(U, \mathcal{O}_{Y \times \mathbb{P}^r})} \Gamma(U, \mathcal{O}_{Y \times \mathbb{P}^r}(n)) \right).$$

This last module is naturally isomorphic to
$$\Gamma(U \cap X, \mathcal{F}) \otimes_{\Gamma(U, \mathcal{O}_{Y \times \mathbb{P}^r})} \Gamma(U, \mathcal{O}_{Y \times \mathbb{P}^r}(n)).$$
Thus $(i_* \mathcal{F})(n) \cong i_*(\mathcal{F}(n))$.

A sheaf $\mathcal{F}$ of $\mathcal{O}_X$-modules on a locally ringed space X is said to be generated by a finite number of global sections if there exist
$$\sigma_1, \ldots, \sigma_n \in \Gamma(X, \mathcal{F})$$
such that $\mathcal{F}_p = \sigma_1 \mathcal{O}_{X,p} + \cdots + \sigma_n \mathcal{O}_{X,p}$ for all $p \in X$.

Theorem 11.45. *Let X be a projective variety, or, more generally, a closed subvariety of $Y \times \mathbb{P}^r$ where Y is an affine variety, with coordinate ring $S = S(X)$, and associated sheaf $\mathcal{O}_X(1) = \widetilde{S(1)}$. Let $\mathcal{F}$ be a coherent sheaf on X. Then there exists an integer m_0 such that for all $n \geq m_0$, the sheaf $\mathcal{F}(n)$ is generated by a finite number of global sections.*

Proof. Write the graded coordinate ring S of X as $S = R[x_0, \ldots, x_r]$ where elements of R have degree 0 and $x_0, \ldots, x_r$ have degree 1, as in (11.19).

Let $B_i = \Gamma(X_{x_i}, \mathcal{O}_X) = R[\frac{x_0}{x_i}, \ldots, \frac{x_r}{x_i}]$ for $0 \leq i \leq r$. Since $\mathcal{F}$ is coherent, there exist finitely generated B_i-modules M_i such that $\mathcal{F}|X_{x_i} \cong \tilde{M}_i$ for each i. Let $\{s_{ij}\}$ be a finite number of generators of M_i for each i. We have that $\mathcal{O}_X(n)|X_{x_i}$ is naturally isomorphic to the module $\widetilde{x_i^n B_i}$ (by (11.21)), so the sheaf $\mathcal{F}(n)|X_{x_i}$ is naturally isomorphic to $M_i \otimes (\widetilde{x_i^n B_i})$.

We will now show that there exists a positive integer n such that $x_i^n s_{ij}$ extends to $t_{ij} \in \Gamma(X, \mathcal{F}(n))$ for all i, j.

It suffices to show that if $n_0 \in M_0$, then $x_0^n n_0$ extends to an element of $\Gamma(X, \mathcal{F}(n))$ for all $n \gg 0$. Let
$$M_{ij} = \Gamma(X_{x_i x_j}, \mathcal{F}) \cong (M_i)_{\frac{x_j}{x_i}}$$
since $\mathcal{F}$ is coherent. There exists $\lambda \in \mathbb{N}$ and $n_i \in M_i$ for $0 < i \leq r$ such that
$$n_0 = \frac{n_i}{\left(\frac{x_0}{x_i}\right)^\lambda}$$
in M_{i0}. Since
$$\frac{n_i}{\left(\frac{x_0}{x_i}\right)^\lambda} = \frac{n_j}{\left(\frac{x_0}{x_j}\right)^\lambda}$$
in $M_{0ij} = (M_{ij})_{\frac{x_0}{x_i}}$, there exists $a \in \mathbb{N}$ such that
$$\left[n_i \left(\frac{x_0}{x_j}\right)^\lambda - n_j \left(\frac{x_0}{x_i}\right)^\lambda\right] \left(\frac{x_0}{x_i}\right)^a = 0$$

11.6. Some theorems about coherent sheaves

in M_{ij}. Let

$$\sigma_i = n_i \left(\frac{x_0}{x_i}\right)^{a+\lambda} x_i^{a+2\lambda} \in \Gamma(X_{x_i}, \mathcal{F}(a+2\lambda)) = M_i x_i^{a+2\lambda}.$$

The differences of the restrictions

$$\begin{aligned}\sigma_i - \sigma_j &= [(\tfrac{x_0}{x_i})^{a+\lambda} n_i - (\tfrac{x_0}{x_i})^{\lambda+a}(\tfrac{x_j}{x_i})^\lambda n_j] x_i^{a+2\lambda} \\ &= (\tfrac{x_0}{x_i})^a (\tfrac{x_j}{x_i})^\lambda [n_i(\tfrac{x_0}{x_j})^\lambda - n_j(\tfrac{x_0}{x_i})^\lambda] x_i^{a+2\lambda} = 0\end{aligned}$$

in $M_{ij} x_i^{a+2j} = \Gamma(X_{x_i x_j}, \mathcal{F}(a+2\lambda))$. Thus $\{\sigma_i\} \in \bigoplus M_i x_i^{a+2\lambda}$ is an element of

$$\Gamma(X, \mathcal{F}(a+2\lambda)),$$

and so $\sigma_0 = n_0 x_0^{a+2\lambda}$ extends to an element of $\Gamma(X, \mathcal{F}(a+2\lambda))$, and so $n_0 x_0^n$ extends to an element of $\Gamma(X, \mathcal{F}(n))$ whenever $n \geq a + 2\lambda$.

The sections $x_i^n s_{ij}$ generate $M_i \otimes (x_i^n B_i)$, and so the t_{ij} generate $\mathcal{F}(n)$ everywhere. $\square$

Theorem 11.46. *Suppose that $\mathcal{F}$ is a quasi-coherent sheaf on a projective variety X, or, more generally, on a closed subvariety X of $Y \times \mathbb{P}^r$ where Y is an affine variety, and X has the coordinate ring $S(X)$. Then there exists a graded $S(X)$-module M such that $\mathcal{F} \cong \tilde{M}$. If $\mathcal{F}$ is coherent, there exists a finitely generated graded $S(X)$-module M such that $\mathcal{F} = \tilde{M}$.*

Proof. Suppose that $\mathcal{F}$ is quasi-coherent. Let $S = S(X) = R[x_0, \ldots, x_r]$ be the coordinate ring of X, and define a graded S-module

$$M = \Gamma_*(\mathcal{F}) = \bigoplus_{n \in \mathbb{Z}} \Gamma(X, \mathcal{F}(n)).$$

We will show that there is a natural isomorphism of $\mathcal{O}_X$-modules $\beta : \tilde{M} \to \mathcal{F}$. It suffices to define isomorphisms over the affine open sets X_{x_i} which agree on the $X_{x_i x_j}$. We have that

$$\Gamma(X_{x_i}, \tilde{M}) = M_{(x_i)} = \left\{\frac{m}{x_i^d} \mid m \in \Gamma(X, \mathcal{F}(d)) \text{ for some } d\right\}.$$

For fixed d, we have that $\mathcal{F} \cong \mathcal{F}(d) \otimes \mathcal{O}_X(-d)$ and $\mathcal{O}_X(-d)|X_{x_i} = \widetilde{x_i^{-d} k[X_{x_i}]}$, so we may define $M_{(x_i)} \to \mathcal{F}(X_{x_i}) \cong [\mathcal{F}(d)(X_{x_i})] \otimes [\mathcal{O}_X(-d)(X_{x_i})]$ by

$$\frac{m}{x_i^d} \mapsto (m|X_{x_i}) \otimes x_i^{-d}.$$

Suppose $m \in \Gamma(X, \mathcal{F}(d))$ and $(m|X_{x_i}) \otimes x_i^{-d} = 0$. Let $m_j = m|X_{x_j}$ for $0 \le j \le r$. We have that $m_i = 0$ so $m_j|X_{x_i x_j} = 0$ in $\Gamma(X_{x_i x_j}, \mathcal{F}(d)) \cong \mathcal{F}(d)(X_{x_j})_{\frac{x_i}{x_j}}$. There exists $\lambda \in \mathbb{N}$ such that for all j, $(\frac{x_i}{x_j})^\lambda m_j = 0$ in $\Gamma(X_{x_i}, \mathcal{F}(d))$. Consider $x_i^\lambda m \in \Gamma(X, \mathcal{F}(d+\lambda))$. Then

$$x_i^\lambda m \mid X_{x_j} = \left(\left(\frac{x_i}{x_j} \right)^\lambda m_j \right) x_j^\lambda = 0 \quad \text{for all } j$$

so $x_i^\lambda m = 0$. Thus $\frac{m}{x_i^d} = 0$ in $M_{(x_i)}$. Thus the homomorphism $M_{(x_i)} \to \mathcal{F}(X_{x_i})$ is injective.

The proof of Theorem 11.45 shows that $M_{(x_i)} \to \mathcal{F}(X_{x_i})$ is surjective and hence is an isomorphism.

Thus we have natural isomorphisms $\beta_i : \widetilde{M}(X_{x_i}) \to \mathcal{F} \mid X_{x_i} = \widetilde{\mathcal{F}(X_{x_i})}$ which agree on $X_{x_i} \cap X_{x_j} = X_{x_i x_j}$. Thus the β_i patch to give an isomorphism of sheaves β.

Now suppose that $\mathcal{F}$ is coherent. By Theorem 11.45, $\mathcal{F}(n)$ is generated by a finite number of global sections for $n \gg 0$. For such an n, let M' be the submodule of M generated by these sections, so that M' is a finitely generated S-module. The inclusion of S-modules $M' \subset M$ induces an inclusion of $\mathcal{O}_X$-modules $\widetilde{M'} \subset \widetilde{M} = \mathcal{F}$. Tensoring with $\mathcal{O}_X(n)$, we have an inclusion $\widetilde{M'}(n) \subset \mathcal{F}(n)$ which is an isomorphism since $\mathcal{F}(n)$ is generated by global sections of $\widetilde{M'}(n)$. After tensoring with $\mathcal{O}_X(-n)$, we obtain that $\widetilde{M'} \cong \mathcal{F}$. □

The proof of the following theorem is similar to that of Theorem 3.35.

Theorem 11.47. *Suppose that X is a projective variety, or, more generally, a closed subvariety of $Y \times \mathbb{P}^r$ where Y is an affine variety. Let $S = S(X)$ be the homogeneous coordinate ring of X for this embedding, and let $\mathcal{O}_X(n)$ be the coherent sheaf $\widetilde{S(n)}$. Let $S' = \bigoplus_{n \ge 0} \Gamma(X, \mathcal{O}_X(n))$. Then S' is a finite S-module and $\Gamma(X, \mathcal{O}_X(n)) = S_n$ for all $n \gg 0$.*

Proof. Write $S = R[x_0, \ldots, x_r]$ where $x_0, \ldots, x_r$ have degree 1. We have that

$$\Gamma(X_{x_i}, \mathcal{O}_X(n)) = S(n)_{(x_i)},$$

the elements of degree 0 in the localization $S(n)_{x_i}$. Now $S(n)$ is the ring S with a different grading, and we can identify $S(n)_{(x_i)}$ with the elements $(S_{x_i})_n$ of degree n in the localization S_{x_i} (by Lemma 11.44). Since $\mathcal{O}_X(n)$ is a sheaf,

$$\Gamma(X, \mathcal{O}_X(n)) = \bigcap_{i=0}^r \Gamma(X_{x_i}, \mathcal{O}_X(n)) = \bigcap_{i=0}^r (S_{x_i})_n$$

where the intersection takes place in the graded ring $\bigcap_{i=0}^{r} S_{x_i}$. Since S_n is certainly contained in $(S_{x_i})_n$ for all n, we have natural graded inclusions

$$S \subset S' \subset \bigcap_{i=0}^{r} S_{x_i}.$$

We will show that S' is integral over S. Let $s' \in S'$ be homogeneous of degree $d \geq 0$. Since $s' \in S_{x_i}$ for each i, there exists an integer m such that $x_i^m s' \in S$ for all i. The monomials of degree m in $x_0, \ldots, x_r$ generate S_m as an R-module for all m. Thus there exists an n_0 such that $S_n s' \subset S$ for all $n \geq n_0$. Thus $S_n s' \subset S_{n+d}$ for all $n \geq n_0$. It follows that $S_n (s')^q \subset S_{n+qd}$ for any $q \geq 1$ and all $n \geq n_0$. Thus $(s')^q \in \frac{1}{x_0^{n_0}} S$ for all $q \geq 1$, and so the ring $S[s']$ is contained in $\frac{1}{x_0^{n_0}} S$ which is a finitely generated S-module. Thus $S[s']$ is a finitely generated S-module and s' is integral over S (Theorem 1.49). Thus S' is contained in the integral closure of S in its quotient field. Since S is a finitely generated k-algebra, the integral closure of S in its quotient field is a finitely generated S-module, by Theorem 1.54. Thus S' is a finitely generated S-module by Lemma 1.55. Let $F_1, \ldots, F_r$ be homogeneous elements in S' which generate S' as an S-module. We showed above that there exists $N > 0$ such that $S_N F_i \subset S$ for all i. Thus $S'_n = S_n$ for all $n \geq \max\{\deg(F_i)\} + N$. $\square$

Proposition 11.48. *Suppose that Y is an affine variety and X is a closed subvariety of $Y \times \mathbb{P}^r$. Let $S(X)$ be the graded coordinate ring of X. Suppose that $\mathcal{I} \subset \mathcal{O}_X$ is an ideal sheaf. Then there exists a graded ideal $I \subset S(X)$ such that $\tilde{I} = \mathcal{I}$.*

Proof. Let $x_0, \ldots, x_r$ be homogeneous coordinates on $\mathbb{P}^r$. For $n \in \mathbb{Z}$ and $p \in X_{x_i}$, $\mathcal{O}_X(n)_p = x_i^n \mathcal{O}_{X,p}$ (by (11.21)), so that $\mathcal{O}_X(n)$ is an invertible sheaf. Thus the natural inclusion $\mathcal{I} \subset \mathcal{O}_X$ induces an inclusion $\mathcal{I}(n) = \mathcal{I} \otimes_{\mathcal{O}_X} \mathcal{O}_X(n) \subset \mathcal{O}_X(n)$, and so $\Gamma_*(\mathcal{I}) = \bigoplus_{n \in \mathbb{Z}} \Gamma(X, \mathcal{I}(n))$ is a graded submodule of $\Gamma_*(\mathcal{O}_X) = \bigoplus_{n \in \mathbb{Z}} \Gamma(X, \mathcal{O}_X(n))$. There exists $n_0 \in \mathbb{N}$ such that $\Gamma(X, \mathcal{O}_X(n)) = S(X)_n$ for $n \geq n_0$ by Theorem 11.47. Thus $I = \bigoplus_{n \geq n_0} \Gamma(X, \mathcal{I}(n))$ is a graded ideal in $S(X)$. We have that $\tilde{I} = \widetilde{\Gamma_*(\mathcal{I})} = \mathcal{I}$ by Theorem 11.46 (and its proof) and Exercise 11.57. $\square$

Theorem 11.49. *Let Y be an affine variety and $X \subset Y \times \mathbb{P}^n$ be a closed subvariety. Let $\mathcal{F}$ be a coherent $\mathcal{O}_X$-module. Then $\Gamma(X, \mathcal{F})$ is a finitely generated $k[Y]$-module. In particular, if $X \subset \mathbb{P}^n$ is a projective variety, then $\Gamma(X, \mathcal{F})$ is a finitely generated k-vector space.*

Proof. Let $A = k[Y]$ and $\mathfrak{p} = I(X)$ which is a graded prime ideal in the graded ring $A[x_0, \ldots, x_n]$. The coordinate ring of X is the graded domain $S = A[x_0, \ldots, x_n]/\mathfrak{p} \cong \bigoplus_{i \geq 0} S_i$, where $S_0 = A/\mathfrak{p} \cap A$. By Theorem 11.46,

there exists a finitely generated S-module M such that $\tilde{M} \cong \mathcal{F}$. By [**73**, Theorem I.7.4], there is a finite filtration
$$0 = M^0 \subset M^1 \subset \cdots \subset M^r = M$$
of M by graded submodules, where for each i, $M^i/M^{i-1} \cong (S/\mathfrak{p}_i)(n_i)$ for some homogeneous prime ideal $\mathfrak{p}_i \subset S$ and some integer n_i. The filtration gives a filtration of $\tilde{M}$ by coherent $\mathcal{O}_X$-modules, and the short exact sequences
$$0 \to \tilde{M}^{i-1} \to \tilde{M}^i \to \widetilde{M^i/M^{i-1}} \to 0$$
induce exact sequences
$$0 \to \Gamma(X, \tilde{M}^{i-1}) \to \Gamma(X, \tilde{M}^i) \to \Gamma(X, \widetilde{M^i/M^{i-1}}).$$
To show that $\Gamma(X, \tilde{M})$ is a finitely generated A-module, it thus suffices to show that each $\Gamma(X, \widetilde{S/\mathfrak{p}_i(n_i)})$ is a finitely generated A-module, which follows from (11.14), Exercise 11.27, (11.25), and Theorem 11.47. $\square$

Theorem 11.50. *Suppose that $\phi : X \to Y$ is a regular map of quasi-projective varieties and $\mathcal{F}$ is a quasi-coherent sheaf on X. Then $\phi_* \mathcal{F}$ is a quasi-coherent sheaf on Y. If Y is an affine variety, X is a closed subvariety of $Y \times \mathbb{P}^r$, $\mathcal{F}$ is coherent on X, and $\pi_1 : Y \times \mathbb{P}^r \to Y$ is the projection with restriction $\overline{\pi}_1$ to X, then $(\overline{\pi}_1)_* \mathcal{F}$ is coherent on Y.*

Proof. Suppose that $\mathcal{F}$ is quasi-coherent. We will show that $\phi_* \mathcal{F}$ is quasi-coherent. First assume that X and Y are affine, so that $\mathcal{F} = \tilde{M}$ for some $k[X]$-module M. Let $A = k[Y]$. Let $f \in A$ and $U = Y_f$. Then
$$(\phi_* \mathcal{F})(U) = \mathcal{F}(\phi^{-1}(U)) = \tilde{M}(X_f) = M_f.$$
Thus writing M_A for M considered as an A-module, we have that $(\tilde{M}_A)(Y_f) = \phi_* \mathcal{F}(Y_f)$. Since any open subset of Y is a union of basic open sets Y_f, we have by the sheaf axioms that $\phi_* \mathcal{F} = \tilde{M}_A$ is quasi-coherent.

Now we prove, more generally, that $\phi_* \mathcal{F}$ is quasi-coherent if X and Y are quasi-projective and $\mathcal{F}$ is quasi-coherent. By Theorem 11.32, we may assume that Y is affine. Cover X with a finite number of open affine sets $U_1, \ldots, U_r$. Let $U_{i,j} = U_i \cap U_j$ which is affine (Exercise 5.21). By the sheaf properties, we have exact sequences of sheaves
$$0 \to \phi_* \mathcal{F} \to \bigoplus_i \phi_*(\lambda_*^i \mathcal{F}|U_i) \to \bigoplus_{i<j} \phi_*(\lambda_*^{i,j} \mathcal{F}|U_{i,j})$$
where $\lambda^i : U_i \to X$ and $\lambda^{i,j} : U_{i,j} \to X$ are the natural inclusions. We have that $\phi_* \lambda_*^i = (\phi \circ \lambda^i)_*$ and $\phi_* \lambda_*^{i,j} = (\phi \circ \lambda^{i,j})_*$, so the sheaves $\phi_* \lambda_*^i(\mathcal{F}|U_i)$ and $\phi_* \lambda_*^{i,j}(\mathcal{F}|U_{ij})$ are quasi-coherent by the first part of this proof. Thus $\phi_* \mathcal{F}$ is quasi-coherent by Exercise 11.35 and Theorem 11.32.

11.6. Some theorems about coherent sheaves

Now suppose that $\mathcal{F}$ is coherent and X is a closed subvariety of $Y \times \mathbb{P}^n$ where Y is affine. We will show that $(\overline{\pi}_1)_*\mathcal{F}$ is coherent. Since $(\overline{\pi}_1)_*\mathcal{F}$ is quasi-coherent, it suffices to show that $\Gamma(Y, (\overline{\pi}_1)_*\mathcal{F})$ is a finitely generated $A = k[Y]$-module. But this follows from Theorem 11.49. $\square$

Theorem 11.51. *Suppose that $\phi : X \to Y$ is a regular map of projective varieties and $\mathcal{F}$ is a coherent sheaf on X. Then $\phi_*\mathcal{F}$ is a coherent sheaf on Y.*

Proof. Let $\Gamma_\phi \subset X \times Y$ be the graph of ϕ, with natural isomorphism $j = (\mathrm{id}, \phi) : X \to \Gamma_\phi$. Let $\pi_2 : X \times Y \to Y$ be the projection, with induced projection $\overline{\pi}_2 : \Gamma_\phi \to Y$. Then $\phi_*\mathcal{F} \cong (\overline{\pi}_2)_*j_*\mathcal{F}$, which is coherent by Theorem 11.50. $\square$

Suppose that X is a variety, with function field $k(X)$. The constant sheaf $k(X)$ satisfies $\Gamma(U, k(X)) = k(X)$ for all open subsets U of X by Proposition 11.14, since all open subsets of X are connected.

Proposition 11.52. *Suppose that $\phi : X \to Y$ is a birational regular map of projective varieties and Y is normal. Then $\phi_*\mathcal{O}_X = \mathcal{O}_Y$.*

Proof. The inclusion $\mathcal{O}_X \to k(X)$ of $\mathcal{O}_X$-modules induces an inclusion $\phi_*\mathcal{O}_X \to \phi_*k(X) = k(X) = k(Y)$ of $\mathcal{O}_Y$-modules. Since $\phi_*\mathcal{O}_X$ is a coherent $\mathcal{O}_Y$-module by Theorem 11.51 under the natural inclusion $\mathcal{O}_Y \to \phi_*\mathcal{O}_X$ and $\mathcal{O}_Y$ is normal, we have that $\phi_*\mathcal{O}_X = \mathcal{O}_Y$. $\square$

Proposition 11.53. *Suppose that X is a variety and $\mathcal{M}$ is a quasi-coherent sheaf on X such that $\mathcal{M}$ is a subsheaf of $\mathcal{O}_X$-modules of the constant sheaf $k(X)$. Suppose that U is an open subset of X. Then*

$$\mathcal{M}(U) = \bigcap_{p \in U} \mathcal{M}_p$$

where the intersection takes place in $k(X)$.

Proof. We have that $\mathcal{M}(U) \subset \Gamma(U, k(X)) = k(X)$ for all open subsets U of X by Proposition 11.14 and Proposition 11.20. Let U be an open subset of X and $\{V_1, \ldots, V_n\}$ be an affine open cover of U. By the sheaf axioms, $\mathcal{M}(U) = \bigcap_{i=1}^n \mathcal{M}(V_i)$, where the intersection is in $k(X)$. The conclusions of the proposition now hold by formula (11.10). $\square$

A coherent sheaf $\mathcal{F}$ on a variety X is locally free if for all $p \in X$, there exists an open neighborhood U of p in X such that $\mathcal{F}|U$ is a free $\mathcal{O}_U$-module (isomorphic to $\mathcal{O}_U^r$ for some r).

Lemma 11.54. *Let X be a quasi-projective variety, let $p \in X$, and let $\mathcal{F}$ be a coherent $\mathcal{O}_X$-module. Let*
$$\lambda(p) = \dim_k \mathcal{F}_p/m_p\mathcal{F}_p$$
where m_p is the maximal ideal of $\mathcal{O}_{X,p}$. Then $\lambda(p)$ is upper semicontinuous on X; that is, for $t \in \mathbb{N}$, the set
$$\{q \in X \mid \lambda(q) \geq t\}$$
is a closed set in X.

Further, there exists an open neighborhood U of p such that $\mathcal{F}|U$ is a free $\mathcal{O}_U$-module (isomorphic to $\mathcal{O}_U^r$ for some r) if and only if λ is constant in a neighborhood of p.

Proof. Let $p \in X$ and $r = \lambda(p)$. Let $k(p) = \mathcal{O}_{X,p}/m_p$. There exist $a_1, \ldots, a_r \in \mathcal{F}_p$ such that their images generate $\mathcal{F}_p \otimes k(p)$. Let U_1 be a neighborhood of p such that the a_i lift to elements of $\Gamma(U_1, \mathcal{F})$. Define an $\mathcal{O}_{U_1}$-module homomorphism $\phi : \mathcal{O}_{U_1}^r \to \mathcal{F}|U_1$ by $\phi(b_1, \ldots, b_r) = \sum a_i b_i$. Let $\mathcal{K}$ be the cokernel of ϕ, which is a coherent $\mathcal{O}_U$-module (Lemma 11.31). We have exact sequences
$$\mathcal{O}_{X,p}^r \xrightarrow{\phi} \mathcal{F}_p \to \mathcal{K}_p \to 0$$
and
$$k(p)^r \xrightarrow{\overline{\phi}} \mathcal{F}_p \otimes k(p) \to \mathcal{K}_p \otimes k(p) \to 0.$$
The homomorphism $\overline{\phi}$ is surjective, so $0 = \mathcal{K}_p \otimes k(p) \cong \mathcal{K}_p/m_p\mathcal{K}_p$. By Nakayama's lemma, Lemma 1.18, $\mathcal{K}_p = 0$. Hence p is in the open set $X \setminus \mathrm{Supp}(\mathcal{K})$ (Exercise 11.36). Let $U_2 = U_1 \cap (X \setminus \mathrm{Supp}(\mathcal{K}))$. We have that $\mathcal{K}|U_2 = 0$, so $a_1, \ldots, a_r$ generate $\mathcal{F}|U_2$. Hence the images of $a_1, \ldots, a_r$ generate $\mathcal{F}_q \otimes k(q)$ for all $q \in U_2$ so that $\lambda(q) \leq r$ if $q \in U_2$.

We now prove the second statement. If $\mathcal{F}$ is a free $\mathcal{O}_X$-module of rank r in a neighborhood U of p, then $\mathcal{F}|U \cong (\mathcal{O}_X|U)^r$ for some r and $\lambda(q) = r$ for all $q \in U$.

Conversely, suppose that U_1 is a nontrivial open subset of X such that $\lambda(p) = r$ for all $p \in U_1$. As in the first part of the proof, there exists a possibly smaller open subset U_2 of X, which we may take to be affine, with a surjective $\mathcal{O}_X$-module homomorphism
$$\psi : \mathcal{O}_{U_2}^r \xrightarrow{\psi} \mathcal{F}|U_2 \to 0.$$
Let $\mathcal{K}$ be the kernel of ψ. We will show that $\mathcal{K} = 0$, establishing that $\mathcal{F}|U_2$ is free, and we will then have completed the proof.

Suppose that $\mathcal{K} \neq 0$. We will derive a contradiction. We have (since U_2 is affine) a short exact sequence
$$0 \to K \to R^r \to M \to 0$$

where $K = \Gamma(U_2, \mathcal{K})$, $R = \Gamma(U_2, \mathcal{O}_X)$, and $M = \Gamma(U_2, \mathcal{F})$. Since $\mathcal{K} = \tilde{K}$, there exists $0 \neq s \in K$. Write $s = (s_1, \ldots, s_r)$ where $s_i \in R$ and some $s_i \neq 0$. Then $Z(s_1, \ldots, s_r) \neq U_2$ (by the nullsellensatz), so there exists $q \in U_2 \setminus Z(s_1, \ldots, s_r)$, giving us that $s \notin m_q \mathcal{O}_{U_2,q}^r$. Thus the leftmost map is nonzero in the exact sequence

$$\mathcal{K} \otimes k(q) \to k(q)^r \to \mathcal{F}_q \otimes k(q) \to 0,$$

giving us that $\lambda(q) < r$, a contradiction. $\square$

The following extension theorem for coherent sheaves will be useful.

Theorem 11.55. *Suppose that X is a variety and U is an open subset of X. Suppose that $\mathcal{F}$ is a coherent sheaf on U and $\mathcal{G}$ is a quasi-coherent sheaf on X such that $\mathcal{F}$ is an $\mathcal{O}_U$-submodule of $\mathcal{G}|U$. Then there exists a coherent sheaf $\mathcal{F}'$ on X such that $\mathcal{F}'$ is an $\mathcal{O}_X$-submodule of $\mathcal{G}$ and $\mathcal{F}'|U = \mathcal{F}$.*

Proof. Let $i : U \to X$ be the inclusion and define $\rho : \mathcal{G} \to i_*(\mathcal{G}|U)$ to be the restriction $\rho(V) : \mathcal{G}(V) \to \mathcal{G}(U \cap V)$ for V an open subset of X. The sheaf $\mathcal{F}$ is a submodule of $\mathcal{G}|U$, so $i_*\mathcal{F}$ is a submodule of $i_*(\mathcal{G}|U)$. Let $\mathcal{H}$ be the submodule $\mathcal{H} = \rho^{-1}(i_*\mathcal{F})$ of $\mathcal{G}$. Since $\rho|U$ is the identity map, we have that $\mathcal{H}|U = \mathcal{F}$. Let V be an affine open subset of X. Then since $i_*\mathcal{F}$ and $\mathcal{G}$ are quasi-coherent and V is affine, there exist $k[V]$-modules A, B, and C with A a submodule of C such that $\mathcal{G}|V \cong \tilde{B}$, $i_*\mathcal{F}|V \cong \tilde{A}$, $i_*(\mathcal{G}|U)|V \cong \tilde{C}$, and there exists a $k[V]$-module homomorphism $\phi : B \to C$ which induces $\rho|V$. Thus $\mathcal{H}|V \cong \tilde{M}$ where M is the $k[V]$-module

$$M = \{(a, b) \in A \oplus B \mid \phi(b) = a\},$$

and so $\mathcal{H}$ is quasi-coherent.

Let $\{\mathcal{M}_i\}$ be the directed system of all finitely generated $\mathcal{O}_X$-submodules of $\mathcal{H}$. We have that $\lim_{\to} \mathcal{M}_i = \bigcup_i \mathcal{M}_i = \mathcal{H}$, so that $\bigcup_i \mathcal{M}_i = \mathcal{H}$, and $\mathcal{F} = \bigcup_i (\mathcal{M}_i|U)$. Each $\mathcal{M}_i|U$ is coherent and $\mathcal{F}$ is coherent, so restricting to a finite affine open cover of X, we see that $\mathcal{F} = \mathcal{M}_i|U$ for some i, establishing the theorem. $\square$

Exercise 11.56. Suppose that $\mathcal{F}$ is a coherent sheaf on a projective variety X with projective coordinate ring S. Show that $\mathcal{F}$ is a quotient sheaf of a finite direct sum $\mathcal{E} = \bigoplus_{i=1}^n \mathcal{O}_X(q_i)$ for some $n \in \mathbb{N}$ and $q_1, \ldots, q_n \in \mathbb{Z}$.

Exercise 11.57. Suppose that X is a closed subvariety of $Y \times \mathbb{P}^r$ where Y is an affine variety and $S(X)$ is the coordinate ring of X. Suppose that M is a graded $S(X)$-module. Suppose that $d_0 \in \mathbb{Z}$. Let $N = \bigoplus_{d \geq d_0} M_d$, which is a graded $S(X)$-module. Show that the quasi-coherent sheaves $\tilde{M}$ and $\tilde{N}$ are isomorphic.

Exercise 11.58. In this problem we consider the question of coherence of the push-forward of a coherent sheaf.

    a) Let p be the origin in $\mathbb{A}^1$, let $U = \mathbb{A}^1 \setminus \{p\}$, and let $i : U \to \mathbb{A}^1$ be the inclusion. Determine if $i_*\mathcal{O}_U$ is a coherent $\mathcal{O}_{\mathbb{A}^1}$-module.

    b) Let q be the origin in $\mathbb{A}^2$, let $V = \mathbb{A}^2 \setminus \{q\}$, and let $j : V \to \mathbb{A}^2$ be the inclusion. Determine if $j_*\mathcal{O}_V$ is a coherent $\mathcal{O}_{\mathbb{A}^2}$-module.

Exercise 11.59. Suppose that $\mathcal{F}$ is a coherent sheaf on a projective variety X with homogeneous coordinate ring $S(X)$. Show that $\bigoplus_{n \geq 0} \Gamma(X, \mathcal{F}(n))$ is a finitely generated $S(X)$-module.

Show that if X is a projective variety, $p \in X$, and $\mathcal{F}$ is the coherent sheaf $\mathcal{F} = \mathcal{O}_X/\mathcal{I}_p$, where $\mathcal{I}_p$ is the ideal sheaf of p in X, then $\Gamma_*(\mathcal{F}) = \bigoplus_{n \in \mathbb{Z}} \Gamma(X, \mathcal{F}(n))$ is not a finitely generated $S(X)$-module.

Chapter 12

Applications to Regular and Rational Maps

In this chapter we define the blow-up of an ideal sheaf and use this to further develop the theory of regular and rational maps.

12.1. Blow-ups of ideal sheaves

In this section, we extend the blow-up of an ideal to the blow-up of an ideal sheaf.

Definition 12.1. Suppose that X is an affine variety and $\mathcal{I}$ is an ideal sheaf on X. From Theorem 11.32, we have that $\mathcal{I} \cong \tilde{I}$ where I is the ideal $\Gamma(X, \mathcal{I})$ in $k[X]$. We define the blow-up of the ideal sheaf $\mathcal{I}$, written as $\pi : B(\mathcal{I}) \to X$, by the construction of Definition 6.2 as $B(\mathcal{I}) = B(I)$.

If X is an affine variety with ideal sheaf $\mathcal{I}$ and U is an affine open subset of X, then a set of generators of $I = \Gamma(X, \mathcal{I})$ is a set of generators of $\Gamma(U, \mathcal{I})$ (by Exercise 1.47 since they generate locally at all local rings of U) so by the construction of Definition 6.2, there is a natural commutative diagram where the horizontal arrows are open embeddings

$$\begin{array}{ccc} B(\mathcal{I}|U) & \longrightarrow & B(\mathcal{I}) \\ \pi_U \downarrow & & \downarrow \pi_X \\ U & \longrightarrow & X \end{array}$$

identifying $B(\mathcal{I} \mid U)$ with $\pi_X^{-1}(U)$.

Definition 12.2. Suppose that X is a projective variety and $\mathcal{I}$ is an ideal sheaf on X. From Proposition 11.48, we can write $\mathcal{I} = \tilde{I}$ where I is a homogeneous ideal in $S(X)$. By Lemma 6.7, we can take I to be generated by homogeneous elements $F_0, \ldots, F_n$ of some common degree d. In Definition 6.8, we defined the blow-up $\pi : B(I) \to X$ of I, where $B(I)$ is a projective variety. We define the blow-up $B(\mathcal{I})$ of $\mathcal{I}$ to be $\pi : B(I) \to X$.

This definition is well-defined by the following lemma, and Proposition 3.39, applied to an affine open cover of X.

Lemma 12.3. *Suppose that X is a projective variety and $\pi : B(I) = B(\mathcal{I}) \to X$ is the blow-up of the ideal sheaf $\mathcal{I}$. Suppose that $U \subset X$ is an open affine subset. Then there is a natural commutative diagram where the horizontal arrows are open embeddings*

$$\begin{array}{ccc} B(\mathcal{I}|U) & \longrightarrow & B(\mathcal{I}) \\ \pi_U \downarrow & & \downarrow \pi_X \\ U & \longrightarrow & X \end{array}$$

identifying $B(\mathcal{I}|U)$ with $\pi_X^{-1}(U)$.

Proof. It is shown after Definition 6.8 that the conclusions of the lemma are true when $U = X_{x_i}$ for homogeneous coordinates $x_0, \ldots, x_n$ on X. For an arbitrary affine open subset U of X, each X_{x_i} is affine, and U_{x_i} is an affine open subset of X_{x_i}. Further, each $U_{x_i} \cap U_{x_j} = U_{x_i x_j}$ is an affine open subset of U_{x_i}. We thus have natural commutative diagrams for all i, j

$$\begin{array}{ccccccc} B(\mathcal{I}|U_{x_i x_j}) & \longrightarrow & B(\mathcal{I}|U_{x_i}) & \longrightarrow & B(\mathcal{I} \mid X_{x_i}) & \longrightarrow & B(\mathcal{I}) \\ \downarrow & & \downarrow & & \downarrow & & \downarrow \pi_X \\ U_{x_i x_j} & \longrightarrow & U_{x_i} & \longrightarrow & X_{x_i} & \longrightarrow & X \end{array}$$

where the horizontal maps are inclusions.

Since $\{U_{x_i}\}$ is an affine cover of U, $B(\mathcal{I}|U) = \bigcup_i B(\mathcal{I}|U_{x_i})$, so there is a commutative diagram of well-defined maps

$$\begin{array}{ccc} B(\mathcal{I}|U) & \to & B(\mathcal{I}) \\ \downarrow & & \downarrow \\ U & \to & X \end{array}$$

which are regular by Proposition 3.39. $\square$

Definition 12.4. Suppose that X is a quasi-projective variety which is an open subset of a projective variety Y and $\mathcal{I}$ is an ideal sheaf on X. By Theorem 11.55, there exists an ideal sheaf $\mathcal{J}$ of Y such that $\mathcal{J}|X = \mathcal{I}$. We

12.1. Blow-ups of ideal sheaves

define the blow-up $B(\mathcal{I})$ of $\mathcal{I}$ to be $B(\mathcal{I}) = \pi^{-1}(X) \to X$ where $\pi : B(\mathcal{J}) \to Y$ is the blow-up of $\mathcal{J}$.

Since an open affine subset of X is also an open affine subset of Y, we have that the conclusions of Lemma 12.3 hold when X is assumed to be quasi-projective, showing that $B(\mathcal{I})$ is well-defined in Definition 12.4. We have that $B(\mathcal{I})$ is a quasi-projective variety since it is an open subset of the projective variety $B(\mathcal{J})$.

If X is a variety and $\mathcal{I}$ is an ideal sheaf of X, then $\mathcal{I}$ is nonzero if and only if $\mathcal{I}_p \neq 0$ for all $p \in X$.

Theorem 12.5 (The universal property of blowing up). *Suppose that Y is a quasi-projective variety and $\mathcal{I}$ is an ideal sheaf on Y. Let $\pi : B(\mathcal{I}) \to Y$ be the blow-up of $\mathcal{I}$. Suppose that $\phi : X \to Y$ is a regular map of quasi-projective varieties such that the $\mathcal{IO}_X$ is a nonzero locally principal ideal sheaf (there is an affine open cover $\{U_i\}$ of X such that the $\Gamma(U_i, \mathcal{IO}_X)$ are nonzero principal ideals). Then there exists a unique regular map $\psi : X \to B(\mathcal{I})$ factoring ϕ.*

A nonzero locally principal ideal sheaf is an example of an invertible sheaf (Definition 11.33).

Proof. Let $p \in X$. Let V be an affine neighborhood of $\phi(p)$ in Y and U be an affine neighborhood of p in X such that $\phi(U) \subset V$. Let $I = \Gamma(V, \mathcal{I}) \subset k[V]$. Since $\mathcal{IO}_X$ is locally principal, we may assume, after possibly replacing U with a smaller affine neighborhood of p, that $Ik[U] = \Gamma(U, \mathcal{IO}_X)$ is a principal ideal.

Let $f_0, \ldots, f_r$ be a set of generators of the $k[V]$-ideal I. Then $B(I) = \pi^{-1}(V) \subset V \times \mathbb{P}^r$ is the graph of the rational map $V \dashrightarrow \mathbb{P}^r$ defined by $(f_0 : \ldots : f_r)$. The projection $\pi : B(I) \to V$ is a regular birational map, which has the inverse rational map $\pi^{-1} : V \dashrightarrow B(I)$ defined by $\pi^{-1} = \text{id} \times (f_0 : \ldots : f_r)$. Thus we have a rational map $\pi^{-1}\phi : U \dashrightarrow B(I)$. Now this rational map is exactly $\phi \times (\phi^*(f_0) : \ldots : \phi^*(f_r))$. The ideal $Ik[U]$ which is generated by $\phi^*(f_0), \ldots, \phi^*(f_r)$ in $k[U]$ is by assumption principal.

We will show that, after possibly replacing U with an affine neighborhood of p in U, we have that there exists an i such that $\phi^*(f_i)$ generates the ideal $Ik[U]$. By our choice of U, $Ik[U] = gk[U]$ for some nonzero $g \in k[U]$. Thus there exists a relation $\sum a_i \phi^*(f_i) = g$ with all $a_i \in k[U]$ and $\phi^*(f_i) = d_i g$ for some $d_i \in k[U]$ for $0 \leq i \leq r$. Thus $(\sum a_i d_i) g = g$ and so $\sum a_i d_i = 1$ since g is nonzero and $k[U]$ is an integral domain. Thus there exists an i such that $d_i \notin I_U(p)$. Thus U_{d_i} is an affine neighborhood of p and $\phi^*(f_i) k[U_{d_i}] = Ik[U_{d_i}]$. Thus after replacing U with U_{d_i}, we have that $Ik[U] = \phi^*(f_i) k[U]$.

Without loss of generality, we may assume that $i = 0$. Thus there exist regular functions $h_i \in k[U]$ such that $\phi^*(f_i) = \phi^*(f_0)h_i$ for $i > 0$. In particular, $\phi \times (\phi^*(f_0) : \ldots : \phi^*(f_r)) = \phi \times (1 : h_1 : \ldots : h_r)$, which is a regular map on U, since the h_i are regular on U and 1 is never zero.

Now the rational map $\pi^{-1}\phi$ is uniquely determined and has at most one regular extension to U. Thus there is a unique regular map $\psi : U \to B(\mathcal{I})$ which factors $\phi|U$.

We can thus construct an affine open cover $\{U_i\}$ of X and unique regular maps $\psi_i : U_i \to B(\mathcal{I})$ factoring $\phi|U_i$. By Proposition 3.39 and since a regular map of a quasi-projective variety is uniquely determined by its restriction to a nontrivial open set, the ψ_i patch to a unique regular map $\psi : X \to B(\mathcal{I})$ which factors ϕ. $\square$

Proposition 12.6. *Suppose that $X \subset \mathbb{P}^m$ is a projective variety, with homogeneous coordinate functions $x_0, \ldots, x_m$ on X. Suppose that $\phi : X \dashrightarrow \mathbb{P}^n$ is a rational map, which is represented by homogeneous elements $F_0, \ldots, F_n \in S(X) = k[x_0, \ldots, x_m]$ of a common degree d (which are necessarily not all zero). Let $I = (F_0, \ldots, F_n) \subset S(X)$ and $\mathcal{I} = \tilde{I}$. Suppose that $p \in X$. Then ϕ is regular at p if and only if $\mathcal{I}_p$ is a principal ideal in $\mathcal{O}_{X,p}$. In particular, ϕ is a regular map if and only if $\tilde{I}$ is a locally principal ideal sheaf.*

Proof. Suppose that $\mathcal{I}_p$ is a principal ideal for some $p \in X$. Without loss of generality, we may suppose that $p \in X_{x_0}$. Then

$$\mathcal{I}(X_{x_0}) = \left(\frac{F_0}{x_0^d}, \ldots, \frac{F_n}{x_0^d}\right) = \left(F_0(1, \frac{x_1}{x_0}, \ldots, \frac{x_m}{x_0}), \ldots, F_n(1, \frac{x_1}{x_0}, \ldots, \frac{x_m}{x_0})\right)$$
$$\subset k[X_{x_0}] = k\left[\frac{x_1}{x_0}, \ldots, \frac{x_m}{x_0}\right].$$

Let $I(p)$ be the ideal of p in $k[X_{x_0}]$. We have by assumption that $\mathcal{I}_p = \mathcal{I}(X_{x_0})_{I(p)}$ is a nonzero principal ideal in the local ring $\mathcal{O}_{X,p} = k[X_{x_0}]_{I(p)}$, so that some $\frac{F_j}{x_0^d}$ generates $\mathcal{I}_p$ (by Exercise 1.12 or as in the proof of Theorem 12.5). Without loss of generality, $\frac{F_0}{x_0^d}$ generates this ideal. Thus there exists an affine neighborhood U of p contained in X_{x_i} such that $\frac{F_0}{x_0^d}$ generates the ideal $\mathcal{I}(U)$ in $k[U]$ (by Exercise 11.42). In particular, there exist regular functions $h_j \in k[U]$ such that

$$\frac{F_j}{x_0^d} = h_j \frac{F_0}{x_0^d} \quad \text{for all } j.$$

Now ϕ is represented by $(\frac{F_0}{x_0^d} : \frac{F_1}{x_0^d} : \ldots : \frac{F_n}{x_0^d})$ which is equivalent to $(1 : h_1 : \ldots : h_n)$, so that ϕ is regular at p.

Now suppose that ϕ is regular at p. Then there exist an affine neighborhood U of p and $h_0, \ldots, h_n \in k[U]$ such that $Z_U(h_0, \ldots, h_n) = \emptyset$, and ϕ is represented by $(h_0 : h_1 : \ldots : h_n)$. Without loss of generality, we may assume that $p \in X_{x_0}$ and that $U \subset X_{x_0}$. Thus

$$\frac{F_i}{x_0^d} h_j = h_i \frac{F_j}{x_0^d} \quad \text{for all } i, j.$$

We must have that some h_i satisfies $h_i(p) \neq 0$. We may suppose that $h_0(p) \neq 0$. After possibly replacing U with a smaller affine neighborhood of p, we may assume that $h_0(q) \neq 0$ for all $q \in U$, so that h_0 is a unit in $k[U]$. We have that

$$\frac{F_i}{x_0^d} = \frac{h_i}{h_0} \frac{F_0}{x_0^d}$$

for all i, so that $\mathcal{I}(U) = \left(\frac{F_0}{x_0^d}, \ldots, \frac{F_n}{x_0^d}\right) k[U]$ is a principal ideal, generated by $\frac{F_0}{x_0^d}$. Thus the localization $\mathcal{I}_p = \mathcal{I}(U)_{I_U(p)}$ is a principal ideal. $\square$

Theorem 12.7 (Resolution of indeterminancy). *Suppose that $X \subset \mathbb{P}^m$ is a projective variety, with homogeneous coordinate functions $x_0, \ldots, x_m$ on X. Suppose that $\phi : X \dashrightarrow \mathbb{P}^n$ is a rational map, which is represented by homogeneous elements $F_0, \ldots, F_n \in S(X) = k[x_0, \ldots, x_m]$ of a common degree d. Let $I = (F_0, \ldots, F_n) \subset S(X)$ and $\mathcal{I} = \tilde{I}$. Let $\pi : B(\mathcal{I}) \to X$ be the blow-up of $\mathcal{I}$, and let $\overline{\phi}$ be the rational map $\overline{\phi} = \phi\pi : B(\mathcal{I}) \dashrightarrow \mathbb{P}^n$. Then $\overline{\phi}$ is a regular map.*

Proof. This follows directly from our definition of the blow-up $B(\mathcal{I})$ as the graph $\Gamma_\phi \subset X \times \mathbb{P}^n$ of the rational map $\phi = (F_0, \ldots, F_n)$. The map $\phi\pi$ is the projection on the second factor $\mathbb{P}^n$. $\square$

12.2. Resolution of singularities

In this section we survey the main results on resolution of singularities. The interested reader is referred to the book *Resolution of Singularities* [**35**] and the article [**38**], which gives an accessible proof of resolution of singularities of 3-folds in characteristic greater than 5, for the proofs of the main results discussed in this section. Reading through Chapter 15 on schemes in this book is adequate preparation for reading [**35**]. Reading [**35**] is a good preparation for reading [**38**].

Definition 12.8. Suppose that X is a quasi-projective variety. A resolution of singularities of X is a closed subvariety Y of $X \times \mathbb{P}^n$ (for some n) such that the projection $\pi : Y \to X$ is birational and Y is nonsingular.

Lemma 12.9. *Suppose that $\pi : Y \to X$ is a resolution of singularities. Then $Y = B(\mathcal{I})$ for some ideal sheaf $\mathcal{I}$ on X.*

Proof. There exists a projective variety $\overline{X}$ such that X is an open subset of $\overline{X}$ and a closed embedding of Y into $X \times \mathbb{P}^n$ for some n. Let $\overline{Y}$ be the Zariski closure of Y in $\overline{X} \times \mathbb{P}^n$. We have that $(X \times \mathbb{P}^n) \cap \overline{Y} = Y$ since Y is closed in $X \times \mathbb{P}^n$. The projection $\overline{Y} \to \overline{X}$ is a birational regular map of projective varieties so $\overline{Y} = B(\mathcal{J})$ for some ideal sheaf $\mathcal{J}$ on $\overline{X}$ by Theorem 6.9. Letting $\mathcal{I} = \mathcal{J}|X$, we have that Y is isomorphic to the blow-up of $\mathcal{I}$ by Lemma 12.3. □

Theorem 12.10. *Suppose that X is a quasi-projective curve (a 1-dimensional variety). Then X has a resolution of singularities.*

Proof. There exists a projective variety $\overline{X}$ such that X is an open subset of $\overline{X}$. Let $\overline{Y}$ be the normalization of $\overline{X}$ in the function field $k(X)$ of X (by Theorem 7.17), with induced regular map $\phi : \overline{Y} \to \overline{X}$. The variety $\overline{Y}$ is a projective variety (by Theorem 7.17), so there is a closed embedding of $\overline{Y}$ in $\mathbb{P}^n$ for some n. The variety $\overline{Y}$ is nonsingular by Theorem 1.87 since $\overline{Y}$ is a curve and $\overline{Y}$ is normal. We have closed embeddings

$$\overline{Y} \cong \Gamma_\phi \subset \overline{Y} \times \overline{X} \subset \mathbb{P}^n \times \overline{X}$$

where Γ_ϕ is the graph of ϕ and the projection onto $\overline{X}$ is the map ϕ. Thus $\overline{Y}$ is a resolution of singularities of $\overline{X}$, and $Y = \Gamma_\phi \cap (\overline{Y} \times X)$ is a resolution of singularities of X. □

Theorem 12.11. *Suppose that X is a quasi-projective curve. Consider the sequence*

(12.1) $$\cdots \to X_n \xrightarrow{\pi_n} \cdots \to X_1 \xrightarrow{\pi_1} X_0 = X,$$

where $X_{n+1} \to X_n$ is obtained by blowing up the (finitely many) singular points on X_n. Then this sequence is finite (there exists an n such that X_n is nonsingular) and is thus a resolution of singularities of X.

Proof. It suffices to prove the theorem when X is projective. Then all X_n are projective. Let $\phi : Y \to X$ be the normalization of X (Theorem 7.17). The variety Y is projective since X is, and Y is a resolution of singularities of X (Theorem 12.10). We have a factorization (by Corollary 10.24)

$$Y \to X_n \xrightarrow{\pi_n} \cdots \to X_1 \xrightarrow{\pi_1} X$$

of regular maps of projective varieties for all n. Since each map in the sequence is a dominant regular map of curves, the preimage of a point by each map is a finite set of points. Thus each map in the sequence is finite by Theorem 9.6. If the sequence (12.1) is infinite, then there exists an affine open subset U of X such that the induced sequence of preimages above U in the sequence (12.1)

$$\cdots \to U_n \to \cdots \to U_1 \to U$$

12.2. Resolution of singularities

is infinite, and each U_n is singular. Let Z be the preimage of U in Y. The sets Z and U_i are all affine open sets since the maps to U are finite. Let $R = R_0 = k[U]$ and let $\overline{R}$ be the integral closure of R in the function field $k(X)$ of X. Then $k[Z] = \overline{R}$. Let $R_i = k[U_i]$ for all i. We have inclusions in $k(X)$

$$R_0 \subset R_1 \subset \cdots \subset R_n \subset \cdots \subset \overline{R}.$$

We have that $R_i \neq R_{i+1}$ for all i since mR_{i+1} is a locally principal ideal if m is a maximal ideal of R_i such that $(R_i)_m$ is not regular (U_{i+1} is the blow-up of all singular points of U_i) and $m(R_i)_m$ is not principal (since $(R_i)_m$ is not regular). Since $\overline{R}$ is finite over R and the sequence (12.1) is assumed infinite, there exists an n such that $R_n = \overline{R}$. But then U_n is nonsingular, a contradiction. □

An approach to resolve surface singularities is by the following algorithm. Suppose that S is a surface. Let S_1 be the normalization of S, so that S_1 has only finitely many singular points (by Theorem 10.17), and let $S_2 \to S_1$ be the blow-up of all singular points on S_2. If S_2 is nonsingular, we have obtained a resolution of singularities of S. Otherwise, we can repeat, computing the normalization S_3 of S_2 and blowing up all singular points of S_3 to obtain S_4. We can repeat as long as S_i is singular, obtaining a sequence

$$(12.2) \qquad \cdots \to S_{2n} \to S_{2n-1} \to \cdots \to S_2 \to S_1 \to S.$$

Theorem 12.12. *Suppose that S is a quasi-projective surface. Then some S_i is a resolution of singularities of S.*

This theorem was proven when the ground field k has characteristic 0 by Zariski in [**149**] and was proven in arbitrary characteristic (in fact over two-dimensional excellent integral schemes) by Lipman [**101**]. Zariski discusses early approaches to resolution of surface singularities in his book [**148**]. Zariski credited Walker's proof [**144**] as the first complete proof of resolution of singularities of complex surfaces.

The first proof of resolution of singularities of surfaces in positive characteristic was given by Abhyankar in [**2**] (by a different method).

Hironaka proved the existence of a resolution of singularities of a variety of any dimension in characteristic 0.

Theorem 12.13 (Hironaka). *Suppose that X is a variety over a field of characteristic 0. Then X has a resolution of singularities.*

Hironaka's first proof is in [**79**]. There have been many simplifications in the proof (some papers and books on this are [**22**], [**24**], [**54**], [**35**], [**90**]).

Abhyankar [**4**] proved resolution of singularities of three-dimensional varieties over fields of positive characteristic greater than 5. Since then, Cossart and Piltant have established the existence of a resolution of singularities for reduced three-dimensional quasi-excellent schemes in [**31**]. All of these proofs are extremely long. A much shorter proof of Abhyankar's result is given in [**38**].

Some recent papers on resolution in higher dimensions and positive characteristic are [**25**], [**75**], [**80**], [**94**], [**87**], [**140**], [**141**].

Exercise 12.14. Resolve the singularities of the curve $C = Z(x^2 - x^4 - y^4) \subset \mathbb{A}^2$ by blowing up points; that is, perform a sequence of blow-ups of points above $\mathbb{A}^2$ so that the strict transform of C is nonsingular.

12.3. Valuations in algebraic geometry

Suppose that K is a field. A valuation ν of K is a map

$$\nu : K^\times \to \Gamma_\nu$$

from the multiplicative group $K^\times$ of nonzero elements of K onto a totally ordered Abelian group Γ_ν, called the value group of ν. The map ν must satisfy two properties:

1. $\nu(fg) = \nu(f) + \nu(g)$ for $f, g \in K^\times$,
2. $\nu(f + g) \geq \min\{\nu(f), \nu(g)\}$ for $f, g \in K^\times$.

The valuation ν extends to K by defining $\nu(0) = \infty$, which is larger than anything in Γ_ν. The valuation ring of ν is

$$V_\nu = \{f \in K \mid \nu(f) \geq 0\}.$$

The basic theory of valuation rings is explained in the paper [**152**], [**160**, Chapter V], [**161**, Chapter VI], and in [**6**]. A quick introduction is given in the section on valuation rings in of [**13**, Chapter 5]. The next two theorems state a couple of basic facts about valuation rings.

Theorem 12.15. *A valuation ring V_ν is Noetherian if and only if $\Gamma_\nu \cong \mathbb{Z}$.*

Proof. [**161**, Theorem 16, page 41]. □

Theorem 12.16. *A valuation ring V_ν is a normal local ring.*

Proof. [**13**, Proposition 5.18]. □

If K is a function field over a field κ, we require a valuation of K to also satisfy the third property that

3. $\nu|\kappa^\times = 0$.

12.3. Valuations in algebraic geometry

A valuation ν of K is called divisorial if V_ν is a localization of a finitely generated algebra over κ. Since such a ring is Noetherian, $\Gamma_\nu \cong \mathbb{Z}$ if ν is divisorial.

If K is the function field of a curve, then all of the valuation rings of K are divisorial. In fact, by [161, Theorem 9, page 17], the valuation rings of K are exactly the local rings of the points on the nonsingular projective curve whose function field is K, and the field K itself (the valuation ring of the trivial valuation).

If K has transcendence degree larger than 1 over its ground field κ (so that K is the function field of a variety of dimension larger than 1), then K admits many valuations whose valuation ring is not Noetherian. The example in [161, pages 102–104] shows that every additive subgroup of the rational numbers must appear as the valuation group of a valuation of K, so K has many valuations whose value group is not finitely generated.

A theory of algebraic geometry built around valuation rings is developed in Zariski's paper [152]. This is the paper where Zariski's main theorem first appears. This theory requires the introduction of more local rings on a variety than we have considered up to now.

Definition 12.17. Suppose that X is a quasi-projective variety and Y is a subvariety of X. Then the local ring $\mathcal{O}_{X,Y}$ is defined to be the localization $\mathcal{O}_{X,Y} = k[U]_{\mathcal{I}_Y(U)}$ where U is any affine open subset of X such that $Y \cap U \neq \emptyset$ and $\mathcal{I}_Y$ is the ideal sheaf of Y in X.

A special case is when Y is a point, in which case the above definition agrees with the definition of the stalk $\mathcal{O}_{X,Y}$ of $\mathcal{O}_X$ at the point Y.

This ring is independent of the choice of U. In fact, we have that if $p \in Y$, then $\mathcal{O}_{X,Y}$ is the localization $\mathcal{O}_{X,Y} = (\mathcal{O}_{X,p})_{\mathcal{I}_{Y,p}}$. We see this as follows. Suppose that U is an affine neighborhood of p in X. Then $\mathcal{O}_{X,p} = k[U]_{I(p)}$ and $\mathcal{I}_{Y,p} = (\mathcal{I}_Y(U))_{I(p)}$ where $I(p)$ is the ideal of p in $k[U]$. Thus

$$(\mathcal{O}_{X,p})_{\mathcal{I}_{Y,p}} \cong k[U]_{\mathcal{I}_Y(U)}$$

by Exercise 1.22 since $\mathcal{I}_Y(U) \subset I(p)$.

If U and V are two affine open subsets of X which intersect Y, then there exists a point p in $U \cap V \cap Y$ since Y is irreducible. Thus

$$k[U]_{\mathcal{I}_Y(U)} = (\mathcal{O}_{X,p})_{\mathcal{I}_{Y,p}} = k[V]_{\mathcal{I}_Y(V)}.$$

If X is affine, then the local rings $\mathcal{O}_{X,Y}$ of X are precisely the local rings $k[X]_P$ for $P \in \text{Spec}(k[X])$ (the spectrum of a ring is defined in Exercise 1.11). If X is projective, then the local rings $\mathcal{O}_{X,Y}$ of X are precisely the local rings $S(X)_{(P)}$ for $P \in \text{Proj}(S(X))$ (the Proj of a graded ring is defined in Exercise 3.7).

Suppose that X is a variety and $K = k(X)$ is the function field of X. Let ν be a valuation of K. We will say that ν dominates a local ring S contained in K if $S \subset V_\nu$ and the maximal ideal of V_ν intersects S in its maximal ideal.

Theorem 12.18 (Zariski). *Suppose that X is a projective variety and ν is a valuation of $k(X)$. Then there is a unique subvariety Y of X such that ν dominates $\mathcal{O}_{X,Y}$.*

Proof. Let $S = k[x_0, \ldots, x_n]$ be the coordinate ring of a projective embedding of X. Suppose that i is such that $\nu(\frac{x_i}{x_0}) \leq \nu(\frac{x_j}{x_0})$ for $0 \leq j \leq n$. Then $\nu(\frac{x_j}{x_i}) = \nu(\frac{x_j}{x_0}) - \nu(\frac{x_i}{x_0}) \geq 0$ for $0 \leq j \leq n$. Thus

$$\mathcal{O}_X(X_{x_i}) = k\left[\frac{x_0}{x_i}, \ldots, \frac{x_n}{x_i}\right] \subset V_\nu.$$

Let $\mathfrak{p} = m_\nu \cap \mathcal{O}_X(X_{x_i})$ where m_ν is the maximal ideal of V_ν. Let Y be the Zariski closure in X of the subvariety $Z(\mathfrak{p})$ of X_{x_i}. Then $\mathcal{O}_{X,Y} = (\mathcal{O}_X(X_{x_i}))_\mathfrak{p}$ is dominated by ν.

Suppose that Z is another closed subvariety of X which has the property that $\mathcal{O}_{X,Z}$ is dominated by ν. There exists a linear form $L \in S$ such that $L \notin I(Y)$ and $L \notin I(Z)$ (by Lemma 8.10). Then $Z \cap X_L \neq \emptyset$ and $Y \cap X_L \neq \emptyset$. Let $R = k[X_L]$, $\mathfrak{p}_1 = I(Y \cap X_L)$, and $\mathfrak{p}_2 = I(Z \cap X_L)$. Here $R_{\mathfrak{p}_1} = \mathcal{O}_{X,Y}$ and $R_{\mathfrak{p}_2} = \mathcal{O}_{X,Z}$. We have that $R \subset V_\nu$ since $R_{\mathfrak{p}_1} \subset V_\nu$. Since V_ν dominates both $R_{\mathfrak{p}_1}$ and $R_{\mathfrak{p}_2}$, we have that $\mathfrak{p}_1 = m_\nu \cap R = \mathfrak{p}_2$. But this is impossible since $I(Y)$ and $I(Z)$ are distinct prime ideals in S. □

We will call the subvariety Y of the conclusions of Theorem 12.18 the center of ν on X.

The following lemma gives a characterization of valuation rings of divisorial valuations.

Lemma 12.19. *Suppose that K is an algebraic function field over a field κ. Suppose ν is a divisorial valuation of K. Then there exists a normal local ring R which is of finite type over κ and a height 1 prime ideal Q in R such that $R_Q = V_\nu$. Conversely, given a subring R of K with these properties and a height 1 prime ideal Q of R, there exists a divisorial valuation ν of K such that $R_Q = V_\nu$.*

Proof. A ring $A = R_Q$ as in the statement of the lemma is a normal Noetherian local ring of dimension 1, so it is a regular local ring of dimension 1 by Theorem 1.87. Let $f \in A$ be a generator of its maximal ideal. Then every nonzero element $h \in K$ has a unique expression $h = uf^n$ where $u \in A$ is a unit and $n \in \mathbb{Z}$. Define $\nu(h) = n \in \mathbb{Z}$. The function ν is a valuation of K as it satisfies the three conditions of a valuation and $A = V_\nu$.

Now suppose that ν is a divisorial valuation of K. The value group of ν is $\mathbb{Z}$ by Theorem 12.15 since V_ν is Noetherian. Let m_ν be the maximal ideal of V_ν. There exists $f \in m_\nu$ such that $\nu(f) = 1$. Suppose $g \in m_\nu$. Then $\nu(g) \geq 1$. Let $h = \frac{g}{f} \in K$. We have that $\nu(h) \geq 0$ so $h \in V_\nu$. Thus $g \in (f)$. We have that $m_\nu = (f)$, so that V_ν is a regular local ring of dimension 1. By assumption, there exist a domain B with quotient field K which is finitely generated over κ and a prime ideal P in B such that $B_P = V_\nu$. Let R be the integral closure of B in K, which is a finitely generated κ-algebra by Theorem 1.54. Then $R \subset B_P$ since V_ν is normal. Let $Q = P_P \cap R$. Then $V_\nu \cong R_Q$. We have that Q is a height 1 prime ideal in R since $\dim V_\nu = 1$. □

Zariski's proof of resolution of singularities of surfaces in Theorem 12.12 ([**149**]) is by assuming that the algorithm described before the statement of the theorem does not terminate. Then we have an infinite sequence of points $p_i \in S_{2i}$ such that p_i maps to p_{i-1} for all i and there is an infinite sequence of distinct local rings

$$R \to R_1 \to \cdots \to R_i \to \cdots$$

where $R_i = \mathcal{O}_{S_{2i}, p_i}$ is a normal but not regular local ring. Let $V = \bigcup_{i=1}^\infty R_i$. Then V is the valuation ring V_ν of a valuation ν of $k(S)$ (this uses the fact that S has dimension 2).

Zariski then shows that ν has a local uniformization (defined below) and then makes a delicate argument to show that this leads to a contradiction to the assumption that the localizations of all of the R_i at the ideal of p_i are not regular.

Definition 12.20. Suppose that K is an algebraic function field and ν is a valuation of K. The valuation ν has a local uniformization if there exists a variety X whose function field is K such that the center of ν on X is a regular local ring.

Zariski proved local uniformization of all valuations on characteristic 0 function fields [**150**] and was able to show from this result that resolution of singularities is true for varieties of dimension ≤ 3 over (algebraically closed) fields of characteristic 0 [**153**]. Hironaka's proof of resolution of singularities of characteristic 0 varieties [**79**] does not use valuations or local uniformization.

The first proof of resolution of singularities of surfaces in characteristic $p > 0$ was by Abhyankar [**2**]. It is by proving local uniformization of all valuations on a two-dimensional algebraic function field. The proof by Lipman [**101**] of resolution of surface singularities does not use local uniformization.

However, all proofs of resolution of singularities of three-dimensional varieties in positive characteristic use valuations and are done by proving local uniformization of all valuations of the function field.

Exercise 12.21. This exercise gives a geometric interpretation of divisorial valuations.

a) Suppose that X is a normal projective variety and ν is a divisorial valuation on $k(X)$. Let Z be the center of ν on X, so that we have an inclusion $\mathcal{O}_{X,Z} \subset V_\nu$. Show that $\mathcal{O}_{X,Z} = V_\nu$ if and only if $\operatorname{codim}_X Z = 1$.

b) Give an example of a divisorial valuation ν on $k(\mathbb{P}^2)$ such that $V_\nu \neq \mathcal{O}_{\mathbb{P}^2,Z}$, where Z is the center of ν on $\mathbb{P}^2$.

c) Suppose that X is a normal projective variety and ν is a divisorial valuation on $k(X)$. Show that there exists a birational regular map $\phi : Y \to X$ of normal projective varieties such that if W is the center of ν on Y, then $\mathcal{O}_{Y,W} = V_\nu$.

12.4. Factorization of birational maps

The blow-ups of a point and of a nonsingular curve in a nonsingular projective 3-fold X (a three-dimensional variety) give examples of birational regular maps of nonsingular projective 3-folds (by Theorem 10.19). Further examples can be found by taking products (sequences) of blow-ups of points and nonsingular curves. In light of the fact that birational maps of nonsingular projective surfaces can always be factored by a product of blow-ups of points (Theorem 10.32), it is natural to ask if every birational regular map of nonsingular projective 3-folds can be factored by a product of blow-ups of points and nonsingular curves. This is, however, not true. Counterexamples have been given by Hironaka [78], Shannon [137], and Sally [129], all in their PhD theses (with Zariski, Abhyankar, and Kaplansky, respectively). The examples of Shannon and Sally do not even factor locally. Here is a simple example of a birational regular map of nonsingular projective 3-folds which does not factor.

Example 12.22. There exists a birational regular map of projective nonsingular 3-folds which cannot be factored by a product of blow-ups of points and nonsingular curves.

The example is constructed as follows. Let $X = \mathbb{P}^3$ with coordinate ring $S(X) = k[x_0, x_1, x_2, x_3]$, where we require that k has characteristic $\neq 3$. Let $C \subset X$ be the curve $C = Z(I)$ where I is the homogeneous prime ideal

$$I = (x_0, x_1 x_2 x_3 + x_1^3 + x_2^3).$$

12.4. Factorization of birational maps

The curve C has an isolated singularity at the point $p = (0:0:0:1)$. Let $\pi_1 : X_1 \to X$ be the blow-up of the curve C (of the ideal sheaf $\tilde{I}$). The variety $X_1 \setminus \pi_1^{-1}(p)$ is nonsingular by Theorem 10.19. Let $U = X_{x_3} \cong \mathbb{A}^3$. The regular functions on U are $k[U] = k[x, y, z]$ where

$$x = \frac{x_0}{x_3}, \quad y = \frac{x_1}{x_3}, \quad z = \frac{x_2}{x_3}.$$

We have that $\tilde{I}(U) = (x, yz + y^3 + z^3)$ and p is the origin in U. We have that $\pi_1^{-1}(U) = B(\tilde{I}(U))$ has an open cover by two affine open sets U_1 and U_2 with

$$k[U_1] = k[U]\left[\frac{x}{yz + y^3 + z^3}\right] \quad \text{and} \quad k[U_2] = k[U]\left[\frac{yz + y^3 + z^3}{x}\right]$$

by Theorem 6.4. Since

$$k[x, y, z, s]/(s(yz + y^3 + z^3) - x) \cong k[y, z, s]$$

is a domain, we have that

$$k[U_1] \cong k[x, y, z, s]/(s(yz + y^3 + z^3) - x) \cong k[y, z, s]$$

by Exercise 1.10, so $U_1 \cong \mathbb{A}^3$ is nonsingular. Since

$$k[x, y, z, t]/(tx - (yz + y^3 + z^3))$$

is a domain, we have that

$$k[U_2] \cong k[x, y, z, t]/(tx - (yz + y^3 + z^3))$$

by Exercise 1.10. Let $f = tx - (yz + y^3 + z^3)$. All singular points of U_2 are in $U_2 \cap \pi_1^{-1}(p) = Z(x, y, z)$. The only point of $U_2 \cap \pi_1^{-1}(p)$ on which $\frac{\partial f}{\partial x} = t$ vanishes is $q := Z(x, y, z, t)$, which is the only singular point of U_2.

Further, we have that $\pi_1^{-1}(C)$ is an irreducible surface E ($yz + y^3 + z^3 = 0$ is a local equation of E in U_1 and $x = 0$ is a local equation of E in U_2) and $\pi_1^{-1}(\alpha) \cong \mathbb{P}^1$ for all $\alpha \in C$.

Let $\pi_2 : X_2 \to X_1$ be the blow-up of the point q. The open set $\pi_2^{-1}(U_2)$ is naturally covered by four affine open sets V_1, V_2, V_3, V_4. We compute their regular functions by taking the strict transform of $f = 0$ in the blow-up of q in $\mathbb{A}^4$ (with $k[\mathbb{A}^4] = k[x, y, z, t]$) to see that X_2 is nonsingular and $\pi_2^{-1}(q)$ is a surface $F \cong \mathbb{P}^1 \times \mathbb{P}^1$ (this is similar to the calculation of Exercise 6.16). Let $\overline{E}$ be the strict transform of E on X_2.

Let $\pi = \pi_1 \pi_2 : X_2 \to X$. The map π is a birational regular map of nonsingular projective 3-folds which is an isomorphism over $X \setminus \pi^{-1}(C)$ and $\pi^{-1}(C) = \overline{E} \cup F$ is a union of two irreducible surfaces.

Suppose that $\pi : X_2 \to X$ factors as a product of blow-ups of points and nonsingular curves. Then the first blow-up must be of a point α contained in C since C is an irreducible singular curve. Let $\mathcal{I}_\alpha$ be the ideal sheaf of

α in X. We have that $\pi^{-1}(\alpha) \cong \mathbb{P}^1$ if $\alpha \neq p$ and $\pi^{-1}(\alpha) = F \cup \gamma$ where $\gamma \cong \mathbb{P}^1$ is a curve which intersects F in a point if $\alpha = p$. In each case, there exists an affine open subset W of X_2 which intersects $\pi^{-1}(\alpha)$ in a curve and such that $\pi(W) \subset T$ for some affine open neighborhood T of α in X. Here $Z_W(I_T(\alpha)) = \pi^{-1}(\alpha) \cap W$ which has codimension > 1 in W so $\mathcal{I}_\alpha \mathcal{O}_W$ cannot be locally principal (by Krull's principal ideal theorem). Since $\mathcal{I}_\alpha \mathcal{O}_{X_2}$ is not locally principal, we have that $\pi : X_2 \to X$ cannot factor through the blow-up $B(\alpha) \to X$ of α (since $\mathcal{I}_\alpha \mathcal{O}_{B(\alpha)}$ is locally principal). This contradiction shows that $\pi : X_2 \to X$ is not a product of blow-ups of points and nonsingular curves.

Hironaka [**78**] and Abhyankar [**7**] proposed the following problem:

Question 12.23. *Suppose that $\phi : X \to Y$ is a birational regular map of nonsingular projective varieties. Does there exist a nonsingular projective variety Z and a commutative diagram of regular maps*

$$\begin{array}{ccc} & Z & \\ \swarrow & & \searrow \\ X & \to & Y \end{array}$$

such that the maps $Z \to X$ and $Z \to Y$ are products of blow-ups of nonsingular subvarieties?

This question is still open, even in dimension 3 and characteristic 0. However, we have the following theorem:

Theorem 12.24 (Abramovich, Karu, Matsuki, Włodarczyk [**9**]). *Suppose that $\phi : X \to Y$ is a birational regular map of nonsingular projective varieties, over a field of characteristic 0. Then there is a factorization (for some n)*

$$\begin{array}{ccccccccc} & Y_n & & & Y_{n-2} & & \cdots & & \\ \swarrow & & \searrow & \swarrow & & \searrow & & \swarrow & \searrow \\ X & & Y_{n-1} & & & Y_{n-3} & & & Y_0 = Y \end{array}$$

where each diagonal arrow is a (finite) product of blow-ups of nonsingular subvarieties.

This theorem is not known in positive characteristic (even in dimension 3).

Theorem 12.24 was proven earlier for toric and toroidal varieties (Włodarczyk [**147**], Abramovich, Matsuki, Rashid [**10**]). This is a category of varieties which is built by only allowing Laurent monomials instead of arbitrary polynomials. A regular map of nonsingular affine toric (or toroidal)

12.4. Factorization of birational maps

varieties is then a monomial map

(12.3) $$\phi : \mathbb{A}^n \to \mathbb{A}^m \quad \text{with } \phi = (M_1, \ldots, M_m)$$

where the M_i are monomials $M_i = \prod_{j=1}^n x_j^{a_{ij}}$ for $1 \le i \le m$ in the coordinate functions $x_1, \ldots, x_n$ of $\mathbb{A}^n$. A regular map of nonsingular toric (or toroidal) varieties is constructed by patching together maps of the form (12.3).

Question 12.25 has been proposed by Oda [123] for the restricted case of toric (toroidal varieties).

Question 12.25. *Suppose that $\phi : X \to Y$ is a birational regular map of nonsingular projective toric (toroidal) varieties. Does there exist a nonsingular projective toric (toroidal) variety Z and a commutative diagram of regular toric maps*

$$\begin{array}{ccc} & Z & \\ \swarrow & & \searrow \\ X & \to & Y \end{array}$$

such that the maps $Z \to X$ and $Z \to Y$ are product of blow-ups of nonsingular toric subvarieties?

This question is open even in dimension 3 and characteristic 0. However, local factorization is true in all dimensions and characteristic 0. The following theorem was conjectured by Abhyankar [8, Section 8].

Theorem 12.26 (Cutkosky [33]). *Suppose that $\phi : X \to Y$ is a birational regular map of nonsingular projective varieties over a field k of characteristic 0 and ν is the valuation of the function field $k(X) = k(Y)$. Then there exists a nonsingular projective variety Z and a commutative diagram of regular maps*

$$\begin{array}{ccc} & Z & \\ \swarrow & & \searrow \\ X & \to & Y \end{array}$$

such that there exist affine neighborhoods U, V, W of the respective centers p, q, r of ν on X, Y, Z such that W is an affine open subset of a sequence of blow-ups of nonsingular subvarieties above U and also W is an affine open subset of a sequence of blow-ups of nonsingular subvarieties above V.

The theorem was first proven by Christensen [29] for the case of a toric valuation in dimension 3. A toric valuation ν of $\mathbb{A}^n$ is obtained by assigning positive real numbers $r_1, \ldots, r_n$ as weights to the coordinate functions $x_1, \ldots, x_n$. The valuation of a polynomial $f = \sum_I a_I x^I$ with $I = (i_1, \ldots, i_n) \in \mathbb{N}^n$ and $a_I \in k$ is then

$$\nu(f) = \min\{i_1 r_1 + \cdots + i_n r_n \mid a_I \ne 0\}.$$

The case of a toric valuation in dimension n is solved by Karu [86] and by Cutkosky and Srinivasan [45]. The case of a general valuation is solved in [33].

12.5. Monomialization of maps

We ask if it is possible to put an arbitrary regular map of varieties $\phi : X \to Y$ into a particularly nice form by performing sequences of blow-ups of nonsingular subvarieties $X_1 \to X$ and $Y_1 \to Y$ to obtain a new map $\phi_1 : X_1 \to Y_1$ which has a simpler form. The simplest such form which is possible is for $X_1 \to Y_1$ to be "locally monomial". This means that for every point p of X_1, there exist uniformizing parameters near p, such that after possibly taking étale covers or using formal coordinates (these concepts will be explained in Chapters 14 and 21), ϕ_1 locally has an expression as a monomial map of the type of (12.3). (Formally, every nonsingular point looks like a point on $\mathbb{A}^n$.)

Specifically, we have the following question.

Question 12.27. *Suppose that $\phi : X \to Y$ is a dominant regular map of characteristic 0 varieties. Does there exist a commutative diagram of regular maps*

(12.4)
$$\begin{array}{ccc} X_1 & \stackrel{\phi_1}{\to} & Y_1 \\ \downarrow & & \downarrow \\ X & \stackrel{\phi}{\to} & Y \end{array}$$

such that X_1 and Y_1 are nonsingular, the vertical arrow are products of blow-ups of nonsingular subvarieties, and $\phi_1 : X_1 \to Y_1$ is locally monomial (or toroidal)?

We will call such a diagram (12.4) a monomialization (toroidalization) of ϕ.

We prove in [34] and [37] that the answer to Question 12.27 is yes if X and Y are three-dimensional varieties over an algebraically closed field k of characteristic 0. (A simpler proof of the results of [34] is given in [39].) As a corollary to this theorem, we obtain a different proof of Theorem 12.24 in dimension 3. This theorem (the solution to Question 12.27 in dimension 3) also shows that the Hironaka-Abhyankar Question 12.23 for factorization of birational maps in dimension 3 and characteristic 0 is implied by the Oda Question 12.25 for factorization of birational maps of toroidal varieties in dimension 3.

Question 12.27 has a negative answer if we allow fields of positive characteristic p. A simple example is the map $\phi : \mathbb{A}^1 \to \mathbb{A}^1$ defined by $t \mapsto t^p + t^{p+1}$. Since $\mathbb{A}^1$ is a nonsingular curve, blow-ups do not change anything, so the

question asks in this case if ϕ itself is locally monomial. The map is given by the expression
$$u = x^p + x^{p+1} = \delta x^p$$
where $\delta = 1 + x$ is a unit in $R = k[x]_x$. To represent u as a monomial, we would have to set $\bar{x} = \delta^{\frac{1}{p}} x$ and then obtain $u = \bar{x}^p$. However, $\delta^{\frac{1}{p}} = 1 + x^{\frac{1}{p}}$ is not in the completion $\hat{R} = k[[x]]$, so such a monomialization is not possible by a formal change of variables.

From this argument, we also see why all maps of nonsingular curves are locally monomial in characteristic 0. We can always represent a map of nonsingular curves $\phi : X \to Y$ locally at a point $q \in X$ by an expression $u = \alpha x^n$ where u is a regular parameter in $R = \mathcal{O}_{Y,\phi(q)}$, x is a regular parameter in $R = \mathcal{O}_{X,q}$, and α is a unit in R. In this case, taking an n-th root of α gives an étale map locally above q on X; in fact, with our assumption that $k = R/m_q$ is algebraically closed, we have that $\alpha^{\frac{1}{n}}$ is in the completion $\hat{R} = k[[x]]$, so we have an expression $u = \bar{x}^n$ where $\bar{x} = \alpha^{\frac{1}{n}} x$ is a regular parameter of $\hat{R}$.

We have a local valuation-theoretic version of Question 12.27.

Question 12.28. *Suppose that $\phi : X \to Y$ is a dominant regular map of projective varieties and ν is a valuation of $k(X)$. Does there exist a commutative diagram of regular maps*

(12.5)
$$\begin{array}{ccc} X_1 & \stackrel{\phi_1}{\to} & Y_1 \\ \downarrow & & \downarrow \\ X & \stackrel{\phi}{\to} & Y \end{array}$$

such that X_1 and Y_1 are nonsingular, there exist affine neighborhoods U, V, W, Z of the respective centers q, r, q', r' of ν on X, Y, X_1, and Y_1, respectively, such that W and Z are affine open subsets of sequences of blow-ups of nonsingular subvarieties above U and V, respectively, and there are regular parameters $x_1, \ldots, x_n$ in $\mathcal{O}_{X_1,q'}$ and $y_1, \ldots, y_m$ in $\mathcal{O}_{Y',r'}$, units $\delta_1, \ldots, \delta_m \in \mathcal{O}_{X_1,q'}$, and a matrix $A = (a_{ij})$ of natural numbers such that

(12.6) $$y_i = \delta_i \prod_{j=1}^n x_j^{a_{ij}} \quad \text{for } 1 \leq i \leq m$$

and

(12.7) $$\mathrm{rank}(A) = m?$$

This question makes sense over all fields (it is true for curves in every characteristic). In characteristic 0, if (12.6) and (12.7) hold, then we can make an étale or formal change of variables to represent the y_i as monomials in regular parameters in an étale extension or the completion of $\mathcal{O}_{X_1,q'}$ (the condition (12.7) is essential!).

Theorem 12.29. *Question 12.28 always has a positive answer over fields of characteristic 0.*

This is proven in [**33**] and [**36**].

Question 12.28 has a negative answer in positive characteristic $p > 0$, even in dimension 2. A counterexample is given in [**40**]. It is shown in [**43**] and most generally in [**41**] that when X and Y have dimension 2 and the valuation has positive residue characteristic, then Question 12.28 has a positive answer for defectless extensions of valued fields. The defect is an interesting invariant of extensions of valuations (in this case from the restriction of ν to the function field of Y to its extension ν on $k(X)$) which can only occur for fairly complicated valuations in characteristic $p > 0$. The presence of defect really says that information about the extension that you should be able to extract from the quotient of value groups is lost. The defect can be viewed as the cause of all of the trouble in local uniformization in positive characteristic [**93**].

Chapter 13

Divisors

We define a divisor on a normal variety X to be a formal sum of prime divisors (codimension 1 irreducible subvarieties of X) in Section 13.1. We associate to a rational function f on X a divisor (f) (or $\mathrm{div}(f)$) which is the difference of the zeros of f and the poles of f, counting multiplicity. The divisor class group $\mathrm{Cl}(X)$ of X (see (13.1)) is the group of equivalence classes of divisors on X, modulo the divisors of rational functions on X. In Section 13.4 we calculate some examples of divisor class groups, and in Section 13.5, we analyze divisors in the most intuitive situation, on a nonsingular projective curve.

Associated to a divisor D on a normal variety X, we have a coherent sheaf $\mathcal{O}_X(D)$, consisting of the rational functions on X whose poles are bounded by D (Section 13.2).

If X is nonsingular, then the sheaf $\mathcal{O}_X(D)$ of a divisor D on X is an invertible sheaf (see (13.2)). Further, we can cover X with affine open subsets U_i such that on each U_i, $D \cap U_i$ is the divisor of a rational function $g_i \in k(X)$ on U_i. The set $\{(g_i, U_i)\}$ is called a Cartier divisor. The concept of a Cartier divisor will be further explored, in the more general situation of schemes, in Section 15.1.

The three concepts of *divisors*, *invertible sheaves*, and *Cartier divisors* are equivalent on a nonsingular variety (see (13.16)). The concepts of invertible sheaves and Cartier divisors agree on varieties, but not on schemes, while the concepts of divisors and invertible sheaves are not the same on a normal (but singular) variety. The concept of invertible sheaf is the most general and is valid on an arbitrary variety (or scheme).

Associated to a divisor D on a normal variety X, we have linear systems on X, parametrizing effective divisors on X which are linearly equivalent to D (Section 13.6). Rational maps from a normal variety can be interpreted through linear systems. The divisors in a linear system (without fixed component) are the pull-backs of linear hyperplane sections in the image. Using this interpretation of rational map, we give criteria for a rational map from a nonsingular variety to be a regular map, injective, and a closed embedding in Section 13.7.

In Sections 13.8 and 13.9, we develop the geometric theory of invertible sheaves on an arbitrary variety, generalizing the theory of divisors on a normal nonsingular variety.

13.1. Divisors and the class group

Suppose that X is a normal quasi-projective variety. A prime divisor on X is an irreducible codimension 1 subvariety E of X. A divisor on X is a finite formal sum $D = \sum a_i E_i$ where the E_i are prime divisors on X and $a_i \in \mathbb{Z}$. Let $\mathrm{Div}(X)$ be the group (under addition) of divisors on X. The group $\mathrm{Div}(X)$ is the free Abelian group on the set of prime divisors of X. The support of a divisor $D = \sum a_i E_i$ is the algebraic set in X,

$$\mathrm{Supp}\, D = \bigcup_{a_i \neq 0} E_i.$$

Suppose that D_1, D_2 are divisors on X. We will say that $D_1 \geq D_2$ if $D_1 - D_2 = \sum a_i E_i$ where all a_i are nonnegative. A divisor D such that $D \geq 0$ is called effective.

Now suppose that E is a prime divisor of X. We associate to E the local ring $\mathcal{O}_{X,E}$, the local ring of a subvariety which we defined in Definition 12.17.

Since X is normal, the singular locus of X has codimension ≥ 2 in X (by Theorem 10.17), so there exists $p \in E$ which is a nonsingular point of X. Let U be an affine neighborhood of p in X. Let I be the ideal of $E \cap U$ in $k[U]$ (that is, let $I = \mathcal{I}_E(U)$ where $\mathcal{I}_E$ is the ideal sheaf of E in X). The ideal I is a height 1 prime ideal in $k[U]$. Let $I(p)$ be the ideal of the point p in U, which is a maximal ideal in $k[U]$. Now $\mathcal{O}_{X,p} = k[U]_{I(p)}$ and the stalk $(\mathcal{I}_E)_p$ is the localization $I_{I(p)}$. Thus $(\mathcal{I}_E)_p = I_{I(p)}$ is a height 1 prime ideal in $\mathcal{O}_{X,p}$. We have that $\mathcal{O}_{X,E}$ is the localization $(\mathcal{O}_{X,p})_{(\mathcal{I}_E)_p}$. Now $\mathcal{O}_{X,p}$ is a regular local ring, since p is a nonsingular point of X, so the localization $\mathcal{O}_{X,E}$ is a regular local ring, by Theorem 1.88. Further, $\mathcal{O}_{X,E}$ has dimension 1, so that the maximal ideal of $\mathcal{O}_{X,E}$ is generated by one element t. Suppose that $0 \neq f \in k(X)$. Then we can (uniquely) write $f = t^n u$ where $u \in \mathcal{O}_{X,E}$ is a unit and $n \in \mathbb{Z}$. This is true for $f \in \mathcal{O}_{X,E}$, and so this is true for

$f \in k(X)$ since every element of $k(X)$ is a quotient of elements of $\mathcal{O}_{X,E}$. We may thus define a map
$$\nu_E : k(X) \setminus \{0\} \to \mathbb{Z}$$
by $\nu_E(f) = n$ if $f = t^n u$ where u is a unit in $\mathcal{O}_{X,E}$. The map ν_E has the properties that $\nu_E(fg) = \nu_E(f) + \nu_E(g)$ and $\nu_E(f+g) \geq \min\{\nu_E(f), \nu_E(g)\}$ for $f, g \in k(X) \setminus \{0\}$. Using the convention that $\nu_E(0) = \infty$, this tells us that ν_E is a valuation of $k(X)$ with valuation ring
$$V_{\nu_E} = \{f \in k(X) \mid \nu_E(f) \geq 0\} = \mathcal{O}_{X,E}.$$

In fact, ν_E is a divisorial valuation (Lemma 12.19). Let V be a nonsingular affine open subset of X such that $V \cap E \neq \emptyset$ and $\mathcal{I}_E(V) = (t)$. Then $t = 0$ is a local equation of $E \cap V$ in V. We see that a nonzero element $f \in k[V]$ has $\nu_E(f) = n$ if and only if t^n divides f and no higher power of t divides f in $k[V]$. Thus n is nonnegative, and n represents the order of vanishing of f along E, that is, the order of E as a "zero" of f. Also, n represents the order of E as a "pole" of $\frac{1}{f}$. Since any element f of $k(X) \setminus \{0\}$ is a quotient of elements of $k[V]$, we can interpret $\nu_E(f)$ as the order of the zero of f along E if $\nu_E(f) > 0$, and $-\nu_E(f)$ as the order of the pole of f along E if $\nu_E(f) < 0$. If $\nu_E(f) = 0$, then E is neither a zero nor a pole of f. This thinking in terms of zeros and poles is most intuitive when X is a curve, so that a divisor is a point, and $\mathcal{O}_{X,E}$ is the local ring of a point.

Lemma 13.1. *Suppose that X is a normal quasi-projective variety and $0 \neq f \in k(X)$. Then there are at most a finite number of prime divisors E on X such that $\nu_E(f) \neq 0$.*

Proof. Since every quasi-projective variety has an open cover by a finite number of affine open sets, we reduce to the case when X is affine. Write $f = \frac{g}{h}$ where $g, h \in k[X]$. Now $Z_X(g)$ has only a finite number of irreducible components, and $\nu_E(g) = 0$ unless E is an irreducible component of $Z_X(g)$. The same statement holds for h, and since $\nu_E(f) = \nu_E(g) - \nu_E(h)$, we have the statement of the lemma. $\square$

We may thus define the divisor of a function $0 \neq f \in k(X)$ (where X is a normal quasi-projective variety) to be
$$(f) = (f)_X = \sum \nu_E(f) E \in \text{Div}(X)$$
where the sum is over the prime divisors E of X. The divisor of zeros of f is
$$(f)_0 = \sum_{E \mid \nu_E(f) > 0} \nu_E(f) E$$

and the divisor of poles of f is
$$(f)_\infty = \sum_{E|\nu_E(f)<0} -\nu_E(f)E,$$
so that
$$(f) = (f)_0 - (f)_\infty.$$
A divisor D is called principal if $D = (f)$ for some $f \in k(X)$.

We define an equivalence relation $\sim$ on $\mathrm{Div}(X)$, called linear equivalence, by $D_1 \sim D_2$ if there exists $0 \neq f \in k(X)$ such that $(f)_X = D_1 - D_2$. Now we define the divisor class group of (a normal quasi-projective variety) X to be

(13.1) $$\mathrm{Cl}(X) = \mathrm{Div}(X)/\sim.$$

We will sometimes write $\mathrm{div}(f)$ or $\mathrm{div}(f)_X$ for $(f) = (f)_X$.

13.2. The sheaf associated to a divisor

Suppose that X is a normal quasi-projective variety. For $D = \sum a_i E_i$ a divisor on X, with E_i prime divisors and $a_i \in \mathbb{Z}$ and U an open subset of X, we define the divisor $D \cap U$ on U to be
$$D \cap U = \sum_{E_i | E_i \cap U \neq \emptyset} a_i(E_i \cap U).$$

We define the presheaf $\mathcal{O}_X(D)$ of "functions whose poles are bounded by D" by
$$\Gamma(U, \mathcal{O}_X(D)) = \{f \in k(X) \mid (f)_U + D \cap U \geq 0\}$$
for U an open subset of X. By the convention that $\nu_E(0) = \infty$ for all prime divisors E, we have $0 \in \Gamma(U, \mathcal{O}_X(D))$ for all D and U.

Each $\Gamma(U, \mathcal{O}_X(D))$ is a group since for $f, g \in k(X)$ and prime divisor E on X, $\nu_E(f+g) \geq \min\{\nu_E(f), \nu_E(g)\}$ and $\nu_E(-f) = \nu_E(f)$.

Lemma 13.2. *Suppose that X is a normal quasi-projective variety and D is a divisor on X. Then $\mathcal{O}_X(D)$ is a sheaf of $\mathcal{O}_X$-modules.*

Proof. To show that $\mathcal{O}_X(D)$ is a sheaf, we will verify that the sheaf axioms hold. Suppose that U is an open subset of X and $\{U_i\}$ is an open cover of U and we have $f_i \in \Gamma(U_i, \mathcal{O}_X(D))$ such that f_i and f_j have the same restriction in $\Gamma(U_i \cap U_j, \mathcal{O}_X(D))$ for all i, j. Now for any open subset V of X we have that $\Gamma(V, \mathcal{O}_X(D))$ is a subset of $k(X)$, so we must have that $f_i = f_j$ (as elements of $k(X)$). Let f be this common element. Then $(f)_U \cap U_i + D \cap U_i \geq 0$ for all i, and the fact that $\{U_i\}$ is an open cover of U, so that each component of $D \cap U$ must intersect some U_i nontrivially, implies $(f)_U + D \cap U \geq 0$. Thus $f \in \Gamma(U, \mathcal{O}_X(D))$.

Suppose that $\{U_i\}$ is an open cover of an open subset U of X and $f \in \Gamma(U, \mathcal{O}_X(D))$ is an element such that the restriction of f is zero in $\Gamma(U_i, \mathcal{O}_X(D))$ for all i. Then we have that $f = 0$ as an element of $k(X)$, so $f = 0$ in $\Gamma(U, \mathcal{O}_X(D))$.

For U an open subset of X and $f \in \Gamma(U, \mathcal{O}_X)$, we have that $\nu_E(f) \geq 0$ for all divisors E on X such that $U \cap E \neq \emptyset$. Thus $\Gamma(U, \mathcal{O}_X(D))$ is a $\Gamma(U, \mathcal{O}_X)$-module. $\square$

Lemma 13.3. *Suppose that X is a normal quasi-projective variety. Let 0 denote the divisor 0. Then $\mathcal{O}_X(0) = \mathcal{O}_X$.*

Proof. Suppose that U is an open subset of X, $f \in \mathcal{O}_X(U)$, and E is a prime divisor such that $U \cap E \neq \emptyset$. Then there exists $p \in U \cap E$. Now $f \in \mathcal{O}_{X,p}$ implies $f \in \mathcal{O}_{X,E}$ (which is a localization of $\mathcal{O}_{X,p}$), so that $\nu_E(f) \geq 0$. Thus $(f)_U \geq 0$, so that $f \in \mathcal{O}_X(0)(U)$. We have established that $\mathcal{O}_X \subset \mathcal{O}_X(0)$, so it suffices by Proposition 11.13 to show that for all $p \in X$, $\mathcal{O}_{X,p} = \mathcal{O}_X(0)_p$. If $f \in \mathcal{O}_X(0)_p$, then $\nu_E(f) \geq 0$ for all prime divisors E of X which contain p. Thus $f \in (\mathcal{O}_{X,p})_\mathfrak{a}$ for all height 1 prime ideals $\mathfrak{a}$ of $\mathcal{O}_{X,p}$. We have that $\mathcal{O}_{X,p}$ is a normal local ring. Thus $\bigcap_\mathfrak{a} (\mathcal{O}_{X,p})_\mathfrak{a} = \mathcal{O}_{X,p}$, where the intersection is over the height 1 prime ideals $\mathfrak{a}$ of $\mathcal{O}_{X,p}$ by Lemma 1.79, so $f \in \mathcal{O}_{X,p}$. $\square$

Let A be a Noetherian local domain with quotient field K. A fractional ideal M of A is an A-submodule of K such that there exists $d \neq 0$ in A such that $dM \subset A$. A fractional ideal is necessarily a finitely generated A-module. If M is a fractional ideal, we have a natural inclusion of A-modules

$$\text{Hom}_A(M, A) = \{f \in K \mid fM \subset A\} \subset K.$$

Since $\text{Hom}_A(M, A)$ is a finitely generated A-module, we have that $\text{Hom}_A(M, A)$ is a fractional ideal.

Given a fractional ideal M, we define its dual to be

$$M^\vee = \text{Hom}_A(M, A).$$

Let $P(A)$ be the set of height 1 prime ideals of A.

Theorem 13.4 ([23, Theorem 1, page 157])**.** *Suppose that M is a fractional ideal of A. Then*

$$M^\vee = \bigcap_{p \in P(A)} (M^\vee)_p.$$

Lemma 13.5. *Suppose that X is a normal quasi-projective variety and D is a divisor on X. Then $\mathcal{O}_X(D)$ is a coherent sheaf of $\mathcal{O}_X$-modules.*

Proof. We first give a proof with the assumption that X is nonsingular. Let $D = a_1 E_1 + \cdots + a_n E_n$. Suppose that $p \in X$. Since p is a nonsingular point on X, every E_i such that $p \in E_i$ has a local equation f_i at p (Lemma 10.18). If $p \notin E_i$, then $f_i = 1$ is a local equation of E_i at p. Let U_p be an affine neighborhood of p such that $\Gamma(U_p, \mathcal{O}_X(-E_i)) = \Gamma(U_p, \mathcal{I}_{E_i}) = (f_i)$ for all i. (Here (f_i) means the ideal in $k[U_p]$.) Let $f = f_1^{a_1} \cdots f_n^{a_n}$. Then $(f)_{U_p} = D \cap U_p$. Suppose that $U \subset U_p$ is an open subset and $g \in \Gamma(U, \mathcal{O}_X(D))$. Then $(g)_U \geq -D \cap U$. We compute $(fg)_U = (f)_U + (g)_U \geq D \cap U - D \cap U = 0$. Thus $fg \in \Gamma(U, \mathcal{O}_X(0)) = \Gamma(U, \mathcal{O}_X)$ by Lemma 13.3, so that $g \in \frac{1}{f}\Gamma(U, \mathcal{O}_X)$. Conversely, if $h \in \Gamma(U, \mathcal{O}_X)$, then $(\frac{h}{f})_U \geq -(f)_U$, so that $\frac{h}{f} \in \Gamma(U, \mathcal{O}_X(D))$. In summary, we have that $\Gamma(U, \mathcal{O}_X(D)) = \frac{1}{f}\Gamma(U, \mathcal{O}_X)$ for all open subsets U of U_p, so that we have equality of sheaves

$$\mathcal{O}_X(D) \mid U_p = \frac{1}{f}\mathcal{O}_{U_p}.$$

We now prove the lemma for normal X. Let Z be the singular locus of X, and let $U = X \setminus Z$. Let $i : U \to X$ be the inclusion. We have that $\operatorname{codim}_X Z > 1$ since X is normal (Theorem 10.17). Let $\mathcal{F} = \mathcal{O}_X(D)|U$, a coherent sheaf on U. We have a natural inclusion of $\mathcal{O}_U$-modules $\mathcal{F} \subset k(U) = k(X)$, where $k(X)$ is the constant sheaf. By Theorem 11.55, there exists a coherent sheaf $\mathcal{H}$ on X such that $\mathcal{H}|U = \mathcal{F}$ and $\mathcal{H}$ is an $\mathcal{O}_X$-submodule of $k(X)$. We will denote the sheaf $Hom_{\mathcal{O}_X}(\mathcal{G}, \mathcal{O}_X)$ (which was defined in Section 11.5) by $\mathcal{G}^*$ for an $\mathcal{O}_X$-module $\mathcal{G}$. Let $\mathcal{A} = (\mathcal{H})^{**}$. The sheaf $\mathcal{H}$ is a coherent $\mathcal{O}_X$-module by Lemma 11.38, and we have natural inclusions of $\mathcal{O}_X$-modules $\mathcal{H} \subset \mathcal{A} \subset k(X)$. Further, since $\mathcal{F}$ is locally isomorphic to $\mathcal{O}_U$ as an $\mathcal{O}_U$-module, we have that $\mathcal{A}|U = \mathcal{F}$.

Suppose that $V \subset X$ is an affine open subset. Let $R = k[V]$. We let $N^\vee = \operatorname{Hom}_R(N, R)$ if N is an R-module. Let $M = \Gamma(V, \mathcal{A})$. By Lemma 11.38 (and its proof) we have that $M = (\Gamma(V, \mathcal{H}))^{\vee\vee}$. By Theorem 13.4, we have that $M = \bigcap_{Q \in P(R)} M_Q$ where $P(R)$ is the set of height 1 prime ideals of R and the intersection is in $k(X)$.

Suppose that $W \subset V$ is a closed set and $\operatorname{codim}_V W \geq 2$. If $Q \in P(R)$, then there exists $p \in Z_V(Q) \setminus W$ (since $\operatorname{codim}_V W \geq 2$ and $\operatorname{codim}_V Z(Q) = 1$) so $Q \subset I(p)$, and hence $M_{I(p)} \subset M_Q$. Thus by Proposition 11.53,

$$M = \Gamma(V, \mathcal{A}) \subset \Gamma(V \setminus W, \mathcal{A}) = \bigcap_{p \in V \setminus W} M_{I(p)} \subset \bigcap_{Q \in P(R)} M_Q = M.$$

Thus $\Gamma(V \setminus W, \mathcal{A}) = \Gamma(V, \mathcal{A})$ for any affine open subset V of X and closed subset W of V of codimension ≥ 2 in V. By the sheaf axioms, we have that

$$\Gamma(V \setminus W, \mathcal{A}) = \Gamma(V, \mathcal{A})$$

13.2. The sheaf associated to a divisor

for any open subset V of X and closed subset W of V of codimension ≥ 2 as we can express V as a union of affine open sets.

Suppose that $V \subset X$ is an open subset. Since $U = X \setminus Z$ does not contain any prime divisor of X, we have that

$$\Gamma(V, \mathcal{O}_X(D)) = \Gamma(V \cap U, \mathcal{O}_X(D)) = \Gamma(V \cap U, \mathcal{A}) = \Gamma(V, \mathcal{A}).$$

Thus $\mathcal{O}_X(D) = \mathcal{A} = i_*(\mathcal{O}_X(D)|U)$ is a coherent $\mathcal{O}_X$-module. $\square$

In the case that X is nonsingular, the fact that there exists an open cover $\{U_i\}$ of X and $f_i \in k(X)$ such that there are expressions

$$(13.2) \qquad \mathcal{O}_X(D) \mid U_i = \frac{1}{f_i} \mathcal{O}_{U_i}$$

for all i tells us that $\mathcal{O}_X(D)$ is an invertible sheaf of $\mathcal{O}_X$-modules (Definition 11.33).

We have that the f_i are "local equations" for D in the sense that $(f_i)_{U_i} = D \cap U_i$ for all i. The set of pairs $\{(f_i, U_i)\}$ determines a Cartier divisor on X (Definition 15.4).

In the appendix to [**126**, Section 1] it is explained that the sheaves $\mathcal{O}_X(D)$ on a normal variety X are the reflexive rank 1 sheaves on X [**23**, Chapter 7, Section 4]. In particular, denoting the sheaf $Hom_{\mathcal{O}_X}(\mathcal{G}, \mathcal{O}_X)$ by $\mathcal{G}^*$ for an $\mathcal{O}_X$-module $\mathcal{G}$, as in the proof of Lemma 13.5, we have that

$$\mathcal{O}_X(-D) \cong \mathcal{O}_X(D)^*$$

and

$$\mathcal{O}_X(D + E) \cong (\mathcal{O}_X(D) \otimes_{\mathcal{O}_X} \mathcal{O}_X(E))^{**}.$$

If $D = \sum_{i=1}^r a_i E_i$ is an effective divisor on a normal variety X and U is an affine open subset of X, then

$$\Gamma(U, \mathcal{O}_X(-D)) = \{f \in k[U] \mid \nu_{E_1}(f) \geq a_1\} \cap \cdots \cap \{f \in k[U] \mid \nu_{E_r}(f) \geq a_r\}$$
$$= \mathfrak{p}_1^{(a_1)} \cap \cdots \cap \mathfrak{p}_r^{(a_r)}$$

where $\mathfrak{p}_i$ is the prime ideal $\mathfrak{p}_i = I_{E_i \cap U} = \Gamma(U, \mathcal{O}_X(-E_i))$ of the codimension 1 subvariety $E_i \cap U$ of U and $\mathfrak{p}_i^{(a_i)}$ is the a_i-th symbolic power of $\mathfrak{p}_i$, defined by $\mathfrak{p}_i^{(a_i)} = k[U] \cap (\mathfrak{p}_i^{a_i} k[U]_{\mathfrak{p}_i})$, since $k[U]_{\mathfrak{p}_i} = \mathcal{O}_{X, E_i}$ is the valuation ring of ν_{E_i}, with maximal ideal $\mathfrak{p}_i \mathcal{O}_{X, E_i}$.

Proposition 13.6. *Suppose that X is a normal quasi-projective variety and D_1 and D_2 are divisors on X. Then $D_1 \sim D_2$ if and only if $\mathcal{O}_X(D_1) \cong \mathcal{O}_X(D_2)$ as sheaves of $\mathcal{O}_X$-modules.*

Proof. We will reduce to the case that X is nonsingular by the argument used in the proof of Lemma 13.5. Let V be the nonsingular locus of X and

$$\lambda : V \to X$$

be the inclusion. We have by Theorem 10.17 that $\operatorname{codim}_X(X \setminus V) \geq 2$ since X is normal. Thus by the definition of $\mathcal{O}_X(D_i)$,

$$\Gamma(U, \mathcal{O}_X(D_i)) = \Gamma(U \cap V, \mathcal{O}_X(D_i))$$

for all open subsets U of X, and so $\lambda_*(\mathcal{O}_X(D_i)|V) \cong \mathcal{O}_X(D_i)$. Thus we may assume that $X = V$ is nonsingular.

Suppose that $D_1 \sim D_2$. Then there exists $g \in k(X)$ such that $(g) = D_1 - D_2$. Multiplication by g thus gives us an $\mathcal{O}_X$-module isomorphism $\mathcal{O}_X(D_1) \to \mathcal{O}_X(D_2)$ since for $f \in k(X)$ and U an open subset of X,

$$(f)_U + D_1 \cap U \geq 0$$

if and only if $(fg)_U + D_2 \cap U \geq 0$.

Suppose that $\phi : \mathcal{O}_X(D_1) \to \mathcal{O}_X(D_2)$ is an $\mathcal{O}_X$-module isomorphism. For $p \in X$, the $\mathcal{O}_{X,p}$-module isomomorphism $\phi_p : \mathcal{O}_X(D_1)_p \to \mathcal{O}_X(D_2)_p$ extends uniquely to a $k(X)$-module isomorphism $\psi_p : k(X) \to k(X)$, since the localizations $(\mathcal{O}_X(D_i)_p)_{\mathfrak{p}} = k(X)$ for $i = 1, 2$ (where $\mathfrak{p}$ is the zero ideal in $\mathcal{O}_{X,p}$). The map ψ_p is defined by $\psi_p(\frac{f}{g}) = \frac{\phi_p(f)}{g}$ for $f \in \mathcal{O}_X(D_1)_p$ and $0 \neq g \in \mathcal{O}_{X,p}$.

Suppose that U is an affine open subset of X such that $\mathcal{O}_X(D_i) \mid U = \frac{1}{f_i}\mathcal{O}_U$ for some $f_i \in k(X)$ for $i = 1, 2$. Then

(13.3) $$\Gamma(U, \mathcal{O}_X(D_i)) = \frac{1}{f_i} k[U] \quad \text{for } i = 1, 2.$$

For $g \in k[U]$, we have that $\phi(U)(\frac{g}{f_1}) = \frac{gu}{f_2}$ for some fixed unit $u \in k[U]$. Localizing at the zero ideal of $k[U]$ gives us a unique extension of $\phi(U)$ to a $k(X)$-module isomorphism $k(X) \to k(X)$. But this extension must agree with our extensions ψ_p for all $p \in U$ (since the $\mathcal{O}_X(D_i)$ are coherent). Since X is connected, all of our extensions ψ_p agree. A nonzero $k(X)$-module homomorphism $k(X) \to k(X)$ is multiplication by a nonzero element of $k(X)$. Thus there exists a nonzero element $g \in k(X)$ such that for all open U in X, the $\Gamma(U, \mathcal{O}_X)$-module isomorphism $\phi(U) : \Gamma(U, \mathcal{O}_X(D_1)) \to \Gamma(U, \mathcal{O}_X(D_2))$ is given by multiplication by g. This tells us that for any $f \in k(X)$ and open subset U of X, $(f)_U + D_1 \cap U \geq 0$ if and only if $(gf)_U + D_2 \cap U \geq 0$.

By consideration of affine open subsets U on which an expression (13.3) holds, we have that $(g)_U = (\frac{f_2}{f_1})_U = (D_2 - D_1)_U$. Thus $(g) = D_2 - D_1$ so that $D_1 \sim D_2$. $\square$

13.2. The sheaf associated to a divisor

In equation (13.2), we associated to a divisor D on a nonsingular variety X an open cover $\{U_i\}$ of X and $g_i \in k(X)$ which have the property that $(g_i) \cap U_i \cap U_j = (g_j) \cap U_i \cap U_j$ for all i, j.

Conversely, given an open cover $\{U_i\}$ of X and $g_i \in k(X)$ such that $(g_i) \cap U_i \cap U_j = (g_j) \cap U_i \cap U_j$ for all i, j, we can associate a divisor D on X, which is defined by the condition that $D \cap U_i = (g_i) \cap U_i$ for all i. As in the first part of the proof of Lemma 13.5, we have that $\mathcal{O}_X(D_i)|U_i = \frac{1}{g_i}\mathcal{O}_{U_i}$.

Theorem 13.7. *Suppose that X is a nonsingular quasi-projective variety and $\mathcal{L}$ is an invertible sheaf on X. Then there exists a divisor D on X such that $\mathcal{L} \cong \mathcal{O}_X(D)$ as $\mathcal{O}_X$-modules.*

Proof. Since X is quasi-compact (Exercise 7.6), X has a finite cover $\{U_1, \ldots, U_r\}$ with $\mathcal{O}_{U_i}$-isomorphisms $\phi_i : \mathcal{O}_{U_i} \to \mathcal{L}|U_i$. Let $\sigma_i = \phi_i(U_i)(1) \in \mathcal{L}(U_i)$, so that $\mathcal{L}|U_i = \mathcal{O}_{U_i}\sigma_i$ is generated as an $\mathcal{O}_{U_i}$-module by σ_i. Thus

$$\mathcal{L}|U_i \cap U_j = \mathcal{O}_{U_i \cap U_j}\sigma_i = \mathcal{O}_{U_i \cap U_j}\sigma_j$$

and there exists a unique unit $g_{ij} \in \mathcal{O}_X(U_i \cap U_j) \subset k(X)$ such that $\sigma_i = g_{ij}\sigma_j$. Thus $\sigma_i = g_{ij}g_{ji}\sigma_i$ implies

$$(13.4) \qquad g_{ji} = g_{ij}^{-1}.$$

The equality $\sigma_i = g_{ij}\sigma_j = g_{ij}g_{jk}\sigma_k$ implies

$$(13.5) \qquad g_{ik} = g_{ij}g_{jk}.$$

We compute

$$(g_{i1})_{U_i \cap U_j} = (g_{ij}g_{j1})_{U_i \cap U_j} = (g_{ij})_{U_i \cap U_j} + (g_{j1})_{U_i \cap U_j} = (g_{j1})_{U_i \cap U_j}$$

since g_{ij} is a unit in $\mathcal{O}_X(U_i \cap U_j)$. Thus there exists a unique divisor D on X such that

$$D \cap U_i = (g_{i1})_{U_i} \quad \text{for } 1 \leq i \leq r,$$

and we have that

$$\mathcal{O}_X(-D)|U_i = g_{i1}\mathcal{O}_{U_i} \quad \text{for } 1 \leq i \leq r.$$

We have a natural inclusion of $\mathcal{O}_X$-modules $\mathcal{L} \subset k(X)\sigma_1$. For all i, we have

$$\mathcal{L}|U_i = \mathcal{O}_{U_i}\sigma_i = g_{i1}\mathcal{O}_{U_i}\sigma_1 = (\mathcal{O}_X(-D)|U_i)\sigma_1.$$

Thus $\mathcal{L} = \mathcal{O}_X(-D)\sigma_1 \cong \mathcal{O}_X(-D)$. $\square$

Lemma 13.8. *Suppose that X is a nonsingular quasi-projective variety and $\mathcal{I}$ is a nonzero ideal sheaf on X. Then there exists an effective divisor D on X and an ideal sheaf $\mathcal{J}$ on X such that the support of $\mathcal{O}_X/\mathcal{J}$ has codimension ≥ 2 in X and $\mathcal{I} = \mathcal{O}_X(-D)\mathcal{J}$.*

Proof. We begin with some remarks about rings. Suppose that R is a local ring of an algebraic variety and I is a nonzero ideal in R. Let d be the dimension of R. Then the following are equivalent:

1. The ring R/I has dimension $d-1$.
2. I has height 1 in R.
3. There exists a height 1 prime $\mathfrak{p}$ in R such that $I \subset \mathfrak{p}$.

If we further have that R is a UFD, then a height 1 prime ideal $\mathfrak{p}$ in R has an expression $\mathfrak{p} = (g)$ where g is an irreducible in R, so that $I \subset \mathfrak{p}$ if and only if $I = gJ$ for some ideal J of R. Thus, assuming that R is a UFD, we have that there exists a nonzero element $f \in R$ and an ideal J in R such that J has height ≥ 2 in R and $I = fJ$.

Now we prove the lemma. We have that $\text{Supp}(\mathcal{O}_X/\mathcal{I}) = E_1 \cup \cdots \cup E_r \cup W$ where $E_1, \ldots, E_r$ are prime divisors and $\text{codim}_X W \geq 2$. Let $\mathcal{I}(E_i)$ be the ideal sheaf of E_i in $\mathcal{O}_X$. Define the localization $\mathcal{I}_{E_i}$ of $\mathcal{I}$ in $\mathcal{O}_{X,E_i}$ by taking any $p \in E_i$ and letting

$$\mathcal{I}_{E_i} = (\mathcal{I}_p)_{\mathfrak{p}} \subset (\mathcal{O}_{X,p})_{\mathfrak{p}} = \mathcal{O}_{X,E_i}$$

where $\mathcal{I}_p$ is the stalk of $\mathcal{I}$ at p, $\mathfrak{p}$ is the height 1 prime ideal $\mathfrak{p} = \mathcal{I}(E_i)_p$ in $\mathcal{O}_{X,p}$. Since $\mathcal{O}_{X,E_i}$ is a one-dimensional regular local ring (a discrete valuation ring), there exists a positive integer a_i such that $\mathcal{I}_{E_i} = m_{E_i}^{a_i}$ where m_{E_i} is the maximal ideal of $\mathcal{O}_{X,E_i}$. Let D be the divisor $D = \sum_{i=1}^{r} a_i E_i$. Define a presheaf $\mathcal{J}$ on X by

$$\mathcal{J}(U) = \{f \in \mathcal{O}_X(U) \mid f\mathcal{O}_X(-D) \subset \mathcal{I}\}.$$

$\mathcal{J}$ is an ideal sheaf on X. The ideal sheaf $\mathcal{O}_X(-D)$ has the properties that

$$\mathcal{O}_X(-D)_{E_i} = m_{E_i}^{a_i} = \mathcal{I}_{E_i} \quad \text{for } 1 \leq i \leq r$$

and

$$\mathcal{O}_X(-D)_G = \mathcal{O}_{X,G} = \mathcal{I}_G$$

if G is a prime divisor on X which is not one of the E_i.

Suppose that $p \in X$. For $1 \leq i \leq r$, let f_i be a local equation of E_i at p (taking $f_i = 1$ if $p \notin E_i$). Then $\mathcal{O}_X(-D)_p = f_1^{a_1} \cdots f_r^{a_r} \mathcal{O}_{X,p}$ and

$$\begin{aligned}\mathcal{I}_p &= f_1^{a_1} \cdots f_r^{a_r}(\mathcal{I}_p : f_1^{a_1} \cdots f_r^{a_r} \mathcal{O}_{X,p}) \\ &= (\mathcal{I}_p : \mathcal{O}_X(-D)_p)\mathcal{O}_X(-D)_p = \mathcal{J}_p \mathcal{O}_X(-D).\end{aligned}$$

Thus $\mathcal{I} = \mathcal{O}_X(-D)\mathcal{J}$. We have that $\text{Supp}(\mathcal{O}_X/\mathcal{J})$ does not contain a prime divisor, since $\dim(\mathcal{O}_{X,p}/\mathcal{J}_p) \leq \dim X - 2$ for all $p \in X$. Thus $\text{codim}_X(\text{Supp}(\mathcal{O}_X/\mathcal{J})) \geq 2$. $\square$

13.3. Divisors associated to forms

Suppose that $X \subset \mathbb{P}^n$ is a normal projective variety. Let $x_0, \ldots, x_n$ be homogeneous coordinates on X, so that $S(X) = k[x_0, \ldots, x_n]$. Suppose that $F \in S(X)$ is a nonzero homogeneous form of degree d. We associate to F an effective divisor $\text{Div}(F)$ on X as follows. Let

$$f_i = \frac{F}{x_i^d} \in k(X) \quad \text{for } 0 \leq i \leq n.$$

We have that

$$f_j = f_i \left(\frac{x_i}{x_j}\right)^d.$$

Let $U_i = X_{x_i}$ for $0 \leq i \leq n$, so that $\{U_0, \ldots, U_n\}$ is an affine open cover of X. We have that

$$f_i \in \mathcal{O}_X(U_i) = k\left[\frac{x_0}{x_i}, \ldots, \frac{x_n}{x_i}\right]$$

for all i and $\frac{x_i}{x_j}$ is a unit in

$$\mathcal{O}_X(U_i \cap U_j) = k\left[\frac{x_0}{x_i}, \ldots, \frac{x_n}{x_i}, \frac{x_i}{x_j}\right]$$

for all i, j. We have that

$$(f_i)_{U_i \cap U_j} = (f_j)_{U_i \cap U_j}$$

for all i, j, so that there is a unique divisor E on X such that

$$E \cap U_i = (f_i)_{U_i} \quad \text{for } 1 \leq i \leq n.$$

We define

$$\text{Div}(F) = E.$$

We have that $\text{Div}(F) \geq 0$ since $(f_i)_{U_i} \geq 0$ for all i, as $f_i \in \mathcal{O}_X(U_i)$ for all i.

13.4. Calculation of some class groups

Example 13.9. $\text{Cl}(\mathbb{A}^n) = (0)$.

Proof. Suppose that E is a prime divisor on $\mathbb{A}^n$. Then the ideal of E, $I(E)$, is a height 1 prime ideal in the polynomial ring $k[\mathbb{A}^n]$. Since $k[\mathbb{A}^n]$ is a UFD, $I(E)$ is a principal ideal, so there exists $f \in k[\mathbb{A}^n]$ such that $(f) = E$. □

Example 13.10. $\mathrm{Cl}(\mathbb{P}^n) \cong \mathbb{Z}$.

Proof. For every irreducible form $F \in S = S(\mathbb{P}^n)$, $Z(F)$ is a prime divisor on $\mathbb{P}^n$. Suppose that E is a prime divisor on $\mathbb{P}^n$. Let $I(E)$ be the homogeneous prime ideal of E in $S = k[x_0, \ldots, x_n]$. Since S is a polynomial ring and $I(E)$ is a height 1 prime ideal in S, we have that $I(E)$ is a principal ideal. Thus there exists an irreducible homogeneous form $F \in S$ such that $I(E) = (F)$. Now another form G in S is a generator of $I(E)$ if and only if $F = \lambda G$ where λ is a unit in S; that is, λ is a nonzero element of k. We can thus associate to E the number $d = \deg(F)$ which is called the degree of E.

Thus we can define a group homomorphism

$$\deg : \mathrm{Div}(\mathbb{P}^n) \to \mathbb{Z}$$

by defining $\deg E = d$ if E is a prime divisor of degree d and

$$\deg\left(\sum a_i E_i\right) = \sum a_i \deg(E_i).$$

Suppose that $f \in k(\mathbb{P}^n) = k(\frac{x_1}{x_0}, \ldots, \frac{x_n}{x_0})$. Then we can write $f = \frac{F}{G}$ where F and G are homogeneous forms of a common degree d and F, G are relatively prime in S (since S is a UFD). This expression is unique up to multiplying F and G by a common nonzero element of k. Since S is a UFD and F, G are homogeneous, there are factorizations

$$F = \prod F_j, \qquad G = \prod G_j$$

where F_j and G_j are irreducible forms in S. Let $D_j = Z(F_j)$ and $E_j = Z(G_j)$ (the D_j may not all be distinct and the E_j may not all be distinct). Let $U_i = X_{x_i} \cong \mathbb{A}^n$ for $0 \le i \le n$. Then

$$(f) \cap U_i = \left(\frac{F}{x_i^d}\right) \cap U_i - \left(\frac{G}{x_i^d}\right) \cap U_i = \sum D_j \cap U_i - \sum E_j \cap U_i.$$

Thus

$$(f) = \sum D_j - \sum E_j,$$

and

$$\deg((f)) = \sum \deg D_j - \sum \deg E_j = d - d = 0.$$

Thus we have a well-defined group homomorphism

$$\deg : \mathrm{Cl}(\mathbb{P}^n) \to \mathbb{Z}.$$

The homomorphism deg is onto since a linear form has degree 1, so the associated prime divisor has degree 1.

13.4. Calculation of some class groups

Suppose that D is a divisor on $\mathbb{P}^n$ which has degree 0. Write
$$D = \sum D_j - \sum E_j$$
where D_j and E_j are prime divisors and $\sum \deg D_j = \sum \deg E_j$. Let d be this common degree.

Let F_j be homogeneous forms such that $I(F_j) = D_j$, and let G_j be homogeneous forms such that $I(G_j) = E_j$. Then $F = \prod F_j$ has degree d and $G = \prod G_j$ has degree d, so $f = \frac{F}{G} \in k(\mathbb{P}^n)$. We have that
$$(f) = \sum Z(F_j) - \sum Z(G_j) = D.$$
Thus the class of D is zero in $\mathrm{Cl}(\mathbb{P}^n)$. □

Example 13.11. $\mathrm{Cl}(\mathbb{P}^m \times \mathbb{P}^n) \cong \mathbb{Z} \times \mathbb{Z}$.

The method of the previous example applied to the bihomogeneous coordinate ring of $\mathbb{P}^m \times \mathbb{P}^n$ proves this result. Let $S = k[x_0, \ldots, x_m, y_0, \ldots, y_n]$ be the bihomogeneous coordinate ring of $\mathbb{P}^m \times \mathbb{P}^n$. Prime divisors on $\mathbb{P}^m \times \mathbb{P}^n$ correspond to irreducible bihomogeneous forms in S. Associated to a form is a bidegree (d, e). Using the fact that S is a UFD, we argue as in the previous example that this bidegree induces the desired isomorphism.

A consequence of this example is that the group $\mathrm{Cl}(\mathbb{P}^m \times \mathbb{P}^n)$ is generated by a prime divisor L_1 of bidegree $(1, 0)$ and a prime divisor L_2 of bidegree $(0, 1)$. The divisor L_1 is equal to $H_1 \times \mathbb{P}^n$, where H_1 is a linear subspace of codimension 1 in $\mathbb{P}^m$, and $L_2 = \mathbb{P}^m \times H_2$, where H_2 is a linear subspace of codimension 1 in $\mathbb{P}^n$.

A particular example is $\mathbb{P}^1 \times \mathbb{P}^1$. Here $L_1 = \{p\} \times \mathbb{P}^1$ and $L_2 = \mathbb{P}^1 \times \{q\}$, where p, q are any points in $\mathbb{P}^1$, are generators of $\mathrm{Cl}(\mathbb{P}^1 \times \mathbb{P}^1)$.

Lemma 13.12. *Suppose that D is a divisor on a nonsingular quasi-projective variety X and $p_1, \ldots, p_m \in X$ are a finite set of points. Then there exists a divisor D' on X such that $D' \sim D$ and $p_i \notin \mathrm{Supp}\, D'$ for $i = 1, \ldots, m$.*

Proof. By assumption, X is an open subset of a projective variety $\overline{X}$. Thus $X = \overline{X} \setminus Z$ for some closed subset Z of $\overline{X}$. Choosing a projective embedding of $\overline{X}$, by the homogeneous nullstellensatz, Theorem 3.12, we find homogeneous forms
$$F_i \in I(Z \cup \{p_1, \ldots, p_{i-1}, p_{i+1}, \ldots, p_m\}) \subset S(\overline{X})$$
for $1 \leq i \leq m$ such that $F_i(p_i) \neq 0$ for $1 \leq i \leq m$. Let $d_i = \deg(F_i)$. Taking $d = d_1 d_2 \cdots d_m$, set
$$F = \sum_{i=1}^m F_i^{\frac{d}{d_i}},$$

a homogeneous form in $S(\overline{X})$ of degree d. Let $U = \overline{X} \setminus Z(F)$. By construction, U is an affine open subset of $X = \overline{X} \setminus Z = X$ which contains $p_1, \ldots, p_m$. We may thus assume that X is an affine variety.

By induction on m, we may assume that $p_1, \ldots, p_i \notin \operatorname{Supp} D$ and $p_{i+1} \in \operatorname{Supp} D$. We must construct a divisor D' such that $D' \sim D$ and $p_1, \ldots, p_{i+1} \notin \operatorname{Supp} D'$. We will prove the inductive statement in the case that D is a prime divisor. The case of a general divisor then follows by applying the statement to each of its components as necessary.

Let π' be a local equation of the prime divisor D in a neighborhood of p_{i+1}. The function π' is regular at p_{i+1}, so if $(\pi')_\infty = \sum k_l G_l$, then $p_{i+1} \notin G_l$ for all l. Then for every l, there exists $f_l \in k[X]$ such that $f_l \in I(G_l)$ and $f_l(p_{i+1}) \neq 0$. The function $\pi = \pi' \prod f_l^{k_l}$ has no poles on X ($\pi \in \Gamma(X, \mathcal{O}_X(0))$) so $\pi \in k[X]$ by Lemma 13.3. Further, $\pi = 0$ is a local equation of D at p_{i+1}. Since $p_j \notin \operatorname{Supp} D \cup \{p_1, \ldots, p_{j-1}, p_{j+1}, \ldots, p_i\}$ for $1 \leq j \leq i$, there exists $g_j \in I(D \cup \{p_1, \ldots, p_{j-1}, p_{j+1}, \ldots, p_i\})$ such that $g_j(p_j) \neq 0$ (by the nullstellensatz, Theorem 2.5).

Let $f = \pi + \sum_{j=1}^{i} \alpha_j g_j^2$, where $\alpha_j \in k$ is chosen so that $\alpha_j \neq -\frac{\pi(p_j)}{g_j(p_j)^2}$. Then

(13.6) $$f(p_j) \neq 0 \quad \text{for } 1 \leq j \leq i.$$

Since $g_j(D) = 0$ for all j, π divides g_j in the regular local ring $\mathcal{O}_{X, p_{i+1}}$ and we have an expression

$$\sum \alpha_j g_j^2 = h\pi^2$$

for some $h \in \mathcal{O}_{X, p_{i+1}}$. Thus $f = \pi(1 + \pi h)$ and since $1 + \pi h$ is a unit in $\mathcal{O}_{X, p_{i+1}}$, $f = 0$ is a local equation of D in a neighborhood of p_{i+1}. Thus we have an expression

$$(f) = D + \sum r_s D_s$$

where none of the prime divisors D_s contain p_{i+1}. Let $D' = D - (f)$. Then $p_{i+1} \notin \operatorname{Supp} D'$. Further, $p_j \notin \operatorname{Supp}(f)$ for $1 \leq j \leq i$ by (13.6). Thus D' satisfies the conclusions of the inductive step. $\square$

Suppose that $\phi : X \to Y$ is a regular map of nonsingular quasi-projective varieties. We will define a natural group homomorphism $\phi^* : \operatorname{Cl}(Y) \to \operatorname{Cl}(X)$.

Let Z be the Zariski closure of $\phi(X)$ in Y. Suppose that D is a divisor on Y. Then by Lemma 13.12, there is a divisor D' on Y such that $D \sim D'$ and Z is not contained in the support of D'. Write $D' = \sum a_i E_i - \sum b_j F_j$, where the a_i, b_j are positive and the E_i, F_j are distinct prime divisors.

We will define a divisor $\phi^*(D')$ on X. Let $\{U_i\}$ be an affine open cover of Y such that there exist $f_i \in k(Y)$ with $(f_i) \cap U_i = D' \cap U_i$ for all i.

After replacing the cover $\{U_i\}$ with a refinement, we may write $f_i = \frac{g_i}{h_i}$ for all i where g_i and h_i are regular on U_i and $(g_i)_{U_i} = \sum a_j(E_j \cap U_i)$ and $(h_i)_{U_i} = \sum b_j(F_j \cap U_i)$. Since Z is not contained in any of these divisors, the restrictions $\overline{g}_i$ and $\overline{h}_i$ of g_i and h_i to $U_i \cap Z$ are nonzero regular functions. Thus we have induced rational functions $\frac{\overline{g}_i}{\overline{h}_i} \in k(Z)$.

Let $V_i = \phi^{-1}(U_i)$, which give an open cover of X. Let $\overline{\phi} : X \to Z$ be the dominant regular map induced by ϕ, with inclusion of function fields $\overline{\phi}^* : k(Z) \to k(X)$. We define $\phi^*(D')$ to be the divisor on X determined by the conditions

$$\phi^*(D') \cap V_i = \left(\overline{\phi}^*\left(\frac{\overline{g}_i}{\overline{h}_i}\right)\right) \cap V_i$$

over the open cover $\{V_i\}$ of X.

If D'' is another divisor whose support does not contain Z such that $D \sim D''$, then there exists $f \in k(Y)$ such that $(f) = D' - D''$. Since the support of (f) contains no component of Z, f restricts to a nonzero element $\overline{f}$ of $k(Z)$. By our construction, we have that $(\overline{\phi}^*(\overline{f})) = \phi^*(D') - \phi^*(D'')$.

Thus we have a well-defined group homomorphism $\phi^* : \mathrm{Cl}(Y) \to \mathrm{Cl}(X)$, defined by taking $\phi^*(D)$ to be the class of $\phi^*(D')$, for any divisor D' on Y which is linearly equivalent to D and whose support does not contain $\phi(X)$. We will write $\phi^*(D) \sim \phi^*(D')$, even though $\phi^*(D)$ may only be defined up to equivalence class. However, the divisor $\phi^*(D')$ is actually a well-defined divisor on X.

A particular case of this construction is the inclusion $i : X \to Y$ of a closed subvariety X of Y into Y. Then the map i^* on class groups can be considered as a restriction map: given a divisor D on Y, find a divisor D' linearly equivalent to D whose support does not contain X, and define $i^*(D)$ to be the class of $D' \cap X$ (of course a prime divisor on Y might intersect X in a sum of prime divisors).

Example 13.13. Suppose that X is a nonsingular surface and $p \in X$ is a point. Let $\pi : B \to X$ be the blow-up of p, with exceptional divisor $\pi^{-1}(p) = E \cong \mathbb{P}^1$. Let $i : E \to B$ be the inclusion. Then $i^*(E) \sim -q$ where q is a point on E.

Proof. Let L be a curve on X which contains p and such that L is nonsingular at p. There exists a divisor D on X such that $L \sim D$ and $p \notin \mathrm{Supp}(D)$ (by Lemma 13.12). Let L' be the strict transform of L on B. Then $\pi^*(L) = L' + E$, and $\pi^*(D) \sim \pi^*(L)$. Thus $i^*(\pi^*(L)) \sim i^*(\pi^*(D)) \sim 0$, since $\mathrm{Supp}(\pi^*(D)) \cap E = \emptyset$. Thus $i^*(E) \sim i^*(-L') = -q$, where q is the intersection point of E and L'. □

Theorem 13.14 (Lefschetz). *Suppose that H is a nonsingular variety which is a codimension 1 closed subvariety of $\mathbb{P}^n$ for some $n \geq 4$. Then the restriction homomorphism $\mathrm{Cl}(\mathbb{P}^n) \cong \mathbb{Z} \to \mathrm{Cl}(H)$ is an isomorphism.*

Proofs of this theorem can be found in [62, pages 156 and 163] over $\mathbb{C}$ and for general fields in [64] and [74]. Since $T = Z(xy - zw) \subset \mathbb{P}^3$ is isomorphic to $\mathbb{P}^1 \times \mathbb{P}^1$, which has class group $\mathbb{Z} \times \mathbb{Z}$, we see that the theorem fails for $n < 4$. The Segre embedding of $\mathbb{P}^1 \times \mathbb{P}^1$ into $\mathbb{P}^3$ is the map $\phi = (x_0 y_0 : x_1 y_0 : x_0 y_1 : x_1 y_1)$. This map is induced by the corresponding homomorphism

$$\phi^* : S(\mathbb{P}^3) = k[x,y,z,w] \to S(\mathbb{P}^1 \times \mathbb{P}^1) = k[x_0, x_1, y_0, y_1]$$

of coordinate rings.

Let L be a form of degree 1 on $\mathbb{P}^3$. Then $\phi^*(L)$ is a bihomogeneous form of bidegree $(1,1)$ in $S(\mathbb{P}^1 \times \mathbb{P}^1)$. Thus the divisor $\phi^*(L)$ has bidegree $(1,1)$ on $\mathbb{P}^1 \times \mathbb{P}^1$. Since the class of L generates $\mathrm{Cl}(\mathbb{P}^3)$, we have that the restriction map

$$\Lambda : \mathrm{Cl}(\mathbb{P}^3) \cong \mathbb{Z} \to \mathrm{Cl}(T) \cong \mathbb{Z}^2$$

is given by $\Lambda(n) = (n,n)$ for $n \in \mathbb{Z}$. So a divisor of degree n on $\mathbb{P}^3$, whose support does not contain T, restricts to a divisor on T which is linearly equivalent to $n(\{p\} \times \mathbb{P}^1 + \mathbb{P}^1 \times \{q\})$ for any points $p, q \in \mathbb{P}^1$.

13.5. The class group of a curve

Suppose that X is a nonsingular projective curve. A divisor D on X is then a sum $D = \sum_{i=1}^r a_i p_i$ where p_i are points on X and $a_i \in \mathbb{Z}$. We define the degree of the divisor D to be

$$\deg(D) = \sum_{i=1}^r a_i.$$

Suppose that X and Y are nonsingular projective curves and $\phi : X \to Y$ is a nonconstant regular map. Then ϕ is finite (by Corollary 10.26). The degree of ϕ is defined to be the degree of the finite field extension

$$\deg(\phi) = [k(X) : k(Y)].$$

Lemma 13.15. *Let $p_1, \ldots, p_r \in X$, and set*

$$R = \bigcap_{i=1}^r \mathcal{O}_{X,p_i},$$

where the intersection is in $k(X)$. Then R is a principal ideal domain. There exist elements $t_i \in R$ such that

(13.7) $$\nu_{p_i}(t_j) = \delta_{ij} \quad \text{for } 1 \leq i \leq j \leq r.$$

13.5. The class group of a curve

If $u \in R$, then

(13.8) $$u = t_1^{l_1} \cdots t_r^{l_r} v$$

where $l_i = \nu_{p_i}(u)$ and v is a unit in R.

Proof. For $1 \leq i \leq r$ we have that

(13.9) $$\mathcal{O}_{X,p_i} = \{f \in k(X) \mid \nu_{p_i}(f) \geq 0\}$$

since X is nonsingular.

Let u_i be a regular parameter in the one-dimensional regular local ring $\mathcal{O}_{X,p_i}$. Then $\nu_{p_i}(u_i) = 1$, so that $(u_i) = p_i + D$ where D is a divisor on X such that p_i does not appear in D. By Lemma 13.12, there exists a divisor D' on X such that the support of D' is disjoint from $\{p_1, \ldots, p_r\}$ and $D' \sim D$ so there exists a rational function $f_i \in k(X)$ such that $(f_i) = D' - D$. Let $t_i = u_i f_i$. Then $(t_i) = p_i + D'$, so that $t_i \in R$ by (13.9) and the equations (13.7) hold.

Let $u \in R$, and set $l_i = \nu_{p_i}(u)$. We necessarily have that $l_i \geq 0$ for all i. Let $v = u t_1^{-l_1} \cdots t_r^{-l_r}$. We have that $\nu_{p_i}(v) = 0$ for all i. By (13.9) we have that v and v^{-1} are in $\mathcal{O}_{X,p_i}$ for all i. Thus $v, v^{-1} \in R$, so that v is a unit in R. We thus have the expression (13.8) for u.

We will now show that R is a principal ideal domain. Let I be an ideal in R. Let

$$l_i = \inf\{\nu_{p_i}(u) \mid u \in I\}$$

for $1 \leq i \leq r$. Set $a = t_1^{l_1} \cdots t_r^{l_r}$, and let $J = aR$. For $u \in I$ we have that $\nu_{p_i}(\frac{u}{a}) \geq 0$ for all i, so that $\frac{u}{a} \in R$, and thus $u \in J$. Thus $I \subset J$. We will now show that $J \subset I$. Let $J' = \{ua^{-1} \mid u \in I\}$. Then J' is an ideal in R, and $\inf\{\nu_{p_i}(v) \mid v \in J'\} = 0$ for $1 \leq i \leq r$. Hence for $1 \leq i \leq r$, there exists $u_i \in J'$ such that $\nu_{p_i}(u_i) = 0$; that is, $u_i(p_i) \neq 0$. Let $c = \sum_{j=1}^r u_j t_1 \cdots \hat{t}_j \cdots t_r \in J'$. Then $c(p_i) \neq 0$ for all i, so that $\nu_{p_i}(c) = 0$ for all i. Thus c is a unit in R so that $J' = R$. Thus $I = aJ' = aR$ is a principal ideal. $\square$

Since $\phi : X \to Y$ is finite, it is surjective, and the preimage of every point of Y in X is a finite set (by Theorems 7.5, 2.57, and 2.56). Suppose that $q \in Y$. Let $\phi^{-1}(q) = \{p_1, \ldots, p_r\}$. Let V be an affine neighborhood of q. Then $U = \phi^{-1}(V)$ is an affine open subset of X and $\phi : U \to V$ is finite (by Theorem 7.5).

Lemma 13.16. *Let $q \in Y$, let V be an affine neighborhood of q in Y, and let $U = \phi^{-1}(V)$. Let $\phi^{-1}(q) = \{p_1, \ldots, p_r\}$, and*

$$R = \bigcap_{i=1}^r \mathcal{O}_{X,p_i}.$$

Then
$$R = k[U]\mathcal{O}_{Y,q} = \left\{\sum a_i b_i \mid a_i \in k[U], b_i \in \mathcal{O}_{Y,q}\right\},$$
where the product is in $k(X)$, and we identify $\mathcal{O}_{Y,q}$ with its image in $k(X)$ by ϕ^*.

Proof. We have that $\mathcal{O}_{Y,q} \subset \mathcal{O}_{X,p_i}$ and $k[U] \subset \mathcal{O}_{X,p_i}$ for all i, since $p_i \in U$ for all i, so that $k[U]\mathcal{O}_{Y,q} \subset R$.

Suppose that $f \in R$. Let $z_1, \ldots, z_s \in X$ be the poles of f on U. Let $w_i = \phi(z_i)$ for $1 \leq i \leq s$. We have that f is regular at the points $\phi^{-1}(q) = \{p_1, \ldots, p_r\}$ since $f \in R$. There exists a function $h \in k[V]$ such that $h(q) \neq 0$ and $h(w_i) = 0$ for all i by Lemma 8.10. Replacing h with a sufficiently high power of h, we then also have that $\nu_{z_i}(hf) \geq 0$ for all i. Thus hf has no poles on U, so that

$$hf \in \bigcap_{q \in U} \mathcal{O}_{X,q} = \mathcal{O}_X(U) = k[U].$$

Since $\frac{1}{h} \in \mathcal{O}_{Y,q}$, we have that $f \in k[U]\mathcal{O}_{Y,q}$. Thus $R \subset k[U]\mathcal{O}_{Y,q}$. □

Theorem 13.17. Let $\phi^{-1}(q) = \{p_1, \ldots, p_r\}$ and $R = \bigcap_{i=1}^r \mathcal{O}_{X,p_i}$. Then R is a free $\mathcal{O}_{Y,q}$-module of rank $n = \deg(\phi)$.

Proof. Since $k[U]$ is a finitely generated $k[V]$-module and $\mathcal{O}_{Y,q}$ is a localization of $k[V]$, we have that $R = k[U]\mathcal{O}_{Y,q}$ is a finitely generated $\mathcal{O}_{Y,q}$-module. By the classification theorem for finitely generated modules over a principal ideal domain [84, Theorem 3.10], the $\mathcal{O}_{Y,q}$-module R is a direct sum of a finite rank free $\mathcal{O}_{Y,q}$-module and a torsion $\mathcal{O}_{Y,q}$-module. Since R and $\mathcal{O}_{Y,q}$ are contained in the field $k(X)$, R has no $\mathcal{O}_{Y,q}$ torsion, so that R is a free $\mathcal{O}_{Y,q}$-module of finite rank.

Suppose that $f_1, \ldots, f_m$ is a basis of the free $\mathcal{O}_{Y,q}$-module R. If $m > n = \deg(\phi)$, then $f_1, \ldots, f_m$ are linearly dependent over $k(Y)$. Clearing denominators in a dependence relation over $k(Y)$ gives a dependence relation over $\mathcal{O}_{Y,q}$. Thus $m \leq n$. Let $h \in k(Y)$. Let l be the maximum of the order of a pole of h at the p_j. Let t be a local parameter at q (a generator of the maximal ideal of $\mathcal{O}_{Y,q}$). Then the function $t^l h$ has no poles in the set $\phi^{-1}(q)$ so $t^l h \in R$. Then there exists an expression $t^l h = \sum_{i=1}^m a_i f_i$ with $a_i \in R$, so

$$h = \sum_{i=1}^m \frac{a_i}{t^l} f_i,$$

showing that $f_1, \ldots, f_m$ span $k(X)$ as a $k(Y)$-vector space. Thus $m \geq n$. □

Theorem 13.18. *Suppose that $\phi : X \to Y$ is a dominant regular map of nonsingular projective curves. Then $\deg \phi^*(q) = \deg \phi$ for every $q \in Y$. In particular,*
$$\deg \phi^*(D) = \deg \phi \deg D$$
for every divisor D on Y.

Proof. Let t be a local parameter at q in Y (a generator of the maximal ideal of $\mathcal{O}_{Y,q}$), and let $\phi^{-1}(q) = \{p_1, \ldots, p_r\}$. Let $R = \bigcap_{i=1}^{r} \mathcal{O}_{X,p_i}$. Let $t_1, \ldots, t_r \in R$ be the generators of the r distinct maximal ideals $\mathfrak{m}_i = (t_i)$ of R, found in Lemma 13.15. Then any two distinct t_i, t_j cannot both be contained in any $\mathfrak{m}_l$. Thus the ideal (t_i, t_j) is not contained in a maximal ideal of R, so that $(t_i, t_j) = R$; that is, the ideals (t_i) and (t_j) are coprime in R.

By Lemma 13.15, we have that $t = t_1^{l_1} \cdots t_r^{l_r} v$ where v is a unit in R and $l_i = \nu_{p_i}(t)$. Thus

$$\phi^*(q) = \sum l_i p_i \quad \text{and} \quad \deg \phi^*(q) = \sum l_i.$$

Since the ideals (t_i) are pairwise coprime, we have by Theorem 1.5 that

$$R/tR \cong \bigoplus_{i=1}^{r} R/t_i^{l_i} R.$$

We will now show that for every i, every $w \in R$ has a unique expression

(13.10) $$w \equiv \alpha_0 + \alpha_1 t_i + \cdots + \alpha_{l_i - 1} t_i^{l_i - 1} \mod (t_i^{l_i})$$

with $\alpha_0, \ldots, \alpha_{l_i - 1} \in k$.

We prove this formula by induction. Suppose that it has been established that w has a unique expression

$$w \equiv \alpha_0 + \alpha_1 t_i + \cdots + \alpha_{s-1} t_i^{s-1} \mod (t_i^s)$$

with $\alpha_0, \ldots, \alpha_{s-1} \in k$. Then $v = t_i^{-s}(w - (\alpha_0 + \alpha_1 t_i + \cdots + \alpha_{s-1} t_i^{s-1})) \in R$. Let $v(p_i) = \alpha_s \in k$. Then $\nu_{p_i}(v - \alpha_s) > 0$, so that $v \equiv \alpha_s \mod (t_i)$. Thus

$$w \equiv \alpha_0 + \alpha_1 t_i + \cdots + \alpha_s t_i^s \mod (t_i^{s+1}),$$

establishing (13.10).

From (13.10), we see that $\dim_k R/(t_i^{l_i}) = l_i$ for all i, so that

$$\dim_k R/(t) = \sum_{i=1}^{r} l_i = \deg \phi^*(q).$$

Now by Theorem 13.17, we have $\mathcal{O}_{Y,q}$-module isomorphisms

$$R/(t) \cong (\mathcal{O}_{Y,q}/(t))^{\deg(\phi)} \cong k^{\deg(\phi)},$$

from which we conclude that $\dim_k R/(t) = \deg \phi$. □

Corollary 13.19. *The degree of a principal divisor on a nonsingular projective curve X is zero.*

Proof. Suppose that $f \in k(X)$ is a nonconstant rational function. The inclusion of function fields $k(\mathbb{P}^1) = k(t) \to k(X)$ defined by mapping t to f determines a nonconstant rational map $\phi : X \to \mathbb{P}^1$, which is a regular map since X is a nonsingular curve (Corollary 10.26). We have a representation $\phi = (f : 1)$.

For $p \in X$, write $f = u_p z_p^{n_p}$ where u_p is a unit in $\mathcal{O}_{X,p}$, z_p is a regular parameter in $\mathcal{O}_{X,p}$, and $n_p = \nu_p(f)$. If $n_p \geq 0$, then $\phi(p) = (f(p) : 1)$, and if $n_p < 0$, then

$$\phi(p) = \left(1 : \left(\frac{1}{f}\right)(p)\right) = \left(1 : \left(\frac{z_p^{-n_p}}{u_p}\right)(p)\right) = (1 : 0).$$

The function t is a local equation of $0 = (0 : 1)$ in $\mathbb{P}^1 \setminus \{(1:0)\}$, so

$$\phi^*(0) = \sum_{p \in \phi^{-1}(0)} \nu_p(\phi^*(t))p = \sum_{p \in \phi^{-1}(0)} \nu_p(f)p = (f)_0.$$

Similarly, $\frac{1}{t}$ is a local equation of $\infty = (1 : 0)$ in $\mathbb{P}^1 \setminus \{(0:1)\}$, so $\phi^*(\infty) = (f)_\infty$.

By Theorem 13.18,

$$\deg(f) = \deg(f)_0 - \deg(f)_\infty = \deg \phi^*(0) - \deg \phi^*(\infty) = \deg \phi - \deg \phi = 0.$$

$\square$

Corollary 13.20. *Suppose that X is a nonsingular projective curve. Then the surjective group homomorphism $\deg : \text{Div}(X) \to \mathbb{Z}$ induces a surjective group homomorphism $\deg : \text{Cl}(X) \to \mathbb{Z}$.*

Let X be a nonsingular projective curve and let $\text{Cl}^0(X)$ be the subgroup of $\text{Cl}(X)$ of classes of divisors of degree 0. By the above corollary, there is a short exact sequence of groups

(13.11) $$0 \to \text{Cl}^0(X) \to \text{Cl}(X) \xrightarrow{\deg} \mathbb{Z} \to 0.$$

Corollary 13.21. *Suppose that X is a nonsingular projective curve. Then $\text{Cl}^0(X) = (0)$ if and only if $X \cong \mathbb{P}^1$. In fact, if X is a nonsingular projective curve and $p, q \in X$ are distinct points such that $p \sim q$, then $X \cong \mathbb{P}^1$.*

Proof. We proved in Example 13.10 that $\text{Cl}^0(\mathbb{P}^1) = (0)$. Suppose that X is a nonsingular projective curve and $p, q \in X$ are distinct points such that $p \sim q$. Then there exists $f \in k(X)$ such that $(f) = p - q$. The inclusion of function fields $k(\mathbb{P}^1) = k(t) \to k(X)$ defined by $t \mapsto f$ determines a regular map $\phi : X \to \mathbb{P}^1$, represented by $(f : 1)$. Now $\phi^*(0) = (f)_0 = p$, so $\deg \phi = 1$

by Theorem 13.18. Thus $\phi^* : k(t) \to k(X)$ is an isomorphism, and so ϕ is birational. Thus ϕ is an isomorphism by Corollary 10.25, since X and $\mathbb{P}^1$ are nonsingular projective curves. $\square$

By Corollary 13.19, the degree, $\deg(\mathcal{L})$, of an invertible sheaf $\mathcal{L}$ on a nonsingular projective curve X is well-defined, since $\mathcal{L} \cong \mathcal{O}_X(D)$ for some divisor D on X by Theorem 13.7.

Exercise 13.22. Let k be an algebraically closed field of characteristic not equal to 2, and let C be an irreducible quadric curve in $\mathbb{P}^2$ ($C = Z(F)$ where $F \in k[x_0, x_1, x_2]$ is an irreducible form of degree 2). Let $p \in C$ be a point and let $\pi : \mathbb{P}^2 \dashrightarrow \mathbb{P}^1$ be the projection from p. Let $\overline{\phi} : C \dashrightarrow \mathbb{P}^1$ be the induced rational map. Show that $\overline{\phi}$ is a regular map and an isomorphism.

13.6. Divisors, rational maps, and linear systems

Let X be a normal quasi-projective variety and let $\phi : X \dashrightarrow \mathbb{P}^n$ be a rational map. We may represent ϕ by an expression

$$\phi = (f_0 : \ldots : f_n)$$

with $f_0, \ldots, f_n \in k(X)$. We assume that none of the f_i are zero. Let

$$D_i = (f_i) = \sum_{j=1}^{m} k_{ij} C_j$$

where the C_j are distinct prime divisors on X.

Define the divisor

$$D = \gcd(D_0, \ldots, D_n) = \sum k_j C_j, \text{ where } k_j = \min_i k_{ij}.$$

The divisor D has the properties that $D_i - D \geq 0$ for all i and

(13.12) \qquad if F is a divisor such that $D_i \geq F$ for all i, then $D \geq F$.

Lemma 13.23. *Suppose that X is a normal variety and $p \in X$. Then ϕ is regular at p if and only if $p \notin \bigcap_{i=0}^{n} \mathrm{Supp}(D_i - D)$.*

Proof. The rational map ϕ is regular at p if and only if there exists $\alpha \in k(X)$ such that $\alpha f_i \in \mathcal{O}_{X,p}$ for all i and there exists j such that $\alpha f_j(p) \neq 0$.

Suppose that such an $\alpha \in k(X)$ exists. Then $\nu_E(\alpha f_i) \geq 0$ for all i and prime divisors E which contain p, and $\nu_E(\alpha f_j) = 0$ for all prime divisors E which contain p. Thus there exists an open neighborhood U of p such that $(\alpha f_i)_U \geq 0$ for all i and $(\alpha f_j)_U = 0$. Thus $(\alpha)_U = -(f_j)_U = -D_j \cap U$. Further, $(D_i - D_j) \cap U = (\alpha f_i)_U \geq 0$, so that $D_i \cap U \geq D_j \cap U$ for all i. This is only possible if $D_j \cap U = D \cap U$, so that $p \notin \bigcap_{i=0}^{n} \mathrm{Supp}(D_i - D)$.

Now suppose that $p \notin \bigcap_{i=0}^n \mathrm{Supp}(D_i - D)$. Then there exists D_s such that $k_{si} = k_i$ for all i such that $p \in C_i$. Thus there exists a neighborhood U of p such that $(f_s) \cap U = D \cap U$. Let $\alpha = \frac{1}{f_s}$. Then α is a local equation for the divisor $-D$ on U. Suppose that E is a prime divisor on X such that $p \in E$. Then

$$\nu_E(\alpha f_i) = \nu_E(\alpha) + \nu_E(f_i) = \begin{cases} 0 & \text{if } E \cap U \neq C_j \cap U \text{ for any } C_j, \\ k_{ij} - k_j & \text{if } E \cap U = C_j \cap U \text{ for some } C_j. \end{cases}$$

In particular, $\alpha f_i \in \mathcal{O}_X(0)_p = \mathcal{O}_{X,p}$ for all i, and αf_s does not vanish at p, so that ϕ is regular at p. $\square$

Suppose that X is a normal projective variety and D is a divisor on X. Then

$$\Gamma(X, \mathcal{O}_X(D)) = \{f \in k(X) \mid (f) + D \geq 0\}$$

is a finite-dimensional vector space over k, by Theorem 11.50, since $\mathcal{O}_X(D)$ is coherent (Lemma 13.5). Suppose that $f_0, \ldots, f_r \in \Gamma(X, \mathcal{O}_X(D))$ are linearly independent over k. For $t_0, \ldots, t_r \in k$ which are not all zero, we have that $t_0 f_0 + \cdots + t_r f_r \in \Gamma(X, \mathcal{O}_X(D))$. We thus have an associated effective divisor $(t_0 f_0 + \cdots + t_r f_r) + D$ on X, which is linearly equivalent to D. If $0 \neq \alpha \in k$, then the divisor $(\alpha t_0 f_0 + \cdots + \alpha t_r f_r) = (t_0 f_0 + \cdots + t_r f_r)$. We define a linear system $L \subset |D|$ by

$$L = \{(t_0 f_0 + \cdots + t_r f_r) + D \mid (t_0 : \ldots : t_r) \in \mathbb{P}^r\}.$$

The linear system L is a family, parameterized by $\mathbb{P}^r$, of effective divisors on X which are linearly equivalent to D. If we take $f_0, \ldots, f_r$ to be a k-basis of $\Gamma(X, \mathcal{O}_X(D))$, then we write $L = |D|$ and say that L is a complete linear system. The linear system $|D|$ is complete in the sense that if G is an effective divisor on X which is linearly equivalent to D, then $G \in |D|$. Observe that the linear system L determines the subspace of $\Gamma(X, \mathcal{O}_X(D))$ spanned by $f_0, \ldots, f_r$, but we cannot recover our specific basis.

Linear systems give us another way to understand rational maps. Suppose that D is a divisor on X, V is a linear subspace of $\Gamma(X, \mathcal{O}_X(D))$, and $L \subset |D|$ is the associated linear system. Then we associate to L a rational map

$$\phi_L = (f_0 : \ldots : f_n) : X \dashrightarrow \mathbb{P}^n,$$

by choosing a basis $f_0, \ldots, f_n$ of V. We will also denote this rational map by ϕ_V. A change of basis of V induces a linear automorphism of $\mathbb{P}^n$, so maps obtained from different choices of bases of V are the same, up to a change of homogeneous coordinates on $\mathbb{P}^n$.

Suppose that $\phi : X \dashrightarrow \mathbb{P}^n$ is a rational map. Then we have a representation $\phi = (f_0 : \ldots : f_n)$ with $f_i \in k(X)$. Let $D_i = (f_i)$ and $D = \gcd(D_0, \ldots, D_n)$. Then $f_i \in \Gamma(X, \mathcal{O}_X(-D))$ for all i. If the f_i are linearly independent, then ϕ is the rational map associated to the linear system $L \subset |-D|$ obtained from the span $V \subset \Gamma(X, \mathcal{O}_X(-D))$ of $f_0, \ldots, f_n$. If some of the f_i are linearly dependent, then the projection $\mathbb{P}^n \dashrightarrow \mathbb{P}^m$ onto an appropriate subspace of $\mathbb{P}^n$ is an isomorphism on the image of ϕ, and the composed rational map $X \dashrightarrow \mathbb{P}^m$ is given by ϕ_L.

Suppose that L is a linear system on a normal projective variety X. The base locus of the linear system L is the closed subset of X:

$$\mathrm{Base}(L) = \bigcap_{F \in L} \mathrm{Supp}(F).$$

We will say that a linear system L is base point free if $\mathrm{Base}(L) = \emptyset$.

Example 13.24. Let $A \in k[x_0, \ldots, x_n]$ be a linear form on $\mathbb{P}^n$. Let $H = \mathrm{Div}(A)$ (a hyperplane of $\mathbb{P}^n$). For $r > 0$, a k-basis of $\Gamma(\mathbb{P}^n, \mathcal{O}_X(rH))$ is

$$\left\{ \frac{x_0^{i_0} \cdots x_n^{i_n}}{A^r} \mid i_0 + \cdots + i_n = r \right\}$$

so the complete linear system

$$|rH| = \left\{ \mathrm{Div}\left(\sum t_{i_0 \ldots i_n} x_0^{i_0} \cdots x_n^{i_n}\right) \mid i_0 + \cdots + i_n = r \text{ and } (t_{i_0 \ldots i_n}) \in \mathbb{P}^{\binom{n+r}{n}-1} \right\}$$

is base point free.

Lemma 13.25. *Suppose that X is a normal projective variety, D is a divisor on X, and $p \in X$. If $\mathcal{O}_X(D)_p$ is not invertible (not isomorphic to $\mathcal{O}_{X,p}$ as an $\mathcal{O}_{X,p}$-module), then $p \in \mathrm{Base}(|D|)$.*

Proof. Suppose that $p \in X$ and $p \notin \mathrm{Base}(|D|)$. We will show that $\mathcal{O}_X(D)_p$ is invertible. There exists $f \in \Gamma(X, \mathcal{O}_X(D))$ such that $p \notin \mathrm{Supp}((f) + D)$. Let $E = (f) + D$. Then $\mathcal{O}_X(D) \xrightarrow{\frac{1}{f}} \mathcal{O}_X(E)$ is an isomorphism of $\mathcal{O}_X$-modules. Since $p \notin \mathrm{Supp}(E)$, $\mathcal{O}_X(E)_p = \mathcal{O}_{X,p}$ so $\mathcal{O}_X(D)_p$ is invertible. $\square$

We point out that for a divisor D on a normal projective variety X and $p \in X$, $\mathcal{O}_X(D)_p$ is invertible if and only if there exists a local equation of D at p; in fact, if h generates $\mathcal{O}_X(D)_p$ as an $\mathcal{O}_{X,p}$-module, then $\frac{1}{h}$ is a local equation of D at p, as we now show. Let U be an affine open neighborhood of p in X and $g \in k(X)$ be such that $D \cap U = (g) \cap U$. Then, as shown in the first part of the proof of Lemma 13.5, $\mathcal{O}_X(D)|U = \frac{1}{g}\mathcal{O}_U$. Further, if V is an open subset of U, then $\mathcal{O}_X(D)|V = \frac{1}{g}\mathcal{O}_V$. Thus $\mathcal{O}_X(D)_p = \frac{1}{g}\mathcal{O}_{X,p}$, so that $h = u\frac{1}{g}$ where u is a unit in $\mathcal{O}_{X,p}$, and so $\frac{1}{h}$ is a local equation of D at p.

Lemma 13.26. *Suppose that X is a normal projective variety and $L \subset |D|$ is the linear system associated to a subspace V of $\Gamma(X, \mathcal{O}_X(D))$. Then*

1) *The rational map*
$$\phi_V : X \dashrightarrow \mathbb{P}^n$$
   associated to V is independent of the divisor G such that $V \subset \Gamma(X, \mathcal{O}_X(G))$.

2) *Suppose that E is a codimension 1 subvariety of X such that $E \subset \text{Base}(L)$. Then $V \subset \Gamma(X, \mathcal{O}_X(D - E))$, and the linear system associated to V, regarded as a subspace of $\Gamma(X, \mathcal{O}_X(D - E))$, is $\{F - E \mid F \in L\}$.*

3) *Suppose that $\text{Base}(L)$ has codimension ≥ 2 in X. Then the locus where the rational map ϕ_L is not a regular map is the closed subset $\text{Base}(L)$.*

A consequence of Lemma 13.26 is that the rational map $\phi_\nu : X \dashrightarrow \mathbb{P}^n$ is defined by a linear system whose base locus has codim ≥ 2 in X.

Proof. The first statement follows since the rational map ϕ_V depends only on V. If $E \subset \text{Base}(L)$ and $f \in V$, then $(f) + D \geq E$. Thus $V \subset \Gamma(X, \mathcal{O}_X(D - E))$. Let $f_0, \ldots, f_n$ be a basis of V. Define $A = -\gcd((f_0), \ldots, (f_n))$ as before Lemma 13.23. We have that $V \subset \Gamma(X, \mathcal{O}_X(A))$ and the base locus of the linear system $\{(f) + A \mid f \in V\}$ has codimension ≥ 2 in X by Lemma 13.23. By (13.12) we have that $A \leq D$. It remains to prove the third statement of the lemma. Since $\text{Base}(L)$ has codimension ≥ 2 in X, we have that $D = A$. By Lemma 13.23, it suffices to show that $\text{Base}(L) = \bigcap_{i=0}^n (\text{Supp}((f_i) + A))$. By Lemma 13.25, we need to show that if $p \in X$ is such that $\mathcal{O}_X(A)_p$ is invertible, $p \in \bigcap_{i=0}^n (\text{Supp}((f_i) + A))$, and $G \in L$, then $p \in \text{Supp}(G)$. With these assumptions, let g be a local equation of A at p. Then $f_i g \in \mathcal{O}_{X,p}$ and $f_i g(p) = 0$ for all i. There exist $t_0, \ldots, t_n \in k$ such that $G = (\sum t_i f_i) + A$ so $\sum t_i f_i g$ is a local equation of G at p. Now $(\sum t_i f_i g_i)(p) = \sum t_i (f_i g)(p) = 0$ so $p \in G$. $\square$

Definition 13.27. A divisor D on a normal projective variety X is called very ample if $\text{Base}|D| = \emptyset$ and the induced regular map $\phi_{|D|} : X \to \mathbb{P}^n$ is a closed embedding. The divisor D is called ample if $\mathcal{O}_X(D)$ is invertible and some positive multiple of D is very ample.

If D is very ample, then $\mathcal{O}_X(D)$ is invertible by Lemma 13.25.

In Example 13.24, rH is very ample if $r \geq 1$ as $\phi_{|rH|}$ is a Veronese embedding.

13.6. Divisors, rational maps, and linear systems

Lemma 13.28. *Suppose that $\phi_L : X \to \mathbb{P}^n$ is a regular map which is given by a base point free linear system L. Then*

$$L = \{\phi_L^*(H) \mid H \text{ is a linear hyperplane on } \mathbb{P}^n \text{ such that } \phi_L(X) \not\subset H\}.$$

Further, if $p \in X$, then the fiber

$$\phi_L^{-1}(\phi_L(p)) = \bigcap_F \mathrm{Supp}(F)$$

where the intersection is over $F \in L$ such that $p \in F$.

The proof of Lemma 13.28 is Exercise 13.40.

Proposition 13.29. *Suppose that D is a divisor on a normal projective variety X and V is a subspace of $\Gamma(X, \mathcal{O}_X(D))$ such that the associated linear system L is base point free. Let t be an indeterminate, and let R be the graded k-algebra $R = k[Vt] \subset k(X)[t]$, where we set t to have degree 1. Then the graded k-algebra $\bigoplus_{n \geq 0} \Gamma(X, \mathcal{O}_X(nD))$ is a finitely generated R-module, where we regard $\bigoplus_{n \geq 0} \Gamma(X, \mathcal{O}_X(nD))$ as a graded R-module by the isomorphism*

$$\bigoplus_{n \geq 0} \Gamma(X, \mathcal{O}_X(nD)) \cong \sum_{n \geq 0} \Gamma(X, \mathcal{O}_X(nD)) t^n \subset k(X)[t].$$

Proof. The rational map $\phi_L : X \dashrightarrow \mathbb{P}^m$ (with $m = \dim_k V - 1$) is a regular map since L is base point free, and $\phi_L^* \mathcal{O}_{\mathbb{P}^m}(1) \cong \mathcal{O}_X(D)$ (by Exercise 13.61). Let $\mathcal{F} = (\phi_L)_* \mathcal{O}_X$. The sheaf $\mathcal{F}$ is a coherent $\mathcal{O}_{\mathbb{P}^m}$-module by Theorem 11.51. We have that

$$\mathcal{F}(n) \cong \mathcal{F} \otimes \mathcal{O}_{\mathbb{P}^m}(n) \cong (\phi_L)_*(\phi_L^* \mathcal{O}_{\mathbb{P}^m}(n)) \cong (\phi_L)_* \mathcal{O}_X(nD)$$

for $n \in \mathbb{Z}$ by Exercise 11.41. Further,

$$\bigoplus_{n \geq 0} \Gamma(X, \mathcal{O}_X(nD)) \cong \bigoplus_{n \geq 0} \Gamma(\mathbb{P}^m, \mathcal{F}(n))$$

is a finitely generated $S = S(\mathbb{P}^m)$-module by Exercise 11.59. Thus

$$\bigoplus_{n \geq 0} \Gamma(X, \mathcal{O}_X(nD))$$

is a finitely generated R-module since R is the image of S in $k(X)[t]$, by the natural graded map which takes S_1 onto V. $\square$

Corollary 13.30. *Suppose that X is a normal projective variety and D is an ample divisor on X. Then there exists $m_0 > 0$ such that mD is very ample for $m > m_0$, and there exists $m_1 \geq m_0$ such that if $m > m_1$ and W is* the image of X in a projective space $\mathbb{P}$ by the closed embedding $\phi_{|mD|}$, and S is the coordinate ring of $W \cong X$ by this embedding, then

$$S \cong \bigoplus_{n\geq 0} \Gamma(X, \mathcal{O}_X(nmD))$$

as graded rings.

However, if $|D|$ is not base point free, then it is possible that the graded k-algebra $\bigoplus_{n\geq 0} \Gamma(X, \mathcal{O}_X(nD))$ is not a finitely generated k-algebra, even if D is an effective divisor on a nonsingular projective surface X. An example is given by Zariski in [**159**, Part 1, Section 2]. We will present this example in Theorem 20.14.

13.7. Criteria for closed embeddings

Suppose that D is a divisor on a nonsingular projective variety X and $V \subset \Gamma(X, \mathcal{O}_X(D))$ is a linear subspace, with associated linear system L, such that $\mathrm{Base}(L)$ has codimension ≥ 2 in X. Let $f_0, \ldots, f_n$ be a k-basis of V, so that the rational map $\phi = \phi_V : X \dashrightarrow \mathbb{P}^n$ is represented by $\phi = (f_0 : \ldots : f_n)$. Then we have seen that for $p \in X$, if $\alpha \in k(X)$ is a local equation of D at p, then $\alpha f_i \in \mathcal{O}_{X,p}$ for all i, and $\phi = (\alpha f_0 : \ldots : \alpha f_n)$ is regular at p if and only if $\alpha f_i(p) \neq 0$ for some i. In this case, $(\alpha f_0 : \ldots : \alpha f_n)$ represents ϕ as a regular map in a neighborhood of p. Let $\mathcal{I}_p$ be the ideal sheaf of p in X.

Consider the commutative diagram:

$$\begin{array}{ccccc} \Gamma(X, \mathcal{O}_X(D)) & \stackrel{\text{restriction}}{\to} & \mathcal{O}_X(D)_p = \frac{1}{\alpha}\mathcal{O}_{X,p} & \stackrel{\alpha}{\to} & \mathcal{O}_{X,p} \\ & & \downarrow & & \downarrow \\ & & \mathcal{O}_X(D)_p/\mathcal{I}_p\mathcal{O}_X(D)_p & \stackrel{\alpha}{\to} & \mathcal{O}_{X,p}/(\mathcal{I}_p)_p \cong k \end{array}$$

where the isomorphism with k is the evaluation map $g \mapsto g(p)$. The horizontal maps, given by multiplication by α, are $\mathcal{O}_{X,p}$-module isomorphisms.

The composed map $\Lambda : \Gamma(X, \mathcal{O}_X(D)) \to \mathcal{O}_X(D)_p/\mathcal{I}_p\mathcal{O}_X(D)_p \cong k$ is given by $f \mapsto (\alpha f)(p)$ for $f \in \Gamma(X, \mathcal{O}_X(D))$.

We see that

(13.13) $\quad\phi$ is regular at p if and only if the natural map of V to $\mathcal{O}_X(D)_p/\mathcal{I}_p\mathcal{O}_X(D)_p \cong k$ is surjective.

Now suppose that ϕ is a regular map on X and p, q are distinct points of X. Let α be a local equation of D at p and let β be a local equation of D at q. Then we have an evaluation map

$$\Lambda' : \Gamma(X, \mathcal{O}_X(D)) \to \mathcal{O}_X(D)_p/\mathcal{I}_p\mathcal{O}_X(D)_p \oplus \mathcal{O}_X(D)_q/\mathcal{I}_q\mathcal{O}_X(D)_q \cong k^2,$$

13.7. Criteria for closed embeddings

defined by $\Lambda'(f) = ((\alpha f)(p), (\beta f)(q))$ for $f \in \Gamma(X, \mathcal{O}_X(D))$. We have that $\phi(p) \neq \phi(q)$ if and only if the vectors

$$(\alpha f_0(p), \ldots, \alpha f_n(p)) \quad \text{and} \quad (\beta f_0(q), \ldots, \beta f_n(q))$$

in k^{n+1} are linearly independent over k. Thus, since the row rank and column rank of the matrix

$$\begin{pmatrix} \alpha f_0(p) & \cdots & \alpha f_n(p) \\ \beta f_0(q) & \cdots & \beta f_n(q) \end{pmatrix}$$

are equal,
(13.14)
> ϕ is regular and injective if and only if the natural map of V to $\mathcal{O}_X(D)_p/\mathcal{I}_p\mathcal{O}_X(D)_p \oplus \mathcal{O}_X(D)_q/\mathcal{I}_q\mathcal{O}_X(D)_q$ is surjective for all $p \neq q \in X$.

Suppose that ϕ is an injective regular map. Then ϕ is a closed embedding if and only if $d\phi_p : T_p(X) \to T_q(\mathbb{P}^n)$ is injective for all $p \in X$ (with $q = \phi(p)$) by Theorem 10.36.

Suppose that $\alpha = 0$ is a local equation of D at p. Then $(\alpha f_i)(p) \neq 0$ for some i, say $(\alpha f_0)(p) \neq 0$. Then $\phi(p)$ is in $U = \mathbb{P}^n_{x_0} \cong \mathbb{A}^n$, which has the regular functions

$$k[U] = k\left[\frac{x_1}{x_0}, \ldots, \frac{x_n}{x_0}\right].$$

Let $\lambda_i = \frac{(\alpha f_i)(p)}{(\alpha f_0)(p)}$ for $1 \leq i \leq n$. Then the maximal ideal $\mathfrak{m}_q = \mathcal{I}_q(U)$ of q in $k[U]$ is

$$\mathfrak{m}_q = \left(\frac{x_1}{x_0} - \lambda_1, \ldots, \frac{x_n}{x_0} - \lambda_n\right).$$

Now $d\phi_p$ is injective if and only if the dual map

$$\phi^* : \mathfrak{m}_q/\mathfrak{m}_q^2 \to \mathfrak{m}_p/\mathfrak{m}_p^2$$

is onto (where $\mathfrak{m}_p = (\mathcal{I}_p)_p$). This map is onto if and only if the classes of

$$\frac{\alpha f_1}{\alpha f_0} - \lambda_1, \ldots, \frac{\alpha f_n}{\alpha f_0} - \lambda_n$$

span $(\mathcal{I}_p)_p/(\mathcal{I}_p)_p^2$ as a k-vector space. Since $(\alpha f_0)(p) \neq 0$, this holds if and only if the classes of

$$\alpha f_1 - \lambda_1(\alpha f_0), \ldots, \alpha f_n - \lambda_n(\alpha f_0)$$

span $(\mathcal{I}_p)_p/(\mathcal{I}_p)_p^2$ as a k-vector space. Now $\{f \in V \mid (\alpha f)(p) = 0\}$ has the basis

$$\{f_1 - \lambda_1 f_0, \ldots, f_n - \lambda_n f_0\}.$$

Thus:

(13.15) Suppose that ϕ is an injective regular map. Let K_p be the kernel of the natural map $V \to \mathcal{O}_X(D)_p/\mathcal{I}_p\mathcal{O}_X(D)_p$. Then ϕ is a closed embedding if and only if the natural map of K_p to $\mathcal{I}_p\mathcal{O}_X(D)_p/\mathcal{I}_p^2\mathcal{O}_X(D)_p$ is surjective for all $p \in X$.

Lemma 13.31. *Suppose that X is a nonsingular quasi-projective variety, D is a divisor on X such that $\mathcal{O}_X(D)$ is invertible, and $p_1, \ldots, p_n \in X$ are distinct points. Then*

1) *The natural $\mathcal{O}_X$-module homomorphism*
$$\mathcal{O}_X(D) \otimes \mathcal{I}_{p_1} \otimes \cdots \otimes \mathcal{I}_{p_n} \to \mathcal{O}_X(D)$$
is injective with image the coherent $\mathcal{O}_X$-submodule $\mathcal{I}_{p_1} \cdots \mathcal{I}_{p_n}\mathcal{O}_X(D)$ of $\mathcal{O}_X(D)$.

2) *If U is an open subset of X, then*

$\Gamma(U, \mathcal{O}_X(D) \otimes \mathcal{I}_{p_1} \otimes \cdots \otimes \mathcal{I}_{p_n})$
$= \{f \in \Gamma(U, \mathcal{O}_X(D)) \mid p_i \in (f)_U + D \cap U \text{ for all } p_i \text{ such that } p_i \in U\}.$

Proof. The first statement follows since $\mathcal{I}_{p_i,q} \cong \mathcal{O}_{X,q}$ if $p \neq q$ and $\mathcal{O}_X(D)_q \cong \mathcal{O}_{X,q}$ for all $q \in X$ and since $\mathcal{O}_{X,q}$ is a flat $\mathcal{O}_{X,q}$-module for all $q \in X$. To prove the second statement, by the sheaf axioms, it suffices to prove the second statement for affine open sets U such that $\mathcal{O}_X(D)|U$ is isomorphic to $\mathcal{O}_U$. Let U be such an open set. Let g be a local equation of D in U. Then $\mathcal{O}_X(D)|U = \frac{1}{g}\mathcal{O}_U$. We may suppose that $p_1, \ldots, p_s \in U$ and $p_{s+1}, \ldots, p_n \notin U$. We compute

$\Gamma(U, \mathcal{O}_X(D) \otimes \mathcal{I}_{p_1} \otimes \cdots \otimes \mathcal{I}_{p_n})$
$= \frac{1}{g}k[U] \otimes_{k[U]} I_U(p_1) \otimes_{k[U]} I_U(p_2) \otimes \cdots \otimes_{k[U]} I_U(p_n)$
$\cong \frac{1}{g} I_U(p_1) I_U(p_2) \cdots I_U(p_s).$

Now $f \in \Gamma(U, \mathcal{O}_X(D) \otimes \mathcal{I}_{p_1} \otimes \cdots \otimes \mathcal{I}_{p_s})$ if and only if $gf \in I_U(p_1) \cdots I_U(p_s)$ which holds if and only if $(gf)_U \geq 0$ and $(gf)(p_i) = 0$ for $1 \leq i \leq s$. These last conditions holds if and only if $(f)_U + D \cap U \geq 0$ and $p_i \in (f)_U + D \cap U$. $\square$

Theorem 13.32. *Suppose that X is a nonsingular projective variety and A and B are divisors on X.*

1) *Suppose that A is very ample and $|B|$ is base point free. Then $A+B$ is very ample.*

2) *Suppose that A is ample and B is an arbitrary divisor on X. Then there exists a positive integer n_0 such that $nA + B$ is very ample for all $n \geq n_0$.*

13.7. Criteria for closed embeddings

Proof. We first establish 1). Since A is very ample, $V = \Gamma(X, \mathcal{O}_X(A))$ satisfies the conditions of (13.13), (13.14), and (13.15). Since $|B|$ is base point free, $\Gamma(X, \mathcal{O}_X(B))$ generates $\mathcal{O}_X(B)_p$ as an $\mathcal{O}_{X,p}$-module for all $p \in X$. Thus we have natural surjections

$$\Lambda : \Gamma(X, \mathcal{O}_X(A)) \otimes \Gamma(X, \mathcal{O}_X(B)) \to (\mathcal{O}_X(A)_p / \mathcal{I}_p \mathcal{O}_X(A)_p) \otimes_{\mathcal{O}_{X_p}} \mathcal{O}_X(B)_p$$
$$\cong \mathcal{O}_X(A+B)_p / \mathcal{I}_p \mathcal{O}_X(A+B)_p$$

for all $p \in X$,

$$\Lambda' : \Gamma(X, \mathcal{O}_X(A)) \otimes \Gamma(X, \mathcal{O}_X(B))$$
$$\to (\mathcal{O}_X(A)_p / \mathcal{I}_p \mathcal{O}_X(A)_p \oplus \mathcal{O}_X(A)_q / \mathcal{I}_q \mathcal{O}_X(A)_q) \otimes_{\mathcal{O}_{X,p}} \mathcal{O}_X(B)_p$$
$$\cong (\mathcal{O}_X(A+B)_p / \mathcal{I}_p \mathcal{O}_X(A+B)_p) \oplus (\mathcal{O}_X(A+B)_q \mathcal{I}_q \mathcal{O}_X(A+B)_q)$$

for all $p, q \in X$ with $p \neq q$, and

$$\Lambda'' : \Gamma(X, \mathcal{O}_X(A) \otimes \mathcal{I}_p) \otimes \Gamma(X, \mathcal{O}_X(B))$$
$$\to (\mathcal{I}_p \mathcal{O}_X(A)_p / \mathcal{I}_p^2 \mathcal{O}_X(A)_p) \otimes \mathcal{O}_X(B)_p$$
$$\cong \mathcal{I}_p \mathcal{O}_X(A+B)_p / \mathcal{I}_p^2 \mathcal{O}_X(A+B)_p.$$

We have a natural k-vector space homomorphism

$$\Gamma(X, \mathcal{O}_X(A)) \otimes \Gamma(X, \mathcal{O}_X(B)) \to \Gamma(X, \mathcal{O}_X(A+B))$$

which factors Λ and Λ' and a natural k-vector space homomorphism

$$\Gamma(X, \mathcal{O}_X(A) \otimes \mathcal{I}_p) \otimes \Gamma(X, \mathcal{O}_X(B)) \to \Gamma(X, \mathcal{O}_X(A+B) \otimes \mathcal{I}_p)$$

which factors Λ''. Thus $A + B$ satisfies the conditions of (13.13), (13.14), and (13.15), so that $A + B$ is very ample.

We now establish 2). There exists a positive integer n_1 such that $n_1 A$ is very ample, so there exists a closed embedding $\phi : X \to \mathbb{P}^r$ such that

$$\mathcal{O}_X(n_1 A) \cong \phi^* \mathcal{O}_{\mathbb{P}^r}(1) = \mathcal{O}_X(1)$$

by Exercise 13.61. By Theorem 11.45, there exists $n_2 \geq n_1$ such that for $0 \leq t < n_1$, $\mathcal{O}_X(tA + nn_1 A)$ is generated by global sections if $n \geq n_2$ and $\mathcal{O}_X(B + nn_1 A)$ is generated by global sections if $n \geq n_2$. Suppose $n > 3n_1 n_2$. Write

$$n - n_1 n_2 = mn_1 + t \quad \text{with } 0 \leq t < n_1.$$

We then have $m > n_2$. Thus

$$nA + B = (m - n_2)n_1 A + [(n_2 n_1 A + tA) + (n_2 n_1 A + B)]$$

is the sum of a very ample divisor and a divisor D such that $|D|$ is base point free. Thus $nA + B$ is very ample by 1) of this theorem. $\square$

Theorem 13.33. *Suppose that X is a nonsingular projective variety and D is a divisor on X such that $\mathrm{Base}(|D|)$ has codimension ≥ 2 in X. Let $\phi_{|D|}$ be the rational map associated to $\Gamma(X, \mathcal{O}_X(D))$. Then*

1) *$\phi_{|D|}$ is a regular map if and only if for all $p \in X$,*

$$\dim_k \Gamma(X, \mathcal{O}_X(D) \otimes \mathcal{I}_p) = \dim_k \Gamma(X, \mathcal{O}_X(D)) - 1.$$

2) *$\phi_{|D|}$ is an injective regular map if and only if for all distinct points p and q in X,*

$$\dim_k \Gamma(X, \mathcal{O}_X(D) \otimes \mathcal{I}_p \otimes \mathcal{I}_q) = \dim_k \Gamma(X, \mathcal{O}_X(D)) - 2.$$

3) *$\phi_{|D|}$ is a closed embedding if and only if 2) holds and for all $p \in X$,*

$$\dim_k \Gamma(X, \mathcal{O}_X(D) \otimes \mathcal{I}_p^2) = \dim_k \Gamma(X, \mathcal{O}_X(D)) - (1 + \dim X).$$

Proof. Consider the short exact sequence for $p \in X$,

$$0 \to \mathcal{I}_p \to \mathcal{O}_X \to \mathcal{O}_X/\mathcal{I}_p \to 0.$$

Since $\mathcal{O}_X(D)$ is locally free, tensoring the sequence with $\mathcal{O}_X(D)$ gives a short exact sequence

$$0 \to \mathcal{O}_X(D) \otimes \mathcal{I}_p \to \mathcal{O}_X(D) \to (\mathcal{O}_X/\mathcal{I}_p) \otimes \mathcal{O}_X(D) \to 0.$$

Now $(\mathcal{O}_X/\mathcal{I}_p) \otimes \mathcal{O}_X(D) \cong \mathcal{O}_X/\mathcal{I}_p$, so taking global sections, we get an exact sequence

$$0 \to \Gamma(X, \mathcal{O}_X(D) \otimes \mathcal{I}_p) \to \Gamma(X, \mathcal{O}_X(D)) \to k,$$

so that the conclusion of 1) follows from (13.13).

Consider the short exact sequence for $p \neq q \in X$, which follows from Theorem 1.5 applied to an affine open subset of X containing p and q,

$$0 \to \mathcal{I}_p \mathcal{I}_q \to \mathcal{O}_X \to \mathcal{O}_X/\mathcal{I}_p\mathcal{I}_q \cong \mathcal{O}_X/\mathcal{I}_p \oplus \mathcal{O}_X/\mathcal{I}_q \to 0.$$

Since the locus of points where $\mathcal{I}_p$ is not locally free and the locus of points where $\mathcal{I}_q$ is not locally free are disjoint, we have that $\mathcal{I}_p \otimes \mathcal{I}_q \cong \mathcal{I}_p\mathcal{I}_q$. Since $\mathcal{O}_X(D)$ is locally free, tensoring the sequence with $\mathcal{O}_X(D)$ gives a short exact sequence

$$0 \to \mathcal{O}_X(D) \otimes \mathcal{I}_p\mathcal{I}_q \to \mathcal{O}_X(D) \to (\mathcal{O}_X/\mathcal{I}_p) \otimes \mathcal{O}_X(D) \oplus (\mathcal{O}_X/\mathcal{I}_q) \otimes \mathcal{O}_X(D) \to 0.$$

Now

$$(\mathcal{O}_X/\mathcal{I}_p\mathcal{I}_q) \otimes \mathcal{O}_X(D) \cong \mathcal{O}_X/\mathcal{I}_p\mathcal{I}_q \cong \mathcal{O}_X/\mathcal{I}_p \oplus \mathcal{O}_X/\mathcal{I}_q,$$

so taking global sections, we get an exact sequence

$$0 \to \Gamma(X, \mathcal{O}_X(D) \otimes \mathcal{I}_p\mathcal{I}_q) \to \Gamma(X, \mathcal{O}_X(D)) \to k^2,$$

so that the conclusion of 2) follows from 1) of this theorem and (13.14).

Finally, 3) of the theorem follows from tensoring the short exact sequence
$$0 \to \mathcal{I}_p^2 \to \mathcal{I}_p \to \mathcal{I}_p/\mathcal{I}_p^2 \to 0$$
with $\mathcal{O}_X(D)$ and taking global sections, applying (13.15), and since
$$\dim_k \mathcal{I}_p/\mathcal{I}_p^2 = \dim X$$
(as p is a nonsingular point on X). □

Corollary 13.34. *Suppose that X is a nonsingular projective curve and D is a divisor on X. Then*

1) *$|D|$ is base point free if and only if*
$$\dim_k \Gamma(X, \mathcal{O}_X(D-p)) = \dim_k \Gamma(X, \mathcal{O}_X(D)) - 1$$
for all $p \in X$.

2) *Suppose that $|D|$ is base point free. Then $\phi_{|D|}$ is injective if and only if*
$$\dim_k \Gamma(X, \mathcal{O}_X(D-p-q)) = \dim_k \Gamma(X, \mathcal{O}_X(D)) - 2$$
for all $p \neq q \in X$.

3) *Suppose that $|D|$ is base point free. Then $\phi_{|D|}$ is a closed embedding (and D is very ample) if and only if*
$$\dim_k \Gamma(X, \mathcal{O}_X(D-p-q)) = \dim_k \Gamma(X, \mathcal{O}_X(D)) - 2$$
for all $p, q \in X$.

13.8. Invertible sheaves

The constructions of $\otimes$ and Hom in Section 11.5 are fairly simple for invertible sheaves. Suppose that $\mathcal{F}$ and $\mathcal{G}$ are invertible sheaves on a quasi-projective variety X. Every point $p \in X$ has an affine neighborhood U such that $\mathcal{F}|U = \sigma \mathcal{O}_U$ where $\sigma \in \Gamma(U, \mathcal{F})$ is a local generator of $\mathcal{F}$ and $\mathcal{G}|U = \tau \mathcal{O}_U$ where $\tau \in \Gamma(U, \mathcal{G})$ is a local generator of $\mathcal{G}$. Thus $\mathcal{F} \otimes \mathcal{G}$ is invertible. In fact, if $\mathcal{F}|U = \sigma \mathcal{O}_U$ and $\mathcal{G}|U = \tau \mathcal{O}_U$, then $\mathcal{F} \otimes_{\mathcal{O}_X} \mathcal{G}|U = \sigma \otimes \tau \mathcal{O}_U$. We usually write $\sigma\tau$ for $\sigma \otimes \tau$.

Suppose that $\mathcal{L}$ is an invertible sheaf on X. Then $\operatorname{Hom}_{\mathcal{O}_X}(\mathcal{L}, \mathcal{O}_X)$ is an invertible sheaf. In fact, if $\mathcal{L}|U = \sigma \mathcal{O}_U$, then $\operatorname{Hom}_{\mathcal{O}_X}(\mathcal{L}, \mathcal{O}_X)|U = \hat{\sigma} \mathcal{O}_U$, where $\hat{\sigma}(\sigma) = 1$. We can write $\operatorname{Hom}_{\mathcal{O}_X}(\mathcal{L}, \mathcal{O}_X)|U = \frac{1}{\sigma} \mathcal{O}_U$. The map $\alpha \otimes \beta \mapsto \beta(\alpha)$ determines an $\mathcal{O}_X$-module isomorphism
$$\mathcal{L} \otimes_{\mathcal{O}_X} \operatorname{Hom}_{\mathcal{O}_X}(\mathcal{L}, \mathcal{O}_X) \cong \mathcal{O}_X.$$
We write $\mathcal{L}^{-1} = \operatorname{Hom}_{\mathcal{O}_X}(\mathcal{L}, \mathcal{O}_X)$, since then
$$\mathcal{L} \otimes \mathcal{L}^{-1} \cong \mathcal{L}^{-1} \otimes \mathcal{L} \cong \mathcal{O}_X.$$

Thus the tensor product and inverse operation make

$$\operatorname{Pic}(X) = \{\mathcal{L} \mid \mathcal{L} \text{ is invertible}\}/\sim$$

into an Abelian group, where two invertible sheaves are equivalent if they are isomorphic as $\mathcal{O}_X$-modules. $\operatorname{Pic}(X)$ is called the Picard group of X.

If X is nonsingular and D, E are divisors on X, then we have an affine cover $\{U_i\}$ of X and $f_i, g_i \in k(X)$ such that

$$\mathcal{O}_X(D)|U_i = \frac{1}{f_i}\mathcal{O}_{U_i}$$

and

$$\mathcal{O}_X(E)|U_i = \frac{1}{g_i}\mathcal{O}_{U_i}.$$

Then

$$\mathcal{O}_X(D) \otimes \mathcal{O}_X(E)|U_i = \frac{1}{f_i g_i}\mathcal{O}_{U_i} = \mathcal{O}_X(D+E)|U_i,$$

so that $\mathcal{O}_X(D+E) \cong \mathcal{O}_X(D) \otimes \mathcal{O}_X(E)$. We also have that $\mathcal{O}_X(D)^{-1}|U_i = f_i \mathcal{O}_X|U_i = \mathcal{O}_X(-D)|U_i$ so that $\mathcal{O}_X(-D) \cong \mathcal{O}_X(D)^{-1}$. It follows from Proposition 13.6 and Theorem 13.7 that (with the assumption that X is nonsingular) the map

(13.16) $$\operatorname{Cl}(X) \to \operatorname{Pic}(X)$$

defined by $D \mapsto \mathcal{O}_X(D)$ is a group isomorphism.

Suppose that $\mathcal{L}$ is an invertible sheaf on a quasi-projective variety X and $\sigma_0, \ldots, \sigma_n \in \Gamma(X, \mathcal{L})$. We have an associated rational map $\phi : X \dashrightarrow \mathbb{P}^n$, which is defined as follows. Suppose that U is an open subset of X such that $\mathcal{L}|U$ is trivial; that is, there exists $\tau \in \Gamma(U, \mathcal{L})$ such that $\mathcal{L}|U = \tau \mathcal{O}_U$. Write $\sigma_i = \tau f_i$ for $0 \leq i \leq n$ where $f_i \in \Gamma(U, \mathcal{O}_X)$. We define $\phi = (f_0 : f_1 : \ldots : f_n)$ on U. This gives a well-defined rational map, which we can write as $\phi = (\sigma_0 : \ldots : \sigma_n)$. The rational map ϕ is a regular map if and only if $\sigma_0, \ldots, \sigma_n$ generate $\mathcal{L}$; that is, for all $p \in X$, the restrictions of $\sigma_0, \ldots, \sigma_n$ to $\mathcal{L}_p$ generate $\mathcal{L}_p$ as an $\mathcal{O}_{X,p}$-module. Every regular map from X to a projective space $\mathbb{P}^n$ can be represented in this way by an appropriate invertible sheaf $\mathcal{L}$ and $\sigma_0, \ldots, \sigma_n \in \Gamma(X, \mathcal{L})$ (take $\mathcal{L} = \phi^* \mathcal{O}_{\mathbb{P}^n}(1)$ and $\sigma_0, \ldots, \sigma_n$ to be the images of $x_0, \ldots, x_n \in \Gamma(\mathbb{P}^n, \mathcal{O}_{\mathbb{P}^n}(1))$ in $\Gamma(X, \phi^* \mathcal{O}_{\mathbb{P}^n}(1))$).

Definition 13.35. An invertible sheaf $\mathcal{L}$ on a projective variety X is called very ample if the global sections $\Gamma(X, \mathcal{L})$ generate $\mathcal{L}$ as an $\mathcal{O}_X$-module and the regular map $\phi = (\sigma_0 : \cdots : \sigma_n) : X \to \mathbb{P}^n$, where $\sigma_0, \ldots, \sigma_n$ is a k-basis of $\Gamma(X, \mathcal{L})$, is a closed embedding. An invertible sheaf $\mathcal{L}$ on a projective variety X is called ample if some positive tensor product $\mathcal{L}^n = \mathcal{L}^{\otimes n}$ is very ample.

Definition 13.35 is consistent with Definition 13.27 of very ample and ample divisors on a normal variety X. A divisor D on a normal projective variety X is very ample (ample) if and only if $\mathcal{O}_X(D)$ is invertible and very ample (ample).

Suppose that $\phi : X \to Y$ is a regular map of quasi-projective varieties and $\mathcal{L}$ is an invertible sheaf on Y. Then $\phi^*\mathcal{L}$ is a coherent sheaf on X (Section 11.5). If $V \subset Y$ is an open subset on which $\mathcal{L}|V$ is trivial, then $\mathcal{L}|V = \tau \mathcal{O}_V$ for some $\tau \in \Gamma(V, \mathcal{O}_Y)$. We have that

$$\phi^*\mathcal{L}|\phi^{-1}(V) = \tau \mathcal{O}_V \otimes_{\mathcal{O}_V} \mathcal{O}_{\phi^{-1}(V)} = (\tau \otimes 1)\mathcal{O}_{\phi^{-1}(V)} = \tau \mathcal{O}_{\phi^{-1}(V)}.$$

Thus $\phi^*\mathcal{L}$ is an invertible sheaf on X. If X and Y are nonsingular and D is a divisor on Y, then $\phi^*(\mathcal{O}_Y(D)) \cong \mathcal{O}_X(\phi^*(D))$. This can be readily seen in the case that ϕ is dominant. We then have an injection $\phi^* : k(Y) \to k(X)$. We cover Y with open subsets V_i such that $\mathcal{O}_Y(D)|V_i = \frac{1}{f_i}\mathcal{O}_{V_i}$ where f_i is a local equation of D on V_i. Then $\phi^*\mathcal{O}_Y(D)|\phi^{-1}(V_i) = \frac{1}{\phi^*(f_i)}\mathcal{O}_{\phi^{-1}(V_i)}$, where $\phi^*(f_i)$ is a local equation of $\phi^*(D)$ on $\phi^{-1}(V_i)$. In the case where ϕ is not dominant we must first shift the support of D as explained after Lemma 13.12.

13.9. Transition functions

Suppose that $\mathcal{L}$ is an invertible sheaf on a variety X. Then there exists an open cover $\{U_i\}$ of X and $\mathcal{O}_{U_i}$-module isomorphisms $\phi_i : \mathcal{O}_X|U_i \xrightarrow{\cong} \mathcal{L}|U_i$ for all i. Consider the $\mathcal{O}_X|U_i \cap U_j$-module isomorphisms $\phi_j^{-1} \circ \phi_i : \mathcal{O}_X|U_i \cap U_j \xrightarrow{\cong} \mathcal{O}_X|U_i \cap U_j$. In the notation of the proof of Theorem 13.7, we have that $\phi_j^{-1} \circ \phi_i$ is multiplication by g_{ij}. We may thus identify $\phi_j^{-1} \circ \phi_i$ with g_{ij}, which is a unit in $\Gamma(U_i \cap U_j, \mathcal{O}_X)$ satisfying the relations (13.4),

$$g_{ij} = g_{ji}^{-1},$$

and (13.5),

(13.17) $$g_{ik} = g_{ij}g_{jk}.$$

We call the g_{ij} transition functions on $U_i \cap U_j$ for $\mathcal{L}$.

Suppose that $\mathcal{M}$ is another invertible sheaf on X and there exist $\mathcal{O}_{U_i}$-module isomorphisms $\psi_i : \mathcal{O}_X|U_i \to \mathcal{M}|U_i$. Let $h_{ij} = \psi_j^{-1} \circ \psi_i$ be transition functions on $U_i \cap U_j$ for $\mathcal{M}$. Then we have that $g_{ij}h_{ij}$ are transition functions for $\mathcal{L} \otimes \mathcal{M}$ on $U_i \cap U_j$ and $\frac{1}{g_{ij}}$ are transition functions for $\mathcal{L}^{-1}$ on $U_i \cap U_j$.

Let $\mathcal{O}_X^*$ be the presheaf of (multiplicative) groups on X, defined for $U \subset X$ an open subset by

$$\Gamma(U, \mathcal{O}_X^*) = \{f \in \Gamma(U, \mathcal{O}_X) \mid f \text{ is a unit}\}.$$

Then $\mathcal{O}_X^*$ is a sheaf of groups on X, and $g_{ij} \in \Gamma(U_i \cap U_j, \mathcal{O}_X^*)$.

Lemma 13.36. *Suppose that $\mathcal{L}$ is an invertible sheaf on a variety X and $\{U_i\}_{i\in I}$ is an open cover of X such that there exist $\mathcal{O}_{U_i}$-module isomorphisms $\phi_i : \mathcal{O}_X|U_i \xrightarrow{\cong} \mathcal{L}|U_i$ for all i. Let $\phi_j^{-1} \circ \phi_i : \mathcal{O}_X|U_i \cap U_j \xrightarrow{\cong} \mathcal{O}_X|U_i \cap U_j$ be the respective associated transition functions, which we identify with elements g_{ij} of $\Gamma(U_i \cap U_j, \mathcal{O}_X^*)$. Then $\mathcal{L}$ is isomorphic to $\mathcal{O}_X$ as an $\mathcal{O}_X$-module if and only if there exist $f_i \in \Gamma(U_i, \mathcal{O}_X^*)$ for all $i \in I$ such that $g_{ij} = f_j f_i^{-1}$ for all i, j.*

Proof. First suppose that there exist $f_i \in \Gamma(U_i, \mathcal{O}_X^*)$ such that $g_{ij} = f_j f_i^{-1}$ for all i, j. Then $\mathcal{L}|U_i = \tau_i \mathcal{O}_{U_i}$ where $\tau_i = \phi_i(f_i) = f_i \phi_i(1)$. For all i, j, on $U_i \cap U_j$, we have $\phi_i = g_{ij} \phi_j$. Thus $\phi_i(1) = g_{ij}\phi_j(1) = f_j f_i^{-1} \phi_j(1)$ so that $\tau_i = \tau_j$ on $U_i \cap U_j$. By the second sheaf axiom, there exists $\tau \in \Gamma(X, \mathcal{L})$ such that $\tau|U_i = \tau_i$ for all i. Define a homomorphism of sheaves of $\mathcal{O}_X$-modules $\Lambda : \mathcal{O}_X \to \mathcal{L}$ by $\Lambda(f) = \tau f$. Here Λ is an isomorphism at all stalks at points of X, so that Λ is an isomorphism by Proposition 11.13.

Now suppose that $\mathcal{L}$ is isomorphic to $\mathcal{O}_X$ as an $\mathcal{O}_X$-module. Then there exists an isomorphism of $\mathcal{O}_X$-modules $\Lambda : \mathcal{O}_X \to \mathcal{L}$. Let $\tau = \Lambda(1)$. Then $\mathcal{L}|U_i = \tau \mathcal{O}_{U_i}$ for all i. Thus there exist $f_i \in \Gamma(U_i, \mathcal{O}_X^*)$ such that $\tau = f_i \phi_i(1) = \phi_i(f_i)$. Thus on $U_i \cap U_j$, we have that $\phi_i(f_i) = \phi_j(f_j)$, so that $f_i \phi_i(1) = f_j \phi_j(1)$. Since $\phi_i(1) = g_{ij} \phi_j(1)$, we have that $g_{ij} = f_j f_i^{-1}$. $\square$

Lemma 13.37. *Suppose that $\mathcal{L}$ and $\mathcal{M}$ are invertible sheaves on a variety X and $\{U_i\}_{i\in I}$ is an open cover of X such that there exist $\mathcal{O}_{U_i}$-module isomorphisms $\phi_i : \mathcal{O}_X|U_i \xrightarrow{\cong} \mathcal{L}|U_i$ and $\psi_i : \mathcal{O}_X|U_i \xrightarrow{\cong} \mathcal{M}|U_i$ for all i. Let $\phi_j^{-1} \circ \phi_i : \mathcal{O}_X|U_i \cap U_j \xrightarrow{\cong} \mathcal{O}_X|U_i \cap U_j$ and $\psi_j^{-1} \circ \psi_i : \mathcal{O}_X|U_i \cap U_j \xrightarrow{\cong} \mathcal{O}_X|U_i \cap U_j$ be the respective associated transition functions, which we identify with elements g_{ij} and h_{ij} of $\Gamma(U_i \cap U_j, \mathcal{O}_X^*)$. Then $\mathcal{L}$ is isomorphic to $\mathcal{M}$ as an $\mathcal{O}_X$-module if and only if there exist $f_i \in \Gamma(U_i, \mathcal{O}_X^*)$ for all $i \in I$ such that $g_{ij} = f_j f_i^{-1} h_{ij}$ for all i, j.*

Proof. We have that $\mathcal{L}$ is isomorphic to $\mathcal{M}$ if and only if $\mathcal{L} \otimes \mathcal{M}^{-1} \cong \mathcal{O}_X$. As commented above Lemma 13.36, $g_{ij} h_{ij}^{-1}$ are the transition functions of $\mathcal{L} \otimes \mathcal{M}^{-1}$ on $U_i \cap U_j$. The conclusions of this lemma now follow from Lemma 13.36. $\square$

A situation where this criterion for isomorphism is very useful is in understanding the pull-back $\phi^* \mathcal{L}$ of an invertible sheaf under a regular map $\phi : Y \to X$ of varieties. Suppose that $\{U_i\}_{i\in I}$ is an open cover of X and we have trivializations $\phi_i : \mathcal{O}_{U_i} \to \mathcal{L}|U_i$ for all i. Let $\phi_j^{-1} \circ \phi_i = g_{ij} \in \Gamma(U_i \cap U_j, \mathcal{O}_X^*)$ be the transition functions on $U_i \cap U_j$ for $\mathcal{L}$. Let $V_i = \phi^{-1}(U_i)$.

For all i, we have an $\mathcal{O}_Y|V_i$-module homomorphism

$$\mathcal{O}_Y|V_i = \mathcal{O}_X \otimes_{\mathcal{O}_X} \mathcal{O}_Y|V_i \xrightarrow{\phi_i \otimes 1} (\mathcal{L} \otimes_{\mathcal{O}_X} \mathcal{O}_Y)|V_i = \phi^*\mathcal{L}|V_i,$$

13.9. Transition functions

which is an isomorphism of sheaves, since it is an isomorphism at stalks of points of V_i. Computing the transition functions of $\phi_i \otimes 1$, we see that

$$(\phi_j \otimes 1)^{-1} \circ (\phi_i \otimes 1) = \overline{g}_{ij} \in \Gamma(V_i \cap V_j, \mathcal{O}_Y^*),$$

where $\overline{g}_{ij} = \phi^*(g_{ij})$ under

$$\phi^* : \Gamma(U_i \cap U_j, \mathcal{O}_X^*) \to \Gamma(V_i \cap V_j, \mathcal{O}_Y^*).$$

We now rework Example 13.13 using this technique.

Example 13.38. Suppose that X is a nonsingular surface and $p \in X$ is a point. Let $\pi : B \to X$ be the blow-up of p. Let $E = \pi^{-1}(p) \cong \mathbb{P}^1$ be the exceptional divisor of π. Let $i : E \to B$ be the inclusion. Then $i^*(\mathcal{O}_B(E)) \cong \mathcal{O}_E(-q)$, where q is a point on E.

Proof. Let x, y be regular parameters in the regular local ring $\mathcal{O}_{X,p}$. Let U be an affine neighborhood of p such that $x = y = 0$ are local equations of p in U. Then $V = \pi^{-1}(U)$ is covered by two affine charts U_1 and U_2 which satisfy $k[U_1] = k[U][\frac{x}{y}]$ and $k[U_2] = k[U][\frac{y}{x}]$. In U_1, $y = 0$ is a local equation of E. Now $(x, y)k[U_1] = yk[U_1]$, so that

$$k[U_1 \cap E] = k[U_1]/(x, y)k[U_1] = k\left[\frac{x}{y}\right].$$

In U_2, $x = 0$ is a local equation of E, so

$$k[U_2 \cap E] = k\left[\frac{y}{x}\right].$$

Now $\mathcal{O}_B(E)|U_1 = \frac{1}{y}\mathcal{O}_B|U_1$ and $\mathcal{O}_B(E)|U_2 = \frac{1}{x}\mathcal{O}_B|U_2$. Thus $g_{12} = \frac{x}{y}$ is a transition function for $\mathcal{O}_B(E)$ on $U_1 \cap U_2$, so that $\overline{g}_{12} = i^*(\frac{x}{y})$, which we can identify with $\frac{x}{y}$, is the transition function for $i^*\mathcal{O}_B(E)$ on $U_1 \cap U_2 \cap E$.

Now $\frac{x}{y}$ is the local equation of a point q in $E \cap U_1$, which is not contained in U_2, so we have that $\mathcal{O}_E(-q)|U_1 \cap E = \frac{x}{y}\mathcal{O}_E|U_1 \cap E$ and $\mathcal{O}_E(-q)|U_2 \cap E = \mathcal{O}_E|U_2 \cap E$. The associated transition function on $U_1 \cap U_2 \cap E$ is $h_{12} = \frac{x}{y}$. Since $h_{12} = \overline{g}_{12}$, we have that $i^*\mathcal{O}_B(E) \cong \mathcal{O}_E(-q)$ by Lemma 13.37. □

Suppose that $\mathcal{L}$ is an invertible sheaf on a nonsingular variety X and $\sigma \in \Gamma(X, \mathcal{L})$ is nonzero. We associate an effective divisor to σ in the following way. Let $\{U_i\}$ be an open cover of X such that there exist $\sigma_i \in \Gamma(U_i, \mathcal{L})$ such that $\mathcal{L}|U_i = \sigma_i \mathcal{O}_{U_i}$. We have expressions $\sigma|U_i = f_i \sigma_i$ where $f_i \in \Gamma(U_i, \mathcal{O}_X)$. We define the divisor (σ) (or $\text{div}(\sigma)$) of σ to be the divisor D on X defined by

$$D \cap U_i = (f_i)_{U_i}.$$

This divisor is well-defined and independent of choice of local trivialization of $\mathcal{L}$. The pairs $\{(f_i, U_i)\}$ determine a Cartier divisor on X (Definition 15.4) and the following discussion.

Proposition 13.39. *Suppose that $\phi : X \to Y$ is a birational regular map of projective varieties such that Y is normal. Suppose that $\mathcal{L}$ is an invertible sheaf on Y. Then $\phi_* \phi^* \mathcal{L} \cong \mathcal{L}$.*

Proof. This follows from Proposition 11.52 and Exercise 11.41 since $\mathcal{L}$ is locally isomorphic to $\mathcal{O}_Y$. □

Exercise 13.40. Prove Lemma 13.28.

Exercise 13.41. Let $x_0, \ldots, x_n$ be homogeneous coordinates on $X = \mathbb{P}^n$, so that $k[x_0, \ldots, x_n] = S(\mathbb{P}^n)$. Let $t_i = \frac{x_i}{x_0}$ for $1 \leq i \leq n$, so that $k(\mathbb{P}^n) = k(t_1, \ldots, t_n)$. Let $E = Z(x_0)$, a prime divisor on X. Suppose that $f(t_1, \ldots, t_n) \in k[t_1, \ldots, t_n] = k[X_{x_0}]$ has degree m. Compute $r = -\nu_E(f)$, and show that $(f) + rE \geq 0$.

Exercise 13.42. Let the notation be as in Exercise 13.41. Suppose that Y is a nonsingular closed subvariety of X such that $Y \not\subset Z(x_0)$. Let $i : Y \to X$ be the inclusion. We have a natural surjection $i^* : k[\mathbb{P}^n_{x_0}] = k[t_1, \ldots, t_n] \to k[Y_{x_0}]$. Suppose that $f \in k[t_1, \ldots, t_n]$ is such that $i^*(f) \neq 0$. Show that $(i^*(f)) + r i^*(E)$ is an effective divisor on Y.

Exercise 13.43. Suppose that m, n are positive integers. Show that $\mathbb{P}^m \times \mathbb{P}^n$ is not isomorphic to $\mathbb{P}^{m+n}$. Show that there is, however, a birational (but not regular) map $\mathbb{P}^m \times \mathbb{P}^n \dashrightarrow \mathbb{P}^{m+n}$.

Exercise 13.44. Suppose that X is a projective variety, with graded coordinate ring $S = S(X)$ and homogeneous coordinates $x_0, \ldots, x_m$. Assume that none of the x_i vanish everywhere on X. Recall that for $n \in \mathbb{Z}$, $S(n)$ is S with the grading $S(n)_t = S_{n+t}$. Recall from (11.21) that $\mathcal{O}_X(n) := \widetilde{S(n)}$ satisfies the following property:

$$\Gamma(X_{x_i}, \mathcal{O}_X(n)) = x_i^n k\left[\frac{x_0}{x_i}, \ldots, \frac{x_m}{x_i}\right] = x_i^n \Gamma(X_{x_i}, \mathcal{O}_X).$$

a) Show that
$$\mathcal{O}_X(n)|X_{x_i} = x_i^n \mathcal{O}_X|X_{x_i}.$$
Conclude that $\mathcal{O}_X(n)$ is an invertible sheaf on X.

b) Show that $S_n \subset \Gamma(X, \mathcal{O}_X(n))$ and that this inclusion is an equality if $X = \mathbb{P}^m$.

c) The following example shows that it is possible for S_n to be strictly smaller than $\Gamma(X, \mathcal{O}_X(n))$. Let s, t be algebraically independent over k and let $R = k[s^4, s^3 t, s t^3, t^4]$ be the ring of Example 1.75. Then R is a standard graded domain with the grading
$$\deg s^4 = \deg s^3 t = \deg s t^3 = \deg t^4 = 1.$$

Thus R is the homogeneous coordinate ring of a projective curve C in $\mathbb{P}^3$ (which is isomorphic to $\mathbb{P}^1$). Show that $s^2t^2 \in \Gamma(C, \mathcal{O}_C(1))$. Show that

$$\bigoplus_{n \geq 0} \Gamma(C, \mathcal{O}_C(n)) = k[s^4, s^3t, s^2t^2, st^3, t^4]$$

where $k[s^4, s^3t, s^2t^2, st^3, t^4]$ is graded by

$$\deg s^4 = \deg s^3 t = \deg s^2 t^2 = \deg st^3 = \deg t^4 = 1.$$

d) Suppose that X is nonsingular. Let $D_i = \mathrm{Div}(x_i)$ be the divisor of x_i on X (Section 13.3). Show that $\mathcal{O}_X(n)$ is isomorphic as an $\mathcal{O}_X$-module to $\mathcal{O}_X(nD_i)$.

e) Suppose that X is nonsingular. Let L be a linear form on X (which does not vanish everywhere on X), and let $D = \mathrm{Div}(L)$ be the divisor associated to L on X. Show that $\mathcal{O}_X(n)$ is isomorphic as an $\mathcal{O}_X$-module to $\mathcal{O}_X(nD)$.

Exercise 13.45. Let X be a projective variety.

a) Suppose that $\mathcal{I} \subset \mathcal{O}_X$ is an ideal sheaf on X, which is not equal to $\mathcal{O}_X$. Show that $\Gamma(X, \mathcal{I}) = (0)$.

b) Suppose that X is nonsingular, and let $D = \sum_{i=1}^r a_i E_i$ be an effective divisor on X, with $a_i > 0$, E_i prime divisors, so that $\mathcal{O}_X(-D)$ is the ideal sheaf $\mathcal{O}_X(-D) = \mathcal{I}(E_1)^{a_1} \cdots \mathcal{I}(E_r)^{a_r}$. Show that the presheaf on X defined by

$$P(U) = \Gamma(U, \mathcal{O}_X(D)) \otimes_{\Gamma(U, \mathcal{O}_X)} \Gamma(U, \mathcal{O}_X(-D))$$

for U an open subset of X is not a sheaf on X; thus this presheaf differs from the sheaf $\mathcal{O}_X(D) \otimes_{\mathcal{O}_X} \mathcal{O}_X(-D)$.

Exercise 13.46. Let X be a nonsingular variety of dimension 3, and let $p \in X$ be a point. Let $\pi : B \to X$ be the blow-up of p. Let $E = \pi^{-1}(p)$. We know that $E \cong \mathbb{P}^2$ and that $\mathrm{Pic}(E) \cong \mathbb{Z}$ is generated by $\mathcal{O}_E(1)$. Let $i : E \to B$ be the inclusion. Compute $i^*\mathcal{O}_B(E)$ in $\mathrm{Pic}(E)$ in terms of the generator $\mathcal{O}_E(1)$.

Exercise 13.47. Show that the regular isomorphisms of $\mathbb{P}^n$ are the linear isomorphisms (Section 4.2). Conclude that the group of automorphisms of $\mathbb{P}^n$ is the variety

$$\mathrm{PGL}(n, k) = \mathbb{P}^{n^2 + 2n} \setminus Z(\mathrm{Det}).$$

Hint: Use Exercise 4.14 and Example 13.10.

Exercise 13.48. Suppose that Y is a nonsingular projective surface and $p \in Y$ is a point. Let $\pi : X \to Y$ be the blow-up of p with exceptional divisor $E = \pi^{-1}(p)$. Show that the group homomorphism

$$\mathrm{Cl}(Y) \oplus \mathbb{Z} \to \mathrm{Cl}(X)$$

defined by $([D], n) \mapsto [\pi^* D + nE]$ is a group isomorphism.

Exercise 13.49. Let H be a hyperplane on $\mathbb{P}^2$ as in Example 13.24. Let $p = (0:0:1)$, and define $V = \Gamma(\mathbb{P}^2, \mathcal{O}_X(H) \otimes \mathcal{I}_p)$. Find a basis of V and compute the linear system $L = \{(f) + H \mid f \in V\}$ as a subsystem of the complete linear system $|H|$ described as the divisors of forms of degree 1 in Example 13.24. Compute the base locus of L. Describe the rational map ϕ_V and the geometry of this map.

Exercise 13.50. Let H be a hyperplane on $\mathbb{P}^2$ as in Example 13.24. Let $P_1 = (0:0:1), P_2 = (0:1:0), P_3 = (1:0:0)$. Let

$$V = \Gamma(\mathbb{P}^2, \mathcal{O}_X(2H) \otimes \mathcal{I}_{P_1} \otimes \mathcal{I}_{P_2} \otimes \mathcal{I}_{P_3}).$$

Find a basis of V and compute the linear system $L = \{(f) + 2H \mid f \in V\}$ as a subsystem of the complete linear system $|2H|$ described as the divisors of forms of degree 2 in Example 13.24. Compute the base locus of L. Describe the rational map ϕ_V and the geometry of this map.

Exercise 13.51. Consider the projections π_1 and π_2 of $\mathbb{P}^1 \times \mathbb{P}^1$ onto the first and second factors. Represent each map as $\phi_{|D|}$ for an appropriate complete linear system $|D|$ on $\mathbb{P}^1 \times \mathbb{P}^1$.

Exercise 13.52. Suppose that Y is a normal variety of dimension ≥ 2.

  a) Suppose that q is a nonsingular point of Y and $\phi : X \to Y$ is the blow-up of q with exceptional divisor E. Show that

$$\phi_* \mathcal{O}_X(nE) = \begin{cases} \mathcal{O}_X & \text{if } n \geq 0, \\ \mathcal{I}_q^{-n} & \text{if } n < 0. \end{cases}$$

  b) Suppose that Y is a normal variety, D is a divisor on Y, and $q_1, \ldots, q_r$ are distinct nonsingular points on Y. Let $\phi : X \to Y$ be the blow-up of $q_1, \ldots, q_r$ with exceptional divisors $E_1, \ldots, E_r$. Show that

$$\Gamma(X, \mathcal{O}_X(\phi^*(D) - E_1 - \cdots - E_r)) = \Gamma(Y, \mathcal{O}_Y(D) \otimes \mathcal{I}_{q_1} \otimes \cdots \otimes \mathcal{I}_{q_r}).$$

Exercise 13.53. Suppose that X is a normal projective variety and D is a divisor on X. Show that the divisor D on X is very ample (Definition 13.27) if and only if $\mathcal{O}_X(D)$ is a very ample invertible sheaf (Definition 13.35).

13.9. Transition functions

Exercise 13.54. Suppose that $\mathcal{L}$ is an invertible sheaf on a projective variety X. Show that $\mathcal{L}$ is very ample if and only if there exists a closed embedding $\psi : X \to \mathbb{P}^n$ such that $\psi^* \mathcal{O}_{\mathbb{P}^n}(1) \cong \mathcal{L}$.

Exercise 13.55. Prove Corollary 13.30.

Exercise 13.56. Extend (and prove) Lemma 13.31 and Theorems 13.33 and 13.32 to invertible sheaves on an arbitrary projective variety.

Exercise 13.57. Let x_0, x_1 be homogeneous coordinates on $X = \mathbb{P}^1$, so that $k(X) = k(\frac{x_0}{x_1})$. Let $p_0 = (1:0)$, $p_1 = (0:1)$, $p_2 = (1:1) \in \mathbb{P}^1$. Suppose that $n \in \mathbb{Z}$. Find a k-basis of each of the following vector spaces:

   a) $\Gamma(X, \mathcal{O}_X(np_0))$,
   b) $\Gamma(X, \mathcal{O}_X(np_2))$,
   c) $\Gamma(X, \mathcal{O}_X(np_0 - p_1))$,
   d) $\Gamma(X, \mathcal{O}_X(np_0 - p_1 - p_2))$.

Exercise 13.58. Let the notation be as in Exercise 13.57, and let V be the subspace of $\Gamma(X, \mathcal{O}_X(3p_0))$ with basis $\{1, (\frac{x_0}{x_1})^2, (\frac{x_0}{x_1})^3\}$. Let L be the linear system $L = \{(f) + 3p_0 \mid f \in V\}$, with associated rational map $\phi = \phi_L : X \dashrightarrow \mathbb{P}^2$.

   a) Show that L is base point free.
   b) Compute the homogeneous coordinate ring of the image of ϕ in $\mathbb{P}^2$.
   c) Is ϕ a closed embedding? Why or why not?

Exercise 13.59. Give an example of a linear system L on a projective variety X such that L is not base point free, but the rational map ϕ_L is a closed embedding of X.

Exercise 13.60. Let $X \subset \mathbb{A}^3$ be the affine surface with regular functions $k[X] = k[\overline{x}, \overline{y}, \overline{z}] = k[x, y, z]/(xy - z^2)$. Define prime divisors on X by $E_1 = Z(x, z)$ and $E_2 = Z(y, z)$.

   a) Show that $\mathcal{O}_X(-E_1)$ and $\mathcal{O}_X(-E_2)$ are not invertible sheaves. Hint: Use a method from the proof of Theorem 10.17.
   b) Show that $E_1 \sim -E_2$.

Exercise 13.61. Suppose that D is a divisor on a normal projective variety and V is a subspace of $\Gamma(X, \mathcal{O}_X(D))$ such that the associated linear system L is base point free. Let $\phi_L : X \to \mathbb{P}^m$ be the associated regular map. Show that $\phi_L^* \mathcal{O}_{\mathbb{P}^m}(n) \cong \mathcal{O}_X(nD)$ for $n \in \mathbb{Z}$.

Chapter 14

Differential Forms and the Canonical Divisor

In Section 14.1, we discuss the algebraic theory of derivations and differentials, which generalizes the notion of 1-forms on a manifold to rings. The results of this section are used extensively in Chapters 21 and 22. In Section 14.2, we define the sheaf of 1-forms on a variety X, and in Section 14.3 we define the sheaf of n-forms, canonical divisors, and the divisor of a rational n-form on an n-dimensional nonsingular variety. We prove the useful adjunction formula (Theorem 14.21) for computing canonical divisors.

14.1. Derivations and Kähler differentials

Definition 14.1. Suppose that A and B are rings and $\lambda : A \to B$ is a ring homomorphism, making B into an A-module by $cx = \lambda(c)x$ for $x \in B$ and $c \in A$. Suppose that M is a B-module. Then a map $D : B \to M$ is an A-derivation from B to M if D satisfies the following three conditions:

1) $D(f + g) = D(f) + D(g)$ for $f, g \in B$,
2) $D(cf) = cD(f)$ if $c \in A$ and $f \in B$,
3) $D(fg) = fD(g) + gD(f)$ for $f, g \in B$.

Let $\mathrm{Der}_A(B, M)$ be the set of all A-derivations from B to M. It is a B-module.

We observe that if $x \in B$ and $n \in \mathbb{N}$, then 3) implies that $D(x^n) = nx^{n-1}D(x)$. Thus $D(1_B) = nD(1_B)$ for all $n > 0$, and so

(14.1) $$D(1_B) = 0.$$

Thus 2) implies that $D(c1_B) = 0$ for all $c \in A$. Conversely, if 3) holds and $D(c1_B) = 0$ for all $c \in A$, then $D(cf) = cD(f)$ for all $c \in A$ and $f \in B$. We thus see that 2) can be replaced with the condition that $D(c1_B) = 0$ for all $c \in A$. A-derivations are often defined in this way ([**161**, page 120] or [**107**, pages 180–181]).

Definition 14.2. Let F be the free B-module on the symbols $\{db|b \in B\}$, and let G be the submodule generated by the relations 1), 2), and 3) in the definition of a derivation. We define the B-module of Kähler differentials of B over A by
$$\Omega_{B/A} = F/G.$$

The map $d = d_{B/A} : B \to \Omega_{B/A}$ defined by letting $d(f)$ be the class of df is an A-derivation of B.

Lemma 14.3. *Suppose that M is a B-module. Then the map*
$$\Phi : \mathrm{Hom}_B(\Omega_{B/A}, M) \to \mathrm{Der}_A(B, M)$$
defined by $\Phi(\tau)(f) = \tau(df)$ for $f \in B$ and $\tau \in \mathrm{Hom}_B(\Omega_{B/A}, M)$ is a B-module isomorphism.

The lemma shows that $d : B \to \Omega_{B/A}$ is a universal derivation: if M is a B-module and $D \in \mathrm{Der}_A(B, M)$, then there is a unique B-module homomorphism $\sigma : \Omega_{B/A} \to M$ such that $D = \sigma d_{B/A}$.

The inverse Ψ to Φ (in Lemma 14.3) is defined as follows. Suppose that $D : B \to M$ is an A-derivation. Define $\Psi(D) = \tau$ by
$$\tau\left(\sum \alpha_i db_i\right) = \sum \alpha_i D(b_i).$$

A detailed proof of Lemma 14.3 is given in [**50**, page 384].

Lemma 14.4. *Define a B-module homomorphism $\delta : B \otimes_A B \to B$ by $\delta(a_1 \otimes a_2) = a_1 a_2$, and let I be the kernel of δ. The quotient I/I^2 is a B-module. Then the map*
$$\Lambda : \Omega_{B/A} \to I/I^2$$
defined by taking $\Lambda(df)$ to be the class of $1 \otimes f - f \otimes 1$ for $f \in B$ is a B-module isomorphism.

This lemma is established in [**50**, Theorem 16.24]. The module $\Omega_{B/A}$ is often defined to be I/I^2 as constructed in Lemma 14.3, with the A-derivation $B \to I/I^2$ defined by mapping f to the class of $1 \otimes f - f \otimes 1$ for $f \in B$.

Example 14.5. Let κ be a ring, and let $B = \kappa[x_1, \ldots, x_n]$ be a polynomial ring over κ. Then $\Omega_{B/\kappa}$ is the free B-module with generators $dx_1, \ldots, dx_n$,
$$\Omega_{B/\kappa} = B dx_1 \oplus \cdots \oplus B dx_n.$$

14.1. Derivations and Kähler differentials

The map $d: B \to \Omega_{B/\kappa}$ is

$$d(f) = \frac{\partial f}{\partial x_1} dx_1 + \cdots + \frac{\partial f}{\partial x_n} dx_n$$

for $f \in B$.

Proof. We first use the three properties of a derivation to prove that

(14.2) $$df = \frac{\partial f}{\partial x_1} dx_1 + \cdots + \frac{\partial f}{\partial x_n} dx_n$$

for $f \in B$. We have that $d(1) = 0$ by (14.1). The properties of a derivation show that (14.2) holds for an arbitrary monomial. Then the formula holds for all f since d and the $\frac{\partial}{\partial x_i}$ are κ-linear.

From equation (14.2) we conclude that $\Omega_{B/\kappa}$ is generated by $dx_1, \ldots, dx_n$ as a B-module. It remains to show that $dx_1, \ldots, dx_n$ are a free basis of $\Omega_{B/\kappa}$. Suppose that we have a relation

(14.3) $$g_1 dx_1 + \cdots + g_n dx_n = 0$$

for some $g_1, \ldots, g_n \in B$. We will show that $g_1 = \cdots = g_n = 0$. We have that $\frac{\partial}{\partial x_i}$ for $1 \leq i \leq n$ are κ-derivations on B. By Lemma 14.3, there exist B-module homomorphisms $\tau_i : \Omega_{B/\kappa} \to B$ for $1 \leq i \leq n$ such that $\tau_i(df) = \frac{\partial f}{\partial x_i}$ for $f \in B$. Thus $\tau_i(dx_j) = \delta_{ij}$ (the Kronecker delta). Applying τ_i to (14.3), we obtain that

$$g_i = \tau_i(g_1 dx_1 + \cdots + g_n dx_n) = 0$$

for $1 \leq i \leq n$. $\square$

The following theorem is proven in [**107**, Theorem 58] or [**50**, Proposition 16.3].

Theorem 14.6. *Suppose that κ is a ring, A is a κ-algebra, J is an ideal of A, and $B = A/J$. Then there is an exact sequence of B-modules*

$$J/J^2 \xrightarrow{\delta} \Omega_{A/\kappa} \otimes_A B \xrightarrow{v} \Omega_{B/\kappa} \to 0$$

where $\delta : J \to \Omega_{A/\kappa} \otimes_A B$ is defined by $x \mapsto d_{A/\kappa}(x) \otimes 1$ and $v : \Omega_{A/\kappa} \otimes_A B \to \Omega_{B/\kappa}$ is defined by $d_{A/\kappa}(a) \otimes b \mapsto b d_{B/\kappa}(a)$.

Example 14.7. Suppose that R is a ring and $A = R[x_1, \ldots, x_n]$ is a polynomial ring over R. Let $I = (f_1, \ldots, f_r) \subset R[x_1, \ldots, x_n]$ be an ideal. Let $B = R[x_1, \ldots, x_n]/I$. Let

$$M = B dx_1 \oplus \cdots \oplus B dx_n.$$

We have that

$$\Omega_{B/R} = M/df_1 B + \cdots + df_r B.$$

Proof. This follows from Theorem 14.6. $\square$

Lemma 14.8. *Suppose that B is an A-algebra and S is a multiplicative set in B. Then we have a natural isomorphism of $S^{-1}B$-modules*

$$\Omega_{S^{-1}B/A} \cong S^{-1}\Omega_{B/A}.$$

This follows from [50, Proposition 16.9].

Theorem 14.9. *Suppose that $f : A \to B$ and $g : B \to C$ are ring homomorphisms. Then there is an exact sequence*

$$\Omega_{B/A} \otimes_B C \xrightarrow{\alpha} \Omega_{C/A} \xrightarrow{\beta} \Omega_{C/B} \to 0$$

of C-modules where α is defined by $\alpha(d_{B/A}b \otimes c) = cd_{C/A}g(b)$ and $\beta(d_{C/A}c) = d_{C/B}c$ for $b \in B$ and $c \in C$.

Proof. [107, Theorem 57] or [106, Theorem 25.1] or [50, Proposition 16.2]. □

Exercise 14.10. Suppose that

$$\begin{array}{ccc} A & \xrightarrow{\phi} & A' \\ \uparrow & & \uparrow \\ \kappa & \xrightarrow{\lambda} & \kappa' \end{array}$$

is a commutative diagram of rings and ring homomorphisms. The A'-module $\Omega_{A'/\kappa'}$ is naturally an A-module by $ax := \phi(a)x$ for $a \in A$ and $x \in \Omega_{A'/\kappa'}$.

Show that there is a natural A-module homomorphism $\delta\phi$ giving a commutative diagram

$$\begin{array}{ccc} \Omega_{A/\kappa} & \xrightarrow{\delta\phi} & \Omega_{A'/\kappa'} \\ d_{A/\kappa} \uparrow & & \uparrow d_{A'/\kappa'} \\ A & \xrightarrow{\phi} & A'. \end{array}$$

Exercise 14.11. If $A' = A \otimes_\kappa \kappa'$ in Exercise 14.10, show that

$$\Omega_{A'/\kappa'} = \Omega_{A/\kappa} \otimes_\kappa \kappa' = \Omega_{A/\kappa} \otimes_A A'.$$

Exercise 14.12. Let k be an (algebraically closed) field of characteristic not equal to 2 or 3. Let X be the affine variety $X = Z(y^2 - x^3) \subset \mathbb{A}^2$. Let $R = k[X]$.

    a) Compute $\Omega_{R/k}$.

    b) Show that $\Omega_{R/k}$ has R-torsion (there exists a nonzero element ω of $\Omega_{R/k}$ and a nonzero element a of R such that $a\omega = 0$).

14.2. Differentials on varieties

Suppose that X is a quasi-projective variety. If X is affine, then we define the coherent sheaf $\Omega_{X/k}$ on X by $\Omega_{X/k} = \widetilde{\Omega_{k[X]/k}}$. We will show that there is a natural isomorphism $\Omega_{X/k}(V) \cong \Omega_{k[V]/k}$ if V is an affine open subset of X. If $V = X_f$ for some $f \in k[X]$, then

$$\Omega_{X/k}(X_f) = (\Omega_{k[X]/k})_f \cong \Omega_{k[X]_f/k} \cong \Omega_{k[X_f]/k}$$

by Lemma 14.8. Suppose that $V \subset X$ is an affine open subset. There exist $f_1, \ldots, f_n \in k[X]$ such that $X_{f_i} \subset V$ for all i and $V = \bigcup_{i=1}^n X_{f_i}$. We then have that $X_{f_i} = V_{f_i}$ for all i, so that $k[X]_{f_i} = k[V]_{f_i}$. Thus

$$\left(\Omega_{k[X]/k}\right)_{f_i} \cong \left(\Omega_{k[V]/k}\right)_{f_i}$$

for all i, and

$$\left(\Omega_{k[X]/k}\right)_{f_i f_j} \cong \left(\Omega_{k[V]/k}\right)_{f_i f_j}$$

for all i, j. We have a commutative diagram

$$\begin{array}{ccccccc}
0 & \to & \Omega_{X/k}(V) & \to & \bigoplus_i \Omega_{X/k}(X_{f_i}) & \to & \bigoplus_{i<j} \Omega_{X/k}(X_{f_i f_j}) \\
& & \downarrow & & \downarrow & & \downarrow \\
0 & \to & \Omega_{k[V]/k} & \to & \bigoplus_i \left(\Omega_{k[V]/k}\right)_{f_i} & \to & \bigoplus_{i<j} \left(\Omega_{k[V]/k}\right)_{f_i f_j}
\end{array}$$

where the horizontal arrows are exact and the vertical arrows are isomorphisms, inducing a natural isomorphism $\Omega_{X/k}(V) \cong \Omega_{k[V]/k}$.

To define $\Omega_{X/k}$ for arbitrary quasi-projective varieties, we cover X with open affine subsets $\{U_i\}$ and define $\Omega_{X/k}|U_i = \Omega_{U_i/k}$. By the universal property of Kähler differentials, $\Omega_{U_i/k}|U_i \cap U_j$ and $\Omega_{U_j/k}|U_i \cap U_j$ are naturally isomorphic ($U_i \cap U_j$ is affine by Exercise 5.21). Thus they patch (Exercise 11.23) to give a coherent sheaf of $\mathcal{O}_X$-modules $\Omega_{X/k}$. If V is an affine open subset of X, we have that $\Omega_{X/k}(V) \cong \Omega_{k[V]/k}$.

Proposition 14.13. *Suppose that X is a variety and $p \in X$. Let m_p be the maximal ideal of $\mathcal{O}_{X,p}$. Let $k(p) = \mathcal{O}_{X,p}/m_p \cong k$ which is naturally an $\mathcal{O}_{X,p}$-module. Then the following are naturally isomorphic k-vector spaces:*

1) *the tangent space $T_p(X) = \mathrm{Hom}_k(m_p/m_p^2, k)$,*
2) *the derivations $\mathrm{Der}_k(\mathcal{O}_{X,p}, k(p))$,*
3) *the module of $\mathcal{O}_{X,p}$-module homomorphisms $\mathrm{Hom}_{\mathcal{O}_{X,p}}((\Omega_{X/k})_p, k(p))$.*

Proof. The vector spaces 1) and 2) are isomorphic since every k-derivation $D : \mathcal{O}_{X,p} \to k(p)$ must vanish on m_p^2 and thus induces a k-linear map $m_p/m_p^2 \to k$, and further, given a k-linear map $\ell : m_p/m_p^2 \to k$, we get a k-derivation $D : \mathcal{O}_{X,p} \to k(p)$ defined by $D(f) = \ell(f - f(p))$.

The vector spaces of 2) and 3) are isomorphic since

$$\mathrm{Hom}_{\mathcal{O}_{X,p}}((\Omega_{X/k})_p, k(p)) \cong \mathrm{Hom}_{\mathcal{O}_{X,p}}(\Omega_{\mathcal{O}_{X,p}/k}, k(p)) \cong \mathrm{Der}_k(\mathcal{O}_{X,p}, k(p))$$

by Lemmas 14.8 and 14.3. $\square$

Theorem 14.14. *Suppose that X is a quasi-projective variety. Then the (nontrivial open) subset U of nonsingular points of X is the largest open subset of X on which $\Omega_{X/k}$ is locally free. The sheaf $\Omega_{X/k}|U$ is locally free of rank equal to the dimension of X.*

Proof. Suppose that $p \in X$. Let $\lambda : (\Omega_{X/k})_p \to k(p) = \mathcal{O}_{X,p}/m_p$ be an $\mathcal{O}_{X,p}$-module homomorphism. Then for $a \in m_p$ and $t \in (\Omega_{X/k})_p$ we have that $\lambda(at) = a\lambda(t) = 0$. Thus we have a natural k-vector space isomorphism

$$\mathrm{Hom}_{\mathcal{O}_{X,p}}((\Omega_{X/k})_p, \mathcal{O}_{X,p}/m_p) \xrightarrow{\cong} \mathrm{Hom}_k((\Omega_{X/k})_p/m_p(\Omega_{X/k})_p, k).$$

This last k-vector space is (noncanonically) isomorphic to $(\Omega_{X/k})_p/m_p(\Omega_{X/k})_p$, and the first k-vector space is isomorphic to $T_p(X)$ by Proposition 14.13. Thus

$$\dim_k (\Omega_{X/k})_p/m_p(\Omega_{X/k})_p = \dim_k T_p(X) \geq \dim X$$

with equality if and only if p is a nonsingular point of X, by Proposition 10.14. The conclusions of the theorem now follow from Lemma 11.54. $\square$

Proposition 14.15. *Suppose that X is a nonsingular quasi-projective variety, $p \in X$, and $x_1, \ldots, x_n$ are regular parameters in $\mathcal{O}_{X,p}$. Then there exists an open neighborhood U of p such that $dx_1, \ldots, dx_n$ is a free basis of $\Omega_{U/k}$.*

This proposition is not true on varieties over nonperfect fields (see Exercise 21.78). The generalization of this proposition to arbitrary fields is given in [154], where a thorough study of the concept of singularity over arbitrary fields is made.

Proof. Let $A = \mathcal{O}_{X,p}$ with maximal ideal $\mathfrak{m} = m_p$. Suppose that $f \in A$. Then $f = c + \sum_{i=1}^n a_i x_i$ with $a_i \in A$ and $c \in k$. Thus $df = \sum a_i dx_i + \sum x_i da_i \in \Omega_{A/k}$, so that $df - \sum a_i dx_i \in \mathfrak{m}\Omega_{A/k}$. By Nakayama's lemma (Lemma 1.18) we have that $dx_1, \ldots, dx_n$ generate $\Omega_{A/k} = (\Omega_{X/k})_p$ as an A-module. Since $\dim X = n$ and p is a nonsingular point of X, by Theorem 14.14, $(\Omega_{X/k})_p$ is a free $\mathcal{O}_{X,p}$-module of rank n. Let N be the kernel of

14.2. Differentials on varieties

the surjection $(dx_1, \ldots, dx_n) : A^n \to (\Omega_{X/k})_p$. Tensoring the short exact sequence

$$0 \to N \to A^n \to (\Omega_{X/k})_p \to 0$$

with $k(X)$, we see that the localization $N \otimes k(X) = 0$ (the sequence is split exact since $(\Omega_{X/k})_p$ is a free A-module) and so N is a torsion submodule of A^n, and thus $N = 0$. Thus $dx_1, \ldots, dx_n$ is a free basis of $(\Omega_{X/k})_p$, and so there exists an open neighborhood U of p on which $dx_1, \ldots, dx_n$ is a free basis of $\Omega_{U/k}$. □

Definition 14.16. Suppose that U is an open subset of a nonsingular variety X. Elements $f_1, \ldots, f_n \in \Gamma(U, \mathcal{O}_X)$ are called uniformizing parameters on U if $df_1, \ldots, df_n$ is a free basis of $\Omega_{X/k}|U$.

Proposition 14.17. *Suppose that U is an open subset of a nonsingular variety X and $f_1, \ldots, f_n \in \Gamma(U, \mathcal{O}_X)$ are uniformizing parameters on U. Suppose that $p \in U$. Then*

$$f_1 - f_1(p), \ldots, f_n - f_n(p)$$

are regular parameters in $\mathcal{O}_{X,p}$.

Proof. Let m_p be the maximal ideal of $\mathcal{O}_{X,p}$. Let $x_1, \ldots, x_n$ be regular parameters in $\mathcal{O}_{X,p}$. We have that $f_1 - f_1(p), \ldots, f_n - f_n(p) \in m_p$. Thus there exist $a_{ij} \in \mathcal{O}_{X,p}$ such that

$$f_i - f_i(p) = \sum_j a_{ij} x_j,$$

so that

$$df_i \equiv \sum a_{ij} dx_j \mod m_p (\Omega_{X/k})_p$$

for all i. Since $\{df_i\}$ and $\{dx_j\}$ are free bases of $(\Omega_{X/k})_p$, we have that $\text{Det}(a_{ij})$ is a unit in $\mathcal{O}_{X,p}$ and the matrix (a_{ij}) is invertible over $\mathcal{O}_{X,p}$, so that $f_1 - f_1(p), \ldots, f_n - f_n(p)$ generate m_p. Since $\mathcal{O}_{X,p}$ has dimension n and $\dim_k m_p/m_p^2 = n$, we have that $f_1 - f_1(p), \ldots, f_n - f_n(p)$ is a regular system of parameters in $\mathcal{O}_{X,p}$. □

Exercise 14.18. Suppose that $f : X \to Y$ is a regular map of varieties over an algebraically closed field k. Show that there are a coherent $\mathcal{O}_X$-module $\Omega_{X/Y}$ and a natural exact sequence of coherent $\mathcal{O}_X$-modules

(14.4) $$f^* \Omega_{Y/k} \to \Omega_{X/k} \to \Omega_{X/Y} \to 0$$

which becomes the sequence of Theorem 14.9 with $A = k$, $B = \Gamma(U, \mathcal{O}_Y)$, and $C = \Gamma(V, \mathcal{O}_X)$ when evaluated at an open affine subset V of Y which maps into an open affine subset U of X.

14.3. n-forms and canonical divisors

In this section, suppose that X is a nonsingular variety of dimension n.

If A is a ring and M is an A-module, then $\wedge^n M$ is the quotient of the tensor product $M \otimes \cdots \otimes M$ with itself n times, by the A-submodule generated by all elements $x_1 \otimes \cdots \otimes x_n$ with all $x_i \in M$ and where $x_i = x_j$ for some $i \neq j$ [**95**, Section 1 of Chapter XIX].

We define $\Omega^n_{X/k}$ to be the coherent sheaf on X associated to the presheaf $U \mapsto \wedge^n \Omega_{X/k}(U)$ for U an open subset of X. If U is an affine open subset of X, then

$$\Omega^n_{X/k}(U) = \wedge^n \Omega_{k[U]/k}.$$

The sheaf $\Omega^n_{X/k}$ is an invertible $\mathcal{O}_X$-module. In fact, if U is an open subset of X such that

$$\Omega_{X/k}|U = \mathcal{O}_U dx_1 \oplus \cdots \oplus \mathcal{O}_U dx_n,$$

then $\Omega^n_{X/k}|U = \mathcal{O}_U dx_1 \wedge \cdots \wedge dx_n$ is a free $\mathcal{O}_U$-module.

The rational differential n-forms on X are

$$\Omega^n_{k(X)/k} = \wedge^n \Omega_{k(X)/k} = k(X) df_1 \wedge \cdots \wedge df_n$$

where $f_1, \ldots, f_n \in k(X)$ are any elements such that $df_1 \wedge \cdots \wedge df_n \neq 0$.

Let $\{U_i\}$ be an affine cover of X, such that there exist $\omega_i \in \Gamma(U_i, \Omega^n_{X/k})$ satisfying $\Omega^n_{X/k}|U_i = \omega_i \mathcal{O}_{U_i}$.

Let ϕ be a nonzero rational differential n-form on X. Then there exist $g_i \in k(X)$ for all i such that $\phi = g_i \omega_i$. We have that

$$(g_i)_{U_i \cap U_j} = (g_j)_{U_i \cap U_j}$$

for all i, j since there exist units $\gamma_{ij} \in \Gamma(U_i \cap U_j, \mathcal{O}_X)$ such that $\omega_i = \gamma_{ij} \omega_j$ as ω_i and ω_j are both generators of $\Omega^n_{X/k}|U_i \cap U_j$. Thus there exists a divisor D on X such that

(14.5) $$D \cap U_i = (g_i)_{U_i} \quad \text{for all } i.$$

The divisor D is independent of our choice of $\{U_i\}$ and $\{\omega_i\}$. We define D to be the divisor of ϕ. We will denote the divisor D of ϕ by (ϕ), $(\phi)_X$, $\text{div}(\phi)$, or $\text{div}(\phi)_X$.

Proposition 14.19. *Suppose that ϕ_1, ϕ_2 are nonzero rational differential n-forms on X. Then $(\phi_1) \sim (\phi_2)$. Let $K_X = (\phi)$ be the divisor of a rational differential n-form. Then $\Omega^n_{X/k} \cong \mathcal{O}_X(K_X)$.*

We call the divisor K_X of a rational differential n-form on X a canonical divisor of X. If X is a normal variety, we can define a canonical divisor on X by defining a canonical divisor on the nonsingular locus U of X (whose complement has codimension ≥ 2 in X) and extending it to a divisor K_X on X. We then have that $i_*\Omega^n_{U/k} \cong \mathcal{O}_X(K_X)$ where $i : U \to X$ is the inclusion (as follows from the proof of Lemma 13.5).

Proof. Let ϕ be a nonzero rational differential n-form on X. Let $D = (\phi)$ and $\mathcal{L} = \phi \mathcal{O}_X$, which is a free $\mathcal{O}_X$-module. We have that
$$\Omega^n_{X/k} \cong \mathcal{O}_X(D) \otimes_{\mathcal{O}_X} \mathcal{L} \cong \mathcal{O}_X(D)$$
by equation (14.5). In particular, if ϕ_1, ϕ_2 are rational differential n-forms, then $(\phi_1) \sim (\phi_2)$ (by Proposition 13.6). $\square$

The following example is established in Exercise 14.23.

Example 14.20. The canonical divisor $K_{\mathbb{P}^n} = -(n+1)L$ where L is a linear hyperplane on $\mathbb{P}^n$.

Theorem 14.21 (Adjunction). *Suppose that V is a nonsingular codimension 1 closed subvariety of a nonsingular variety W, so that V is a prime divisor on W, and let $i : V \to W$ be the inclusion. Then $K_V = i^*(K_W + V)$; that is,*
$$\mathcal{O}_V(K_V) \cong \mathcal{O}_W(K_W + V) \otimes_{\mathcal{O}_W} \mathcal{O}_V.$$

Proof. Let $n = \dim(W)$. There exist an affine open cover $\{U_i\}$ of a neighborhood of V in W and uniformizing parameters $x_1(i), \ldots, x_n(i) \in \Gamma(U_i, \mathcal{O}_W)$, such that $x_1(i) = 0$ is a local equation of V in U_i. Thus there are units $g_{ij} \in \Gamma(U_i \cap U_j, \mathcal{O}_W^*)$ such that

(14.6) $$x_1(i) = g_{ij} x_1(j)$$

for all i, j. Since $\{dx_1(i), \ldots, dx_n(i)\}$ and $\{dx_1(j), \ldots, dx_n(j)\}$ are free bases of $\Gamma(U_i \cap U_j, \Omega^n_{W/k})$, there exists $a_{l,m}(i,j) \in \Gamma(U_i \cap U_j, \mathcal{O}_W)$ such that
$$d(x_l(i)) = \sum a_{l,m}(i,j) dx_m(j)$$
and $h_{ij} = \mathrm{Det}(a_{l,m}(i,j))$ is in $\Gamma(U_i \cap U_j, \mathcal{O}_W^*)$. Now we have that
$$dx_1(i) \wedge \cdots \wedge dx_n(i) = h_{ij} dx_1(j) \wedge \cdots \wedge dx_n(j).$$
Let
$$c_{ij} = \mathrm{Det} \begin{pmatrix} a_{22}(i,j) & \cdots & a_{2n}(i,j) \\ \vdots & & \vdots \\ a_{n2}(i,j) & \cdots & a_{nn}(i,j) \end{pmatrix}.$$
The ideal of V in W is $\mathcal{I}_V \cong \mathcal{O}_W(-V)$. Taking d of (14.6), we have that for all i,j,
$$h_{ij} \equiv g_{ij} c_{ij} \mod \Gamma(U_i \cap U_j, \mathcal{O}_W(-V)).$$
Thus the transition functions of $\Omega^n_{W/k} \otimes \mathcal{O}_V$ on $U_i \cap U_j$ are $\bar{g}_{ij} \bar{c}_{ij}$, where $\bar{g}_{ij}$ is the image of g_{ij} in $\Gamma(U_i \cap U_j, \mathcal{O}_V^*)$ and $\bar{c}_{ij}$ is the image of c_{ij} in $\Gamma(U_i \cap U_j, \mathcal{O}_V^*)$.

We have that the $\bar{g}_{ij}$ are the transition functions of $\mathcal{O}_W(-V) \otimes \mathcal{O}_V$ on $U_i \cap U_j$.

The images $\bar{x}_2(i), \ldots, \bar{x}_n(i)$ of $x_2(i), \ldots, x_n(i)$ in $\Gamma(U_i, \mathcal{O}_V)$ are uniformizing parameters on $U_i \cap V$ since $x_1(i) = 0$ is a local equation of V on U_i and so $d\bar{x}_2(i), \ldots, d\bar{x}_n(i)$ is a free basis of $\Gamma(U_i, \Omega_{V/k})$ by Proposition 14.15. Thus the $\bar{c}_{ij}$ are transition functions for $\Omega_{V/k}^{n-1}$ on $U_i \cap U_j$.

Since $\Omega_{W/k}^n \otimes \mathcal{O}_V$ and $\Omega_{V/k}^{n-1} \otimes \mathcal{O}_W(-V)$ have the same transition functions on $U_i \cap U_j$, $\Omega_{W/k}^n \otimes \mathcal{O}_V$ and $\Omega_{V/k}^{n-1} \otimes \mathcal{O}_W(-V)$ are isomorphic by Lemma 13.37. Thus $\Omega_{V/k}^{n-1} \cong \Omega_{W/k}^n \otimes \mathcal{O}_W(V) \otimes \mathcal{O}_V$. □

Corollary 14.22. *Suppose that C is a nonsingular cubic curve in $\mathbb{P}^2$. Then $K_C = 0$ (so that $\mathcal{O}_C(K_C) \cong \mathcal{O}_C$).*

Proof. This follows from adjunction since $C \sim 3L$ and $K_{\mathbb{P}^2} = -3L$, where L is a linear hyperplane on $\mathbb{P}^2$. □

Exercise 14.23. Prove Example 14.20.

Exercise 14.24. Suppose that S is a nonsingular surface and $\pi : X \to S$ is the blow-up of a point. Let E be the exceptional divisor of π. Show that $K_X \sim \pi^*(K_S) + E$.

Chapter 15

Schemes

In this chapter we discuss some generalizations of quasi-projective varieties, beginning with subschemes of quasi-projective varieties. This concept will be important in later chapters. Then we will give a quick survey of some more general spaces. All of our constructions will be locally ringed spaces, which were defined in Definition 11.15.

15.1. Subschemes of varieties, schemes, and Cartier divisors

Definition 15.1. A closed subscheme Z of a quasi-projective variety X is a locally ringed space, consisting of the pair of a closed subset Y of X and a sheaf of rings $\mathcal{O}_Z = \mathcal{O}_X/\mathcal{I}$ on Y where $\mathcal{I}$ is an ideal sheaf on X such that $\mathrm{Supp}(\mathcal{O}_X/\mathcal{I}) = Y$. We call Y the associated topological space of Z. The ideal sheaf of Z in X is $\mathcal{I}_Z = \mathcal{I}$. We define the closed subscheme Z_{red} of X to be the subscheme associated to Y with ideal sheaf $\sqrt{\mathcal{I}}$.

An open subscheme of a closed subscheme Z is an open subset U of a closed subsheme Z, with the sheaf $\mathcal{O}_U = \mathcal{O}_Z|U$.

We will sometimes write "subscheme" or "scheme" to mean an open or closed subscheme.

A subscheme is called affine if it is a closed subscheme of an affine variety, projective if it is a closed subscheme of a projective variety, and quasi-projective if it is an open subscheme of a projective subscheme.

Definition 15.2. A scheme X is a locally ringed space such that every point $p \in X$ has an open neighborhood U such that the open subset U with the sheaf $\mathcal{O}_X|U$ is an affine scheme.

If Y is a closed subvariety of a quasi-projective variety X, then we can regard Y as the scheme with associated topological space Y and sheaf of rings $\mathcal{O}_Y \cong \mathcal{O}_X/\mathcal{I}_Y$; in particular, we have $Y = Y_{\text{red}}$. More generally, if Y is a (closed) algebraic set in a quasi-projective variety X, then we can regard Y as a closed subscheme of X, with the structure such that $Y = Y_{\text{red}}$.

If Z is a subscheme, then the natural inclusion $\mathcal{I}_Z \subset \sqrt{\mathcal{I}_Z}$ induces a surjection $\mathcal{O}_Z \to \mathcal{O}_{Z_{\text{red}}}$, and so we have an inclusion of subschemes $Z_{\text{red}} \subset Z$.

We say that a scheme Z is irreducible if the associated topological space of Z is irreducible. A scheme Z is said to be reduced if $\mathcal{O}_Z$ is reduced; that is, $Z = Z_{\text{red}}$ so that $\mathcal{I}_Z$ is a reduced ideal sheaf. A scheme Z is said to be integral if Z is both reduced and irreducible. Thus a quasi-projective scheme Z is integral if and only if Z is a variety and Z is reduced if and only if Z is an algebraic set.

Example 15.3. Let $X = \mathbb{A}^2$ with regular functions $k[X] = k[x,y]$ and let Y be the closed subset $Y = Z(y) \subset X$. For n a positive integer, let I_n be the ideal $I_n = (y^n)$ in $k[X]$, and let $\mathcal{I}_n$ be the ideal sheaf $\mathcal{I}_n = \tilde{I}_n$ on X. Let Y_n be the locally ringed space

$$Y_n = (Y, \mathcal{O}_{\mathbb{A}^2}/\mathcal{I}_n).$$

The schemes Y_n are different (nonisomorphic) closed subschemes of X with the same underlying topological space Y. We have that $(Y_n)_{\text{red}} = Y_1 = Y$ for all n. The subscheme $Y_1 = Y$ is a subvariety of X.

We consider some situations where schemes appear naturally.

Suppose that $\phi : X \to Y$ is a regular map of quasi-projective varieties and Z is a closed subscheme of Y. We define the scheme-theoretic fiber X_Z over Z to be the closed subscheme of X with associated topological space the subset $\phi^{-1}(Z)$ of X with the ideal sheaf $\mathcal{I}_{X_Z} = \mathcal{I}_Z \mathcal{O}_X$, where $\mathcal{I}_Z$ is the ideal sheaf of Z in Y. A particularly important case is of the fiber X_p over a point $p \in Y$. We will continue to write $\phi^{-1}(Z)$ for the algebraic set $(X_Z)_{\text{red}}$.

Suppose that Z_1 and Z_2 are closed subschemes of a quasi-projective variety X, with respective ideal sheaves $\mathcal{I}_{Z_1}$ and $\mathcal{I}_{Z_2}$. Then the scheme-theoretic intersection $Z_1 \cap Z_2$ is the closed subscheme of X with ideal sheaf $\mathcal{I}_{Z_1} + \mathcal{I}_{Z_2}$. Earlier, we encountered the set-theoretic intersection $Z_1 \cap Z_2$, which is the algebraic set with ideal sheaf $\sqrt{\mathcal{I}_{Z_1} + \mathcal{I}_{Z_2}}$.

An effective divisor D on a normal variety X can be regarded as a closed subscheme of X. The sheaf $\mathcal{O}_D = \mathcal{O}_X/\mathcal{O}_X(-D)$.

We define sheaves, quasi-coherent sheaves, and coherent sheaves on a scheme as in Chapter 11.

The definition of invertible sheaf and analysis in Section 13.9 are valid on an arbitrary scheme. If $\mathcal{L}$ is an invertible sheaf on a scheme X, then

15.1. Subschemes of varieties, schemes, and Cartier divisors

there exists an affine cover $\{U_i\}$ of X with $\mathcal{O}_{U_i}$-module isomorphisms $\phi_i : \mathcal{O}_X|U_i \to \mathcal{L}|U_i$ and transition functions $g_{ij} \in \Gamma(U_i \cap U_j, \mathcal{O}_X^*)$, the units in $\Gamma(U_i, \cap U_j, \mathcal{O}_X)$.

The total quotient ring of a ring A is defined to be $\mathrm{QR}(A) = S^{-1}A$ where S is the multiplicative set of nonzero divisors in A.

Let X be a scheme. There is a sheaf $\mathcal{K}$ on X which has the property that $\mathcal{K}(U) = \mathrm{QR}(\mathcal{O}_X(U))$ whenever U is affine. The sheaf $\mathcal{K}$ is called the sheaf of total quotient rings of $\mathcal{O}_X$. The sheaf of multiplicative groups $\mathcal{K}^*$ is defined to be the group of invertible elements of $\mathcal{K}$ and $\mathcal{O}_X^*$ is the sheaf of invertible elements of $\mathcal{O}_X$. We have inclusions $\mathcal{O}_X \subset \mathcal{K}$ and $\mathcal{O}_X^* \subset \mathcal{K}^*$.

Definition 15.4. A Cartier divisor D on a scheme X is a collection $\{(U_i, f_i)\}$ where $\{U_i\}$ is an open cover of X and $f_i \in \Gamma(U_i, \mathcal{K}^*)$ are such that for each i, j, $\frac{f_i}{f_j} \in \Gamma(U_i \cap U_j, \mathcal{O}_X^*)$.

We have encountered Cartier divisors on varieties earlier, in (13.2). Cartier divisors are developed in more detail in [**118**, Section 9] and [**73**, Section II.6].

Let X be a scheme. Then for D a Cartier divisor on X, $\mathcal{O}_X(D)$ is the invertible sheaf on X defined by

$$\mathcal{O}_X(D)|U_i = \frac{1}{f_i}\mathcal{O}_X|U_i.$$

We have that $\mathcal{O}_X(D_1 - D_2) \cong \mathcal{O}_X(D_1) \otimes \mathcal{O}_X(D_2)^{-1}$ if D_1 and D_2 are Cartier divisors. We say that D_1 is linearly equivalent to D_2 if $\mathcal{O}_X(D_1) \cong \mathcal{O}_X(D_2)$. A Cartier divisor on a scheme X is effective if it can be represented by $\{(U_i, f_i)\}$ where each $f_i \in \mathcal{O}_X(U_i)$. In this case, we have an associated subscheme D of X defined by $\mathcal{I}_D|U_i = f_i\mathcal{O}_{U_i}$.

Suppose that $\mathcal{L}$ is an invertible sheaf on a scheme X. A global section $s \in \Gamma(X, \mathcal{L})$ is called a nonzero divisor if the annihilator of the stalk s_p by $\mathcal{O}_{X,p}$ is zero for all $p \in X$. This is equivalent to the annihilator of $s|U_i$ as a $\Gamma(U_i, \mathcal{O}_X)$-module being zero for all U_i in an affine cover of X.

Suppose that $s \in \Gamma(X, \mathcal{L})$ is a nonzero divisor. Then there is associated to s a subscheme $\mathrm{div}(s)$ (or (s)) of X which is an effective Cartier divisor. The closed subscheme $\mathrm{div}(s)$ has the ideal sheaf $\mathcal{I}_{\mathrm{div}(s)}$ which is defined on each U_i by

$$\mathcal{I}_{\mathrm{div}(s)}|U_i = \phi_i^{-1}(s)\mathcal{O}_{U_i}$$

where $\phi_i : \mathcal{O}_{U_i} \to \mathcal{L}|U_i$ are local trivializations of $\mathcal{L}$.

The effective divisor $\mathrm{div}(s)$ associated to s is well-defined; it is independent of choice of local trivialization. This generalizes the construction of $\mathrm{div}(s)$ on a nonsingular variety, given after Example 13.38.

It is possible for $\mathcal{I}_{\mathrm{div}(s),p}$ to have embedded primes, where $p \in \mathrm{div}(s)$, even when X is a variety. This is illustrated by Example 1.75.

Suppose that $\mathcal{I}$ is an ideal sheaf on a projective variety X. Let $x_0, \ldots, x_n$ be homogeneous coordinates on X, and let $S(X) = k[x_0, \ldots, x_n]$ be the coordinate ring of X. Let $\mathfrak{m} = (x_0, \ldots, x_n)$ be the graded maximal ideal. There exists a homogeneous ideal I in $S(X)$ such that $\mathcal{I}$ is the sheafification $\tilde{I}$ of I (Proposition 11.48). Let $I = Q_0 \cap Q_1 \cap \cdots \cap Q_s$ be a homogeneous primary decomposition of I, where Q_0 is the $\mathfrak{m}$-primary component of I if $\mathfrak{m}$ is an associated prime of I and $Q_0 = S(X)$ otherwise. Let $P_i = \sqrt{Q_i}$ be the associated (homogeneous) prime ideals to Q_i. We have that

$$I^{\mathrm{sat}} = Q_1 \cap \cdots \cap Q_s.$$

Let $Y_i = Z(P_i)$. Then

$$\mathcal{I} = \tilde{I} = \widetilde{(I^{\mathrm{sat}})} = \tilde{Q}_1 \cap \cdots \cap \tilde{Q}_s$$

is a primary decomposition of $\mathcal{I}$, where the ideal sheaf $\tilde{Q}_i$ is $\mathcal{I}_{Y_i} = \tilde{P}_i$ primary.

Let Z be the subscheme of X with ideal sheaf $\mathcal{I}_Z = \mathcal{I}$. We define the homogeneous ideal of Z to be

$$I(Z) = I^{\mathrm{sat}}.$$

The homogeneous ideal $I(Z)$ is uniquely determined by Z.

Definition 15.5. Suppose that $\mathcal{L}$ is an invertible sheaf on a projective scheme V, $\mathcal{F}$ is a coherent sheaf on V, and $s \in \Gamma(V, \mathcal{L})$. We will say that s is not a zero divisor on $\mathcal{F}$ if for all $p \in V$, $\phi^{-1}(s)$ is not a zero divisor on $\mathcal{F}_p$, where $\phi : \mathcal{O}_U \to \mathcal{L}|U$ is a trivialization of $\mathcal{L}$ in a neighborhood U of p.

Recall (after Definition 15.4) that for $s \in \Gamma(V, \mathcal{L})$ a nonzero divisor, $\mathrm{div}(s)$ is the subscheme of V defined by

$$\mathcal{I}_{\mathrm{div}(s)} \mid U = \phi^{-1}(s)\mathcal{O}_U,$$

where U is an open neighorhood of p in V admitting a local isomorphism $\phi : \mathcal{O}_U \to \mathcal{L}|U$.

As V is a closed subscheme of a projective space $\mathbb{P}^n$, we have that $\mathcal{O}_V = \widetilde{S(V)}$ where $S(V) = S(\mathbb{P}^n)/J$ for some homogeneous ideal J in $S(\mathbb{P}^n)$. We can realize $\mathcal{F} = \tilde{F}$ where F is a graded $S(V)$-module. Let I be the saturation of the annihilator of F as a $S(V)$-module. Let

$$I = Q_1 \cap \cdots \cap Q_t$$

be an irredundant primary decomposition of I, and let P_i be the prime ideals associated to Q_i. Suppose that $s \in \Gamma(V, \mathcal{L})$ is a nonzero divisor. Then s is a nonzero divisor on $\mathcal{F}$ if and only if the subvarieties $V_i = Z(P_i)$ of V satisfy

$V_i \not\subset \mathrm{Supp}(\mathrm{div}(s))$ for $1 \le i \le t$. We will call the V_i the associated varieties of $\mathcal{F}$.

Suppose that U is an affine open subset of V and ϕ is a local trivialization $\phi : \mathcal{O}_U \to \mathcal{L}|U$. Let $S \subset \{1, \ldots, t\}$ be the indices such that $U \cap V_i \ne \emptyset$. Then the annihilator of the module $\Gamma(U, \mathcal{F})$ is the ideal $\tilde{I}(U)$ which has the irredundant primary decomposition

$$\tilde{I}(U) = \bigcap_{i \in S} \tilde{Q}_i(U)$$

with $\sqrt{\tilde{Q}_i(U)} = I(U \cap V_i)$. The element s is a nonzero divisor of $\mathcal{F}$ on U if and only if $\phi^{-1}(s) \notin I(U \cap V_i)$ for $i \in S$.

Exercise 15.6. Suppose that X is a normal affine variety with regular functions $R = k[X]$. Suppose that D is a divisor on X such that $\mathcal{O}_X(D)$ is invertible and $f \in \Gamma(X, \mathcal{O}_X(D))$ is nonzero. Suppose that $(f) + D = \sum a_i E_i$ with E_i prime divisors and $a_i \in \mathbb{Z}_+$. Show that, regarding f as an element of $\Gamma(X, \mathcal{O}_X(D))$, we have

$$I_{\mathrm{div}(f)} = I_{E_1}^{(a_1)} \cap \cdots \cap I_{E_r}^{(a_r)},$$

where I_{E_i} are the prime ideals of E_i in R and $I_{E_i}^{(a_i)} = (I_{E_i}^{a_i} R_{I_{E_i}}) \cap R$ is the a_i-th symbolic power of I_{E_i}. Hint: Use Theorem 1.79.

Exercise 15.7. Suppose that X is nonsingular in Exercise 15.6. Show that we then have

$$I_{\mathrm{div}(f)} = I_{E_1}^{a_1} I_{E_2}^{a_2} \cdots I_{E_r}^{a_r}.$$

Exercise 15.8. Use Example 1.75 to give an example of a projective variety X with coordinate ring S such that if $L = \mathrm{Div}(F)$ for some nonzero homogeneous linear form $F \in S$, then $I(L) \ne (F)$.

15.2. Blow-ups of ideals and associated graded rings of ideals

Suppose that X is an affine variety with regular functions $R = k[X]$ and $J \subset R$ is an ideal. Let $\pi : B(J) \to X$ be the blow-up of J. From Theorem 6.4, we know that a coordinate ring of $B(J)$ is the graded ring $S(B(J)) \cong \bigoplus_{i \ge 0} J^i$. Let Z be the closed affine subscheme of X with ideal sheaf $\mathcal{I}_Z = \tilde{J}$. From Theorem 6.4, we see that the coordinate ring of the fiber $B(J)_Z$ of Z in $B(J)$ is

$$S(B(J)_Z) \cong S(B(J))/JS(B(J)) \cong \bigoplus_{i \ge 0} J^i/J^{i+1}.$$

This last ring is (by definition) the associated graded ring $\mathrm{gr}_J R$ of J. In the case that $J = \mathfrak{m}$ is a maximal ideal of R (which corresponds to a point p of X) we have that

$$S(B(J)_Z) \cong \bigoplus_{i \geq 0} \mathfrak{m}^i/\mathfrak{m}^{i+1} = \mathrm{gr}_\mathfrak{m}(R).$$

When $J = \mathfrak{m}$, the scheme $B(J)_Z$ (and the ring $S(B(J)_Z)$) is often called the tangent cone of p (the tangent cone of $R_\mathfrak{m}$).

There is a nice way to compute the tangent cone which we now indicate. Let $k[x_1, \ldots, x_n]$ be a polynomial ring. Given $0 \neq f \in k[x_1, \ldots, x_n]$, there is an expression $f = f_d + f_{d+1} + \cdots + f_r$ where the f_i are homogeneous forms of degree i for all i and $f_d \neq 0$. The initial form of f is defined to be $\mathrm{in}(f) = f_d$. Given an ideal I in $k[x_1, \ldots, x_n]$, the initial ideal of I is $\mathrm{in}(I) = (\mathrm{in}(f) \mid 0 \neq f \in I)$, the homogeneous ideal in $k[x_1, \ldots, x_n]$ generated by the initial forms of elements of I.

Suppose that $R \cong k[x_1, \ldots, x_n]/I$ and $\mathfrak{m} = (x_1, \ldots, x_n)R$. Then (as explained on the bottom of page 249 through the top of page 250 of [**161**]), we have a graded isomorphism

$$\bigoplus_{i \geq 0} \mathfrak{m}^i/\mathfrak{m}^{i+1} \cong k[x_1, \ldots, x_n]/\mathrm{in}(I).$$

We have that $\mathrm{in}(I) = I$ if I is a homogeneous ideal and $\mathrm{in}(I) = (\mathrm{in}(f))$ if $I = (f)$ is a principal ideal. However, if I is generated by $f_1, \ldots, f_m$, it is not true in general that $\mathrm{in}(I)$ is generated by $\mathrm{in}(f_1), \ldots, \mathrm{in}(f_m)$. In fact, this can fail even if $f_1, \ldots, f_m$ is a complete intersection.

Suppose that $p \in Z(J) \subset X$, with corresponding maximal ideal $\mathfrak{n}$ in R. Then the fiber $B(J)_p$ of p in $B(J)$ has the coordinate ring

$$S(B(J)_p) \cong \bigoplus_{i \geq 0} J^i/\mathfrak{n} J^i.$$

This ring is sometimes called the fiber cone.

Exercise 15.9. Let X be the reduced subscheme $X = Z(x_1 x_2)$ of $\mathbb{P}^2$ (a union of two lines). Compute the blow-up $\pi : \overline{X} \to X$ of the line $Z(x_1)$ in X. Show that $\overline{X}$ is isomorphic to $\mathbb{P}^1$.

Exercise 15.10. Suppose that k has characteristic > 3 and let

$$f = x^2 + y^2 + z^2 + x^3 + y^3 + z^3,$$

an irreducible polynomial in the polynomial ring $k[x, y, z]$. Let $S = Z(f) \subset \mathbb{A}^3$. Let $\pi : \overline{S} \to S$ be the blow-up of the origin p of S. Show that the fiber $\overline{S}_p$ over p is isomorphic to $Z(F) \subset \mathbb{P}^2$, where $F = x^2 + y^2 + z^2 \in k[x, y, z] = S(\mathbb{P}^2)$.

15.3. Abstract algebraic varieties

Definition 15.11. An abstract prevariety is an irreducible scheme (Definition 15.2).

A topological space X is called Noetherian if it satisfies the descending chain condition for closed subsets. A prevariety X is a Noetherian topological space [**116**, Proposition 1 and Proposition 2, pages 48–49]. The function field $k(X)$ of a prevariety X is the quotient field of $\mathcal{O}_X(U)$ for any affine open subset U of X

If X is a prevariety, then the product $X \times X$ is naturally a prevariety (it is covered by the affine varieties $U_i \times U_j$ where $\{U_i\}$ is an affine cover of X).

Definition 15.12. An abstract prevariety X is an abstract variety if

$$\Delta(X) = \{(p,p) \mid p \in X\}$$

is closed in $X \times X$.

Example 15.13. Let p be a point in $U_1 = \mathbb{A}^1$ and let q be a point in $U_2 = \mathbb{A}^1$. Define a prevariety X by gluing U_1 to U_2 by identifying the open subsets $U_1 \setminus \{p\}$ and $U_2 \setminus \{q\}$. Then X is a prevariety which is not a variety.

The following theorem follows from the valuative criterion of separatedness [**73**, Theorem II.4.3].

Theorem 15.14. *An abstract prevariety X is an abstract variety if and only if for $p, q \in X$ the stalks $\mathcal{O}_{X,p} \subset k(X)$ and $\mathcal{O}_{Y,q} \subset k(X)$ are equal if and only if $p = q$.*

A quasi-projective variety is an abstract variety since it is separated by Proposition 3.36.

Definition 15.15. An abstract variety X is complete if for all abstract varieties Y the projection $\pi : X \times Y \to Y$ is a closed map.

A projective variety is complete. This follows from Corollary 5.13 or the valuative criterion of properness [**73**, Theorem II.4.7]. An example by Hironaka of a nonsingular, complete abstract variety which is not projective is given in [**73**, Example 3.4.1 of Appendix B].

Theorem 15.16. *Suppose that X is an abstract prevariety over $\mathbb{C}$. Then:*

1) *The abstract prevariety X is Hausdorff in the Euclidean topology if and only if X is an abstract variety.*

2) *The abstract prevariety X is compact (and Hausdorff) in the Euclidean topology if and only if X is a complete abstract variety.*

Proof. This is proven in [**116**, Section 10 of Chapter 1]. Part 1) is established for quasi-projective varieties over $\mathbb{C}$ in Theorem 10.41. □

Exercise 15.17. Show that an affine variety X is complete if and only if X is a point.

Exercise 15.18. Show that the prevariety constructed in Example 15.13 is not an abstract variety.

15.4. Varieties over nonclosed fields

Suppose that $V \subset \mathbb{A}^n$ is an affine variety over our algebraically closed field k. Then $k[V] = k[x_1, \ldots, x_n]/P$ where P is the prime ideal $I(V)$ of V. If k_0 is a subfield of k such that the prime ideal $P_0 = P \cap k_0[x_1, \ldots, x_n]$ satisfies $P = P_0 k[x_1, \ldots, x_n]$, then we say that V is defined over k_0 or that k_0 is a field of definition of V, and we define $k_0[V] = k_0[x_1, \ldots, x_n]/P_0$. We define the k_0-rational points $V(k_0)$ of V to be the set of points $p = (\alpha_1, \ldots, \alpha_n) \in k_0^n$ such that $p \in V$. These points correspond to the maximal ideals in $k_0[V]$ whose residue field is k_0. The rational function field of V, regarded as a variety over k_0, written as $k_0(V)$, is the quotient field of $k_0[V]$. We can also consider field extensions k_1 of k_0 in k. Then we can regard V as a variety over k_1 with $k_1[V] = k_1[x_1, \ldots, x_n]/Pk_1[x_1, \ldots, x_n] \cong k[V] \otimes_k k_1$.

If $V \subset \mathbb{P}^n$, we can extend these notions to define the concept of a field of definition k_0 of V. We then have the set of k_0-rational points $V(k_0)$ of V, which are the set of points $(\alpha_0 : \ldots : \alpha_n) \in \mathbb{P}^n$ such that $\alpha_0, \ldots, \alpha_n \in k_0$.

This philosophy, including the notions of generic points, independent generic points, and specialization of points is developed in [**146**]. An excellent discussion on this topic is in [**116**, Section 4 of Chapter 2].

An inherent difficulty with this approach is that if we start with a prime ideal P in a polynomial ring $k_0[x_1, \ldots, x_n]$, giving a variety X_{k_0} in $\mathbb{A}^n_{k_0}$, we might have that $Pk[x_1, \ldots, x_n]$ is no longer a prime ideal if k is an algebraic closure of k_0. In this case there is not a corresponding variety X_k in $\mathbb{A}^n_k$. If $Pk[x_1, \ldots, x_n]$ is a prime ideal, then X_{k_0} is said to be "absolutely irreducible".

15.5. General schemes

A very general definition of schemes is given in [**65**] and [**69**]. Some sections in books giving good introductions to this are [**73**, Section 2 of Chapter II], [**116**, Chapter II], and [**53**].

An affine scheme is defined to be the spectrum $\mathrm{Spec}(R)$ of a commutative ring R (Exercise 1.11). The points of $\mathrm{Spec}(R)$ are the prime ideals of R,

15.5. General schemes

and it is given the Zariski topology. The closed sets are
$$Z(I) = \{P \in \mathrm{Spec}(R) \mid I \subset P\}$$
where I is an ideal of R. The closed points are then the maximal ideals of R. An affine scheme $X = \mathrm{Spec}(A)$ has a structure sheaf $\mathcal{O}_X$ which generalizes the notion of regular functions on an affine variety. The structure sheaf has the property that the stalk $\mathcal{O}_{X,\mathfrak{p}} = A_\mathfrak{p}$ for all $\mathfrak{p} \in X$.

A homomorphism $\phi : A \to B$ of commutative rings determines a continuous map $f : \mathrm{Spec}(B) \to \mathrm{Spec}(A)$ by $f(\mathfrak{p}) = \phi^{-1}(\mathfrak{p})$ for $\mathfrak{p} \in \mathrm{Spec}(B)$. These maps ϕ are the morphisms (as locally ringed spaces) from $\mathrm{Spec}(B)$ to $\mathrm{Spec}(A)$, with $f^\#$ induced by ϕ.

If R is the ring of regular functions $k[V]$ of an affine variety V, then the closed points of $\mathrm{Spec}(k[V])$ are the ideals $I(p)$ of the points $p \in V$.

A scheme is a topological space X with a sheaf of rings $\mathcal{O}_X$ such that there exists an open cover $\{U_\alpha\}$ of X such that
$$(U_\alpha, \mathcal{O}_X|U_\alpha) \cong (\mathrm{Spec}(R_\alpha), \mathcal{O}_{\mathrm{Spec}(R_\alpha)})$$
for some commutative rings R_α. A morphism of schemes is a morphism of locally ringed spaces. A scheme X is separated if the image of the diagonal map $\Delta : X \to X \times X$ is closed. An affine scheme is separated.

Let $S = \bigoplus_{n \geq 0} S_n$ be a graded ring where $S_0 = R$ is a commutative ring. The projective scheme $\mathrm{Proj}(S)$ is the topological space which consists of the graded prime ideals of S which do not contain $S_+ = \bigoplus_{n > 0} S_n$. The closed sets are
$$Z(I) = \{P \in \mathrm{Proj}(S) \mid I \subset P\}$$
where I is a graded ideal of R. For $F \in S$ homogeneous, $\mathrm{Spec}(S_{(F)})$ is an affine open subset of $\mathrm{Proj}(S)$, and such open subsets are a basis for the topology.

Our definition of a scheme follows the definitions in the later, second version of EGA I [**69**] and in [**73**]. In the earlier edition of EGA I [**65**] and in [**116**] a scheme is called a prescheme and a scheme is a separated prescheme.

Chapter 16

The Degree of a Projective Variety

In this chapter we define the degree of a projective variety X embedded in a projective space $\mathbb{P}^n$. Classically, the degree is defined to be the number of intersection points of X with a general linear subvariety of $\mathbb{P}^n$ of codimension equal to the dimension of X. In a more algebraic approach, the degree is defined from the Hilbert polynomial of Y. We will indicate why these two definitions are in fact equal and derive a classical bound on the degree of a nondegenerate variety (a variety which is not contained in a linear hyperplane of $\mathbb{P}^n$).

Let $\mathbb{P}^n$ be projective space over an algebraically closed field k. Let $x_0, \ldots, x_n$ be homogeneous coordinates on $\mathbb{P}^n$, so that the coordinate ring of $\mathbb{P}^n$ is $S = S(\mathbb{P}^n) = k[x_0, \ldots, x_n]$.

The linear hyperplanes in $\mathbb{P}^n$ are parametrized by the projective space

$$V = \mathbb{P}^n$$

by associating to $A = (a_0 : \ldots : a_n) \in V$ the hyperplane L_A which is the subscheme of $\mathbb{P}^n$ with underlying topological space $Z(\sum_{i=0}^n a_i x_i)$ and ideal sheaf

$$\mathcal{I}_{L_A} = \widetilde{\left(\sum a_i x_i\right)}.$$

We say that a property holds for a general linear hyperplane if it holds for all L_A with A in a dense open subset of V. A hyperplane L_A with A in this dense open set is called a general hyperplane.

The intersection $Y \cap Z$ of two closed subschemes will be the scheme-theoretic intersection defined in Section 15.1.

The proof of the next theorem follows directly from [55, Corollary 3.4.14 and Theorem 3.4.10].

Theorem 16.1. *Suppose that $Y \subset \mathbb{P}^n$ is a closed algebraic set (a reduced subscheme). Then for a general hyperplane $H \subset \mathbb{P}^n$, we have that the scheme-theoretic intersection $Y \cap H$ does not contain any irreducible component of Y and $Y \cap H$ is reduced (an algebraic set). If all irreducible components of Y have dimension m, then all irreducible components of $Y \cap H$ have dimension $m - 1$. Further, if Y is nonsingular, then $Y \cap H$ is nonsingular, and if Y is normal, then $Y \cap H$ is normal. If Y is a variety (integral) of dimension ≥ 2, then $Y \cap H$ is a variety (integral).*

Definition 16.2. Suppose that Y is a closed algebraic set in $\mathbb{P}^n$ (a reduced closed subscheme) such that all irreducible components of Y have a common dimension m. A linear subvariety L of $\mathbb{P}^n$ of codimension m is said to be general for Y if there exist linear hyperplanes $L_1, \ldots, L_m$ of $\mathbb{P}^n$ such that $L = L_1 \cap \cdots \cap L_m$ and for $1 \leq r \leq m$, $X \cap L_1 \cap \cdots \cap L_r$ is a closed algebraic set in $\mathbb{P}^n$ such that all irreducible components have dimension $m - r$. If Y is a variety, we further require that for $1 \leq r \leq m - 1$, $X \cap L_1 \cap \cdots \cap L_r$ is a variety.

The next corollary follows from successive application of Theorem 16.1.

Corollary 16.3. *Suppose that Y is a closed algebraic set in $\mathbb{P}^n$ (a reduced closed subscheme) such that all irreducible components of Y have dimension $m > 0$. Then there exists a linear subvariety L of $\mathbb{P}^n$ of codimension m which is general for Y.*

Definition 16.4. A numerical polynomial is a polynomial $P(z) \in \mathbb{Q}[z]$ such that $P(n) \in \mathbb{Z}$ for all integers $n \gg 0$.

For $n \in \mathbb{N}$, define $\binom{z}{n} \in \mathbb{Q}[z]$ by $\binom{z}{0} = 1$ and

$$\binom{z}{n} = \frac{1}{n!} z(z-1) \cdots (z-n+1)$$

for $n \geq 1$.

Lemma 16.5. *Suppose that $P \in \mathbb{Q}[z]$ is a numerical polynomial of degree r. Then there are integers $c_0, c_1, \ldots, c_r \in \mathbb{Z}$ such that*

$$(16.1) \qquad P(z) = c_0 \binom{z}{r} + c_1 \binom{z}{r-1} + \cdots + c_r.$$

Proof. We prove the lemma by induction on $r = \deg P$, the case $r = 0$ certainly being true. Since $\deg \binom{z}{r} = r$, we can express any numerical polynomial $P(z)$ in the form (16.1), with $c_0, \ldots, c_r \in \mathbb{Q}$. Define the difference polynomial $\Delta P(z) = P(z+1) - P(z)$. Since $\Delta \binom{z}{r} = \binom{z}{r-1}$,

$$\Delta P = c_0 \binom{z}{r-1} + c_1 \binom{z}{r-2} + \cdots + c_{r-1}.$$

By induction, $c_0, \ldots, c_{r-1} \in \mathbb{Z}$. But then $c_r \in \mathbb{Z}$ since $P(n) \in \mathbb{Z}$ for $n \gg 0$. □

Theorem 16.6 (Hilbert, Serre). *Let $M = \bigoplus_{n \in \mathbb{Z}} M_n$ be a finitely generated graded $S = k[x_0, \ldots, x_n]$-module. Then there is a unique polynomial $P_M(z) \in \mathbb{Q}[z]$ such that*

$$P_M(n) = \dim_k M_n$$

for $n \gg 0$. The degree of the polynomial $P_M(z)$ is $\dim Z(\mathrm{Ann}(M))$ (viewed as an algebraic set in $\mathbb{P}^n$), where $\mathrm{Ann}(M) = \{f \in S \mid fM = 0\}$.

Proof. [161, Theorem 41 on page 232 and Theorem 42′ on page 235] or [50, Section 12.1]. □

The polynomial P_M is called the Hilbert polynomial of M. We define the degree of M to be $r!$ times the leading coefficient of P_M, where $r = \dim Z(\mathrm{Ann}(M))$ is the degree of P_M.

Suppose that I, J are homogeneous ideals in S such that $I^{\mathrm{sat}} = J^{\mathrm{sat}}$. Then

$$P_{S/I} = P_{S/J}$$

by Exercise 16.13.

Definition 16.7. Suppose that $Y \subset \mathbb{P}^n$ is a closed subscheme of dimension r. We define the Hilbert polynomial P_Y of Y to be the Hilbert polynomial $P_{S/I(Y)}$. By Theorem 16.6, this polynomial has degree r. We define the degree $\deg(Y)$ of Y to be $r!$ times the leading coefficient of P_Y.

Proposition 16.8. *The following are true.*

1) *The degree of a nonempty closed subsheme Y of $\mathbb{P}^n$ is a positive integer.*

2) *Suppose that Y is a closed algebraic set (a closed reduced subscheme) in $\mathbb{P}^n$ and $Y = Y_1 \cup Y_2$ is the union of closed algebraic sets of the same dimension r such that $\dim Y_1 \cap Y_2 < r$. Then $\deg(Y) = \deg(Y_1) + \deg(Y_2)$.*

3) $\deg(\mathbb{P}^n) = 1$.

4) *If $H \subset \mathbb{P}^n$ is a hypersurface whose ideal is generated by a homogeneous polynomial of degree d, then $\deg(H) = d$.*

Proof. 1) The polynomial P_Y is nonzero of degree $r = \dim Y$ by Theorem 16.6. By Lemma 16.5, we have a unique expression

$$P_Y(x) = c_0 \binom{x}{r} + c_1 \binom{x}{r-1} + \cdots + c_r$$

where $c_0, \ldots, c_r \in \mathbb{Z}$ and $c_0 \neq 0$. Thus $\deg(Y) = c_0$ is a nonzero integer. We have that $c_0 > 0$ since $P_Y(n) = \dim_k S(Y)_n > 0$ for $n \gg 0$.

2) Let $I_1 = I(Y_1)$, $I_2 = I(Y_2)$, and $I = I(Y)$, so that $I = I(Y_1 \cup Y_2) = I_1 \cap I_2$. We have an exact sequence of graded $S = S(\mathbb{P}^r)$-modules

$$0 \to S/I \xrightarrow{h \mapsto (h,-h)} S/I_1 \oplus S/I_2 \xrightarrow{(f,g) \mapsto f+g} S/(I_1 + I_2) \to 0.$$

The algebraic set $Z(I_1 + I_2) = Y_1 \cap Y_2$ has dimension $< r$. Hence $P_{S/(I_1+I_2)}$ has degree $< r$, and the leading coefficient of $P_{S/I}$ is the sum of the leading coefficients of P_{Y_1} and P_{Y_2}.

3) Let $S = S(\mathbb{P}^n) = k[x_0, \ldots, x_n]$. For $m > 0$, $\dim_k S_m = \binom{m+n}{n}$ so $P_S(x) = \binom{x+n}{n}$. The leading coefficient of this polynomial is $\frac{1}{n!}$, so $\deg \mathbb{P}^n = 1$.

4) Let $F \in S$ be a degree d homogeneous polynomial which generates the ideal of the scheme H. We have an exact sequence of graded S-modules

$$0 \to S(-d) \xrightarrow{F} S \to S/(F) \to 0.$$

Thus

$$\dim_k S(H)_m = \dim_k S_m - \dim_k S_{m-d},$$

so

$$P_H(x) = \binom{x+n}{n} - \binom{x+n-d}{n} = \frac{d}{(n-1)!} x^{n-1} + \text{lower-order terms}.$$

Thus $\deg H = d$. $\square$

Theorem 16.9. *Suppose that Y is a closed algebraic set in $\mathbb{P}^n$ such that all irreducible components of Y have dimension m and L is a linear subvariety of $\mathbb{P}^n$ of codimension m which is general for Y. Let d be the number of irreducible components (points) of the reduced zero-dimensional scheme $L \cap Y$. Then*

$$d = \deg(Y).$$

Proof. We first show that

$$\deg(Y) = \deg(Y \cap L).$$

By induction on m, we need only verify that $\deg(Y) = \deg(Y \cap L_A)$ where L_A is a hyperplane section of $\mathbb{P}^n$ such that all irreducible components of $L_A \cap Y$ have dimension $m - 1$ and $L_A \cap Y$ is reduced. With these assumptions, the radical ideal $I(L_A \cap Y) = (I(Y) + I(L_A))^{\text{sat}}$.

We have that $I(Y) = \bigcap Q_i$ where the Q_i are the homogeneous prime ideals of the irreducible components of Y. Let $T = S/I(Y)$, a graded S-module where $S = S(\mathbb{P}^n)$. Let $F_A = \sum_{i=0}^n a_i x_i$. By our assumptions on A, $F_A \notin Q_i$ for any i, so we have a short exact sequence of graded S-modules

$$0 \to T(-1) \xrightarrow{F_A} T \to T/F_A T \to 0.$$

Thus

$$P_{T/F_A T}(n) = P_T(n) - P_T(n-1) = \frac{\deg(Y)}{(m-1)!} n^{m-1} + \text{lower-order terms in } n.$$

Since

$$I(Y \cap L_A)_n = (I(Y) + I(L_A))_n^{\text{sat}} = (I(Y) + (F_A))_n^{\text{sat}} = (I(Y) + (F_A))_n$$

for $n \gg 0$ and $S/(I(Y) + (F_A)) \cong T/F_A T$, we have that

$$P_{S/I(Y \cap L_A)} = P_{S/(I(Y)+(F_A))} = P_{T/F_A T}$$

and thus $\deg(Y \cap L_A) = \deg(Y)$. By induction, we have that $\deg(Y \cap L) = \deg(Y)$.

By our assumption, $I(Y \cap L) = \bigcap_{i=1}^d P_i$ where the P_i are the homogeneous ideals of points. Thus

$$S/P_i \cong k[x]$$

is a standard graded polynomial ring in one variable. Thus $P_{S/P_i}(n)$ is the constant polynomial 1. By 2) of Proposition 16.8,

$$\deg(Y \cap L) = d. \qquad \square$$

Theorem 16.10. *Suppose that X is a projective subvariety of $\mathbb{P}^n$ and X is not contained in a linear hyperplane (X is nondegenerate). Then*

$$\deg(X) \geq \operatorname{codim}(X) + 1.$$

Proof. Let $m = \operatorname{codim}(X)$. By Corollary 16.3, we can construct a linear subspace L of $\mathbb{P}^n$ of dimension m such that $I(L) = (F_1, \ldots, F_m)$ with $F_1, \ldots, F_m$ in $S = S(\mathbb{P}^n)$ linear forms such that the ideals $I_0 = I(X)$ and $I_i = (I_{i-1} + (F_i))^{\text{sat}}$ for $i < m$ are prime ideals and

$$I_m = (I_{m-1} + (F_m))^{\text{sat}} = (I(L) + I(X))^{\text{sat}} = I(X \cap L).$$

Let $Y_i = Z(I_i)$ and $L_i = Z(F_1, \ldots, F_i) \cong \mathbb{P}^{n-i}$. The variety $Y_0 = X$, Y_i is a subvariety of L_i for $i < m$, and $Y_m = X \cap L$ is a closed algebraic set in $L = L_m$ (a union of d points). We will prove that Y_i is nondegenerate in L_i for all i by induction on i.

Suppose that $i < m$. Sheafify the graded short exact sequence of $S(L_i) = S/(F_1, \ldots, F_i)$-modules

$$0 \to (S/I_i)(-1) \xrightarrow{F_{i+1}} S/I_i \to S/(I_i + (F_{i+1})) \to 0$$

to obtain a short exact sequence of sheaves of $\mathcal{O}_{L_i}$-modules

$$0 \to \mathcal{O}_{Y_i}(-1) \to \mathcal{O}_{Y_i} \to \mathcal{O}_{Y_{i+1}} \to 0,$$

and obtain, after tensoring with $\mathcal{O}_{Y_i}(1)$, a commutative diagram of sheaves of $\mathcal{O}_{L_i}$-modules with exact rows

$$\begin{array}{ccccccccc} 0 & \to & \mathcal{O}_{L_i} & \to & \mathcal{O}_{L_i}(1) & \to & \mathcal{O}_{L_{i+1}}(1) & \to & 0 \\ & & \downarrow & & \downarrow & & \downarrow & & \\ 0 & \to & \mathcal{O}_{Y_i} & \to & \mathcal{O}_{Y_i}(1) & \to & \mathcal{O}_{Y_{i+1}}(1) & \to & 0. \end{array}$$

Taking global sections, we obtain a commutative diagram of k-vector spaces

$$\begin{array}{ccccccccc} 0 & \to & \Gamma(L_i, \mathcal{O}_{L_i}) & \to & \Gamma(L_i, \mathcal{O}_{L_i}(1)) & \to & \Gamma(L_{i+1}, \mathcal{O}_{L_{i+1}}(1)) & \to & 0 \\ & & \alpha \downarrow & & \beta \downarrow & & \gamma \downarrow & & \\ 0 & \to & \Gamma(Y_i, \mathcal{O}_{Y_i}) & \to & \Gamma(Y_i, \mathcal{O}_{Y_i}(1)) & \to & \Gamma(Y_{i+1}, \mathcal{O}_{Y_{i+1}}(1)) & & \end{array}$$

where the top and bottom rows are exact. The map α is an isomorphism since Y_i is a variety (and k is algebraically closed) by Theorem 3.35. By the induction hypothesis, Y_i is nondegenerate in L_i; that is, β is injective. After a diagram chase, we see that γ must also be injective, so that Y_{i+1} is nondegenerate in L_{i+1}.

We have established that the algebraic set Y_m, which is a union of d points in $L_m \cong \mathbb{P}^{\mathrm{codim}(X)}$, is nondegenerate. Thus Y_m must contain at least $\mathrm{codim}(X) + 1$ points, so

$$d = \deg(X) \geq \mathrm{codim}(X) + 1. \qquad \square$$

A nondegenerate subvariety X of $\mathbb{P}^n$ such that $\deg(X) = \mathrm{codim}(X) + 1$ is called a variety of minimal degree. There is a beautiful classification of varieties of minimal degree by Del Pezzo [48] and Bertini [19]. A modern proof is given by Eisenbud and Harris in [52].

Exercise 16.11. Let X be a proper subvariety of $\mathbb{P}^n$. Give a simple direct proof that a general linear hyperplane of $\mathbb{P}^n$ does not contain X.

Exercise 16.12. Suppose that $f : \mathbb{Z} \to \mathbb{Z}$ is a function such that the first difference function $\Delta f(n) = f(n+1) - f(n)$ is a numerical polynomial. Show that there exists a numerical polynomial $P(z)$ such that $f(n) = P(n)$ for $n \gg 0$.

Exercise 16.13. Suppose that $S = k[x_0, \ldots, x_n]$ and $I \subset S$ is a homogeneous ideal. Show that $P_{S/I} = P_{S/I^{\mathrm{sat}}}$.

16. The Degree of a Projective Variety

Exercise 16.14. Suppose that X is a subvariety of $\mathbb{P}^n$ which has degree 1. Show that X is a linear subvariety.

Exercise 16.15. Show that the degree of the d-th Veronese embedding of $\mathbb{P}^n$ in $\mathbb{P}^N$ is d^n.

Exercise 16.16. Show that the degree of the Segre embedding of $\mathbb{P}^r \times \mathbb{P}^s$ in $\mathbb{P}^N$ is $\binom{r+s}{r}$.

Exercise 16.17. Let $Y \subset \mathbb{P}^n$ be an r-dimensional variety of degree 2. Show that Y is contained in a linear subvariety of dimension $r+1$ in $\mathbb{P}^n$, and that Y is isomorphic to a quadric hypersurface in $\mathbb{P}^{r+1}$.

Chapter 17

Cohomology

In this chapter we study basic properties of sheaf and Čech cohomology and discuss some applications.

17.1. Complexes

We begin with some preliminaries on homological algebra. A complex A^* is a sequence of homomorphisms of Abelian groups (or modules over a ring R):

$$\cdots \to A^i \xrightarrow{d^i} A^{i+1} \xrightarrow{d^{i+1}} \cdots$$

for $i \in \mathbb{Z}$ with $d^{i+1}d^i = 0$ for all i. The homomorphisms d^i are called differentials or coboundary maps.

Associated to a complex A^* are cohomology groups (modules)

$$H^i(A^*) = \mathrm{Kernel}(d^i)/\mathrm{Image}(d^{i-1}).$$

Definition 17.1. Suppose that A^* and B^* are complexes. A chain map (or map of complexes) $f : A^* \to B^*$ is a sequence of homomorphisms $f^i : A^i \to B^i$ for all i such that the following diagram commutes:

$$\begin{array}{ccccccccc} \cdots \to & A^i & \xrightarrow{d^i} & A^{i+1} & \xrightarrow{d^{i+1}} & A^{i+2} & \to & \cdots \\ & \downarrow f^i & & \downarrow f^{i+1} & & \downarrow f^{i+2} & & \\ \cdots \to & B^i & \xrightarrow{e^i} & B^{i+1} & \xrightarrow{e^{i+1}} & B^{i+2} & \to & \cdots. \end{array}$$

A chain map $f : A^* \to B^*$ is nullhomotopic if there are homomorphisms $s^i : A^i \to B^{i-1}$ such that

$$f^i = e^{i-1}s^i + s^{i+1}d^i$$

for all i. The homomorphisms s^i are called a homotopy. If f and g are two chain maps from A^* to B^*, then f is homotopic to g if $f - g$ is nullhomotopic.

Suppose that f is a chain map between complexes A^* and B^*. Then there are induced homomorphisms $f^* : H^i(A^*) \to H^i(B^*)$.

Proposition 17.2. *Suppose that f and g are homotopic chain maps between complexes A^* and B^*. Then $f^* = g^* : H^n(A^*) \to H^n(B^*)$ for all $n \in \mathbb{Z}$.*

Proof. Let s^i be the homotopy. Suppose that $z \in \text{Kernel}(d^i)$. Then
$$fz - gz = e^{i-1} s^i z + s^{i+1} d^i z = e^{i-1} s^i z \in \text{Image}(e^{i-1})$$
so $f^* = g^*$. □

17.2. Sheaf cohomology

In this section, we summarize some material from [73, Chapter III]. Associated to a sheaf of Abelian groups F on a topological space X are sheaf cohomology groups $H^i(X, F)$ for all nonnegative integers i, which have the properties that we have a natural isomorphism $H^0(X, F) \cong \Gamma(X, F)$, and if

(17.1) $$0 \to A \to A' \to A'' \to 0$$

is a short exact sequence of sheaves of Abelian groups on X, then there is a long exact cohomology sequence of Abelian groups
(17.2)
$$0 \to H^0(X, A) \to H^0(X, A') \to H^0(X, A'') \to H^1(X, A) \to H^1(X, A')$$
$$\to H^1(X, A'') \to H^2(X, A) \to \cdots .$$

Further, given a commutative diagram of homomorphisms of sheaves of Abelian groups on X,

(17.3)
$$\begin{array}{ccccccccc} 0 & \to & A & \to & A' & \to & A'' & \to & 0 \\ & & \downarrow & & \downarrow & & \downarrow & & \\ 0 & \to & B & \to & B' & \to & B'' & \to & 0 \end{array}$$

where the horizontal diagrams are short exact, we have an induced commutative diagram
(17.4)
$$\begin{array}{ccccccccc} 0 & \to & H^0(X, A) & \to & H^0(X, A') & \to & H^0(X, A'') & \to & H^1(X, A) & \to & \cdots \\ & & \downarrow & & \downarrow & & \downarrow & & \downarrow & & \\ 0 & \to & H^0(X, B) & \to & H^0(X, B') & \to & H^0(X, B'') & \to & H^1(X, B) & \to & \cdots . \end{array}$$

The sheaf cohomology groups are constructed by an injective resolution. A sheaf of Abelian groups I on X is called injective if the functor $\text{Hom}(\cdot, I)$ is exact on exact sequences of sheaves of Abelian groups on X. An injective resolution of a sheaf of Abelian groups F on X is an exact sequence

(17.5) $$0 \to F \to I^0 \to I^1 \to \cdots$$

17.2. Sheaf cohomology

of sheaves of Abelian groups on X such that each I^i is injective. It follows from [**73**, Corollary III.2.3] that every sheaf of Abelian groups F on X has an injective resolution. The sheaf cohomology groups $H^i(X, F)$ are computed by choosing an injective resolution (17.5) and taking the cohomology of the associated complex ($d^{i+1}d^i = 0$ for all i)

$$(17.6) \qquad \Gamma(X, I^0) \xrightarrow{d^0} \Gamma(X, I^1) \xrightarrow{d^1} \cdots.$$

The sheaf cohomology groups $H^i(X, F)$ are defined by

$$(17.7) \qquad H^i(X, F) = \operatorname{Kernel}(d^i)/\operatorname{Image}(d^{i-1})$$

for $i \geq 0$. These cohomology groups are independent of choice of injective resolution [**128**, Theorem 6.14]. Given a commutative diagram (17.3), we have an associated diagram (17.4) by [**128**, Theorem 6.26].

Lemma 17.3. *Let X be a Noetherian topological space, and let $\mathcal{F} = \bigoplus_j \mathcal{F}_j$ be a direct sum of sheaves of Abelian groups $\mathcal{F}_j$ on X. Then*

$$H^i(X, \mathcal{F}) \cong \bigoplus_j H^i(X, \mathcal{F}_j)$$

for all i.

This is established in [**73**, Proposition III.2.9 and Remark III.2.9.1].

If X is a locally ringed space, then we can construct corresponding cohomology groups for sheaves of $\mathcal{O}_X$-modules. In this case, the cohomology groups are constructed by taking an injective resolution (17.5) where the I^i are injective $\mathcal{O}_X$-modules, taking the complex of global sections (17.6), and then taking the cohomology (17.7) of this complex. The fact that every sheaf of $\mathcal{O}_X$-modules has an injective resolution follows from [**73**, Proposition III.2.2]. The cohomology groups of an $\mathcal{O}_X$-module F are in fact the same as those that we compute regarding F as a sheaf of groups, by [**73**, Proposition III.2.6].

Lemma 17.4. *Let Y be a closed subscheme of a scheme X, let $\mathcal{F}$ be a sheaf of Abelian groups on Y, and let $j : Y \to X$ be the inclusion. Then $H^i(Y, \mathcal{F}) = H^i(X, j_*\mathcal{F})$ for all i.*

This is proved in [**73**, Lemma III.2.10].

Suppose that Y is a closed subscheme of a scheme X and $\mathcal{F}$ is a sheaf of Abelian groups on X, such that the support of $\mathcal{F}$ is contained in the underlying topological space of Y. Then

$$(17.8) \qquad H^i(X, \mathcal{F}) = H^i(Y, \mathcal{F}|Y)$$

for all i. This follows, since letting $j : Y \to X$ be the inclusion, we have that $j_*(\mathcal{F}|Y) = \mathcal{F}$.

Theorem 17.5. *Suppose that X is a Noetherian topological space of dimension n. Then for all $i > n$ and all sheaves of Abelian groups F on X, we have that $H^i(X, F) = 0$.*

This is proven in [**73**, Theorem III.2.7].

Corollary 17.6. *Suppose that X is a scheme and $\mathcal{F}$ is a coherent sheaf on X such that the support of $\mathcal{F}$ (which is a Zariski closed subset of X by Exercise 11.36) has dimension n. Then $H^i(X, \mathcal{F}) = 0$ for all $i > n$.*

Theorem 17.7. *Suppose that X is an affine scheme and $\mathcal{F}$ is a quasi-coherent sheaf on X. Then $H^i(X, \mathcal{F}) = 0$ for all $i > 0$.*

This follows from [**73**, Theorem III.3.5]. In fact, if $\mathcal{F} = \tilde{M}$ where M is an $A = k[X]$-module and

$$0 \to M \to I^0 \xrightarrow{d^0} I^1 \xrightarrow{d^1} \cdots$$

is an injective resolution of M as an A-module, then

$$H^i(X, \mathcal{F}) = \text{Kernel}(d^i)/\text{Image}(d^{i-1}).$$

17.3. Čech cohomology

In practice, the most effective way to compute cohomology is by Čech cohomology.

Suppose that X is a topological space and $\mathcal{F}$ is a sheaf of Abelian groups on X.

Let $\underline{U} = \{U_i\}_{i \in I}$ be an open cover of X. Fix a well-ordering on I. For a finite set $i_0, \ldots, i_p \in I$, let $U_{i_0,\ldots,i_p} = U_{i_0} \cap \cdots \cap U_{i_p}$. We define a complex of Abelian groups $C^* = C^*(\underline{U}, \mathcal{F})$ on X by

$$C^p(\underline{U}, \mathcal{F}) = \prod_{i_0 < \cdots < i_p} \mathcal{F}(U_{i_0,\ldots,i_p}).$$

Define the coboundary map $d = d^p : C^p(\underline{U}, \mathcal{F}) \to C^{p+1}(\underline{U}, \mathcal{F})$ by

$$(17.9) \qquad (d^p(\alpha))_{i_0,\ldots,i_{p+1}} = \sum_{k=0}^{p+1} (-1)^k \alpha_{i_0,\ldots,\hat{i}_k,\ldots,i_{p+1}} |U_{i_0,\ldots,i_{p+1}}$$

for $\alpha = (\alpha_{i_0,\ldots,i_p}) \in C^p(\underline{U}, \mathcal{F})$. The notation $\hat{i}_k$ means omit i_k. We have that $d^2 = 0$, so that C^* is a complex.

We write out the first part of the complex as

$$(17.10) \qquad \prod_i \mathcal{F}(U_i) \xrightarrow{d^0} \prod_{j<k} \mathcal{F}(U_j \cap U_k) \xrightarrow{d^1} \prod_{l<m<n} \mathcal{F}(U_l \cap U_m \cap U_n)$$

17.3. Čech cohomology

where the indicies $i, j < k$ and $l < m < n$ range over I. If $\alpha = (\alpha_i) \in C^0(\underline{U}, \mathcal{F})$, then
$$d^0(\alpha)_{j,k} = \alpha_k - \alpha_j,$$
and if $\alpha = (\alpha_{j,k}) \in C^1(\underline{U}, \mathcal{F})$, then
$$d^1(\alpha)_{l,m,n} = \alpha_{l,m} - \alpha_{l,n} + \alpha_{m,n}.$$

Remark 17.8. We will sometimes find it useful to extend the symbol $\alpha_{i_0,i_1,\ldots,i_p}$ to be defined for all $(p+1)$-tuples of elements of I. To do this, we define $\alpha_{i_0,\ldots,i_p} = 0$ if any of the i_j are equal, and if the i_j are all distinct, define
$$\alpha_{i_0,\ldots,i_p} = (-1)^{\text{sign}(\sigma)} \alpha_{\sigma(i_0),\ldots,\sigma(i_p)}$$
where σ is the permutation such that $\sigma(i_0) < \cdots < \sigma(i_p)$. With this convention, the formula (17.9) holds for any $(p+1)$-tuple $i_0, \ldots, i_{p+1}$ of elements of I.

Definition 17.9. We define the p-th Čech cohomology group of $\mathcal{F}$ with respect to the covering $\underline{U}$ to be
$$\check{H}^p(\underline{U}, \mathcal{F}) = H^p(C^*(\underline{U}, \mathcal{F})) = \text{Kernel}(d^p)/\text{Image}(d^{p-1}).$$

From the sheaf axioms, we have the following lemma.

Lemma 17.10. *We have that*
$$\check{H}^0(\underline{U}, \mathcal{F}) = \Gamma(X, \mathcal{F}) = H^0(X, \mathcal{F}).$$

Let $\underline{V} = \{V_j\}_{j \in J}$ be another open cover of X. The cover $\underline{V}$ is a refinement of $\underline{U}$ if there is an order-preserving map of index sets $\lambda : J \to I$ such that $V_j \subset U_{\lambda(j)}$ for all $j \in J$. If $\underline{V}$ is a refinement of $\underline{U}$, then there is a natural map of complexes
$$\phi : C^*(\underline{U}, \mathcal{F}) \to C^*(\underline{V}, \mathcal{F})$$
defined by
$$\phi^p(\alpha)_{j_0,\ldots,j_p} = \alpha_{\lambda(j_0),\ldots,\lambda(j_p)} | V_{j_0,\ldots,j_p}$$
for $\alpha \in C^p(\underline{U}, \mathcal{F})$, where d^* is the differential of $C^*(\underline{U}, \mathcal{F})$ and e^* is the differential of $C^*(\underline{V}, \mathcal{F})$. The map ϕ is a map of complexes since $e^p \phi^p = \phi^{p+1} d^p$.

Thus we have natural homomorphisms of cohomology groups
$$\check{H}^p(\underline{U}, \mathcal{F}) \to \check{H}^p(\underline{V}, \mathcal{F}).$$

This map is independent of choice of function λ, as the maps of complexes ϕ and ψ are homotopic (Proposition 17.2) if ψ is the induced map of complexes obtained from another order-preserving map $\tau : J \to I$ such that $V_j \subset U_{\tau(j)}$.

As the coverings of X form a partially ordered set under refinement, we may make the following definition.

Definition 17.11. The p-th Čech cohomology group of $\mathcal{F}$ is
$$\check{H}^p(X, \mathcal{F}) = \varinjlim \check{H}^p(\underline{U}, \mathcal{F})$$
where the limit is over the (ordered) open covers $\underline{U}$ of X.

Theorem 17.12. *There are natural homomorphisms $\check{H}^i(X, \mathcal{F}) \to H^i(X, \mathcal{F})$, which are isomorphisms if $i \leq 1$.*

Proof. [59, Corollary, page 227] or [73, Exercise III.4.4] (which gives a sketch of the proof). $\square$

Theorem 17.13. *Suppose that X is a scheme, $\underline{U}$ is an affine cover of X, and $\mathcal{F}$ is a quasi-coherent sheaf on X. Then*
$$\check{H}^p(\underline{U}, \mathcal{F}) \cong \check{H}^p(X, \mathcal{F}) \cong H^p(X, \mathcal{F})$$
for all $p \geq 0$.

Theorem 17.13 is proven in [73, Theorem III.4.5]. The key point of the proof in comparing the cohomologies is that for $p > 0$, the cohomology groups $H^p(U, \mathcal{F})$ for $p > 0$ of a quasi-coherent sheaf on an affine scheme U vanish (by Theorem 17.7).

17.4. Applications

Now we give some applications of cohomology.

Theorem 17.14. *Let $X = \mathbb{P}^r$ and $h^i(n) = \dim_k H^i(X, \mathcal{O}_X(n))$. Then*
$$h^i(n) = \begin{cases} \binom{r+n}{r} & \text{for } i = 0 \text{ and } n \geq 0, \\ 0 & \text{for } i = 0 \text{ and } n < 0, \\ 0 & \text{for } 0 < i < r \text{ and all } n \in \mathbb{Z}, \\ h^0(-n-r-1) & \text{if } i = r. \end{cases}$$

Proof. Let $\mathcal{F}$ be the quasi-coherent sheaf $\mathcal{F} = \bigoplus_{n \in \mathbb{Z}} \mathcal{O}_X(n)$. Cohomology commutes with direct sums (by Lemma 17.3) so
$$H^i(X, \mathcal{F}) \cong \bigoplus_{n \in \mathbb{Z}} H^i(X, \mathcal{O}_X(n)).$$

Let $U_i = X_{x_i} = D(x_i)$ for $0 \leq i \leq r$. Here $\underline{U} = \{U_i\}$ is an affine cover of X so $H^i(X, \mathcal{F}) = \check{H}^i(\underline{U}, \mathcal{F})$ (by Theorem 17.13). We have that
$$\mathcal{F}(U_{i_0 \ldots i_p}) \cong \bigoplus_{n \in \mathbb{Z}} \Gamma(U_{i_0 \ldots i_p}, \mathcal{O}_X(n)).$$

Let S be the graded ring $S = S(X) = k[x_0, \ldots, x_r]$.

17.4. Applications

We have that $\Gamma(U_{i_p}, \mathcal{O}_X(n)) = S(n)_{(x_{i_p})}$, the elements of degree 0 in the localization $S(n)_{x_{i_p}}$, which is equal to the elements of degree n in $S_{x_{i_p}}$. Writing

$$\begin{aligned} U_{i_0\ldots i_p} &= U_{i_0} \cap \cdots \cap U_{i_p} \\ &= (U_{i_p})_{\frac{x_{i_0}}{x_{i_p}}} \cap \cdots \cap (U_{i_p})_{\frac{x_{i_{p-1}}}{x_{i_p}}} \\ &= (U_{i_p})_{\frac{x_{i_0}}{x_{i_p}} \ldots \frac{x_{i_{p-1}}}{x_{i_p}}}, \end{aligned}$$

we have that

$$\Gamma(U_{i_0\ldots i_p}, \mathcal{O}_X(n)) = [S(n)_{(x_{i_p})}]_{\frac{x_{i_0}}{x_{i_p}} \ldots \frac{x_{i_{p-1}}}{x_{i_p}}}$$

which is the set of elements of degree n in $S_{x_{i_0}\ldots x_{i_p}}$. We thus have that the Čech complex $C^*(\underline{U}, \mathcal{F})$ is isomorphic as a graded S-module to the complex C^* :

$$\bigoplus_{i_0} S_{x_{i_0}} \xrightarrow{d^0} \bigoplus_{i_0 < i_1} S_{x_{i_0} x_{i_1}} \xrightarrow{d^1} \cdots \xrightarrow{d^{r-1}} S_{x_0 x_1 \cdots x_r}$$

where

$$d^s(\alpha)_{i_0\ldots i_{s+1}} = \sum_{k=0}^{s+1}(-1)^k \alpha_{i_0\ldots \hat{i}_k\ldots i_s}|U_{i_0\ldots i_{s+1}}$$

for $\alpha \in C^s(\underline{U}, \mathcal{F})$. The kernel of d^0 can be identified with the intersection $\bigcap_i S_{x_i}$ in the quotient field of S, and this intersection is just S since $S = k[x_0, \ldots, x_r]$ is a polynomial ring. Thus $H^0(X, \mathcal{F}) = S$, which establishes the theorem for $i = 0$.

The group $H^r(X, \mathcal{F})$ is the cokernel of

$$d^{r-1} : \bigoplus_k S_{x_0 \cdots \hat{x}_k \cdots x_r} \to S_{x_0 \cdots x_r}.$$

The k-vector space $S_{x_0 \cdots x_r}$ has the basis $x_0^{l_0} \cdots x_r^{l_r}$ with $l_i \in \mathbb{Z}$. The image of d^{r-1} is generated by the monomials $x_0^{l_0} \cdots x_r^{l_r}$ such that at least one $l_i \geq 0$. Thus the (classes) of the monomials

$$\{x_0^{l_0} \cdots x_r^{l_r} | l_i < 0 \text{ for all } i\}$$

is a k-basis of $H^r(X, \mathcal{F})$. We then have that a basis of $H^r(X, \mathcal{O}_X(n))$ is

$$\{x_0^{-1} \cdots x_r^{-1} x_0^{m_0} \cdots x_r^{m_r} | m_i \leq 0 \text{ for all } i \text{ and } m_0 + \cdots + m_r = n + r + 1\}$$

so

$$\dim_k H^r(X, \mathcal{O}_X(n)) = \dim_k S_{-n-r-1} = \dim_k H^0(X, \mathcal{O}_X(-n-r-1))$$

establishing the theorem for $i = r$.

We now prove the theorem when $0 < i < r$ by induction on r. This statement is vacuously true for $r = 1$, so we may assume that $r > 1$. Localizing the complex C^* with respect to x_r, we obtain the Čech complex for the sheaf

$\mathcal{F}|U_r$ on U_r with respect to the affine open cover $\{U_i \cap U_r | i = 0, \ldots, r\}$. By Theorem 17.13, the cohomology of the complex $C^*_{x_r}$ is the sheaf cohomology of $\mathcal{F}|U_r$ on U_r, which is zero for $i > 0$ by Theorem 17.7. Since localization is exact, we conclude that $H^i(X, \mathcal{F})_{x_r} = 0$ for $i > 0$. Thus every element of $H^i(X, \mathcal{F})$ for $i > 0$ is annihilated by some power of x_r. To complete the proof that $H^i(X, \mathcal{F}) = 0$ for $0 < i < r$, we will show that multiplication by x_r induces a bijection of $H^i(X, \mathcal{F})$ into itself, from which it follows that $H^i(X, \mathcal{F}) = 0$.

We have an exact sequence of graded S-modules
$$0 \to S(-1) \xrightarrow{x_r} S \to S/x_r S \to 0.$$

Sheafifying gives an exact sequence of sheaves
$$0 \to \mathcal{O}_X(-1) \to \mathcal{O}_X \to \mathcal{O}_H \to 0$$
where $H \cong \mathbb{P}^{r-1}$ is the hyperplane $Z(x_r)$. Tensoring this last sequence with $\mathcal{O}_X(n)$ and taking the direct sum over $n \in \mathbb{Z}$ gives us a short exact sequence of quasi-coherent $\mathcal{O}_H$-modules
$$0 \to \mathcal{F}(-1) \to \mathcal{F} \to \mathcal{F}_H \to 0$$
where $\mathcal{F}_H = \bigoplus_{n \in \mathbb{Z}} \mathcal{O}_H(n)$. Taking cohomology, we obtain a long exact sequence
$$\cdots \to H^i(X, \mathcal{F}(-1)) \xrightarrow{x_r} H^i(X, \mathcal{F}) \to H^i(X, \mathcal{F}_H) \to \cdots.$$

Now
$$H^i(X, \mathcal{F}_H) = H^i\left(H, \bigoplus \mathcal{O}_H(n)\right) \cong \bigoplus H^i(H, \mathcal{O}_H(n)),$$
so by induction on r, $H^i(X, \mathcal{F}_H) = 0$ for $0 < i < r - 1$.

For $i = 0$ and $n \in \mathbb{Z}$, the left exact sequence
$$0 \to H^0(X, \mathcal{O}_X(n-1)) \to H^0(X, \mathcal{O}_X(n)) \to H^0(H, \mathcal{O}_H(n)) \to 0$$
is actually exact (the rightmost map is a surjection) since from our calculations earlier in this proof it is the short exact sequence
$$0 \to S_{n-1} \xrightarrow{x_r} S_n \to (S/x_r S)_n \to 0.$$

We also have at the end of the long exact sequence the right exact sequence
$$H^{r-1}(X, \mathcal{O}_H(n)) \xrightarrow{\delta} H^r(X, \mathcal{O}_X(n)) \xrightarrow{x_r} H^r(X, \mathcal{O}_X(n)) \to 0.$$

From earlier in this proof, we know that $H^r(X, \mathcal{O}_X(n))$ has the k-basis of (classes) of monomials
$$\{x_0^{l_0} \cdots x_r^{l_r} \mid l_i < 0 \text{ for all } i \text{ and } l_0 + \cdots + l_r = n\}.$$

The kernel K_n of $x_r : H^r(X, \mathcal{O}_X(n-1)) \to H^r(X, \mathcal{O}_X(n))$ has the k-basis
$$\{x_0^{m_0} \cdots x_{r-1}^{m_{r-1}} x_r^{-1} \mid m_i < 0 \text{ for all } i \text{ and } m_0 + \cdots + m_{r-1} - 1 = n - 1\}.$$

Thus $\dim_k K_n = \dim_k H^{r-1}(H, \mathcal{O}_H(n))$ so δ is injective. Thus multiplication by $x_r : H^i(X, \mathcal{F}(-1)) \to H^i(X, \mathcal{F})$ is bijective for $0 < i < r$ and so $H^i(X, \mathcal{F}) = 0$ for $0 < i < r$. $\square$

Lemma 17.15. *Suppose that X is a variety and $k(X)$ is the function field of X. Let F be the presheaf defined by $F(U) = k(X)$ for U an open subset of X. Then F is a sheaf. We will also write the sheaf F as the constant sheaf $k(X)$. We then have that*

$$H^0(X, k(X)) = k(X) \quad and \quad H^1(X, k(X)) = 0.$$

Proof. F is a sheaf by Proposition 11.14, since all open subsets of the variety X are irreducible and hence connected.

We have that $H^0(X, k(X)) = F(X) = k(X)$ since the presheaf F is a sheaf.

Let $\underline{U} = \{U_i\}_{i \in I}$ be an open cover of X. We form an augmented complex

$$G^* : 0 \to G^{-1} \to G^0 \to G^1 \to \cdots$$

of the complex $C^*(\underline{U}, F)$ by

$$G^* : 0 \to k(X) \xrightarrow{\epsilon} C^0(\underline{U}, F) \xrightarrow{d_0} C^1(\underline{U}, F) \xrightarrow{d_1} \cdots$$

where ϵ is the product of the identifications of $k(X)$ with $F(U_i, F)$, for $i \in I$.

Fix $j \in I$. For $p \geq 1$, define $\lambda : C^p(\underline{U}, F) \to C^{p-1}(\underline{U}, F)$ by $(\lambda \alpha)_{i_0, \ldots, i_{p-1}} = \alpha_{j, i_0, \ldots, i_{p-1}}$, using the convention of Remark 17.8. We define $\lambda : C^0(\underline{U}, F) \to k(X)$ by $(\lambda \alpha) = \alpha_j$.

Suppose $\alpha \in C^p(\underline{U}, F)$ with $p \geq 1$. We compute

$$d\lambda(\alpha)_{i_0, \ldots, i_p} = \sum_{k=0}^{p}(-1)^k \lambda(\alpha)_{i_0, \ldots, \hat{i}_k, \ldots, i_p} = \sum_{k=0}^{p}(-1)^k \alpha_{j, i_0, \ldots, \hat{i}_k, \ldots, i_p}$$

and

$$\lambda d(\alpha)_{i_0, \ldots, i_p} = d(\alpha)_{j, i_0, \ldots, i_p} = \alpha_{i_0, \ldots, i_p} - \sum_{k=0}^{p}(-1)^k \alpha_{j, i_0, \ldots, \hat{i}_k, \ldots, i_p}$$

so $(d\lambda + \lambda d)(\alpha) = \alpha$.

Suppose $\alpha \in C^0(\underline{U}, F)$. We compute

$$d\lambda(\alpha)_i = (d\alpha_j)_i = (\alpha_j)_i = \alpha_j$$

and

$$\lambda d(\alpha)_i = d(\alpha)_{j,i} = \alpha_i - \alpha_j$$

so we again have that $(d\lambda + \lambda d)(\alpha) = \alpha$.

Thus λ is a homotopy operator for the complex G^*, and we have shown that the identity map is homotopic to the zero map. Thus the identity map and the zero map are the same maps $H^i(G^*) \to H^i(G^*)$ (by Proposition 17.2), so $H^i(G^*) = 0$ for all i and G^* is an exact complex. We conclude that $H^0(\underline{U}, F) = k(X)$ and $H^i(\underline{U}, F) = 0$ for $i > 0$. Thus

$$\check{H}^1(X, F) = \varinjlim H^1(\underline{U}, F) = 0$$

and by Theorem 17.12, we have that

$$H^1(X, F) \cong \check{H}^1(X, F) = 0. \qquad \square$$

Suppose that X is a variety and $\underline{U} = \{U_i\}_{i \in I}$ is an open cover of X. The sheaf $\mathcal{O}_X^*$ of units in $\mathcal{O}_X$ is a group under multiplication. Thus the Čech complex $C^*(\underline{U}, \mathcal{O}_X^*)$ is a complex of Abelian groups under multiplication. Taking $\mathcal{F} = \mathcal{O}_X^*$ in (17.10), we have

$$d^0(\alpha)_{j,k} = \alpha_k \alpha_j^{-1} \quad \text{for } \alpha = (\alpha_i) \in C^0(\underline{U}, \mathcal{O}_X^*)$$

and

$$d^1(\alpha)_{l,m,n} = \alpha_{l,m} \alpha_{l,n}^{-1} \alpha_{m,n} \quad \text{for } \alpha = (\alpha_{j,k}) \in C^1(\underline{U}, \mathcal{O}_X^*)$$

(recall the convention on indexing of Remark 17.8).

Theorem 17.16. *Suppose that X is a variety. Then $\mathrm{Pic}(X) \cong H^1(X, \mathcal{O}_X^*)$.*

Proof. Let $\underline{U} = \{U_i\}$ be an open cover of X. Let

$$P(\underline{U}) = \{\mathcal{L} \in \mathrm{Pic}(X) \mid \underline{U} \text{ is a trivializing open cover of } X \text{ for } \mathcal{L}\}.$$

Suppose that $[\mathcal{L}] \in P(\underline{U})$ is a class. Let $\phi_i : \mathcal{O}_X|U_i \to \mathcal{L}|U_i$ be a trivialization of $\mathcal{L}$. Let $g_{ij} = \phi_j^{-1} \phi_i$. Then by formula (13.17), we have

$$g_{ij} g_{ik}^{-1} g_{jk} = 1$$

for all i, j, k so $\{g_{ij}\}$ is a cocycle in $C^1(\underline{U}, \mathcal{O}_X^*)$, giving a class $[\{g_{ij}\}]$ in $\check{H}^1(\underline{U}, \mathcal{O}_X^*)$. Taking $\mathcal{M} \cong \mathcal{L}$ in Lemma 13.37, we see that for a trivialization $\psi_i : \mathcal{O}_X|U_i \to \mathcal{M}|U_i$, the transition functions $h_{ij} = \psi_j^{-1} \psi_i$ differ from the g_{ij} by a coboundary. Thus the map $\Psi_{\underline{U}} : P(\underline{U}) \to \check{H}^1(\underline{U}, \mathcal{O}_X^*)$ given by $\Psi_{\underline{U}}([\mathcal{L}]) = [\{g_{ij}\}]$ is well-defined.

We now establish that $\Psi_{\underline{U}}$ is onto. Suppose that $\{g_{ij}\} \in C^1(\underline{U}, \mathcal{O}_X^*)$ is a cocycle, so that $g_{ij} g_{ik}^{-1} g_{jk} = 1$ by (17.10). Fix $k \in I$ and define $\mathcal{L}$ by $\mathcal{L}|U_i = g_{ik} \mathcal{O}_{U_i}$. We compute

$$g_{ik} \mathcal{O}_{U_i \cap U_j} = g_{ij} g_{jk} \mathcal{O}_{U_i \cap U_j} = g_{jk} \mathcal{O}_{U_i \cap U_j}$$

since $g_{ij} \in \Gamma(U_i \cap U_j, \mathcal{O}_X^*)$ is a unit on $U_i \cap U_j$, so $\mathcal{L}$ is a well-defined invertible sheaf and $[\mathcal{L}] \in P(\underline{U})$. We have $\phi_i = g_{ik} : \mathcal{O}_X|U_i \to \mathcal{L}|U_i$ is an isomorphism for $i \in I$, so

$$\Psi_{\underline{U}}([\mathcal{L}]) = [\{\phi_j^{-1} \phi_i\}] = [\{g_{jk}^{-1} g_{ik}\}] = [\{g_{ij}\}].$$

Thus $\Psi_{\underline{U}}$ is onto. The map $\Psi_{\underline{U}}$ is injective by Lemma 13.37.

17.4. Applications

If $[\mathcal{L}], [\mathcal{M}] \in P(\underline{U})$ with transition functions g_{ij} and h_{ij}, then the g_{ij}^{-1} are transition functions of $\mathcal{L}^{-1}$ and the $g_{ij}h_{ij}$ are transition functions of $\mathcal{L} \otimes \mathcal{M}$. Thus $\Psi_{\underline{U}}$ is a group isomorphism.

Suppose that $[\mathcal{L}] \in \text{Pic}(X)$ and $\underline{U} = \{U_i\}_{i \in I}$ is a trivializing open cover of X for $\mathcal{L}$, with isomorphisms $\phi_i : \mathcal{O}_X|U_i \to \mathcal{L}|U_i$. Let $\underline{V} = \{V_j\}_{j \in J}$ be a refinement of $\underline{U}$, so there is an order-preserving map of index sets $\lambda : J \to I$ such that $V_j \subset U_{\lambda(j)}$ for all $j \in J$. The map

$$\phi : C^*(\underline{U}, \mathcal{O}_X^*) \to C^*(\underline{V}, \mathcal{O}_X^*)$$

of complexes defined after Lemma 17.10 takes a 1-cycle $\{g_{kl}\}$ to $\{g_{\lambda(k)\lambda(l)}\}$. Now $\psi_i = \phi_{\lambda(i)}|V_i$ gives a trivialization of $\mathcal{L}|V_i$ with transition functions $\psi_j^{-1}\psi_i = g_{\lambda(i)\lambda(j)}$. Since $\text{Pic}(X) = \bigcup P(\underline{U}) = \lim_{\to} P(\underline{U})$, the $\Psi_{\underline{U}}$ patch by Theorem 17.12 to a group isomorphism

$$\text{Pic}(X) \to \lim_{\to} \check{H}^1(\underline{U}, \mathcal{O}_X^*) = H^1(X, \mathcal{O}_X^*). \qquad \square$$

Remark 17.17. Theorem 17.16 is true when X is an arbitrary scheme and in fact for an arbitrary locally ringed space X. The construction of an invertible sheaf $\mathcal{L}$ with a given cocycle $\{g_{ij}\}$ as its transition functions is a little more delicate in this case because of the possibility of zero divisors in $\mathcal{O}_X$. Using the cocycle $\{g_{ij}\}$, we realize $\mathcal{L}$ as an inverse limit of sheaves $(\lambda_i)_*\mathcal{O}_{U_i}$ and $(\lambda_{ij})_*\mathcal{O}_{U_i \cap U_j}$ where $\lambda_i : U_i \to X$ and $\lambda_{ij} : U_i \cap U_j \to X$ are the inclusions.

Theorem 17.18. Let $X \subset \mathbb{P}^r$ be a projective scheme and let $\mathcal{O}_X(n) = \mathcal{O}_{\mathbb{P}^r}(n) \otimes_{\mathcal{O}_{\mathbb{P}^r}} \mathcal{O}_X$ for $n \in \mathbb{Z}$. Let $\mathcal{F}$ be a coherent sheaf on X. Then:

1) For each $i \geq 0$, $H^i(X, \mathcal{F})$ is a finite-dimensional k-vector space.

2) There is an integer n_0, depending only on $\mathcal{F}$, such that for each $i > 0$ and each $n \geq n_0$, $H^i(X, \mathcal{F}(n)) = 0$ where $\mathcal{F}(n) = \mathcal{F} \otimes \mathcal{O}_X(n)$.

Proof. Let $i : X \to \mathbb{P}^r$ be the given closed embedding. Let $\mathcal{I}_X$ be the ideal sheaf of the closed subscheme X in $\mathcal{O}_{\mathbb{P}^r}$. Then $i_*\mathcal{O}_X = \mathcal{O}_{\mathbb{P}^r}/\mathcal{I}_X$ is a coherent $\mathcal{O}_{\mathbb{P}^r}$-module, so $i_*\mathcal{F}$ is a coherent $\mathcal{O}_{\mathbb{P}^r}$-module. For U an open subset of $\mathbb{P}^r$, we have that

$$\Gamma(U, i_*\mathcal{F}) = \Gamma(U \cap X, \mathcal{F}).$$

Computing cohomology by Čech cohomology, we see that

$$H^i(X, \mathcal{F}) = H^i(\mathbb{P}^r, i_*\mathcal{F})$$

for all i (alternatively, we get this isomorphism directly from Lemma 17.4). Thus we may assume that $X = \mathbb{P}^r$. Now conclusions 1) and 2) of Theorem 17.18 are true when $\mathcal{F} = \mathcal{O}_X(m)$ for some $m \in \mathbb{Z}$ by Theorem 17.14.

We prove the theorem for arbitrary coherent sheaves $\mathcal{F}$ on X by descending induction on i. For $i > r$ we have that $H^i(X, \mathcal{F}(n)) = 0$ for all $n \in \mathbb{Z}$ by Corollary 17.6 or by Theorem 17.13, since X can be covered by $r + 1$ open affine subschemes. The theorem is thus true in this case.

We can write $\mathcal{F}$ as a quotient sheaf of a sheaf $\mathcal{E} = \bigoplus_{i=0}^{m} \mathcal{O}_X(q_i)$ for suitable $q_i \in \mathbb{Z}$ by Exercise 11.56. Let $\mathcal{K}$ be the kernel of this quotient, giving an exact sequence

(17.11) $$0 \to \mathcal{K} \to \mathcal{E} \to \mathcal{F} \to 0.$$

Then $\mathcal{K}$ is also coherent. For instance, letting S be the coordinate ring of X, we can realize $\mathcal{E} \to \mathcal{F}$ as the sheafification of a surjection of finitely generated graded S-modules $\Psi : \bigoplus S(q_i) \to M$. Then the kernel K of Ψ is a finitely generated graded S-module since S is Noetherian. Thus $\mathcal{K} = \tilde{K}$ is coherent.

We have an exact sequence of k-vector spaces
$$\cdots \to H^i(X, \mathcal{E}) \to H^i(X, \mathcal{F}) \to H^{i+1}(X, \mathcal{K}) \to \cdots.$$

The module on the left is a finite-dimensional k-vector space since $\mathcal{E}$ is a finite direct sum of sheaves $\mathcal{O}_X(q_i)$ and cohomology commutes with direct sums. The vector space on the right is finite dimensional by induction. Thus $H^i(X, \mathcal{F})$ is a finite-dimensional vector space, establishing 1).

To prove 2), tensor (17.11) with $\mathcal{O}_X(n)$ and then take the long exact cohomology sequence
$$\cdots \to H^i(X, \mathcal{E}(n)) \to H^i(X, \mathcal{F}(n)) \to H^{i+1}(X, \mathcal{K}(n)) \to \cdots.$$

For $i > 0$ and $n \gg 0$, the vector space on the left is zero by Theorem 17.14 and the vector space on the right is zero by induction. Thus $H^i(X, \mathcal{F}(n)) = 0$ for $n \gg 0$. □

Remark 17.19. Suppose that Y is an affine variety, $X \subset Y \times \mathbb{P}^r$ is a closed subscheme, and $\mathcal{F}$ is a coherent sheaf on X. Then the proof of Theorem 17.18 extends to show that $H^i(X, \mathcal{F})$ is a finitely generated $k[Y]$-module for all i and $H^i(X, \mathcal{F}(n)) = 0$ for all $i > 0$ and $n \gg 0$.

Corollary 17.20. *Suppose that $\mathcal{F}$ is a coherent sheaf on a projective scheme X. Then $\mathcal{F} \otimes \mathcal{O}_X(n)$ is generated by global sections for all $n \gg 0$.*

Proof. For $p \in X$, tensor the short exact sequence
$$0 \to \mathcal{I}_p \mathcal{F} \to \mathcal{F} \to \mathcal{F}/\mathcal{I}_p \mathcal{F} \to 0$$
with $\mathcal{O}_X(n)$ for $n \gg 0$ and take cohomology to conclude that $\Gamma(X, \mathcal{F} \otimes \mathcal{O}_X(n))$ surjects onto $\mathcal{F}_p/\mathcal{I}_p \mathcal{F}_p$. The stalk $\mathcal{F}_p$ is thus generated by $\Gamma(X, \mathcal{F} \otimes \mathcal{O}_X(n))$ as an $\mathcal{O}_{X,p}$-module by Nakayama's lemma. The conclusions of the corollary then follow from the facts that X is a Noetherian

17.4. Applications

space and the closed set of points where $\mathcal{F}(n)$ is not generated by global sections contains the closed set of points where $\mathcal{F}(n+1)$ is not generated by global sections. □

We write $h^i(X, \mathcal{F}) = \dim_k H^i(X, \mathcal{F})$ if $\mathcal{F}$ is a coherent sheaf on a projective scheme X.

The Euler characteristic of a coherent sheaf $\mathcal{F}$ on an n-dimensional projective scheme X is

$$\chi(\mathcal{F}) = \sum_{i=0}^{n}(-1)^i h^i(X, \mathcal{F}).$$

Corollary 17.21. *Let Z be a projective scheme. Then*

$$\chi(\mathcal{O}_Z(n)) = P_Z(n)$$

for $n \gg 0$, where $P_Z(n)$ is the Hilbert polynomial of the homogeneous coordinate ring $S(Z)$ of Z.

Proof. The statement of Theorem 11.47 is valid for arbitrary projective schemes (although the proof requires a little modification). Thus for $n \gg 0$, we have

$$P_Z(n) = \dim_k S(Z)_n = \dim_k \Gamma(Z, \mathcal{O}_Z(n)) = \chi(\mathcal{O}_Z(n))$$

by Theorems 16.6, 11.47, and 17.18. □

Theorem 17.22 (Serre duality). *Let X be a nonsingular projective variety of dimension n, let D be a divisor on X, and let K_X be a canonical divisor on X. Then for all i,*

$$\dim_k H^i(X, \mathcal{O}_X(D)) = \dim_k H^{n-i}(X, \mathcal{O}_X(-D + K_X)).$$

This follows from [**73**, Corollary III.7.7]. We will present a proof for curves (by Serre in [**134**]) in Section 18.2.

Theorem 17.23 (Künneth formula). *Suppose X and Y are schemes and $\mathcal{F}$ and $\mathcal{G}$ are quasi-coherent sheaves on X and Y, respectively. Let $\pi_1 : X \times Y \to X$ and $\pi_2 : X \times Y \to Y$ be the projections. Then*

$$\bigoplus_{p+q=n} H^p(X, \mathcal{F}) \otimes_k H^q(Y, \mathcal{G}) \cong H^n(X \times Y, \pi_1^* \mathcal{F} \otimes_{\mathcal{O}_{X \times Y}} \pi_2^* \mathcal{G}).$$

Proof. Let $\underline{U}$ be an affine open cover of X and $\underline{V}$ be an affine open cover of Y. Then $\underline{W} = \{U_i \times V_j\}$ is an affine open cover of $X \times Y$. Define complexes $A_n = C^{-n}(\underline{U}, \mathcal{F})$ and $C_n = C^{-n}(\underline{V}, \mathcal{G})$ for $n \in \mathbb{Z}$ (where $C^*(\underline{U}, \mathcal{F})$

and $C^*(\underline{V}, \mathcal{G})$ are the Čech complexes). Define B to be the complex $B = A \otimes_k C$, so that $B_n = \bigoplus_{p+q=n} A_p \otimes C_q$ for $n \in \mathbb{Z}$. Taking homology of these complexes, we have that for $p \in \mathbb{Z}$

$$H_p(A_*) \cong H^{-p}(C^*(\underline{U}, \mathcal{F})) \cong H^{-p}(X, \mathcal{F}),$$

$$H_p(C_*) \cong H^{-p}(C^*(\underline{V}, \mathcal{G})) \cong H^{-p}(Y, \mathcal{G}),$$

and

$$H_p(B_*) \cong H^{-p}(C^*(\underline{W}, \pi_1^* \mathcal{F} \otimes \pi_2^* \mathcal{G})).$$

By [**128**, Corollary 11.29 on page 340], we have

$$\bigoplus_{p+q=n} H_p(A_*) \otimes H_q(C_*) \cong H_n(A_* \otimes C_*)$$

for $n \in \mathbb{Z}$, giving us the conclusions of the theorem. □

Exercise 17.24. Suppose that X is a projective scheme and

$$0 \to \mathcal{M}_0 \to \mathcal{M}_1 \to \cdots \to \mathcal{M}_n \to 0$$

is an exact sequence of coherent $\mathcal{O}_X$-modules. Show that

$$\sum_{i=0}^n (-1)^i \chi(\mathcal{M}_i) = 0.$$

Exercise 17.25. Let $X = \mathbb{P}^m \times \mathbb{P}^n$ with projections π_1 and π_2 onto the first and second factors, respectively. For $a, b \in \mathbb{Z}$ let $\mathcal{O}_X(a, b) = \pi_1^* \mathcal{O}_{\mathbb{P}^m}(a) \otimes \pi_2^* \mathcal{O}_{\mathbb{P}^n}(b)$. Compute $h^i(X, \mathcal{O}_X(a, b))$ for $a, b \in \mathbb{Z}$ and $i \in \mathbb{N}$.

17.5. Higher direct images of sheaves

Definition 17.26. Suppose that $\phi : X \to Y$ is a continuous map of topological spaces and $\mathcal{F}$ is a sheaf of groups on X. For $i \geq 0$, the i-th direct image sheaf of $\mathcal{F}$ is the sheaf $R^i \phi_* \mathcal{F}$ on Y associated to the presheaf

$$U \mapsto H^i(\phi^{-1}(U), \mathcal{F})$$

for U an open subset of Y.

The sheaf $R^0 \phi_* \mathcal{F}$ is equal to the sheaf $\phi_* \mathcal{F}$.

Given an open cover $\underline{U}$ of a topological space X and a sheaf $\mathcal{F}$ of Abelian groups on X, define a complex of sheaves of Abelian groups $\mathcal{C}^* = \mathcal{C}^*(\underline{U}, \mathcal{F})$ on X by

(17.12) $$\mathcal{C}^p(\underline{U}, \mathcal{F})|V = \prod_{i_0 < \cdots < i_p} \mathcal{F}|U_{i_0,\ldots,i_p} \cap V$$

for V an open subset of X, with coboundary map $d : \mathcal{C}^p \to \mathcal{C}^{p+1}$ as defined by (17.9).

17.5. Higher direct images of sheaves

Proposition 17.27. *Suppose that $\phi : X \to Y$ is a regular map of varieties and $\mathcal{F}$ is a quasi-coherent sheaf on X. Then the sheaves $R^i\phi_*\mathcal{F}$ are quasi-coherent on Y. In particular, for V an affine open subset of Y,*

$$R^i\phi_*\mathcal{F} \mid V = \tilde{M}$$

where $M = H^i(\phi^{-1}(V), \mathcal{F})$. If ϕ is the composition of a closed embedding $X \to Y \times \mathbb{P}^n$ for some n, followed by projection onto the first factor, then the sheaves $R^i\phi_*\mathcal{F}$ are coherent on Y.

Proof. Let $\underline{U} = \{U_i\}$ be an affine open cover of X. The sheaves $\mathcal{C}^p(\underline{U}, \mathcal{F})$ defined in (17.12) are quasi-coherent on X so the complex $\phi_*\mathcal{C}^*(\underline{U}, \mathcal{F})$ of $\mathcal{O}_Y$-modules is quasi-coherent by Theorem 11.50, and thus the sheaves $H^p(\phi_*\mathcal{C}^*(\underline{U}, \mathcal{F}))$ are quasi-coherent on Y by Exercise 11.35.

Suppose $V \subset Y$ is an affine open set. Then the open subset $U_i \times V$ of $X \times Y$ is affine, so $\Gamma_\phi \cap (U_i \times V) \cong U_i \cap \phi^{-1}(V)$ is an open affine subset of $X \cong \Gamma_\phi$. Let $W_i = U_i \cap \phi^{-1}(V)$. Then $\underline{W} = \{W_i\}$ is an affine open cover of $\phi^{-1}(V)$.

We have that

$$\begin{aligned} H^p(\phi_*\mathcal{C}^*(\underline{U}, \mathcal{F}))(V) &= H^p(\Gamma(V, \phi_*\mathcal{C}^*(\underline{U}, \mathcal{F}))) \\ &= H^p(C^*(\underline{W}, \mathcal{F})) = H^p(\phi^{-1}(V), \mathcal{F}) \end{aligned}$$

by Exercise 11.35 and Theorem 17.13, and so

$$R^p\phi_*\mathcal{F}|V = \tilde{M}$$

where $M = H^p(\phi^{-1}(V), \mathcal{F})$ by Theorem 11.32.

The coherence of $R^p\phi_*\mathcal{F}$ in the case when X is a closed subvariety of $Y \times \mathbb{P}^n$ follows from Remark 17.19. $\square$

Lemma 17.28. *Suppose that $\phi : X \to Y$ is a regular map of varieties, $\mathcal{F}$ is a quasi-coherent $\mathcal{O}_X$-module, and $\mathcal{E}$ is a locally free $\mathcal{O}_Y$-module of finite rank. Then*

$$R^i\phi_*(\mathcal{F} \otimes \phi^*\mathcal{E}) \cong R^i\phi_*\mathcal{F} \otimes \mathcal{E}.$$

Proof. Let notation be as in Proposition 17.27. Since $\mathcal{E}$ is locally free,

$$\phi_*(\mathcal{C}^*(\underline{U}, \mathcal{F} \otimes \phi^*\mathcal{E})) \cong \phi_*(\mathcal{C}^*(\underline{U}, \mathcal{F}) \otimes \phi^*\mathcal{E}) \cong (\phi_*\mathcal{C}^*(\underline{U}, \mathcal{F})) \otimes \mathcal{E}$$

by Exercise 11.41. Since $\mathcal{E}$ is locally free,

$$H^p((\phi_*\mathcal{C}^*(\underline{U}, \mathcal{F})) \otimes \mathcal{E}) \cong H^p(\phi_*\mathcal{C}^*(\underline{U}, \mathcal{F})) \otimes \mathcal{E},$$

so if V is an affine open subset of Y,

$$\begin{aligned} \Gamma(V, R^p\phi_*(\mathcal{F} \otimes \phi^*\mathcal{E})) &= H^p(\phi_*\mathcal{C}^*(\underline{U}, \mathcal{F} \otimes \phi^*\mathcal{E}))(V) \\ &\cong H^p(\phi_*\mathcal{C}^*(\underline{U}, \mathcal{F}))(V) \otimes_{\mathcal{O}_Y(V)} \mathcal{E}(V) \\ &\cong \Gamma(V, (R^p\phi_*\mathcal{F}) \otimes \mathcal{E}). \end{aligned}$$
$\square$

The Leray spectral sequence [**59**, II.4.7.1]

(17.13) $$^2E^{i,j} = H^i(Y, R^j\phi_*\mathcal{F}) \Rightarrow_i H^{i+j}(X, \mathcal{F})$$

relates the sheaf cohomology of $\mathcal{F}$ on X and the sheaf cohomology of the higher direct image sheaves of $\mathcal{F}$ on Y. The method of using a double complex to compute this spectral sequence using Čech cohomology is given a lucid explanation in [**105**, Section 3].

Proposition 17.29. *Suppose that X and Y are projective varieties and $\phi : X \to Y$ is a regular map. Let*

$$Y_i = \{p \in Y \mid \dim \phi^{-1}(p) \geq i\}.$$

The Y_i are closed subsets of Y by Corollary 8.14. Suppose that $\mathcal{F}$ is a coherent sheaf on X. Then

$$\operatorname{Supp}(R^i\phi_*\mathcal{F}) \subset Y_i$$

for all i.

Proof. [**105**, Proposition 4.3]. $\square$

Proposition 17.30. *Suppose that Y is a nonsingular variety and W is a nonsingular closed subvariety. Let $\pi : X = B(W) \to Y$ be the blow-up of W with exceptional divisor $E = X_W$. Then E is a codimension 1 subvariety of X, X and E are nonsingular,*

(17.14) $$R^i\pi_*\mathcal{O}_X(mE) = 0 \quad \text{for } i > 0 \text{ and } m \leq 0,$$

and

(17.15) $$\pi_*\mathcal{O}_X(mE) = \begin{cases} \mathcal{O}_Y & \text{if } m \geq 0, \\ \mathcal{I}_W^{-m} & \text{if } m < 0. \end{cases}$$

Proof. The assertions that E is a codimension 1 subvariety of X and X and E are nonsingular in Proposition 17.30 were proven earlier in Theorem 10.19.

By Proposition 17.27, it suffices to show that there exists an affine cover $\{V\}$ of Y such that for all V,

(17.16) $$H^i(\pi^{-1}(V), \mathcal{O}_X(mE)) = 0 \quad \text{for } i > 0 \text{ and } m \leq 0$$

and

(17.17) $$H^0(\pi^{-1}(V), \mathcal{O}_X(mE)) = \begin{cases} k[V] & \text{if } m \geq 0, \\ I_V(W \cap V)^{-m} & \text{if } m < 0. \end{cases}$$

Suppose that $q \in Y \setminus W$. Let V be an affine neighborhood of q in $Y \setminus W$. Then $\pi^{-1}(V) \cong V$ since $W \cap V = \emptyset$, so $\pi^{-1}(V)$ is affine, and

17.5. Higher direct images of sheaves

thus (17.16) holds on V by Theorem 17.7. Since $W \cap V = \emptyset$, we have that $I_V(W \cap V) = k[V]$. Further, $E \cap \pi^{-1}(V) = \emptyset$, so

$$\mathcal{O}_X(mE)|\pi^{-1}(V) = \mathcal{O}_{\pi^{-1}(V)} \cong \mathcal{O}_V$$

for all m. Thus (17.17) holds on V.

Suppose that $q \in W$. Since Y and W are nonsingular, Lemma 10.3 implies that there exist $f_0, \ldots, f_r \in \mathcal{O}_{Y,q}$ with $r = \text{codim}_Y X - 1$, such that $\mathcal{I}_{W,q} = (f_0, \ldots, f_r)$ and $f_0, \ldots, f_r$ is an $\mathcal{O}_{Y,q}$-regular sequence. There then exists an affine neighborhood V of q in Y such that $f_0, \ldots, f_r \in k[V]$, $(f_0, \ldots, f_r) = I_V(W \cap V)$, and $f_0, \ldots, f_r$ is a $k[V]$-regular sequence. Let $J = (f_0, \ldots, f_r) \subset k[V]$, $T = W \cap V$, $U = \pi^{-1}(V)$, and $F = E \cap U$. We have that the coordinate ring

$$S(U) = \bigoplus_{n \geq 0} J^n \cong k[V][tf_0, \ldots, tf_r]$$

where t is an indeterminate with $\deg t = 1$ by Theorem 6.4, and so for $m \in \mathbb{Z}$ and $0 \leq i \leq r$,

$$\mathcal{O}_U(m)|U_{tf_i} = \widetilde{S(U)(m)}|U_{tf_i} = t^m f_i^m \mathcal{O}_{U_{tf_i}} = t^m \mathcal{O}_U(-mF)|U_{tf_i}.$$

Thus

$$\mathcal{O}_X(-mE)|U = \mathcal{O}_U(-mF) \cong \mathcal{O}_U(m)$$

for $m \in \mathbb{Z}$. Now

$$S(U)/JS(U) = \bigoplus_{i \geq 0} J^i/J^{i+1} = k[T][\overline{f}_0, \ldots, \overline{f}_r]$$

where $\overline{f}_i$ is the class of f_i in J/J^2 and is a polynomial ring over $k[T] = k[V]/J$ in $\overline{f}_0, \ldots, \overline{f}_r$ by Theorem 1.76. Thus $JS(U)$ is a prime ideal in $S(U)$, and

$$S(F) = S(U)/JS(U)$$

and so $F \cong W \times \mathbb{P}^r$. Since $S(F)$ is integrally closed, we have by Theorem 11.47 that

(17.18) $$H^0(F, \mathcal{O}_F(n)) = S(F)_n = J^n/J^{n+1}$$

for $n \geq 0$.

Since

$$\mathcal{O}_F(m) \cong \pi_2^* \mathcal{O}_{\mathbb{P}^r}(m) \cong \pi_1^* \mathcal{O}_T \otimes \pi_2^* \mathcal{O}_{\mathbb{P}^r}(m)$$

where $\pi_1 : F \cong W \times \mathbb{P}^r \to W$ and $\pi_2 : F \cong W \times \mathbb{P}^r \to \mathbb{P}^r$ are the natural projections, we have that for all $i \geq 0$,

$$H^i(F, \mathcal{O}_F(m)) \cong \bigoplus_{\alpha+\beta=i} H^\alpha(T, \mathcal{O}_T) \otimes H^\beta(\mathbb{P}^r, \mathcal{O}_{\mathbb{P}^r}(m))$$

by Theorem 17.23, so

(17.19) $$H^i(F, \mathcal{O}_F(m)) = 0 \quad \text{for } i > 0 \text{ and } m \geq 0$$

since $H^\beta(\mathbb{P}^r, \mathcal{O}_{\mathbb{P}^r}(m)) = 0$ for $\beta > 0$ and $m \geq 0$ and $H^\alpha(T, \mathcal{O}_T) = 0$ for $\alpha > 0$ since T is affine, and by Theorem 17.7. Further,

(17.20) $$H^0(F, \mathcal{O}_F(n)) = 0 \quad \text{for } n < 0.$$

We have short exact sequences

(17.21) $$0 \to \mathcal{O}_U(m+1) \to \mathcal{O}_U(m) \to \mathcal{O}_F(m) \to 0$$

for all $m \in \mathbb{Z}$, giving surjections, by (17.19),

$$H^i(U, \mathcal{O}_U(m+1)) \to H^i(U, \mathcal{O}_U(m)) \quad \text{for } i > 0 \text{ and } m \geq 0.$$

Since $\mathcal{O}_U(1)$ is ample on U, $H^i(U, \mathcal{O}_U(m)) = 0$ for $m \gg 0$ by Remark 17.19. Thus (17.16) holds on $U = \pi^{-1}(V)$.

By (17.16) and (17.18), we have short exact sequences

$$0 \to H^0(U, \mathcal{O}_U(m+1)) \to H^0(U, \mathcal{O}_U(m)) \to J^m/J^{m+1} \to 0$$

for $m \geq 0$. By Theorem 11.47, $H^0(U, \mathcal{O}_U(m)) = J^m$ for $m \gg 0$. Thus $H^0(U, \mathcal{O}_U(m)) = J^m$ for all $m \geq 0$. By (17.20) and (17.21), we have isomorphisms $H^0(U, \mathcal{O}_U(m+1)) \to H^0(U, \mathcal{O}_U(m))$ for $m < 0$. Since $H^0(U, \mathcal{O}_U) = k[V]$, we have that $H^0(U, \mathcal{O}_U(m)) = k[V]$ for $m \leq 0$. Thus (17.17) holds for $U = \pi^{-1}(V)$.

Since equations (17.16) and (17.17) hold on an affine cover $\{V\}$ of Y, the conclusions (17.14) and (17.15) of this proposition hold on Y. $\square$

Suppose that Y is a nonsingular affine variety and W is a nonsingular subvariety of Y. Let $R = k[Y]$ and $\mathfrak{p} = I(W)$. Let $\pi : X \to Y$ be the blow-up of W and let $E = X_W$. Then $\mathcal{O}_{X,E}$ is a valuation ring, since E is a codimension 1 subvariety of the nonsingular variety X. We then have that

(17.22) $$\Gamma(X, \mathcal{O}_X(-nE)) = \{f \in R \mid \nu_E(f) \geq n\} = (\mathfrak{p}^n R_\mathfrak{p}) \cap R = \mathfrak{p}^{(n)}$$

is the n-th symbolic power of $\mathfrak{p}$. Comparing (17.15) and (17.22), we have the formula, with our assumption that $\mathfrak{p}$ is a "regular prime" ($R/\mathfrak{p}$ is a regular ring),

(17.23) $$\mathfrak{p}^{(n)} = \mathfrak{p}^n.$$

This formula (17.23) was previously derived in Proposition 10.4.

For many prime ideals $\mathfrak{p}$ in a regular local ring, this formula is not true and we do not have equality of ordinary and symbolic powers of prime ideals. In fact, the symbolic algebra $\bigoplus_{n \geq 0} \mathfrak{p}^{(n)}$ is in general not Noetherian. An example of this is given by Roberts [**127**], using an earlier example of Nagata [**122**]. A non-Noetherian example with $\mathfrak{p}$ being a rational monomial curve is given by Goto, Nishida, and Watanabe in [**60**].

We now give some more useful formulas on cohomology.

17.6. Local cohomology and regularity

Theorem 17.31. *Suppose that Y is a nonsingular variety and $\phi : X \to Y$ is the blow-up of a nonsingular subvariety W of Y and $\mathcal{L}$ is an invertible sheaf on Y. Then $H^i(X, \phi^*\mathcal{L}) = H^i(Y, \mathcal{L})$ for all i.*

Proof. We have that $R^i\phi_*(\phi^*\mathcal{L}) = 0$ for $i > 0$ and $\phi_*(\phi^*\mathcal{L}) \cong \mathcal{L}$, by Proposition 17.30 and Lemma 17.28. Thus the Leray spectral sequence

$$^2E^{i,j} = H^i(Y, R^j\phi_*\phi^*\mathcal{L}) \Rightarrow_i H^{i+j}(X, \phi^*\mathcal{L})$$

degenerates at the ^2E level, so that $H^i(Y, \mathcal{L}) \cong H^i(X, \phi^*\mathcal{L})$ as desired. $\square$

Theorem 17.32. *Suppose that $\phi : X \to Y$ is a birational regular map of nonsingular projective varieties over a field of characteristic 0 and $\mathcal{L}$ is an invertible sheaf on Y. Then $H^i(X, \phi^*\mathcal{L}) = H^i(Y, \mathcal{L})$ for all i.*

Proof. By resolution of indeterminancy ([**79**] or [**35**, Theorem 6.39]), there exists a regular birational map $f : Z \to X$ from a nonsingular projective variety Z such that $g = \phi \circ f$ factors as a product of blow-ups of nonsingular subvarieties

$$g : Z = Z_n \overset{g_n}{\to} Z_{n-1} \overset{g_{n-1}}{\to} \cdots \overset{g_2}{\to} Z_1 \overset{g_1}{\to} Z_0 = Y.$$

By Theorem 17.31 we have that $g^* : H^i(Y, \mathcal{L}) \cong H^i(Z, g^*\mathcal{L})$ for all i. The isomorphisms $g^* : H^*(Y, \mathcal{L}) \to H^i(Z, g^*\mathcal{L})$ factor as

$$H^i(Y, \mathcal{L}) \overset{\phi^*}{\to} H^i(X, \phi^*\mathcal{L}) \overset{f^*}{\to} H^i(Z, g^*\mathcal{L}).$$

Thus ϕ^* is injective. By resolution of indeterminancy [**79**] there exists a projective variety W and a birational regular map $\gamma : W \to Z$ such that $\beta = f \circ \gamma$ is a product of blow-ups of nonsingular subvarieties, so we have isomorphisms $\beta^* : H^i(X, \phi^*\mathcal{L}) \cong H^i(W, \beta^*\mathcal{L})$ for all i. Thus f^* is also injective, and the theorem follows. $\square$

17.6. Local cohomology and regularity

Suppose that R is a Noetherian ring and M is an R-module. Suppose that I is an ideal in R, with generators $I = (f_1, \ldots, f_n)$. Consider the modified Čech complex

$$C^* : 0 \to C^0 \to C^1 \to \cdots \to C^d \to 0$$

where $C^0 = R$ and

$$C^t = \bigoplus_{1 \leq i_1 < i_2 < \cdots < i_t \leq n} R_{f_{i_1} f_{i_2} \cdots f_{i_t}}.$$

The local cohomology of M is

(17.24) $$H_I^i(M) = H^i(M \otimes_R C^*).$$

We have that
$$H_I^0(M) = \{f \in M \mid I^k f = 0 \text{ for some } k \geq 0\} = \Gamma_I(M),$$
the set of elements of M which have support in I.

$H_I^i(M)$ does not depend on the choice of generators of I. Further,
$$H_I^i(M) = H_{\sqrt{I}}^i(M).$$

If $0 \to A \to B \to C \to 0$ is a short exact sequence of R-modules, then there is a long exact sequence
$$0 \to H_I^0(A) \to H_I^0(B) \to H_I^0(C) \to H_I^1(A) \to \cdots.$$

An important use of local cohomology is to compute depth (Definition 1.67).

We have the following interpretation of depth in terms of local cohomology ([**27**], [**72**], [**73**, Exercise III.3.4], [**50**, Theorem A4.3]).

Proposition 17.33. *Suppose that M is a finitely generated R-module and $n \geq 0$. Then the following are equivalent:*

1) $\operatorname{depth}_I M \geq n$.
2) $H_I^i(M) = 0$ for all $i < n$.

If R is a regular local ring of dimension d and I is the maximal ideal of R, then
$$\operatorname{depth}_I R = d$$
since a regular system of parameters in R is a maximal R-regular sequence in R [**107**, Theorem 36, page 121].

Suppose that X is a Noetherian affine scheme with $R = \Gamma(X, \mathcal{O}_X)$ and M is an R-module. Let $\mathcal{M} = \tilde{M}$ be the quasi-coherent sheaf on X associated to M. Suppose that $I = (f_1, \ldots, f_n)$ is an ideal in R.

Consider the Čech complex
$$F^* : F^0 \to F^1 \to \cdots \to F^{n-1},$$
where
$$F^t = \bigoplus_{1 \leq i_1 < i_2 < \cdots < i_{t+1} \leq n} R_{f_{i_1} f_{i_2} \cdots f_{i_{t+1}}}.$$

The sheaf cohomology of $\tilde{M}$ on $U = X \setminus Z(I)$ is
$$H^i(U, \tilde{M}) = H^i(M \otimes_R F^*).$$

17.6. Local cohomology and regularity

The modified Čech complex C^*, used to compute local cohomology, is obtained from the Čech complex F^* by shifting the Čech complex one to the right and setting $C^0 = R$. From this we see that there is an exact sequence

(17.25) $\quad 0 \to H_I^0(M) \to M \to H^0(U, \tilde{M}) \to H_I^1(M) \to 0$

and isomorphisms

(17.26) $\quad H^i(U, \tilde{M}) \cong H_I^{i+1}(M) \quad \text{for } i \geq 1,$

where $U = X \setminus Z(I)$. We have the interpretation of

$$H^0(U, \tilde{M}) \cong \varinjlim \operatorname{Hom}_R(I^n, M)$$

as an "ideal transform".

A particularly important case of this is when $X = \mathbb{A}^{n+1}$ so that $\kappa[X] = R = \kappa[x_0, \ldots, x_n]$ is a polynomial ring over a field κ. We give R the standard grading. Let $\mathfrak{m} = (x_0, \ldots, x_n)$. Let Q be the point $Z(\mathfrak{m})$ in $\mathbb{A}^{n+1}$. Suppose that M is a graded module over R. Let $\tilde{M}$ be the sheafification of M on the affine variety X. Then the local and sheaf cohomology modules $H^i(\mathbb{A}^{n+1} \setminus Q, \tilde{M})$ and $H_{\mathfrak{m}}^i(R)$ are graded, and the maps of equations (17.25) and (17.26) are graded. From the natural surjection of the affine cone $\mathbb{A}^{n+1} \setminus Q$ onto the projective space $\mathbb{P}^n$, we obtain graded isomorphisms

$$H^i(\mathbb{A}^{n+1} \setminus Q, \tilde{M}) \cong \bigoplus_{j \in \mathbb{Z}} H^i(\mathbb{P}^n, \tilde{M}(j)).$$

In the first cohomology module, $\tilde{M}$ is the sheaf associated to M on $\mathbb{A}^{n+1}$. In the second cohomology module, $\tilde{M}(j)$ is the sheaf associated to $M(j)$ on $\mathbb{P}^n$ ($M(j)_d = M_{j+d}$ for $d \in \mathbb{Z}$).

We thus have a degree-preserving exact sequence of graded R-modules

(17.27) $\quad 0 \to H_{\mathfrak{m}}^0(M) \to M \to \bigoplus_{j \in \mathbb{Z}} H^0(\mathbb{P}^n, \tilde{M}(j)) \to H_{\mathfrak{m}}^1(M) \to 0$

and isomorphisms

(17.28) $\quad \bigoplus_{j \in \mathbb{Z}} H^i(\mathbb{P}^n, \tilde{M}(j)) \cong H_{\mathfrak{m}}^{i+1}(M) \quad \text{for } i \geq 1.$

We have the interpretation

$$\bigoplus_{j \in \mathbb{Z}} H^0(\mathbb{P}^n, \tilde{M}(j)) \cong \varinjlim \operatorname{Hom}_R(\mathfrak{m}^n, M)$$

as an ideal transform.

We continue to study the graded polynomial ring $R = \kappa[x_0, \ldots, x_n]$ and assume that M is a finitely generated graded R-module. Define

$$a^i(M) = \begin{cases} \sup\{j \mid H_{\mathfrak{m}}^i(M)_j \neq 0\} & \text{if } H_{\mathfrak{m}}^i(M) \neq 0, \\ -\infty & \text{otherwise.} \end{cases}$$

The regularity of M is defined to be

$$\operatorname{reg}(M) = \max_i \{a^i(M) + i\}.$$

Interpreting R as the coordinate ring $S(\mathbb{P}^n)$ of $\mathbb{P}^n$ and considering the sheaf $\tilde{M}$ on $\mathbb{P}^n$ associated to M, we can define the regularity of $\tilde{M}$ to be

$$\begin{aligned}\operatorname{reg}(\tilde{M}) &= \max\{m \mid H^i(\mathbb{P}^n, \tilde{M}(m-i-1)) \neq 0 \text{ for some } i \geq 1\} \\ &= \max_{i \geq 2}\{a^i(M) + i\}.\end{aligned}$$

Thus

$$\operatorname{reg}(\tilde{M}) \leq \operatorname{reg}(M).$$

The classical interpretations of regularity are for sections of invertible sheaves on a projective variety X.

Definition 17.34. Suppose that $\mathcal{F}$ is a coherent sheaf on $\mathbb{P}^n$. The sheaf $\mathcal{F}$ is said to be m-regular if $H^i(\mathbb{P}^n, \mathcal{F}(m-i)) = 0$ for all $i > 0$.

Thus $\operatorname{reg}(\mathcal{F})$ is the smallest m such that $\mathcal{F}$ is m-regular. The following theorem by Mumford ([**118**, page 99]) generalizes a classical result of Castelnuovo.

Theorem 17.35 (Geometric regularity theorem). *Suppose that $\mathcal{F}$ is an m-regular coherent sheaf on $\mathbb{P}^n$. Then:*

a) $H^0(\mathbb{P}^n, \mathcal{F}(k))$ *is spanned by* $H^0(\mathbb{P}^n, \mathcal{F}(k-1)) \otimes H^0(\mathbb{P}^n, \mathcal{O}(1))$ *if* $k > m$.

b) $H^i(\mathbb{P}^n, \mathcal{F}(k)) = 0$ *whenever* $i > 0$, $k + i \geq m$.

c) $\mathcal{F}(k)$ *is generated by global sections if* $k \geq m$.

d)
$$\bigoplus_{d \in \mathbb{Z}} H^0(\mathbb{P}^n, \mathcal{F}(d))$$

is generated as an $R = \bigoplus_{d \geq 0} H^0(\mathbb{P}^n, \mathcal{O}_{\mathbb{P}^n}(d))$-*module in degrees* $\leq m$.

The following proof is based on [**118**].

Proof. The proof is by induction on n. If $n = 0$, the result is immediate. For $n > 0$, let H be a hyperplane of $\mathbb{P}^n$ not containing any of the associated varieties of $\mathcal{F}$ (defined after Definition 15.5 in Section 15.1). Tensor the exact sequence

$$0 \to \mathcal{O}_{\mathbb{P}^n}(-H) \cong \mathcal{O}_{\mathbb{P}^n}(-1) \to \mathcal{O}_{\mathbb{P}^n} \to \mathcal{O}_H \to 0$$

17.6. Local cohomology and regularity

with $\mathcal{F}(k)$. For all $p \in \mathbb{P}^n$, if f is a local equation of H at p, then multiplication by f is injective on $\mathcal{F}_p$. Thus the sequence

(17.29) $\qquad 0 \to \mathcal{F}(k-1) \to \mathcal{F}(k) \to \mathcal{F}_H(k) \to 0,$

where $\mathcal{F}_H(k) = (\mathcal{F} \otimes \mathcal{O}_H)(k)$ is short exact. Taking cohomology, we obtain an exact sequence

$$H^i(\mathbb{P}^n, \mathcal{F}(m-i)) \to H^i(H, \mathcal{F}_H(m-i)) \to H^{i+1}(\mathbb{P}^n, \mathcal{F}(m-i-1)).$$

Thus if $\mathcal{F}$ is m-regular, then the sheaf $\mathcal{F}_H$ on $H \cong \mathbb{P}^{n-1}$ is m-regular. Thus the induction hypothesis gives us the conclusions of the theorem for $\mathcal{F}_H$.

From the short exact sequence (17.29), we obtain an exact sequence

$$H^{i+1}(\mathbb{P}^n, \mathcal{F}(m-i-1)) \to H^{i+1}(\mathbb{P}^n, \mathcal{F}(m-i)) \to H^{i+1}(H, \mathcal{F}_H(m-i)).$$

If $i \geq 0$, by b) for $\mathcal{F}_H$, the last group is (0). By m-regularity, the first group is zero. Thus the middle group is (0) and $\mathcal{F}$ is $(m+1)$-regular. By induction on $k \geq m - i$, we obtain the conclusion b) for $\mathcal{F}$.

We now prove a). Consider the commutative diagram

$$\begin{array}{c}
 & & H^0(\mathbb{P}^n, \mathcal{F}(k-1)) \\
 & & \downarrow \\
H^0(\mathbb{P}^n, \mathcal{F}(k-1)) \otimes H^0(\mathbb{P}^n, \mathcal{O}_{\mathbb{P}^n}(1)) & \xrightarrow{\mu} & H^0(\mathbb{P}^n, \mathcal{F}(k)) \\
\downarrow \lambda & & \downarrow \nu \\
H^0(H, \mathcal{F}_H(k-1)) \otimes H^0(H, \mathcal{O}_H(1)) & \xrightarrow{\tau} & H^0(H, \mathcal{F}_H(k))
\end{array}$$

where the bottom row is exact. The map λ is surjective. Further, τ is surjective if $k > m$ by conclusion a) for $\mathcal{F}_H$. Thus $\nu(\text{Image}(\mu)) = H^0(H, \mathcal{F}_H(k))$; that is, $H^0(\mathbb{P}^n, \mathcal{F}(k))$ is spanned by $\text{Image}(\mu)$ and $H^0(\mathbb{P}^n, \mathcal{F}(k-1))$. Let $\sigma \in H^0(\mathbb{P}^n, \mathcal{O}_{\mathbb{P}^n}(1))$ be such that $\text{div}(\sigma) = H$. Then the image of $H^0(\mathbb{P}^n, \mathcal{F}(k-1))$ in $H^0(\mathbb{P}^n, \mathcal{F}(k))$ is equal to $\sigma \otimes H^0(\mathbb{P}^n, \mathcal{F}(k-1))$ which is contained in $\text{Image}(\mu)$. Thus μ is surjective and a) is proven for $\mathcal{F}$.

By Corollary 17.20, $\mathcal{F}(k)$ is generated by its global sections if k is sufficiently large. Thus by a),

(17.30)

$H^0(\mathbb{P}^n, \mathcal{F}(m)) \otimes H^0(\mathbb{P}^n, \mathcal{O}_{\mathbb{P}^n}(k-m))$ generates the sheaf $\mathcal{F}(k)$ if $k \gg 0$.

Let $p \in \mathbb{P}^n$, and fix a local isomorphism of $\mathcal{O}_{\mathbb{P}^n}(1)$ and $\mathcal{O}_{\mathbb{P}^n}$ in a neighborhood of p. For $k \geq m$, this identifies $\mathcal{O}_{\mathbb{P}^n}(k-m)$ with $\mathcal{O}_{\mathbb{P}^n}$ in a neighborhood of p. Then $H^0(\mathbb{P}^n, \mathcal{O}_{\mathbb{P}^n}(k-m))$ is identified with a vector space of elements of the local ring $\mathcal{O}_{\mathbb{P}^n, p}$ which generate $\mathcal{O}_{\mathbb{P}^n, p}$, and (17.30) tells us that $H^0(\mathbb{P}^n, \mathcal{F}(m))$ generates $\mathcal{F}(m)_p$ as an $\mathcal{O}_{\mathbb{P}^n, p}$-module; that is, $\mathcal{F}(m)$ is generated by its global sections. $\square$

Continuing to assume that M is a finitely generated graded module over the polynomial ring $R = \kappa[x_0, \ldots, x_n]$, with maximal ideal $\mathfrak{m} = (x_0, \ldots, x_n)$, let
$$F_* : 0 \to \cdots \to F_j \to \cdots \to F_1 \to F_0 \to M \to 0$$
be a minimal free resolution of M as a graded R-module. Let b_j be the maximum degree of the generators of F_j. Then
$$\mathrm{reg}(M) = \max\{b_j - j \mid j \geq 0\}.$$
In fact, we have (Eisenbud and Goto [**51**], Bayer and Mumford [**17**]) that
$$\begin{aligned}
\mathrm{reg}(M) &= \max\{b_j - j \mid j \geq 0\} \\
&= \max\{n \mid \exists j \text{ such that } \mathrm{Tor}^R_j(\kappa, M)_{n+j} \neq 0\} \\
&= \max\{n \mid \exists j \text{ such that } H^j_{\mathfrak{m}}(M)_{n-j} \neq 0\}.
\end{aligned}$$
The equality of the first and second of the right-hand sides of these equations follows since $\mathrm{Tor}^R_j(\kappa, M) = H_j(F_* \otimes R/\mathfrak{m})$, and as F_* is minimal, we have that the maps of the complex $F_* \otimes R/\mathfrak{m}$ are all zero. To obtain the equality of the first and third conditions, we take the cohomology of the dual of F_*, to compute $\mathrm{Ext}^j_R(M, R)$, and then apply graded local duality. The right-hand side of the third equation is equal to $\mathrm{reg}(M)$ by the definition of regularity.

We may also define local cohomology for sheaves of Abelian groups on a topological space. Let X be a topological space, Y be a closed subset, and $\mathcal{F}$ be a sheaf of Abelian groups on X. Let $\Gamma_Y(X, \mathcal{F})$ be the subgroup of $\Gamma(X, \mathcal{F})$ consisting of all sections whose support is contained in Y. If
$$0 \to \mathcal{A} \to \mathcal{B} \to \mathcal{C} \to 0$$
is a short exact sequence of sheaves of Abelian groups on X, then
$$0 \to \Gamma_Y(X, \mathcal{A}) \to \Gamma_Y(X, \mathcal{B}) \to \Gamma_Y(X, \mathcal{C})$$
is exact, so we can define local cohomology groups $H^i_Y(X, \mathcal{F})$ by taking a resolution
$$0 \to \mathcal{F} \to I^0 \to I^1 \to \cdots$$
by injective sheaves of groups I^i, taking the associated complex
$$\Gamma_Y(X, I^0) \xrightarrow{d^0} \Gamma_Y(X, I^1) \xrightarrow{d^1} \cdots$$
and defining
$$H^i_Y(X, \mathcal{F}) = \mathrm{Kernel}(d^i)/\mathrm{Image}(d^{i+1})$$
for $i \geq 0$.

We summarize some properties of local cohomology, whose proofs can be found in [**72**] and are derived in exercises in [**73**, Chapter III]. If X is a locally ringed space and $\mathcal{F}$ is a sheaf of $\mathcal{O}_X$-modules, then $H^i_Y(X, \mathcal{F})$ can be computed by taking an injective resolution by $\mathcal{O}_X$-modules.

17.6. Local cohomology and regularity

Continuing to suppose that $\mathcal{F}$ is a sheaf of Abelian groups on a topological space, let U be the open subset $U = X \setminus Y$ of X. Then we have a long exact sequence of cohomology groups

$$0 \to H^0_Y(X, \mathcal{F}) \to H^0(X, \mathcal{F}) \to H^0(U, \mathcal{F}|U) \to H^1_Y(X, \mathcal{F})$$
$$\to H^1(X, \mathcal{F}) \to H^1(U, \mathcal{F}|U) \to \cdots.$$

Proposition 17.36 (Excision). *Let V be an open subset of X containing Y. Then for all $i \geq 0$ there are natural isomorphisms*

$$H^i_Y(X, \mathcal{F}) \cong H^i_Y(V, \mathcal{F}|V).$$

Proposition 17.37 (Mayer-Vietoris sequence). *Let Y_1, Y_2 be two closed subsets of X. Then there is a long exact sequence*

$$\cdots \to H^i_{Y_1 \cap Y_2}(X, \mathcal{F}) \to H^i_{Y_1}(X, \mathcal{F}) \oplus H^i_{Y_2}(X, \mathcal{F}) \to H^i_{Y_1 \cup Y_2}(X, \mathcal{F})$$
$$\to H^{i+1}_{Y_1 \cap Y_2}(X, \mathcal{F}) \to \cdots.$$

Now suppose that X is a Noetherian affine scheme with $R = \Gamma(X, \mathcal{O}_X)$ and $\mathcal{F} = \tilde{M}$ where M is an R-module. Let $I \subset R$ be an ideal, and let Y be the subscheme of X with $\Gamma(X, \mathcal{O}_Y) = R/I$. Then (by [72] and [27] or [73, Exercise III.3.3] and [50, Appendix 4]) we have that

$$H^i_Y(X, \mathcal{F}) = H^i_I(M)$$

as defined in equation (17.24).

Chapter 18

Curves

In this chapter we consider the geometry of nonsingular projective curves.

In Sections 18.1–18.3, we prove the Riemann-Roch theorem on a nonsingular projective curve X, Theorem 18.13, which gives a formula for the dimension of the vector space $\Gamma(X, \mathcal{O}_X(D))$ of functions whose poles are bounded by a given divisor D on X in terms of the genus g of X, the degree $\deg D$ of D, and the dimension $h^0(X, \mathcal{O}_X(K_X - D))$, where K_X is a canonical divisor on X. The Riemann-Roch theorem follows from the Riemann-Roch inequality, Theorem 18.2 and Corollary 18.3, proven in Section 18.1 and from Serre duality, Corollary 18.10, proven in Section 18.2.

The Riemann-Roch inequality, Theorem 18.2 and Corollary 18.3, give a lower bound for $h^0(X, \mathcal{O}_X(D))$, which is only in terms of g and $\deg D$ and which is an equality if and only if $h^1(X, \mathcal{O}_X(D)) = 0$. Clifford's theorem, Theorem 18.20, gives an upper bound for $h^0(X, \mathcal{O}_X(D))$ which only depends on $\deg D$ if $h^1(X, \mathcal{O}_X(D)) > 0$.

As a consequence of the Riemann-Roch theorem, we show in Theorem 18.21 that if $\deg D \geq 2g + 1$, then D is very ample and the complete linear system $|D|$ induces a closed embedding of X into a projective space. We deduce in Theorem 18.22 a subdivision of curves by Kodaira dimension: the curves of genus larger than 1, for which K_X is ample; the elliptic curves (genus 1), for which $K_X \sim 0$; and $\mathbb{P}^1$ (genus 0), for which $-K_X$ is ample. The theory of Kodaira dimension generalizes to higher-dimensional varieties [18], [91], [112], [113], [114], and [20].

In Section 18.4, we consider the Riemann-Roch problem, which is the problem of computing the function $h^0(X, \mathcal{O}_X(nD))$ for large n, where X is a nonsingular projective variety and D is a divisor on X.

In Sections 18.5 and Section 18.6, we consider regular maps $f : X \to Y$ of nonsingular projective curves and find formulas relating the genus of X, the genus of Y, and the ramification of f.

We work out the basic geometric theory of elliptic curves in Section 18.7, study the topology of complex curves in Section 18.8, and introduce the theory of Abelian varieties and Jacobians of curves in Section 18.9.

18.1. The Riemann-Roch inequality

Suppose that X is a nonsingular projective curve. The genus of X is

$$g = g(X) = h^0(X, \mathcal{O}_X(K_X)).$$

We have that

(18.1) $$g(X) = h^1(X, \mathcal{O}_X)$$

as follows from Serre duality (Corollary 18.10) which will be established in Section 18.2. Recall the definition of the degree of a divisor on a curve from Section 13.5. If D_1 and D_2 are linearly equivalent divisors on X, then $\deg D_1 = \deg D_2$ by Corollary 13.19.

Lemma 18.1. *Let D be a divisor on X. If $h^0(X, \mathcal{O}_X(D)) > 0$, then $\deg(D) \geq 0$. If $h^0(X, \mathcal{O}_X(D)) > 0$ and $\deg D = 0$, then $D \sim 0$.*

Proof. If $h^0(X, \mathcal{O}_X(D)) > 0$, then there exists $0 \neq f \in \Gamma(X, \mathcal{O}_X(D))$. Then $E = (f) + D$ is an effective divisor, so that $\deg E \geq 0$. We have $\deg D = \deg E \geq 0$ by Corollary 13.19. If $\deg D = 0$, then D is linearly equivalent to an effective divisor of degree 0. The only such divisor is 0. □

For a coherent sheaf $\mathcal{F}$ on X, we have

$$\chi(\mathcal{F}) = h^0(X, \mathcal{F}) - h^1(X, \mathcal{F})$$

by Theorem 17.5.

Theorem 18.2. *Let D be a divisor on a nonsingular projective curve X of genus g. Then*

$$\chi(\mathcal{O}_X(D)) = h^0(X, \mathcal{O}_X(D)) - h^1(X, \mathcal{O}_X(D)) = \deg D + 1 - g.$$

Proof. We must show that

(18.2) $$\chi(\mathcal{O}_X(D)) = \deg D + 1 - g$$

for every divisor D on X. The formula is true for $D = 0$ by Theorem 3.35, the definition of genus, and equation (18.1).

Let D be any divisor, and let $p \in X$ be a point. We will show that the formula is true for D if and only if it is true for $D + p$. Since any divisor on X can be obtained by a finite sequence of addition and subtraction of points, this will establish the formula (18.2) and prove the theorem.

Let $\mathcal{I}(p)$ be the ideal sheaf of the point $p \in X$. Using the fact that $\mathcal{I}(p) = \mathcal{O}_X(-p)$ (a point is a divisor on a curve), we have a short exact sequence of sheaves of $\mathcal{O}_X$-modules

$$0 \to \mathcal{O}_X(-p) \to \mathcal{O}_X \to \mathcal{O}_X/\mathcal{I}(p) \to 0.$$

Now tensor with $\mathcal{O}_X(D+p)$ to get a short exact sequence

(18.3) $$0 \to \mathcal{O}_X(D) \to \mathcal{O}_X(D+p) \to \mathcal{O}_X/\mathcal{I}(p) \to 0.$$

The sequence is short exact since $\mathcal{O}_X(D+p)$ is a locally free (and thus flat) $\mathcal{O}_X$-module (in fact, locally, this is just like tensoring with $\mathcal{O}_X$). The support of $\mathcal{O}_X/\mathcal{I}(p)$ is just the point p, so that $(\mathcal{O}_X/\mathcal{I}(p)) \otimes_{\mathcal{O}_X} \mathcal{O}_X(D+p) \cong \mathcal{O}_X/\mathcal{I}(p)$. Taking the long exact cohomology sequence associated to (18.3) and using Corollary 17.6, we get an exact sequence

$$0 \to H^0(X, \mathcal{O}_X(D)) \to H^0(X, \mathcal{O}_X(D+p)) \to H^0(X, \mathcal{O}_X/\mathcal{I}(p)) \cong k$$
$$\to H^1(X, \mathcal{O}_X(D)) \to H^1(X, \mathcal{O}_X(D+p)) \to H^1(X, \mathcal{O}_X/\mathcal{I}(p)) = 0.$$

Thus

$$\chi(\mathcal{O}_X(D+p)) = \chi(\mathcal{O}_X(D)) + 1.$$

Since $\deg(D+p) = \deg(D) + 1$, we obtain the formula (18.2). $\square$

Corollary 18.3 (The Riemann-Roch inequality)**.** *Suppose that D is a divisor on a nonsingular projective curve X of genus g. Then*

$$h^0(X, \mathcal{O}_X(D)) \geq \deg(D) + 1 - g.$$

18.2. Serre duality

The proof in this section follows Serre [**134**].

We continue to assume that X is a nonsingular projective curve. For a divisor $D = \sum a_i p_i$ on X where p_i are distinct points of X and $a_i \in \mathbb{Z}$, we define for a point $q \in X$

$$\nu_q(D) = \begin{cases} a_i & \text{if } q = p_i, \\ 0 & \text{if } q \neq p_i \text{ for all } i. \end{cases}$$

A répartition r is a family $\{r_p\}_{p \in X}$ of elements of $k(X)$ such that $r_p \in \mathcal{O}_{X,p}$ for all but finitely many $p \in X$. The set of all répartitions is an algebra R over k. Suppose that D is a divisor on X. Then define $R(D)$ to be the k-subspace of R consisting of all $r = \{r_p\}$ such that $\nu_p(r_p) \geq -\nu_p(D)$ for all $p \in X$.

To every $f \in k(X)$, we associate the répartition $\{r_p\}$ such that $r_p = f$ for every $p \in X$, giving an injection of $k(X)$ into R. We may then view $k(X)$ as a subring of R.

Proposition 18.4. *Suppose that D is a divisor on X. Then the k-vector space $I(D) = H^1(X, \mathcal{O}_X(D))$ is canonically isomorphic to $R/(R(D)+k(X))$.*

Proof. For $p \in X$ and $r \in R$, let $[r_p]$ be the class of r_p in $k(X)/\mathcal{O}_X(D)_p$. Define a k-vector space homomorphism $\Lambda : R \to \bigoplus_{p \in X} k(X)/\mathcal{O}_X(D)_p$ by $r \mapsto \{[r_p]\}$. Here Λ is well-defined since $r_p \in \mathcal{O}_X(D)_p$ for all but finitely many $p \in X$. We have that $\Lambda(r) = 0$ if and only if $r_p \in \mathcal{O}_X(D)_p$ for all $p \in X$ which holds if and only if $\nu_p(r_p) \geq -\nu_p(D)$ for all $p \in X$. Thus Λ induces an isomorphism

$$(18.4) \qquad R/R(D) \cong \bigoplus_{p \in X} k(X)/\mathcal{O}_X(D)_p.$$

Let $\mathcal{A}$ be the sheaf $k(X)/\mathcal{O}_X(D)$. By Lemma 17.15, we have natural exact sequences

$$0 \to \Gamma(V, \mathcal{O}_X(D)) \to \Gamma(V, k(X)) = k(X) \to \Gamma(V, \mathcal{A})$$

for all open subsets V of X.

Suppose that U is a neighborhood of a point $p \in X$ and $s \in \mathcal{A}(U)$. There exists $t \in k(X)$ such that the image of t in $\mathcal{A}_p$ is equal to s_p. Let t' be the image of t in $\mathcal{A}(U)$. Then the germ of $s - t'$ in $\mathcal{A}_p$ is zero. Since $\mathcal{A}_p$ is the limit of $\mathcal{A}(V)$ over open sets V containing p, we have that there exists an open neighborhood V of p in U such that the restriction of $s - t'$ in $\mathcal{A}(V)$ is zero.

Then replacing U with V and s with its restriction to V, we have that s is the class of $t \in k(X)$, which is necessarily in $\mathcal{O}_X(D)_q$ for all but finitely many $q \in U$. Thus there exists a neighborhood U' of p such that $s = 0$ on $U' \setminus \{p\}$. In particular, every $s \in H^0(X, \mathcal{A})$ has finite support, so

$$(18.5) \qquad \Phi : H^0(X, \mathcal{A}) \to \bigoplus_{p \in X} \mathcal{A}_p$$

defined by $s \mapsto \{s_p\}$ is a well-defined homomorphism. By the sheaf axioms, every element $\{\alpha_p\} \in \bigoplus \mathcal{A}_p$ lifts to a section of $H^0(X, \mathcal{A})$, and the kernel of Φ is zero. Thus Φ is an isomorphism.

We have that $\mathcal{A}_p = k(X)/\mathcal{O}_X(D)_p$ for $p \in X$ so (18.4) and (18.5) give us an isomorphism

$$(18.6) \qquad R/R(D) \cong H^0(X, \mathcal{A}).$$

The sheaf $\mathcal{O}_X(D)$ is a subsheaf of the constant sheaf $k(X)$, so there is an exact sequence

$$0 \to \mathcal{O}_X(D) \to k(X) \to k(X)/\mathcal{O}_X(D) \to 0.$$

18.2. Serre duality

By Lemma 17.15, we have that $H^0(X, k(X)) = k(X)$ and $H^1(X, k(X)) = 0$ so we have an exact sequence of cohomology modules

$$k(X) \to H^0(X, \mathcal{A}) \to H^1(X, \mathcal{O}_X(D)) \to 0.$$

Now using the isomorphism (18.6), we have the desired isomorphism

$$H^1(X, \mathcal{O}_X(D)) \cong R/(R(D) + k(X)). \qquad \square$$

From now on, we identify $H^1(X, \mathcal{O}_X(D))$ and $R/(R(D) + k(X))$ which we will denote by $I(D)$.

Let $J(D)$ be the dual of the k-vector space $I(D) = R/(R(D) + k(X))$. An element of $J(D)$ is thus a linear form on R which vanishes on $k(X)$ and $R(D)$. Suppose that $D' \geq D$. Then $R(D') \supset R(D)$ so that $J(D) \supset J(D')$. The union of the $J(D)$ for D running through the divisors of X will be denoted by J.

Let $f \in k(X)$ and $\alpha \in J$. The map $r \mapsto \alpha(fr)$ is a linear form on R vanishing on $k(X)$, which we will denote by $f\alpha$. If $\alpha \in J$, then $f\alpha \in J$. This follows since if $\alpha \in J(D)$ and $f \in \Gamma(X, \mathcal{O}_X(\Delta))$, then the linear form $f\alpha$ vanishes on $R(D - \Delta)$ and thus belongs to $J(D - \Delta)$. The operator $(f, \alpha) \mapsto f\alpha$ gives J the structure of a vector space over $k(X)$.

Proposition 18.5. *The dimension of J as a $k(X)$-vector space is ≤ 1.*

Proof. Suppose that $\alpha, \alpha' \in J$ are linearly independent over $k(X)$. There exists a divisor D such that $\alpha \in J(D)$ and $\alpha' \in J(D)$. Let $d = \deg(D)$. For every integer $n \geq 0$, let Δ_n be a divisor of degree n (for example, $\Delta_n = np$, where p is a fixed point of X).

Suppose that $f, g \in \Gamma(X, \mathcal{O}_X(\Delta_n))$. Then $f\alpha, g\alpha' \in J(D - \Delta_n)$. Since α, α' are linearly independent over $k(X)$, any relation $f\alpha + g\alpha' = 0$ implies $f = g = 0$. Thus the map $(f, g) \mapsto f\alpha + g\alpha'$ is an injective k-vector space homomorphism

$$\Gamma(X, \mathcal{O}_X(\Delta_n)) \oplus \Gamma(X, \mathcal{O}_X(\Delta_n)) \to J(D - \Delta_n),$$

so we have the inequality

(18.7) $$\dim_k J(D - \Delta_n) \geq 2 \dim_k \Gamma(X, \mathcal{O}_X(\Delta_n))$$

for all n. We will now show that (18.7) leads to a contradiction as $n \to \infty$. The left-hand side is

$$\begin{aligned} \dim_k I(D - \Delta_n) &= h^1(X, \mathcal{O}_X(D - \Delta_n)) \\ &= -\deg(D - \Delta_n) + g - 1 + h^0(X, \mathcal{O}_X(D - \Delta_n)) \\ &= n + (g - 1 - d) + h^0(D, \mathcal{O}_X(D - \Delta_n)) \end{aligned}$$

by Theorem 18.2.

When $n > d$, $\deg(D-\Delta_n) < 0$ so that $h^0(X, \mathcal{O}_X(D-\Delta_n)) = 0$ by Lemma 18.1. Thus for large n, the left-hand side of (18.7) is equal to $n + A_0$, A_0 a constant. The right-hand side of (18.7) is equal to $2h^0(X, \mathcal{O}_X(\Delta_n))$. By Theorem 18.2,
$$h^0(X, \mathcal{O}_X(\Delta_n)) \geq \deg(\Delta_n) + 1 - g = n + 1 - g.$$
Thus the right-hand side of (18.7) is $\geq 2n + A_1$ for some constant A_1, giving a contradiction for large n. □

The sheaf $\Omega_{X/k}$ is a subsheaf of $\Omega_{k(X)/k}$. If $p \in X$ and t is a regular parameter in $\mathcal{O}_{X,p}$, then $\Omega_{X/k,p} = \mathcal{O}_{X,p} dt$ by Proposition 14.15. We further have that $\Omega_{k(X)/k} = k(X)dt$. We define $\nu_p(\omega) = \nu_p(f)$ if $\omega = fdt \in k(X)dt$. Recall (Section 14.3) that the divisor (ω) of $\omega \in \Omega_{k(X)/k}$ is
$$(\omega) = \sum_{p \in X} \nu_p(\omega) p.$$
Thus $\nu_p(\omega) = \nu_p(K)$ where $(\omega) = K$ is the divisor of ω. The quotient field of $\hat{\mathcal{O}}_{X,p} = k[[t]]$ (Proposition 21.41) is the field of Laurent series $k((t))$. Identifying f with its image in $k((t))$ by the inclusion $k(X) \to k((t))$ induced by the inclusion $\mathcal{O}_{X,p} \to k[[t]]$, we have an expression
$$f = \sum_{n \gg -\infty} a_n t^n$$
with all $a_n \in k$ ($n \gg -\infty$ in the summation means that $a_n = 0$ for $n \ll 0$). The coefficient a_{-1} of t^{-1} in f is called the residue of $\omega = fdt$ at p, denoted by $\text{Res}_p(\omega)$. The following proposition shows that the definition is well-defined.

Proposition 18.6 (Invariance of the residue). *The preceding definition is independent of the choice of regular parameter t in $\mathcal{O}_{X,p}$.*

Proposition 18.6 is proven in [**134**, Section 11 of Chapter II].

Proposition 18.7 (Residue formula). *For every $\omega \in \Omega_{k(X)/k}$,*
$$\sum_{p \in X} \text{Res}_p(\omega) = 0.$$

Proposition 18.7 is proven in [**134**, Sections 12 and 13 of Chapter II]. The proof is by taking a projection to $\mathbb{P}^1$ and showing that it reduces to verifying the formula for $\mathbb{P}^1$.

Given a divisor D on X, let $\Omega_{X/k}(D)$ be the subsheaf of $\Omega_{k(X)/k}$ defined by
$$\Gamma(U, \Omega_{X/k}(D)) = \{\omega \in \Omega_{k(X)/k} \mid (\omega) \cap U \geq D \cap U\}$$
for U an open subset of X.

18.2. Serre duality

Let ω_0 be a nonzero rational differential form, and let $K = (\omega_0)$. Every rational differential form ω can be written as $\omega = f\omega_0$ for some $f \in k(X)$ and $(\omega) \cap U \geq D \cap U$ if and only if $(f) \cap U + (\omega_0) \cap U \geq D \cap U$, which holds if and only if $f \in \Gamma(U, \mathcal{O}_X(K - D))$. Thus

$$\Omega_{X/k}(D) \cong \Omega_X(K - D) \cong \Omega_{X/k} \otimes \mathcal{O}_X(-D).$$

Let $\Omega(D) = \Gamma(X, \Omega_{X/k}(D))$.

We define a product $\langle \omega, \rangle$ of differentials $\omega \in \Omega_{k(X)/k}$ and répartitions $r \in R$ by the following formula:

$$\langle \omega, r \rangle = \sum_{p \in X} \mathrm{Res}_p(r_p \omega).$$

This formula is well-defined since $r_p \omega \in (\Omega_{X/k})_p$ for all but finitely many $p \in X$. The product has the following properties:

a) $\langle \omega, r \rangle = 0$ if $r \in k(X)$.
b) $\langle \omega, r \rangle = 0$ if $r \in R(D)$ and $\omega \in \Omega(D)$.
c) If $f \in k(X)$, then $\langle f\omega, r \rangle = \langle \omega, fr \rangle$.

Property a) follows from the residue formula (Proposition 18.7) and property b) follows since then $r_p \omega \in (\Omega_{X/k})_p$ for all $p \in X$.

For every $\omega \in \Omega_{k(X)/k}$, let $\theta(\omega)$ be the linear form on R defined by

$$\theta(\omega)(r) = \langle \omega, r \rangle.$$

If $\omega \in \Omega(D)$, then $\theta(\omega) \in J(D)$ by properties a) and b) since $J(D)$ is the dual of $R/(R(D) + k(X))$.

Lemma 18.8. *Suppose that $\omega \in \Omega_{k(X)/k}$ is such that $\theta(\omega) \in J(D)$. Then $\omega \in \Omega(D)$.*

Proof. Suppose that $\omega \notin \Omega(D)$. Then there is a point $p \in X$ such that $\nu_p(\omega) < \nu_p(D)$. Set $n = \nu_p(\omega) + 1$, and let r be the répartition defined by

$$r_q = \begin{cases} 0 & \text{if } q \neq p, \\ \frac{1}{t^n} & \text{where } t \text{ is a regular parameter at } p \text{ if } q = p. \end{cases}$$

We have $\nu_p(r_p \omega) = -1$ so that $\mathrm{Res}_p(r_p \omega) \neq 0$ and $\langle \omega, r \rangle \neq 0$. But $n \leq \nu_p(D)$ so $r \in R(D)$ ($\nu_q(0) = \infty$). This is a contradiction since $\theta(\omega)$ is assumed to vanish on $R(D)$. □

Theorem 18.9 (Serre duality). *For every divisor D, the map θ is a k-vector space isomorphism from $\Omega(D)$ to $J(D)$.*

Proof. Suppose that $\omega \in \Omega(D)$ is such that $\theta(\omega) = 0$ in $J(D)$. Then $\theta(\omega) \in J(\Delta)$ for all divisors Δ so $\omega \in \Omega(\Delta)$ for all divisors Δ by Lemma 18.8 so that $\omega = 0$. Hence θ is injective.

By property c), θ is a $k(X)$-linear map from $\Omega_{k(X)/k}$ to J. As $\Omega_{k(X)/k}$ has dimension 1 and J has dimension ≤ 1 as $k(X)$-vector spaces by Proposition 18.5, θ maps $\Omega_{k(X)/k}$ onto J. Thus if $\alpha \in J(D)$, there exists $\omega \in \Omega_{k(X)/k}$ such that $\theta(\omega) = \alpha$ and Lemma 18.8 then shows that $\omega \in \Omega(D)$. □

Corollary 18.10. *Suppose that D is a divisor on X. Then*

$$h^1(X, \mathcal{O}_X(D)) = h^0(X, \mathcal{O}_X(K_X - D))$$

where K_X is a canonical divisor of X.

Exercise 18.11. Prove the residue formula of Proposition 18.7 for $X = \mathbb{P}^1$.

Exercise 18.12. Strengthen the conclusions of Proposition 18.5 to show that $\dim_{k(X)} J = 1$.

18.3. The Riemann-Roch theorem

Theorem 18.13 (Riemann-Roch theorem). *Let D be a divisor on a nonsingular projective curve X of genus g. Then*

$$h^0(X, \mathcal{O}_X(D)) = h^0(X, \mathcal{O}_X(K_X - D)) + \deg D + 1 - g.$$

Proof. The theorem follows from Theorem 18.2 and Serre duality (Corollary 18.10). □

Corollary 18.14. *Suppose that X is a nonsingular projective curve of genus g. Then the degree of the canonical divisor is $\deg K_X = 2g - 2$.*

Proof. Take $D = K_X$ in the Riemann-Roch theorem. □

Corollary 18.15. *Suppose that D is a divisor on a nonsingular projective curve X of genus g such that $\deg D > 2g - 2$. Then*

$$h^0(X, \mathcal{O}_X(D)) = \deg D + 1 - g.$$

Proof. Since $\deg(K_X - D) < 0$, we have that $h^0(X, \mathcal{O}_X(K_X - D)) = 0$ by Lemma 18.1. □

Corollary 18.16. *Suppose that D is a divisor on a nonsingular projective curve X of genus g such that $\deg(D) > 0$. Then*

$$h^0(X, \mathcal{O}_X(nD)) = n \deg(D) + 1 - g$$

for $n > \frac{2g-2}{\deg(D)}$.

18.3. The Riemann-Roch theorem

Theorem 18.17. *Suppose that X is a nonsingular projective curve. Then $X \cong \mathbb{P}^1$ if and only if $g(X) = 0$.*

Proof. Theorem 17.14 implies that $g(\mathbb{P}^1) = h^1(\mathbb{P}^1, \mathcal{O}_{\mathbb{P}^1}) = 0$. Suppose $g(X) = 0$ and $p \in X$ is a point. Then $h^0(X, \mathcal{O}_X(p)) = 2$ by the Riemann-Roch theorem (Theorem 18.13), Corollary 18.14, and Lemma 18.1. Now the complete linear system $|p|$ consists of effective divisors of degree equal to $1 = \deg p$ by Corollary 13.19 and so $X \cong \mathbb{P}^1$ by Corollary 13.20. □

A nonsingular projective curve X is called an elliptic curve if $g(X) = 1$.

Corollary 18.18. *A nonsingular projective curve X is an elliptic curve if and only if $K_X \sim 0$.*

Proof. If $g(X) = 1$, then $\deg K_X = 0$ by Corollary 18.14. Since
$$h^0(X, \mathcal{O}_X(K_X)) = 1,$$
we have $K_X \sim 0$ by Lemma 18.1.

If $K_X \sim 0$, then $g = 1$ by Corollary 18.14 □

Theorem 18.19. *Suppose that X is an elliptic curve and $p_0 \in X$ is a point. Then the map $X \to \mathrm{Cl}^0(X)$ defined by $p \mapsto [p - p_0]$ is a bijection.*

Proof. Suppose that D is a divisor of degree 0 on X. Then
$$h^0(X, \mathcal{O}_X(K_X - D - p_0)) = 0$$
since $\deg(K_X - D - p_0) = -1$. By the Riemann-Roch theorem, we then have that
$$h^0(X, \mathcal{O}_X(D + p_0)) = 1.$$
Thus there is a unique effective divisor linearly equivalent to $D + p_0$ which must be a single point p, since $\deg(D + p_0) = 1$. In particular, there exists a unique point $p \in X$ such that $p - p_0 \sim D$ from which the theorem follows. □

If D is a divisor on a nonsingular projective curve X and
$$h^1(X, \mathcal{O}_X(D)) = 0,$$
then the Riemann-Roch theorem gives the dimension of $h^0(X, \mathcal{O}_X(D))$ but only gives a lower bound if $h^1(X, \mathcal{O}_X(D)) \neq 0$. The following theorem gives an upper bound for $h^0(X, \mathcal{O}_X(D))$ when D is effective and $h^1(X, \mathcal{O}_X(D)) \neq 0$. The bound is sharp, and in fact the curves X and divisors D for which the upper bound is achieved are extremely special and are completely characterized (this is part of Clifford's original theorem). A proof of this characterization is given in [**73**, Theorem IV.5.4].

Theorem 18.20 (Clifford's theorem). *Suppose that D is a divisor on a nonsingular projective curve X such that*

$$h^0(X, \mathcal{O}_X(D)) > 0 \quad \text{and} \quad h^1(X, \mathcal{O}_X(D)) > 0.$$

Then

$$h^0(X, \mathcal{O}_X(D)) \leq \frac{1}{2} \deg(D) + 1.$$

Proof. Let $g = g(x)$. After possibly replacing D with a divisor linearly equivalent to D and K_X with a divisor linearly equivalent to K_X, we may assume that $D \geq 0$ and $D' = K_X - D \geq 0$. Further, we may assume that $h^0(X, \mathcal{O}_X(D-p)) \neq h^0(X, \mathcal{O}_X(D))$ for all $p \in X$ since otherwise we can replace D with $D - p$ and get a stronger inequality. We can then choose

$$g \in \Gamma(X, \mathcal{O}_X(D)) = \{f \in k(X) \mid (f) + D \geq 0\}$$

such that $g \notin \Gamma(X, \mathcal{O}_X(D-p))$ for all $p \in \mathrm{Supp}(D')$.

Consider the k-linear map

$$\phi : \Gamma(X, \mathcal{O}_X(D'))/\Gamma(X, \mathcal{O}_X) \to \Gamma(X, \mathcal{O}_X(K_X))/\Gamma(X, \mathcal{O}_X(D))$$

defined by $\phi(\overline{f}) = \overline{fg}$, where bar denotes residue. The map ϕ is well-defined, since for $f \in \Gamma(X, \mathcal{O}_X(D'))$, $(fg) \geq -D - D' = -K_X$ and since

$$k = \Gamma(X, \mathcal{O}_X) = \{f \in k(X) \mid (f) \geq 0\}$$

by Theorem 3.35 and Lemma 13.3, so we have that $(gf) + D \geq 0$ if $f \in \Gamma(X, \mathcal{O}_X)$.

Suppose $\phi(\overline{f}) = 0$ for some $f \in \Gamma(X, \mathcal{O}_X(D'))$. Then $(f) + D' \geq 0$ so if $p \notin \mathrm{Supp}(D')$, then $\nu_p(f) \geq 0$. Suppose $p \in \mathrm{Supp}(D')$. Then $\nu_p(g) = -\nu_p(D)$ by our choice of g. Since $(gf) + D \geq 0$, we have $\nu_p(fg) \geq -\nu_p(D)$ and so

$$\nu_p(f) \geq -\nu_p(D) - \nu_p(g) = 0.$$

Thus $\nu_p(f) \geq 0$ for all $p \in X$ and so $f \in \Gamma(X, \mathcal{O}_X)$ and we have that ϕ is injective. Thus

(18.8) $$h^0(X, \mathcal{O}_X(D')) - 1 \leq g - h^0(X, \mathcal{O}_X(D)).$$

By the Riemann-Roch theorem,

(18.9) $$\begin{aligned} h^0(X, \mathcal{O}_X(D')) &= \deg(D') + 1 - g + h^0(X, \mathcal{O}_X(K_X - D')) \\ &= g - 1 - \deg(D) + h^0(X, \mathcal{O}_X(D)) \end{aligned}$$

since

$$2g - 2 = \deg K_X = \deg D + \deg D'.$$

Combining equations (18.8) and (18.9), we obtain the conclusions of the theorem. $\square$

Theorem 18.21. *Let D be a divisor on a nonsingular projective curve X of genus g. Then:*

1) *If $\deg D \geq 2g$, then $|D|$ is base point free.*
2) *If $\deg D \geq 2g + 1$, then D is very ample, so that the regular map*

$$\phi_{|D|} : X \to \mathbb{P}^{h^0(X,\mathcal{O}_X(D))}$$

is a closed embedding.

Proof. Conclusion 1) of this theorem follows from Corollary 18.15, which tells us that

$$h^0(X, \mathcal{O}_X(D - p)) = h^0(X, \mathcal{O}_X(D)) - 1$$

for all $p \in X$, and 1) of Corollary 13.34. Conclusion 2) follows from Corollary 18.15, which shows that

$$h^0(X, \mathcal{O}_X(D - p - q)) = h^0(X, \mathcal{O}_X(D)) - 2$$

for all $p, q \in X$, and 3) of Corollary 13.34. □

Theorem 18.22. *Suppose that X is a nonsingular projective curve. Then K_X is ample if $g(X) > 1$, $K_X \sim 0$ if X is an elliptic curve ($g(X) = 1$), and $-K_X$ is ample if $X \cong \mathbb{P}^1$ ($g(X) = 0$).*

Proof. This follows from Corollary 18.14, Theorem 18.21, Theorem 18.17, and Corollary 18.18. □

Theorem 18.22 generalizes to the theory of Kodaira dimension for higher-dimensional varieties. This is especially worked out in the classification of surfaces [18]. Some papers on the theory in higher dimensions are [91], [112], [113], [114], and [20].

18.4. The Riemann-Roch problem on varieties

From Theorems 18.21, 17.18, and 17.35 we obtain the following theorem.

Theorem 18.23. *Suppose that D is a divisor on a nonsingular projective curve X such that $\deg D > 0$. Then*

$$R[D] = \bigoplus_{n \geq 0} \Gamma(X, \mathcal{O}_X(nD))$$

is a finitely generated k-algebra.

Thus Corollary 18.16 is not so surprising, since $h^0(X, \mathcal{O}_X(nD))$ is the Hilbert function of $R[D]$. However, it may be that $R[D]$ is not generated in degree 1, so just knowing that $R[D]$ is a finitely generated k-algebra is not enough to conclude that its Hilbert function is eventually a polynomial. We do have that the Hilbert function of a finitely generated graded k-algebra is eventually a quasi-polynomial, which has an expression $P(n) = a_d(n)n^d + a_{d-1}(n)n^{d-1} + \cdots + a_0(n)$ where the coefficients $a_i(n)$ are periodic functions.

The Riemann-Roch problem is to compute the function

$$P_D(n) = h^0(X, \mathcal{O}_X(nD))$$

for large n where D is a divisor on a nonsingular projective variety X.

It will follow from Theorem 19.1 that $\chi(\mathcal{O}_X(nD))$ is a polynomial in n. Thus if D is ample, we have that $P_D(n)$ is a polynomial for $n \gg 0$, as $P_D(n) = \chi(\mathcal{O}_X(nD))$ for $n \gg 0$ by Theorem 17.18.

If D is a divisor of degree 0 on a nonsingular projective curve X, then $h^0(X, \mathcal{O}_X(nD)) > 0$ if and only if $nD \sim 0$ by Lemma 18.1. We thus have the following complete solution to the Riemann-Roch problem on a curve.

Theorem 18.24. *Suppose that X is a nonsingular projective curve and D is a divisor on X. Then for $n \gg 0$,*

$$h^0(X, \mathcal{O}_X(nD)) = \begin{cases} n \deg D + 1 - g(X) & \text{if } \deg D > 0, \\ \text{a periodic function in } n & \text{if } \deg D = 0, \\ 0 & \text{if } \deg D < 0. \end{cases}$$

There are examples of effective divisors D on a nonsingular projective surface S such that $R[D] = \bigoplus_{n \geq 0} \Gamma(X, \mathcal{O}_X(nD))$ is not a finitely generated k-algebra. This was shown by Zariski in [**159**]; we will construct Zariski's example in Theorem 20.14. It may thus be expected that (the sometimes not finitely generated k-algebra) $R = \bigoplus_{n \geq 0} \Gamma(S, \mathcal{O}_S(nD))$ will not always have a good Hilbert function, that is, that $h^0(S, \mathcal{O}_S(nD))$ will not be polynomial-like. However, Zariski showed in [**159**] that this function is almost a polynomial on a surface.

Theorem 18.25 (Zariski). *Let D be an effective divisor on a nonsingular projective surface S over an algebraically closed field k. Then there exists a quadratic polynomial $P(n)$ and a bounded function $\lambda(n)$ such that*

$$h^0(S, \mathcal{O}_S(nD)) = P(n) + \lambda(n).$$

for $n \geq 0$.

In this same paper, Zariski asked if $\lambda(n)$ is always eventually a periodic function of n (a periodic function in n for $n \gg 0$). This question is answered in [**44**].

Theorem 18.26. *Let D be an effective divisor on a nonsingular projective surface S. Let $\lambda(n)$ be the function of Theorem 18.25. Then:*

1. *If k has characteristic 0 or is the algebraic closure of a finite field, then $\lambda(n)$ is eventually a periodic function.*

2. *There are examples where $\lambda(n)$ is not eventually periodic if k is of positive characteristic and is not the algebraic closure of a finite field.*

Proof. Cutkosky and Srinivas [**44**, Theorems 2 and 3 and Example 3]. $\square$

While the function $h^0(X, \mathcal{O}_X(nD))$ is almost a polynomial function when X is a surface, the behavior of the function $h^0(X, \mathcal{O}_X(nD))$ in higher dimensions can be much more complicated.

Example 18.27. Over any algebraically closed field k, there exists a nonsingular projective 3-fold X and an effective divisor D on X such that

$$\lim_{n \to \infty} \frac{h^0(X, \mathcal{O}_X(nD))}{n^3}$$

is an irrational number. In particular, $h^0(X, \mathcal{O}_X(nD))$ is not eventually a polynomial-like function.

Proof. Cutkosky and Srinivas [**44**, Example 4]. $\square$

The volume of an invertible sheaf $\mathcal{L}$ on a d-dimensional projective variety X is defined as

$$\operatorname{Vol}(\mathcal{L}) = \lim_{n \to \infty} \sup \frac{h^0(X, \mathcal{L}^n)}{n^d/d!}.$$

The volume always exists as a limit over an algebraically closed field k (by Lazarsfeld [**98**] and Lazarsfeld and Mustaţă [**99**]) and over an arbitrary field (by Cutkosky [**42**]) but can be an irrational number (by Example 18.27).

Exercise 18.28. Find an example of a divisor D on a nonsingular projective curve such that $|D|$ is not base point free but $|n_0 D|$ is base point free for some positive multiple n_0.

Exercise 18.29. Give an example of a finitely generated graded k-algebra such that its Hilbert function is not eventually a polynomial.

18.5. The Hurwitz theorem

Suppose that $\phi : X \to Y$ is a dominant regular map of nonsingular projective curves. Recall that ϕ is then finite (Corollary 10.26). Suppose that $P \in X$.

The ramification index e_P of ϕ at P is defined as follows. Let $Q = \phi(P)$. Recall that the valuation ν_Q is a valuation of $k(Y)$ whose valuation ring is

$\mathcal{O}_{Y,Q}$ and ν_P is a valuation of $k(X)$ whose valuation ring is $\mathcal{O}_{X,P}$. Thus ν_P is an extension of ν_Q to $k(X)$. Let x be a regular parameter in $\mathcal{O}_{X,P}$ and y be a regular parameter in $\mathcal{O}_{Y,Q}$. Then

$$y = u x^{e_P}$$

for some unit $u \in \mathcal{O}_{X,P}$ and positive integer e_P. The number e_P is called the ramification index of ν_P over ν_Q or the ramification index of P over Q. Since $y = 0$ is a local equation for the divisor Q on Y, we have that

$$\phi^*(Q) = \sum_{P \in \phi^{-1}(Q)} e_P P,$$

and by Theorem 13.18,

$$\sum_{P \in \phi^{-1}(Q)} e_P = \deg(\phi^*(Q)) = \deg(\phi) = [k(X) : k(Y)]$$

does not depend on Q.

We will say that ϕ is ramified at P if $e_P > 1$, tamely ramified at P if the characteristic p of k does not divide e_P, and wildly ramified at P if p divides e_P. We can then consider the set of all ramification points of ϕ in X.

We will say that a dominant regular map $\phi : X \to Y$ of varieties is separable if the induced extension of fields $k(Y) \to k(X)$ is finite and separable.

Proposition 18.30. *Suppose that $\phi : X \to Y$ is a finite regular map of nonsingular curves and that ϕ is separable. Then there is an exact sequence of $\mathcal{O}_X$-modules*

(18.10) $$0 \to \phi^* \Omega_{Y/k} \to \Omega_{X/k} \to \Omega_{X/Y} \to 0.$$

Proof. By formula (14.4), the sequence (18.10) is right exact, so we need only show that the map

(18.11) $$\phi^* \Omega_{Y/k} \to \Omega_{X/k}$$

is injective. Since $\phi^* \Omega_{Y/k}$ and $\Omega_{X/k}$ are invertible sheaves of $\mathcal{O}_X$-modules, we need only show that the map (18.11) is nonzero. Tensoring over $\mathcal{O}_X$ with $k(X)$, we reduce by Lemma 14.8 to showing that the natural map $\Omega_{k(Y)/k} \otimes_{k(Y)} k(X) \to \Omega_{k(X)/k}$ is nonzero, which will follow if the natural map

(18.12) $$\Omega_{k(Y)/k} \to \Omega_{k(X)/k}$$

is nonzero. The field $k(Y)$ is separably generated over the algebraically closed field k (by Theorem 1.14). Let $z \in k(Y)$ be a transcendental element over k such that $k(Y)$ is separable over $k(z)$. Then $k(X)$ is separable over

18.5. The Hurwitz theorem

$k(z)$, so z is also a separable transcendence basis of $k(X)$ over k. By Theorem 21.75, $d_{K(Y)/k}(z)$ is a generator of $\Omega_{k(Y)/k}$ and $d_{k(X)/k}(z)$ is a generator of $\Omega_{k(X)/k}$. Thus (18.12) is an injection, and $\Omega_{k(Y)/k} \otimes_{k(Y)} k(X) \to \Omega_{k(X)/k}$ is nonzero. $\square$

Suppose that $P \in X$. Let $Q = \phi(P)$. Let x be a regular parameter in $\mathcal{O}_{X,P}$ and y be a regular parameter in $\mathcal{O}_{Y,Q}$. Taking stalks at P in (18.10) gives us (by Proposition 14.15) the short exact sequence

(18.13) $\quad 0 \to \mathcal{O}_{X,P} dy \to \mathcal{O}_{X,P} dx \to (\Omega_{X/Y})_P \to 0.$

We define
$$\frac{dy}{dx} \in \mathcal{O}_{X,P}$$
by
$$dy = \frac{dy}{dx} dx.$$

We have that $y = u x^{e_P}$ where u is a unit in $\mathcal{O}_{X,P}$. Since d is a derivation, we have that
$$dy = e_P u x^{e_P - 1} dx + x^{e_P} du.$$
Now $du = a dx$ for some $a \in \mathcal{O}_{X,P}$, so $dy = (e_P u x^{e_P - 1} + a x^{e_P}) dx$. We thus obtain the following proposition.

Proposition 18.31. *Let $\phi : X \to Y$ be a separable finite regular map of nonsingular curves. Then:*

1. *The support of $\Omega_{X/Y}$ is the finite set of ramification points of ϕ in X.*

2. *For each $P \in X$, $(\Omega_{X/Y})_P$ is a cyclic $\mathcal{O}_{X,P}$-module (generated by one element) of k-dimension equal to $\nu_P(\frac{dy}{dx})$.*

3. *If ϕ is tamely ramified at P, then*
$$\dim_k (\Omega_{X/Y})_P = e_P - 1.$$

4. *If ϕ is wildly ramified at P, then*
$$\dim_k (\Omega_{X/Y})_P > e_P - 1.$$

Let $\mathcal{D}_{X/Y}$ be the ideal sheaf in $\mathcal{O}_X$ which is the annihilator of $\Omega_{X/Y}$. Let R be the effective divisor such that $\mathcal{D}_{X/Y} = \mathcal{O}_X(-R)$. We then have that $\mathcal{D}_{X/Y}$ is the annihilator of $\Omega_{X/Y} \otimes \Omega_{X/k}^{-1}$. Tensoring (18.10) with the invertible sheaf $\Omega_{X/k}^{-1}$, we obtain the short exact sequence
$$0 \to \phi^* \Omega_{Y/k} \otimes \Omega_{X/k}^{-1} \to \mathcal{O}_X \to \Omega_{X/Y} \otimes \Omega_{X/k}^{-1} \to 0$$
so that
$$\phi^* \Omega_{Y/k} \otimes \Omega_{X/k}^{-1} \cong \mathcal{O}_X(-R)$$

and
$$\mathcal{O}_R = \mathcal{O}_R/\mathcal{O}_X(-R) \cong \Omega_{X/Y} \otimes \Omega_{X/k}^{-1} \cong \Omega_{X/Y}$$
since $\Omega_{X/Y}$ has finite support. Thus

(18.14)
$$R = \sum_{p \in X} \dim_k(\Omega_{X/Y})_P P.$$

Taking degrees of divisors in
$$\mathcal{O}_X(-R) \cong \phi^*\Omega_{Y/k} \otimes \Omega_{X/k}^{-1} \cong \mathcal{O}_X(\phi^*(K_Y) - K_X),$$
we have that
$$\begin{aligned} \deg R &= \deg K_X - \deg \phi^*(K_Y) \\ &= \deg K_X - \deg(\phi)\deg K_Y \\ &= (2g(X) - 2) - \deg(\phi)(2g(Y) - 2) \end{aligned}$$
by Theorem 13.18 and Corollary 18.14. We thus have the following theorem.

Theorem 18.32 (Hurwitz). *Let $\phi : X \to Y$ be a dominant separable regular map of nonsingular projective curves. Then*
$$2g(X) - 2 = \deg(\phi)(2g(Y) - 2) + \deg(R),$$
where R is the ramification divisor (18.14). If ϕ has only tame ramification, then
$$\deg(R) = \sum_{P \in X}(e_P - 1).$$

In the case that X and Y are affine, with coordinate rings $A = k[Y]$ and $B = k[X]$, the annihilator $\Gamma(X, \mathcal{D}_{X/Y}) = \Gamma(X, \mathcal{O}_X(-R))$ of $\Gamma(X, \Omega_{X/Y}) = \Omega_{B/A}$ is the different $\mathfrak{D}_{B/A}$ ([**135**, Proposition 14]). The different is defined in [**135**, Chapter III] and [**160**, Section 11 of Chapter V], using the trace of the quotient field of B over A. Proposition 18.31 is proven in [**160**, Theorem 28 of Section 11, Chapter V] and [**135**, Proposition 13]. On [**160**, page 312], a derivation of "Hilbert's formula" is given to compute $\deg(R)$ in the case of a Galois extension, even in the presence of wild ramification.

18.6. Inseparable maps of curves

Recall that a dominant regular map $\phi : X \to Y$ of varieties is separable if the induced field extension $k(Y) \to k(X)$ is finite and separable. We will say that $\phi : X \to Y$ is inseparable if $k(Y) \to k(X)$ is not separable and that $\phi : X \to Y$ is purely inseparable if $k(Y) \to k(X)$ is purely inseparable.

If $K \to L$ is a finite field extension, then there exists a (unique) intermediate field M (called the separable closure of K in L) such that L is purely inseparable over M and M is separable over K (Theorem 1.15). It

18.6. Inseparable maps of curves

follows that any dominant finite regular map of algebraic varieties factors as a purely inseparable finite map, followed by a separable finite map.

Suppose that κ is a perfect field of characteristic $p > 0$ and R is a κ-algebra. Let $Fr : R \to R$ be the Frobenius homomorphism, defined by $Fr(x) = x^p$ for $x \in R$. The map Fr is a ring homomorphism but it is not a κ-algebra homomorphism. Let R_p be the ring R with the κ-algebra structure $\cdot$ given by $a \cdot x = a^p x$ for $a \in \kappa$ and $x \in R$. Then $Fr : R \to R_p$ is a κ-algebra homomorphism. Since R_p is equal to R as a ring, R_p is a domain if and only if R is a domain, R_p is normal if and only if R is normal, and R_p is regular if and only if R is regular.

Now suppose that R is also a domain with quotient field K. We can express R as a κ-algebra by $R = \kappa[S]$ for some subset S of K. Let Ω be an algebraic closure of K. Define $\Lambda : R_p \to \Omega$ by $\Lambda(f) = f^{\frac{1}{p}}$. For $a \in \kappa$ and $x \in R_p$, we have that

$$\Lambda(a \cdot x) = \Lambda(a^p x) = a x^{\frac{1}{p}} = a\Lambda(x),$$

so Λ is a κ-algebra homomorphism, which identifies R_p with the κ-subalgebra $\Lambda(R_p) = \kappa[S^{\frac{1}{p}}]$ of Ω (we have that $\Lambda(\kappa) = \kappa$ as κ is perfect). The composition $\Lambda Fr(x) = x$ for $x \in R$, so $Fr : R \to R_p$ is identified with the natural inclusion of κ-algebras

(18.15) $$\kappa[S] \to \kappa[S^{\frac{1}{p}}].$$

In particular, the quotient field of R_p is identified with $K^{\frac{1}{p}}$ as a κ-algebra.

Now suppose that X is an affine variety over an algebraically closed field k of characteristic $p > 0$. Let $R = k[X]$. The above construction gives us a finitely generated k-algebra R_p which is a domain and a k-algebra homomorphism $Fr : R \to R_p$. Thus there is an affine variety X_p and a regular map $F : X_p \to X$ such that $F^* = Fr$ (by Proposition 2.40).

If X is a quasi-projective variety, we can apply the above construction on an affine open cover of X to obtain by Proposition 3.39 a quasi-projective variety X_p with regular map $F : X_p \to X$. (If X is embedded in $\mathbb{P}^n$, then X_p is embedded in $\mathbb{P}^n_p$ which is isomorphic to $\mathbb{P}^n$ as a variety over k.) Applying the construction (18.15) to $Fr : k(X) \to k(X)$, we see that $k(X_p) \cong k(X)^{\frac{1}{p}}$ and $F^* : k(X) \to k(X_p)$ is the natural inclusion $k(X) \subset k(X)^{\frac{1}{p}}$. The regular map $F : X_p \to X$ is called the k-linear Frobenius map. If X is normal, then X_p is also normal, since X_p has an affine cover by normal varieties.

Theorem 18.33. *Suppose that X is a variety of dimension n. Then*

$$[k(X)^{\frac{1}{p}} : k(X)] = p^n.$$

Proof. An algebraic function field over a perfect field k is separably generated over k (by Theorem 1.14). The theorem then follows from 2) of Theorem 21.76 since $\operatorname{trdeg}_k k(X) = \dim X = n$. □

Theorem 18.34. *Suppose that $f : X \to Y$ is a finite purely inseparable regular map of nonsingular projective curves. Then f is a composition of k-linear Frobenius maps. In particular, $g(X) = g(Y)$.*

Proof. Let the degree of f be $[k(X) : k(Y)] = p^r$. Suppose that $g \in k(X)$. The minimal polynomial of g over $k(Y)$ is $z^{p^i} - h$ for some $h \in k(Y)$ and $i \in \mathbb{N}$ with $i \le r$ since g is purely inseparable over $k(Y)$ and $[k(Y)[g] : k(Y)]$ divides p^r. Thus $k(X)^{p^r} \subset k(Y)$ so $k(X) \subset k(Y)^{\frac{1}{p^r}}$. Let F' be the composition of k-linear Frobenius maps

$$Y_{p^r} \xrightarrow{F} Y_{p^{r-1}} \to \cdots \to Y_p \xrightarrow{F} Y$$

where $Y_{p^i} = (Y_{p^{i-1}})_p$. Here F' has degree p^r by Theorem 18.33. Since $k(X) \subset k(Y)^{\frac{1}{p^r}}$ and both $k(X)$ and $k(Y)^{\frac{1}{p^r}}$ have the same degree over $k(Y)$, we have that $k(X) = k(Y)^{\frac{1}{p^r}}$. Since a nonsingular projective curve is uniquely determined by its function field (by Corollary 10.25), we have that $X \cong Y_{p^r}$, and thus $f = F'$.

Let $\underline{U} = \{U_i\}$ be an affine open cover of Y with corresponding affine open cover $\underline{V} = \{V_i\}$ of Y_{p^r}. Now each $\Gamma(U_i, \mathcal{O}_Y)$ is isomorphic to $\Gamma(V_i, \mathcal{O}_{Y_{p^r}})$ as a ring (but not as a k-algebra) and the Čech complexes $C^*(\underline{U}, \mathcal{O}_Y)$ and $C^*(\underline{V}, \mathcal{O}_{Y_{p^r}})$ are isomorphic complexes of rings. Thus the cohomology is isomorphic, so $H^1(Y_{p^r}, \mathcal{O}_{Y_{p^r}})$ is $H^1(Y, \mathcal{O}_Y)$ with the vector space operation $a \cdot v = a^{p^r} v$ for $a \in k$. Since k is perfect, we have that

$$h^1(Y, \mathcal{O}_Y) = \dim_k H^1(Y, \mathcal{O}_Y) = \dim_k H^1(Y_{p^r}, \mathcal{O}_{Y_{p^r}}) = h^1(Y_{p^r}, \mathcal{O}_{Y_{p^r}})$$

and $g(Y) = g(Y_{p^r})$. □

Exercise 18.35. Let $\mathbb{P}^n$ be projective space over an algebraically closed field k of characteristic $p > 0$. Show that $(\mathbb{P}^n)_p$ is isomorphic to $\mathbb{P}^n$ (as varieties over k).

Exercise 18.36. Suppose that $f : X \to Y$ is a finite regular map of nonsingular projective curves. Show that $g(X) \ge g(Y)$.

Exercise 18.37 (Lüroth's theorem). Suppose that k is an algebraically closed field and L is a subfield of a one-dimensional rational function field $k(t)$ over k such that L contains k and is not equal to k. Show that L is a one-dimensional rational function field over k.

Exercise 18.38. Give an example of a finite purely inseparable regular map of nonsingular projective surfaces which is not a composition of Frobenius maps.

18.7. Elliptic curves

Recall that a nonsingular projective curve X is called an elliptic curve if it has genus $g(X) = 1$. An elliptic curve is characterized by $K_X \sim 0$ by Corollary 18.18. The theory of elliptic curves is particularly remarkable and extensive. We give a brief introduction here. The group of regular isomorphisms of a variety X will be denoted by $\operatorname{Aut}(X)$.

Every nonsingular cubic curve X in $\mathbb{P}^2$ is an elliptic curve. This follows since by adjunction, Theorem 14.21, $\mathcal{O}_X(K_X) \cong \mathcal{O}_{\mathbb{P}^2}(K_{\mathbb{P}^2} + X) \otimes \mathcal{O}_X$ and since $K_{\mathbb{P}^2} = -X$ by Example 14.20.

The reader should peruse the definitions and statements of Section 21.7 on the Galois theory of varieties before reading the proofs of this section.

Lemma 18.39. *Suppose that X is an elliptic curve and P, Q are two not necessarily distinct points in X. Then there exists a regular automorphism $\sigma : X \to X$ such that $\sigma^2 = \mathrm{id}$, $\sigma(P) = Q$, and for any $R \in X$, $R + \sigma(R) \sim P + Q$.*

Proof. We have that $h^0(X, \mathcal{O}_X(P+Q)) = 2$ by Corollary 18.15 and $|P+Q|$ is base point free by Theorem 18.21. We thus have a regular map $\phi = \phi_{|P+Q|} : X \to \mathbb{P}^1$. A linear hyperplane section H on $\mathbb{P}^1$ is a point and $\phi^*(H)$ is an effective divisor linearly equivalent to $P + Q$ by Lemma 13.28. Thus $\deg(\phi) = [k(X) : k(\mathbb{P}^1)] = 2$ by Theorem 13.18. The field extension $k(X)/k(\mathbb{P}^1)$ is separable by Theorem 18.34, since otherwise $g(x) = g(\mathbb{P}^1) = 0$. Thus $k(X)$ is a Galois extension of $k(\mathbb{P}^1)$, so X is Galois over $\mathbb{P}^1$ by Theorem 21.69, with $G(X/\mathbb{P}^1) \cong \mathbb{Z}_2$ by Proposition 21.67. Let $\sigma \in G(X/\mathbb{P}^1)$ be a generator. Since X is Galois over $\mathbb{P}^1$, σ interchanges the two points of a fiber. There exists $S \in \mathbb{P}^1$ such that $\phi^*(S) = P + Q$ by Lemma 13.28, so $\sigma(P) = Q$. We have that

$$\bigcup_{F \in |P+Q|} F = \bigcup_{S \in \mathbb{P}^1} \phi^*(S) = X,$$

so for any $R \in X$, $R + \sigma(R) \in |P + Q|$, and thus $R + \sigma(R) \sim P + Q$. $\square$

Corollary 18.40. *The group $\operatorname{Aut}(X)$ of regular automorphisms of an elliptic curve X is transitive on X.*

Lemma 18.41. *Suppose that $\phi_1 : X \to \mathbb{P}^1$ and $\phi_2 : X \to \mathbb{P}^1$ are two regular maps of degree 2 from an elliptic curve X to $\mathbb{P}^1$. Then there exist automorphisms $\sigma \in \operatorname{Aut}|(X)$ and $\tau \in \operatorname{Aut}(\mathbb{P}^1)$ such that $\phi_2 \sigma = \tau \phi_1$.*

Proof. Let $P_1 \in X$ be a ramification point of ϕ_1 and $P_2 \in X$ be a ramification point of ϕ_2 (which exist by Theorem 18.32). By Corollary 18.40 there is $\sigma \in \operatorname{Aut}(X)$ such that $\sigma(P_1) = P_2$. Since P_1 is a ramification point of the degree 2 map ϕ_1 and $h^0(X, \mathcal{O}_X(2P_1)) = 2$, we have that $\phi_1 = \phi_{|2P_1|}$, and since P_2 is a ramification point of the degree 2 map ϕ_2, $\phi_2 = \phi_{|2P_2|}$. Since σ takes P_1 to P_2, ϕ_1 and $\phi_2\sigma$ are induced by base point free linear systems which are contained in $|2P_1|$. But $|2P_1|$ is the only such linear system. Thus ϕ_1 and $\phi_2\sigma$ are induced by the same linear system, so they differ only by an automorphism of $\mathbb{P}^1$. □

Proposition 18.42. *Suppose that X is an elliptic curve over an algebraically closed field k of characteristic $\neq 2$ and let $P_0 \in X$ be a point. Then there is a closed embedding $\phi : X \to \mathbb{P}^2$ such that the image is the curve with the homogeneous equation*

(18.16) $$x_2 x_1^2 - x_0(x_0 - x_2)(x_0 - \lambda x_2) = 0$$

for some $\lambda \in k \setminus \{0, 1\}$ and $\phi(P_0) = (0 : 1 : 0)$.

The affine equation of the image $\phi(X) \setminus \{(0 : 1 : 0)\}$ of $X \setminus P_0$ in $\mathbb{P}^2_{x_2} \cong \mathbb{A}^2$ is

(18.17) $$y^2 = x(x-1)(x-\lambda),$$

where $x = \frac{x_0}{x_2}$, $y = \frac{x_1}{x_2}$. We think of P_0 as being the "point at infinity" on X under this embedding, since $\phi(P_0) = (0 : 1 : 0) = \phi(X) \cap Z(x_2)$.

Proof. We have that $h^0(X, \mathcal{O}_X(nP_0)) = n$ for $n > 0$ by the Riemann-Roch theorem. The linear system $|3P_0|$ gives a closed embedding $\phi = \phi_{|3P_0|}$ of X into $\mathbb{P}^2$ by Theorem 18.21. Within the function field $k(X)$, we have inclusions

$$k = \Gamma(X, \mathcal{O}_X) = \Gamma(X, \mathcal{O}_X(P_0)) \subset \Gamma(X, \mathcal{O}_X(2P_0)) \subset \cdots.$$

Choose $x \in k(X)$ so that $1, x$ are a basis of $\Gamma(X, \mathcal{O}_X(2P_0))$ and choose $y \in k(X)$ so that $1, x, y$ are a basis of $\Gamma(X, \mathcal{O}_X(3P_0))$. Since $h^0(X, \mathcal{O}_X(6P_0)) = 6$, there is a linear relation between the seven functions $1, x, y, x^2, xy, x^3, y^2 \in \Gamma(X, \mathcal{O}_X(6P_0))$. Further, x^3 and y^2 must both have nonzero coefficients in this relation, since otherwise the relation will have a pole of finite order at P_0, as x has a pole of order 2 at P_0 and y has a pole of order 3 at P_0. (1 has no pole at P_0 and xy has a pole of order 5 at P_0.) Replacing x and y by suitable scalar multiples, we may assume that we have a relation

$$y^2 + b_1 xy + b_2 y = f(x)$$

where $f(x)$ is a degree 3 monic polynomial in x and $b_1, b_2 \in k$. Completing the square in y by replacing y with

$$y' = y + \frac{1}{2}(b_1 x + b_2),$$

we obtain the relation

(18.18) $$y^2 = g(x)$$

where $g(x) = x^3 + a_1 x^2 + a_2 x + a_3$ for some $a_1, a_2, a_3 \in k$ (this is where we need characteristic $\neq 2$).

We represent the closed embedding $\phi: X \to \mathbb{P}^2$ by $\phi = (x : y : 1) = (\frac{x_0}{x_2} : \frac{x_1}{x_2} : 1)$. The relation (18.18) becomes

(18.19) $$x_1^2 x_2 = x_0^3 + a_1 x_0^2 x_2 + a_2 x_0 x_2^2 + a_3 x_2^3.$$

Thus $\phi(X) \subset Z(F)$ where

$$F = x_1^2 x_2 - x_0^3 - a_1 x_0^2 x_2^2 - a_2 x_0 x_2 - a_3 x_2^3.$$

The image of ϕ is a closed irreducible curve $\phi(X)$, which has codimension 1 in $\mathbb{P}^2$. Since F is irreducible in the coordinate ring $k[x_0, x_1, x_2]$ of $\mathbb{P}^2$, $(F) = I(\phi(X))$.

The regular functions on the affine open subset $\mathbb{P}^2_{x_2} \cong \mathbb{A}^2$ of $\mathbb{P}^2$ are $k[\mathbb{P}^2_{x_2}] = k[\bar{x}, \bar{y}]$ where $\bar{x} = \frac{x_0}{x_2}, \bar{y} = \frac{x_1}{x_2}$. The ideal of $\phi(X) \cap \mathbb{P}^2_{x_2}$ is generated by $f = \bar{y}^2 - g(\bar{x})$.

Since $\phi(X)$ is nonsingular, $g(\bar{x})$ can have no multiple roots (by the Jacobian criterion of Proposition 10.14). Thus we can make a change of variables $x' = \alpha x + \beta$ for some $\alpha \neq 0, \beta \in k$, and replace y with a scalar multiple of y to obtain an expression (18.18) with $g(x) = x(x-1)(x-\lambda)$ for some $\lambda \in k$ with $\lambda \neq 0$ or 1. Finally, we see that the set of points at infinity on $\phi(X)$ is the algebraic set $Z(x_2) \cap \phi(X) = Z(x_2, x_0^3) = \{(0 : 1 : 0)\}$. Since x has a pole of order 2 at P_0 and y has a pole of order 3 at P_0,

$$\phi(P_0) = \left(\frac{x}{y}(P_0) : 1 : \frac{1}{y}(P_0)\right) = (0 : 1 : 0). \qquad \square$$

We can regard $\mathbb{P}^1$ as $\mathbb{P}^1 = \mathbb{A}^1 \cup \{\infty\}$, with $k[\mathbb{A}^1] = k[z]$ and $k(\mathbb{P}^1) = k(z)$. The group of automorphisms of $\mathbb{P}^1$ consists of the linear automorphisms (Exercise 13.47), so they can be represented as fractional linear transformations

$$\frac{az + b}{cz + d}$$

with $a, b, c, d \in k$ and $ad - bc \neq 0$. The corresponding transformation in homogeneous coordinates is

$$(ax_0 + bx_1 : cx_0 + dx_1).$$

Let G be the subgroup of $\mathrm{Aut}(\mathbb{P}^1)$ consisting of the automorphisms which permute $\{0, 1, \infty\}$. Then $G \cong S_3$ with

(18.20) $$G = \left\{z, \frac{1}{z}, 1-z, \frac{1}{1-z}, \frac{z}{z-1}, \frac{z-1}{z}\right\}.$$

We have that the group G is generated by $\frac{1}{z}$ and $1-z$.

Suppose that X is an elliptic curve with char $k \neq 2$ and $P_0 \in X$. Consider the linear system $|2P_0|$ which gives a regular map $\Psi = \phi_{|2P_0|} : X \to \mathbb{P}^1$ which is Galois of degree 2 (as we saw in the proof of Lemma 18.39). By Hurwitz's theorem (Ψ is tamely ramified since char $k \neq 2$), Ψ is ramified over four points: $a, b, c, d \in \mathbb{P}^1$ with $\Psi(P_0) = d$. There exists a unique linear automorphism τ of $\mathbb{P}^1$ which takes d to ∞, a to 0, and b to 1. Let λ be the image of c. Then $\tau\Psi$ is ramified over $0, 1, \lambda, \infty$.

Define the j invariant of X as

$$j(\lambda) = 2^8 \frac{(\lambda^2 - \lambda + 1)^3}{\lambda^2(\lambda - 1)^2}.$$

We have (we need only check the generators $\frac{1}{z}$ and $1-z$ of G in (18.20)) that

(18.21) $$j(\sigma(\lambda)) = j(\lambda) \quad \text{for } \sigma \in G.$$

Write $\mathbb{P}^2 = \mathbb{A}^2 \cup H$ where $H = Z(x_2)$ is "the hyperplane at infinity". The closed embedding $\phi = \phi_{|3P_0|}$ of Proposition 18.42 gives an expression of X as (isomorphic) to the union of the affine curve

$$C = Z(y^2 - x(x-1)(x-\lambda)) \subset \mathbb{A}^2$$

and the point "at infinity" $P_0 = (0 : 1 : 0)$. The degree 2 map $\Psi = \phi_{|2P_0|}$ is the linear projection to $\mathbb{P}^1$ which takes $(a, b) \in C$ to $a \in \mathbb{A}^1$ and P_0 to ∞.

Theorem 18.43. *Suppose that k is an algebraically closed field of characteristic not equal to 2. For $\lambda_1, \lambda_2 \in k \setminus \{0, 1\}$, if X_1 is an elliptic curve which gives λ_1 in the above construction of λ and X_2 is an elliptic curve which gives λ_2, then X_1 is isomorphic to X_2 if and only if $j(\lambda_1) = j(\lambda_2)$. Further, every element of k is the j invariant of some elliptic curve X.*

Proof. We will first show that $j(\lambda)$ is uniquely determined by an elliptic curve X. Suppose $P_1, P_2 \in X$ and $\Psi_1 : X \to \mathbb{P}^1$ is induced by $|2P_1|$ so that the ramification points of Ψ_1 in $\mathbb{P}^1$ are $0, 1, \lambda_1, \infty$ with $\Psi_1(P_1) = \infty$ and $\Psi_2 : X \to \mathbb{P}^1$ is induced by $|2P_2|$ so that the ramification points of Ψ_2 in $\mathbb{P}^1$ are $0, 1, \lambda_2, \infty$ with $\Psi_2(P_2) = \infty$. By Lemma 18.41 and its proof, there exist automorphisms $\sigma \in \mathrm{Aut}(X)$ and $\tau \in \mathrm{Aut}(\mathbb{P}^1)$ such that $\Psi_2 \sigma = \tau \Psi_1$ with $\sigma(P_1) = P_2$ so that $\tau(\infty) = \infty$ and τ sends the other ramification

18.7. Elliptic curves

points $\{0, 1, \lambda_1\}$ to $\{0, 1, \lambda_2\}$ in some order. Let $\gamma(z)$ be the fractional linear transformation of $\mathbb{P}^1$ defined by

$$\gamma(z) = \frac{z - \tau(0)}{\tau(1) - \tau(0)}.$$

Then $\gamma\tau(0) = 0$, $\gamma\tau(1) = 1$, and $\gamma\tau(\infty) = \infty$, so $\gamma\tau$ is the identity map. Thus

$$\lambda_1 = \gamma\tau(\lambda_1) = \frac{\tau(\lambda_1) - \tau(0)}{\tau(1) - \tau(0)}.$$

Since the sets $\{\tau(0), \tau(1), \tau(\lambda_1)\}$ and $\{0, 1, \lambda_2\}$ are equal, we have that

$$\lambda_2 \in \left\{\lambda_1, \frac{1}{\lambda_1}, 1 - \lambda_1, \frac{1}{1 - \lambda_1}, \frac{\lambda_1}{\lambda_1 - 1}, \frac{\lambda_1 - 1}{\lambda_1}\right\},$$

and so $j(\lambda_1) = j(\lambda_2)$ by (18.21).

Now suppose that X_1 and X_2 are two elliptic curves, giving λ_1 and λ_2, respectively, and such that $j(\lambda_1) = j(\lambda_2)$. The regular map $j : \mathbb{A}^1 \setminus \{0, 1\} \to \mathbb{A}^1$ extends to a regular map $j = (j : 1) : \mathbb{P}^1 \to \mathbb{P}^1$ with $j^{-1}(\infty) = \{0, 1, \infty\}$, which induces

$$j^* : k(\mathbb{P}^1) = k(j) \to k(\mathbb{P}^1) = k(\lambda)$$

defined by

$$j \mapsto 2^8 \frac{(\lambda^2 - \lambda + 1)^3}{\lambda^2(\lambda - 1)^2}.$$

For $j_0 \in \mathbb{A}^1$,

$$2^8(\lambda^2 - \lambda + 1)^3 - j_0\lambda^2(\lambda - 1)^2 = 0$$

is an equation of degree 6 in λ, so counting multiplicities, it has six roots. Thus

$$6 = \deg(j^*(j_0)) = [k(\lambda) : k(j)]$$

by Theorem 13.18. By (18.21), substituting λ for z, G acts on $k(\lambda)$ by k-automorphisms which leave $k(j)$ invariant. Since $[k(\lambda) : k(j)] = 6$ and $|G| = 6$, we have that $k(\lambda)$ is Galois over $k(j)$ with Galois group G. Thus $j : \mathbb{P}^1 \to \mathbb{P}^1$ is Galois with Galois group G (Theorem 21.69), and so

$$j(\lambda_1) = j(\lambda_2)$$

if and only if there exists $\tau \in G$ such that $\tau(\lambda_1) = \lambda_2$.

By Proposition 18.42, X_1 and X_2 can be embedded in $\mathbb{P}^2$ with respective affine equations

(18.22) $$y^2 = x(x-1)(x-\lambda_1)$$

and

(18.23) $$y^2 = x(x-1)(x-\lambda_2).$$

Since $j(\lambda_1) = j(\lambda_2)$, there exists $\tau \in G$ such that $\tau(\lambda_1) = \lambda_2$, and thus τ permutes $0, 1, \infty$. Let $\Psi_1 : X_1 \to \mathbb{P}^1$ be the 2-1 cover induced by projection onto the x-axis in (18.22). Then $\tau\Psi_1 : X_1 \to \mathbb{P}^1$ is a degree 2 map which is ramified over $0, 1, \lambda_2, \infty$ in $\mathbb{P}^1$. Let $Q = (\tau\Psi_1)^{-1}(\infty)$. Then $x \in H^0(X_1, \mathcal{O}_{X_1}(2Q))$ where $\tau\Psi_1 = (x : 1)$. Proceeding as in the proof of Proposition 18.42, we find $y \in H^0(X_1, \mathcal{O}_{X_1}(3Q))$, giving the relation $y^2 = g(x)$ of (18.18) and such that $\phi_{|3Q|} = (x : y : 1)$ is a closed embedding of X_1 into $\mathbb{P}^2$. The points in $X_1 \subset \mathbb{P}^2$ where $\tau\Psi_1 : X_1 \to \mathbb{P}^1$ is ramified are Q and the points in $X_1 \cap \mathbb{A}^2$ where $y = 0$. Since $\tau\Psi_1$ is ramified over $0, 1, \lambda_2$, and ∞ and g is monic of degree 3, we see that $g(x) = x(x-1)(x-\lambda_2)$. Thus X_1 is isomorphic to the cubic curve with affine equation (18.23), so X_1 is isomorphic to X_2.

Now given $j_0 \in k$, we can solve the equation
$$2^8(\lambda^2 - \lambda + 1)^3 - j_0\lambda^2(\lambda - 1)^2 = 0$$
to find a solution $\lambda_0 \in k$, which cannot be 0 or 1. The elliptic curve with affine equation
$$y^2 = x(x-1)(x-\lambda_0)$$
defines a nonsingular cubic curve of degree 3 in $\mathbb{P}^2$ which is an elliptic curve that has j_0 as its j invariant. $\square$

Let X be an elliptic curve with a fixed point $P_0 \in X$. By Theorem 18.19, the map $P \mapsto [P - P_0]$ is a bijection from X to $\mathrm{Cl}^0(X)$. This induces a group structure on X with P_0 as the zero element and with addition $\oplus$ defined by $P \oplus Q = R$ if and only if $P + Q \sim R + P_0$ as divisors on X.

Proposition 18.44. *Suppose that X is an elliptic curve with the group structure given as above by the choice of a point $P_0 \in X$. Then the addition map $X \times X \to X$ and the inverse map $X \to X$ are regular maps.*

Proof. We will denote the addition of P and Q in X by $P \oplus Q$ and the inverse of P by $\ominus P$.

By Lemma 18.39, taking $P = Q = P_0$, there is an automorphism σ of X such that for any $R \in X$, $R + \sigma(R) \sim 2P_0$. Thus $\ominus R = \sigma(R)$, and so the inverse map $\ominus$ is a regular map.

Let $P \in X$. By Lemma 18.39, there is an automorphism τ of X such that $R + \tau(R) \sim P + P_0$ for all $R \in X$. Thus $P \ominus R = \tau(R)$, and since $\ominus$ is a regular map, we have that translation $R \mapsto R \oplus P$ is a regular map for fixed $P \in X$.

Embed X into $\mathbb{P}^2$ by $\phi_{|3P_0|}$. Let $F = 0$ be the homogeneous cubic equation of X in $\mathbb{P}^2$. If L is a linear form on $\mathbb{P}^2$, then L intersects X in three points with multiplicity, considering the restriction of F to L as a degree

18.7. Elliptic curves

3 form on $L \cong \mathbb{P}^1$. (This is a special case of Bézout's theorem, which we will prove later in Theorem 19.20.) Thus we have a map $\lambda : X \times X \to X$ obtained by letting the image of (P, Q) be the third point of intersection of the line through P and Q with X (if $P = Q$, the line through P is required to be tangent to X at p).

We will establish that this map is in fact regular everywhere. It will follow that addition, $(P, Q) \mapsto P \oplus Q$, is a regular map, since $P + Q + \lambda(P, Q) \sim 3P_0$ and $P + Q \sim (P \oplus Q) + P_0$, so $(P \oplus Q) \oplus \lambda(P, Q) = P_0$ and thus $P \oplus Q = \ominus \lambda(P, Q)$.

Given $P, Q \in X$, there exists a linear form H of $\mathbb{P}^2$ such that all three intersection points of the line through P and Q with X lie in $\mathbb{P}_H^2$. Thus we are reduced to showing that if $f = 0$ is the equation of $C = X \cap \mathbb{P}_H^2$ in $\mathbb{P}_H^2 \cong \mathbb{A}^2$ and if P, Q are points of C such that the line through P and Q in C has three intersections with C in $\mathbb{A}^2$ (counting multiplicity), then the rational map λ is regular near (P, Q).

We now consider $P = (\alpha, \beta)$ and $Q = (\gamma, \delta)$ as variable points in $\mathbb{A}^2$. The line through P and Q in $\mathbb{A}^2$ can be parameterized as

$$x = \alpha + t(\gamma - \alpha), \qquad y = \beta + t(\delta - \beta).$$

The intersection points of this line with C are obtained from the solutions in t to

$$g(t) = f(\alpha + t(\gamma - \alpha), \beta + t(\delta - \beta)) = 0.$$

Write $g(t) = at^3 + bt^2 + ct + d$ with a, b, c, d in the polynomial ring $k[\alpha, \beta, \gamma, \delta]$. We now constrain $P = (\alpha, \beta)$ and $Q = (\gamma, \delta)$ to lie on C. Thus we consider the residues $\bar{a}, \bar{b}, \bar{c}, \bar{d}$ of a, b, c, d in

$$R = k[\alpha, \beta, \gamma, \delta]/(f(\alpha, \beta), f(\gamma, \delta))$$

(which is a domain by Proposition 5.7). Let $\bar{\alpha}, \bar{\beta}, \bar{\gamma}, \bar{\delta}$ be the residues of $\alpha, \beta, \gamma, \delta$ in R, so that $R = k[\bar{\alpha}, \bar{\beta}, \bar{\gamma}, \bar{\delta}]$. Let $\bar{g}(t) = \bar{a}t^3 + \bar{b}t^2 + \bar{c}t + \bar{d}$ be the residue of $g(t)$ in $R[t]$. We have that $\bar{g}(0) = f(\bar{\alpha}, \bar{\beta}) = 0$ and $\bar{g}(1) = f(\bar{\gamma}, \bar{\delta}) = 0$. Thus $\bar{d} = 0$ and $\bar{a} + \bar{b} + \bar{c} = 0$, and we have a factorization

$$\bar{g}(t) = t(t - 1)(\bar{a}t + (\bar{a} + \bar{b})).$$

We see that if $a \neq 0$, then λ is the regular map defined by

$$\lambda((u_1, v_1) \times (u_2, v_2))$$
$$= \left(u_1 - \frac{a+b}{a}(u_1, v_1, u_2, v_2)(u_2 - u_1), v_1 - \frac{a+b}{a}(u_1, v_1, u_2, v_2)(v_2 - v_1) \right).$$

$\square$

In the language of [146], (α, β) and (γ, δ) in the above proof are "independent generic points".

Lemma 18.45 (Rigidity lemma). *Let X be a projective variety, Y and Z be quasi-projective varieties, and $f : X \times Y \to Z$ be a regular map such that for some $Q \in Y$, $f(X \times \{Q\}) = P$ is a single point of Z. Then there is a regular map $g : Y \to Z$ such that if $\pi_2 : X \times Y \to Y$ is the projection, we have that $f = g \circ \pi_2$.*

Proof. Let $R \in X$ be a point and define $g : Y \to Z$ by $g(y) = f(R, y)$. Since two regular maps on a variety are equal if they agree on a nontrivial open set, we need only show that f and $g \circ \pi_2$ agree on some open subset of $X \times Y$. Let U be an affine open neighborhood of P in Z, $F = Z \setminus U$, and $G = \pi_2(f^{-1}(F))$. The set G is closed in Y since $f^{-1}(F)$ is closed in $X \times Y$ and X is projective, and hence π_2 is a closed map (Corollary 5.13). We have that $Q \notin G$ since $f(X \times \{Q\}) = P \notin F$. Thus $V = Y \setminus G$ is a nonempty open neighborhood of Q in Y. For each $y \in V$, the projective variety $X \times \{y\}$ is mapped by f into the affine variety U and hence to a single point of U (by Corollary 5.16). Thus for any $x \in X$ and $y \in V$, we have that

$$f(x, y) = f(R, y) = g \circ \pi_2(x, y),$$

proving the lemma. □

Corollary 18.46. *Let X be an elliptic curve with group structure defined by a point $P_0 \in X$ and let Y be an elliptic curve with group structure defined by $Q_0 \in Y$. Suppose that $\Phi : X \to Y$ is a regular map such that $\Phi(P_0) = Q_0$. Then Φ is a group homomorphism.*

Proof. Consider the regular map $\Psi : X \times X \to Y$ defined by

$$\Psi(x, y) = \Phi(x \oplus y) \ominus \Phi(x) \ominus \Phi(y).$$

Then $\Psi(X \times \{P_0\}) = \Psi(\{P_0\} \times X) = Q_0$, so $\Psi(x, y) = Q_0$ for all $x, y \in X$ by Lemma 18.45. □

18.8. Complex curves

A nonsingular projective curve X over $k = \mathbb{C}$ has the structure of a Riemann surface, and $g = g(X)$ is the topological genus of X (X is topologically a sphere with g handles). This is discussed, for instance, in [115] and [62]. Now such X has the Euclidean topology. We proved that when G is an Abelian group, then $\Gamma(U, G) \cong G^r$ where r is the number of connected components of U (by Proposition 11.14). This is the same as the first singular cohomology $H^0_{\text{Sing}}(U, G)$. Now the Čech complex computes singular cohomology of X, since X can be triangulated ([49, Section 9 of Chapter X]

18.8. Complex curves

or [**62**]) and computes sheaf cohomology, so we obtain that the sheaf cohomology $H^i(X, \mathbb{Z}_{an})$ is isomorphic to $H^i_{Sing}(X, \mathbb{Z})$. We write $\mathbb{Z}_{an}$ to indicate that we are in the Euclidean topology. Now we regard X as a compact two-dimensional oriented real manifold, and then we have (for instance by [**103**]) that

$$H^i_{Sing}(X, \mathbb{Z}) = \begin{cases} \mathbb{Z} & \text{if } i = 0, \\ \mathbb{Z}^{2g} & \text{if } i = 1, \\ \mathbb{Z} & \text{if } i = 2, \\ 0 & \text{if } i > 2. \end{cases}$$

Let $\mathcal{O}_X^{an}$ be the sheaf of analytic functions on X and $(\mathcal{O}_X^{an})^*$ be the sheaf of nonvanishing analytic functions. Then we have a short exact sequence of sheaves

$$0 \to \mathbb{Z} \to \mathcal{O}_X^{an} \xrightarrow{e} (\mathcal{O}_X^{an})^* \to 0,$$

where e denotes the exponential map $f \mapsto e^f$.

It follows from GAGA [**133**] that if Y is a complex projective variety and $\mathcal{F}$ is a coherent sheaf on Y, then the cohomology of the extension $\mathcal{F}^{an}$ of $\mathcal{F}$ to an analytic sheaf is the same as the cohomology of $\mathcal{F}$. Thus $H^i(X, \mathcal{O}_X^{an}) \cong H^i(X, \mathcal{O}_X)$ for all i. Taking sheaf cohomology, we get the long exact sequence

$$0 \to \mathbb{Z} \to \mathbb{C} \xrightarrow{e} \mathbb{C}^* \to H^1(X, \mathbb{Z}) \to H^1(X, \mathcal{O}_X)$$
$$\to H^1(X, (\mathcal{O}_X^{an})^*) \xrightarrow{c} H^2(X, \mathbb{Z}) \to H^2(X, \mathcal{O}_X).$$

Now X has genus g and dimension 1, so that $H^2(X, \mathcal{O}_X) = 0$. Further, $e : \mathbb{C} \to \mathbb{C}^*$ is onto. Thus from our above exact sequence, we deduce that we have an exact sequence of groups

$$0 \to \mathbb{C}^g/\mathbb{Z}^{2g} \to H^1(X, (\mathcal{O}_X^{an})^*) \xrightarrow{c} \mathbb{Z} \to 0,$$

since

$$H^1(X, \mathcal{O}_X)/H^1(X, \mathbb{Z}) \cong \mathbb{C}^g/\mathbb{Z}^{2g}.$$

From the argument of Theorem 17.16, we have that

$$H^1(X, (\mathcal{O}_X^{an})^*) \cong \text{Pic}^{an}(X),$$

the group of invertible analytic sheaves on X, modulo isomorphism. Now again by GAGA, we know that all global analytic sheaves on X are isomorphic to algebraic sheaves, and this isomorphism takes global analytic homomorphisms to algebraic homomorphisms. Thus the natural map

$$\text{Pic}(X) \to \text{Pic}^{an}(X)$$

is an isomorphism. In conclusion, we have obtained the following theorem:

Theorem 18.47. *Suppose that X is a nonsingular projective curve of genus g over the complex numbers. Then there is a short exact sequence of groups*
$$0 \to G \to \text{Pic}(X) \xrightarrow{c} \mathbb{Z} \to 0,$$
where G is a group $\mathbb{C}^g/\mathbb{Z}^{2g}$.

The subset $H^1(X, \mathbb{Z}) \cong \mathbb{Z}^{2g}$ of $H^1(X, \mathcal{O}_X) \cong \mathbb{C}^g$ is in fact a lattice, if we regard $\mathbb{C}^g$ as a $2g$-dimensional real vector space. Thus in the Euclidean topology, $G \cong (S^1)^{2g}$ where S is the circle $\mathbb{R}/\mathbb{Z}$ and G is a "torus". This group G naturally has the structure of an analytic manifold (of complex dimension g), and it is even an algebraic variety (of dimension g). The group structure on G is algebraic. The map c is just the degree map, and the exact sequence of the theorem is just the exact sequence
$$0 \to \text{Cl}^0(X) \to \text{Cl}(X) \xrightarrow{\deg} \mathbb{Z} \to 0$$
of (13.11).

Using our natural isomorphism of $\text{Pic}(X)$ with $\text{Cl}(X)$, the map c (for Chern) is actually the degree of a divisor which we studied on a curve earlier. We can thus identify the algebraic group G with the group $\text{Cl}^0(X)$ of linear equivalence classes of divisors of degree 0 on X. This group G is called the Jacobian of X (in honor of Jacobi). Fixing a point $P_0 \in X$, we obtain a map $X \to J$ defined by mapping a point P to the class of $P - P_0$. This map is a regular map and is a closed embedding if $g > 0$.

18.9. Abelian varieties and Jacobians of curves

In this section we discuss the algebraic construction of the Jacobian. We need to introduce a couple of new concepts first.

An Abelian variety A (in honor of Abel) is a projective variety with a group structure such that the multiplication $m : A \times A \to A$ is a regular map and the inverse map $i : A \to A$ is a regular map. There is an extensive literature on these remarkable varieties. A few references are [**146**], [**96**], [**119**], and [**110**]. The elliptic curves are the one-dimensional Abelian varieties. An Abelian variety is commutative and nonsingular (as is shown in any of these references). A g-dimensional Abelian variety over the complex numbers is isomorphic by an analytic isomorphism to a complex torus $\mathbb{C}^g/\Lambda$, where Λ is a lattice in $\mathbb{C}^g$.

Suppose that X is a variety and r is a positive integer. The symmetric group S_r acts on the product X^r by permuting factors. There exists a variety $X^{(r)}$ whose function field is $k(X^r)^{S_r}$ which is a quotient X^r/S_r [**119**, II, Section 7 and III Section 11]. In the case when X is a nonsingular projective curve, $X^{(r)}$ is nonsingular [**111**, Proposition 3.2]. The points of $X^{(r)}$ can be considered as effective divisors $p_1 + p_2 + \cdots + p_r$ of degree r on X.

We have the following theorem.

Theorem 18.48. *Suppose that X is a nonsingular projective curve of genus g. Then there exists an Abelian variety J of dimension g and a regular map $\phi : X \to J$ such that:*

1. *ϕ is a closed embedding.*
2. *ϕ induces a birational regular map $X^{(g)} \to J$ by*

$$p_1 + \cdots + p_g \mapsto \sum_{i=1}^{g} \phi(p_i).$$

3. *ϕ induces a group isomorphism $\mathrm{Cl}^0(X) \to J$ by $[D] \mapsto \sum n_i \phi(p_i)$ if $D = \sum n_i p_i$.*

The variety J of Theorem 18.48 is called the Jacobian of X.

An Abelian variety A of positive dimension $n > 0$ over the complex numbers has lots of points of infinite order (under the group law of A). This follows from the fact that there is an analytic isomorphism of A with the quotient of $\mathbb{C}^g$ by a lattice of $\mathbb{C}^g$. However, if A is an Abelian variety over the algebraic closure of a finite field, then every element of A has finite order. We see this as follows. Suppose that k is the algebraic closure of a finite field and A is an Abelian variety over k. Then A is a subvariety of a projective space $\mathbb{P}^n_k$. Suppose that $x \in A$. Then there exists a finite field k' such that the embedding of A into $\mathbb{P}^n$ is defined over k, x is a rational point over k', and the addition on A is defined over k'. There are only finitely many points of $\mathbb{P}^n$ which are rational over k' so there are only finitely many points of A which are rational over k'. All multiples of x are rational over k' since the multiplication is defined over k' and x is rational over k'. Thus x has finite order in the group A.

Let A be an Abelian variety over an algebraically closed field k. Let $A(k)$ be the group of points of A, so that $A(k)$ is a $\mathbb{Z}$-module. The rank of $A(k)$ is $\mathrm{rank}(A(k)) = \dim_\mathbb{Q} A(k) \otimes \mathbb{Q}$. We have seen that if A is an Abelian variety over the algebraic closure k of a finite field, then $\mathrm{rank}(A(k)) = 0$. However, we have the following theorem ensuring us that there are lots of points of infinite order on an Abelian variety of positive dimension over any other algebraically closed field.

Theorem 18.49. *Suppose that A is a positive-dimensional Abelian variety defined over an algebraically closed field k which is not the algebraic closure of a finite field. Then the rank of $A(k)$ is equal to the cardinality of k.*

Proof. [**56**, Theorem 10.1]. □

We also have the following proposition describing the points of finite order in an Abelian variety.

Proposition 18.50. *Let A be an Abelian variety of dimension g over an algebraically closed field k. For $n \in \mathbb{Z}_{\geq 0}$, let*
$$A_n(k) = \{x \in A \mid nx = 0\}.$$
Suppose that the characteristic p of k does not divide n. Then $A_n(k) \cong (\mathbb{Z}/n\mathbb{Z})^{2g}$.

Proof. [**119**, Proposition, page 64]. □

In the case when $k = \mathbb{C}$, so that there is an analytic isomorphism $A \cong \mathbb{C}^g/\Lambda$ where Λ is a lattice in $\mathbb{C}^g$, the proposition follows since $A_n(\mathbb{C}) \cong (\frac{1}{n}\Lambda)/\Lambda$.

The history of Abelian varieties and Jacobian varieties is outlined at the end of Milne's article [**111**]. Milne proves many interesting facts about the Jacobian in [**111**], including giving in [**111**, Section 7] a proof in modern language of Weil's proof in [**146**] of Theorem 18.48, the original proof using the language of *Foundations of Algebraic Geometry* [**145**]. Milne refers to Section 2 of Artin [**12**] for a proof in modern language of Weil's theorem that a "birational group" is isomorphic to an algebraic group [**146**].

Throughout these exercises C will denote a nonsingular projective curve of genus g.

Exercise 18.51. Show that $|K_C|$ is base point free if $g \geq 1$.

Exercise 18.52. Show that mK_C is very ample if $g \geq 3$ and $m \geq 2$.

Exercise 18.53. If $g = 2$, show that mK_C is very ample for $m \geq 3$ and $\phi_{|2K_C|} : C \to \mathbb{P}^2$ is a degree 2 regular map onto a quadric curve in $\mathbb{P}^2$ (which is isomorphic to $\mathbb{P}^1$).

Exercise 18.54. Suppose that C is defined over an algebraically closed field k of characteristic 0. Suppose that $0 \neq f \in k(C)$ and that the regular map $\phi = (f : 1) : C \to \mathbb{P}^1$ has degree n. Show that
$$g = \frac{1}{2}\left(\sum_{p \in C}(e_p - 1)\right) - n + 1.$$

Exercise 18.55. A curve C is called a hyperelliptic curve if there exists a degree 2 regular map $\phi : C \to \mathbb{P}^1$. Suppose that k is an algebraically closed field of characteristic 0 and $a_1, \ldots, a_l \in k$ are distinct. Let γ be the affine

18.9. Abelian varieties and Jacobians of curves

curve $\gamma = Z(y^2 - f(x)) \subset \mathbb{A}^2$ where $f(x) = \prod_{i=1}^{l}(x - a_i)$. Let C be the resolution of singularities of the Zariski closure of γ in $\mathbb{P}^2$, and let $\pi : C \to \mathbb{P}^1$ be the regular map which when restricted to γ is the projection of γ onto the x-axis. Show that π is a degree 2 map. Compute the ramification of π and show that C has genus $g = l - 1$.

Exercise 18.56. Suppose that C is a plane curve of degree 4.

  a) Show that the effective canonical divisors on C are the divisors $i^*(L)$ where $i : C \to \mathbb{P}^2$ is inclusion and L is a line on $\mathbb{P}^2$.

  b) Show that the genus of C is 3.

  c) If D is any effective divisor of degree 2 on C, show that $h^0(C, \mathcal{O}_C(D)) = 1$.

  d) Conclude that C is not hyperelliptic.

Exercise 18.57. Suppose that C is not a hyperelliptic curve. Show that K_C is very ample.

Exercise 18.58. Suppose that C is a hyperelliptic curve with degree 2 regular map $\phi : C \to \mathbb{P}^1$ and $p \in C$ is a ramification point. Show that

$$h^0(C, \mathcal{O}_C(mp)) = \begin{cases} i+1 & \text{if } m = 2i, \ 1 \leq i \leq g, \\ i+1 & \text{if } m = 2i+1, \ 1 \leq i \leq g, \\ m+1-g & \text{if } m \geq 2g. \end{cases}$$

Exercise 18.59. Suppose that $g \geq 2$ and $\phi : C \to C$ is a dominant regular map. Show that ϕ is an isomorphism.

Chapter 19

An Introduction to Intersection Theory

We give a treatment of intersection theory, based on the Snapper polynomial ([**139**]). Let V be a d-dimensional projective scheme (over an algebraically closed field k). The results in this chapter are mostly from [**89**, Section 1]. More general intersection theories are presented in [**57**]. We conclude this chapter with some examples and applications.

In Theorem 19.20, we give a proof of Bézout's theorem, showing that two projective plane curves intersect in m points counting multiplicity, where m is the product of the degrees of the two curves. In Theorem 19.21, Corollary 19.22, and Theorem 19.23, we give formulas relating the degrees of projective varieties W and $\pi(W)$ under a projection π.

Suppose that V is a nonsingular projective variety and W is a t-dimensional closed subvariety of V. Associated to divisors $D_1, \ldots, D_t$ on V we have the intersection product $I(D_1 \cdots D_t \cdot W)$ of (19.9). This product satisfies the conditions that

(19.1) \qquad it is multilinear in divisors on V

by Proposition 19.6, and if $D'_1, \ldots, D'_t$ are divisors on V such that

$$D'_1 \sim D_1, \ldots, D'_t \sim D_t,$$

then

(19.2) \qquad $I(D'_1 \cdots D'_t \cdot W) = I(D_1 \cdots D_t \cdot W)$

by the comment after Definition 19.2.

By Theorem 16.1, if $H_1, \ldots, H_t$ are very ample divisors on V, then there exist effective divisors $H_1', \ldots, H_t'$ on V such that $H_1' \sim H_1, \ldots, H_t' \sim H_t$ and the scheme $H_1' \cap \cdots \cap H_t' \cap Y$ is reduced and finite (a zero-dimensional algebraic set). Then if d is the number of points in $H_1' \cap \cdots \cap H_t' \cap Y$, we have by Propositions 19.8 and 19.5 that

(19.3) $$I(H_1' \cdots H_t') = d.$$

Using the three properties (19.1), (19.2), and (19.3), we can compute the intersection product $I(D_1 \cdots D_t \cdot W)$ for any divisors $D_1, \ldots, D_t$ on V, since any divisor on V is the difference of very ample divisors by 2) of Theorem 13.32.

Alternatively, we can use the three properties (19.1), (19.2), and (19.3) to define the intersection product $I(D_1 \cdots D_t \cdot W)$. This is the way that intersections products were first defined. Of course we must then prove that our construction is well-defined. A first step on this is given by Theorem 16.9, which shows that the number of points d in the reduced scheme

$$H_1' \cap \cdots \cap H_t' \cap Y$$

does not depend on the choice of general $H_1', \ldots, H_t'$ which are linearly equivalent to $H_1, \ldots, H_t$.

19.1. Definition, properties, and some examples of intersection numbers

Suppose that V is a projective scheme (over an algebraically closed field k).

Theorem 19.1 (Snapper). *Let $\mathcal{F}$ be a coherent sheaf on V and let $s = \dim \operatorname{Supp} \mathcal{F}$. Let $\mathcal{L}_1, \ldots, \mathcal{L}_t$ be t invertible sheaves on V. Then the Euler characteristic*

$$\chi(\mathcal{F} \otimes \mathcal{L}_1^{n_1} \otimes \cdots \otimes \mathcal{L}_t^{n_t}) = \sum_{i=0}^{\infty}(-1)^i h^i(V, \mathcal{F} \otimes \mathcal{L}_1^{n_1} \otimes \cdots \otimes \mathcal{L}_t^{n_t})$$

is a numerical polynomial in $n_1, \ldots, n_t$ of total degree s.

A numerical polynomial in $n_1, \ldots, n_t$ is a polynomial $f(n_1, \ldots, n_t)$ in $n_1, \ldots, n_t$ with rational coefficients such that $f(\overline{n}_1, \ldots, \overline{n}_t)$ is an integer whenever $\overline{n}_1, \ldots, \overline{n}_t$ are integers.

A coherent sheaf $\mathcal{G}$ of $\mathcal{O}_V$-modules is a torsion sheaf if for all $p \in V$ and $a \in \mathcal{G}_p$, there exists a non-zerodivisor $b \in \mathcal{O}_{V,p}$ such that $ab = 0$.

Proof. We will prove the theorem by induction on s. The theorem is trivial when $s = -1$. Assume $s \geq 0$. We may replace V with $\operatorname{Supp}(\mathcal{F})$, given the subscheme structure defined by $\operatorname{Ann}(\mathcal{F})$. We then have reduced to the case that the theorem holds for torsion sheaves on V.

19.1. Definition, properties, and examples of intersection numbers

Let K' be the set of $(\mathcal{F}, \mathcal{L}_1, \ldots, \mathcal{L}_t)$ on V such that the theorem holds. Since χ is exact on exact sequences of coherent modules, by Grothendieck's theory of dévissage (unscrewing) in [**67**, Section 3.1], we need only prove the theorem for $\mathcal{F} = \mathcal{O}_V$ when V is a projective variety and if we assume that all torsion sheaves $\mathcal{G}$ on V satisfy the conclusions of the theorem. We proceed by induction on t, the case $t = 0$ being trivial. Since V is a projective variety, there exists $a > 0$ such that the sheaf $\mathcal{L}_1 \otimes \mathcal{O}_V(a)$ is generated by global sections by Theorem 11.45. Since $\mathcal{L}_1 \otimes \mathcal{O}_V(a)$ is generated by global sections, there exists $0 \neq \sigma \in \Gamma(V, \mathcal{L}_1 \otimes \mathcal{O}_V(a))$. We thus have a short exact sequence

$$(19.4) \qquad 0 \to \mathcal{O}_V(-a) \otimes \mathcal{L}_1^{-1} \xrightarrow{\sigma} \mathcal{O}_V \to \mathcal{B} \to 0$$

where $\mathcal{B} = \mathcal{O}_V / \sigma \mathcal{O}_V(-a) \otimes \mathcal{L}_1^{-1}$. We can assume that a is sufficiently large that there exists $0 \neq \tau \in \Gamma(V, \mathcal{O}_V(a))$ giving a short exact sequence

$$(19.5) \qquad 0 \to \mathcal{O}_V(-a) \xrightarrow{\tau} \mathcal{O}_V \to \mathcal{C} \to 0$$

where $\mathcal{C} = \mathcal{O}_V / \tau \mathcal{O}_V(-1)$. Tensor (19.4) with $\mathcal{L}_1^{n_1+1} \otimes \mathcal{L}_2^{n_2} \otimes \cdots \otimes \mathcal{L}_t^{n_t}$ and tensor (19.5) with $\mathcal{L}_1^{n_1} \otimes \mathcal{L}_2^{n_2} \otimes \cdots \otimes \mathcal{L}_t^{n_t}$ and take the long exact cohomology sequences to get

$$\chi(\mathcal{L}_1^{n_1+1} \otimes \mathcal{L}_2^{n_2} \otimes \cdots \otimes \mathcal{L}_t^{n_t})$$
$$= \chi(\mathcal{O}_V(-a) \otimes \mathcal{L}_1^{n_1} \otimes \mathcal{L}_2^{n_2} \otimes \cdots \otimes \mathcal{L}_t^{n_t}) + \chi(\mathcal{B} \otimes \mathcal{L}_1^{n_1+1} \otimes \mathcal{L}_2^{n_2} \otimes \cdots \otimes \mathcal{L}_t^{n_t})$$

and

$$\chi(\mathcal{L}_1^{n_1} \otimes \mathcal{L}_2^{n_2} \otimes \cdots \otimes \mathcal{L}_t^{n_t})$$
$$= \chi(\mathcal{O}_V(-a) \otimes \mathcal{L}_1^{n_1} \otimes \mathcal{L}_2^{n_2} \otimes \cdots \otimes \mathcal{L}_t^{n_t}) + \chi(\mathcal{C} \otimes \mathcal{L}_1^{n_1} \otimes \mathcal{L}_2^{n_2} \otimes \cdots \otimes \mathcal{L}_t^{n_t}).$$

Since $\dim \operatorname{Supp} \mathcal{B} < \dim V$ and $\dim \operatorname{Supp} \mathcal{C} < \dim V$, we have that

$$Q(n_1, \ldots, n_t)$$
$$= \chi(\mathcal{L}_1^{n_1+1} \otimes \mathcal{L}_2^{n_2} \otimes \cdots \otimes \mathcal{L}_t^{n_t}) - \chi(\mathcal{L}_1^{n_1} \otimes \mathcal{L}_2^{n_2} \otimes \cdots \otimes \mathcal{L}_t^{n_t})$$
$$= \chi(\mathcal{B} \otimes \mathcal{L}_1^{n_1+1} \otimes \mathcal{L}_2^{n_2} \otimes \cdots \otimes \mathcal{L}_t^{n_t}) - \chi(\mathcal{C} \otimes \mathcal{L}_1^{n_1} \otimes \mathcal{L}_2^{n_2} \otimes \cdots \otimes \mathcal{L}_t^{n_t})$$

is a numerical polynomial in $n_1, \ldots, n_t$ of total degree $< s$. Expand $Q \in \mathbb{Q}[x_1, \ldots, x_t]$ as

$$Q = c_0 \binom{x_1}{r} + c_1 \binom{x_1}{r-1} + \cdots + c_r$$

where $c_i \in \mathbb{Q}[x_2, \ldots, x_t]$ and $r < s$ is the degree of Q in x_1. Set

$$P = c_0 \binom{x_1}{r+1} + \cdots + c_r \binom{x_1}{1}.$$

The first difference
$$\Delta P := P(n_1+1, n_2, \ldots, n_t) - P(n_1, n_2, \ldots, n_t) = Q(n_1, \ldots, n_t)$$
since $\Delta\binom{x_1}{j} = \binom{x_1}{j-1}$. Expand

$$\begin{aligned}
&\chi(\mathcal{L}_1^{n_1} \otimes \mathcal{L}_2^{n_2} \otimes \cdots \otimes \mathcal{L}_t^{n_t}) - P(n_1, \ldots, n_t) \\
&= \sum_{i=0}^{n_1-1} \Delta\left(\chi(\mathcal{L}_1^i \otimes \mathcal{L}_2^{n_2} \otimes \cdots \otimes \mathcal{L}_t^{n_t}) - P(i, n_2, \ldots, n_t)\right) \\
&\quad + \chi(\mathcal{L}_2^{n_2} \otimes \cdots \otimes \mathcal{L}_t^{n_t}) - P(0, n_2, \ldots, n_t) \\
&= \chi(\mathcal{L}_2^{n_2} \otimes \cdots \otimes \mathcal{L}_t^{n_t}) - P(0, n_2, \ldots, n_t).
\end{aligned}$$

Since $\chi(\mathcal{L}_2^{n_2} \otimes \cdots \otimes \mathcal{L}_t^{n_t})$ is a numerical polynomial of total degree s by induction on t, we have that $\chi(\mathcal{L}_1^{n_1} \otimes \mathcal{L}_2^{n_2} \otimes \cdots \otimes \mathcal{L}_t^{n_t})$ is a numerical polynomial of total degree s. □

Definition 19.2. Let $\mathcal{L}_1, \ldots, \mathcal{L}_t$ be t invertible sheaves on V, and let $\mathcal{F}$ be a coherent sheaf on V such that $\dim \operatorname{Supp} \mathcal{F} \leq t$. The intersection number

$$(\mathcal{L}_1 \cdots \mathcal{L}_t; \mathcal{F})_V$$

of $\mathcal{L}_1, \ldots, \mathcal{L}_t$ with $\mathcal{F}$ is the coefficient of the monomial $n_1 \cdots n_t$ in

$$\chi(\mathcal{F} \otimes \mathcal{L}_1^{n_1} \otimes \cdots \otimes \mathcal{L}_t^{n_t}).$$

Observe that $(\mathcal{L}_1 \cdots \mathcal{L}_t; \mathcal{F})_V$ is independent of isomorphism class of $\mathcal{L}_1, \ldots, \mathcal{L}_t$ and $\mathcal{F}$.

Lemma 19.3. *Suppose that $\mathcal{F}$ is a coherent sheaf on V and $\mathcal{L}_1, \ldots, \mathcal{L}_t$ are invertible sheaves on V with $\dim \operatorname{Supp} \mathcal{F} \leq t$. Then*

$$\begin{aligned}
(\mathcal{L}_1 \cdots \mathcal{L}_t; \mathcal{F})_V &= \chi(\mathcal{F}) - \sum_{i=1}^t \chi(\mathcal{F} \otimes \mathcal{L}_i^{-1}) + \sum_{i<j} \chi(\mathcal{F} \otimes \mathcal{L}_i^{-1} \otimes \mathcal{L}_j^{-1}) \\
&\quad - \cdots + (-1)^t \chi(\mathcal{F} \otimes \mathcal{L}_1^{-1} \otimes \cdots \otimes \mathcal{L}_t^{-1}).
\end{aligned}$$

Proof. The polynomial $P(n_1, \ldots, n_t) = \chi(\mathcal{F} \otimes \mathcal{L}_1^{n_1} \otimes \cdots \otimes \mathcal{L}_t^{n_t})$ has an expansion

$$P(n_1, n_2, \ldots, n_t) = \sum_{j_1 + \cdots + j_t \leq t} a_{j_1, \ldots, j_t} n_1^{j_1} \cdots n_t^{j_t}$$

with $a_{j_1, \ldots, j_t} \in \mathbb{Q}$. Now for $i_1 < i_2 < \cdots < i_s \leq t$,

$$\begin{aligned}
&\chi(\mathcal{F} \otimes \mathcal{L}_{i_1}^{-1} \otimes \mathcal{L}_{i_2}^{-1} \cdots \otimes \mathcal{L}_{i_s}^{-1}) \\
&= \sum_{j_1 + \cdots + j_s \leq t} (-1)^{j_1 + \cdots + j_s} a_{0, \ldots, 0, j_1, 0, \ldots, 0, j_2, 0, \ldots, 0, j_s, 0, \ldots, 0},
\end{aligned}$$

from which the conclusion of the lemma follows, using the identity

$$0 = (1-1)^d = \sum_{i=0}^d \binom{d}{i} (-1)^i. \qquad \square$$

Proposition 19.4. *The intersection number $(\mathcal{L}_1 \cdots \mathcal{L}_t; \mathcal{F})_V$ is an integer.*

19.1. Definition, properties, and examples of intersection numbers

Proof. This follows from Lemma 19.3, since an Euler characteristic is an integer. $\square$

Proposition 19.5. *The intersection number*
$$(\mathcal{L}_1 \cdots \mathcal{L}_t; \mathcal{F})_V = \begin{cases} 0 & \text{if } \dim \operatorname{Supp} \mathcal{F} < t, \\ h^0(V, \mathcal{F}) & \text{if } \dim \operatorname{Supp} \mathcal{F} = t = 0. \end{cases}$$

Proof. If $\dim \operatorname{Supp} \mathcal{F} < t$, then $\chi(\mathcal{F} \otimes \mathcal{L}_1^{n_1} \otimes \cdots \otimes \mathcal{L}_t^{n_t})$ is polynomial of degree less than t, and if $\dim \operatorname{Supp} \mathcal{F} = t = 0$, then
$$(\mathcal{F})_V = \chi(\mathcal{F}) = h^0(\mathcal{F}). \qquad \square$$

Suppose that $\mathcal{F}$ is a coherent sheaf on V whose support has dimension 0. Let $\operatorname{Supp}(\mathcal{F}) = \{Q_1, \ldots, Q_m\}$. Then the intersection multiplicity

(19.6) $$(\mathcal{F})_V = h^0(V, \mathcal{F}) = \sum_{i=1}^m \dim_k \mathcal{F}_{Q_i}.$$

Proposition 19.6. *The intersection number $(\mathcal{L}_1 \cdots \mathcal{L}_t; \mathcal{F})_V$ is a symmetric t-linear form in $\mathcal{L}_1, \ldots, \mathcal{L}_t$.*

Proof. Let $\mathcal{M}$ and $\mathcal{N}$ be invertible sheaves on V. Taking successively $n = 0$ and $m = 0$, we have that
$$\chi(\mathcal{F} \otimes \mathcal{M}^m \otimes (\mathcal{N}^{-1})^n \otimes \mathcal{L}_2^{n_2} \otimes \cdots \otimes \mathcal{L}_t^{n_t})$$
$$= (\mathcal{M} \cdot \mathcal{L}_2 \cdots \mathcal{L}_t \cdot \mathcal{F})_V m n_2 \cdots n_t - (\mathcal{N} \cdot \mathcal{L}_2 \cdots \mathcal{L}_t \cdot \mathcal{F})_V n n_2 \cdots n_t + \cdots.$$
Now taking $m = n = n_1$, we obtain
$$\chi(\mathcal{F} \otimes (\mathcal{M} \otimes \mathcal{N}^{-1})^{n_1} \otimes \mathcal{L}_2^{n_2} \otimes \cdots \otimes \mathcal{L}_t^{n_t})$$
$$= ((\mathcal{M} \cdot \mathcal{L}_2 \cdots \mathcal{L}_t \cdot \mathcal{F})_V - (\mathcal{N} \cdot \mathcal{L}_2 \cdots \mathcal{L}_t \cdot \mathcal{F})_V) n_1 n_2 \cdots n_t + \cdots,$$
establishing linearity. $\square$

Proposition 19.7. *If*
$$0 \to \mathcal{F}' \to \mathcal{F} \to \mathcal{F}'' \to 0$$
is a short exact sequence of coherent $\mathcal{O}_V$-modules, then
$$(\mathcal{L}_1 \cdots \mathcal{L}_t; \mathcal{F})_V = (\mathcal{L}_1 \cdots \mathcal{L}_t; \mathcal{F}')_V + (\mathcal{L}_1 \cdots \mathcal{L}_t; \mathcal{F}'')_V.$$

Proof. This follows from the additivity of the Euler characteristic on short exact sequences. $\square$

Let W be a closed subscheme of V of dimension less than or equal to t. Then we define
$$(\mathcal{L}_1 \cdots \mathcal{L}_t \cdot W) = (\mathcal{L}_1 \cdots \mathcal{L}_t; \mathcal{O}_W)_V.$$
We define
$$(\mathcal{L}_1 \cdots \mathcal{L}_t) = (\mathcal{L}_1 \cdots \mathcal{L}_t \cdot V).$$

In Proposition 19.8, we use notation introduced in Section 15.1.

Proposition 19.8. *Suppose that $s \in \Gamma(V, \mathcal{L}_1)$ is a nonzero divisor and also that s is not a zero divisor on $\mathcal{F}$. Then*

$$(\mathcal{L}_1 \cdots \mathcal{L}_t; \mathcal{F})_V = (\mathcal{L}_2 \cdots \mathcal{L}_t; \mathcal{F} \otimes \mathcal{O}_\Delta)_V$$

where $\Delta = \mathrm{div}(s)$ and $\mathcal{O}_\Delta = \mathcal{O}_V/\mathcal{I}_{\mathrm{div}(s)}$. In particular, if $\dim V \leq t$, then

$$(\mathcal{L}_1 \cdots \mathcal{L}_t) = (\mathcal{L}_2 \cdots \mathcal{L}_t \cdot \Delta).$$

Proof. With our assumptions on s, we have short exact sequences

$$0 \to \mathcal{F} \otimes \mathcal{L}_1^{n_1-1} \otimes \mathcal{L}_2^{n_2} \otimes \cdots \otimes \mathcal{L}_t^{n_t} \xrightarrow{s} \mathcal{F} \otimes \mathcal{L}_1^{n_1} \otimes \cdots \otimes \mathcal{L}_t^{n_t} \to (\mathcal{F} \otimes \mathcal{O}_\Delta) \otimes \mathcal{L}_1^{n_1} \otimes \cdots \otimes \mathcal{L}_t^{n_t} \to 0.$$

Hence

$$\begin{aligned}
\chi(\mathcal{F} \otimes \mathcal{O}_\Delta &\otimes \mathcal{L}_1^{n_1} \otimes \cdots \otimes \mathcal{L}_t^{n_t}) \\
&= \chi(\mathcal{F} \otimes \mathcal{L}_1^{n_1} \otimes \cdots \otimes \mathcal{L}_t^{n_t}) - \chi(\mathcal{F} \otimes \mathcal{L}_1^{n_1-1} \otimes \cdots \otimes \mathcal{L}_t^{n_t}) \\
&= (\mathcal{L}_1 \cdots \mathcal{L}_t \cdot \mathcal{F})_V n_1 n_2 \cdots n_t \\
&\quad + \cdots - (\mathcal{L}_1 \cdots \mathcal{L}_t \cdot \mathcal{F})_V (n_1 - 1) n_2 \cdots n_t + \cdots \\
&= (\mathcal{L}_1 \cdots \mathcal{L}_t \cdot \mathcal{F})_V n_2 \cdots n_t + \cdots.
\end{aligned}$$

Further, taking $n_1 = 0$, we have that

$$\chi((\mathcal{F} \otimes \mathcal{O}_\Delta) \otimes \mathcal{L}_2^{n_2} \otimes \cdots \otimes \mathcal{L}_t^{n_t}) = (\mathcal{L}_2 \cdots \mathcal{L}_t \cdot \mathcal{F}_\Delta)_V n_2 \cdots n_t + \cdots. \quad \square$$

Proposition 19.9. *Suppose that W is a closed subscheme of V which contains the subscheme $X = \mathrm{Supp}(\mathcal{F})$, where X is provided with the subscheme structure defined by the annihilator of $\mathcal{F}$; that is, there is a natural surjection $\mathcal{O}_W \to \mathcal{O}_X = \mathcal{O}_V/\mathrm{Ann}(\mathcal{F})$. Then $\mathcal{F}$ may be considered as an $\mathcal{O}_W$-module, and*

$$(\mathcal{L}_1 \cdots \mathcal{L}_t; \mathcal{F})_V = (\mathcal{L}_1 \otimes \mathcal{O}_W \cdots \mathcal{L}_t \otimes \mathcal{O}_W; \mathcal{F})_W.$$

In particular,

$$(\mathcal{L}_1 \cdots \mathcal{L}_t \cdot W) = (\mathcal{L}_1 \otimes \mathcal{O}_W \cdots \mathcal{L}_t \otimes \mathcal{O}_W).$$

Proof. If $\mathcal{L}$ is any invertible sheaf on V, then

$$\mathcal{F} \otimes \mathcal{L} \cong \mathcal{F} \otimes \mathcal{O}_W \otimes \mathcal{L}.$$

Now set $\mathcal{L} = \mathcal{L}_1^{n_1} \otimes \cdots \otimes \mathcal{L}_t^{n_t}$, and take Euler characteristics to obtain the conclusions of the proposition. $\square$

Corollary 19.10. *Suppose that $V = \mathrm{Supp}(\mathcal{F})$. Let $V_1, \ldots, V_s$ be the irreducible components of V (which might not be reduced), and let $\mathcal{F}_i = \mathcal{F} \otimes \mathcal{O}_{V_i}$ for $1 \leq i \leq s$. Then*

$$(\mathcal{L}_1 \cdots \mathcal{L}_t; \mathcal{F})_V = (\mathcal{L}_1 \cdots \mathcal{L}_t; \mathcal{F}_1)_V + \cdots + (\mathcal{L}_1 \cdots \mathcal{L}_t; \mathcal{F}_s)_V.$$

Proof. The canonical homomorphisms $\mathcal{O}_V \to \mathcal{O}_{V_j}$ for $j = 1, \ldots, s$ give an exact sequence of coherent $\mathcal{O}_V$-modules

$$0 \to \mathcal{A} \to \mathcal{F} \to \bigoplus_j \mathcal{F}_j \to \mathcal{B} \to 0$$

with Supp $\mathcal{A}$, Supp $\mathcal{B} \subset \bigcup_{i \neq j}(V_i \cap V_j)$. The corollary now follows from Propositions 19.5 and 19.7. □

If V is a scheme and W is a subvariety, then we compute the local ring $\mathcal{O}_{V,W}$ as in Definition 12.17. If U is an open affine subset of V which intersects W, then $\mathcal{O}_{V,W} = \Gamma(U, \mathcal{O}_V)_P$ where P is the prime ideal of W in $\Gamma(U, \mathcal{O}_V)$. If $\mathcal{F}$ is a coherent sheaf on V and $\mathcal{F}|U = \tilde{M}$ for some $\Gamma(U, \mathcal{O}_V)$-module M, then we define $\mathcal{F}_W = M_P$.

Corollary 19.11. *Suppose that V is irreducible and $\dim V \leq t$. Let $\ell = \ell_{\mathcal{O}_{V,V_{\text{red}}}}(\mathcal{F}_{V_{\text{red}}})$. Then*

$$(\mathcal{L}_1 \cdots \mathcal{L}_t; \mathcal{F})_V = \ell(\mathcal{L}_1 \cdots \mathcal{L}_t; \mathcal{O}_{V_{\text{red}}}).$$

Proof. We have that $\mathcal{O}_{V,V_{\text{red}}}$ is an Artin local ring and $\mathcal{F}_{V_{\text{red}}}$ is a finite $\mathcal{O}_{V,V_{\text{red}}}$-module. Let $\mathcal{K}'$ be the set of $\mathcal{F}$ for which the corollary is true. The corollary is certainly true when $\mathcal{F} = \mathcal{O}_{V_{\text{red}}}$ and when $\dim \text{Supp } \mathcal{F} < t$. The conclusions of the corollary now follow from dévissage ([67, Section 3.1]). □

We define the degree of a regular map $\phi : V \to V'$ of varieties (Definition 21.25) as follows:

$$\deg(\phi) = \begin{cases} [k(V') : k(V)] & \text{if } \dim V = \dim V' \text{ and } \phi \text{ is dominant,} \\ 0 & \text{otherwise.} \end{cases}$$

Proposition 19.12. *Let $\phi : V' \to V$ be a regular map of projective varieties and assume that $t \geq \max\{\dim V, \dim V'\}$. Let $\mathcal{L}_1, \ldots, \mathcal{L}_t$ be invertible sheaves on V and set $\mathcal{L}'_i = \phi^* \mathcal{L}_i$ for $1 \leq i \leq t$. Then*

$$(\mathcal{L}'_1 \cdots \mathcal{L}'_t) = \deg(\phi) (\mathcal{L}_1 \cdots \mathcal{L}_t).$$

Proof. [89, Proposition 6]. □

If $D_1, D_2, \ldots, D_n$ are divisors on an n-dimensional nonsingular projective variety V, then we define

(19.7) $\quad (D_1 \cdot D_2 \cdots D_n) = (\mathcal{O}_V(D_1) \cdot \mathcal{O}_V(D_2) \cdots \mathcal{O}_V(D_n)).$

Observe that if $D'_i \sim D_i$ for $1 \leq i \leq n$, then

(19.8) $\quad (D'_1 \cdot D'_2 \cdots D'_n) = (D_1 \cdot D_2 \cdots D_n).$

Suppose that W is a t-dimensional subvariety of V. Then we define
$$(19.9) \qquad (D_1 \cdots D_t \cdot W) = (\mathcal{O}_V(D_1) \cdots \mathcal{O}_V(D_t) \cdot \mathcal{O}_W).$$

Example 19.13. Suppose that $\mathcal{L} \cong \mathcal{O}_X(D)$ is an invertible sheaf on a nonsingular projective curve X. Then the intersection multiplicity
$$(\mathcal{L}) = (\mathcal{L} \cdot X) = \deg(D).$$

If D is effective, this follows since
$$(\mathcal{L}) = (\mathcal{O}_X(D); \mathcal{O}_X)_X = (\mathcal{O}_D)_X = h^0(X, \mathcal{O}_D)$$
by Propositions 19.8 and 19.5. For general D, write $D = D_1 - D_2$ where D_1 and D_2 are effective. Then
$$(\mathcal{O}_X(D)) = (\mathcal{O}_X(D_1 - D_2)) = (\mathcal{O}_X(D_1)) - (\mathcal{O}_X(D_2))$$
$$= \deg(D_1) - \deg(D_2) = \deg(D).$$
Alternatively, $(\mathcal{L} \cdot X)$ is the linear term of
$$\chi(\mathcal{O}_X(nD)) = n \deg(D) + 1 - g(X)$$
by the Riemann-Roch theorem.

Example 19.14. Suppose that C is a nonsingular curve on a nonsingular projective surface X and D is a divisor on X. Then by Propositions 19.8 and 19.9 and Example 19.13,
$$(D \cdot C) = (\mathcal{O}_X(D) \otimes \mathcal{O}_C) = \deg(\mathcal{O}_X(D) \otimes \mathcal{O}_C).$$

We will denote the self-intersection number $(\mathcal{L} \cdots \mathcal{L})$ of an invertible sheaf $\mathcal{L}$ on a d-dimensional scheme V, d times with itself by $(\mathcal{L}^d)$.

Example 19.15. The intersection product
$$(\mathcal{O}_{\mathbb{P}^d}(1)^d) = 1$$
for $d \geq 1$.

We prove this formula by induction on d. When $d = 1$, we have
$$(\mathcal{O}_{\mathbb{P}^1}(1)) = \deg \mathcal{O}_{\mathbb{P}^1}(1) = 1$$
by Example 19.13. If $d > 1$, by Propositions 19.8 and 19.9 we have
$$(\mathcal{O}_{\mathbb{P}^d}(1)^d) = (\mathcal{O}_{\mathbb{P}^d}(1)^{d-1} \cdot \mathcal{O}_{\mathbb{P}^d}(1)) = (\mathcal{O}_{\mathbb{P}^d}(1)^{d-1} \cdot L) = (\mathcal{O}_{\mathbb{P}^{d-1}}(1)^{d-1})$$
where $L \cong \mathbb{P}^{d-1}$ is a hyperplane section of $\mathbb{P}^d$.

Theorem 19.16. *Let Z be a projective scheme of dimension d and let $\mathcal{L}$ be an invertible sheaf on Z. Then*
$$\chi(\mathcal{L}^n) = \frac{(\mathcal{L}^d)}{d!} n^d + Q(n)$$
where $Q(n)$ is a polynomial with rational coefficients of degree $\leq d - 1$.

19.1. Definition, properties, and examples of intersection numbers

Proof. There exists $b \in \mathbb{Q}$ and a polynomial $Q(n)$ with rational coefficients of degree $\leq d-1$ such that

(19.10) $$\chi(\mathcal{L}^n) = bn^d + Q(n)$$

by Theorem 19.1. We have

(19.11)
$$(-1)^d(\mathcal{L}^d)n^d = (\mathcal{L}^{-n} \cdots \mathcal{L}^{-n})$$
$$= \chi(\mathcal{O}_Z) - d\chi(\mathcal{L}^n) + \binom{d}{2}\chi(\mathcal{L}^{2n}) - \cdots + (-1)^d\chi(\mathcal{L}^{dn})$$

by multilinearity of the intersection product, Proposition 19.6, and Lemma 19.3. Substituting (19.10) into (19.11), we obtain

$$(-1)^d(\mathcal{L}^d)n^d = \left(-d + \binom{d}{2}2^d - \cdots + (-1)^d d^d\right)bn^d + \sum_{i=0}^{d}(-1)^i\binom{d}{i}Q(in).$$

Thus

(19.12) $$b = \left(-d + \binom{d}{2}2^d - \cdots + (-1)^d d^d\right)^{-1}(-1)^d(\mathcal{L}^d).$$

In the case that $Z = \mathbb{P}^d$ and $\mathcal{L} = \mathcal{O}_{\mathbb{P}^d}(1)$, we have that $(\mathcal{O}_{\mathbb{P}^d}(1)^d) = 1$ by Example 19.15 and

$$\chi(\mathcal{L}^n) = \binom{d+n-1}{n-1} = \frac{n^d}{d!} + \text{lower-order terms},$$

so that $b = \frac{1}{d!}$ in this case. We then see from (19.12) that

$$\left(-d + \binom{d}{2}2^d + \cdots + (-1)^d d^d\right)^{-1}(-1)^d = \frac{1}{d!},$$

so that for general $\mathcal{L}$,

$$b = \frac{(\mathcal{L}^d)}{d!}$$

in (19.10). $\square$

Suppose that $\phi: X \to Y$ is a regular map of normal varieties such that $\phi = \phi_{|D|}$ where D is a divisor on X such that $|D|$ is base point free. We have that $D = \phi^*(H)$ where H is a hyperplane section of Y. Since the linear system $|D|$ has no base points, we have that $(C \cdot D) \geq 0$ for all closed curves C on X. An invertible sheaf $\mathcal{L}$ such that $(\mathcal{L} \cdot C) \geq 0$ for all closed curves C on X is called numerically effective (nef). For a closed curve C on X, we have that $\phi(C)$ is contracted to a point on Y if and only if $(C \cdot D) = 0$. A nef divisor F does not always have the property that some positive multiple mF gives a base point free linear system $|mF|$. We will give an example in Theorem 20.14 (and Exercise 20.17). More examples and explanations of this can be found in [**159**], [**44**], [**98**].

Theorem 19.17. *Suppose that Z is an arbitrary closed subscheme of $\mathbb{P}^n$ of dimension r. Then*

(19.13) $$\deg(Z) = (\mathcal{O}_{\mathbb{P}^n}(1)^r \cdot Z) = (\mathcal{O}_Z(1)^r).$$

Proof. The quickest proof of the theorem follows from Corollary 17.21, which tells us that $\chi(\mathcal{O}_Z(n))$ is the Hilbert polynomial $P_Z(n)$ of Z, and from Theorem 19.16.

We also give an alternate, more conceptual, proof. Let $I \subset S = S(\mathbb{P}^n)$ be the saturated homogeneous ideal such that $\mathcal{I}_Z = \tilde{I}$. By prime avoidance and since k is infinite ([**28**, Lemma 1.5.10–Proposition 1.5.12]), there exists a linear form $F \in S_1$ such that F_1 is not contained in any associated prime ideals of I. Thus the sequence

$$0 \to (S/I)(-1) \xrightarrow{F_1} S/I \to S/(I+(F_1)) \to 0$$

is short exact. Let $L_1 = Z(F_1) \subset \mathbb{P}^n$. Comparing Hilbert functions of S/I and $S/(I+(F_1))$ as in the proof of Theorem 16.9, we have that

$$P_{S/I}(n) - P_{S/I}(n-1) = P_{S/(I+(F_1))}(n)$$

so that $\deg Z = \deg Z \cap L_1$. With our assumptions, we have that $F_1 \in \Gamma(\mathbb{P}^n, \mathcal{O}_{\mathbb{P}^n}(1))$ is not a zero divisor on $\mathcal{O}_Z$. Thus $(\mathcal{O}_{\mathbb{P}^n}(1)^r \cdot Z) = (\mathcal{O}_Z(1)^{r-1} \cdot Z \cap L_1)$ by Propositions 19.8 and 19.9. By induction, we can choose $F_1, \ldots, F_r \in S$ with $r = \dim Z$ such that if $L_i = Z(F_i)$, we have that

$$\deg Z = \deg Z \cap L_1 \cap \cdots \cap L_r$$

and

$$(\mathcal{O}_{\mathbb{P}^n}(1)^r \cdot Z) = (\mathcal{O}_{Z \cap L_1 \cap \cdots \cap L_r})_{\mathbb{P}^r}.$$

The Hilbert polynomial of the zero-dimensional scheme $Z \cap L_1 \cap \cdots \cap L_r$ is the constant dimension

$$\dim_k (S/I + (F_1) + \cdots + (F_r))_m \quad \text{for } m \gg 0,$$

which is

$$h^0(\mathbb{P}^n, \mathcal{O}_{Z \cap L_1 \cap \cdots \cap L_r}(m)) = h^0(\mathbb{P}^n, \mathcal{O}_{Z \cap L_1 \cap \cdots \cap L_r})$$

by Theorem 11.47, which is valid for arbitrary projective schemes. Since $Z \cap L_1 \cap \cdots \cap L_r$ is a zero-dimensional scheme, we also have that

$$(\mathcal{O}_{Z \cap L_1 \cap \cdots \cap L_r})_{\mathbb{P}^r} = h^0(\mathbb{P}^n, \mathcal{O}_{Z \cap L_1 \cap \cdots \cap L_r})$$

by Proposition 19.5, from which the theorem follows. $\square$

Corollary 19.18. *Suppose that X is a nonsingular projective variety of dimension d, H is an ample divisor on X, and D is a nonzero effective divisor on X. Then $(H^{d-1} \cdot D) > 0$.*

Proof. There exists a positive integer n_0 such that $n_0 H$ is very ample on X, and so there is a closed embedding $X \subset \mathbb{P}^r$ such that $\mathcal{O}_{\mathbb{P}^r}(1) \otimes \mathcal{O}_X \cong \mathcal{O}_X(n_0 H)$. Let $A = n_0 H$ and suppose that E is a codimension 1 subvariety of X. Then

$$\begin{aligned}
(A^{d-1} \cdot E) &= (\mathcal{O}_X(A)^{d-1} \cdot \mathcal{O}_X(E)) \\
&= (\mathcal{O}_X(A)^{d-1} \cdot \mathcal{O}_E) \quad \text{by Proposition 19.8} \\
&= (\mathcal{O}_{\mathbb{P}^r}(1)^{d-1} \cdot \mathcal{O}_E) \quad \text{by Proposition 19.9} \\
&= (\mathcal{O}_{\mathbb{P}^r}(1)^{d-1} \cdot E) \\
&= \deg(E) \quad \text{by Theorem 19.17.}
\end{aligned}$$

Writing $D = \sum a_i E_i$ with E_i prime divisors on X and $a_i > 0$, we have

$$\begin{aligned}
(H^{d-1} \cdot D) &= \tfrac{1}{n_0^{d-1}}(A^{d-1} \cdot D) \\
&= \tfrac{1}{n_0^{d-1}}\left(\sum a_i \deg(E_i)\right) > 0
\end{aligned}$$

by Proposition 19.6. $\square$

Remark 19.19. Suppose that X is a nonsingular projective variety of dimension d and $D_1, \cdots, D_d$ are divisors on X with $D_d \sim 0$. Then $(D_1 \cdot \cdots \cdot D_d) = 0$.

The remark follows since

$$\chi(\mathcal{O}_X(D_1)^{n_1} \otimes \cdots \otimes \mathcal{O}_X(D_d)^{n_d}) = \chi(\mathcal{O}_X(n_1 D_1 + \cdots + n_{d-1} D_{d-1}))$$

does not depend on n_d, so

$$(\mathcal{O}_X(D_1) \cdot \cdots \cdot \mathcal{O}_X(D_d)) = (\mathcal{O}_X(D_1) \cdot \cdots \cdot \mathcal{O}_X(D_d); \mathcal{O}_X)_X = 0$$

by Definition 19.2 of the intersection product.

19.2. Applications to degree and multiplicity

Theorem 19.20 (Bézout's theorem). *Let Y and Z be distinct irreducible, closed curves in $\mathbb{P}^2$, having degrees d and e, respectively. Let the intersection points of Y and Z be $\{Q_1, \ldots, Q_s\}$. Then*

$$\sum_{i=1}^{s} \dim_k \mathcal{O}_{Y \cap Z, Q_i} = de$$

where $Y \cap Z$ is the scheme-theoretic intersection. In particular, if Q_i are nonsingular points of both Y and Z and if Y and Z have distinct tangent spaces at these points (Y and Z intersect transversally), then $s = de$.

Proof. We have that

$$de = de(\mathcal{O}_{\mathbb{P}^2}(1) \cdot \mathcal{O}_{\mathbb{P}^2}(1)) = (\mathcal{O}_{\mathbb{P}^2}(d) \cdot \mathcal{O}_{\mathbb{P}^2}(e))$$

by Example 19.15 and Proposition 19.6. We also calculate

$$\begin{aligned}(\mathcal{O}_{\mathbb{P}^2}(d) \cdot \mathcal{O}_{\mathbb{P}^2}(e)) &= (\mathcal{O}_{\mathbb{P}^2}(Y) \cdot Z) = (\mathcal{O}_{Y \cap Z})_{\mathbb{P}^2} \\ &= h^0(\mathbb{P}^2, \mathcal{O}_{Y \cap Z}) = \sum_{i=1}^{s} \dim_k \mathcal{O}_{Y \cap Z, Q_i}\end{aligned}$$

by Propositions 19.8, 19.9, and 19.5. □

Theorem 19.21. *Suppose that W is an m-dimensional closed subvariety of $\mathbb{P}^n$. Let $P \in \mathbb{P}^n$ be a point not in W and let $\pi : \mathbb{P}^n \dashrightarrow \mathbb{P}^{n-1}$ be projection from the point P. Let $W_1 = \pi(W)$ and let $\phi : W \to W_1$ be the induced regular map. Suppose that $\dim W_1 = \dim W$. Then*

$$\deg(W_1) = \deg(\phi)\deg(W).$$

Proof. Since $\phi^*\mathcal{O}_{W_1}(1) \cong \mathcal{O}_W(1)$, we have by Proposition 19.12 and Theorem 19.17 that

$$\begin{aligned}\deg(W_1) &= (\mathcal{O}_{\mathbb{P}^{n-1}}(1)^m \cdot W_1) = (\mathcal{O}_{W_1}(1)^m) = \deg(\phi)(\mathcal{O}_W(1)^m) \\ &= \deg(\phi)(\mathcal{O}_{\mathbb{P}^n}(1)^m \cdot W) = \deg(\phi)\deg(W).\end{aligned}$$ □

Corollary 19.22. *Suppose that W is a closed subvariety of $\mathbb{P}^n$ of dimension m and L is a linear subspace of $\mathbb{P}^n$ of dimension $n-m-1$ such that $W \cap L = \emptyset$ and the projection W_1 of W from L to $\mathbb{P}^{m+1}$ is birational. Then $\deg(W_1) = \deg(W)$. In particular, the degree of the homogeneous form defining W_1 in $\mathbb{P}^{m+1}$ is degree $d = \deg W$.*

Suppose that R is a d-dimensional local ring with maximal ideal $\mathfrak{m}$. Then the length $\ell_R(R/\mathfrak{m}^{t+1})$ is a polynomial in t of degree d for $t \gg 0$, called the Hilbert-Samuel polynomial. The leading coefficient times $d!$ is an integer, called the mutiplicity $e(R)$ of R. This theory is explained in [**161**, Chapter VIII].

Theorem 19.23. *Suppose that W is a projective variety of dimension d which is a closed subvariety of $\mathbb{P}^n$. Suppose that $P \in W$. Let $\pi : \mathbb{P}^n \dashrightarrow \mathbb{P}^{n-1}$ be the rational map which is the projection from the point P. Let W_1 be the projective subvariety of $\mathbb{P}^{n-1}$ which is the closure of $\pi(W \setminus \{P\})$. Let μ be the multiplicity of the local ring $\mathcal{O}_{W,P}$. Then:*

1. *$\mu \leq \deg(W)$.*

2. *Suppose that $\mu < \deg(W)$. Then $\dim W = \dim W_1$ and*

(19.14) $$[k(W) : k(W_1)]\deg(W_1) = \deg(W) - \mu.$$

3. *Suppose that $\mu = \deg(W)$. Then $\dim W > \dim W_1$ and W is a cone over W_1.*

Proof. Let $\sigma : Z \to \mathbb{P}^n$ be the blow-up of P with exceptional divisor $E = Z_P$. Let $\lambda : Z \to \mathbb{P}^{n-1}$ be the induced regular map (a resolution of

19.2. Applications to degree and multiplicity

indeterminacy of π). Let H_0 be a hyperplane of $\mathbb{P}^n$ and H_1 be a hyperplane of $\mathbb{P}^{n-1}$. We have a linear equivalence of divisors

(19.15) $$\sigma^*(H_0) - E \sim \lambda^*(H_1).$$

Let $\overline{W}$ be the strict transform of W on Z. By Proposition 19.6, we have that

(19.16) $$(\lambda^*(H_1)^d \cdot \overline{W}) = (\sigma^*(H_0)^d \cdot \overline{W}) + ((-E)^d \cdot \overline{W}).$$

Since $\deg(\sigma|\overline{W}) = 1$, by Proposition 19.12 and Theorem 19.17, we have that

$$(\sigma^*(H_0)^d \cdot \overline{W}) = (H_0^d \cdot W) = \deg(W).$$

Let $y_0, \ldots, y_n$ be homogeneous coordinates on $\mathbb{P}^n$. After a linear change of variables, we may assume that $P = (0, \ldots, 0) \in U = \mathbb{A}^n = \mathbb{P}^n \setminus Z(y_0)$. Let $Z' = \sigma^{-1}(U)$, $W' = W \cap U$, and $\overline{W}' = \overline{W} \cap \sigma^{-1}(U)$. The variables $y_1, \ldots, y_n$ are homogeneous coordinates on $\mathbb{P}^{n-1}$ and $\pi = (y_1 : \cdots : y_n)$. Let m_P be the maximal ideal of P in $k[U]$.

We have an embedding of Z' in $U \times \mathbb{P}^{n-1}$. The variety $U \times \mathbb{P}^{n-1}$ has the homogeneous coordinate ring $k[U][y_1, \ldots, y_n]$. The fiber $Z'_P = E \cong \mathbb{P}^{n-1}$ has the homogeneous ideal $I_E = m_P k[U][y_1, \ldots, y_n]$ and coordinate ring $S(E) = k(P)[y_1, \ldots, y_n]$.

Let $\overline{m}_P = m_P k[W']$. Then the coordinate ring of $\overline{W}'$ is the coordinate ring of the blow-up of $\overline{m}_P$, which is $S(\overline{W}') = \bigoplus_{t \geq 0} \overline{m}_P^t$ by Theorem 6.4 and Proposition 6.6. The homogeneous ideal of $\overline{W}'_P = E \cap \overline{W}'$ is $\overline{m}_P S(\overline{W}')$, so the homogeneous coordinate ring of $E \cap \overline{W}' = E \cap \overline{W}$ is

$$S(E \cap \overline{W}) = \bigoplus_{t \geq 0} \overline{m}_P^t / \overline{m}_P^{t+1}.$$

The embedding of $E \cap \overline{W}$ in E is realized by the natural surjection

$$k(P)[y_1, \ldots, y_n] \to \bigoplus_{t \geq 0} \overline{m}_P^t / \overline{m}_P^{t+1}$$

which maps the y_i to the corresponding generators of $\overline{m}_P$.

Under this embedding, we have that $\deg(\overline{W} \cap E)$ is $(d-1)!$ times the leading coefficient of the Hilbert polynomial of $S(E \cap \overline{W})$, since $\dim \overline{W} \cap E = \dim \overline{W}_P = d - 1$.

Let $R = \mathcal{O}_{W,P}$ with maximal ideal m_R. Since
$$\ell(R/m_R^{t+1}) = \sum_{s=0}^{t} \dim_k \overline{m}_P^s/\overline{m}_P^{s+1},$$
we have that $\deg(\overline{W} \cap E) = e(R)$.

As in Example 13.38, we calculate that $\mathcal{O}_E \otimes \mathcal{O}_Z(-E) \cong \mathcal{O}_{\mathbb{P}^{n-1}}(1)$. Thus by Proposition 19.8 and (19.13),
$$((-E)^d \cdot \overline{W}) = -((-E)^{d-1} \cdot (\overline{W} \cap E)) = -\deg(\overline{W} \cap E) = -e(R) = -\mu.$$

We can thus rewrite (19.16) as
$$(19.17) \qquad (\lambda^*(H_1)^d \cdot \overline{W}) = \deg(W) - \mu.$$

By Proposition 19.12, we have that
$$(\lambda^*(H_1)^d \cdot \overline{W}) = \begin{cases} [k(W):k(W_1)]\deg(W_1) & \text{if } \dim W = \dim W_1, \\ 0 & \text{if } \dim W > \dim W_1. \end{cases}$$

Substituting into (19.17), we conclude that $\mu \leq \deg(W)$ and $\mu = \deg(W)$ if and only if W_1 has dimension $< d$, which holds if and only if W is a cone with vertex P. If $\mu < \deg(W)$, then we obtain (19.14). $\square$

Exercise 19.24. Suppose that X is an r-dimensional variety of degree 2 in $\mathbb{P}^n$. Show that X is birationally equivalent to $\mathbb{P}^r$. Hint: We know that X is a quadric hypersurface in a linear subvariety $L \cong \mathbb{P}^{r+1}$ of $\mathbb{P}^n$ by Exercise 16.17. Let $p \in X$ be a nonsingular point, and consider the projection of L from p to $\mathbb{P}^r$ and its effect on X. Is X necessarily isomorphic to $\mathbb{P}^r$?

Chapter 20

Surfaces

In this chapter we will derive some basic properties of nonsingular projective surfaces. For further reading on this topic, the books [**18**], [**16**], and [**118**] are recommended.

20.1. The Riemann-Roch theorem and the Hodge index theorem on a surface

Associated to divisors D_1 and D_2 on a nonsingular projective surface S, we have the intersection product $(D_1 \cdot D_2)$ defined in (19.7). The product $(D_1 \cdot D_2)$ is symmetric and bilinear by Proposition 19.6. If $D_1' \sim D_1$ and $D_2' \sim D_2$, then $(D_1' \cdot D_2') = (D_1 \cdot D_2)$ by (19.8). If $D_1 = \sum a_i E_i$ and $D_2 = \sum b_j E_j$, where E_i are prime divisors, then

$$(D_1 \cdot D_2) = \sum_{i,j} a_i b_j (E_i \cdot E_j).$$

Thus the computation of $(D_1 \cdot D_2)$ reduces to the case when D_1 and D_2 are prime divisors. In the case when D_1, D_2 are distinct prime divisors, the intersection product has a nice interpretation, analogous to Bézout's theorem on $\mathbb{P}^2$.

Lemma 20.1. *Suppose that D_1, D_2 are distinct prime divisors on the non-singular projective surface S. Then*

$$(D_1 \cdot D_2) = \sum_{p \in D_1 \cap D_2} \dim_k(\mathcal{O}_{D_1 \cap D_2, p})$$

where $D_1 \cap D_2$ is the scheme-theoretic intersection.

Proof. By Propositions 19.8 and 19.5, we have that

$$\begin{aligned}(D_1 \cdot D_2) &= (\mathcal{O}_S(D_1); \mathcal{O}_{D_2})_S = (\mathcal{O}_{D_1 \cap D_2})_S \\ &= h^0(S, \mathcal{O}_{D_1 \cap D_2}) \\ &= \sum_{p \in D_1 \cap D_2} \dim_k(\mathcal{O}_{D_1 \cap D_2, p}).\end{aligned}$$
□

Theorem 20.2 (Adjunction). *Suppose that C is a nonsingular projective curve of genus g on a nonsingular projective surface X with canonical divisor K_X. Then*

$$(C \cdot (C + K_X)) = 2g - 2.$$

Proof. We have

$$(C \cdot (C + K_X)) = \deg(\mathcal{O}_X(C + K_X) \otimes \mathcal{O}_C) = \deg \mathcal{O}_C(K_C) = 2g - 2$$

by Theorem 14.21, Corollary 18.14, and Example 19.14. □

The following theorem is the Riemann-Roch theorem on a a surface. It is often used with Serre duality (Theorem 17.22) which implies that

$$h^2(X, \mathcal{O}_X(D)) = h^0(X, \mathcal{O}_X(K_X - D)).$$

Theorem 20.3. *Suppose that X is a nonsingular projective surface and D is a divisor on X. Then*

$$\begin{aligned}\chi(\mathcal{O}_X(D)) &= h^0(X, \mathcal{O}_X(D)) - h^1(X, \mathcal{O}_X(D)) + h^2(X, \mathcal{O}_X(D)) \\ &= \tfrac{1}{2}(D \cdot (D - K_X)) + \chi(\mathcal{O}_X)\end{aligned}$$

where K_X is a canonical divisor on X.

Proof. We have that $h^i(X, \mathcal{O}_X(D)) = 0$ for $i > 2$ by Theorem 17.5.

Write $D = D_2 - D_1$ where D_1 and D_2 are effective divisors. Let H be an ample divisor on X. By Theorem 13.32, there exists a positive integer n_0 such that $n_0 H + D_1$ and $n_0 H + D_2$ are very ample. By Theorem 16.1, there exist nonsingular curves C_1 and C_2 such that $n_0 H + D_1 \sim C_1$ and $n_0 H + D_2 \sim C_2$ so $D \sim C_2 - C_1$. We have short exact sequences

$$0 \to \mathcal{O}_X(C_2 - C_1) \to \mathcal{O}_X(C_2) \to \mathcal{O}_X(C_2) \otimes \mathcal{O}_{C_1} \to 0$$

and

$$0 \to \mathcal{O}_X \to \mathcal{O}_X(C_2) \to \mathcal{O}_X(C_2) \otimes \mathcal{O}_{C_2} \to 0.$$

Since χ is additive on short exact sequences (Exercise 17.24), we have

$$\begin{aligned}(20.1) \quad \chi(\mathcal{O}_X(D)) &= \chi(\mathcal{O}_X(C_2 - C_1)) \\ &= \chi(\mathcal{O}_X) + \chi(\mathcal{O}_X(C_2) \otimes \mathcal{O}_{C_2}) - \chi(\mathcal{O}_X(C_2) \otimes \mathcal{O}_{C_1}).\end{aligned}$$

20.1. The Riemann-Roch theorem and the Hodge index theorem

By the Riemann-Roch theorem for curves (Theorem 18.2) and Example 19.14, we have

(20.2) $$\chi(\mathcal{O}_X(C_2) \otimes \mathcal{O}_{C_2}) = (C_2)^2 + 1 - g(C_2)$$

and

(20.3) $$\chi(\mathcal{O}_X(C_2) \otimes \mathcal{O}_{C_1}) = (C_1 \cdot C_2) + 1 - g(C_1).$$

By Theorem 20.2, we have

(20.4) $$g(C_1) = \frac{1}{2}(C_1 \cdot (C_1 + K_X)) + 1$$

and

(20.5) $$g(C_2) = \frac{1}{2}(C_2 \cdot (C_2 + K_X)) + 1.$$

The theorem now follows from formulas (20.1)–(20.5). □

Corollary 20.4. *Suppose that D is a divisor on a nonsingular projective surface S. Then*

$$h^0(X, \mathcal{O}_X(D)) \geq \frac{1}{2}(D \cdot (D - K_X)) + \chi(\mathcal{O}_X) - h^0(X, \mathcal{O}_X(K_X - D)).$$

This follows by combining Theorems 20.3 and 17.22.

Lemma 20.5. *Let H be an ample divisor on a nonsingular projective surface X. Then there is an integer n_0 such that for any divisor D on X, if $(D \cdot H) > n_0$, then $H^2(X, \mathcal{O}_X(D)) = 0$.*

Proof. Let $n_0 = (K_X \cdot H)$. By Serre duality,

$$h^2(X, \mathcal{O}_X(D)) = h^0(X, \mathcal{O}_X(K_X - D)).$$

If $(D \cdot H) > n_0$ and $h^0(X, \mathcal{O}_X(K_X - D)) > 0$, then $K_X - D$ is linearly equivalent to an effective divisor and thus $((K_X - D) \cdot H) \geq 0$ by Corollary 19.18 and Remark 19.19. But

$$((K_X - D) \cdot H) = (K_X \cdot H) - (D \cdot H) < 0$$

if $(D \cdot H) > n_0 = (K_X \cdot H)$, giving a contradiction. □

Corollary 20.6. *Let H be an ample divisor on a nonsingular projective surface X and let D be a divisor on X such that $(D \cdot H) > 0$ and $(D^2) > 0$. Then for all $n \gg 0$, nD is linearly equivalent to an effective divisor.*

Proof. Let $n_0 = (K_X \cdot H)$ be the constant of Lemma 20.5. Then there exists $n_1 > 0$ such that for $n \geq n_1$, we have

$$(nD \cdot H) = n(D \cdot H) > n_0$$

so $h^0(X, \mathcal{O}_X(K_X - nD)) = 0$. By Corollary 20.4, we have
$$h^0(X, \mathcal{O}_X(nD)) \geq \frac{1}{2}n^2(D^2) - \frac{1}{2}n(D \cdot K_X) + \chi(\mathcal{O}_X)$$
for $n \geq n_1$. Since $(D^2) > 0$, $h^0(X, \mathcal{O}_X(nD)) > 0$ for $n \gg 0$. $\square$

Definition 20.7. A divisor D on a nonsingular projective surface X is said to be numerically equivalent to zero, written $D \equiv 0$, if $(D \cdot E) = 0$ for all divisors E on X. We say that divisors D and E are numerically equivalent, written as $D \equiv E$ if $D - E \equiv 0$.

If D and E are linearly equivalent divisors, then $\mathcal{O}_X(D - E) \cong \mathcal{O}_X$, so that $D \equiv E$ by Remark 19.19.

Theorem 20.8 (Hodge index theorem). *Let H be an ample divisor on a nonsingular projective surface X and suppose that D is a divisor on X such that $D \not\equiv 0$ and $(D \cdot H) = 0$. Then $(D^2) < 0$.*

Proof. Suppose that $(D^2) \geq 0$. We will derive a contradiction.

First suppose that $(D^2) > 0$. Let $H' = D + nH$. For $n \gg 0$, H' is ample by Theorem 13.32. Since
$$(D \cdot H') = (D^2) > 0,$$
we have that mD is linearly equivalent to a nonzero effective divisor for $m \gg 0$ by Corollary 20.6 and Remark 19.19. Then
$$(mD \cdot H) = m(D \cdot H) > 0$$
by Corollary 19.18. Hence $(D \cdot H) > 0$, giving a contradiction.

Now suppose that $(D^2) = 0$. Since $D \not\equiv 0$, there exists a divisor E such that $(D \cdot E) \neq 0$. Let
$$E_1 = (H^2)E - (E \cdot H)H.$$
Then $(D \cdot E_1) \neq 0$ $((H^2) > 0$ by Corollary 19.18) and $(E_1 \cdot H) = 0$.

Let $D_1 = nD + E_1$ for $n \in \mathbb{Z}$. Then $(D_1 \cdot H) = 0$ and
$$(D_1^2) = 2n(D \cdot E_1) + (E_1^2).$$
Since $(D \cdot E_1) \neq 0$, there exists $n \in \mathbb{Z}$ such that $(D_1^2) > 0$. Now the first case of the proof applied to D_1 gives a contradiction. $\square$

Let X be a nonsingular projective surface. Define
$$\text{Num}(X) = \text{Pic}(X)/\equiv.$$
The group $\text{Num}(X)$ is a finitely generated group [**97**] without torsion, so $\text{Num}(X) \cong \mathbb{Z}^\rho$ for some ρ. Let $N(X) = \text{Num}(X) \otimes \mathbb{R}$, which is a finite-dimensional vector space. The intersection pairing is a nondegenerate bilinear form on $N(X)$. By Sylvester's theorem [**95**, Theorem 4.1, page 577],

the form can be diagonalized with ±1's on the diagonal, and the difference of the number of +1's minus the number of −1's is an invariant called the signature of the form. The Hodge index theorem tells us that the diagonalized intersection form on a surface has exactly one +1. In particular, the signature of the form is $2 - \rho$.

20.2. Contractions and linear systems

The self-intersection number $(C^2) = (C \cdot C)$ of a curve C on a nonsingular projective surface can by negative, as is shown by the following example.

Example 20.9. Let S be a nonsingular projective surface, and let $p \in S$ be a point. Let $\pi : S_1 \to S$ be the blow-up of p with exceptional divisor E. Then
$$(E^2) = (E \cdot E) = -1.$$

Proof. We have that $(E \cdot E) = \deg(\mathcal{O}_{S_1}(E) \otimes \mathcal{O}_E)$ by Example 19.14. Now $E \cong \mathbb{P}^1$ and $\mathcal{O}_{S_1}(E) \otimes \mathcal{O}_E \cong \mathcal{O}_{\mathbb{P}^1}(-q)$ where q is a point on $\mathbb{P}^1$ by Example 13.38, so $(E \cdot E) = -1$. $\square$

A remarkable fact is that there is a converse to this example.

Suppose that X is a normal projective surface and $\mathcal{C} = \{C_1, \ldots, C_n\}$ is a finite set of closed curves on X. A contraction of $\mathcal{C}$ is a regular birational map $\phi : X \to Y$ such that Y is normal, there exists a point $q \in Y$ such that $\phi(C_i) = q$ for all i, and $\phi : X \setminus \mathcal{C} \to Y \setminus \{q\}$ is an isomorphism.

The contraction $\phi : X \to Y$ of $\mathcal{C}$, if it exists, must be unique. To see this, suppose that $\phi_1 : X \to Y_1$ and $\phi_2 : X \to Y_2$ are two contractions of $\mathcal{C}$. Let $\Psi : Y_1 \dashrightarrow Y_2$ be the induced birational map and let $\Gamma_\Psi \subset Y_1 \times Y_2$ be the graph. Let $\pi_1 : \Gamma_\Psi \to Y_1$ and $\pi_2 : \Gamma_\Psi \to Y_2$ be the projections. Let $q_1 = \phi_1(\mathcal{C})$ and $q_2 = \phi_2(\mathcal{C})$. Then $\Psi : Y_1 \setminus \{q_1\} \to Y_2 \setminus \{q_2\}$ is an isomorphism, so $\pi_1^{-1}(q_1) = \pi_2^{-1}(q_2) = \{(q_1, q_2)\}$. By Zariski's main theorem, Theorem 9.3, both projections π_1 and π_2 are isomorphisms, so $Y_1 \cong Y_2$.

By Zariski's connectedness theorem, Theorem 9.7, a set of curves $\mathcal{C}$ must have a connected union to be contractible.

Theorem 20.10 (Castelnuovo's contraction theorem). *Suppose that S is a nonsingular projective surface and $E \cong \mathbb{P}^1$ is a curve on S such that $(E^2) = -1$. Then there exists a birational regular map $\phi : S \to T$ where T is a nonsingular projective surface such that $\phi(E) = p$ is a point on T and $\phi : S \to T$ is isomorphic to the blow-up $\pi : B(p) \to T$ of p.*

Proof. We will first prove the existence of a contraction T of E. Let H be a very ample divisor on S. After possibly replacing H with a positive multiple of H, we may assume that $H^1(S, \mathcal{O}_S(H)) = 0$ (by Theorem 17.18).

Let $m = (H \cdot E) > 0$ by Corollary 19.18. Then $(E \cdot (H + mE)) = 0$. We have an exact sequence
$$0 \to \mathcal{O}_S(-mE) \to \mathcal{O}_S \to \mathcal{O}_{mE} \to 0,$$
where mE is the subscheme of S with $\mathcal{O}_{mE} = \mathcal{O}_S/\mathcal{I}_E^m$. Tensoring with $\mathcal{O}_S(H + mE)$ and taking global sections, we have an exact sequence

(20.6)
$$H^0(S, \mathcal{O}_S(H + mE)) \to H^0(S, \mathcal{O}_S(H + mE) \otimes \mathcal{O}_{mE})$$
$$\to H^1(S, \mathcal{O}_S(H)) = 0.$$

Let $\mathcal{L} = \mathcal{O}_S(H + mE)$. For all $n \geq 1$, we have short exact sequences
$$0 \to \mathcal{O}_E \otimes \mathcal{O}_S(-nE) \to \mathcal{O}_{(n+1)E} \to \mathcal{O}_{nE} \to 0$$
since
$$\mathcal{O}_S(-nE)/\mathcal{O}_S(-(n+1)E) \cong \mathcal{O}_E \otimes \mathcal{O}_S(-nE).$$

Tensoring with $\mathcal{L}$ and taking global sections, we have exact sequences

(20.7)
$$H^0(S, \mathcal{L} \otimes \mathcal{O}_{(n+1)E}) \to H^0(S, \mathcal{L} \otimes \mathcal{O}_{nE})$$
$$\to H^1(E, \mathcal{O}_E \otimes \mathcal{O}_S(-nE) \otimes \mathcal{L}).$$

Now $(E \cdot (-nE)) = n$ and $(\mathcal{O}_S(E) \cdot \mathcal{L}) = 0$ so
$$\mathcal{O}_E \otimes \mathcal{O}_S(-nE) \otimes \mathcal{L} \cong \mathcal{O}_E(n) = \mathcal{O}_{\mathbb{P}^1}(n).$$
Thus $H^1(E, \mathcal{O}_S(-nE) \otimes \mathcal{L} \otimes \mathcal{O}_E) = 0$ for $n \geq 1$ by Theorem 17.14.

Combining (20.6) and (20.7), we have a surjection
$$H^0(S, \mathcal{O}_S(H + mE)) \to H^0(E, \mathcal{O}_S(H + mE) \otimes \mathcal{O}_E) \cong H^0(E, \mathcal{O}_E).$$

Thus $\text{Base}(|H + mE|) \cap E = \emptyset$, and so $\text{Base}(|H + mE|) = \emptyset$ since H is very ample. Let $\phi : S \to \mathbb{P}^r$ be the regular map induced by $|H + mE|$. Suppose C is a curve on S. Then $\phi(C)$ is a point if and only if $((H + mE) \cdot C) = 0$, so the only curve contracted by ϕ is E. Let T be the normalization of the image $\phi(S)$ in the function field of S (Theorem 7.17). Then we have a factorization $\Psi : S \to T$ by Exercise 9.9 since S is normal. Let $p = \Psi(E)$. Then $S \backslash E \to T \backslash \{p\}$ is an isomorphism by Zariski's main theorem (Theorem 9.3).

We have that $\Psi_* \mathcal{O}_S \cong \mathcal{O}_T$ be Proposition 11.52. By the theorem on formal functions ([**156**] or [**73**, Theorem III.11.1])
$$\hat{\mathcal{O}}_{T,p} \cong \varprojlim H^0(E_n, \mathcal{O}_{E_n})$$
where E_n is the closed subscheme of S with the ideal sheaf $m_p^n \mathcal{O}_S$ (the completion $\hat{A}$ of a local ring A is defined in Section 21.5). Since $\Psi^{-1}(p) = E$, the sequence of ideal sheaves $m_p^n \mathcal{O}_S$ is confinal with respect to $\mathcal{I}_E^n$, so
$$\hat{\mathcal{O}}_{T,p} \cong \varprojlim H^0(S, \mathcal{O}_S/\mathcal{I}_E^n).$$

20.2. Contractions and linear systems

We will show that $H^0(S, \mathcal{O}_{nE}) \cong A_n = k[x,y]/(x,y)^n$ so
$$\hat{\mathcal{O}}_{T,p} \cong \varprojlim A_n \cong k[[x,y]].$$

We then have that $\mathcal{O}_{T,p}$ is a regular local ring (by Theorem 21.36). For $n = 1$, $H^0(E, \mathcal{O}_E) \cong k$. For $n > 1$, we have short exact sequences
$$0 \to \mathcal{I}_E^n/\mathcal{I}_E^{n+1} \to \mathcal{O}_{E_{n+1}} \to \mathcal{O}_{E_n} \to 0,$$
with
$$\mathcal{I}_E^n/\mathcal{I}_E^{n+1} \cong \mathcal{O}_S(-nE) \otimes \mathcal{O}_E \cong \mathcal{O}_{\mathbb{P}^1}(n).$$
For $n = 1$, $H^0(\mathbb{P}^1, \mathcal{O}_{\mathbb{P}^1}(1))$ is a two-dimensional vector space. Let x, y be a basis. Then $H^0(S, \mathcal{O}_{E_2}) \cong A_2$. Inductively, if $H^0(S, \mathcal{O}_{E_n})$ is isomorphic to A_n, lift x, y to $H^0(S, \mathcal{O}_{E_{n+1}})$. Since $H^0(\mathbb{P}^1, \mathcal{O}_{\mathbb{P}^1}(n))$ is the vector space with basis $x^n, x^{n-1}y, \ldots, y^n$, we have that $H^0(S, \mathcal{O}_{E_{n+1}}) \cong A_{n+1}$.

The contraction $S \to T$ is then the blow-up of p by Theorem 10.32. $\square$

The above theorem may lead one to hope that any curve with negative self-intersection number on a nonsingular projective surface can be contracted by a regular map, but this is not true, as we show in the following example.

Example 20.11. Suppose that the algebraically closed field k is not an algebraic closure of a finite field. Then there exists a nonsingular projective surface X over k and a nonsingular closed curve C on X with $(C^2) = -1$ such that C is not contractible.

Proof. Let γ be a nonsingular cubic curve in $S = \mathbb{P}^2_{\mathbb{C}}$ and L be a line in S such that $\gamma \cap L$ is reduced. Since $(\gamma \cdot L) = 3$, $\gamma \cap L$ is a divisor $\gamma \cap L = q_1 + q_2 + q_3$ for distinct points q_1, q_2, q_3 on γ.

Since the curve γ is nonsingular of degree 3, its canonical divisor is $K_\gamma = 0$ by Corollary 14.22. Thus γ has genus 1 by Corollary 18.14. The elliptic curve γ is isomorphic as a group to $\text{Cl}^0(\gamma)$, by Theorem 18.19, and as explained before Proposition 18.44.

We will show that we can find a point $p_0 \in \gamma$ such that $q_1 + q_2 + q_3 - 3p_0$ has infinite order in $\text{Cl}^0(\gamma)$. Let p be a point in γ. If $q_1 + q_2 + q_3 - 3p$ has infinite order, then take $p_0 = p$. Otherwise, there exists a positive integer m such that $m(q_1 + q_2 + q_3 - 3p) \sim 0$. By Theorem 18.49, there exists $p_0 \in \gamma$ such that $p_0 - p$ has infinite order. We will show that $q_1 + q_2 + q_3 - 3p_0$ has infinite order. Suppose it does not. Then there eixsts a positive integer n such that $n(q_1 + q_2 + q_3 - 3p_0) \sim 0$, so
$$0 \sim mn(q_1+q_2+q_3-3p_0) = nm(q_1+q_2+q_3-3p) + mn3(p-p_0) \sim mn3(p-p_0),$$
a contradiction since $p_0 - p$ is assumed to have infinite order. By Theorem 18.49, there exist distinct points $p_1, \ldots, p_{10} \in \gamma$ such that the classes of

$q_1+q_2+q_3-3p_0, p_1-p_0, \ldots, p_{10}-p_0$ are linearly independent in the rational vector space $\text{Cl}^0(\gamma) \otimes \mathbb{Q}$. After possibly replacing L with a different line linearly equivalent to L, we may assume that $q_1, q_2, q_3, p_1, \ldots, p_{10}$ are distinct points of γ.

Suppose that there exist $n, m_1, \ldots, m_{10} \in \mathbb{Z}$ such that $n(q_1 + q_2 + q_3) \sim m_1 p_1 + \cdots + m_{10} p_{10}$ on γ. Then from $m_1 + \cdots + m_{10} = \deg(n(q_1+q_2+q_3)) = 3n$ we have that $m_1(p_1 - p_0) + \cdots + m_{10}(p_{10} - p_0) \sim n(q_1 + q_2 + q_3 - 3p_0)$ so that $m_1 = \cdots = m_{10} = n = 0$.

Let $\pi : X \to S$ be the blow-up of the ten points $p_1, \ldots, p_{10}$. Let $H = \pi^*(L)$ and E_i be the rational curves $E_i = \pi^{-1}(p_i)$ for $1 \le i \le 10$. We have that

(20.8) $$\text{Cl}(X) = [H]\mathbb{Z} \oplus [E_1]\mathbb{Z} \oplus \cdots \oplus [E_{10}]\mathbb{Z}$$

by Exercise 13.48 and Example 13.10. Let C be the strict transform of γ on X. Since γ is a nonsingular curve, $\pi : C \to \gamma$ is an isomorphism (for instance since γ is nonsingular and $\pi : C \to \gamma$ is birational). This allows us to identify C with γ.

Since $p_1, \ldots, p_{10} \not\in L$, we can identify $H \cap C = L \cap \gamma = q_1 + q_2 + q_3$. We also have $E_i \cap C = p_i$ for $1 \le i \le 10$. We have that

$$C = \pi^*(\gamma) - E_1 - \cdots - E_{10} \sim \pi^*(3L) - E_1 - \cdots - E_{10} \sim 3H - E_1 - \cdots - E_{10}.$$

We compute

$$(C^2) = (C \cdot (3H - E_1 - \cdots - E_{10})) = -1.$$

Suppose that there exists a contraction $\phi : X \to Y$ of C. Let $q = \phi(C)$. Let A be a very ample effective divisor on Y which does not contain q. Let $\overline{A} = \phi^*(A)$. Then $\overline{A} \cap C = \emptyset$ so $\mathcal{O}_X(\overline{A}) \otimes \mathcal{O}_C \cong \mathcal{O}_C$. There exist $m_1, \ldots, m_{10}, n \in \mathbb{Z}$ such that $\overline{A} \sim nH + m_1 E_1 + \cdots + m_{10} E_{10}$ by (20.8). Thus

$$\begin{aligned}\mathcal{O}_C &\cong \mathcal{O}_X(\overline{A}) \otimes \mathcal{O}_C \cong \mathcal{O}_X(nH + m_1 E_1 + \cdots + m_{10} E_{10}) \otimes \mathcal{O}_C \\ &\cong \mathcal{O}_\gamma(n(q_1 + q_2 + q_3) + m_1 p_1 + \cdots + m_{10} p_{10})\end{aligned}$$

so $n(q_1 + q_2 + q_3) + m_1 p_1 + \cdots + m_{10} p_{10} \sim 0$ on γ. Thus $m_1 = \cdots = m_{10} = n = 0$, so that $\overline{A} = 0$. Since $\phi^* : \text{Cl}(Y) \to \text{Cl}(X)$ is injective (by Proposition 13.39), we have that $A \sim 0$ on Y. But this is impossible since A is a hyperplane section of Y. $\square$

We have the following necessary condition for the contractibility of a union of curves.

Theorem 20.12. *Suppose that S is a nonsingular projective surface and $\mathcal{C} = \{C_1, \ldots, C_n\}$ is a finite set of curves on S such that $\mathcal{C}$ is contractible. Then the intersection matrix $A = ((C_i \cdot C_j))$ is negative definite.*

Proof. Let $\phi : S \to S'$ be the contraction of $\mathcal{C}$ and let H be a very ample divisor on S', which we may assume does not contain the point which is the image of the C_i (Lemma 13.12). Let $D = \phi^*(H)$. Then $(D^2) = (H^2) > 0$ (Proposition 19.12 and Corollary 19.18) and $(D \cdot C_i) = 0$ for all i. With the notation following the Hodge index theorem, Theorem 20.8, let

$$L = \{v \in \text{Num}(S) \mid (v \cdot D) = 0\}.$$

The restriction of the intersection form to L is negative definite as commented after the proof of Theorem 20.8. $\square$

If $\mathcal{C} = \{C_1, \ldots, C_n\}$ is a finite set of curves on a nonsingular projective surface S over $\mathbb{C}$ such that the union of the curves in $\mathcal{C}$ is connected and the intersection matrix $(C_i \cdot C_j)$ is negative definite, then although there may not be a contraction of these curves by a regular map, there does exist an analytic map $\phi : S \to T$ where T is a complex analytic (but not necessarily algebraic) normal surface such that ϕ is an analytic contraction of the $\mathcal{C}$. This is proven by Grauert in [**61**]. We also have [**11**] that $\mathcal{C}$ is contractible on a nonsingular projective surface S over a field k which is an algebraic closure of a finite field if and only if the intersection matrix $(C_i \cdot C_j)$ is negative definite (and the union of the curves in $\mathcal{C}$ is connected), showing that the assumption that k is not an algebraic closure of a finite field is necessary in Example 20.11.

Theorem 20.10 is the first result in a general philosophy that curves with negative intersection number with the canonical divisor play a major role in geometry; a nonsingular closed rational curve E on a nonsingular projective surface S with $(E^2) = -1$ satisfies $(E \cdot K_S) = -1$ by adjunction (Theorem 20.2). This philosophy has been realized to a remarkable degree. The theory for projective surfaces is classical. In higher dimensions, much of the theory has been developed, although many questions still remain (especially in positive characteristic). A few papers and articles on this subject, which contain detailed references, are Mori's papers [**112**], [**113**], and [**114**], the book [**91**] by Kollár and Mori, and the article [**20**] by Birkar, Cascini, Hacon, and McKernan.

Let D be an effective divisor on a normal projective variety X. The fixed component B_n of the complete linear system $|nD|$ is the largest effective divisor $B_n \le nD$ such that $\Gamma(X, \mathcal{O}_X(nD - B_n)) = \Gamma(X, \mathcal{O}_X(nD))$.

Lemma 20.13. *Suppose that D is an effective divisor on a normal projective variety X. Let $R[D] = \bigoplus_{n \ge 0} \Gamma(X, \mathcal{O}_X(nD))$. Then $R[D]$ is not a finitely generated k-algebra if*

1) $B_n \ne 0$ *for all* $n > 0$,

2) B_n *is bounded from above.*

Proof. Assume that 1) and 2) hold and that $R[D]$ is a finitely generated k-algebra. Then there exists a positive integer N, $\lambda(i) \in \mathbb{Z}_+$ and $u_i \in \Gamma(X, \mathcal{O}_X(\lambda(i)D))$ for $1 \leq i \leq N$ such that $\Gamma(X, \mathcal{O}_X(nD))$ is spanned by the products

$$\left\{ \prod_{i=1}^{N} u_i^{\nu_i} \mid \nu_i \geq 0 \text{ for all } i \text{ and } \sum \lambda(i)\nu_i = n \right\}.$$

Thus

$$B_n \geq \text{glb} \left\{ \sum_{i=1}^{N} \nu_i B_i \mid \sum \lambda(i)\nu_i = n \right\}$$

(glb means "greatest lower bound"). We have $mB_n \geq B_{mn}$ for all $m, n \in \mathbb{N}$. Thus $\frac{N!}{i} B_i \geq B_{N!}$ for $i = 1, 2, \ldots, N$. The divisor $B_{N!}$ is nonzero by 1). Thus $B_1, B_2, \ldots, B_N$ have at least one prime divisor as a common component, so condition 2) cannot be satisfied, as $\sum \nu_i \mapsto \infty$ as $n \mapsto \infty$. $\square$

The following example is given by Zariski in [**159**].

Theorem 20.14 (Zariski). *Suppose that k is a field which is not an algebraic closure of a finite field. Then there is an effective divisor D on a nonsingular projective surface X over k such that $R[D] = \bigoplus_{n \geq 0} \Gamma(X, \mathcal{O}_X(nD))$ is not a finitely generated k-algebra.*

Proof. Let $X' = \mathbb{P}_k^2$, let E' be a nonsingular cubic curve on X', and let H' be a line on X'. Let α be a divisor on E' such that $\mathcal{O}_{E'}(\alpha) \cong \mathcal{O}_{X'}(H' + E') \otimes \mathcal{O}_{E'}$. The elliptic curve E' is isomorphic to $\text{Cl}^0(E')$ as a group by Theorem 18.19 and is explained before Proposition 18.44. By Theorem 18.49, there exists a divisor β on E' of degree 0 such that the class of β has infinite order in $\text{Cl}^0(E')$. The genus g of E' is 1, as commented before Lemma 18.39. We compute $\deg \alpha = ((H' + E') \cdot E') = 12$ (by Example 19.14), so

$$\deg(\alpha - \beta) = 12 > 2g + 1 = 5,$$

so $\alpha - \beta$ is a very ample divisor on E' by Theorem 18.21. Thus there exist distinct points $p'_1, \ldots, p'_{12} \in E'$ such that $\alpha - \beta \sim p'_1 + \cdots + p'_{12}$ by Theorem 16.1. Now $n(p'_1 + \cdots + p'_{12}) - n\alpha \sim -n\beta$ for all $n \in \mathbb{Z}$, so

(20.9) $$n(p'_1 + \cdots + p'_{12}) \not\sim n\alpha$$

for all $0 \neq n \in \mathbb{Z}$.

Let $\pi : X \to X'$ be the blow-up of the points $p'_1, \ldots, p'_{12}$, with exceptional divisors $F_1, \ldots, F_{12}$. Let $\Gamma' \in |H'|$ be an irreducible curve which does not pass through any of the points $p'_1, \ldots, p'_{12}$ (Theorem 16.1 and since $|H'|$ is base point free). Let $\Gamma = \pi^*(\Gamma')$ and let E be the strict transform of E' on X, so that $\pi^*(E') = E + F_1 + \cdots + F_{12}$. Set $D = \Gamma + E$. We will show that the fixed component B_n of $|nD|$ is precisely E (for $n \in \mathbb{Z}_+$). The theorem will then follow from Lemma 20.13.

20.2. Contractions and linear systems

The restriction of π to E is an isomorphism onto E'. We compute
$$(E \cdot E) = ((3\Gamma - F_1 - \cdots - F_{12}) \cdot E) = 3(H' \cdot E') - 12 = -3,$$
so for $m, n \in \mathbb{Z}$, $((n\Gamma + mE) \cdot E) = 3n - 3m$. The canonical divisor K_E is zero, since E is an elliptic curve (Corollary 18.18). By Serre duality, Corollary 18.10,
$$h^1(E, \mathcal{O}_X(n\Gamma + mE) \otimes \mathcal{O}_E) = 0 \quad \text{if } n \geq 1 \text{ and } 0 \leq m \leq n-1$$
and
(20.10)
$\mathcal{O}_X(n\Gamma + mE) \otimes \mathcal{O}_E$ is generated by global sections if $n \geq 1$ and $0 \leq m \leq n-1$
by Theorem 18.21. However, when $0 \neq m = n$, we have
$$\begin{aligned} \mathcal{O}_X(n(\Gamma + E)) \otimes \mathcal{O}_E &\cong \mathcal{O}_X(n(\Gamma + \pi^*(E')) - F_1 - \cdots - F_{12})) \otimes \mathcal{O}_E \\ &\cong \mathcal{O}_{X'}(n(H' + E')) \otimes \mathcal{O}_{E'}(-n(p'_1 + \cdots + p'_{12})) \\ &\cong \mathcal{O}_{E'}(n(\alpha - p'_1 - \cdots - p'_{12})). \end{aligned}$$
Thus
$$h^0(E, \mathcal{O}_X(n\Gamma + nE) \otimes \mathcal{O}_E) = 0$$
by (20.9) and
$$h^1(E, \mathcal{O}_X(n\Gamma + nE) \otimes \mathcal{O}_E) = 0$$
by Theorem 18.2.

We have
(20.11) $$H^1(X, \mathcal{O}_X(n\Gamma)) = H^1(X', \mathcal{O}_{X'}(n\Gamma')) = 0$$
and
(20.12) $$H^0(X, \mathcal{O}_X(n\Gamma)) = H^0(X', \mathcal{O}_{X'}(n\Gamma'))$$
for all $n \geq 0$ by Theorem 17.32 and Theorem 17.14.

We have a short exact sequence of $\mathcal{O}_X$-modules
$$0 \to \mathcal{O}_X(-E) \to \mathcal{O}_X \to \mathcal{O}_E \to 0.$$
Tensor this short exact sequence with $\mathcal{O}_X(n\Gamma + mE)$ and take cohomology to obtain long exact sequences
$$\begin{aligned} 0 \to H^0(X, \mathcal{O}_X(n\Gamma + (m-1)E)) &\to H^0(X, \mathcal{O}_X(n\Gamma + mE)) \\ \to H^0(E, \mathcal{O}_X(n\Gamma + mE) \otimes \mathcal{O}_E) &\to H^1(X, \mathcal{O}_X(n\Gamma + (m-1)E)) \\ \to H^1(X, \mathcal{O}_X(n\Gamma + mE)) &\to H^1(E, \mathcal{O}_X(n\Gamma + mE) \otimes \mathcal{O}_E) = 0 \end{aligned}$$
for $n \geq 1$ and $0 \leq m \leq n$. By induction on m and (20.11), we have that
$$H^1(X, \mathcal{O}_X(n\Gamma + mE)) = 0 \quad \text{for } n \geq 1 \text{ and } 0 \leq m \leq n.$$
The $\mathcal{O}_X$-module $\mathcal{O}_X(n\Gamma)$ is generated by global sections by (20.12), so $\mathcal{O}_X(n\Gamma + E)$ is generated by global sections except possibly at points of E. But $H^0(X, \mathcal{O}_X(n\Gamma + E))$ surjects onto $H^0(E, \mathcal{O}_X(n\Gamma + E) \otimes \mathcal{O}_E)$ and

$\mathcal{O}_X(n\Gamma + E) \otimes \mathcal{O}_E$ is generated by global sections by (20.10) if $n \geq 2$, so $\mathcal{O}_X(n\Gamma + E)$ is generated by global sections (if $n \geq 2$). By induction and (20.10), $\mathcal{O}_X(n\Gamma + (n-1)E)$ is generated by global sections for all $n \geq 1$. Now $H^0(X, \mathcal{O}_X(n\Gamma + nE) \otimes \mathcal{O}_E) = 0$, so the fixed component B_n of $|nD| = |n\Gamma + nE|$ is E. □

This type of example cannot occur if k is an algebraic closure of a finite field. If D is an effective divisor on a nonsingular projective surface X over a field k which is the algebraic closure of a finite field, then

$$R[D] = \bigoplus_{n \geq 0} \Gamma(X, \mathcal{O}_X(nD))$$

is a finitely generated k-algebra by [44, Theorem 3].

Exercise 20.15. Suppose that C is a nonsingular curve of degree d in $\mathbb{P}^2$. Show that

$$g(C) = \frac{1}{2}(d-1)(d-2).$$

Exercise 20.16. Let C be a nonsingular curve and $S = C \times \mathbb{P}^1$. Let $\pi : S \to C$ be the projection onto C.

a) Suppose that $p, q \in C$. Show that $\pi^*(p) \equiv \pi^*(q)$.

b) Let C be an elliptic curve over an algebraically closed field k which is not the algebraic closure of a finite field, so there exist $p, q \in C$ such that the class of $p - q$ in $\mathrm{Cl}^0(C)$ has infinite order (by Theorem 18.49). Show that $n\pi^*(p) \not\sim m\pi^*(q)$ for all $m, n \in \mathbb{Z}$.

Exercise 20.17. Let D be the divisor constructed in Theorem 20.14. Show that D is numerically effective; that is, $(D \cdot C) \geq 0$ for all curves C on X.

Exercise 20.18. Compute the functions $h^i(X, \mathcal{O}_X(nD))$ for $i = 0, 1, 2$ and $n \in \mathbb{N}$ for the divisor D constructed in Theorem 20.14.

Exercise 20.19. Modify the proof of Theorem 20.14 by starting with a divisor β on E' which has degree 0 and finite order r in $\mathrm{Cl}^0(E')$. (We can always find such a β if r is relatively prime to the characteristic of k by Proposition 18.50). Let X and D be the surface and divisor which we construct. Is $R[D]$ a finitely generated k-algebra? Is it generated in degree 1? Compute the functions $h^i(X, \mathcal{O}_X(nD))$ for $i = 0, 1, 2$ and $n \in \mathbb{N}$.

Chapter 21

Ramification and Étale Maps

In this chapter, we consider algebraic analogues of the topological covering spaces and branched (ramified) covers in the Euclidean topology. Finite maps correlate with branched covers and the algebraic counterpart of the topological covering spaces are the étale morphisms (Definition 21.79). A regular map $\phi : X \to Y$ of nonsingular complex varieties is a topological covering space in the Euclidean topology if and only if ϕ is étale, by the analytic implicit function theorem (Exercise 21.87). The concept of a covering space in the Zariski topology is much too restrictive a notion; there are many finite maps of nonsingular complex varieties which are étale but are not covering spaces in the Zariski topology (Exercise 21.88).

A related concept to étale is the more classical notion of an unramified map (Section 21.4). If $\phi : X \to Y$ is a finite regular map of varieties and Y is normal, then ϕ is étale if and only if ϕ is unramified (Proposition 21.84).

We develop various characterizations of the concepts of étale and unramified maps. We define Galois maps of varieties (Section 21.7). We develop the concept of completion (Section 21.5) and prove the local form of Zariski's main theorem (Proposition 21.54) which we earlier used in the proof of global forms of Zariski's main theorem (Chapter 9) and prove Zariski's subspace theorem (Proposition 21.61).

We prove the purity of the branch locus, Theorem 21.92, which tells us that if $\phi : X \to Y$ is a finite separable regular map, Y is nonsingular, and X is normal, then the ramification locus of ϕ has pure codimension 1 (all irreducible components of the ramification locus have codimension 1). We also prove the Abhyankar-Jung theorem (Theorem 21.93) showing that tamely ramified covers have a very simple form.

Most of the results in this chapter are due to Zariski and Abhyankar.

21.1. Norms and Traces

We summarize some properties of norms and traces which will be useful in this chapter.

Suppose that L is a finite extension of a field K and $\overline{K}$ is an algebraic closure of K. Let $\sigma_1, \ldots, \sigma_r$ be the distinct embeddings of L into $\overline{K}$ by K-homomorphisms. For $\alpha \in L$, the norm of α over K is defined to be

$$(21.1) \qquad N_{L/K}(\alpha) = \left(\prod_{i=1}^{r} \sigma_i(\alpha) \right)^{[L:K]_i}.$$

The trace of α over K is defined to be

$$(21.2) \qquad \operatorname{Tr}_{L/K}(\alpha) = [L:K]_i \left(\sum_{i=1}^{r} \sigma_i(\alpha) \right).$$

The inseparability index $[L:K]$ is defined in formula (1.2).

Theorem 21.1. *Let L/K be a finite extension. Then the norm $N_{L/K}$ is a multiplicative homomorphism of $L^\times$ into $K^\times$ (where $L^\times$ and $K^\times$ are the respective multiplicative groups of nonzero elements of L and K) and the trace is an additive homomorphism of L into K. If $K \subset F \subset L$ is a tower of fields, then the two maps are transitive; in other words,*

$$N_{L/K} = N_{F/K} N_{L/F} \quad \text{and} \quad \operatorname{Tr}_{L/K} = \operatorname{Tr}_{F/K} \operatorname{Tr}_{L/F}.$$

If $L = K(\alpha)$ and $f(t) = t^n + a_{n-1} t^{n-1} + \cdots + a_0 \in K[t]$ is the minimal polynomial of α over K, then

$$N_{L/K}(\alpha) = (-1)^n a_0 \quad \text{and} \quad \operatorname{Tr}_{L/K}(\alpha) = -a_{n-1}.$$

Proof. [95, Theorem 5.1, page 285]. □

We now give an alternate construction of the trace and norm. Suppose that L is a finite field extension of K and $\omega_1, \ldots, \omega_n$ is a basis of L over K. For $\alpha \in L$, we have expressions

$$(21.3) \qquad \alpha \omega_i = \sum_{j=1}^{n} a_{ij} \omega_j$$

with $a_{ij} \in K$. Let $A = A_\alpha = (a_{ij})$. From (21.3), we have $\operatorname{Det}(\alpha I_n - A) = 0$. Letting X be an indeterminate, we define the field polynomial $P_{L/K,\alpha}(X) \in K[X]$ as

$$P_{L/K,\alpha}(X) = \operatorname{Det}(X I_n - A) \in K[X].$$

By its construction, we see that the field polynomial is independent of choice of basis of L. It will only be the minimal polynomial of α over K if $L = K(\alpha)$.

Proposition 21.2. *For $\alpha \in L$,*

$$N_{L/K}(\alpha) = \mathrm{Det}(A_\alpha) \quad \text{and} \quad \mathrm{Tr}_{L/K}(\alpha) = \mathrm{trace}(A_\alpha) = \sum_{i=1}^{n} a_{ii}.$$

Proof. This follows from [**160**, formulas (6) and (7) on page 87 and (19) and (20) on page 91]. $\square$

21.2. Integral extensions

Suppose that K is a field and K^* is a finite extension field. Suppose that A is a subring of K whose quotient field is K and B is a subring of K^* whose quotient field is K^* and that B is integral over A.

If A is Noetherian and normal, B is the integral closure of A in K^*, and K^* is finite separable over K, then B is a finite A-algebra [**160**, Corollary 1, page 265]. However, there exist examples of Noetherian domains A such that the normalization of A (in K) is not Noetherian and examples of normal Noetherian domains A whose integral closure in a finite (necessarily inseparable) field extension is not Noetherian [**121**, Example 5, page 207]. This type of pathology cannot occur when A is a localization of a finite type algebra over a field (Theorem 1.54) and, more generally, when A is excellent [**107**, Chapter 13].

Suppose that P and Q are respective prime ideals of A and B. We say that Q lies over P or P lies below Q if $Q \cap A = P$.

We have the following useful lemmas.

Lemma 21.3 (Going up theorem). *Let P and P^* be prime ideals in A such that $P^* \subset P$ and let Q^* be a prime ideal in B lying over P^*. Then there exists a prime ideal Q in B lying over P such that $Q^* \subset Q$.*

Proof. [**160**, Corollary, page 259]. $\square$

Corollary 21.4. *Let notation be as above. Then:*

1) *Two distinct prime ideals P' and Q' of B such that $P' \subset Q'$ cannot lie over the same prime ideal of A.*

2) *Let P' be a prime ideal of B lying over P. For P' to be a maximal ideal of B it is necessary and sufficient that P be a maximal ideal of A.*

Proof. [**160**, Complements 1) and 2) on page 259]. $\square$

Lemma 21.5 (Going down theorem). *Suppose that A is normal. Let $P \subset Q$ be prime ideals in A and let Q^* be a prime ideal in B lying over Q. Then there exists a prime ideal P^* in B lying over P with $P^* \subset Q^*$.*

Proof. [**160**, Theorem 6, page 262]. □

The example on page 32 [**107**] and the remark on page 37 [**107**] show that the assumption that A is normal is necessary in Lemma 21.5.

Definition 21.6. An extension L of a field K is a normal extension of K (L is normal over K) if L is an algebraic extension of K and if every irreducible polynomial $f(x) \in K[x]$ which has a root in L factors completely in $L[x]$ into linear factors.

Lemma 21.7. *A finite extension L of a field K is normal over K if and only if it satisfies the following condition: If M is any extension of L, then any K-homomorphism of L into M is necessarily a K-automorphism of L.*

Proof. [**160**, Theorem 15, page 77]. □

Lemma 21.8. *Suppose that R is an integrally closed local domain with quotient field K and T is the integral closure of R in a finite normal extension L of K. Let $\{m_i\}$ be the prime ideals of T lying over m_R. Then the m_i are the maximal ideals of T and $\mathrm{Aut}(L/K)$ acts transitively on the set $\{m_i\}$, so that $\{m_i\}$ is a finite set.*

Proof. The fact that the m_i are the maximal ideals of T follows from 2) of Corollary 21.4. Let $G = \mathrm{Aut}(L/K)$. We have that $\sigma(T) = T$ for $\sigma \in G$ since T is the integral closure of R. Thus σ permutes the m_i. Suppose there exists an m_j such that m_j is not a conjugate $\sigma(m_1)$ of m_1. We will derive a contradiction. By Theorem 1.5, there exists $x \in m_j$ such that $x \notin \sigma(m_1)$ for all $\sigma \in G$. Thus none of the conjugates $\sigma(x)$ of x are in m_1, and thus no power of a product of conjugates of x is in the prime ideal m_1. By (21.1), $y = N_{L/K}(x) \in K$ is of this form (a power of a product of conjugates of x). Thus $y \in K \cap T = R$ since R is integrally closed. Finally, $y \in m_j \cap R = m_R$. Since $m_R \subset m_1$ and m_1 is a prime ideal, this is a contradiction. □

With the notation introduced in the first paragraph of this section, suppose that P is a prime ideal of A. Then there exists a prime ideal Q of B lying over P by Proposition 1.56, and if A is normal, then there are only a finite number of prime ideals in B lying over P (by Lemma 21.8, applied to the integral closure of A_P in a finite normal extension of K containing K^*).

Let R be a normal (not necessarily Noetherian) local domain with maximal ideal m_R and quotient field K, and let R^* be the integral closure of R in a finite extension field K^* of K. Let $m_1^*, \ldots, m_r^*$ be the prime ideals of R^* which lie over m_R. The m_i^* are the maximal ideals of R^* by Corollary 21.4. Let $S_i = R^*_{m_i^*}$ for $1 \leq i \leq r$. We shall say that the S_i are the local rings of K^* lying above (or over) R and that R is the local ring of K lying below S_i.

21.2. Integral extensions

Lemma 21.9. *With the notation introduced in the above paragraph, we have that $S_i \cap K = R$ for all i.*

Proof. Let $\tilde{K}$ be a finite normal extension of K containing K^*. Let $\tilde{R}$ be the integral closure of R in $\tilde{K}$ and let $\tilde{m}_1, \ldots, \tilde{m}_t$ be the maximal ideals of $\tilde{R}$ lying over m_R. We may assume that $\tilde{m}_1 \cap R^* = m_i^*$. Let $\tilde{S}_j = \tilde{R}_{\tilde{m}_j}$ for $1 \leq j \leq t$. Let $G = \text{Aut}(\tilde{K}/K)$. We have that $R \subset S_i \cap K \subset \tilde{S}_1 \cap K$ so it suffices to show that $\tilde{S}_1 \cap K = R$. Let $u \in \tilde{S}_1 \cap K$. Given $\tilde{S}_j$, there exists $g \in G$ such that $g(\tilde{S}_1) = \tilde{S}_j$, by Lemma 21.8, and hence $g(u) \in \tilde{S}_j$. But $u \in K$ implies $g(u) = u$. Hence $u \in \tilde{S}_j$ for $i = 1, \ldots, t$ so that $u \in \bigcap_{j=1}^{t} \tilde{S}_j = \tilde{R}$ by Lemma 1.77. Thus $u \in \tilde{R} \cap K = R$, so that $\tilde{S}_1 \cap K = R$, and hence $S_i \cap K = R$. $\square$

Now, with the notation introduced in the first paragraph of this section, suppose that A is normal (not necessarily Noetherian) and B is the integral closure of A in K^*. Then $B \cap K = A$. Further, if K' is an intermediate field between K and K^*, then $B \cap K'$ is the integral closure of A in K'. We saw above that if P^* is a prime ideal of B and $P = P^* \cap A$, then $B_{P^*} \cap K = A_P$.

Suppose that C is a normal subring of K^* whose quotient field is K^*. Then $C \cap K$ is normal, but it can happen that the quotient field of $C \cap K$ is not equal to K. Here is a simple example which was constructed by Bill Heinzer. Let x and y be algebraically independent over a field κ, and let $S^* = \kappa[x^3, x^2y]_{(x^3, x^2y)}$. Consider the automorphism of the quotient field $K^* = \kappa(x^3, x^2y, xy^2, y^3)$ of the regular local ring S^* over κ which interchanges x and y. The image of S^* by this automorphism is the two-dimensional regular local ring $S' = \kappa[y^3, y^2x]_{(y^3, y^2x)}$. Regarding S^* and S' as subrings of the formal power series ring $\kappa[[x, y]]$, we see that the intersection of S^* and S' is κ. Hence if K is the fixed field of this automorphism of K^*, we have $S^* \cap K = S' \cap K = \kappa$.

However, every valuation ring V of K^* has the property that $K \cap V$ is a valuation ring of K, and hence its quotient field is K. This follows since the restriction ν of a valuation ν^* of K^* to K is a valuation of K, and $V \cap K$ is the set of all element in K whose value is nonnegative.

A valuation ν of a field K is called a discrete valuation if its value group is $\mathbb{Z}$. A local domain R is called a discrete valuation ring (dvr) if R is the valuation ring of the quotient field of R. The divisorial valuations, defined in Section 12.3, are examples of discrete valuation rings.

Theorem 21.10. *A domain R is a discrete valuation ring if and only if R is a one-dimensional regular local ring.*

Proof. Let K be the quotient field of R. First suppose that R is a one-dimensional regular local ring. If $0 \neq f \in R$, then there exists $n \in \mathbb{N}$ such that $0 \neq f \in m_R^n \setminus m_R^{n+1}$. Since $m_R = (t)$ for some $t \in R$, $f = t^n v$ with $v \in R \setminus m_R$. Thus v is a unit in R. Since K is the quotient field of R, every nonzero element $f \in K$ has a unique expression

(21.4) $$f = t^n v$$

with $n \in \mathbb{Z}$ and $v \in R$ a unit. Define $\nu : K^\times \to \mathbb{Z}$ by $\nu(f) = n$ if f has the expression (21.4). From (21.4), it follows that ν is a valuation of K and that R is the valuation ring of ν.

Now suppose that $R = V_\nu$ where ν is a discrete valuation of K. Then $R = \{f \in K \mid \nu(f) \geq 0\}$ (with the convention that $\nu(0) = \infty$). We have that R is a Noetherian local domain by Theorems 12.15 and 12.16. There exists $t \in R$ such that $\nu(t) = 1$. Suppose that $f \in R$ is nonzero. Let $\nu(f) = n \geq 0$ and let $u = \frac{f}{t^n}$. Then $\nu(u) = 0$ so $u \in R$. Further, $\nu(\frac{1}{u}) = -\nu(u) = 0$ so $\frac{1}{u} \in R$ and thus u is a unit in R. It follows that $m_R = (t)$ and so $\dim_{R/m_R} m_R/m_R^2 = 1$. Now $R \neq K$ since $\frac{1}{t} \notin R$. Thus

$$1 \leq \dim R \leq \dim_{R/m_R} m_R/m_R^2 = 1$$

by Theorem 1.81, so R is a one-dimensional regular local ring. □

From the above proof, we have the following remark.

Remark 21.11. The maximal ideal m_R in a discrete valuation ring R is a principal ideal, $m_R = (t)$, and the nonzero ideals in R are the principal ideals (t^n) for $n \in \mathbb{N}$.

Proposition 21.12. *Let K be a field, ν a valuation of K, K^* an algebraic extension of K, and $R_1^*, R_2^*, \ldots$ the local rings in K^* lying over V_ν. Then R_i^* is the valuation ring $V_{\nu_i^*}$ of a valuation ν_i^* of K^* where $\nu_1^*, \nu_2^*, \ldots$ are exactly the extensions of ν to K^*.*

Proof. [6, Proposition 2.38] □

Let R be a one-dimensional regular local ring with quotient field K and let L be a finite extension of K. Let $S_1, \ldots, S_g$ be the local rings of L which lie over R. The S_i are one-dimensional regular local rings by [160, Theorem 19 on page 281], [160, Theorem 13, page 275], and Theorem 1.87. Since the maximal ideals m_R of R and m_{S_i} of the S_i are principal, there exist positive integers n_i such that $m_R S_i = m_{S_i}^{n_i}$ for $1 \leq i \leq g$. We define

(21.5) $\quad e(S_i/R) = n_i \quad \text{and} \quad f(S_i/R) = [S_i/m_{S_i} : R/m_R]$.

The index $f(S_i/R)$ is finite by [160, Lemma 1 on page 284]. The index $e(S_i/R)$ is called the reduced ramification index of S_i over R and $f(S_i/R)$ is called the relative degree of S_i over R. The reduced ramification index has already been encountered in Section 18.5.

We define

(21.6) $\quad e(\nu^*/\nu) = e(S_i/R) \quad \text{and} \quad f(\nu^*/\nu) = f(S_i/R)$

if $R = V_\nu$ and $S_i = V_{\nu^*}$ where ν is a discrete valuation of K and ν^* is an extension of ν to K^* (which is necessarily a discrete valuation).

Lemma 21.13 (Kronecker). *Let A be a normal domain with quotient field K. Let $f(t), g(t)$ be monic polynomials in $K[t]$ and let $h(t) = f(t)g(t)$. Then $h(t) \in A[t]$ implies $f(t), g(t) \in A[t]$.*

Let K^* be an overfield of K. If $u \in K^*$ is such that u is integral over A, then the minimal polynomial of u over K is in $A[t]$.

Proof. The second statement of the lemma follows from the first. We will prove the first statement. Let L be a splitting field of $h(t)$ over K. Let

$$f(t) = \sum_{i=0}^{n} f_{n-i} t^i = \prod_{i=1}^{n} (t - u_i)$$

where $f_i \in K$, $f_0 = 1$, and $u_i \in L$. Let $p_i(Y_1, \ldots, Y_n)$ be $(-1)^i$ times the i-th elementary symmetric function in $Y_1, \ldots, Y_n$, so that $f_i = p_i(u_1, \ldots, u_n)$. Now $f(u_i) = 0$ implies $h(u_i) = 0$ which implies that the u_i are integral over A. Since the integral closure of A in L is a domain, the fact that $u_1, \ldots, u_n$ are integral over A implies $f_i = p_i(u_1, \ldots, u_n)$ are integral over A. Since A is normal and $f_i \in K$, we must have $f_i \in A$ for $i = 1, \ldots, n$, so $f(t) \in A[t]$. Similarily, $g(t) \in A[t]$. $\square$

Theorem 21.14. *Let A be an integral domain, let K be its quotient field, and let x be an element of an extension of K. Suppose that x is integral over A. Then x is algebraic over K, and the coefficients of the minimal polynomial $f(t)$ of x over K, in particular the norm and trace of x over K, are elements of K which are integral over A. If A is integrally closed, these coefficients are in A.*

Proof. Let L be the algebraic closure of K. Since x is integral over A, it is necessarily algebraic over K. Let n be the degree of the minimal polynomial $f(x)$ of x over K. Then $f(t) = \prod(t - x_i) \in L[t]$ where the x_i are conjugates of x in L over K. Since an equation of integral dependence of x over A is satisfied by all the conjugates x_i of x over K, the coefficients of $f(t)$ are integral over A by Corollary 1.50. In particular, the norm $N_{K(x)/K}(x)$ and trace $T_{K(x)/K}(x)$ are in K and are integral over A by Theorem 21.1. It follows from (21.1) and (21.2) that if L is any finite extension of K containing x, then $N_{L/K}(x)$ and $T_{L/K}(x)$ are in K and are integral over A. $\square$

Exercise 21.15. Let $f(t) \in tk[[t]]$ be transcendental over $k(t)$. Define an injective k-algebra homomorphism $\phi : k[x, y] \to k((t))$, where $k((t))$ is the quotient field of $k[[t]]$, by $\phi(g(x,y)) = g(t, f(t))$ for $g \in k[x, y]$, with induced inclusion $\phi : k(x, y) \to k((t))$. Define a valuation ν of $k(x, y)$ by

$$\nu(g) = \mathrm{ord}_t(g(t, f(t)))$$

for $g \in k(x, y)$. Show that $\nu|k^\times = 0$ so that ν is a k-valuation and ν is a dvr, but ν is not a divisorial valuation.

21.3. Discriminants and ramification

We first define the discriminant ideal of a normal domain. The construction in this section follows [**92**]. It generalizes the classical construction of discriminants of field extensions ([**131**], [**160**, Section 11 of Chapter 2]). Our treatment is based on the section "The discriminant of an ideal" in [**6**].

Let K be a field and L be a finite field extension of K. Let $\{w_1, \ldots, w_n\}$ be a basis of L/K. For $a \in L$, let $T_a : L \to L$ be the K-linear transformation defined by $T_a(b) = ab$ for $b \in L$. For $a \in L$, the trace of a relative to L/K (Section 1.2) is

$$\mathrm{Tr}_{L/K}(a) = \sum_{i=1}^{n} k_{ii}$$

where

$$T_a(w_i) = \sum_{j=1}^{n} k_{ij} w_j \quad \text{with } k_{ij} \in K.$$

For $a_1, \ldots, a_n \in L$, the discriminant of $(a_1, \ldots, a_n)$ relative to L/K is defined to be

$$D_{L/K}(a_1, \ldots, a_n) = \mathrm{Det}(\mathrm{Tr}_{L/K}(a_i a_j)).$$

Basic properties of the discriminant are derived in [**160**, Section 11 of Chapter 2]. The condition of the discriminant $D_{L/K}(a_1, \ldots, a_n)$ being zero or not zero is independent of choice of basis $\{a_1, \ldots, a_n\}$ of L over K, so it makes sense to say that the discriminant $D_{L/K}(a_1, \ldots, a_n)$ of a basis is zero or not zero ([**160**, Corollary, page 93]). The discriminant $D_{L/K}(a_1, \ldots, a_n)$ of a basis $\{a_1, \ldots, a_n\}$ of L over K is nonzero if and only if L is separable over K ([**160**, Theorem 22, page 95] or equation (21.2)). If L/K is separable, then L has a primitive element b over K (Theorem 1.16), and then if $n = [L : K]$,

$$(21.7) \qquad D_{L/K}(1, b, b^2, \ldots, b^{n-1}) = \prod_{i<j}(b_i - b_j)^2$$

where the b_i are the n distinct roots of the minimal polynomial $f(x)$ of b over K in an algebraic closure of K containing L by [**160**, formula (7) on page 95]. Let $f(x) = x^n + a_1 x^{n-1} + \cdots + a_n$ be the minimal polynomial of b over K. We can then compute the discriminant $D_{L/K}(1, b, b^2, \ldots, b^{n-1})$ as

$$(21.8) \qquad D_{L/K}(1, b, b^2, \ldots, b^{n-1}) = (-1)^{\frac{n(n-1)}{2}} \mathrm{Res}(f, f')$$

21.3. Discriminants and ramification

by [**95**, Proposition 8.5, page 204], where the resultant $\mathrm{Res}(f, f')$ is the determinant of the $(2n-1) \times (2n-1)$ matrix

$$\begin{pmatrix} 1 & a_1 & a_2 & \cdots & a_{n-1} & a_n & 0 & \cdots & & \cdots & 0 \\ 0 & 1 & a_1 & a_2 & \cdots & a_{n-1} & a_n & & 0 & \cdots & 0 \\ \vdots & & & & & & & & & & \\ 0 & \cdots & & 0 & 1 & a_1 & a_2 & \cdots & & a_{n-1} & a_n \\ n & (n-1)a_1 & (n-2)a_2 & \cdots & & a_{n-1} & 0 & & \cdots & & 0 \\ 0 & n & (n-1)a_1 & (n-2)a_2 & \cdots & & a_{n-1} & & 0 & \cdots & 0 \\ \vdots & & & & & & & & & & \\ 0 & 0 & & \cdots & & 0 & n & (n-1)a_1 & (n-2)a_2 & \cdots & a_{n-1} \end{pmatrix}.$$

We have that

$$(21.9) \qquad \mathrm{Res}(f, f') = \prod_{i=1}^{n} f'(b_i)$$

by [**95**, formula (4), page 204]. The formula (21.8) is also called the discriminant of f.

Let A be a normal domain with quotient field K, and let L be a finite extension of K with $[L:K] = n$. Let B be a domain with quotient field L which contains A and is integral over A. If $a_1, \ldots, a_n \in B$, then $\mathrm{Tr}_{L/K}(a_i a_j) \in A$ for all i, j by Theorem 21.14, so that $D_{L/K}(a_1, \ldots, a_n) \in A$. We define the discriminant ideal $D(B/A)$ of B over A to be the ideal in A generated by the discriminants $D_{L/K}(w_1, \ldots, w_n)$ of all the bases $\{w_1, \ldots, w_n\}$ of L/K which are in B. If B is the integral closure of A in L, we will sometimes write $D(L/A)$ to denote $D(B/A)$.

The following formula is useful in computing discriminants. Suppose that $\{a_1, \ldots, a_n\}$ and $\{b_1, \ldots, b_n\}$ are bases of L/K. Define $k_{ij} \in K$ by $b_i = \sum_j k_{ij} a_j$. Then ([**160**, equation (2), page 93] or [**6**, equation (5), page 26])

$$(21.10) \qquad D_{L/K}(b_1, \ldots, b_n) = \mathrm{Det}(k_{ij})^2 D_{L/K}(a_1, \ldots, a_n).$$

We thus have that $D(B/A)$ localizes; that is, if S is a multiplicative set in A, then

$$(21.11) \qquad D(S^{-1}B/S^{-1}A) = S^{-1}D(B/A).$$

Definition 21.16. Suppose R is a local domain and S is a local domain which dominates R ($R \subset S$ and $m_S \cap R = m_R$). We say that the extension $R \to S$ is unramified if $m_R S = m_S$ and S/m_S is finite and separable over R/m_R. Otherwise, we say that the extension is ramified.

This definition generalizes our definition of ramification for a regular map of nonsingular curves in Section 18.5. A dominant regular map of nonsingular projective curves $\phi : X \to Y$ is ramified at $P \in X$ if and only if the extension $\mathcal{O}_{Y,\phi(P)} \to \mathcal{O}_{X,P}$ is ramified.

Let A be a normal local domain with quotient field K. Let L be a finite extension of K, and let B be a domain with quotient field L which contains A and is integral over A. Let $Q_1, \ldots, Q_t$ be the prime ideals in B lying over the maximal ideal m_A. If the B_{Q_i} are unramified over A for $1 \le i \le t$, then we say that B is unramified over A. Otherwise, we will say that B is ramified over A. We will say that A is unramified in L if the integral closure B of A in L is unramified over A.

Suppose that L is a finite extension field of a field K. Let M be the maximal separable extension field of K in L. The separable degree of L over K is $[L:K]_s = [M:K]$ (Section 1.2).

Let A be a ring containing a field κ such that A is a finite-dimensional vector space over κ. We can extend the definitions of the trace and discriminant to this situation to define the discriminant $D_{A/\kappa}(a_1, \ldots, a_n)$ of a basis $a_1, \ldots, a_n$ of A over κ. The formula (21.10) holds in this situation, so the question of whether a discriminant of A over κ is zero or nonzero is independent of the choice of a basis of A over κ. Since A is an Artinian ring,

$$A \cong A_1 \oplus \cdots \oplus A_m$$

where the A_i are Artin local rings by [**13**, Theorem 8.7].

Proposition 21.17. *With the above assumptions on A and κ, a discriminant of A over κ is nonzero if and only if each A_i is a field which is a separable field extension of κ.*

Proof. Let $e_i = (0, \ldots, 0, 1_{A_i}, 0, \ldots, 0) \in A$ (where 1_{A_i} is in the i-th position) for $1 \le i \le m$. Let $\{w_{ij} \mid 1 \le j \le n_i\}$ be a κ-basis of A_i. Then $\{e_i w_{ij} \mid 1 \le i \le m, 1 \le j \le n_i\}$ is a κ-basis of A. Let $a \in A$ and expand

$$a = ae_1 + \cdots + ae_m \quad \text{with } a_i \in A_i.$$

Then

$$ae_i w_{ij} = \left(\sum_l a_l e_l \right) e_i w_{ij} = a w_{ij} e_i = \left(\sum_{u=1}^{n_i} k_{ju}^{(i)} w_{iu} \right) e_i \quad \text{for some } k_{ju}^{(i)} \in \kappa.$$

Thus

$$\operatorname{Tr}_{A/\kappa}(a) = \operatorname{Tr}_{A_1/\kappa}(a_1) + \cdots + \operatorname{Tr}_{A_m/\kappa}(a_m)$$

21.3. Discriminants and ramification

and $i \neq u$ implies $e_i w_{ij} e_u w_{uv} = 0$, so that $\text{Tr}_{A/\kappa}(e_i w_{ij} e_u w_{uv}) = 0$. Thus

$$D_{A/k}(e_1 w_{11}, e_1 w_{12}, \ldots, e_1 w_{1n_1}, e_2 w_{21}, \ldots, e_m w_{mn_m})$$
$$= \text{Det}(\text{Tr}_{A/\kappa}(e_i w_{ij} e_u w_{uv}))$$
$$= \prod_{i=1}^{m} \text{Det}(\text{Tr}_{A/\kappa}(e_i w_{ij} e_i w_{iv}))$$
$$= \prod_{i=1}^{m} D_{A_i/\kappa}(w_{i1}, \ldots, w_{in_i}),$$

so that

(21.12)   the discriminant of A over κ is nonzero if and only if the discriminant of A_i over κ_i is nonzero for $i = 1, \ldots, m$.

Now suppose that the radical $N = \sqrt{(0)}$ of A is nonzero, and let $s > 1$ be the integer which satisfies $N^{s-1} \neq 0$ and $N^s = 0$. For $0 \leq i \leq s-1$, N^i is a finite-dimensional vector space over κ, so there exist $p_{i1}, \ldots p_{iq_i}$ in N^i whose residue classes mod N^{i+1} form a κ-basis of N^i/N^{i+1}. Then $\{p_{ij} \mid 0 \leq i \leq s-1, 1 \leq j \leq q_i\}$ is a κ-basis of A. Now $a \in N$ implies

$$a p_{ij} = \sum_{v > i} k_{ijvu} p_{vu}$$

for some $k_{ijvu} \in \kappa$, which implies $\text{Tr}_{A/\kappa}(a) = 0$. Thus since there exists a basis $\{w_1, \ldots, w_n\}$ of A over κ with at least the last element $w_n \in N$ if $N \neq 0$,

(21.13) $N \neq 0$ implies the discriminant of A/κ is zero.

The proposition now follows from (21.12), (21.13), and the fact that the discriminant of a finite field extension L over κ is nonzero if and only if L is separable over κ [**160**, Theorem 22, page 95]. □

We will say that $a \in A$ is a primitive element of A over κ if $A = \kappa[a]$.

Lemma 21.18. *Suppose that κ is an infinite field and $A \cong A_1 \oplus \cdots \oplus A_m$ where the A_i are finite separable field extensions of κ. Then A has a primitive element over κ.*

Proof. Let $e_i = (0, \ldots, 0, 1_{A_i}, 0, \ldots, 0) \in A$ (where 1_{A_i} is in the i-th position) for $1 \leq i \leq m$. Let a_j be primitive elements of A_j over κ (Theorem 1.16) and let $g_j(x) \in \kappa[x]$ be the respective minimal polynomials of the a_j over κ. Since κ is assumed to be infinite and two polynomials in $\kappa[x]$ are coprime if and only if they have no common roots in an algebraic closure of κ, after possibly multiplying the a_j by suitable nonzero elements of κ, we may assume that the $g_j(x)$ are pairwise coprime in $\kappa[x]$. Let $a = e_1 a_1 + \cdots + e_m a_m$. Suppose $f(x) = b_0 x^q + b_1 x^{q-1} + \cdots + b_q \in \kappa[x]$ satisfies $f(a) = 0$. Then

$0 = e_j f(a) = e_j f(a_j)$ so that $f(a_j) = 0$ for $1 \leq j \leq m$ and thus $g_j(x)$ divides $f(x)$ in $\kappa[x]$ for $1 \leq j \leq m$, so that the minimal polynomial $g(x)$ of a over κ is divisible by $\prod_{j=1}^m g_j(x)$. Since $\dim_\kappa A = \sum_{i=1}^m \deg(g_i(x))$, we have that a is a primitive element of A over κ. $\square$

Lemma 21.19. *Suppose that the ring A is a finite-dimensional vector space over an infinite field κ, so that $A = A_1 \oplus \cdots \oplus A_m$ where the A_i are Artin local rings. Suppose that some A_i is not a separable field extension of κ. Let N_i be the maximal ideal of A_i. Then there exists $a \in A$ such that the minimal polynomial of a in $\kappa[x]$ has degree $> \sum_{i=1}^m [A_i/N_i : \kappa]_s$.*

Proof. We will first find $a_i \in A_i$ such that the minimal polynomial of a_i over κ has degree $\geq [A_i/N_i : \kappa]_s$, with a strict inequality if A_i/N_i is not a separable field extension of κ. Let $b_i \in A_i$ be such that the minimal polynomial of the residue $\bar{b}_i$ of b_i in A_i/N_i has degree $[A_i/N_i : \kappa]$ if A_i/N_i is separable over κ and has degree $\geq [A_i/N_i : \kappa]_s p$, where p is the characteristic of κ if A_i/N_i is not separable over κ (by the primitive element theorem in [**142**, page 139]). If $N_i = 0$, then we take $a_i = b_i$. Suppose $N_i \neq 0$ and A_i/N_i is separable over κ. Let $h_i(x)$ be the minimal polynomial of $\bar{b}_i$ over κ. If $h_i(b_i) \neq 0$, then we can take $a_i = b_i$, so suppose $h_i(b_i) = 0$. There exists $0 \neq c_i \in N_i$. Let $a_i = b_i + c_i$. Then $h_i'(b_i)^r \neq 0$ for all r, where $h_i'(x)$ is the derivative of $h_i(x)$, so that $h_i'(b_i) \notin N_i$. We have $h_i(a_i) = c_i[h_i'(b_i) + c_i f]$ with $f \in A_i$. Thus $h_i'(b_i) + c_i f \notin N_i$ which implies that $h_i'(b_i) + c_i f$ is a nonzero divisor in A_i, so that $h_i(a_i) \neq 0$. Since $h_i(\bar{a}_i) = h_i(\bar{b}_i) = 0$, $h_i(a_i)^r = 0$ and $h_i(a_i)^{r-1} \neq 0$ for some $r > 1$ and so $h_i^r(x)$ is the minimal polynomial of a_i over κ, since $h_i(x)$ is irreducible in $\kappa[x]$. $\square$

Now by the argument of the proof of Lemma 21.18, we can find $a \in A$ satisfying the conclusions of the lemma.

Theorem 21.20. *Let R be a normal local domain with quotient field K and K^* be a finite algebraic extension of K. Let R^* be a domain with quotient field K^* such that R^* contains and is integral over R. Then $D(R^*/R) = R$ implies R^* is unramified over R.*

Proof. Let $N^* = m_R R^*, \kappa = R/m_R$ and $A = R^*/N^*$. We will denote the class of the residue of an element $x \in R^*$ in A by $\bar{x}$.

Since $D(R^*/R) = R$, there exists a basis $w_1, \ldots, w_n$ of K^* over K in R^* such that $D(w_1, \ldots, w_n)$ is a unit in R. Given $w \in K^*$, we have that $w = a_1 w_1 + \cdots + a_n w_n$ with $a_j \in K$. If $w \notin w_1 R + \cdots + w_n R$, then some $a_j \notin R$, say $a_1 \notin R$. It follows that $a_1^2 \notin R$, since if $a_1^2 \in R$, then a_1 is integral over R, which is impossible since R is normal. Then

$$D(w, w_2, \ldots, w_n) = a_1^2 D(w_1, w_2, \ldots, w_n) \notin R$$

implies w is not integral over R, since if w is integral over R, then
$$D(w, w_2, \ldots, w_n) \in \overline{R} \cap K = R,$$
where $\overline{R}$ is the integral closure of R in K^* (by Theorem 21.14). Since
$$w_1 R + \cdots + w_n R \subset R^*,$$
we have that $R^* = w_1 R + \cdots + w_n R$ is the integral closure of R in K^*. Thus $w_1, \ldots, w_n$ is a free R-basis of R^*, and so
$$N^* = m_R R^* = w_1 m_R + \cdots + w_n m_R$$
and A is a free κ-module of rank n with basis $\overline{w}_1, \ldots, \overline{w}_n$. Let
$$(w_i w_j) w_p = \sum_{q=1}^n a_{ijpq} w_q$$
with $a_{ijpq} \in R$. Then
$$(\overline{w}_i \overline{w}_j) \overline{w}_p = \sum_{q=1}^n \overline{a}_{ijpq} \overline{w}_q.$$
Thus $D(\overline{w}_1, \ldots, \overline{w}_n)$ is the discriminant of $D(w_1, \ldots, w_n)$ modulo m_R. Hence
$$D(\overline{w}_1, \ldots, \overline{w}_n) \neq 0,$$
since $D(w_1, \ldots, w_n)$ is a unit in R. Thus R^* is unramified over R by Proposition 21.17. $\square$

Theorem 21.21. *Let R be a Noetherian normal local domain and let K be the quotient field of R. Let K^* be a finite extension of K, and let R^* be the integral closure of R in K^*. Suppose that R^* is a finitely generated R-module. Then R^* is unramified over R if and only if the discriminant ideal $D(R^*/R) = R$.*

The separability index $[L : K]_s$ of a finite extension of fields L/K is defined in (1.2).

Proof. The "if" direction follows from Theorem 21.20. We will prove the "only if" direction. With our assumptions, R^* is a finite R-module. Let $\kappa = R/m_R$ and $A = R^*/m_R R^*$. The ring A is a finite κ-vector space, so $A = \bigoplus_{i=1}^r A_i$ where the A_i are Artin local rings. Since R^* is unramified over R, we have that each A_i is a separable field extension of κ. Let $u_1, \ldots, u_n \in R^*$ be such that their residues $\overline{u}_1, \ldots, \overline{u}_n$ in A are a κ-basis of A. By Nakayama's lemma (Lemma 1.18),

(21.14) $$R^* = u_1 R + \cdots + u_n R.$$

Let $L = u_1 K + \cdots + u_n K$, a subring of K^* by (21.14). Since L is a finite-dimensional K-vector space, L is a subfield of K^*, so that $L = K^*$. Thus

$$[K^* : K] \leq n = [A : \kappa] = \sum_{i=1}^{t} [A_i : k']_s.$$

The proof for general κ now follows from [**6**, Theorem 1.42] (referring to [**92**] in the case when κ is finite). We will present here a proof with the assumption that κ is an infinite field. Since the A_i are separable over the infinite field κ, there exists a primitive element $\overline{v} \in A$ of A over κ by Lemma 21.18. Let v be a lift of $\overline{v}$ to R^*. Let $f(x) \in K[x]$ be the minimal polynomial of v over K. Then $f(x) \in R[x]$ by Theorem 21.14 since R is normal. Let $h(x) \in \kappa[x]$ be the minimal polynomial of $\overline{v}$ over κ. Then h divides the reduction $\overline{f}$ of f in $\kappa[x]$, so

(21.15) $\qquad [K^* : K] \geq \deg(\overline{f}) \geq \deg(h) = [A : \kappa] = n.$

Thus $[K^* : K] = [A : \kappa] = n$. Now Proposition 21.17 implies that the discriminant $D(\overline{u}_1, \ldots, \overline{u}_n)$ of A over κ is nonzero. Thus $D(u_1, \ldots, u_n) \not\in m_R$, which we have shown is in the discriminant ideal of R^* over R, so $D(R^*/R) = R$. $\qquad\square$

Theorem 21.22. *Let A be a normal domain with quotient field K. Let L be a finite extension of K, and let B be the integral closure of A in L. Let P be a prime ideal in A. Let $Q_1, \ldots, Q_t$ be the prime ideals in B lying over P. Let $\kappa = A_P/PA_P$ and $\kappa_i = B_{Q_i}/Q_i B_{Q_i}$ for $1 \leq i \leq t$. Then*

(21.16) $\qquad \sum_{i=1}^{t} [\kappa_i : \kappa]_s \leq [L : K],$

and equality holds if and only if the discriminant ideal $D(B/A) \not\subset P$.

Proof. We will prove the theorem with the assumptions that A is Noetherian, κ is infinite, and B is finite over A, referring to [**6**, Theorem 1.45] for the general case. Write $PB_P = I_1 \cap \cdots \cap I_t$ where the I_i are Q_i-primary ideals by Lemma 21.8. There exists $r > 0$ such that $Q_i^r \subset I_i$ for $1 \leq i \leq t$. We have that

$$\begin{aligned} B_P/PB_P &\cong \bigoplus_{i=1}^{t} B_P/(Q_i^r B_P + PB_P) \quad \text{by Theorem 1.5} \\ &\cong \bigoplus_{i=1}^{t} B_{Q_i}/(Q_i^r B_{Q_i} + PB_{Q_i}) \quad \text{by Lemma 1.28} \\ &\cong \bigoplus_{i=1}^{t} B_{Q_i}/I_i B_{Q_i}. \end{aligned}$$

We have a natural surjection $B_P/PB_P \to \bigoplus_{i=1}^{t} \kappa_i$. By the theorem of the primitive element (Theorem 1.16) there exist $\overline{u}_i \in \kappa_i$ such that the minimal polynomial $g_i(x) \in \kappa[x]$ of $\overline{u}_i$ has degree equal to $[\kappa_i : \kappa]_s$. Since κ is assumed to be infinite, we can replace the $\overline{u}_i$ with suitable products of the $\overline{u}_i$ with elements of κ to get that the minimal polynomials $g_i(x)$ are pairwise coprime in $\kappa[x]$. Let $\overline{v} = \overline{u}_1 + \cdots + \overline{u}_t$. Then the minimal polynomial $g(x) \in \kappa[x]$ of $\overline{v}$ is divisible by $\prod_{i=1}^{t} g_i(x)$. Let v be a lift of $\overline{v}$ to B_P. Let $f(x) \in K[x]$ be the minimal polynomial of v over K. Then $f(x) \in A_P[x]$ by Theorem 21.14. Let $\overline{f}(x)$ be the reduction of $f(x)$ in $\kappa[x]$. Then $\overline{f}(\overline{v}) = 0$, so that $g(x)$ divides $\overline{f}(x)$ and thus

$$[L:K] \geq [K(v):K] \geq \deg g(x) \geq \sum_{i=1}^{t} [\kappa_i : \kappa]_s.$$

Let $R = A_P$ and S be the integral closure of R in L. We have that $D(B/A)_P = D(S/R)$ by (21.11) since $S = T^{-1}B$ where $T = A \setminus P$. Thus $D(S/R) = R$ if and only if $D(B/A) \not\subset P$. By Theorem 21.21, $D(S/R) = R$ implies $R \to S$ is unramified, and it is shown in the proof of Theorem 21.21 that $R \to S$ unramified implies equality in (21.16). Assume equality holds in (21.16). Then $B_P/PB_P \cong \bigoplus_{i=1}^{t} \kappa_i$ and each κ_i is a separable extension of κ by Lemma 21.19. Thus B_P/PB_P has a primitive element $\overline{u}$ over κ by Lemma 21.18, and the discriminant $D(1, \overline{u}, \ldots, \overline{u}^{n-1}) \neq 0$ by Proposition 21.17. Let u be a lift of $\overline{u}$ to B_P. Then the residue of $D(1, u, \ldots, u^{n-1}) \in A_P$ in κ is $D(1, \overline{u}, \ldots, \overline{u}^{n-1})$, so that $D(1, u, \ldots, u^{n-1})$ is a unit in A_P. Now u is a primitive element of L over K since $1, u, \ldots, u^{n-1}$ are linearly independent over κ, since $D(1, u, \ldots, u^{n-1}) \neq 0$ and by (21.10), so that $L = K(u)$ since $[L:K] = n$. Thus $D(S/R) = R$. □

A particularly strong form of Theorem 21.22 holds if A is a Dedekind domain, as we have seen in the case of curves, Theorem 13.18, and more generally as shown in [**160**, Section 9 of Chapter V].

Proposition 21.23. *Suppose that R is a one-dimensional regular local ring with quotient field K and L is a finite separable extension of K. Let $S_1, \ldots, S_g$ be the normal local rings of L which lie over R. Then the S_i are one-dimensional regular local rings, and the indices $e(S_i/R)$ and $f(S_i/R)$ defined in equation (21.5) satisfy*

$$\sum_{i=1}^{g} e(S_i/R) f(S_i/R) = [L:K],$$

and if L is a Galois extension of K, then the $e(S_i/R)$ all have a common value e and the $f(S_i/R)$ all have a common value f, so that $efg = [L:K]$.

Proof. [**160**, Corollary, page 287] and [**160**, Theorem 22, page 289]. □

Suppose that R is a one-dimensional regular local ring with quotient field K and L is a finite extension of K. Suppose that S is a normal local ring of L which lies over R. Then $R \to S$ is unramified if and only if $e(S/R) = 1$ and S/m_S is a separable field extension of R/m_R.

21.4. Ramification of regular maps of varieties

Proposition 21.24. *Let A be a normal domain with quotient field K and let L be a finite extension field of K. Suppose that the integral closure of A in L is a finite A-module. Let*

$$U = \{P \in \mathrm{Spec}(A) \mid A_P \text{ is unramified in } L\}.$$

Then $U = \mathrm{Spec}(A) \setminus Z(D(B/A))$ is an open subset of $\mathrm{Spec}(A)$.

Proof. By Theorem 21.21, for $P \in \mathrm{Spec}(A)$, A_P is unramified in L if and only if $D(B_P/A_P) = A_P$. By (21.11), $D(B_P/A_P) = D(B/A)_P$. Thus A_P is unramified in L if and only if $D(B/A) \not\subset P$. □

If R is a normal local ring and L is a finite extension field of the quotient field of R such that the integral closure of R in L is a finite R-module and R is unramified in L, then R_P is unramified in L for all $P \in \mathrm{Spec}(R)$. We may thus define a normal domain A to be unramified in a finite extension L of the quotient field of A if A_P is unramified in L for all $P \in \mathrm{Spec}(A)$.

Definition 21.25. Let X and Y be varieties, and let $\phi : X \to Y$ be a regular map. The degree of ϕ is

$$\deg(\phi) = \begin{cases} [k(X) : k(Y)] & \text{if } \dim X = \dim Y \text{ and } \phi \text{ is dominant,} \\ 0 & \text{otherwise.} \end{cases}$$

This definition was anticipated in Chapter 19.

Theorem 21.26. *Suppose that $\phi : X \to Y$ is a dominant finite map of varieties and Y is normal. Then the number of points above any point $y \in Y$ is less than or equal to $\deg(\phi)$.*

Proof. Let U be an affine neighborhood of y in Y and let $V = \phi^{-1}(U)$. Then V is affine and $\phi : V \to U$ is finite (by Theorem 7.5). Let W be the normalization of V (Theorem 7.17). The variety W is also affine and the induced map $\psi : W \to U$ is finite and factors through V. Further, $\deg(\psi) = \deg(\phi)$ since $k(W) = k(V) = k(X)$. It thus suffices to prove the theorem with Y replaced by U and X replaced by W. The theorem now follows from Theorem 21.22, taking P to be the prime ideal of y in $A = k[U]$. □

Definition 21.27. Suppose that $\phi : X \to Y$ is a dominant finite map of normal varieties and $y \in Y$. We will say that ϕ is ramified at y if $\mathcal{O}_{Y,y} \to B$ is ramified, where B is the integral closure of $\mathcal{O}_{Y,y}$ in $k(X)$. The map ϕ is said to be unramified at y if this extension is not ramified. We will say that the map ϕ is unramified if ϕ is unramified at y for all $y \in Y$. We will say that ϕ is unramified at $x \in X$ if $\mathcal{O}_{Y,\phi(x)} \to \mathcal{O}_{X,x}$ is unramified.

Theorem 21.28. *Suppose that $\phi : X \to Y$ is a dominant finite map of normal varieties and $y \in Y$. Then ϕ is unramified at y if and only if the number of points in the preimage $\phi^{-1}(y)$ is equal to the degree $\deg(\phi)$.*

Proof. As in the proof of Theorem 21.26, we may assume that Y and X are affine. The theorem then follows from Theorems 21.22 and 21.21, with the observation that in the language of these theorems, $k \cong A/P \cong B/Q_i$ for all Q_i, since P is the ideal of y in $A = k[Y]$ and the Q_i are the ideals in $B = k[X]$ of the points in the preimage of y by ϕ. □

Theorem 21.29. *Suppose that $\phi : X \to Y$ is a dominant finite map of normal varieties. Then the set of points in Y at which ϕ is unramified is open and is nonempty if and only if $k(X)$ is a separable extension of $k(Y)$.*

Proof. It suffices to prove this in the case when Y and X are affine. Let $A = k[Y]$ and $B = k[X]$. Then ϕ is unramified at $y \in Y$ if and only if the ideal $P = I(y)$ of y in A does not contain the discriminant ideal $D(B/A)$ by Theorem 21.21, that is, if and only if $y \notin Z(D(B/A))$. Thus the set of unramified points is open.

By Theorem 21.21 we have that the set of points in Y at which ϕ is unramified is nonempty if and only if $Z(D(B/A)) \neq Y$, which holds if and only if $D(B/A) \neq (0)$. By Theorem 21.22, taking P to be the prime ideal (0) in A, we have that $D(B/A) \neq (0)$ if and only if

$$[k(X) : k(Y)]_s = [k(X) : k(Y)].$$ □

Definition 21.30. *Suppose that $\phi : X \to Y$ is a dominant finite map of normal varieties. The closed set of points $p \in Y$ at which ϕ is ramified is called the ramification locus of ϕ in Y.*

Exercise 21.31. Suppose that $R = \kappa[x_1, \ldots, x_m]$ is a polynomial ring over a field κ and $f = z^n + a_1 z^{n-1} + \cdots + a_n$ with $a_i \in R$ is an irreducible polynomial in $R[z]$. Let $S = R[z]/(f) = R[\bar{z}]$ where $\bar{z}$ is the residue of z in S. Then S is a domain which is finite over R. Let K be the quotient field of R and L be the quotient field of S. Using the definition of $D(S/R)$, show that $D(S/R)$ is generated by $D_{L/K}(1, \bar{z}, \ldots, \bar{z}^{n-1})$.

Exercise 21.32. Let $R = k[x]$ be a polynomial ring over an algebraically closed field k of characteristic $\neq 2$ or 3 and let $S = R[y]/(f)$. Let X be the affine variety with regular functions S and induced regular map $\Phi : X \to \mathbb{A}^1$. In each of the following problems, show that S is normal and finite over R. Compute the ideal $D(S/R)$ in terms of $\text{Res}(f, f')$ using formula (21.8). Compute the ramification locus of the map ϕ in $\mathbb{A}^1$.

a) $f = y^2 + 3xy + (x^2 + 3)$.

b) $f = y^3 + x^3 + 1$.

21.5. Completion

In this section, we will give a brief survey of the topic of completion of a ring, referring to [**161**, Chapter VIII], [**107**, Chapter 9], and [**50**, Chapter 7] for more details.

Let A be a ring with a given topology. The ring A is called a topological ring if the ring operations are continuous. Let A be a topological ring. An A-module E, with a given topology, is said to be a topological A-module if the module operations on E are continuous.

Let E be a topological A-module and let $\Sigma(E)$ be a system of open sets in E which contain the zero 0 of E and such that any open set in E containing 0 contains a set of the system $\Sigma(E)$. Then the system of sets of the form $x + U$ where $x \in E$ and $U \in \Sigma(E)$ is a basis for the topology of E. Such a set $\Sigma(E)$ is called a basis of neighborhoods of 0 for the topological module E. It follows that if $\Sigma(E)$ is a basis of neighborhoods of the zero of a topological A-module E, then E is a Hausdorff (separated) space if and only if

$$\bigcap_{U \in \Sigma(E)} U = \{0\}.$$

Let I be an ideal of a ring R. The system $\{I^n \mid n \in \mathbb{N}\}$ is a basis of neighborhoods of 0 of a topology of R called the I-adic topology. If E is an R-module, then the I-adic topology of E is defined by taking $\{I^n E \mid n \in \mathbb{N}\}$ as a basis of neighborhoods of 0. A submodule F of E has the I-adic topology and the induced topology which has $\{(I^n E) \cap F \mid n \in \mathbb{N}\}$ as a basis of neighborhoods of 0. Since $I^n F \subset (I^n E) \cap F$ for all $n \geq 0$, the inclusion map $F \to E$ is continuous, with the respective I-adic topologies.

Theorem 21.33. *Suppose that A is a Noetherian ring, I is an ideal of A, and E is a finite A-module. Then for every submodule F of E, the I-adic topology of F is induced by the I-adic topology of E.*

Proof. Since E is a finite A-module, the Artin-Rees lemma, [**161**, Theorem $4'$, page 255], tells us that there exists an integer k such that

$$I^n E \cap F = I^{n-k}(I^k E \cap F) \subset I^{n-k} F$$

for every $n \geq k$. $\square$

Suppose that E is an A-module with the I-adic topology. A sequence (x_n) in E is called a Cauchy sequence in E if $x_n - x_{n+i} \in I^{N(n)} E$ for all $i \geq 0$, where $N(n) \mapsto \infty$ as $n \mapsto \infty$. A limit of a Cauchy sequence (x_n) in E is an element y of E such that given $m \in \mathbb{N}$, there exists $n_0 \in \mathbb{N}$ such that $x_n - y \in I^m E$ whenever $n \geq n_0$. If E is separated, a limit, if it exists, is

21.5. Completion

unique. The module E is said to be complete if every Cauchy sequence in E converges to a limit in E.

Theorem 21.34. *Let A be a ring with an ideal I and M be an A-module with the I-adic topology. Then there exists an A-module $\hat{M}$ which is complete and separated for the I-adic topology, with a continuous homomorphism $\phi : M \to \hat{M}$, which satisfies the following universal property: for every A-module M' which is complete and separated for the I-adic topology, and for any continuous homorphism $f : M \to M'$, there exists a unique continuous homomorphism $\hat{f} : \hat{M} \to M'$ satisfying $\hat{f} \circ \phi = f$.*

It follows from the universal property that $\hat{M}$ is unique up to isomorphism. The module $\hat{M}$ is called the completion of M. It is a topological $\hat{A}$-module. Several proofs of the theorem are given in [**107**, Section (23.H), page 165]. The proof shows that the kernel of $\phi : M \to \hat{M}$ is $\bigcap_{n \geq 0} I^n M$, so ϕ is injective if and only if M is separated. One of the constructions shows that

$$\hat{M} \cong \varprojlim M/I^n M.$$

The topology on $\hat{M}$ is defined as follows. By (11.2), $\hat{M}$ is naturally a submodule of $\prod_n M/I^n M$ which has the product topology, and $\hat{M}$ has the subspace topology. We have the following useful properties.

Theorem 21.35. *Suppose that R is a Noetherian ring and I is an ideal in R. Let $\hat{R}$ be the I-adic completion of R. Then $\hat{R}$ is Noetherian.*

Proof. [**13**, Theorem 10.26]. □

Theorem 21.36. *Suppose that R is a Noetherian local ring and I is a proper ideal of R. Let $\hat{R}$ be the I-adic completion of R. Then $\dim \hat{R} = \dim R$.*

Proof. [**106**, Theorem 15.7]. □

Lemma 21.37. *Suppose that A is a Noetherian local ring and $\hat{A}$ is the m_A-adic completion of A. Let I be an ideal of A. Then $I\hat{A} \cap A = I$.*

Proof. The map $A \to \hat{A}$ is faithfully flat by [**107**, (24.B), page 173, and Theorem 56 (5), page 172]. The conclusions of the lemma now follow from [**107**, (4.C)(ii), page 28]. □

If R is a local ring, then $\hat{R}$ will denote the m_R-adic completion of R, unless we explicitly say otherwise.

Proposition 21.38. *A Noetherian local ring R is separated in its m_R-adic topology. Thus the natural map $R \to \hat{R}$ is an inclusion.*

Proof. [**161**, Theorem 9, page 262]. □

A local ring R is said to be equicharacteristic if R and its residue field have the same characteristic. The ring R is equicharacteristic if and only if R contains a field. A coefficient field of a local ring R is a subfield L of R which is mapped onto the residue field R/m_R. A coefficient field of R is thus isomorphic to R/m_R.

A theorem of fundamental importance is the following theorem of Cohen. Proofs can be found in [30] and [161, Theorem 27].

Theorem 21.39. *An equicharacteristic complete Noetherian local ring R has a coefficient field.*

If R has equicharacteristic 0, then the existence of a coefficient field follows from Hensel's lemma [161, Corollary 2, page 280].

We have the following corollary of Theorem 21.39 ([30] or [161, Corollary on page 307]).

Corollary 21.40. *An equicharacteristic complete regular local ring R is isomorphic to a formal power series ring over a field. If K is a coefficient field and $x_1, \ldots, x_d$ is a regular system of parameters in R, then $R = K[[x_1, \ldots, x_d]]$ is a d-dimensional power series ring over K.*

If R is a regular local ring of mixed characteristic (the characteristic of R is 0 but the characteristic of its residue field is $p > 0$), then $\hat{R}$ is a power series ring over a complete discrete valuation ring if R is unramified ($p \notin m_R^2$). However, if R is ramified ($p \in m_R^2$), this may not be true. Examples of this are given on [30, pages 93–94].

The following two propositions are very helpful in computing completions. The following proposition follows from Corollary 21.40, as we always have that our algebraically closed field k is a coefficient field of the local ring of a point on a variety.

Proposition 21.41. *Suppose that X is a variety and $p \in X$ is a nonsingular point. Let $x_1, \ldots, x_n$ be regular parameters in $\mathcal{O}_{X,p}$. Then $\hat{\mathcal{O}}_{X,p} = k[[x_1, \ldots, x_n]]$ is a power series ring over k in $x_1, \ldots, x_n$.*

The following proposition follows from [161, Theorem 6, page 257].

Proposition 21.42. *Suppose that R is a Noetherian ring and I and J are ideals in R. Then $\widehat{R/J} \cong \hat{R}/J\hat{R}$ where the completion is the I-adic completion.*

Lemma 21.43. *Suppose that R and S are local rings and $\phi : R \to S$ is a homomorphism such that $\phi(m_R) \subset m_S$. Then there exists a unique homomorphism $\hat{\phi} : \hat{R} \to \hat{S}$ such that $\hat{\phi}(m_{\hat{R}}) \subset m_{\hat{S}}$ and $\hat{\phi}(x) = \phi(x)$ for $x \in R$.*

Proof. We first prove existence. Suppose that $y \in \hat{R}$. Let (y_n) be a Cauchy sequence in R which has y as its limit. Then $(\phi(y_n))$ is a Cauchy sequence in S since $\phi(m_R) \subset m_S$. Thus there exists $z \in \hat{S}$ which is the limit of $(\phi(y_n))$. We have that z does not depend on the choice of Cauchy sequence (y_n) which has y as its limit. We may thus define $\hat{\phi}(y) = z$. The map $\hat{\phi} : \hat{R} \to \hat{S}$ is a homomorphism such that $\hat{\phi}(m_{\hat{R}}) \subset m_{\hat{S}}$ and $\hat{\phi}(x) = \phi(x)$ for $x \in R$.

We now prove uniqueness. Let $\psi : \hat{R} \to \hat{S}$ be a homomorphism such that $\psi(m_{\hat{R}}) \subset m_{\hat{S}}$ and $\psi(x) = \phi(x)$ for $x \in R$. Let $y \in \hat{R}$, and suppose that (y_n) is a Cauchy sequence in R which has y as its limit. Then 0 is the limit of the Cauchy sequence $(y - y_n)$. Since $\psi(m_{\hat{R}}) \subset m_{\hat{S}}$, we have that the Cauchy sequence $(\psi(y - y_n))$ has 0 as its limit. Now

$$\psi(y - y_n) = \psi(y) - \psi(y_n) = \psi(y) - \phi(y_n)$$

so that the Cauchy sequence $(\phi(y_n))$ has $\psi(y)$ as its limit, so that $\psi(y) = \hat{\phi}(y)$ by our construction of $\hat{\phi}$. □

Lemma 21.44. *Suppose that R and S are equicharacteristic regular local rings and $\phi : R \to S$ is a homomorphism such that $\phi(m_R) \subset m_S$. Suppose that $x_1, \ldots, x_m$ are regular parameters in R and $y_1, \ldots y_n$ are regular parameters in S and S/m_S is finite and separable over R/m_R. Then there exist coefficient fields κ_1 of $\hat{R}$ and κ_2 of $\hat{S}$ such that $\hat{R} = \kappa_1[[x_1, \ldots, x_n]]$ and $\hat{S} = \kappa_2[[y_1, \ldots, y_n]]$ are power series rings and $\hat{\phi}(\kappa_1) \subset \kappa_2$.*

Proof. This follows from Hensel's lemma ([**161**, Theorem 17, page 279]) and Corollary 21.40. □

The conclusions of Lemma 21.44 may be false if S/m_S is not separable over R/m_R; that is, there may not exist coefficient fields κ_1 of $\hat{R}$ and κ_2 of $\hat{S}$, respectively, such that $\hat{\phi}(\kappa_1) \subset \kappa_2$. An example is given in [**35**, page 23].

A semilocal ring is a ring with a finite number of maximal ideals.

Theorem 21.45 (Chevalley). *Let A be a complete Noetherian semilocal ring, let $\mathfrak{m}$ be the intersection of its maximal ideals, and let (a_n) be a descending sequence of ideals in A such that $\bigcap_{n=0}^{\infty} a_n = (0)$. Then there exists an integral-valued function $s(n)$ which tends to infinity with n, such that $a_n \subset \mathfrak{m}^{s(n)}$ for all $n \geq 0$.*

Proof. [**161**, Theorem 13, page 270]. □

Suppose that R and S are Noetherian local rings such that R is a subring of S. We say that R is a subspace of S if R, with its m_R-adic topology, is a subspace of S with its m_S-adic topology. This is so if and only if S dominates R and there exists a sequence of nonnegative integers $a(n)$ such that $a(n) \mapsto \infty$ as $n \mapsto \infty$ and $R \cap m_S^n \subset m_R^{a(n)}$ for all $n \gg 0$. It follows from Theorem 21.45 that if R is a complete local ring and S is local ring dominating R, then R is a subspace of S.

Lemma 21.46. *Suppose that R and S are Noetherian local rings such that S dominates R, with inclusion $f : R \to S$. Let $\hat{R}$ and $\hat{S}$ be the respective completions of R and S and let $\hat{f} : \hat{R} \to \hat{S}$ be the induced homomorphism. Then R is a subspace of S if and only if $\hat{f}$ is an injection.*

Proof. First assume that R is a subspace of S. Suppose that $y \in \hat{R}$ and $\hat{f}(y) = 0$. There exists a Cauchy sequence (y_n) in R such that y is the limit of (y_n). Then $(f(y_n))$ is a Cauchy sequence in S whose limit is $\hat{f}(y) = 0$. Since R is a subspace of S, given $m > 0$, there exists t_0 such that $R \cap m_S^t \subset m_R^m$ if $t \geq t_0$, and there exists m_0 such that $n > m_0$ implies $f(y_n) \in m_S^{t_0}$ so that $n > m_0$ implies $y_n \in m_R^m$. Thus $y = 0$ is the limit of the Cauchy sequence (y_n) and so $\hat{f}$ is an injection.

Now suppose that $\hat{f}$ is injective. Then in $\hat{R}$,
$$(0) = \text{Kernel } \hat{f} = \bigcap_{n=0}^{\infty} a_n$$
where $a_n = \hat{R} \cap m_{\hat{S}}^n$. By Theorem 21.45, there exists an integer-valued function $s(n)$ which tends to ∞ with n, such that $a_n \subset m_{\hat{R}}^{s(n)}$ for all $n \gg 0$. Thus
$$R \cap m_S^n = R \cap (S \cap m_{\hat{S}^n}) = R \cap a_n \subset R \cap m_{\hat{R}}^{s(n)} = m_R^{s(n)}$$
by Lemma 21.37, and so R is a subspace of S. □

Exercise 21.47. In this exercise, consider two power series rings
$$\kappa[[x_1, \ldots, x_m]] \quad \text{and} \quad \kappa[[y_1, \ldots, y_n]]$$
over a field κ.

  a) Suppose that $h_1, \ldots, h_m$ are in the maximal ideal of $\kappa[[y_1, \ldots, y_n]]$. Show that there is a unique local κ-algebra homomorphism
  $$\phi : \kappa[[x_1, \ldots, x_m]] \to \kappa[[y_1, \ldots, y_n]]$$
  such that $\phi(x_i) = h_i$. Explain why all local κ-algebra homomorphisms $\phi : \kappa[[x_1, \ldots, x_m]] \to \kappa[[y_1, \ldots, y_n]]$ have this form. Hint: Let $R = \kappa[x_1, \ldots, x_m]$. Use the universal property of polynomial rings to define a (unique) κ-algebra homomorphism $\psi : R \to \kappa[[y_1, \ldots, y_n]]$ such that $\psi(x_i) = h_i$. Let I be the maximal ideal $I = (x_1, \ldots, x_n)$ of R. Explain why ψ extends to a homomorphism of local rings $\psi : R_I \to \kappa[[y_1, \ldots, y_n]]$, which has the property that $\psi(IR_I) \subset (y_1, \ldots, y_n)$. By Lemma 21.43, ψ extends to a local homomorphism $\hat{\psi} : \widehat{R}_I = \kappa[[x_1, \ldots, x_m]] \to \kappa[[y_1, \ldots, y_n]]$.

b) Can there be local homomorphisms

$$\phi : \kappa[[x_1,\ldots,x_m]] \to \kappa[[y_1,\ldots,y_n]]$$

which are not κ-algebra homomorphisms? Can there be ring homomorphisms $\phi : \kappa[[x_1,\ldots,x_m]] \to \kappa[[y_1,\ldots,y_n]]$ such that ϕ is not a local homomorphism (ϕ does not map the maximal ideal of $\kappa[[x_1,\ldots,x_m]]$ into the maximal ideal of $\kappa[[y_1,\ldots,y_n]]$)?

c) Suppose that $\phi : \kappa[[x_1,\ldots,x_n]] \to \kappa[[x_1,\ldots,x_n]]$ is a local κ-algebra homomorphism. Show that ϕ is an isomorphism if and only if $\mathrm{Det}(a_{ij}) \neq 0$ where

$$\phi(x_i) = a_{i1}y_1 + \cdots + a_{in}y_n + \sum_{i_1+\cdots+i_n>1} a(i)_{i_1,\ldots,i_n} y_1^{i_1}\cdots y_n^{i_n}$$

with $a_{ij}, a(i)_{i_1,\ldots,i_n} \in \kappa$ is the series expansion of $\phi(x_i)$ for $q \le i \le n$.

21.6. Zariski's main theorem and Zariski's subspace theorem

In this section we give some generalizations by Abhyankar in [4, Section 10] of some theorems of Zariski in [155].

Proposition 21.48. *Suppose that R and S are complete Noetherian local rings such that S dominates R, S/m_S is finite algebraic over R/m_R, and $m_R S$ is m_S-primary. Then S is a finite R-module. If $S/m_S = R/m_R$ and $m_R S = m_S$, then $R = S$.*

Proof. The quotient S/m_S is a finite length S-module since $m_R S$ is m_S-primary. Since S/m_S is a finite R/m_R-vector space, we have that $S/m_R S$ is a finitely generated R-module.

By Proposition 21.38 and since S is a Noetherian local ring with $m_R S \subset m_S$, we have that

$$\bigcap_{n=1}^{\infty} m_R^n S = (0).$$

Let $u_1,\ldots,u_s \in S$ generate $S/m_R S$ as an R-module. Let $N = \sum_{i=1}^{s} R u_i$.

Suppose $a \in S$. We will show that there exists a sequence of elements $a_1,\ldots,a_n,\ldots$ in N such that

$$a_n = \sum_{i=1}^{s} m_{ni} u_i$$

with $m_{ni} \in m_R^{n-1}$ and $a - \sum_{j=1}^{n} a_j \in m_R^n S$.

We will prove this by induction on n. The case $n = 1$ is immediate from our assumptions. Assume that $a_1, \ldots, a_n$ are defined. Then

$$a - \sum_{j=1}^{n} a_j = \sum m_i b_i$$

with $m_i \in m_R^n$, $b_i \in S$. Let c_i be elements of N such that $b_i - c_i \in m_R S$ and set

$$a_{n+1} = \sum m_i c_i = \sum m_{n+1,i} u_i.$$

Then a_{n+1} is the required element and the sequence in well-defined.

Set

$$m_i^* = \sum_{n=1}^{\infty} m_{ni} \in R$$

and $a^* = \sum m_i^* u_i \in N$. Then $a - a^* \in m_R^n S$ for all n, so that $a = a^*$ and thus $S = N$. □

A local ring R is said to be analytically irreducible if $\hat{R}$ is a domain.

Proposition 21.49. *Let R and S be analytically irreducible Noetherian local domains such that S dominates R. Assume that there exists a subring T of S with $R \subset T$ such that T is a finite R-module and $S = T_{T \cap m_S}$. Also assume that R is a subspace of S. Let $\hat{R}$ and $\hat{S}$ be the completions of R and S. Let K, L, K^*, and L^* be the quotient fields of R, S, $\hat{R}$, and $\hat{S}$, respectively, where K^* is identified with a subfield of L^*. Then $\hat{S} = \hat{R}[T]$, $\hat{S}$ is a finite $\hat{R}$-module, and $L^* = K^*(L)$.*

Proof. We have that $\hat{R}[T]$ is a finite $\hat{R}$-module and hence is a complete Noetherian local domain such that $\hat{R}[T]$ dominates $\hat{R}$ and $m_{\hat{R}[T]} = \sqrt{m_{\hat{R}} \hat{R}[T]}$ by [**161**, Theorem 15, page 276] and [**161**, Corollary 2 on page 283]. Thus $\hat{S}$ dominates $\hat{R}[T]$. In particular,

$$T \cap m_{\hat{R}[T]} = T \cap m_{\hat{S}} = T \cap m_S$$

and hence $\hat{R}[T]$ dominates S. Consequently $\hat{S}$ and $\hat{R}[T]$ have isomorphic residue fields and $m_{\hat{R}[T]} \hat{S} = m_{\hat{S}}$. By Proposition 21.48 we have that $\hat{S} = \hat{R}[T]$. Thus $\hat{S}$ is a finite $\hat{R}$-module and $L^* = K^*(L)$. □

Proposition 21.50. *Suppose that R is a normal local ring which is essentially of finite type over a field. Then its completion $\hat{R}$ is a normal local ring.*

Proof. [**121**, Theorem 37.5] or [**107**, Theorem 79, page 258]. □

Proposition 21.51. *Let R be an analytically irreducible Noetherian local domain, let $\hat{R}$ be the completion of R, let K and $\hat{K}$ be the respective quotient fields of R and $\hat{R}$, let V be a local (not necessarily Noetherian) domain with quotient field K such that V dominates R, and let H be the smallest subring of $\hat{K}$ such that H contains V and $\hat{R}$. Then $m_V H \neq H$ and there exists a valuation ring V^* of $\hat{K}$ such that V^* dominates V and $\hat{R}$.*

Proof. Suppose that $m_V H = H$. We will derive a contradiction. Since H is the set of all finite sums $\sum a_i b_i$ with $a_i \in V$ and $b_i \in \hat{R}$, we have an expression $1 = x_1 y_1 + \cdots + x_n y_n$ where $x_1, \ldots, x_n \in m_V$ and $y_1, \ldots, y_n \in \hat{R}$. Since R and V have the same quotient field, we have expressions $x_i = \frac{z_i}{z}$, where $z, z_1, \ldots, z_n \in R$ with $z \neq 0$. Then
$$z = z_1 y_1 + \cdots + z_n y_n \in R \cap \left((z_1, \ldots, z_n) \hat{R} \right) = (z_1, \ldots, z_n) R$$
by Lemma 21.37. Hence $z = z_1 r_1 + \cdots + z_n r_n$ with $r_1, \ldots, r_n \in R$. Then
$$1 = x_1 r_1 + \cdots + x_n r_n \in m_V,$$
a contradiction. Thus $m_V H \neq H$, and so by the existence theorem of valuations [**161**, Theorem 4, page 11] there exists a valuation ring V^* of $\hat{K}$ such that $H \subset V^*$ and $m_V H \subset m_{V^*}$. The valuation ring V^* thus dominates V and $\hat{R}$. □

Theorem 21.52. *Let A be a subring of a field K and let P be a prime ideal of A. Then*
$$\bigcap_{V \in N} V = \text{the integral closure of } A_P \text{ in } K,$$
where N is the set of all valuation rings V of K such that V dominates A_P.

Proof. [**161**, Theorem 8, page 17]. □

Proposition 21.53. *Let R and S be Noetherian local domains such that R is analytically irreducible, S dominates R, $\dim R = \dim S$, S/m_S is finite over R/m_R, and $m_R S$ is m_S-primary. Let $\hat{R}$ and $\hat{S}$ be the respective completions of R and S. Let $f : R \to S$ be the inclusion, with induced homomorphism $\hat{f} : \hat{R} \to \hat{S}$. Then $\hat{f}$ is an injection (and hence R is a subspace of S by Lemma 21.46). If R is normal and R and S have the same quotient fields, then $R = S$.*

Proof. The ring $\hat{f}(\hat{R})$ is a complete local ring since it is isomorphic to a quotient of $\hat{R}$. By Proposition 21.48, $\hat{S}$ is a finite $\hat{f}(\hat{R})$-module as $\hat{S}/m_{\hat{S}} \cong S/m_S$ is finite over $\hat{f}(\hat{R})/m_{\hat{f}(\hat{R})} \cong R/m_R$ and $m_{\hat{f}(\hat{R})} \hat{S} = m_R \hat{S}$ is $m_{\hat{S}}$-primary. Hence $\dim \hat{S} = \dim \hat{f}(\hat{R})$ by Theorem 1.62. Thus $\dim \hat{f}(\hat{R}) = \dim \hat{R}$, as $\dim \hat{R} = \dim R$ and $\dim \hat{S} = \dim S$ by Theorem 21.36. Since $\hat{R}$ is a domain, $\hat{f}$ is then a monomorphism.

Now assume that R is normal and the quotient fields of R and S coincide. Let K and $\hat{K}$ be the respective quotient fields of R and $\hat{R}$. Since R is normal, we have by Theorem 21.52 that R is the intersection of all valuation rings of K which dominate R. Thus it suffices to show that if V is any valuation ring of K which dominates R and z is an element of S, then $z \in V$. Since K is the quotient field of R, we have an expression $z = \frac{x}{y}$ where $x, y \in R$ and $y \neq 0$. Since $z \in S \subset \hat{S}$ and $\hat{S}$ is integral over $\hat{R}$ (by Proposition 21.48), there exist $z_1, \ldots, z_n \in \hat{R}$ such that $z^n + z_1 z^{n-1} + \cdots + z_n = 0$. By Proposition 21.51, there exists a valuation ring V^* of $\hat{K}$ such that V^* dominates V and $\hat{R}$. Since V^* is normal, $\hat{R} \subset V^*$, and z is integral over $\hat{R}$, we have that $z \in V^*$. Now $V = V^* \cap K$ so $z \in V$. □

Proposition 21.54 (Zariski's main theorem). *Let κ be a field, let R be a normal Noetherian local domain which is a localization of a finite type κ-algebra, and let S be a Noetherian local domain such that S dominates R, $\dim R = \dim S$, S/m_S is finite over R/m_R, $m_R S$ is m_S-primary, and R and S have a common quotient field. Then $R = S$.*

Proof. The ring R is analytically irreducible by Proposition 21.50. The proposition now follows from Proposition 21.53. □

Proposition 21.55. *Let R be a Noetherian normal local ring which is essentially of finite type over a field κ, let T be the integral closure of R in a finite algebraic extension L of the quotient field K of R, let P be a prime ideal in T with $R \cap P = m_R$, and let $S = T_P$. Then S is normal and a localization of a finite type κ-algebra, S dominates R, $\dim R = \dim S$, S/m_S is finite algebraic over R/m_R, $m_R S$ is primary for m_S, the completions $\hat{R}$ and $\hat{S}$ are normal domains, R is a subspace of S, and upon identifying the quotient field $\hat{K}$ of $\hat{R}$ with a subfield of the quotient field $\hat{L}$ of $\hat{S}$, we have that $\hat{S} = \hat{R}[T]$, $\hat{S}$ is a finite $\hat{R}$-module, and $\hat{L} = \hat{K}(T)$.*

Proof. The ring T is a finitely generated R-module and S is a localization of a finite type κ-algebra by Theorem 1.54. Thus S/m_S is finite algebraic over R/m_R and $m_R S$ is primary for m_S. We have that $\dim S = \dim R$ by Theorem 1.62. The completions $\hat{R}$ and $\hat{S}$ are normal domains by Proposition 21.50. Proposition 21.53 implies that R is a subspace of S. The facts that $\hat{S} = \hat{R}[T]$, $\hat{S}$ is a finite $\hat{R}$-module, and $\hat{L} = \hat{K}(L)$ now follow from Proposition 21.49. □

Proposition 21.56. *Let R be a Noetherian local domain with quotient field K and let V be a valuation ring of K such that $V \neq K$, V dominates R, and $\mathrm{trdeg}_{R/m_R} V/m_V \geq \dim R - 1$. Then $\mathrm{trdeg}_{R/m_R} V/m_V = \dim R - 1$ and V is a one-dimensional regular local ring.*

Proof. We have that the value group of V is isomorphic to $\mathbb{Z}$ and
$$\operatorname{trdeg}_{R/m_R} V/m_V = \dim R - 1$$
by [**161**, Proposition 2, page 331] and [**161**, Proposition 3, page 335] (which are generalizations of "Abhyankar's inequality" in [**3**]). Thus V is a one-dimensional regular local ring by Theorem 21.10. $\square$

The following proposition generalizes Theorem 10.19.

Proposition 21.57. *Let R be an n-dimensional Noetherian local domain with $n > 1$. Let $x_1, \ldots, x_n \in R$ be a system of parameters in R and let $Q = (x_1, \ldots, x_n)$. Let $A = R[\frac{x_2}{x_1}, \ldots, \frac{x_n}{x_1}]$ and let $\pi : A \to A/m_R A$ be the natural quotient map. Then $m_R A$ is a prime ideal, $\dim A_{m_R} = 1$, $m_R A = \sqrt{QA}$, $R \cap m_R A = m_R$, and $\pi(\frac{x_2}{x_1}), \ldots, \pi(\frac{x_n}{x_1})$ are algebraically independent over $\pi(R) \cong R/m_R$.*

Proof. We have that $QA = x_1 A$. There exists a positive integer e such that $m_R^e \subset Q$ since Q is m_R-primary, so that $(m_R A)^e \subset x_1 A$. Let $X_1, \ldots, X_n$ be indeterminates. Suppose that $R \cap m_R A \ne m_R$. We will derive a contradiction. Then $m_R A = A$ and thus $x_1 A = A$, so there exists a nonzero element $y \in A$ such that $x_1 y = 1$. There thus exists a nonzero polynomial $f(X_2, \ldots, X_n)$ of some degree d in $X_2, \ldots, X_n$ with coefficients in R such that $y = f(\frac{x_2}{x_1}, \ldots, \frac{x_n}{x_1})$. Thus
$$x_1^d = x_1^{d+1} y = x_1 g(x_1, \ldots, x_n)$$
where $g(X_1, \ldots, X_n)$ is a nonzero homogeneous polynomial of degree d in $X_1, \ldots, X_n$ with coefficients in R. In particular, $x_1^d \in m_R Q^d$, which is a contradiction by [**161**, Theorem 21, page 292], since $x_1, \ldots, x_n$ is a system of parameters in R. Thus $R \cap m_R A = m_R$ and $\pi(R) \cong R/m_R$.

Suppose that $\pi(\frac{x_2}{x_1}), \ldots, \pi(\frac{x_n}{x_1})$ are algebraically dependent over R/m_R. We will derive a contradiction. By our assumption that they are algebraically dependent, there exists a nonzero polynomial $F(X_2, \ldots, X_n)$ of some degree u in $X_2, \ldots, X_n$ with coefficients in R at least one of which is not in m_R such that $F(\frac{x_2}{x_1}, \ldots, \frac{x_n}{x_1}) \in m_R A$. Thus there exists a polynomial $G(X_2, \ldots, X_n)$ in $X_2, \ldots, X_n$ with coefficients in m_R such that
$$F\left(\frac{x_2}{x_1}, \ldots, \frac{x_n}{x_1}\right) = G\left(\frac{x_2}{x_1}, \ldots, \frac{x_n}{x_1}\right).$$
After multiplying both sides of this equation by x_1^v for a suitable integer $v \ge u$, we obtain that
$$U(x_1, \ldots, x_n) = V(x_1, \ldots, x_n)$$
where $U(X_1, \ldots, X_n)$ is a nonzero homogeneous polynomial of degree v in $X_1, \ldots, X_n$ with coefficients in R, at least one of which is not in m_R,

and $V(X_1,\ldots,X_n)$ is either the zero polynomial or a nonzero homogeneous polynomial of degree v in $X_1,\ldots,X_n$ with coefficients in m_R. Thus $U(x_1,\ldots,x_n) \in m_R Q^v$, which is a contradiction by [**161**, Theorem 21, page 292], since $x_1,\ldots,x_n$ is a system of parameters. Thus $\pi(\frac{x_2}{x_1}),\ldots,\pi(\frac{x_n}{x_1})$ are algebraically independent over $\pi(R)$. Since

$$\pi(A) = \pi(R)\left[\pi\left(\frac{x_2}{x_1}\right),\ldots,\pi\left(\frac{x_n}{x_1}\right)\right],$$

we have that $\pi(A)$ is a domain and thus $m_R A$ is a prime ideal in A.

Now $(m_R A)^e \subset x_1 A = QA$, which implies that $x_1 A_{m_R A}$ is $m_R A_{m_R A}$-primary, which implies that $\dim A_{m_R A} = 1$ by Krull's principal ideal theorem, Theorem 1.65. $\square$

Proposition 21.58. *Let R be an n-dimensional Noetherian local domain with $n > 0$, and let K be the quotient field of R. Then there exists a one-dimensional regular local ring V with quotient field K such that V dominates R and $\operatorname{trdeg}_{R/m_R} V/m_V = n - 1$.*

Proof. By Theorem 8.12, there exists a system of parameters $x_1,\ldots,x_n$ in R, so that $(x_1,\ldots,x_n)$ is m_R-primary. Let $A = R[\frac{x_2}{x_1},\ldots,\frac{x_n}{x_1}]$. By Proposition 21.57, $m_R A$ is a prime ideal in A, and upon letting $S = A_{m_R A}$ we have that S is a one-dimensional Noetherian local domain with $\operatorname{trdeg}_{R/m_R} S/m_S = n-1$. Let T be the integral closure of S in K. By the Krull-Akizuki theorem, [**121**, Theorem 33.2] (or by Theorem 1.54 if R is a localization of a finite type algebra over a field), we have that T is Noetherian and $\dim T = 1$. Let P be a maximal ideal of T, and let $V = T_P$. Then V dominates S and V/m_V is algebraic over S/m_S. We have that V is a regular local ring by Theorem 1.87, since V is normal by Exercise 1.58. $\square$

Proposition 21.59. *Suppose that A and B are local domains which are localizations of finitely generated algebras over a field, A and B have respective quotient fields K and L, and B dominates A. Then*

$$\dim A + \operatorname{trdeg}_K L = \dim B + \operatorname{trdeg}_{A/m_A} B/m_B.$$

Proof. [**107**, Corollary, page 86, and Corollary 3, page 92]. $\square$

Proposition 21.60. *Let R and S be Noetherian local domains, with respective quotient fields K and L, such that R is analytically irreducible, S dominates R, $\operatorname{trdeg}_K L < \infty$, and*

$$\dim R + \operatorname{trdeg}_K L - \dim S + \operatorname{trdeg}_{R/m_R} S/m_S.$$

Then R is a subspace of S.

Proof. Let $\hat{K}$ be the quotient field of the completion $\hat{R}$ of R. If $\dim R = 0$, then the conclusions of the proposition follow trivially. Assume $\dim R > 0$.

21.6. Zariski's main theorem and Zariski's subspace theorem

Then $\dim S > 0$, and by Proposition 21.58, there exists a one-dimensional regular local ring W with quotient field L such that W dominates S and $\operatorname{trdeg}_{S/m_S} W/m_W = \dim S - 1$. Let $V = K \cap W$. Since W is a valuation ring of L, we have that V is a valuation ring of K, W dominates V, and V dominates R. In particular, $R \cap m_V = m_R \ne (0)$, and hence $m_V \ne (0)$. Now W is the valuation ring of a discrete valuation ω of L, so we have that V is the valuation ring of the discrete valuation $\nu = \omega|K$ and V is a one-dimensional regular local ring by Theorem 21.10. Thus the principal ideals $m_V W = m_W^u$ where u is a positive integer, and $K \cap m_W^{ui} = m_V^i$ for every nonnegative integer i (by Remark 21.11).

We will now establish that

$$\operatorname{trdeg}_{V/m_V} W/m_W \le \operatorname{trdeg}_K L.$$

Suppose that $\bar{t}_1, \ldots, \bar{t}_r \in W/m_W$ are algebraically independent over V/m_V. Let $t_1, \ldots, t_r \in W$ be lifts of $\bar{t}_1, \ldots, \bar{t}_r$. Suppose that $t_1, \ldots, t_r$ are algebraically dependent over K. We will derive a contradiction. With this assumption, there exists a relation

$$(21.17) \qquad \sum_{i_1, \ldots, i_r} a_{i_1, \ldots, i_r} t_1^{i_1} \cdots t_r^{i_r} = 0$$

with the finitely many coefficients $a_{i_1, \ldots, i_r} \in K$ not all zero. Let $j_1, \ldots, j_r$ be such that

$$\nu(a_{j_1, \ldots, j_r}) = \min\{\nu(a_{i_1, \ldots, i_r})\}.$$

Dividing the relation (21.17) by $a_{j_1, \ldots, j_r}$, we may assume that $a_{j_1, \ldots, j_r} = 1$ and $\nu(a_{i_1, \ldots, i_r}) \ge 0$ for all $i_1, \ldots, i_r$. Let $\bar{a}_{i_1, \ldots, i_r}$ be the residue of $a_{i_1, \ldots, i_r}$ in V/m_V. Then

$$\sum_{i_1, \ldots, i_r} \bar{a}_{i_1, \ldots, i_r} \bar{t}_1^{i_1} \cdots \bar{t}_r^{i_r} = 0$$

is a nontrivial relation, contradicting our assumption that $\bar{t}_1, \ldots, \bar{t}_r$ are algebraically independent over V/m_V.

We have that

$$\operatorname{trdeg}_{S/m_S} W/m_W = \dim S - 1,$$

$$\operatorname{trdeg}_{V/m_V} W/m_W + \operatorname{trdeg}_{R/m_R} V/m_V = \operatorname{trdeg}_{S/m_S} W/m_W + \operatorname{trdeg}_{R/m_R} S/m_S,$$

and, by assumption,

$$\dim R + \operatorname{trdeg}_K L = \dim S + \operatorname{trdeg}_{R/m_R} S/m_S.$$

Thus $\mathrm{trdeg}_{R/m_R} V/m_V \geq \dim R - 1$. By Proposition 21.51, there exists a valuation ring V^* of $\hat{K}$ such that V^* dominates V and $\hat{R}$. Since $\dim \hat{R} = \dim R$ (by Theorem 21.36) and $\hat{R}/m_{\hat{R}} = R/m_R$, we have that $\mathrm{trdeg}_{\hat{R}/m_{\hat{R}}} V^*/m_{V^*} \geq \dim \hat{R} - 1$. Thus V^* is a one-dimensional regular local ring by Proposition 21.56. In particular, $\bigcap_{i=0}^{\infty} m_{V^*}^i = (0)$, and hence

$$\bigcap_{i=0}^{\infty}(\hat{R} \cap m_{V^*}^i) = (0).$$

Thus by Theorem 21.45, there exists a sequence of nonnegative integers $a(i)$ which tend to infinity with i such that $\hat{R} \cap (m_{V^*})^i \subset m_{\hat{R}}^{a(i)}$ for every nonnegative integer i. We thus have

$$R \cap m_S^{ui} \subset R \cap m_W^{ui} = R \cap m_V^i \subset R \cap m_{V^*}^i \subset R \cap (\hat{R} \cap m_{V^*}^i) \subset R \cap m_{\hat{R}}^{a(i)} = m_R^{a(i)}$$

by Lemma 21.37. Thus there exists a sequence of nonnegative integers $b(i)$ which tend to infinity with i such that $R \cap m_S^i \subset m_R^{b(i)}$ for every nonnegative integer i. Thus R is a subspace of S. □

Proposition 21.61 (Zariski's subspace theorem). *Let R and S be local domains which are localizations of finite type algebras over a field such that R is analytically irreducible and S dominates R. Then R is a subspace of S so that the natural map $\hat{R} \to \hat{S}$ is an inclusion.*

Proof. This follows from Propositions 21.59 and 21.60. □

Proposition 21.62. *Suppose that $\phi : X \to Y$ is a dominant regular map of varieties, $p \in X$, and $q = \phi(p)$. Assume that $\mathcal{O}_{Y,q}$ is analytically irreducible (which holds if $\mathcal{O}_{Y,q}$ is normal by Proposition 21.50) and that $m_q \mathcal{O}_{X,p} = m_p$ is the maximal ideal of $\mathcal{O}_{X,p}$. Then $\hat{\phi}^* : \hat{\mathcal{O}}_{Y,q} \to \hat{\mathcal{O}}_{X,p}$ is an isomorphism.*

Proof. Let $R = \mathcal{O}_{Y,q}$ and $S = \mathcal{O}_{X,p}$. By Proposition 21.61, we have that $\hat{\phi} : \hat{R} \to \hat{S}$ is an injection. Since $R/m_R = S/m_S = k$ and $m_R S = S$, we have that $\hat{\phi}^*$ is an isomorphism by Proposition 21.48. □

Corollary 21.63. *Suppose that $\phi : X \to Y$ is a finite map of normal varieties. Suppose that ϕ is unramified at $p \in X$. Then $\hat{\phi}^* : \hat{\mathcal{O}}_{Y,\phi(p)} \to \hat{\mathcal{O}}_{X,p}$ is an isomorphism.*

The subspace theorem is not true in complex analytic geometry.

Example 21.64 (Gabrièlov, [58]). *There exists an injective local $\mathbb{C}$-algebra homomorphism $R \to S$ of rings of germs of convergent power series, such that the induced map $\hat{R} \to \hat{S}$ of formal power series rings is not an injection.*

21.7. Galois theory of varieties

Exercise 21.65. In this exercise, we show that the assumption that $\mathcal{O}_{Y,q}$ is analytically irreducible is necessary in Proposition 21.62.

  a) Let Y be the nodal curve $Y = Z(x^2 - y^2 - y^3) \subset \mathbb{A}_k^2$ where k is an algebraically closed field of characteristic $\neq 2$ or 3. Show that Y is a variety with an isolated singularity at the origin q.

  b) Let $\phi : X \to Y$ be the blow-up of the ideal $I(q)$ of Y. Show that X is an affine variety which is the normalization of Y and the ring of regular functions on X is
  $$k[X] = k[x_1, y_1]/(1 - y_1^2 - x_1 y_1^3),$$
  with the inclusion $k[Y] \to k[X]$ defined by the substitutions $x = x_1, y = x_1 y_1$.

  c) Let $p \in \phi^{-1}(q)$. Show that $\mathcal{O}_{Y,q} \to \mathcal{O}_{X,p}$ is unramified but the induced map on completions $\hat{\phi}^* : \hat{\mathcal{O}}_{Y,q} \to \hat{\mathcal{O}}_{X,p}$ is not an isomorphism.

Exercise 21.66. Let $f_1 = xy$, $f_2 = x$, $f_3 = y$ in the polynomial ring $k[x, y]$. Let $R = k[x, y]_{(x,y)}$ and $Q = (f_1, f_2, f_3)R = m_R$. Let $A = R[\frac{f_2}{f_1}, \frac{f_3}{f_1}]$. Show that $m_R A = A$.

21.7. Galois theory of varieties

Suppose that $\phi : Y \to X$ is a dominant finite regular map of normal varieties. Let $G(Y/X)$ be the group of all regular isomorphisms of Y/X, that is, the group of all regular isomorphisms $\tau : Y \to Y$ such that there is a commutative diagram

The group of $k(X)$-algebra isomorphisms of $k(Y)$ is denoted by $\text{Aut}(k(Y)/k(X))$ (Section 1.2).

Proposition 21.67. *The map*
$$\Phi : G(Y/X) \to \text{Aut}(k(Y)/k(X))^{\text{op}}$$
(where $\text{Aut}(k(Y)/k(X))^{\text{op}}$ is the opposite group) defined by $\tau \mapsto \tau^*$ is a group isomorphism.

Proof. Suppose that $\tau \in G(Y/X)$. Then τ^* gives an isomorphism of $k(Y)$ which fixes $k(X)$, so $\tau^* \in \text{Aut}(k(Y)/k(X))$.

Now suppose that $\sigma \in \text{Aut}(k(Y)/k(X))$. Then there exists a unique birational map $\tau : Y \dashrightarrow Y$ such that $\tau^* = \sigma$. Suppose that $U \subset X$ is an affine open subset. Let $V = \phi^{-1}(U)$ which is an affine open subset of Y (by Theorem 7.5). Let $A = k[U]$ and $B = k[V]$. Here B is the integral closure of A in $k(Y)$ (by $\phi^* : k(X) \to k(Y)$). Suppose $f \in B$. Then f is integral over A and $\sigma(f) \in k(Y)$ must satisfy the same equation of integrality over A as f since $A \subset k(X)$ is fixed by σ. Thus $\sigma(f) \in B$ and $\sigma(B) \subset B$. Since σ is an isomorphism, we have $\sigma(B) = B$. Hence σ is an A-algebra isomorphism of B so $\tau|V \in G(V/U)$. Since this is true for all members of an affine cover $\{U_i\}$ of X, we have that $\tau \in G(Y/X)$ by Proposition 3.39.

Finally, we observe that the group structure is preserved since $(\tau_1\tau_2)^* = \tau_2^*\tau_1^*$ for $\tau_1, \tau_2 \in G(Y/X)$. $\square$

Suppose that H is a subgroup of $G(Y/X)$. Then we can define (by Theorem 7.17) a normal variety Y^H by taking Y^H to be the normalization of X in the fixed field $k(Y)^H = k(Y)^{\Phi(H)}$ ($\sigma \in H$ acts as σ^*). We call Y^H the quotient of Y by H.

Definition 21.68. Suppose that $\phi : Y \to X$ is a dominant finite regular map of normal varieties and $k(Y)$ is a separable extension of $k(X)$. The map ϕ is said to be *Galois* and Y is said to be Galois over X if for every $p \in X$ and $q_1, q_2 \in \phi^{-1}(p)$ there exists $\tau \in G(Y/X)$ such that $\tau(q_1) = q_2$.

Theorem 21.69. *Suppose that $\phi : Y \to X$ is a dominant finite regular map of normal varieties. Then Y is Galois over X if and only if $k(Y)$ is Galois over $k(X)$.*

Proof. Suppose that Y/X is Galois. Let $Z = Y^{G(Y/X)}$ and let

$$Y \xrightarrow{\alpha} Z \xrightarrow{\beta} X$$

be the regular maps factoring ϕ. By Theorems 21.29 and 21.28, there exists $p \in X$ such that $\#\{\phi^{-1}(p)\} = [k(Y) : k(X)]$. Thus β is unramified above p and α is unramified above all points of $\beta^{-1}(p)$ by Theorem 21.26 and Theorem 21.28. Suppose that $Z \neq X$ so that $k(Z) \neq k(X)$. Then there exist $a_1, a_2 \in \beta^{-1}(p)$ which are not equal. Let $q_1 \in \alpha^{-1}(a_1)$ and $q_2 \in \alpha^{-1}(a_2)$. Since Y is Galois over X, there exists $\tau \in G(Y/X)$ such that $\tau(q_1) = q_2$. But $G(Y/Z) = G(Y/X)$ (since $\tau \in G(Y/X)$ implies $\tau^* : k(Z) \to k(Z)$ is the identity) so $a_1 = \alpha(\tau(q_1)) = \alpha(q_2) = a_2$, a contradiction. Thus $Z = X$ and so $k(Y)^{G(Y/X)} = k(X)$, so that $k(Y)$ is Galois over $k(X)$.

Now suppose that $k(Y)$ is Galois over $k(X)$. Suppose that $q \in X$ and $\phi^{-1}(q) = \{p_1, \ldots, p_r\}$. Let T be the integral closure of $\mathcal{O}_{X,q}$ in $k(Y)$, and let $m_1, \ldots, m_r$ be the maximal ideals of T, with $T_{m_i} = \mathcal{O}_{Y,p_i}$. By Lemma 21.8, if $i \neq j$, then there exists $\sigma \in G(k(Y)/k(X))$ such that $\sigma(m_i) = m_j$. If $\tau \in G(Y/X)$ corresponds to σ, then we have $\tau(p_j) = p_i$. Thus Y is Galois over X. $\square$

21.7. Galois theory of varieties

We now summarize some results on quotients that we found in this section. Suppose that $\phi : Y \to X$ is a dominant finite map of normal varieties and $U \subset X$ is affine. Then $V = \phi^{-1}(U)$ is affine (since ϕ is finite) and $k[V]$ is the integral closure of $k[U]$ in $k(Y)$. Further, if Y is Galois over X, then $G = G(Y/X)$ acts naturally on $k[V]$ and the ring of invariants is $k[V]^G = k[U]$.

Exercise 21.70. Suppose that X and Y are normal varieties over an algebraically closed field of characteristic $\neq 2$ and $\phi : Y \to X$ is a finite map with $\deg(\phi) = 2$. Show that ϕ is Galois.

Exercise 21.71. Suppose that $\phi : Y \to X$ is a Galois map of nonsingular curves. Suppose that $p \in X$ and $\phi^{-1}(p) = \{q_1, \ldots, q_t\}$. Show that the divisor $\phi^*(p) = eq_1 + \cdots + eq_t$ where $et = \deg(\phi)$. Hint: Use Theorem 13.18.

Exercise 21.72. Suppose that k is an algebraically closed field of characteristic $\neq 2$. Let $\phi : \mathbb{A}_k^1 \to \mathbb{A}_k^1$ be defined by $\phi(z) = (z^2 + 1)^2$. Show that ϕ is not Galois. Hint: Use Exercise 21.71.

Exercise 21.73. Suppose that $\phi : Y \to X$ is a finite regular Galois map of varieties and H is a subgroup of $G(Y/X)$. Let $Z = Y^H$ with natural regular maps

$$Y \xrightarrow{\alpha} Z \xrightarrow{\beta} X$$

factoring ϕ. Suppose that U is an affine open subset of X. Let $V = \phi^{-1}(U)$ and $W = \beta^{-1}(U)$. Here V and W are affine open subsets of Y and Z, respectively, by Theorem 7.5.

a) Show that

$$G(V/U) = \{(\sigma|V) \mid \sigma \in G(Y/X)\}.$$

b) Let

$$k[V]^H = \{f \in k[V] \mid \sigma^*(f) = f \text{ for all } \sigma \in H\}.$$

Show that $k[W] = k[V]^H$.

c) Show that Y is Galois over Z.

Exercise 21.74. Let k be an algebraically closed field of characteristic $\neq 3$ and let $\phi : Y = \mathbb{A}_k^1 \to X = \mathbb{A}_k^1$ be the finite map defined by $\phi(t) = t^3 - 3t$. Compute $G(Y/X)$ and show that ϕ is not Galois. Hint: Use the fact that every automorphism of $\mathbb{A}^1$ extends to an automorphism of $\mathbb{P}^1$.

21.8. Derivations and Kähler differentials redux

We require some more results on derivations and differentials.

Theorem 21.75. *Suppose that κ is a field and K is a finitely generated extension field, of transcendence degree n over k. Then $\Omega_{K/\kappa}$ is a vector space of dimension $\geq n$ over K. Suppose that $x_1, \ldots, x_n \in K$. Then $dx_1, \ldots, dx_n$ is a K-basis of $\Omega_{K/\kappa}$ if and only if $x_1, \ldots, x_n$ is a separating transcendence basis of K over κ.*

Theorem 21.75 follows from [50, Theorem 16.4 and Corollary 16.17] or the material in [160, Section 17, Chapter II] on derivations, along with the isomorphism of K-vector spaces $\mathrm{Der}_\kappa(K, K) \cong \mathrm{Hom}_K(\Omega_{K/\kappa}, K)$ of Lemma 14.3.

Suppose that K is a finitely generated extension field of an algebraically closed field k and $x_1, \ldots, x_n$ is a separating transcendence basis of K over k (which exists by Theorem 1.14). Let $L = k(x_1, \ldots, x_n)$. By Example 14.5 and Lemma 14.8, $\Omega_{L/k} \cong \bigoplus_{i=1}^n L dx_i$. Since $\mathrm{Der}_k(L, L) \cong \mathrm{Hom}_L(\Omega_{L/k}, L)$ by Lemma 14.3, $\frac{\partial}{\partial x_1}, \ldots, \frac{\partial}{\partial x_n}$ is an L-basis of $\mathrm{Der}_k(L, L)$, and by Theorem 21.75 and since $\mathrm{Der}_k(K, K) \cong \mathrm{Hom}_K(\Omega_{K/k}, K)$ by Lemma 14.3, $\frac{\partial}{\partial x_1}, \ldots, \frac{\partial}{\partial x_n}$ extend uniquely to a K-basis of $\mathrm{Der}_k(K, K)$. A direct proof of this is given in Theorem 39 and its corollaries in [160, Section 17, Chapter II].

Suppose that $0 \neq \alpha \in K$ and $\delta \in \mathrm{Der}_k(L, L)$. Let $g(t) \in L[t]$ be the minimal polynomial of α over L. Write $g(t) = t^d + a_{d-1}t^{d-1} + \cdots + a_0$ with $a_i \in L$. Let
$$g^\delta(t) = \delta(a_{d-1})t^{d-1} + \delta(a_{d-2})t^{d-2} + \cdots + \delta(a_0) \in L[t].$$
By the properties of derivations, we have that
$$0 = \delta(g(\alpha)) = g^\delta(\alpha) + g'(\alpha)\delta(\alpha)$$
where
$$g'(t) = \frac{dg}{dt} = dt^{d-1} + (d-1)a_{d-1}t^{d-1} + \cdots + a_1$$
is the formal derivative of $g(t)$. Since α is separable over L, we have that $g'(\alpha) \neq 0$. Thus
$$\delta(\alpha) = -\frac{g^\delta(\alpha)}{g'(\alpha)}.$$

If $\delta(\alpha) = 0$, then $g^\delta(\alpha) = 0$. Since $g^\delta(t)$ has smaller degree in t than the minimal polynomial $g(t)$ of α, we have that $g^\delta(t) = 0$.

Suppose that k has characteristic 0. If $\delta(\alpha) = 0$ for all $\delta \in \mathrm{Der}_k(L, L)$, then $g(t) \in k[t]$, so that $\alpha \in k$ (as k is algebraically closed). Thus if k has characteristic 0,

(21.18) $\qquad k = \{f \in K \mid \delta(f) = 0 \text{ for all } \delta \in \mathrm{Der}_k(K, K)\}.$

21.8. Derivations and Kähler differentials redux

Suppose that K is a field of characteristic $p > 0$. A finite set of elements $x_1, \ldots, x_n$ in K are said to be p-independent if the n^p monomials $x_1^{i_1} x_2^{i_2} \cdots x_n^{i_n}$ with $0 \le i_q < p$ for $1 \le q \le n$ are linearly independent over K^p. If we also have that the set $S = \{x_1, \ldots, x_n\}$ satisfies $K = K^p(S)$, then we say that S is a p-basis of K.

Theorem 21.76. *Suppose that K is a finitely generated extension field of an algebraically closed field k of characteristic $p > 0$ and $x_1, \ldots, x_n$ is a separating transcendence basis of K over k. Then:*

1) *The $k(x_1, \ldots, x_n)$-basis*
$$\frac{\partial}{\partial x_1}, \ldots, \frac{\partial}{\partial x_n}$$
of $\mathrm{Der}_k(k(x_1, \ldots, x_n), k(x_1, \ldots, x_n))$ extends uniquely to a K-basis of
$$\mathrm{Der}_k(K, K) = \mathrm{Der}_{K^p}(K, K).$$

2) $[K : K^p] = p^n$ *and* $x_1, \ldots, x_n$ *is a p-basis of K.*

3) $K^p = \{f \in K \mid \delta(f) = 0 \text{ for all } \delta \in \mathrm{Der}_k(K, K)\}.$

Proof. Statement 1) follows from Theorem 21.75 since
$$\mathrm{Der}_k(K, K) \cong \mathrm{Hom}_k(\Omega_{K/k}, K)$$
by Lemma 14.3.

Suppose that $f \in K^p(x_1, \ldots, x_n)$. Then f has an expression

(21.19) $$f = \sum_I a_I x_1^{i_1} x_2^{i_2} \cdots x_n^{i_n}$$

where the sum is over $I = (i_1, \ldots, i_n) \in \mathbb{N}^n$ with $0 \le i_q < p$ for $1 \le q \le n$ and all $a_I \in K^p$. Let I be such that $i_1 + \cdots + i_n$ is maximal for $a_I \ne 0$. Suppose that $i_1 + \cdots + i_n > 0$. Without loss of generality, $i_1 > 0$. Then
$$\frac{\partial f}{\partial x_n^{i_n} \cdots \partial x_1^{i_1}} = i_1! i_2! \cdots i_n! a_I \ne 0$$

so $\frac{\partial f}{\partial x_1} \ne 0$. Thus $f \ne 0$, and so $x_1, \ldots, x_n$ are p-independent, and we have that
$$[K^p(x_1, \ldots, x_n) : K^p] = p^n.$$

We have that $[K : K^p] = p^n$ by [**160**, Theorem 41, Section 17, Chapter II], so $x_1, \ldots, x_n$ is a p-basis of K.

Now 3) follows from 2) and the above calculation showing that if an element f with an expansion (21.19) has the property that all derivations vanish on f, then $f \in K^p$. □

Theorem 21.77. *Let K be a field and F be a finitely generated extension field of K. Then*
$$\dim_K \mathrm{Der}_K(F,F) \geq \mathrm{trdeg}_K F$$
and F is separably generated over K if and only if
$$\dim_K \mathrm{Der}_K(F,F) = \mathrm{trdeg}_K F.$$

Proof. [160, Theorem 41, page 127]. □

Exercise 21.78. Let $\kappa = F_p(t)$, and let $R = \kappa[x,y]/(x^p + y^p - t)$ be the ring considered in Exercise 10.21. Let K be the quotient field of R. It was shown in Exercise 10.21 that R is a regular ring of dimension 1.

a) Compute $\Omega_{R/\kappa}$.

b) Show that K is not separably generated over κ.

21.9. Étale maps and uniformizing parameters

Definition 21.79. A regular map of varieties $\phi : X \to Y$ is said to be étale if for all $p \in X$ there are open neighborhoods $U \subset X$ of p and $V \subset Y$ of $\phi(p)$ such that $\phi(U) \subset V$ and there exists a commutative diagram

$$\begin{array}{ccc} U & \xrightarrow{\text{open embedding}} & Z \\ \phi \downarrow & & \downarrow \\ V & \xrightarrow{\text{open embedding}} & W \end{array}$$

where Z and W are affine varieties, and
$$k[Z] = R[x_1, \ldots, x_n]/(f_1, \ldots, f_n)$$
with $R = k[W]$, and the rank of the $n \times n$ matrix $(\frac{\partial f_i}{\partial x_j}(p))$ over k is n.

This definition of étale is equivalent to the definition of étale in [109, page 20] and [73, Exercise III.10.3], as will be explained after Definition 22.8. The definition is valid with $\phi : X \to Y$ a map of schemes (and Z, W affine schemes). A refinement of Definition 21.79 is given in Exercise 21.86.

The following is a version of Hensel's lemma.

Lemma 21.80. *Let R be a complete Noetherian local ring and let $f_1, \ldots, f_n$ be elements of the polynomial ring $R[x_1, \ldots, x_n]$. Assume $a_1, \ldots, a_n \in R$ are such that $f_1(a_1, \ldots, a_n), \ldots, f_n(a_1, \ldots, a_n) \in m_R$ and*
$$\mathrm{Det}\left(\frac{\partial f_i}{\partial x_j}\right)(a_1, \ldots, a_n) \notin m_R.$$
Then there exist $\alpha_1, \ldots, \alpha_n \in R$ such that $\alpha_i - a_i \in m_R$ and
$$f_1(\alpha_1, \ldots, \alpha_n) = \cdots = f_n(\alpha_1, \ldots, \alpha_n) = 0.$$

21.9. Étale maps and uniformizing parameters

Proof. We inductively define approximate roots $a_1^{(r)}, \ldots, a_n^{(r)} \in R$, with $a_i^{(1)} = a_i$, such that

(21.20) $\qquad a_i^{(r)} \equiv a_i^{(r-1)} \mod m_R^{r-1} \quad \text{for } 1 \leq i \leq n \text{ and } r \geq 2$

and

(21.21) $\qquad f_1(a_1^{(r)}, \ldots, a_n^{(r)}) \equiv \cdots \equiv f_n(a_1^{(r)}, \ldots, a_n^{(r)}) \equiv 0 \mod m_R^r.$

Suppose $a_1^{(r)}, \ldots, a_n^{(r)} \in R$ satisfy (21.20) and (21.21). Let $\epsilon_1, \ldots, \epsilon_2 \in m_R^r$. Then

$$f_i(a_1^{(r)}+\epsilon_1, \ldots, a_n^{(r)}+\epsilon_n) \equiv f_i(a_1^{(r)}, \ldots, a_n^{(r)}) + \sum_{j=1}^n \frac{\partial f_i}{\partial x_j}(a_1, \ldots, a_n)\epsilon_j \mod m_R^{r+1}.$$

Let

$$B = (b_{ij}) = \left(\left(\frac{\partial f_i}{\partial x_j}\right)(a_1, \ldots, a_n)\right)^{-1},$$

a matrix with coefficients in R. Set

$$\epsilon_i = -\sum_{j=1}^n b_{ij} f_j(a_1^{(r)}, \ldots, a_n^{(r)}) \quad \text{for } 1 \leq i \leq n$$

and $a_i^{(r+1)} = a_i^{(r)} + \epsilon_i$. Then

$$f_i(a_1^{(r)} + \epsilon_1, \ldots, a_n^{(r)} + \epsilon_n) \equiv 0 \mod m_R^{r+1}$$

for all i. Setting α_i to be the limit of the Cauchy sequence $(a_i^{(r)})$ for $1 \leq i \leq n$, we have that

$$f_i(\alpha_1, \ldots, \alpha_n) \in \bigcap_{i=1}^\infty m_R^i = (0)$$

by Proposition 21.38. $\qquad\square$

Theorem 21.81. *Suppose that $\phi : X \to Y$ is a regular map of varieties and $p \in X$. Then ϕ is étale in some neighborhood of p if and only if the induced map on complete local rings*

$$\hat{\phi}^* : \hat{\mathcal{O}}_{Y,\phi(p)} \to \hat{\mathcal{O}}_{X,p}$$

is an isomorphism.

Proof. First suppose that $\phi : X \to Y$ is étale in some neighborhood of $p \in X$. Let notation be as in Definition 21.79. Let $y_1, \ldots, y_m$ be generators of the maximal ideal $\mathfrak{n}$ of $q = \phi(p)$ in R. Let $\mathfrak{m}$ be the maximal ideal of p in $R[x_1, \ldots, x_n]$. Observe that every $h \in R[x_1, \ldots, x_n]$ has a unique expression $h = \lambda + f$ with $\lambda \in k$ and $f \in \mathfrak{m}$. We have that $f_1, \ldots, f_n \in \mathfrak{m}$.

Replacing the x_i with $x_i - x_i(p)$ for $1 \le i \le n$, we may assume that $\mathfrak{m} = (y_1, \ldots, y_m, x_1, \ldots, x_n)$. We have that

$$f_i - \sum_j \frac{\partial f_i}{\partial x_j}(p) x_j \in (y_1, \ldots, y_m) + \mathfrak{m}^2$$

for $1 \le i \le n$. Define an $n \times n$ matrix $A = (a_{ij})$ by $A = \left(\frac{\partial f_i}{\partial x_j}(p)\right)^{-1}$. Then for $1 \le i \le n$,

(21.22) $$x_i = \sum_{j=1}^n a_{ij} f_j + \sum_{j=1}^m c_{ij} y_j + h_i$$

for some $c_{ij} \in k$ and with $h_i \in \mathfrak{m}^2$. Now substitute the n expressions (21.22) into h_i in (21.22) to obtain an expression

$$x_i = \sum_j a_{ij} f_j + \sum_j c_{ij} y_j + \sum_{j,k} d_{ijk} f_j f_k + \sum_{j,k} g_{ijk} f_j y_k + \sum_{j,k} h_{ijk} y_j y_k + \Omega_i$$

with $d_{ijk}, g_{ijk}, h_{ijk} \in k$ and $\Omega_i \in \mathfrak{m}^3$. Iterating, we obtain Cauchy sequences in $R[x_1, \ldots, x_n]$ which converge to series

$$x_i = \sum a_{i_1 \ldots i_n j_1 \ldots j_m} f_1^{i_1} \cdots f_n^{i_n} y_1^{j_1} \cdots y_m^{j_m}$$

in $\hat{R}[[x_1, \ldots, x_n]]$ with $a_{i_1 \ldots i_n j_1 \ldots j_m} \in k$. Thus we have expansions for $1 \le i \le n$,

(21.23) $$x_i = \psi_i(f_1, \ldots, f_n),$$

with $\psi_i(z_1, \ldots, z_n) \in \hat{R}[[z_1, \ldots, z_n]]$ (a power series ring over $\hat{R}$). By Lemma 21.80, there exist $\alpha_1, \ldots, \alpha_n \in m_{\hat{R}}$ such that

$$f_1(\alpha_1, \ldots, \alpha_n) = \cdots = f_n(\alpha_1, \ldots, \alpha_n) = 0.$$

Let $\Lambda : \hat{R}[[x_1, \ldots, x_n]] \to \hat{R}$ be the homomorphism defined by $\Lambda(g) = g(\alpha_1, \ldots, \alpha_n)$ for $g \in \hat{R}[[x_1, \ldots, x_n]]$ (Lemma 21.43), which has the kernel $(x_1 - \alpha_1, \ldots, x_n - \alpha_n)$, so that $(f_1, \ldots, f_n) \subset (x_1 - \alpha_1, \ldots, x_n - \alpha_n)$.

Evaluating (21.23) at $(\alpha_1, \ldots, \alpha_n)$, we have that

$$\psi_i(f_1, \ldots, f_n)(\alpha_1, \ldots, \alpha_n) = \alpha_i,$$

so that $(f_1, \ldots, f_n) = (x_1 - \alpha_1, \ldots, x_n - \alpha_n)$. Thus

$$\hat{\mathcal{O}}_{X,p} \cong \hat{R}[[x_1, \ldots, x_n]]/(x_1 - \alpha_1, \ldots, x_n - \alpha_n) \cong \hat{R} \cong \hat{\mathcal{O}}_{Y,q}.$$

Now suppose that $\hat{\phi}^* : \hat{\mathcal{O}}_{Y,q} \to \hat{\mathcal{O}}_{X,p}$ is an isomorphism. We may assume that X and Y are affine. Let $A = k[Y]$ and $B = k[X]$ be the respective rings of regular functions. For any ideal $I \subset \mathcal{O}_{X,p}$, we have that $(I\hat{\mathcal{O}}_{X,p}) \cap \mathcal{O}_{X,p} = I$ by Lemma 21.37. We have that

(21.24) $$m_q \mathcal{O}_{X,p} = (m_q \hat{\mathcal{O}}_{X,p}) \cap \mathcal{O}_{X,p} = (m_p \hat{\mathcal{O}}_{X,p}) \cap \mathcal{O}_{X,p} = m_p.$$

Let $\mathfrak{n}$ be the ideal of q in A. We have a representation $B = A[x_1, \ldots, x_n]/I$, where I is an ideal in a polynomial ring $A[x_1, \ldots, x_n]$ over A and $\mathfrak{m} = \mathfrak{n}A[x_1, \ldots, x_n] + (x_1, \ldots, x_n)$ is the ideal of p in $A[x_1, \ldots, x_n]$. We have that

$$\Omega_{A[x_1,\ldots,x_n]_{\mathfrak{m}}/A} \cong \Omega_{A[x_1,\ldots,x_n]/A} \otimes_{A[x_1,\ldots,x_n]} A[x_1,\ldots,x_n]_{\mathfrak{m}} \quad \text{by Lemma 14.8}$$

$$\cong \bigoplus_{i=1}^{n} A[x_1, \ldots, x_n]_{\mathfrak{m}} dx_i$$

by Example 14.5. Let $N = \{df \mid f \in I_{\mathfrak{m}}\}$, so that

$$\Omega_{B_{\mathfrak{m}}/A} \cong \left(\bigoplus_{i=1}^{n} A[x_1, \ldots, x_n]_{\mathfrak{m}} dx_i \right) / N$$

by Example 14.7. For $1 \leq i \leq n$, there exists $a_i \in A$, $b_i \in I$, and $c_i \in \mathfrak{m}^2$ such that $x_i = a_i + b_i + c_i$ by (21.24). Thus $dx_i \in N = \mathfrak{m}\Omega_{A[x_1,\ldots,x_n]_{\mathfrak{m}}/A}$ and so $(\Omega_{B/A}) \otimes_B B_{\mathfrak{m}} \cong \Omega_{B_{\mathfrak{m}}/A} = 0$ by Nakayama's lemma. Thus there exist $f_1, \ldots, f_n \in I$ such that $\text{Det}\left(\frac{\partial f_i}{\partial x_j}(p)\right) \neq 0$.

Define a ring C by $C = A[x_1, \ldots, x_n]/(f_1, \ldots, f_n)$. By the first part of this proof, the completion $\hat{C}$ of C at the maximal ideal $\mathfrak{m}C$ is equal to $\hat{A}$. Thus the natural maps $A \to C \to B$ induce isomorphisms

$$\hat{\mathcal{O}}_{Y,q} \cong \hat{A} \cong \hat{C} \cong \hat{B} \cong \hat{\mathcal{O}}_{X,p}.$$

Since C is a subring of $\hat{C}$ and B is a subring of $\hat{B}$ (by Proposition 21.38), we have that $C_{\mathfrak{m}C}$ is a subring of $B_{\mathfrak{m}B}$. Since B is a quotient of C, we have that $B_{\mathfrak{m}B} = C_{\mathfrak{m}C}$. Thus $I_{\mathfrak{m}} = (f_1, \ldots, f_m)_{\mathfrak{m}}$ in $A[x_1, \ldots, x_n]_{\mathfrak{m}}$, from which it follows that ϕ is étale near p. □

Theorem 21.82. *Suppose that X is an n-dimensional variety and U is an open subset of X. Suppose that $f_1, \ldots, f_n \in \Gamma(U, \mathcal{O}_X)$. Let $\phi = (f_1, \ldots, f_n) : U \to \mathbb{A}^n$ be the induced regular map. Then the following conditions are equivalent:*

1) *ϕ is étale.*

2) *For all $p \in U$, $t_1 = f_1 - f_1(p), \ldots, t_n = f_n - f_n(p)$ generate $\mathfrak{m}_p/\mathfrak{m}_p^2$.*

3) *For all $p \in U$, the k-algebra homomorphism*

$$k[[T_1, \ldots, T_n]] \to \hat{\mathcal{O}}_{X,p}$$

   defined by $T_i \mapsto t_i$ is an isomorphism.

4) *$\Omega_{X/k}|U = \bigoplus_{i=1}^{n} \mathcal{O}_U df_i$ ($f_1, \ldots, f_n$ are uniformizing parameters on U).*

5) *$\Omega_{U/\mathbb{A}^n} = 0$.*

Proof. We observe that 3) is equivalent to the statement that

$$\hat{\phi}^* : \hat{\mathcal{O}}_{\mathbb{A}^n, \phi(p)} \to \hat{\mathcal{O}}_{U,p}$$

is an isomorphism for all $p \in U$. The equivalence of 1) and 3) follows from Theorem 21.81. The equivalence of 4) and 5) follows from the exact sequence (14.4) and Example 14.5. If we assume 5) and N is the kernel of the surjection $\phi^*\Omega_{\mathbb{A}^n/k} \to \Omega_{U/k}$, then we have that $N \otimes k(X) = 0$ since both $\phi^*\Omega_{\mathbb{A}^n/k}$ and $\Omega_{U/k}$ have rank n, so that $N = 0$ since it is a torsion submodule of the locally free sheaf $\phi^*\Omega_{\mathbb{A}^n/k}$. The equivalence of 2) and 3) is by Proposition 21.62.

It remains to establish that 2) is equivalent to 4). Condition 2) implies U is nonsingular by Definition 10.12, and condition 4) implies U is nonsingular by Theorem 14.14, so we may assume that U is nonsingular.

Condition 2) is the statement that for all $p \in U$, $t_1, \ldots, t_n$ generate m_p/m_p^2, which is equivalent to the statement that $df_1, \ldots, df_n$ generate $(\Omega_{X/k,p}) \otimes k(p)$ by Proposition 14.13. This is equivalent to the statement that $df_1, \ldots, df_n$ generate $\Omega_{X/k}$ in some neighborhood of p (by Nakayama's lemma). So condition 2) is equivalent to the statement that $\Omega_{U/k}$ is a quotient of $\bigoplus_{i=1}^n \mathcal{O}_U df_i$. Since $\Omega_{U/k}$ is locally free of rank n (as U is nonsingular) this is equivalent to statement 4). $\square$

Ramification can also be expressed in terms of Kähler differentials.

Theorem 21.83. *Suppose that $\phi : X \to Y$ is a dominant regular map of varieties, $p \in X$, and $q = \phi(p)$. Then*

$$\mathcal{O}_{Y,q} \to \mathcal{O}_{X,p}$$

is unramified if and only if $(\Omega_{X/Y})_p = \Omega_{\mathcal{O}_{X,p}/\mathcal{O}_{Y,q}} = 0$.

Proof. We may suppose that X, Y are affine. We express $k[X]$ as a quotient of a polynomial ring $k[x_1, \ldots, x_n, \ldots, x_m]$ so that the subring $k[Y]$ is a quotient of the polynomial ring $k[x_1, \ldots, x_n]$. There exist $g_1, \ldots, g_N \in k[x_1, \ldots, x_m]$ such that $k[X] = k[x_1, \ldots, x_m]/(g_1, \ldots, g_N)$. Let $\bar{x}_i$ be the residues of x_i in $k[X]$. The point $p \in X$ has a maximal ideal $\alpha = I(p)$ in $k[x_1, \ldots, x_m]$. Under the corresponding embedding $X \subset \mathbb{A}^m$, suppose that p is the point $(\xi_1, \ldots, \xi_m)$.

We will first establish that the following condition (21.25) holds if and only if $\mathcal{O}_{Y,q} \to \mathcal{O}_{X,p}$ is unramified:

(21.25) the Jacobian matrix $\frac{\partial(g_1,\ldots,g_N)}{\partial(x_{n+1},\ldots,x_m)}(p)$ has rank $m - n$.

First suppose that $\mathcal{O}_{Y,q} \to \mathcal{O}_{X,p}$ is unramified. Suppose that $D \in \mathrm{Der}_k(\mathcal{O}_{X,p}, k(p))$ and $D|\mathcal{O}_{Y,q} = 0$. Suppose that $f \in \mathcal{O}_{X,p}$. There exists

21.9. Étale maps and uniformizing parameters

$g \in \mathcal{O}_{Y,q}$ such that $f - g \in m_p^2$ since $m_p = m_q \mathcal{O}_{X,p}$ and $\mathcal{O}_{X,p}/m_p \cong \mathcal{O}_{Y,q}/m_q$. Thus $D(f) = D(g) = 0$, so $D = 0$.

Suppose that $u_1, \ldots, u_m \in k$. A necessary and sufficient condition that there exists $D \in \mathrm{Der}_k(\mathcal{O}_{X,p}, k(p))$ such that $D(\overline{x}_i) = u_i$ is that

$$\sum_{j=1}^{m} \frac{\partial g_i}{\partial x_j}(p) u_j = 0 \quad \text{for } 1 \leq i \leq N.$$

This follows since

$$\mathrm{Der}_k(\mathcal{O}_{X,p}, k(p)) \cong \mathrm{Hom}_{\mathcal{O}_{X,p}}(\Omega_{\mathcal{O}_{X,p}/k}, k(p))$$

by Proposition 14.13 and

$$\Omega_{\mathcal{O}_{X,p}/k} = \mathcal{O}_{X,p} dx_1 \oplus \cdots \oplus \mathcal{O}_{X,p} dx_m / \left(\sum_{j=1}^{m} \frac{\partial g_i}{\partial x_j} dx_j \mid 1 \leq i \leq N \right).$$

We have earlier shown that if $u_1 = u_2 = \cdots = u_n = 0$, then necessarily $u_{n+1} = \cdots = u_m = 0$. Thus (21.25) holds.

Now suppose that (21.25) holds. We may assume after reindexing the g_i that

$$\frac{\partial(g_1, \ldots, g_{m-n})}{\partial(x_{n+1}, \ldots, x_m)}(p) \text{ has rank } m - n.$$

Thus $g_1(\xi_1, \ldots, \xi_n, x_{n+1}, \ldots, x_n), \ldots, g_{m-n}(\xi_1, \ldots, \xi_n, x_{n+1}, \ldots, x_n)$ are uniformizing parameters at the point $(\xi_{n+1}, \ldots, \xi_m)$ in the affine space $\mathbb{A}^{m-n}$ with coordinate ring $k(p)[x_{n+1}, \ldots, x_m]$ by Theorem 21.82. Thus $x_1 - \xi_1, \ldots, x_n - \xi_n, g_1, \ldots, g_{m-n}$ are uniformizing parameters in $\mathbb{A}^m$ at p (they are in fact a k-basis of α/α^2). Thus

$$m_q \mathcal{O}_{X,p} = (x_1 - \xi_1, \ldots, x_n - \xi_n) \mathcal{O}_{X,p} = \alpha \mathcal{O}_{X,p} = m_p$$

and so $\mathcal{O}_{Y,q} \to \mathcal{O}_{X,p}$ is unramified.

It remains to show that equation (21.25) holds if and only if $\Omega_{\mathcal{O}_{X,p}/\mathcal{O}_{Y,q}} = 0$. From the surjection

$$(\mathcal{O}_{Y,q}[x_{n+1}, \ldots, x_m])_\alpha \to (\mathcal{O}_{Y,q}[x_{n+1}, \ldots, x_m]/(g_1, \ldots, g_N))_\alpha = \mathcal{O}_{X,p},$$

we have that

$$\Omega_{\mathcal{O}_{X,p}/\mathcal{O}_{Y,q}} = \mathcal{O}_{X,p} dx_{n+1} \oplus \cdots \oplus \mathcal{O}_{X,p} dx_m / \left(\sum_{j=n+1}^{m} \frac{\partial g_i}{\partial x_j} dx_j \mid 1 \leq i \leq N \right).$$

Thus $\left(\Omega_{\mathcal{O}_{X,p}/\mathcal{O}_{Y,q}} \right) \otimes k(p) = 0$ if and only if (21.25) holds, and this condition is equivalent to $\Omega_{\mathcal{O}_{X,p}/\mathcal{O}_{Y,q}} = 0$ by Nakayama's lemma, Lemma 1.18. $\square$

Proposition 21.84. *Suppose that $\phi : X \to Y$ is a dominant regular map of varieties such that ϕ is étale. Then ϕ is unramified.*

Proof. The proof of Theorem 21.81 shows that $\Omega_{X/Y} = 0$, which implies ϕ is unramified by Theorem 21.83. □

The converse of Proposition 21.84 is false. An example is given in Exercise 21.89. However, the converse is true if Y is normal, as shown by the following proposition, whose proof is Exercise 21.91.

Proposition 21.85. *Suppose that $\phi : X \to Y$ is a dominant regular map of varieties and Y is normal. Then ϕ is unramified if and only if ϕ is étale.*

Exercise 21.86. Show that a regular map of varieties $\phi : X \to Y$ is étale if and only if for every $p \in X$, there exist open affine neighborhoods A of p and B of $q = \phi(p)$ such that $k[A]$ is a quotient of a polynomial ring over $k[B]$, $k[A] = k[B][x_1, \ldots, x_n]/(f_1, \ldots, f_n)$ where $\mathrm{Det}(\frac{\partial f_i}{\partial x_j})$ is a unit in $k[A]$. Hint: Use Exercise 1.7.

Exercise 21.87. Suppose that X is a variety over the complex numbers and U is an open subset of X with uniformizing parameters $f_1, \ldots, f_n$ on U. Let $\phi = (f_1, \ldots, f_n) : U \to \mathbb{A}^n$. Suppose that $p \in U$. Use the implicit function theorem (Theorem 10.42) to show that there exists an open neighborhood V of p in the Euclidean topology, which is contained in U such that $\phi : V \to \phi(V)$ is an analytic isomorphism.

Exercise 21.88. Show that the implicit function theorem (Theorem 10.42) is false in the Zariski topology by considering the following example. Suppose that k has characteristic $\neq 2$. Let $X = Z(x_1^2 - x_2) \subset \mathbb{A}^2$ and let $\phi : X \to \mathbb{A}^1$ be defined by $\phi(a_1, a_2) = a_2$. At $p = (1, 1)$, we have that $\frac{\partial}{\partial x_1}(x_1^2 - x_2) \neq 0$. Show that ϕ is not 1-1 in any Zariski open subset U of X.

Exercise 21.89. Let $\phi : X \to Y$ be the regular map of Exercise 21.65. Show that $\Omega_{X/Y} = 0$ (so that ϕ is unramified) but that ϕ is not étale.

Exercise 21.90. Suppose that $\phi : X \to Y$ is a finite regular map of varieties. The ideal sheaf $\mathrm{Ann}(\Omega_{X/Y})$ is defined (in Exercise 11.43) by

$$\mathrm{Ann}(\Omega_{X/Y})(U) = \{f \in \mathcal{O}_X(U) \mid f\Omega_{X/Y}(U) = 0\}$$

for U an open subset of X. Show that $\phi(\mathrm{Supp}(\mathcal{O}_X/\mathrm{Ann}(\Omega_{X/Y})))$ is the locus in Y above which ϕ is ramified.

Conclude that

$$\{p \in X \mid \phi \text{ is unramified at } p\}$$

is an open subset of X which is nonempty if and only if $k(X)$ is separable over $k(Y)$.

Exercise 21.91. Prove Proposition 21.85.

21.10. Purity of the branch locus and the Abhyankar-Jung theorem

Suppose that $\phi : X \to Y$ is a dominant finite regular map of normal varieties. Let

(21.26) $\qquad \Delta = \{p \in X \mid \phi^* : \mathcal{O}_{Y,\phi(p)} \to \mathcal{O}_{X,p}$ is ramified$\}$

be the locus of points in X where ϕ is ramified. We will call Δ the ramification locus of ϕ in X. By Theorem 21.83, we have that $\Delta = \operatorname{Supp}(\Omega_{X/Y})$, so Δ is a closed subset of X. If we also assume that $\phi^* : k(Y) \to k(X)$ is separable, then Δ is a proper subset of X (for instance by Theorem 21.29). We have that $\phi(\Delta)$ is the ramification locus of ϕ in Y (or the branch locus of ϕ). Since ϕ is finite, if Δ has pure codimension 1 in X (all irreducible components have codimension 1), then the ramification locus $\phi(\Delta)$ has pure codimension 1 in Y.

The proof of the following theorem is based on the proof by Zariski in [**157**, Proposition 2]. Stronger forms of Theorem 21.92 are by Nagata [**121**, Theorem 41.1], Auslander [**14**], Grothendieck [**64**], and Bhatt, Carvajal-Rojas, Grant, Schwede, and Tucker [**21**], although Y cannot be too far from being a nonsingular variety for purity of the branch locus to hold.

Theorem 21.92 (Purity of the branch locus). *Suppose that X is a normal variety, Y is a nonsingular variety, and $\phi : X \to Y$ is a dominant finite regular map such that $k(Y) \to k(X)$ is separable. Then the closed set of points of X at which ϕ is ramified has pure codimension 1 in X (all irreducible components of the ramification locus have codimension 1).*

Proof. Suppose that $a \in X$ and a is not contained in an irreducible component of Δ which has codimension 1 in X. Let $q = \phi(a)$ and let $x_1, \ldots, x_n$ be regular parameters in $\mathcal{O}_{Y,q}$. There exists, by Proposition 14.15, an affine open neighborhood U of q in Y such that $x_1, \ldots, x_n$ are uniformizing parameters on U. Let $\alpha : U \to \mathbb{A}^n$ be the corresponding étale map (Theorem 21.82). Let V be an affine neighborhood of a such that $\phi(V) \subset U$ and $\Delta \cap V$ has codimension ≥ 2 in V.

The elements $x_1, \ldots, x_n$ are a separating transcendence basis of $k(X)$ over k since they are a separating transcendence basis of $k(Y)$ over k by Theorem 21.75 and $k(X)$ is finite and separable over $k(Y)$. Thus the derivatives $\frac{\partial}{\partial x_i}$ on the rational function field $k(x_1, \ldots, x_n)$ extend uniquely to a $k(Y)$-basis of $\operatorname{Der}_k(k(Y), k(Y))$ and to a $k(X)$-basis of $\operatorname{Der}_k(k(X), k(X))$, as explained after Theorem 21.75. Suppose that E is a prime divisor on V. Then there exists $p' \in E \setminus \Delta$. By Theorem 21.83, we have that

$$\Omega_{\mathcal{O}_{X,p'}/\mathcal{O}_{Y,q'}} = 0$$

where $q' = \phi(p')$, so
$$\Omega_{\mathcal{O}_{Y,q'}/k} \otimes_{\mathcal{O}_{Y,q'}} \mathcal{O}_{X,p'} \cong \Omega_{\mathcal{O}_{X,p'}/k}$$
by (14.4), since $\Omega_{\mathcal{O}_{Y,q'}/k}$ is a free module of rank equal to the dimension of X by Proposition 14.15. Now $\mathrm{Der}_k(\mathcal{O}_{Y,q'}, \mathcal{O}_{Y,q'})$ is a free $\mathcal{O}_{Y,q'}$-module with basis $\frac{\partial}{\partial x_1}, \ldots, \frac{\partial}{\partial x_n}$, so
$$\begin{aligned}
\mathrm{Der}_k(\mathcal{O}_{X,p'}, \mathcal{O}_{X,p'}) &\cong \mathrm{Hom}_{\mathcal{O}_{X,p'}}(\Omega_{\mathcal{O}_{X,p'}/k}, \mathcal{O}_{X,p'}) \\
&\cong \mathrm{Hom}_{\mathcal{O}_{Y,q'}}(\Omega_{\mathcal{O}_{Y,q'}/k}, \mathcal{O}_{Y,q'}) \otimes_{\mathcal{O}_{Y,q'}} \mathcal{O}_{X,p'} \\
&\cong \mathrm{Der}_k(\mathcal{O}_{Y,q'}, \mathcal{O}_{Y,q'}) \otimes_{\mathcal{O}_{Y,q'}} \mathcal{O}_{X,p'}
\end{aligned}$$
by Lemma 14.3 and since $\Omega_{\mathcal{O}_{Y,q'}/k}$ is a free $\mathcal{O}_{Y,q'}$-module.

Thus $\frac{\partial}{\partial x_1}, \ldots, \frac{\partial}{\partial x_n}$ is a free basis of $\mathrm{Der}_k(\mathcal{O}_{X,p'}, \mathcal{O}_{X,p'})$ as an $\mathcal{O}_{X,p'}$-module. In particular, the derivations $\frac{\partial}{\partial x_i} : k(X) \to k(X)$ map $\mathcal{O}_{X,p'}$ into $\mathcal{O}_{X,p'}$ for all i, and so $\frac{\partial}{\partial x_i} : \mathcal{O}_{X,E} \to \mathcal{O}_{X,E}$ for all i, since $\mathcal{O}_{X,E}$ is a localization of $\mathcal{O}_{X,p'}$.

Now, by Theorem 1.79, $\mathcal{O}_{X,a} = \bigcap_{a \in E} \mathcal{O}_{X,E}$, where the intersection is over all prime divisors E of V which contain a since $\mathcal{O}_{X,a}$ is integrally closed. Thus the derivations

$$(21.27) \qquad \frac{\partial}{\partial x_i} : \mathcal{O}_{X,a} \to \mathcal{O}_{X,a} \quad \text{for all } i.$$

Suppose that k has characteristic 0. Then we have a natural k-algebra homomorphism $\psi : \mathcal{O}_{X,a} \to k[[x_1, \ldots, x_n]]$ defined by
$$\psi(f) = \sum c_{i_1,\ldots,i_n} x_1^{i_1} \cdots x_n^{i_n}$$
where
$$c_{i_1,\ldots,i_n} = \frac{1}{i_1! \cdots i_n!} \left(\frac{\partial^{i_1 + \cdots + i_n} f}{\partial x_1^{i_1} \cdots \partial x_n^{i_n}} \right)(a).$$
Now fibers of the composed map $\alpha \circ \phi : V \to \mathbb{A}^n$ are finite sets, so the ideal $(x_1, \ldots, x_n)\mathcal{O}_{X,a}$ contains a power m_a^r of the maximal ideal m_a of $\mathcal{O}_{X,a}$. Now $\psi(m_a^r) \subset (x_1, \ldots, x_n)$ implies $\psi(m_a) \subset (x_1, \ldots, x_n)$ since $(x_1, \ldots, x_n)$ is a prime ideal in $k[[x_1, \ldots, x_n]]$. Thus ψ extends uniquely to a k-algebra homomorphism $\overline{\psi} : \hat{\mathcal{O}}_{X,a} \to k[[x_1, \ldots, x_n]]$ such that $\overline{\psi}(\hat{m}_a) \subset (x_1, \ldots, x_n)$ and $\overline{\psi}(f) = \psi(f)$ for $f \in \mathcal{O}_{X,a}$ by Lemma 21.43. The natural inclusions of normal domains
$$k[x_1, \ldots, x_n]_{(x_1,\ldots,x_n)} = \mathcal{O}_{\mathbb{A}^n, \alpha(q)} \to \mathcal{O}_{Y,q} \to \mathcal{O}_{X,a}$$
induce k-algebra homomorphisms of integral domains, by Proposition 21.50, of the same dimension n by Theorem 21.36,
$$k[[x_1, \ldots, x_n]] = \hat{\mathcal{O}}_{\mathbb{A}^n, \alpha(q)} \to \hat{\mathcal{O}}_{Y,q} \to \hat{\mathcal{O}}_{X,a} \xrightarrow{\overline{\psi}} k[[x_1, \ldots, x_n]]$$

whose composite is the identity map. Thus each of these maps must be an equality, and in particular, $\hat{\mathcal{O}}_{Y,q} = \hat{\mathcal{O}}_{X,a}$, so that ϕ is unramified at a. Thus $a \notin \Delta$, completing the proof of Theorem 21.92 in the case that k has characteristic 0.

Now suppose that k has characteristic $p > 0$. Then $x_1, \ldots, x_n$ is a p-basis of $k(X)$ by Theorem 21.76. Let $f \in \mathcal{O}_{X,a}$. Since $x_1, \ldots, x_n$ is a p-basis, we can write f uniquely in the form

$$f = \sum A_{i_1,\ldots,i_n}^p x_1^{i_1} \cdots x_n^{i_n}$$

with all $A_{i_1,\ldots,x_n} \in k(X)$ and $0 \leq i_1 + \cdots + i_n < p$ for all $i_1, \ldots, i_n$ in the sum. Since all partials

$$\frac{\partial^{i_1+\cdots+i_n} f}{\partial x_1^{i_1} \cdots \partial x_n^{i_n}} \in \mathcal{O}_{X,a}$$

by (21.27), we have that all $A_{i_1,\ldots,i_n}^p \in \mathcal{O}_{X,a}$, and thus all $A_{i_1,\ldots,i_n} \in \mathcal{O}_{X,a}$ since $\mathcal{O}_{X,a}$ is integrally closed. If f is in the maximal ideal m_a of $\mathcal{O}_{X,a}$, then we have that $A_{0,\ldots,0} \in m_a$, so $A_{0,\ldots,0}^p \in m_a^p \subset m_a^2$. Thus $m_a \subset \sum_{i=1}^n x_i \mathcal{O}_{X,a} + m_a^2$, so

$$m_a = (x_1, \ldots, x_n)\mathcal{O}_{X,a} + m_a^2.$$

Thus $m_a = (x_1, \ldots, x_n)\mathcal{O}_{X,a}$ by Nakayama's lemma, Lemma 1.18. Thus $m_a = m_q \mathcal{O}_{X,a}$, where m_q is the maximal ideal of $\mathcal{O}_{Y,q}$ and so $a \notin \Delta$, completing the proof of Theorem 21.92 in the case that k has positive characteristic p. $\square$

The following theorem is the Abhyankar-Jung theorem, which generalizes a topological proof in the complex analytic case by Jung [85]. A proof based on Abhyankar's original proof in [1] will be given in Section 21.12. The above theorem on the purity of the branch locus is an important ingredient in the proof. Generalizations of the Abhyankar-Jung theorem can be found in [70, Section XII], [71], and other references.

Let $K \to K^*$ be a finite separable field extension of algebraic function fields, R be a normal algebraic local ring of K, and S be a normal local ring of K^* which lies over R. Let $J(S/R) = \sqrt{\text{Ann}(\Omega_{S/R})}$, which is an ideal in S defining the locus in S where $R \to S$ is ramified (Theorem 21.83 and Exercise 21.90). That is,

$$Z(J(S/R)) = \{Q \in \text{Spec}(S) \mid R_{Q \cap R} \to S_Q \text{ is ramified}\}.$$

If S is finite over R, define

$$I(S/R) = \sqrt{J(S/R) \cap R}$$

which defines by Exercise 21.90 the locus in R where $R \to S$ is ramified. That is, if S is finite over R,

$Z(I(S/R))$
$= \{P \in \mathrm{Spec}(R) \mid \text{there exists } Q \in \mathrm{Spec}(S) \text{ such that } R_P \to S_Q \text{ is ramified}\}$.

Theorem 21.93 (Abhyankar-Jung theorem). *Suppose that Y is a nonsingular variety and X is a normal variety, $\phi : X \to Y$ is a dominant finite regular map such that $k(Y) \to k(X)$ is separable, and if the characteristic of the ground field k is $p > 0$, then p does not divide the index $[K' : k(Y)]$ where K' is a Galois closure of $k(X)/k(Y)$. Suppose that $p \in X$ and $q = \phi(p) \in Y$ are such that there exist regular parameters $x_1, \ldots, x_n \in \mathcal{O}_{Y,q}$ such that $\prod_{i=1}^t x_i \in J(\mathcal{O}_{X,p}/\mathcal{O}_{Y,q})$ for some $t \leq n$. Let $d = [K' : k(Y)]$. Then there exists a subgroup Γ of $\mathbb{Z}_d^t$ such that $\mathcal{O}_{X,p} \cong k[[x_1^{\frac{1}{d}}, \ldots, x_t^{\frac{1}{d}}, x_{t+1}, \ldots, x_n]]^\Gamma$, where the basis element $e_i \in \mathbb{Z}_d^t$ acts on $x_i^{\frac{1}{d}}$ by multiplication by a d-th root of unity (in the ground field k) and is the identity on $x_j^{\frac{1}{d}}$ if $j \neq i$.*

Corollary 21.94. *Let k be an algebraically closed field of characteristic 0, let $R = k[[x_1, \ldots, x_n]]$ be a power series ring over k, and let $f \in R[z]$ be an irreducible monic polynomial. Suppose that the discriminant (21.8) of f is a unit in R times a monomial in $x_1, \ldots, x_n$. Then $f(z)$ has a root $\alpha \in k[[x_1^{\frac{1}{d}}, \ldots, x_n^{\frac{1}{d}}]]$ (a fractional power series) for some $d \in \mathbb{Z}_{>0}$. We in fact have a factorization $f(z) = \prod_{i=1}^{\deg(f)}(z - \alpha_i)$ where $\alpha_i \in k[[x_1^{\frac{1}{d}}, \ldots, x_n^{\frac{1}{d}}]]$ for all i.*

Proof. Let L be the quotient field of $S = R[z]/(f) = R[\bar{z}]$. By Theorem 21.20, S_P is unramified over R_P if P is a prime ideal in R which does not contain the monomial $x_1 \cdots x_n$. Let T be the integral closure of S in L. We have that $S_Q \to T_Q$ is an isomorphism if Q is a prime ideal in S such that S_Q is a regular local ring. Thus T_Q is unramified over S_Q if S_Q is regular, and so T_P is unramified over R_P if P is a prime ideal in R which does not contain $x_1 \cdots x_n$. Now the proof of Theorem 21.93 generalizes to prove the Abhyankar-Jung theorem in the situation of this corollary, if we use Nagata's theorem [**121**, Theorem 41.1], which proves the purity of the branch locus over an arbitrary regular local ring, giving us inclusions

$$R \to S \to T \to k[[x_1^{\frac{1}{d}}, \ldots, x_n^{\frac{1}{d}}]],$$

showing that the root $\alpha = \bar{z}$ of f is in $k[[x_1^{\frac{1}{d}}, \ldots, x_n^{\frac{1}{d}}]]$. The last assertion of the corollary follows since the quotient field E of $k[[x_1^{\frac{1}{d}}, \ldots, x_n^{\frac{1}{d}}]]$ is Galois over the quotient field of R, and the irreducible polynomial f has a root in E, so all roots of f must be contained in the integral closure $k[[x_1^{\frac{1}{d}}, \ldots, x_n^{\frac{1}{d}}]]$ of R in E. □

21.10. Purity of the branch locus and the Abhyankar-Jung theorem

Corollary 21.95 (Newton). *Suppose that k is an algebraically closed field of characteristic 0 and $R = k[[x]]$ is a power series ring over k. Suppose that $f \in R[z]$ is irreducible and monic. Then $f(z)$ has a root $\alpha \in k[[x^{\frac{1}{d}}]]$ (a fractional power series) for some $d \in \mathbb{Z}_{>0}$.*

It follows that when k is algebraically closed of characteristic 0, the Laurent field of fractional power series expansions (all series $\sum_{i=m}^{\infty} a_i x^{\frac{1}{d}}$ for some $d \in \mathbb{Z}_{>0}$, $m \in \mathbb{Z}$, and $a_i \in k$) is an algebraic closure of the Laurent field $k((x))$ of formal power series in x over k.

Newton's proof is constructive. His algorithm is explained in [35, Section 2.1] and in the book [26]. Some letters of Newton presenting this algorithm are translated in [26].

This result is no longer true when k has characteristic $p > 0$. An example showing that this fails is given in Exercise 21.98. The example computed in the exercise is essentially the worst thing that can happen. A construction of an algebraic closure of $k((x))$ when k is algebraically closed of positive characteristic is given in [88]

The following example shows that Corollary 21.95 does not extend to power series rings of dimension greater than 1.

Example 21.96. Let k be an algebraically closed field of characteristic 0 or $p > 5$ and let $n \geq 2$ be a positive integer which is prime to p. Let $R = k[[x,y]]$ and $f = z^n + x^2 - y^3$. Then there does not exist a fractional power series solution $z = \alpha(x^{\frac{1}{d}}, y^{\frac{1}{d}}) \in k[[x^{\frac{1}{d}}, y^{\frac{1}{d}}]]$ to $f(x,y,z) = 0$ for any $d \in \mathbb{Z}_{>0}$. Further, this statement is true for any regular system of parameters y_1, y_2 in R (so that $R = k[[y_1, y_2]]$).

Proof. If there were such a fractional power series solution α, then we would have an inclusion $S = R[z]/(f) \to T = k[[x^{\frac{1}{d}}, y^{\frac{1}{d}}]]$. The finite extension $R \to T$ is then unramified above primes that do not contain xy and thus $R \to S$ is unramified above primes that do not contain xy. This is a contradiction since $R \to S$ is ramified above the prime ideal $(y^3 - x^2)$. □

Exercise 21.97. This exercise shows that the conclusions of Theorem 21.92 may fail if Y is normal but not nonsingular. Consider the finite map $\phi : X \to Y$ of affine varieties where

$$\phi^* : k[Y] = k[x^2, xy, y^2] \to k[X] = k[x,y]$$

is the natural inclusion and the characteristic of k is $\neq 2$. Show that X and Y are normal and compute the locus of points in X at which ϕ is ramified. Conclude that the ramification locus of ϕ in X does not have pure codimension 1 in X, so the conclusions of Theorem 21.92 do not hold.

Exercise 21.98. (This is [**1**, Example 1].) Consider the finite map $\phi : X \to Y$ of affine varieties over a field k of characteristic $p > 2$ where $\phi^* : k[Y] = k[x,y] \to k[X] = k[x,y,z]/(z^p - x^{p-1}z - y^{p-1})$ is the natural inclusion. Show that X is normal and Y is nonsingular and that $k(X)$ is a degree p Galois extension of $k(Y)$ (an Artin-Schreier extension). Compute the locus of points in X at which ϕ is ramified. Show that all of the hypotheses of Theorem 21.8 hold, except the assumption that p does not divide the index $[k(Y) : k(X)]$. Show that the conclusions of Theorem 21.8 do not hold.

Exercise 21.99. Let k be a field of positive characteristic $p > 0$. Show that

$$\sigma = \sum_{i=1}^{\infty} x^{1 - \frac{1}{p^i}}$$

is algebraic over the rational function field $k(x)$ with minimal polynomial $f(z) = z^p - x^{p-1}z - x^{p-1}$. Show that we have a factorization

$$f(z) = \prod_{i=1}^{p}(z - (\sigma + ix)).$$

Since this factorization holds in the field $k(x^{\mathbb{Q}})$ of all series with exponents being well-ordered subsets of $\mathbb{Q}$ and coefficients elements of k, we see that $f(z)$ cannot have a root which is a fractional power series in x (with bounded denominator). The field $k(x)(\sigma)$ is an example of an Artin-Schreier extension of $k(x)$. This type of extension is ultimately responsible for all of the problems which arise in ramification in positive characteristic.

21.11. Galois theory of local rings

We introduce in this section some material from the section "Galois theory of local rings" in [**6**], which we will need in the proof of the Abhyankar-Jung theorem in Section 21.12.

Let $K \to K^*$ be a finite Galois field extension with Galois group $G = G(K^*/K)$, let R be a normal local ring with quotient field K, and let S be a normal local ring with quotient field K^* which lies over R. The splitting group $G^s(S/R)$ is defined to be

$$G^s(S/R) = \{\sigma \in G \mid \sigma(S) = S\}.$$

The splitting field $K^s = K^s(S/R)$ is defined to be $K^s = (K^*)^{G^s(S/R)}$.

Lemma 21.100. *The field K^s is the smallest field K' lying between K and K^* for which S is the only normal local ring in K^* lying above $S \cap K'$.*

21.11. Galois theory of local rings

Proof. Let T be the integral closure of R in K^*. Let $n_1, \ldots, n_u$ be the maximal ideals of T, indexed so that $S = T_{n_1}$. Let $T^s = T \cap K^s$, $m_i^s = n_i \cap T_s$ for $1 \le i \le u$. By the Chinese remainder theorem, Theorem 1.5, there exists $a \in T$ such that $a \equiv 0 \bmod n_1$ and $a \equiv 1 \bmod n_i$ if $i > 1$. Then the norm

$$N_{K^*/K^s}(a) = \prod_{\sigma \in G^s(S/R)} \sigma(a) \in n_1 \cap K^s = m_1^s$$

and

$$N_{K^*/K^s}(a) \equiv 1 \bmod n_i \quad \text{for } i > 1.$$

Thus for $i > 1$, $m_1^s \not\subset n_i$ so that n_i does not lie above m_1^s.

Hence it is enough to show that if K' satisfies the assumption of the lemma, then $K^s \subset K'$. Suppose that $\sigma \in G(K^*/K')$. If $\sigma(n_j) = n_1$ for some $j > 1$, then $n_j \cap K' = \sigma(n_j \cap K') = n_1 \cap K'$, which is a contradiction to our assumptions. Thus $G^s(K^*/K') \subset G^s$, and so $K^s \subset K'$. $\square$

Lemma 21.101. *Let $R^s = K^s \cap S$. Then $m_R R^s = m_{R^s}$ and $R^s/m_{R^s} \cong R/m_R$ so that $R \to R^s$ is unramified.*

Proof. Let T be the integral closure of R in K^*. Let $n_1, \ldots, n_u$ be the maximal ideals of T, indexed so that $S = T_{n_1}$. Let $T^s = T \cap K^s$, $m_i^s = n_i \cap T^s$ for $1 \le i \le u$.

Suppose that $a \in T^s$. By Lemma 21.100 and the Chinese remainder theorem (Theorem 1.5), there exists $b \in T^s$ such that $b \equiv a \bmod m_1^s$ and $b \equiv 1 \bmod m_i^s$ for $i > 1$, so that $b \equiv a \bmod n_1$ and $b \equiv 1 \bmod n_i$ for all $i > 1$. Let $\sigma_1 = \mathrm{id}, \sigma_2, \ldots, \sigma_q$ be a complete set of representatives of the cosets of $G^s(S/R)$ in G. Then $\sigma_1(b), \ldots, \sigma_q(b)$ are the K^s/K conjugates of b and hence

$$c = N_{K^s/K}(b) = \prod_{t=1}^{q} \sigma_t(b) \in R.$$

We have that $\sigma_1(b) = b \equiv a \bmod n_1$, and if $t > 1$, then there exists an $i > 1$ such that $\sigma_t(n_i) = n_1$, so that $\sigma_t(b) \equiv 1 \bmod n_1$. Thus $\sigma_1(b) \equiv a \bmod n_1$ and for $t > 1$, $\sigma_t(b) \equiv 1 \bmod n_1$. Hence $c \equiv a \bmod n_1$ and $c - a \in n_1 \cap K^s = m_1^s$ so that $c \equiv a \bmod m_1^s$. Thus $R^s/m_{R^s} \cong T^s/m_1^s \cong R/m_R$.

Let $X_1 = m_1^s$ and let $X_2, \ldots, X_v$ be the other distinct maximal ideals in T^s, so that given $i > 1$, $m_i^s = X_t$ for some $t > 1$ and given $t > 1$, $X_t = m_i^s$ for some $i > 1$. We have a primary decomposition

$$m_R T^s = Y_1 \cap \cdots \cap Y_v = Y_1 \cdots Y_v$$

where the Y_t are primary for X_t by Theorem 1.5. Let

$$Z = X_1 \cap \cdots \cap X_v = X_1 \cdots X_v.$$

Suppose that $a \in X_1$ is such that $a \notin X_t$ for $t > 1$. We will show that $a \in Y_1$. To see this, let
$$a^* = \prod_{i=2}^{q} \sigma_i(a) \in \frac{1}{a}K \subset K^s.$$
Since a^* is integral over R, $a^* \in T^s$. Since $a^* \notin n_1$, we have that $a^* \notin X_1$. Now $aa^* \in m_R \subset Y_1$. Thus $a \in Y_1$, since Y_1 is X_1-primary.

By the Chinese remainder theorem (Theorem 1.5), there exists $e \in T^s$ such that $e \equiv 0 \mod X_1$ and $e \equiv 1 \mod X_t$ for $t > 1$. By the above observation, $e \in Y_1$. Suppose that $f \in Z$. Then $f+e \in X_1$ and $f+e \notin X_t$ if $t > 1$. Thus $f+e \in Y_1$ (again by the observation) and so $f \in Y_1$. Thus $Z \subset Y_1$ and hence $X_1 \subset Y_1$ so that $X_1 = Y_1$, so that $m_R R^s = X_1 R^s = m_{R_s}$. $\square$

The inertia group $G^i(S/R)$ is defined to be
$$G^i(S/R) = \{\sigma \in G^s(S/R) \mid \sigma(u) \equiv u \mod m_S \text{ for all } u \in S\}.$$

The inertia field $K^i = K^i(S/R)$ is defined to be $K^i = (K^*)^{G^i(S/R)}$.

Lemma 21.102. *Let $R^i = S \cap K^i$. Then $m_R R^i = m_{R^s} R^i = m_{R^i}$. Further, R^i/m_{R^i} is isomorphic to the separable closure of R/m_R in S/m_S, so that $R \to R^i$ and $R^s \to R^i$ are unramified. The group $G^i(S/R)$ is a normal subgroup of $G^s(S/R)$ and R^i/m_{R^i} is a Galois extension of R/m_R whose Galois group is isomorphic to $G^s(S/R)/G^i(S/R)$.*

Proof. Let $\kappa = R/m_R = R^s/m_{R^s}$, $\kappa^i = R^i/m_{R^i}$, and $\kappa^* = S/m_S$.

For $a \in S$, let $\bar{a}$ denote the residue of a in κ^*. For $g \in G^s(S/R)$, define a κ-algebra automorphism $\Phi(g) : \kappa^* \to \kappa^*$ by $\Phi(g)(\bar{a}) = \overline{g(a)}$. We have that $\Phi : G^s(S/R) \to \text{Aut}(\kappa^*/\kappa)$ is a group homomorphism. Let $f_a(t)$ be the minimal polynomial of a over K^s and let $f_{\bar{a}}(t)$ be the minimal polynomial of $\bar{a}$ over κ. We have that $f_a(t) \in R^s[t]$ by Theorem 21.14. Let $\overline{f}_a(t) \in \kappa[t]$ be obtained from $f_a(t)$ by reducing its coefficients modulo m_R.

Since K^* is Galois over K^s, we have a factorization
$$f_a(t) = \prod_{j=1}^{\deg f_a} (t - a_j) \quad \text{with } a_j \in K^*.$$

By Lemma 21.13, all $a_j \in S$. Since $f_{\bar{a}}(t)$ divides $\overline{f}_a(t)$ in $\kappa[t]$, we have that
$$f_{\bar{a}}(t) = \prod_{u=1}^{\deg f_{\bar{a}}} (t - \bar{a}_{t_u}) \quad \text{with } \bar{a}_{t_u} \in \kappa^*.$$

Thus κ^* is a normal field extension of κ. Let κ' be the separable closure of κ in κ^*. The field κ' is finite algebraic over κ by Theorem 21.22. Thus κ' is finite Galois over κ.

Let $a \in S$ be such that $\bar{a}$ is a primitive element of κ' over κ. Then there exists $\sigma_u \in G^s(K^*/K)$ such that $\sigma_u(a) = a_{t_u}$. Thus $\Phi(\sigma_u)(\bar{a}) = \bar{a}_{t_u}$, and so Φ is surjective onto $G(\kappa'/\kappa)$. We have that Kernel(Φ) = $G^i(S/R)$. In particular, $G^i(S/R)$ is a normal subgroup of $G^s(S/R)$. Let $G = G^s(S/R)/G^i(S/R)$. We have that
$$\mathrm{Aut}(\kappa^*/\kappa) = G(\kappa'/\kappa) \cong G.$$
Further,
$$[K^i : K^s] = [G^s(K^*/K) : G^i(K^*/K)] = [\kappa' : \kappa]$$
and K^i is a Galois extension of K^s, with Galois group G.

Since S is the only local ring of K^* lying over R^i,
$$G^s(S/R^i) = G(K^*/K^i) = G^i(S/R) = G^i(S/R^i).$$
Thus by the first part of the proof,
$$\mathrm{Aut}(\kappa^*/\kappa^i) \cong G^s(S/R^i)/G^i(S/R^i) = (1).$$
Thus κ^* is purely inseparable over κ^i since κ^* is a normal extension of κ. Hence $\kappa' \subset \kappa^i$. Thus
$$[\kappa^i : \kappa]_s = [\kappa' : \kappa] = [K^i : K^s].$$
Now R^i is the unique local ring of K^i lying over R^s, so R^i is the integral closure of R^s in K^i. By Theorem 21.22, $D(R^i/R^s) = R^s$ and so by Theorem 21.20, $R^s \to R^i$ is unramified. Thus $\kappa^i = \kappa'$ and $m_{R^s} R^i = m_{R^i}$, and so $R \to R^i$ is unramified. $\square$

21.12. A proof of the Abhyankar-Jung theorem

In this section we give a proof of Theorem 21.93, based on Abhyankar's original proof in [**1**].

If K is an algebraic function field over a field κ, we will say that a local ring R with quotient field K is an algebraic local ring of K if R is a localization of a finite type κ-algebra.

Suppose that $R \to S$ is an extension of dvrs (discrete valuation rings are discussed at the end of Section 21.2). We have that $e(S/R) = n$ if a generator t of m_R has an expansion $t = u^n v$ where u is a generator of m_S and v is a unit in S (Section 21.2). Thus the extension $R \to S$ is unramified if and only if $e(S/R) = 1$ and S/m_S is a finite separable extension of R/m_R.

If ν is a valuation of a field K, we will denote the valuation ring of ν by V_ν. Thus
$$V_\nu = \{f \in K \mid \nu(f) \geq 0\}.$$
We denote the maximal ideal of V_ν by m_ν. Suppose that $K \to K^*$ is a finite field extension. If $S = V_\nu$ is the valuation ring of a valuation ν of K^* and

$R = V_\nu \cap K = V_\mu$ is the valuation ring of the restriction μ of ν to K, then we will write $e(\nu/\mu) = e(V_\nu/V_\mu)$.

Proposition 21.103. *Let R be a normal local domain which is essentially of finite type over a field κ. Let K be the quotient field of R. Let K^* be a finite separable extension of K and let $n = [K^* : K]$. Let $x \in R$ be a primitive element of K^* over K, with minimal polynomial $f(t) \in R[t]$ (such an x exists by Theorem 21.14). Let S_i for $1 \le i \le h$ be the local rings in K^* lying over R and let S be the integral closure of R in K^*. Then $\hat{R}$ and $\hat{S}_i$ are normal local domains, the natural homomorphisms $\hat{R} \to \hat{S}_i$ are injective for all i, and we have a natural isomorphism*

$$S \otimes_R \hat{R} \cong \bigoplus_{i=1}^{h} \hat{S}_i.$$

Let E be the quotient field of $\hat{R}$ and E_i be the quotient field of $\hat{S}_i$ for $1 \le i \le h$. Let $e_i = [E_i : E]$. Then $n = e_1 + \cdots + e_h$. Further, x is a primitive element of E_i over E for all i with minimal polynomial $f_i(t) \in E[t]$ and there is a factorization $f(t) = f_1(t) \cdots f_h(t)$ in $E[t]$.

Proof. We have that $\hat{R}$ is a normal local domain by Proposition 21.50. By [161, Theorem 16, page 277] and [161, Corollary 2, page 283], we have a natural isomorphism

$$S \otimes_R \hat{R} \cong \bigoplus \hat{S}_i.$$

We have that the $\hat{S}_i$ are normal local domains by Proposition 21.50. We have that

$$K^* \otimes_K E \cong E[x] \cong E[t]/(f(t)).$$

Let A be a ring. The total quotient ring of A is $\mathrm{QR}(A) = S^{-1}A$ where S is the multiplicative set of all nonzero divisors of A. Let

$$g \in K^* \otimes_K E = E[x].$$

Then $g = \frac{a}{b}$ where $a \in \hat{R}[x]$ and $b \in \hat{R} \setminus \{0\}$. Since $\hat{R}$ is a domain, b is not a zero divisor in $S \otimes_R \hat{R}$ by [161, Theorem 16, page 277]. Thus $K^* \otimes_K E \subset \mathrm{QR}(S \otimes_R \hat{R})$. Since the reduced ring $S \otimes_R \hat{R}$ is naturally a subring of $QR(S \otimes_R \hat{R})$, we have a natural inclusion $S \otimes_R \hat{R} \subset K^* \otimes_K E$. Now $f(t)$ is reduced in $E[t]$ since K^* is separable over K. We have that $K^* \otimes_K E \cong \bigoplus E[t]/(f_i(t))$ is a direct sum of fields, where the $f_i(t)$ are the irreducible factors of $f(t)$ in $E[t]$ by the Chinese remainder theorem, Theorem 1.5. Thus $\mathrm{QR}(S \otimes_R \hat{R}) = K^* \otimes_K E$. Now $S \otimes_R \hat{R}$ is reduced, so $\mathrm{QR}(S \otimes_R \hat{R}) \cong \bigoplus_i E_i$. Thus, after reindexing, we have that $E_i \cong E[t]/(f_i(t))$ and we have that

$$\sum [E_i : E] = \sum \deg(f_i) = \deg(f) = [K^* : K]. \qquad \square$$

21.12. A proof of the Abhyankar-Jung theorem

Suppose that K is an algebraic function field over a field κ and K' is a finite separable extension of K. Suppose that R is a normal, algebraic local ring of K and R' is a normal local ring of K' which lies over R. Let E be the quotient field of $\hat{R}$ and let E' be the quotient field of $\hat{R}'$. Define

$$d(R' : R) = [E' : E], \qquad g(R' : R) = [R'/m_{R'} : R/m_R]_s,$$

and

$$r(R' : R) = d(R' : R)/g(R' : R).$$

We have that $d(R' : R), g(R' : R), r(R' : R)$ are all multiplicative in towers of fields.

Now suppose that K^* is a finite Galois extension of K and that R^* is a normal local ring of K^* which lies over R. Let $R^s = R^* \cap K^s$ where $K^s = K^s(R^*/R)$ is the splitting field of R^* over R. Let $R^i = R^* \cap K^i$ where K^i is the inertia field $K^i = K^i(R^*/R)$ of R^* over R. Let E, E^s, E^i, E^* be the respective quotient fields of $\hat{R}, \hat{R}^s, \hat{R}^i$, and $\hat{R}^*$.

Since R^* is the unique local ring of K^* which dominates R^s by Lemma 21.100, we have by Proposition 21.103 that

(21.28) $$[K^* : K^s] = d(R^* : R^s) \quad \text{and} \quad [K^* : K^i] = d(R^* : R^i).$$

By Proposition 21.103 and Lemma 21.102 we have that

(21.29) $$[K^i : K^s] = d(R^i : R^s) = g(R^* : R).$$

We have that $\hat{R}^s \cong \hat{R}$ by Lemma 21.101, Proposition 21.49, and Proposition 21.48, and so $E^s \cong E$. Thus

(21.30) $$d(R^s : R) = 1.$$

Lemma 21.104. *Let notation be as above. Then $r(R' : R)$ is a positive integer.*

Proof. Let K^* be a finite Galois extension of K which contains K', and let R^* be a normal local ring of K^* such that R^* lies over R'. We have that

$$G^i(R^*/R') = G^i(R^*/R) \cap G(K^*/K')$$

and

$$G^s(R^*/R') = G^s(R^*/R) \cap G(K^*/K'),$$

so $(K')^s = K'K^s$ and $(K')^i = K'K^i$. Thus we have a commutative diagram of fields

$$\begin{array}{ccc}
& K^* & \\
\nearrow & & \nwarrow \\
K^i & \to & (K')^i \\
\uparrow & & \uparrow \\
K^s & \to & (K')^s \\
\uparrow & & \uparrow \\
K & \to & K'
\end{array}$$

where $K^s = K^s(R^*/R)$, $K^i = K^i(R^*/R)$, $(K')^s = (K')^s(R^*/R')$, and $(K')^i = (K')^i(R^*/R')$. Considering the induced commutative diagram of fields

$$\begin{array}{ccc}
& E^* & \\
\nearrow & & \nwarrow \\
E^i & \to & (E')^i \\
\uparrow & & \uparrow \\
E^s & \to & (E')^s \\
\uparrow & & \uparrow \\
E & \to & E'
\end{array}$$

where $E, E', E^*, E^s, E^i, (E')^s, (E')^i$ are the respective quotient fields of

$$\hat{R}, \quad \widehat{R'}, \quad \widehat{R^*}, \quad \widehat{R^s}, \quad \widehat{R^i}, \quad \widehat{(R')^s}, \quad \widehat{(R')^i},$$

we see from (21.29) and (21.30) that

$$d(R':R)g(R^*:R') = g(R^*:R)d((R')^i:R^i).$$

Thus $d(R':R) = g(R':R)d((R')^i:R^i)$. $\square$

We also read off the following formulas from the commutative diagrams of the proof of Lemma 21.104 and from (21.28), (21.29), and (21.30):

$$[K^*:K^i][R^*/m_{R^*}:R/m_R]_s = d(R':R)[K^*:(K')^i][R^*/m_{R^*}:R'/m_{R'}]_s,$$

so that

(21.31) $$[K^*:K^i] = r(R':R)[K^*:(K')^i]$$

and

(21.32) $$[K^*:K^s] = d(R':R)[K^*:(K')^s].$$

Lemma 21.105. *Let notation be as above. Then:*

1) $G^i(R^*/R) \subset G(K^*/K')$ *if and only if* $r(R':R) = 1$.
2) $G^s(R^*/R) \subset G(K^*/K')$ *if and only if* $d(R':R) = 1$.

21.12. A proof of the Abhyankar-Jung theorem

Proof. The lemma follows from equations (21.31) and (21.32) and the observations that $G^i(R^*/R) \subset G(K^*/K')$ if and only if $(K')^i = K^i$ and $G^s(R^*/R) \subset G(K^*/K')$ if and only if $(K')^s = K^s$. □

Proposition 21.106. *Suppose that R is an algebraic normal local ring with quotient field K, K' is a finite field extension of K, and R' is a normal local ring of K' which lies over R. Then $R \to R'$ is unramified if and only if $r(R'/R) = 1$.*

Proof. The extension $R \to R'$ is unramified if and only if $\hat{R} \to \hat{R}'$ is unramified which holds if and only if $D(\hat{R}'/\hat{R}) = \hat{R}$ by Theorem 21.21. This holds if and only if
$$[R'/m_{R'} : R/m_R]_s = [E' : E],$$
where E' is the quotient field of $\hat{R}'$ and E is the quotient field of $\hat{R}$ by Theorem 21.22. But this is equivalent to the condition that $r(R'/R) = 1$. □

Proposition 21.107. *Let R be a normal algebraic local ring with quotient field K. Let K' be a finite separable extension of K and let K^* be a least Galois extension of K containing K'. Let I be an ideal defining the ramification locus in R of $R \to K'$ (for a prime $P \in \mathrm{Spec}(R)$, $R_P \to K'$ is ramified if and only if $I \subset P$). Then I is an ideal defining the ramification locus in R of $R \to K^*$.*

Proof. We must show that for $P \in \mathrm{Spec}(R)$, $R_P \to K'$ is ramified if and only if $R_P \to K^*$ is ramified. It follows from the definition of ramification that $R_P \to K^*$ unramified implies $R_P \to K'$ is unramified.

Suppose that $R_P \to K'$ is unramified. Let R_j^* with $1 \leq j \leq t$ be the local rings of K^* lying over R_P, $G = G(K^*/K)$, $G' = (K^*/K')$, and $G_j = G^i(R_j^*/R)$. Then $G_j \subset G'$ for $1 \leq j \leq t$ by Lemma 21.105 and Proposition 21.106. Let $\overline{G}$ be the smallest subgroup of G which contains $G_1, \ldots, G_t$. Since the G_j are conjugate subgroups, $\overline{G}$ is a normal subgroup of G and $\overline{G} \subset G'$. Thus the fixed field $\overline{K}$ of $\overline{G}$ is a Galois extension of K containing K', and hence by minimality of K^*, we have that $\overline{K} = K^*$; that is, $\overline{G} = (1)$. Thus $G_1 = G_2 = \cdots = G_t = (1)$. Defining K_j^i to be the inertial field $K_j^i = (K^*)^{G_i}$, this implies that $K^* = K_j^i$. Thus R is unramified in K^* by Lemma 21.102. □

Lemma 21.108. *Let R be a normal algebraic local ring with quotient field K, which is an algebraic function field over an algebraically closed field, and suppose that K^* is a finite Galois extension of K. Let p be the characteristic of K and let $y_1, \ldots, y_h$ be a finite number of elements in R such that for all j, $N_j = y_j R$ is a height 1 prime ideal in R and hence that R_{N_j} is the valuation ring of a discrete valuation ω_j of K. Let ω_j^* be an extension of ω_j to K^**

and let $n_j = e(\omega_j^*/\omega_j)$, which is defined in (21.6) (n_j, the residue degree of ω_j^* over ω_j, and the inseparable degree of the residue field of ω_j^* over ω_j only depend on j since $K \to K^*$ is Galois, as shown in [**160**, Section 10, Chapter V]). Assume that if $p \neq 0$, then p does not divide n_j for $j = 1, \ldots, h$. Let $y_j^{\frac{1}{n_j}}$ be an n_j-th root of y_j. Let

$$\overline{K} = K(y_1^{\frac{1}{n_1}}, \ldots, y_h^{\frac{1}{n_h}}) \quad \text{and} \quad \overline{K}^* = K^*(y_1^{\frac{1}{n_1}}, \ldots, y_h^{\frac{1}{n_h}}).$$

Let N_0 be a height 1 prime ideal in R different from $N_1, \ldots, N_h$ and let ω_0 be the discrete valuation of K with valuation ring R_{N_0}. Let $\overline{\omega}_j^*$ be an extension of ω_j^* to $\overline{K}^*$ and let $\overline{\omega}_j$ be the restriction of $\overline{\omega}_j^*$ to $\overline{K}$. Let V_{ω_j}, $V_{\overline{\omega}_j}$, $V_{\omega_j^*}$, and $V_{\overline{\omega}_j}$ be the respective valuation rings. Then:

a) $\overline{K}^*/K$, $\overline{K}/K$, $\overline{K}^*/K^*$, and $\overline{K}^*/\overline{K}$ are Galois extensions.

b) For $1 \leq j \leq h$, we have that $e(\overline{\omega}_j^*/\omega_j^*) = 1$ and $V_{\overline{\omega}_j^*}/m_{\overline{\omega}_j^*}$ is separable over $V_{\omega_j^*}/m_{\omega_j^*}$. Further, $e(\overline{\omega}_j/\omega_j) = n_j$ and $V_{\overline{\omega}_j}/m_{\overline{\omega}_j}$ is separable over $V_{\omega_j}/m_{\omega_j}$.

c) $e(\overline{\omega}_0^*/\omega_0^*) = 1$ and $V_{\overline{\omega}_0^*}/m_{\overline{\omega}_0^*}$ is separable over $V_{\omega_0^*}/m_{\omega_0^*}$. Further, $e(\overline{\omega}_0^*/\overline{\omega}_0) = e(\omega_0^*/\omega_0)$ and the inseparability indices

$$[V_{\overline{\omega}_0^*}/m_{\overline{\omega}_0^*} : V_{\overline{\omega}_0}/m_{\overline{\omega}_0}]_i = [V_{\omega_0^*}/m_{\omega_0^*} : V_{\omega_0}/m_{\omega_0}]_i.$$

Proof. We assume that $h = 1$. The general case then follows by induction on h, using the fact that e and the inseparable degree of residue field extensions are multiplicative for extensions of dvrs in towers of fields. Statement a) follows since the composition of two Galois extensions is again Galois (Theorem 1.14, page 267 of [**95**]).

Let $n = [\overline{K}^* : K^*]$. Then $n \mid n_1$. Let $x = y_1^{\frac{1}{n_1}}$ which is a generator of $\overline{K}^*$ over K^* and let $z = x^n$. Then $f(X) = X^n - z$ is the minimal polynomial of x over K^*. Now z^{n-1} (times the unit $(-1)^{\frac{n(n-1)}{2}} n^n$) is the discriminant of $f(X)$ (as follows from (21.7) and (21.9)). We have that z^{n-1} is a unit in $V_{\omega_0^*}$, so $V_{\omega_0^*} = D(\overline{K}^*/V_{\omega_0^*})$ and thus $V_{\omega_0^*} \to V_{\overline{\omega}_0^*}$ is unramified, so $e(\overline{\omega}_0^*/\omega_0^*) = 1$ and $V_{\omega_0^*}/m_{\omega_0^*} \to V_{\overline{\omega}_0^*}/m_{\overline{\omega}_0^*}$ is a finite separable extension.

Fix $u \in V_{\omega_1^*}$ which is a generator of $m_{\omega_1^*}$. Now $\frac{x}{u}$ is a primitive element of $\overline{K}^*$ over K^* and $g(X) = X^n - zu^{-n}$ is the minimal polynomial of $\frac{x}{u}$ over K^*. Now $\omega_1^*(z) = n\omega_1^*(x)$ and $\omega_1^*(y_1) = e(\omega_1^*/\omega_1) = n_1$, so $\omega_1^*(x) = 1$ and thus $\omega_1^*(\frac{z}{u^n}) = n - n = 0$. We thus have that $\frac{z}{u^n}$ is a unit in $V_{\omega_1^*}$ and since $(\frac{z}{u^n})^{n-1}$ is the discriminant of $g(X)$, we have that $D(\overline{K}^*/V_{\omega_1^*}) = V_{\omega_1^*}$ and thus $V_{\omega_1^*} \to V_{\overline{\omega}_1^*}$ is unramified and so $e(\overline{\omega}_1^*/\omega_1^*) = 1$ and $V_{\omega_1^*}/m_{\omega_1^*} \to V_{\overline{\omega}_1^*}/m_{\overline{\omega}_1^*}$ is a finite separable extension.

21.12. A proof of the Abhyankar-Jung theorem

We have that $n_1\overline{\omega}_1(x) = \overline{\omega}_1(y_1)$, so $e(\overline{\omega}_1/\omega_1) \geq n_1$. But
$$[V_{\overline{\omega}_1}/m_{\overline{\omega}_1} : V_{\omega_1}/m_{\omega_1}]e(\overline{\omega}_1/\omega_1) \leq [\overline{K} : K] = n_1$$
by [**160**, Theorem 22, page 289]. Thus $e(\overline{\omega}_1/\omega_1) = n_1$ and $V_{\overline{\omega}_1}/m_{\overline{\omega}_1} = V_{\omega_1}/m_{\omega_1}$.

By a similar calculation to the analysis of $V_{\omega_0^*} \to V_{\overline{\omega}_0^*}$, we obtain that $e(\overline{\omega}_0/\omega_0) = 1$ and $V_{\omega_0}/m_{\omega_0} \to V_{\overline{\omega}_0}/m_{\overline{\omega}_0}$ is finite and separable.

From the identities
$$e(\overline{\omega}_0^*/\overline{\omega}_0)e(\overline{\omega}_0/\omega_0) = e(\overline{\omega}_0^*/\omega_0^*)e(\omega_0^*/\omega_0)$$
and
$$[V_{\overline{\omega}_0^*}/m_{\overline{\omega}_0^*} : V_{\omega_0^*}/m_{\omega_0^*}]_i[V_{\omega_0^*}/m_{\omega_0^*} : V_{\omega_0}/m_{\omega_0}]_i$$
$$= [V_{\overline{\omega}_0^*}/m_{\overline{\omega}_0^*} : V_{\overline{\omega}_0}/m_{\overline{\omega}_0}]_i[V_{\overline{\omega}_0}/m_{\overline{\omega}_0} : V_{\omega_0}/m_{\omega_0}]_i,$$
we obtain that $e(\overline{\omega}_0^*/\overline{\omega}_0) = e(\omega_0^*/\omega_0)$ and
$$[V_{\overline{\omega}_0^*}/m_{\overline{\omega}_0^*} : V_{\overline{\omega}_0}/m_{\overline{\omega}_0}]_i = [V_{\omega_0^*}/m_{\omega_0^*} : V_{\omega_0}/m_{\omega_0}]_i. \qquad \square$$

Lemma 21.109. *Let K be an algebraic function field and let K^* be a finite Galois extension of K. Let R be a normal algebraic local ring of K and let R^* be a normal algebraic local ring of K^* such that R^* lies over R. Let E be the quotient field of $\hat{R}$ and let E^* be the quotient field of $\hat{R}^*$. Then E^* is Galois over E with Galois group isomorphic to $G^s(R^*/R)$ by restriction in the commutative diagram*
$$\begin{array}{ccc} E & \to & E^* \\ \uparrow & & \uparrow \\ K & \to & K^*. \end{array}$$

Proof. Let K^s be the splitting field of R^* over R. The field K^* is Galois over K^s with Galois group $\overline{G} = G^s(R^*/R)$. Let $R^s = R^* \cap K^s$. Then $\widehat{R^s} = \hat{R}$ by Lemma 21.101, and so the quotient field E^s of $\widehat{R^s}$ is isomorphic to E. Since R^* is the unique local ring of K^* lying over R^s, by Lemma 21.100, we have that

(21.33) $$[E^* : E] = [K^s : K]$$

by Proposition 21.103.

Suppose that $\sigma \in G^s(R^*/R)$. Then there is a commutative diagram
$$\begin{array}{ccc} R & \to & R^* \\ & \searrow & \downarrow \sigma \\ & & R^*. \end{array}$$

Taking completions and quotient fields, we obtain an E-automorphism of E^* which extends σ, and thus we have an inclusion of $G^s(R^*/R)$ into $\text{Aut}(E^*/E)$. Thus by (21.33), we have that E^* is Galois over E with Galois group $G^s(R^*/R)$. $\qquad \square$

We now give the proof of Theorem 21.93. Let $K = k(Y)$, $K^* = k(X)$, $R = \mathcal{O}_{Y,q}$, and $R^* = \mathcal{O}_{X,p}$, so that R^* lies over R. We are given the assumption that $\prod_{i=1}^{t} x_i \in J(R^*/R)$ for some regular system of parameters $x_1, \ldots, x_t, \ldots, x_n$ in R. Let K' be a least Galois extension of K containing K^*. By assumption, the characteristic p of $k(X)$ does not divide $[K' : K]$. Let R' be an algebraic local ring of K' which lies over R^*. Let $K^s = (K')^{G^s(R'/R)}$ and let $\overline{K}'$ be the composite of K^* and K^s. Then $\overline{K}' = (K')^{G^s(R'/R^*)}$ since

$$G^s(R'/R^*) = G^s(R'/R) \cap G(K'/K^*).$$

Let R^s be the normal local ring of K^s such that R' lies over R^s and let $\overline{R}'$ be the normal local ring of $\overline{K}'$ such that R' lies over $\overline{R}'$.

Now $R \to R^s$ and $R^* \to \overline{R}'$ are unramified by Lemma 21.101, so that

$$J(\overline{R}'/R^s) = J(\overline{R}'/R) \supset J(R^*/R)$$

by Theorem 21.83 and Theorem 14.9 and thus $\prod_{i=1}^{t} x_i \in I(\overline{R}'/R^s)$, which defines the ramification locus of $R^s \to \overline{K}'$ since $R^s \to \overline{R}'$ is finite as $\overline{R}'$ is the unique local ring of $\overline{K}'$ lying over R^s by Lemma 21.100. Let $\tilde{K}$ be a least Galois extension over K^s containing $\overline{K}'$ in K'. Let $\tilde{R}$ be the normal local ring of $\tilde{K}$ such that R' lies over $\tilde{R}$. By Proposition 21.107 and since $\tilde{R}$ is the unique local ring of K' lying over R^s, $\prod_{i=1}^{t} x_i \in I(\overline{R}'/R^s) = I(\tilde{R}'/R^s)$ defines the ramification locus of R^s in $\tilde{K}$.

Let E be the quotient field of $\hat{R}$, let E^s be the quotient field of $\hat{R}^s$, let $\overline{E}'$ be the quotient field of $\widehat{\overline{R}'}$, let E^* be the quotient field of $\hat{R}^*$, and let $\tilde{E}$ be the quotient field of $\hat{\tilde{R}}$. Then $\hat{R} \cong \hat{R}^s$ and $\hat{R}^* \cong \widehat{\overline{R}'}$, by Lemma 21.101, so $E \cong E^s$ and $E^* \cong \overline{E}'$. Since $\tilde{K}$ is Galois over $\overline{K}'$, we have that $\tilde{E}$ is Galois over $\overline{E}'$ by Lemma 21.109. Thus it suffices to prove the theorem with R replaced with R^s and R^* replaced with $\overline{R}'$, K replaced with K^s and K^* replaced with $\overline{K}'$. In summary, we have reduced to the situation where we have a tower of fields

$$K \to K^* \to \tilde{K}$$

where $\tilde{K}$ is Galois over K, p does not divide $[\tilde{K} : K]$, and (after possibly permuting $x_1, \ldots, x_t$ and decreasing t) $\sqrt{D(\tilde{K}/R)} = (\prod_{i=1}^{t} x_i)$ where $x_1, \ldots, x_n$ is a suitable regular system of parameters in R, and $\tilde{R}$ is the unique local ring of $\tilde{K}$ which lies over R.

Let w_i be the valuation of K with valuation ring $V_{\omega_i} = R_{(x_i)}$ for $1 \leq i \leq t$, and let $\tilde{\omega}_i$ be an extension of ω_i to $\tilde{K}$ for $1 \leq i \leq t$. Let $n_i = e(\tilde{\omega}_i/\omega_i)$ for $1 \leq i \leq t$ (n_i does not depend on the extension $\tilde{\omega}_i$ of ω_i, as explained in Proposition 21.23). Now n_i divides $[\tilde{K} : K]$ by Proposition 21.23, so p

21.12. A proof of the Abhyankar-Jung theorem

does not divide n_i for $1 \leq i \leq t$. Let $K_1 = K(x_1^{\frac{1}{n_1}}, \ldots, x_t^{\frac{1}{n_t}})$ and $K_2 = \tilde{K}(x_1^{\frac{1}{n_1}}, \ldots, x_t^{\frac{1}{n_t}})$. Let R_2 be a normal local ring of K_2 which lies over $\tilde{R}$ and let $R_1 = R_2 \cap K_1$. All extensions in

$$\begin{array}{ccc} K & \to & \tilde{K} \\ \downarrow & & \downarrow \\ K_1 & \to & K_2 \end{array}$$

are Galois by a) of Lemma 21.108.

We have that $\hat{R} \cong k[[x_1, \ldots, x_n]]$ and $\hat{R}_1$ is the integral closure of $\hat{R}$ in

$$k((x_1^{\frac{1}{n_1}}, \ldots, x_t^{\frac{1}{n_t}}, x_{t+1}, \ldots, x_n))$$

by Proposition 21.55. Thus $\hat{R}_1 \cong k[[x_1^{\frac{1}{n_1}}, \ldots, x_t^{\frac{1}{n_t}}, x_{t+1}, \ldots, x_n]]$. Since $\hat{R}_1$ is regular, we have that R_1 is a regular local ring.

Suppose that ν is a valuation of K whose valuation ring is R_P where P is a height 1 prime ideal in R. Let ν_2 be an extension of ν to K_2, let $\tilde{\nu}$ be the restriction of ν_2 to $\tilde{K}$, and let ν_1 be the restriction of ν_2 to K_1. We have that p does not divide $[V_{\tilde{\nu}}/m_{\tilde{\nu}} : V_\nu/m_\nu]$ since p does not divide $[\tilde{K} : K]$ and by Proposition 21.23. Hence $V_{\tilde{\nu}}/m_{\tilde{\nu}}$ is separable over V_ν/m_ν. If P is a height 1 prime ideal of R such that $P \neq (x_i)$ for some i with $1 \leq i \leq t$, then we have that $D(\tilde{K}/K) \not\subset P$, so $R_P \to \tilde{K}$ is unramified. Thus if ν is a valuation of K whose valuation ring is R_P, we have that $e(\tilde{\nu}/\nu) = 1$. Thus by Lemma 21.108, $R_1 \to R_2$ is unramified in codimension 1. By the purity of the branch locus, Theorem 21.92, $R_1 \to R_2$ is unramified since R_1 is regular. Thus by Proposition 21.62,

$$\hat{R}_1 \cong \hat{R}_2 \cong k[[x_1^{\frac{1}{n_1}}, \ldots, x_t^{\frac{1}{n_t}}, x_{t+1}, \ldots, x_n]].$$

Let E be the quotient field of $\hat{R}$, let E^* be the quotient field of $\hat{R}^*$, let E_1 be the quotient field of $\hat{R}_1$, and let E_2 be the quotient field of $\hat{R}_2$.

Now E_2 is Galois over E^* with Galois group $G(E_2/E^*) = G^s(R_2/R^*)$ and E_2 is Galois over E with Galois group $G(E_2/E) = G^s(R_2/R)$ by Lemma 21.109. We further have that $E_2 \cong E_1$ is Galois over E with Galois group $G(E_2/E) \cong \bigoplus_{i=1}^{t} \mathbb{Z}^{n_i}$, which acts by multiplication of roots of unity on the $x_i^{\frac{1}{n_i}}$. Now $G(E_2/E^*) \cong G^s(R_2/R^*)$ is a subgroup of $G(E_2/E)$ and thus $\hat{R}^* \cong \hat{R}_2^{G^s(R_2/R^*)}$ has the desired form.

Chapter 22

Bertini's Theorems and General Fibers of Maps

We first consider the question of when a variety X over a not necessarily algebraically closed field k_0 satisfies the property that the scheme X' obtained from X by extending the field k_0 to a larger field k' is always reduced, irreducible, or integral (reduced and irreducible). If this property holds for all extension fields k' of k_0, then X is said to be geometrically reduced, geometrically irreducible, or geometrically integral. In Corollary 22.3 we characterize these conditions in terms of properties of the extension field $k_0(X)$ of k_0, and we say that the extension $k_0 \to k_0(X)$ is geometrically reduced, geometrically irreducible, or geometrically integral if the variety X has this property over k_0. In Proposition 22.14 and more generally in Theorem 22.18, it is shown that the general fiber of a dominant regular map $\phi: X \to Y$ of varieties has one of the properties geometrically reduced, geometrically irreducible, or geometrically integral if and only if the extension of function fields $k(Y) \to k(X)$ has this property.

We establish Theorem 22.4, showing that the general fiber of a map from a nonsingular characteristic 0 variety is nonsingular. This property does not hold in positive characteristic, as shown by Exercise 22.20. In Definition 22.8 and Theorem 22.9 we introduce the general notion of a smooth morphism of schemes.

We derive the two theorems of Bertini over algebraically closed fields k. The first theorem of Bertini, Theorem 22.12, states that if X is a normal variety over an algebraically closed field k of characteristic $p \geq 0$ and L is a linear system on X without fixed component which is not composite with a pencil (L does not induce a rational map to a curve), then there is

451

a power p^e such that a general member of L has the form $p^e D$ where D is a prime divisor. Classical algebraic proofs of the first theorem are given by Zariski in [**151**] (when k has characteristic 0), Matsusaka [**104**], and Zariski in [**158**, Section I.6] (in any characteristic). Our proof of the first theorem is based on the proof in [**158**, Section 1.6].

The second theorem of Bertini, Theorem 22.11, states that if X is a nonsingular variety over an algebraically closed field k of characteristic 0 and L is a linear system without fixed component on X, then a general member of L is nonsingular outside the base locus of L. The second theorem is only true in characteristic 0, as shown by Exercise 22.19.

22.1. Geometric integrality

A ring A containing a field κ is called *geometrically irreducible* over κ if the nilradical of $A \otimes_\kappa k'$ is a prime ideal for all extension fields k' of κ. The ring A is called *geometrically reduced* over κ if $A \otimes_\kappa k'$ is reduced for all extension fields k' of κ. The ring A is called *geometrically integral* over κ if $A \otimes_\kappa k'$ is a domain for all extension fields k' of κ.

A scheme X over a field κ (there is a natural inclusion $\kappa \to \mathcal{O}_X(U)$ for all open subsets U of X) is called *geometrically irreducible* over κ if the nilradical of $\mathcal{O}_X(U) \otimes_\kappa k'$ is a prime ideal for all open affine subsets U of X and extension fields k' of κ. The scheme X is called *geometrically reduced* over κ if $\mathcal{O}_X(U) \otimes_\kappa k'$ is reduced for all open affine subsets U of X and extension fields k' of κ. The scheme X is called *geometrically integral* over κ if $\mathcal{O}_X(U) \otimes_\kappa k'$ is a domain for all open affine subsets U of X and extension fields k' of κ.

In the language of *Foundations of Algebraic Geometry* [**145**], a geometrically integral variety is called "absolutely irreducible".

Definition 22.1. If K is an extension field of a field κ of characteristic p, then K is said to be a separable extension of κ if K and $\kappa^{p^{-1}}$ are linearly disjoint over κ.

If K is an algebraic extension of κ, then this definition of separability is equivalent to the definition of separability for algebraic extensions, by [**160**, Theorem 34, page 109]. If K is a finitely generated extension of κ, then K is separably generated over κ if and only if K is a separable extension of κ, by [**160**, Theorem 35, page 111].

A dominant regular map $\varphi : X \to Y$ of varieties is said to be separable if the induced extension of fields $k(Y) \to k(X)$ is separable.

Proposition 22.2. *Suppose that F is a finitely generated extension field of a field K. Then:*

1) *F is geometrically reduced over K if and only if F is separable over K.*

2) *F is geometrically irreducible over K if and only if K is separably closed in F (if $\alpha \in F$ is separably algebraic over K, then $\alpha \in K$).*

3) *F is geometrically integral over K if and only if F is separable over K and K is separably closed in F.*

Proof. The "if" statement of 1) follows from [**160**, Theorem 39, Chapter 3, page 195]. For the "only if" statement of 1), suppose that F is not separable over K. Then F and $K^{p^{-1}}$ (where p is the characteristic of K) are not linearly disjoint over K (Definition 22.1), so by Theorem 5.3, the kernel $\mathfrak{p}$ of the canonical homomorphism Λ of $F \otimes_K K^{p^{-1}}$ onto the subring S of an algebraic closure of F generated by F and $K^{p^{-1}}$ is nontrivial. Now $f \in \mathfrak{p}$ implies $f^p \in (F \otimes 1) \cap \text{kernel}(\Lambda) = (0)$ so all elements of $\mathfrak{p}$ are nilpotent. We now prove the "if" statement of 2). In the case that K has characteristic 0, this follows from [**160**, Corollary 2 to Theorem 40, page 198]. Suppose that K has characteristic $p > 0$. Since K is separably closed in F, we have that F and K' are quasi-linearly disjoint over K by [**160**, Theorem 40, page 197], and thus all free joins of F/K and K'/K are equivalent by [**160**, Theorem 38 page 192], and so the zero ideal of $F \otimes_K K'$ is primary by [**160**, Corollary 2, page 195]. For the "only if" statement of 2), suppose that K is not separably closed in F. There exists $\alpha \in F \setminus K$ which is separably algebraic over K. Let $f(x) \subset K[x]$ be the minimal polynomial of α. Let $L = K(\alpha)$. We have that

$$L \otimes_K L \cong K(\alpha)[x]/(f(x)).$$

Since we have a factorization $f(x) = (x - \alpha)g(x)$ in $K(\alpha)[x]$ where $x - \alpha$ and $g(x)$ are reduced and relatively prime, the zero ideal of $L \otimes_K L$ is not primary. Since we have a natural inclusion $L \otimes_K L \subset F \otimes_K L$, the zero ideal of $F \otimes_K L$ is not primary. Conclusion 3) follows from 1) and 2). □

Corollary 22.3. *Suppose that X is a variety over a (not necessarily algebraically closed) field k_0 (as defined in Section 15.4). Then:*

1) *X is geometrically reduced over k_0 if and only if $k_0(X)$ is separable over k_0.*

2) *X is geometrically irreducible over k_0 if and only if k_0 is separably closed in $k_0(X)$.*

3) *X is geometrically integral over k_0 if and only if $k_0(X)$ is separable over k_0 and k_0 is separably closed in $k_0(X)$.*

Proof. By considering an affine cover of X, we reduce to the case when X is an affine variety. Suppose that X is affine. Let S be the multiplicative set $S = k_0[X] \setminus \{0\}$, and let k' be an extension field of k_0. By consideration of the natural inclusion

$$k_0[X] \otimes_{k_0} k' \to k_0(X) \otimes_{k_0} k' = S^{-1}(k_0[X] \otimes_{k_0} k'),$$

we see that $k_0[X]$ is geometrically irreducible, geometrically reduced, or geometrically integral over k_0 if and only if $k_0(X)$ has this property over k_0. $\square$

22.2. Nonsingularity of the general fiber

In this section, we prove the following theorem.

Theorem 22.4. *Let $\phi : X \to Y$ be a dominant regular map of varieties over an (algebraically closed) field k of characteristic 0, and suppose that X is nonsingular. Then there exists a nonempty open subset U of Y such that the fiber X_b is nonsingular for all $b \in U$; that is, $\mathcal{O}_{X_b,a} = \mathcal{O}_{X,a}/m_b\mathcal{O}_{X,a}$, where m_b is the maximal ideal of $\mathcal{O}_{Y,b}$, is a regular local ring for all $a \in \phi^{-1}(b)$.*

The conclusions of Theorem 22.4 are false in positive characteristic. Exercise 22.20 gives a positive characteristic example where $k(X)$ is geometrically integral over $k(Y)$ and all fibers of ϕ are singular.

Throughout this section we will assume that the assumptions of Theorem 22.4 hold. By Theorems 8.13 and 10.16, after replacing Y with an open subset, we may assume that for all $b \in Y$, every irreducible component of $\phi^{-1}(b)$ has dimension $n - m$ where $m = \dim Y$ and $n = \dim X$ and Y is nonsingular.

Lemma 22.5. *With the above assumptions, suppose that $b \in Y$ and $T_a(X) \to T_b(Y)$ is surjective for all $a \in \phi^{-1}(b)$. Then X_b is nonsingular.*

Proof. Suppose that $a \in \phi^{-1}(b)$. Let $A = \mathcal{O}_{X_b,a} = \mathcal{O}_{X,a}/m_b\mathcal{O}_{X,a}$, and let n be the maximal ideal of A. We have that $A/n \cong k(a) \cong k$. The natural surjection of local rings $\mathcal{O}_{X,a} \to A$ induces a surjection of k-vector spaces

(22.1) $$m_a/m_a^2 \to n/n^2 \to 0$$

where m_a is the maximal ideal of $\mathcal{O}_{X,a}$. Taking the dual of (22.1), we have an injection of k-vector spaces,

$$0 \to T(A) = \mathrm{Hom}_k(n/n^2, k) \to T_a(X).$$

The argument above (10.5) (after Definition 10.8) is applicable to our slightly more general situation and shows that

$$T(A) \subset \mathrm{Kernel}(d\phi_a).$$

22.2. Nonsingularity of the general fiber

Thus

$$\dim_{A/n}(n/n^2) = \dim_k(T(A)) \leq \dim T_a(X) - \dim T_b(Y) = \dim X - \dim Y$$

since a is a nonsingular point of X and b is a nonsingular point of Y. We have that

$$\dim_{A/n} n/n^2 \geq \dim A = \dim \phi^{-1}(b) = \dim X - \dim Y.$$

Thus $\dim A = \dim_{A/n} n/n^2$, and A is a regular local ring. $\square$

Lemma 22.6. *Let assumptions be as above. Then there exists a nonempty open subset V of X such that $d\phi_a$ is a surjection for all $a \in V$.*

Proof. Let $b \in Y$ and let $u_1, \ldots, u_m$ be regular parameters in $\mathcal{O}_{Y,b}$. Then $du_1, \ldots, du_m$ is a free $\mathcal{O}_{Y,b}$-basis of $(\Omega_{Y/k})_b$ by Proposition 14.15, and since $\Omega_{k(Y)/k}$ is a localization of $(\Omega_{Y/k})_b$, it is also a $k(Y)$-basis of $\Omega_{k(Y)/k}$. Thus $u_1, \ldots, u_m$ is a transcendence basis of $k(Y)/k$ by Theorem 21.75. We can extend $u_1, \ldots, u_m$ to a transcendence basis $u_1, \ldots, u_n$ of $k(X)/k$. Since k has characteristic 0, $u_1, \ldots, u_n$ is a separating transcendence basis of $k(X)$ over k, so by Theorem 21.75, $du_1, \ldots, du_n$ is a $k(X)$-basis of $\Omega_{k(X)/k}$, and thus there exists a nonempty open subset U of X such that $du_1, \ldots, du_n$ is a free basis of $\Omega_{U/k}$. By Proposition 14.17, $u_1 - u_1(a), \ldots, u_n - u_n(a)$ are regular parameters in $\mathcal{O}_{X,a}$ for all $a \in U$. (Here we are making the usual abuse of notation, identifying $\phi^*(u_i)$ with u_i for $1 \leq i \leq m$; in particular, by $u_i(a)$ for $1 \leq i \leq m$ we mean $\phi^*(u_i)(a) = u_i(\phi(a))$). By Propositions 14.15 and 14.17, there exists an open neighborhood W of b in Y such that $u_1 - u_1(c), \ldots, u_m - u_m(c)$ are regular parameters in $\mathcal{O}_{Y,c}$ for all $c \in W$.

Suppose that $a \in U \cap \phi^{-1}(W)$. Then $u_1 - u_1(a), \ldots, u_m - u_m(a)$ are regular parameters in $\mathcal{O}_{Y,\phi(a)}$ and $u_1 - u_1(a), \ldots, u_n - u_n(a)$ are regular parameters in $\mathcal{O}_{X,a}$.

The classes of $u_1 - u_1(a), \ldots, u_m - u_m(a)$ form a k-basis of $m_{\phi(a)}/m_{\phi(a)}^2$ and the classes of $u_1 - u_1(a), \ldots, u_n - u_n(a)$ form a k-basis of m_a/m_a^2. Thus

$$m_{\phi(a)}/m_{\phi(a)}^2 \to m_a/m_a^2$$

is an inclusion of k-vector spaces, so

$$d\phi_a : T_a(X) \to T_{\phi(a)}(Y)$$

is surjective. $\square$

Lemma 22.7. *Let assumptions be as above. Suppose that $r \in \mathbb{N}$, and let*

$$X_r = \{a \in X \mid \operatorname{rank} d\phi_a \leq r\}.$$

Then $\dim \overline{\phi(X_r)} \leq r$.

Proof. Let Y' be an irreducible component of the Zariski closure $\overline{\phi(X_r)}$ of $\phi(X_r)$ in Y, and let X' be an irreducible component of the Zariski closure $\overline{X}_r$ of X_r in X which dominates Y'. Let $\phi' : X' \to Y'$ be the induced dominant regular map. By Lemma 22.6, there exists a nonempty open subset V of X' such that $d\phi'_a$ is surjective for $a \in V$. Let $a \in V \cap X_r$. We then have, by (10.5) (after Definition 10.8), a commutative diagram of k-vector spaces,

$$\begin{array}{ccc} T_a(X') & \to & T_a(X) \\ d\phi'_a \downarrow & & \downarrow d\phi_a \\ T_{\phi'(a)}(Y') & \to & T_{\phi(a)}(Y). \end{array}$$

The horizontal arrows are injections since X' is a subvariety of X and Y' is a subvariety of Y. Since $\operatorname{rank} d\phi_a \leq r$, we have that

$$\dim Y' \leq \dim_k T_{\phi'(a)}(Y') \leq r. \qquad \square$$

Now we give the proof of Theorem 22.4. Let

$$X_{m-1} = \{a \in X \mid \operatorname{rank} d\phi_a \leq m - 1\}.$$

Then $\dim \overline{\phi(X_{m-1})} \leq m - 1$ by Lemma 22.7. Recall that we have made the reduction that Y is nonsingular. Let $U = Y \setminus \overline{\phi(X_{m-1})}$, a nonempty open subset of Y. Suppose that $b \in U$ and $a \in \phi^{-1}(b)$. Then $a \notin X_{m-1}$, so $\operatorname{rank} d\phi_a \geq m$, and since b is a nonsingular point of Y, $d\phi_a$ is surjective. By Lemma 22.5, X_b is nonsingular.

We end this section by giving the definition of a smooth morphism. A morphism of schemes is defined in Section 15.5. The k-morphisms of varieties are the regular maps.

Definition 22.8. A morphism of schemes $\phi : X \to Y$ is said to be smooth of relative dimension m if for all $p \in X$ there are open neighborhoods $U \subset X$ of p and $V \subset Y$ of $\phi(p)$ such that $\phi(U) \subset V$, and there exists a commutative diagram

$$\begin{array}{ccc} U & \xrightarrow{\text{open embedding}} & Z \\ \phi \downarrow & & \downarrow \\ V & \xrightarrow{\text{open embedding}} & W \end{array}$$

where Z and W are affine schemes, and if $R = \mathcal{O}_W(W)$, then

$$\mathcal{O}_Z(Z) = R[x_1, \ldots, x_{n+m}]/(f_1, \ldots, f_n)$$

and the rank of the $(n+m) \times n$ matrix $(\frac{\partial f_i}{\partial x_j}(p))$ over $\kappa(p) = \mathcal{O}_{X,p}/m_p$ is n.

This is the definition given in [**116**, Definition 3 on pages 436–437]. An étale morphism (Definition 21.79) is a smooth morphism of relative dimension 0. A refinement of Definition 22.8 is given in Exercise 22.10.

Using the general notion of a scheme of Section 15.5 (which has a larger topological space including nonclosed points) we have the following.

Theorem 22.9. *A morphism of schemes $\phi : X \to Y$ is smooth of relative dimension m if and only if the following three conditions hold.*

1. *f is flat,*
2. *if $X' \subset X$ and $Y' \subset Y$ are irreducible components such that $f(X') \subset Y'$, then $\dim X' = \dim Y' + m$,*
3. *for each point $x \in X$ (closed or not)*

$$\dim_{\kappa(x)}(\Omega_{X/Y} \otimes \kappa(x)) = m,$$

where $\kappa(x)$ is the residue field of $\mathcal{O}_{X,x}$.

The proof of Theorem 22.9 follows from [116, Theorem 3', page 437] and [73, Theorem III.10.2]. The criterion of Theorem 22.9 is the definition of a smooth morphism given in Section 10 of Chapter III of [73].

Exercise 22.10. Show that a regular map of varieties $\phi : X \to Y$ is smooth of relative dimension m if and only if for every $p \in X$, there exist open affine neighborhoods A of p and B of $q = \phi(p)$ such that $k[A]$ is a quotient of a polynomial ring over $k[B]$ of the form

$$k[A] = k[B][x_1, \ldots, x_{n+m}]/(f_1, \ldots, f_n)$$

where the ideal $I_n(\frac{\partial f_i}{\partial x_j})$ generated by the $n \times n$ minors of the matrix $(\frac{\partial f_i}{\partial x_j})$ is equal to $k[A]$. Hint: Use Exercise 1.7 and Definition 22.8.

22.3. Bertini's second theorem

Suppose that X is a normal variety and L is a linear system on X without fixed component ($\text{Base}(L)$ has codimension ≥ 2 in X). Then there exists an effective D on X and linearly independent sections $s_0, \ldots, s_n \in \Gamma(X, \mathcal{O}_X(D))$ such that

$$L = \{L_a = \text{div}(a_0 s_0 + a_1 s_1 + \cdots + a_n s_n) + D \mid a = (a_0 : \ldots : a_n) \in V_L = \mathbb{P}^n\}.$$

The projective space V_L parametrizes the linear system L. The linear system L induces a rational map

$$\phi_L = (s_0 : \ldots : s_n) : X \dashrightarrow \mathbb{P}^n.$$

Let X' be the projective variety which is the Zariski closure of $\phi_L(X)$ in $\mathbb{P}^n$. The function field of X' is

$$k(X') = k\left(\frac{s_1}{s_0}, \ldots, \frac{s_n}{s_0}\right) \subset k(X).$$

We now assume that X is nonsingular, and we associate to the linear system L a closed subvariety Z_L of $X \times V_L$. Let $x_0, \ldots, x_n$ be homogeneous coordinates on V_L. Suppose that $U \subset X$ is an affine open subset such that $D \cap U$ has a local equation $g_U = 0$ on U. Then $\mathcal{O}_X(D) \mid U = \frac{1}{g_U} \mathcal{O}_U$. Let $s_j \mid U = \frac{h_{U,j}}{g_U}$ with $h_{U,j} \in \Gamma(U, \mathcal{O}_X)$ for $0 \le j \le n$. Define the closed subscheme Z_U of $U \times V_L$ so that the homogeneous ideal of Z_U in the graded ring $k[U][x_0, \ldots, x_n]$ is generated by $\sum_{j=0}^n h_{U,j} x_j$. Then the coordinate ring of Z_U is

$$S(Z_U) = k[U] \otimes_k k[x_0, \ldots, x_n] / \left(\sum_{j=0}^n h_{U,j} x_j \right).$$

The Z_U patch to determine a closed subscheme Z of $X \times V_L$. This is since $\frac{g_U}{g_V}$ is a unit on $U \cap V$ if U, V are affine open subsets of X such that $g_U = 0$ is a local equation of D on U and g_V is a local equation of D on V.

Let $\pi : Z \to V_L$ be the natural projection. The scheme-theoretic fiber Z_a by π of $a \in V_L$ is isomorphic to the divisor D_a for all $a \in V_L$.

Theorem 22.11 (the second theorem of Bertini). *Suppose that X is a nonsingular variety over an algebraically closed field k of characteristic 0, and L is a linear system on X without fixed component. Let W be the base locus of L. Then there exists a dense open subset U of V_L such that $L_a \setminus W$ is nonsingular for $a \in U$.*

This theorem is not true in positive characteristic. A counterexample is given in Exercise 22.19.

Proof. Let $Y = X \setminus W$. Then $L|Y$ is a base point free linear system. We construct the family $\pi : Z \to V_L$ for $L|Y$ as above. The fibers Z_a are equal to $L_a \setminus W$ for $a \in V_L$. Here Z is locally a hypersurface in homogeneous coordinates of V_L. Since $L|Y$ is base point free, at every point of Z one of the coefficients of the hypersurface is a unit. Thus Z is nonsingular by the Jacobian criterion. The theorem now follows from Theorem 22.4. $\square$

22.4. Bertini's first theorem

Suppose that X is a normal variety and L is a linear system on X without fixed component. Let notation be as in Section 22.3, with $V_L \cong \mathbb{P}^m$. The linear system L is said to be composite with a pencil if $\dim X' = 1$; that is,

$$\operatorname{trdeg}_k k \left(\frac{s_1}{s_0}, \ldots, \frac{s_m}{s_0} \right) = 1.$$

In this section, we prove the following theorem. Our proof is based on the proof by Zariski in [**158**, Section 1.6].

22.4. Bertini's first theorem

Theorem 22.12 (the first theorem of Bertini). *Suppose that X is a normal variety over an algebraically closed field k of characteristic $p \geq 0$ and L is a linear system on X without fixed component. Suppose that L is not composite with a pencil.*

Define p^e by $p^e = 1$ if K has characteristic 0, and so that p^e is the largest exponent such that $k(X') \subset k(X)^{p^e}$ if k has characteristic $p > 0$. Then there is a linear system L' on X such that $L = p^e L'$ and there exists a Zariski open subset C of V_L such that $L_a = p^e L'_{\frac{1}{a^{p^e}}}$ where $L'_{\frac{1}{a^{p^e}}}$ is a prime divisor for $a \in C$.

Suppose that κ is a field and $f(x) = f(x_1, \ldots, x_n)$ is in the polynomial ring $\kappa[x_1, \ldots, x_n]$. We say that f is geometrically irreducible over κ if $f(x)$ is irreducible in $F[x_1, \ldots, x_n]$ for all extension fields F of K; that is, $f = f_1 f_2$ with $f_1, f_2 \in F[x_1, \ldots, x_n]$ implies f_1 or f_2 has degree 0.

Remark 22.13. By Propositions 1.31 and 22.2, the following are equivalent:

1) f is geometrically irreducible over κ,
2) $F[x_1, \ldots, x_n]/(f)$ is a domain for all extension fields F of κ,
3) $\kappa[x_1, \ldots, x_n]/(f)$ is geometrically integral over κ,
4) $\kappa[x_1, \ldots, x_n]/(f)$ is a domain, and letting L be the quotient field of $\kappa[x_1, \ldots, x_n]/(f)$, we have that L is separable over κ and κ is separably closed in L.

Proposition 22.14. *Let $A = k[X]$ be the ring of regular functions on an affine variety X, and suppose that a nonzero element F in the polynomial ring $A[T_1, \ldots, T_n]$ over A has degree d. If R is an A-algebra, let $F_{(R)}$ be the image of F in $R[T_1, \ldots, T_n] \cong A[T_1, \ldots, T_n] \otimes_A R$. Let $K = k(X)$ be the quotient field of A. Assume that $F_{(K)}$ is geometrically irreducible. Then there exists a nonzero element $f \in A$ such that for all $r \in X_f$, $F_{(k(r))}$ is irreducible.*

Proof. Write $F = \sum c_\alpha T^\alpha$ with $c_\alpha \in A$ and where the indexing is over multi-indices $\alpha = (\alpha_1, \ldots, \alpha_n)$ with $T^\alpha = T_1^{\alpha_1} \cdots T_n^{\alpha_n}$. Define $|\alpha| = \alpha_1 + \cdots + \alpha_n$.

Let p, q be positive integers such that $p + q = d$. Let T'_β and T''_γ be indeterminates indexed by multi-indices $\beta = (i_1, \ldots, i_n)$ and $\gamma = (j_1, \ldots, j_n)$ such that $|\beta| \leq p$ and $|\gamma| \leq q$. Let $B_{(p,q)}$ be the polynomial ring

$$B_{(p,q)} = A[\{T'_\beta \mid |\beta| \leq p\}, \{T''_\gamma \mid |\gamma| \leq q\}],$$

which we will denote by $A[T''_\beta, T''_\gamma]$. Let $I_{(p,q)}$ be the ideal in $B_{(p,q)}$ generated by $\{P_\alpha \mid |\alpha| \leq d\}$ where

$$P_\alpha = \sum_{\beta+\gamma=\alpha} T'_\beta T''_\gamma - c_\alpha.$$

Let Ω be an algebraically closed field containing K. Suppose that we have a factorization

(22.2) $$F_{(\Omega)} = F_1 F_2$$

for some $F_1, F_2 \in \Omega[T_1, \ldots, T_n]$ with $\deg F_1 = p$ and $\deg F_2 = q$. Then writing $F_1 = \sum t'_\beta T^\beta$ and $F_2 = \sum t''_\gamma T^\gamma$ with $t'_\beta, t''_\gamma \in \Omega$, we have that $P_\alpha(t'_\beta, t''_\gamma) = 0$ for all α with $|\alpha| \leq d$. In fact, we have that there exists a factorization (22.2) if and only if the equations $P_\alpha(T'_\beta, T''_\gamma) = 0$ have a common solution in Ω for all α with $|\alpha| \leq d$, so $F_{(\Omega)}$ does not have a factorization (22.2) if and only if $Z(I_{(p,q)}) = \emptyset$ in $\mathbb{A}^n_\Omega$. By the nullstellensatz, this is equivalent to the statement that

$$I_{(p,q)} \Omega[T'_\beta, T''_\gamma] = \Omega[T'_\beta, T''_\gamma],$$

which is equivalent to the statement that

$$I_{(p,q)} K[T'_\beta, T''_\gamma] = K[T'_\beta, T'_{\gamma'}].$$

Thus if $F_{(\Omega)}$ does not have a factorization (22.2), then there exists $0 \neq f_{(p,q)} \in A$ such that

$$I_{(p,q)} A_f[T'_\beta, T''_\gamma] = A_{f_{(p,q)}}[T'_\beta, T''_\gamma].$$

Suppose that $r \in X$ has the associated maximal ideal $m = I(r) \subset A$. We have (since $k(r) \cong k$ is algebraically closed) that $F_{(k(r))}$ has a factorization $F_{(k(r))} = \overline{F}_1 \overline{F}_2$ for some $\overline{F}_1, \overline{F}_2 \in k(r)[T_1, \ldots, T_n]$ with $\deg \overline{F}_1 = p$ and $\deg \overline{F}_2 = q$ if and only if

$$I_{(p,q)} k(r)[T'_\beta, T''_\gamma] = k(r)[T'_\beta, T''_\gamma],$$

which holds if and only if the closed set in $\mathbb{A}^n_{k(r)}$

$$Z(I_{(p,q)} k(r)[T'_\beta, T''_\gamma]) = \emptyset.$$

Thus, letting $f = \prod_{p+q=d} f_{(p,q)}$ (with the restriction that p and q are positive), we have that $F_{(k(r))}$ is irreducible for all $r \in X_f$. □

Lemma 22.15. *Suppose that P is a field which is algebraically closed in a field Σ and $\Sigma^* = \Sigma(x_1, \ldots, x_m)$ is a pure transcendental extension of Σ. Then $P(x_1, \ldots, x_m)$ is algebraically closed in Σ^*.*

Proof. This is [**160**, Lemma, page 196] or [**151**, Lemma 2]. □

The following proposition and its proof are based on [**158**, Proposition 1.6.1].

22.4. Bertini's first theorem

Proposition 22.16. *Let K/F be a field of algebraic functions, of transcendence degree $r \geq 2$. Let $z_1, \ldots, z_m$ be elements of K such that the field $F(z) = F(z_1, \ldots, z_m)$ has transcendence degree $s \geq 2$ over F, let $u_1, \ldots, u_m$ be algebraically independent elements over K, and let z_u denote the linear form $u_1 z_1 + \cdots + u_m z_m$. If F is separably closed in K, then the field $F(z_u, u) = F(z_u, u_1, \ldots, u_m)$ is separably closed in $K(u_1, \ldots, u_m)$.*

Proof. Step a). We will first reduce to the case $s = r = 2$ and F algebraically closed in K. Fix a transcendence basis $\{x_{r-s+1}, x_{r-s+2}, \ldots, x_r\}$ of $F(z)$ over F and extend it to a transcendence basis $\{x_1, x_2, \ldots, x_r\}$ of K over F. Let F' be the algebraic closure of $F(x_1, \ldots, x_{r-2})$ in K. Then $\mathrm{trdeg}_{F'} K = \mathrm{trdeg}_{F'} F'(z) = 2$. Suppose the proposition is true if $s = r = 2$. Then $F'(z_u, u)$ is separably closed in $K(u)$.

We will now show that $u_1, \ldots, u_m$ and z_u are algebraically independent over F'. Assume that this is not the case. We will derive a contradiction. At least one of the z_i is transcendental over F'. Without loss of generality, we may assume that z_m is transcendental over F'. Since $u_1, \ldots, u_m$ are transcendental over K, and hence over F', we then have a relation

$$a_0 z_u^n + a_1 z_u^{n-1} + \cdots + a_n = 0$$

with $a_i \in F'[u_1, \ldots, u_m]$ for all i and the a_i not all zero. After possibly dividing out a common power of u_1 from all of the a_i, we may assume that u_1 does not divide a_i for some i. Then setting $u_1 = 0$, we have a relation $\bar{a}_0 \bar{z}_u^n + \cdots + \bar{a}_n = 0$, with $\bar{z}_n = u_2 z_2 + \cdots + u_m z_m$ and $\bar{a}_i = a_i(0, u_2, \ldots, u_n)$ with the $\bar{a}_i$ not all zero. Now, after dividing this new relation by the largest power of u_2 which divides all of the $\bar{a}_i$, we may repeat this argument, eventually getting a relation

(22.3) $$b_0 z_m^l + b_1 z_m^{l-1} + \cdots + b_0 = 0$$

with all $b_i \in F'[u_m]$ and the b_i not all zero. Now $z_m \in K$ and u_m is transcendental over K, so expanding (22.3) as a polynomial in $K[u_m]$, the coefficients must all be zero. Now these coefficients are of the form $h_i(z_m)$ with $h_i(y)$ in the polynomial ring $F'[y]$ not all zero. Since $h_i(z_m) = 0$ for all i, we have that z_m is algebraic over F', a contradiction. Thus $u_1, \ldots, u_m, z_u$ are algebraically independent over F'.

Since F is assumed to be separably closed in K (and therefore also in F') it follows from Lemma 22.15 (taking $\Sigma = F'$, $\Sigma^* = F'(z_u, u)$, and P to be the algebraic closure of F in F') that $P(z_u, u)$ is algebraically closed in $F'(z_u, u)$ and thus is separably closed in $K(u)$. Now P is purely inseparable over F, so $F(z_u, u)$ is separably closed in $P(z_u, u)$. Thus $F(z_u, u)$ is separably closed in $K(u)$.

Step b). We now assume that $s = r = 2$ and that F is algebraically closed in K. We will next find a reduction to the case $m = 2$. Let v_i, w_i for $1 \le i \le m$ and t_1, t_2 be $m+2$ elements which are algebraically independent over K. Set

(22.4) $$u_i = t_1 v_i + t_2 w_i \quad \text{for } i = 1, 2, \ldots, m,$$

(22.5) $$z_v = v_1 z_1 + v_2 z_2 + \cdots + v_m z_m,$$

(22.6) $$z_w = w_1 z_1 + w_2 z_2 + \cdots + w_m z_m,$$

(22.7) $$z_u = t_1 z_v + t_2 z_w = u_1 z_1 + u_2 z_2 + \cdots + u_m z_m.$$

By (22.4), we have that

(22.8) $$K(t_1, t_2, v, w) = K(t_1, t_2, v, u),$$

and thus the $2m+2$ elements t_1, t_2, v_i, u_i are algebraically independent over K. In particular, the u_i are algebraically independent over K.

Let $F' = F(v, w) = F(\{v_i\}, \{w_j\})$ and $K' = K(v, w) = K(\{v_i\}, \{w_j\})$. We have that $\operatorname{trdeg}_{F'} K' = 2$, and F' is algebraically closed in K' by Lemma 22.15. By an argument as in the first step of the proof, regarding v_i and w_j as indeterminates, we see that z_v and z_w are algebraically independent over F'. Since t_1 and t_2 are algebraically independent over K' and since $z_u = t_1 z_v + t_2 z_w$, we conclude, since we are assuming the proposition is true if $s = m = r = 2$, that $F'(z_u, t_1, t_2)$ is separably closed in $K'(t_1, t_2)$.

By (22.4), we have that the field $F'(z_u, t_1, t_2)$ is generated over $F(z_u, u)$ by the $m+2$ element t_1, t_2, v_i. By (22.8), these elements are algebraically independent over $K(u)$, and hence over $F(z_u, u)$. Thus $F(z_u, u)$ is algebraically closed in $F'(z_u, t_1, t_2)$ and so $F(z_u, u)$ is separably closed in $K'(t_1, t_2)$, and thus also in $K(u)$, since $K(u)$ is contained in $K'(t_1, t_2)$.

Step c). Now assume that $s = r = m = 2$. We will next reduce to the case that K is a Galois extension of $F(z_1, z_2)$. We have that K is an algebraic extension of $F(z_1, z_2)$. Let K_0 be the separable closure of $F(z_1, z_2)$ in K. Then $F(z_u, u)$ is separably closed in $K(u)$ if it is separably closed in $K_0(u)$. Hence we may replace the field K with K_0 in the proof, and so we may assume that K is a separable extension of $F(z_1, z_2)$.

Let K' be the smallest Galois extension of $F(z_1, z_2)$ containing K. Then the u_i are also algebraically independent over the finite extension K' of K. Let F' be the algebraic closure of F in K'. Then, assuming that the proposition is true in the case that K is a Galois extension of $F(z_1, z_2)$, we have that $F'(z_u, u)$ is separably closed in $K'(u)$. We will now show that $F'(z_u, u) \cap K(u) = F(Z_u, u)$, which will show that $F(z_u, u)$ is separably closed in $K(u)$.

Let $t \in F'(z_u, u) \cap K(u)$. The elements u_1, u_2 and $u_z = u_1 z_1 + u_2 z_2$ are algebraically independent over F, hence also over the finite extension F' of F. Thus the expression of t as a quotient

$$t = \frac{f(u_1, u_2, z_u)}{g(u_1, u_2, z_u)}$$

with $f, g \in F'[u_1, u_2, z_u]$ is uniquely determined if we normalize one of these polynomials by imposing the condition that a preassigned nonzero coefficient of the polynomial is 1.

Suppose $\sigma \in G(K'/K)$. Then σ extends naturally to an automorphism of $K'(u)$ over K such that $\sigma(u) = u$. Hence $\sigma(t) = t$ since $t \in K(u)$, and thus the coefficients of the polynomials must be invariant under σ. Thus these coefficients must be in K, and since they are algebraic over F, they must be in F. Hence $t \in F(z_u, u)$, proving our assertion.

Step d). We now further assume that K is a Galois extension of $F(z_1, z_2)$. Let $z_v = v_1 z_1 + v_2 z_2$ be a second linear form with indeterminate coefficients v_1, v_2, which are assumed to be algebraically independent over $K(u)$. The automorphisms in $G(K/F(z_1, z_2))$ extend uniquely to automorphisms in $G(K(u, v)/F(z_1, z_2, u, v))$ and $K(u, v)$ is Galois over $F(z_1, z_2, u, v)$ with Galois group $G(K/F(z_1, z_2))$. Let H_u be the algebraic closure of $F(z_u, u)$ in $K(u)$ and let H_v be the algebraic closure of $F(z_v, v)$ in $K(v)$. We shall prove that

(22.9) $$H_u(z_v, v) = H_v(z_u, u).$$

Since both fields in (22.9) contain the field $F(u, v, z_1, z_2)$, it will suffice to show that we have equality of Galois groups $G_1 = G_2$, where $G_1 = G(K(u, v)/H_u(z_v, v))$ and $G_2 = G(K(u, v)/H_v(z_u, u))$. The field $K(u, v)$ has an automorphism σ which interchanges u_i and v_i for $i = 1, 2$ and which is the identity on K. Then $\sigma(H_u(z_v, v)) = H_v(z_u, u)$, and $G_2 = \sigma^{-1} G_1 \sigma$. Now we have that σ commutes with each element of the Galois group $G(K(u, v)/F(u, v, z_1, z_2))$ since these automorphisms are extensions of elements of $G(K/F(z_1, z_2))$. Thus $G_1 = G_2$.

Since F is algebraically closed in K, we have that F is algebraically closed in $K(v)$, and thus also in H_v. We have that z_u, u_1, and u_2 are algebraically independent over $F(z_v, v)$ since z_u and z_v are algebraically independent over $F(u, v)$. Hence z_u, u_1, u_2 are also algebraically independent over H_v. Thus $F(z_u, u)$ is algebraically closed in $H_v(z_u, u)$, and thus by (22.9), we have that $F(z_u, u) = H_u$, completing the proof of the proposition. $\square$

We now give the proof of Theorem 22.12.

Define an effective divisor D_0 by $(s_0) + D = D_0$. Then $1, \frac{s_1}{s_0}, \ldots, \frac{s_m}{s_0} \in \Gamma(X, \mathcal{O}_X(D_0))$ and

$$L = \left\{ \operatorname{div}\left(a_0 + a_1\left(\frac{s_1}{s_0}\right) + \cdots + a_m\left(\frac{s_m}{s_0}\right)\right) + D_0 \mid a = (a_0 : \ldots : a_m) \in V_L \right\}.$$

Thus after replacing D with D_0 and the s_i with $\frac{s_i}{s_0}$, we may assume that $s_0 = 1$. If $p^e > 1$, we construct another linear system L' as follows. Let $s_i' \in k(X)$ be such that $(s_i')^{p^e} = s_i$ for $0 \leq i \leq m$. Write $D = \sum n_i E_i$ where the E_i are the integral components of D. The linear system L has no fixed component, so for each i, there exists $a_0, \ldots, a_m \in k$ such that E_i is not a component of $\operatorname{div}(a_0 s_0 + \cdots + a_m s_m) + D$. So $\nu_{E_i}(a_0 s_0 + \cdots + a_m s_m) = -n_i$. Now

$$\nu_{E_i}(a_0 s_0 + \cdots + a_m s_m) = p^e \nu_{E_i}(a_0^{\frac{1}{p^e}} s_0' + \cdots + a_m^{\frac{1}{p^e}} s_m')$$

so $p^e \mid n_i$ for all i. Let $D' = \frac{1}{p^e} D$. Then

$$L' = \{\operatorname{div}(b_0 s_0' + \cdots + b_m s_m') + D'\}$$

is a linear system such that $L = p^e L'$. We have thus reduced to the case that $p^e = 1$, which we will assume for the rest of the proof.

Let $S(V_L) = k[u_0, \ldots, u_m]$ be the homogeneous coordinate ring of V_L (a graded polynomial ring over k with $\deg u_i = 1$ for all i).

Let U be a nonsingular affine open subset of X such that $U \cap \operatorname{Supp} D = \emptyset$. Then $s_i \in k[U]$ for all i by Lemma 13.3. Let $F_1, \ldots, F_s$ be the irreducible components of $X \setminus U$ which have codimension 1 in X. Since L has no fixed component, for each i there exists $p_i \in F_i$ which is not in the base locus of L. Let

$$W_i = \{a \in V_L \mid p_i \in L_a\}.$$

W_i is a proper linear subspace of V_L since for each i there exists $G_i \in L$ such that $p_i \notin G_i$. Hence $V_L \setminus \bigcup_{i=0}^{s} W_i$ is a nonempty open subset of V_L. For $a \in V_L \setminus \bigcup_{i=0}^{s} W_i$, L_a is a prime divisor if and only if $L_a \cap U$ is a prime divisor, since L_a cannot contain a codim 1 component of $X \setminus U$.

The coordinate ring of $U \times V_L$ is

$$S(U \times V_L) = R \otimes_k k[u_0, \ldots, u_m] = R[u_0, \ldots, u_m]$$

where $R = k[U]$. The ideal $(\sum_{i=0}^{m} s_i u_i) \subset R[u_0, \ldots, u_m]$ is a homogeneous prime ideal ($s_0 = 1$). Let $Z = Z(\sum_{i=0}^{m} s_i u_i) \subset U \times V_L$. Here Z is a subvariety of $U \times V_L$ with coordinate ring

$$S(Z) = R[u_0, \ldots, u_m] / \left(\sum_{i=0}^{m} s_i u_i\right) \cong R[u_1, \ldots, u_m].$$

22.4. Bertini's first theorem

Let $k^* = k(u_0, \ldots, u_m)$ and let K^* be the quotient field of $S(Z)$. We have natural inclusions $k^* \subset K^*$ and $k(X) \subset K^*$ such that $u_1, \ldots, u_m$ are algebraically independent over $k(X)$.

Let $F = k$, $z_i = s_i$ for $1 \leq i \leq m$, $z_u = -u_0$, and $K = k(X)$. Then $K^* = K(u_1, \ldots, u_m)$ and $k^* = F(z_u, u_1, \ldots, u_m)$. Since $F(z) = F(z_1, \ldots, z_m) = k(X')$ has transcendence degree ≥ 2 over $F = k$ by assumption, we conclude from Proposition 22.16 that k^* is separably closed in K^*.

Observing that $k(V_L) = k(\frac{u_1}{u_0}, \ldots, \frac{u_m}{u_0})$ and that $k(Z)$ is the quotient field of

$$R\left[\frac{u_1}{u_0}, \ldots, \frac{u_m}{u_0}\right] / \left(1 + \sum_{i=1}^{m} s_i \frac{u_i}{u_0}\right),$$

we see that we have a commutative diagram of inclusions of fields

$$\begin{array}{ccc} k(V_L) & \to & k(Z) \\ \downarrow & & \downarrow \\ k^* & \to & K^*. \end{array}$$

Since k^* is separably closed in K^*, the separable closure of $k(V_L)$ in $k(Z)$ is contained in k^*. Now $k^* = k(V_L)(u_0)$ is a transcendental extension of $k(V_L)$, so $k(V_L)$ is separably closed in $k(Z)$.

Since k is algebraically closed, there exist $x_1, \ldots, x_r \in K$ which are a separating transcendence basis of K/k (by Theorem 1.14). After possibly replacing U with a smaller affine open subset of X, we may assume that $x_1, \ldots, x_r$ are uniformizing parameters in $R = \Gamma(U, \mathcal{O}_X)$ (Theorem 21.75 and Definition 14.16), so

$$\Omega_{R/k} = R dx_1 \oplus \cdots \oplus R dx_r$$

and $d_{R/k} : R \to \Omega_{R/k}$ is the map

$$f \mapsto D_1(f) dx_1 + \cdots + D_r(f) dx_r$$

where $\{D_1, \ldots, D_r\}$ is the basis in $\text{Der}_k(k(X), k(X))$ which is defined by $D_i(x_j) = \delta_{ij}$ for $1 \leq i, j \leq r$.

Let $S = R \otimes_k k(V_l)$. We have that

$$\Omega_{S/k(V_L)} \cong \Omega_{R/k} \otimes_k k(V_L) \cong S dx_1 \oplus \cdots \oplus S dx_r$$

by Exercise 14.11 and $D_1, \ldots, D_r$ extend naturally to a basis of $\text{Der}_{k(V_L)}(S, S)$. Let

$$A = k[Z \cap (U \times (V_L)_{u_0})] \otimes_{k[\frac{u_1}{u_0}, \ldots, \frac{u_m}{u_0}]} k(V_L) \cong S / \left(1 + \sum_{i=1}^{m} s_i \frac{u_i}{u_0}\right).$$

By Theorem 14.6, we have a right exact sequence of A-modules

$$A \xrightarrow{\delta} A dx_1 \oplus \cdots \oplus A dx_r \to \Omega_{A/k(V_L)} \to 0$$

where

$$\delta(1) = \sum_{i=1}^{r} \sum_{j=1}^{m} \frac{u_j}{u_0} D_i(s_j) dx_i.$$

Localizing at the quotient field $k(Z)$ of A, we have, by Lemma 14.8, a right exact sequence of $k(Z)$-vector spaces

$$k(Z) \xrightarrow{\delta} k(Z)dx_1 \oplus \cdots \oplus k(Z)dx_r \to \Omega_{k(Z)/k(V_L)} \to 0.$$

Since at least one of the s_j is not contained in $k(X)^p$ (not contained in k if k has characteristic 0), we have for this s_j that $D_i(s_j) \neq 0$ for some i by Theorem 21.76 or (21.18). Further, since $\mathrm{Der}_{k(V_L)}(k(Z), k(Z)) \cong \mathrm{Hom}_{k(Z)}(\Omega_{k(Z)/k(V_L)}, k(Z))$ by Lemma 14.3, we have that

(22.10) $\qquad \dim_{k(Z)} \mathrm{Der}_{k(V_L)}(k(Z), k(Z)) = r - 1 = \mathrm{trdeg}_{k(V_L)} k(Z)$

since $\dim Z = \dim V_L + \dim X - 1$ and

$$\mathrm{trdeg}_k k(Z) = \mathrm{trdeg}_{k(V_L)} k(Z) + \mathrm{trdeg}_k k(V_L)$$

by (1.1).

We conclude from Theorem 21.77 and the equality (22.10) that $k(Z)$ is separably generated over $k(V_L)$. Since $k(Z)$ is an algebraic function field over $k(V_L)$, we have that $k(Z)$ is a separable extension of $k(V_L)$ as observed before Proposition 22.2. Thus $k(Z)$ is geometrically integral over $k(V_L)$ by Proposition 22.2, since we earlier showed that $k(V_L)$ is separably closed in $k(Z)$.

Let $Y = Z \cap (U \times (V_L)_{u_0})$ with projection $\pi : Y \to (V_L)_{u_0}$. We must find an open subset A of $(V_L)_{u_0} \setminus (\bigcup_{i=0}^{s} W_i)$ such that for $a \in A$, the scheme Y_a has only one irreducible component γ and $\mathcal{O}_{Y_a,\gamma} \cong \mathcal{O}_\gamma$. For any $a \in (V_L)_{u_0}$ we have that the irreducible components of Y_a all have dimension $r - 1$ (where $r = \dim X$).

The field $k(Z)$ is separably generated over $k(V_L)$, so $k(Z)$ has a separating transcendence basis $y_1, \ldots, y_{r-1}$ over $k(V_L)$. Let $L = k(V_L)(y_1, \ldots, y_{r-1})$. By the theorem of the primitive element, there exists a primitive element t of $k(Z)$ over L. Let $g_1(y_r) \in L[y_r]$ be the minimal polynomial of t over L. Then $k(Z) \cong L[y_r]/(g_1)$. By Proposition 1.31, there exists $g \in k[(V_L)_{u_0}][y_1, \ldots, y_r]$ which is irreducible and $gL[y_r] = (g_1)$.

Let B be the affine variety with

$$k[B] = k[(V_L)_{u_0}][y_1, \ldots, y_r]/(g).$$

Then B is birationally equivalent to Z, so there exist affine open subsets C of Y and E of B such that $C \cong E$ and $C_a \cong E_a$ for all $a \in (V_L)_{u_0}$.

22.4. Bertini's first theorem

Proposition 22.14 implies that there exists an affine open subset F of $(V_L)_{u_0}$ such that B_a is a variety for all $a \in F$.

Let $F_1, \ldots, F_e$ be the irreducible components of $Y \setminus C$. By Theorem 8.13, there exists an affine open subset G of $(V_L)_{u_0}$ such that for $a \in G$, $(F_i)_a = \emptyset$ if $\pi(F_i)$ is not dense in $(V_L)_{u_0}$ and $\dim(F_i)_a < r - 1$ if $\pi(F_i)$ is dense in $(V_L)_{u_0}$. Then for $a \in (F \cap G) \setminus (\bigcup_{i=0}^s W_i)$, Y_a is a variety, and Y_a is dense in Z_a, and so L_a is a prime divisor (since we have reduced to $L = L'$), and the conclusions of the theorem follow.

Remark 22.17. If the assumptions of Theorem 22.12 hold, except that L is composite with a pencil, then we can still conclude that $L = p^e L'_{a^{\frac{1}{p^e}}}$ where $L'_{a^{\frac{1}{p^e}}}$ is a reduced divisor for a in a suitable open subset C of $V_{L'_{a^{\frac{1}{p^e}}}}$. If L is composite with a pencil, then (after replacing L with L') Proposition 22.16 is not applicable and it may be that $k(V_L)$ is not separably closed in $k(Z)$. However, the argument showing that $k(Z)$ is a separable extension of $k(V_L)$ is still valid, allowing us to conclude that $k(Z)$ is geometrically reduced over $k(V_L)$. Now a modification of Proposition 22.14 and the arguments at the end of the proof of Theorem 22.12 (or Theorem 22.18 below) allow us to conclude that $L'_{a^{\frac{1}{p^e}}}$ is reduced for a in a suitable open subset C of V_L.

We mention a general result, which can be deduced from the methods of this chapter. The proof follows directly from [68, Theorem IV.9.7.7].

Theorem 22.18. *Suppose that $\phi : X \to Y$ is a dominant regular map of varieties over an algebraically closed field k. Then there exists a nonempty open subset $U \subset Y$ such that for all $p \in U$,*

1) *the fiber X_p is irreducible if and only if $k(X)/k(Y)$ is geometrically irreducible,*
2) *the fiber X_p is reduced if and only if $k(X)/k(Y)$ is geometrically reduced,*
3) *the fiber X_p is integral if and only if $k(X)/k(Y)$ is geometrically integral.*

Proof. In this proof, we regard X and Y as general schemes (Section 15.5). Let $\{V_i\}$ be an affine cover of X such that there exist affine open subsets U_i of Y such that $f(V_i) \subset U_i$ for all i. Let η be the generic point of Y, so that $\mathcal{O}_{Y,\eta} = k(Y)$. Let k' be an extension field of $k(Y)$. The ring $k[V_i] \otimes_{k[U_i]} k(Y)$ is the localization of $k[V_i]$ with respect to the multiplicative set $T = k[U_i] \setminus \{0\}$, so it is a subring of $k(X)$. Thus $(k[V_i] \otimes_{k[U_i]} k(Y)) \otimes_{k(Y)} k'$ is a subring of $k(X) \otimes_{k(Y)} k'$ and $k[V_i] \otimes_{k[U_i]} k'$ is irreducible (respectively, reduced, integral) if and only if $k(X) \otimes_{k(Y)} k'$ is irreducible (respectively, reduced, integral) and so X_η is geometrically irreducible (respectively, reduced, integral) over

$k(X)$ if and only if $k(Y)$ is geometrically irreducible (respectively, reduced, integral) over $k(X)$. The theorem now follows from [**68**, Theorem IV.9.7.7]. □

Exercise 22.19. This exercise (from an example of Serre in [**73**, Exercise III.10.7]) shows that the second theorem of Bertini, Theorem 22.11, is not true in characteristic $p > 0$, even when the linear system is complete, base point free, irreducible, and not composite with a pencil. Let k be an algebraically closed field of characteristic 2. Let $p_1, \ldots, p_7 \in X = \mathbb{P}^2_k$ be the seven points with coefficients in $\mathbb{Z}_2$. Let L be the linear system of all cubic curves passing through $p_1, \ldots, p_7$. Prove the following statements.

   a) L is a linear system with the base points $p_1, \ldots, p_7$, and ϕ_L is an inseparable regular map of degree 2 from $X \setminus \{p_1, \ldots, p_7\} \to \mathbb{P}^2$.

   b) Every curve $C \in L$ is singular. More precisely, either C consists of three lines all passing through one of the p_i or C is an irreducible cubic curve with its only singular point some $p \neq p_i$. Furthermore, the correspondence $C \mapsto$ the singular point of C is a 1-1 correspondence between L and $\mathbb{P}^2$.

   c) Let H be a cubic curve on $\mathbb{P}^2_k$, and let $\phi : Y \to X$ be the blow-up of the seven points $p_1, \ldots, p_7$ with exceptional divisors $E_1, \ldots, E_7$. Show that the complete linear system $|\phi^*(H) - E_1 - \cdots - E_7|$ has no base points, is not composite with a pencil, all but finitely many members of the linear system are irreducible (integral), but every member of the linear system is singular.

Exercise 22.20. This exercise shows that Theorem 22.4 is not true in characteristic $p > 0$, even when $k(X)$ is geometrically integral over $k(Y)$. Let $D = \phi^*(H) - E_1 - \cdots - E_{12}$ be the divisor on Y constructed in Exercise 22.19. Let $\pi : Z \to V_{|D|}$ be the regular map constructed before Theorem 22.11 for the linear system $L = |D|$.

   a) Show that the fiber Z_p is a singular plane cubic for all $p \in V_{|D|}$.

   b) Show that the field $k(Z)$ is geometrically integral over $k(V_{|D|})$.

Exercise 22.21. Give an example of a linear system L on $\mathbb{P}^n$ which does not have a fixed component, is composite with a pencil, and such that there is a nontrivial open subset U of V_L such that L_a is not integral for all $a \in U$.

Bibliography

[1] S. Abhyankar, *On the ramification of algebraic functions*, Amer. J. Math. **77** (1955), 575–592, DOI 10.2307/2372643. MR0071851

[2] S. Abhyankar, *Local uniformization on algebraic surfaces over ground fields of characteristic $p \neq 0$*, Ann. of Math. (2) **63** (1956), 491–526, DOI 10.2307/1970014. MR0078017

[3] S. Abhyankar, *On the valuations centered in a local domain*, Amer. J. Math. **78** (1956), 321–348, DOI 10.2307/2372519. MR0082477

[4] S. S. Abhyankar, *Resolution of singularities of embedded algebraic surfaces*, 2nd ed., Springer Monographs in Mathematics, Springer-Verlag, Berlin, 1998. MR1617523

[5] S. Abhyankar, *On Macaulay's examples*, Notes by A. Sathaye, Conference on Commutative Algebra (Univ. Kansas, Lawrence, Kan., 1972), pp. 1–16. Lecture Notes in Math., Vol. 311, Springer, Berlin, 1973. MR0466156

[6] S. Abhyankar, *Ramification theoretic methods in algebraic geometry*, Annals of Mathematics Studies, no. 43, Princeton University Press, Princeton, N.J., 1959. MR0105416

[7] S. S. Abhyankar, *Algebraic geometry for scientists and engineers*, Mathematical Surveys and Monographs, vol. 35, American Mathematical Society, Providence, RI, 1990. MR1075991

[8] S. S. Abhyankar, *Resolution of singularities and modular Galois theory*, Bull. Amer. Math. Soc. (N.S.) **38** (2001), no. 2, 131–169, DOI 10.1090/S0273-0979-00-00892-2. MR1816069

[9] D. Abramovich, K. Karu, K. Matsuki, and J. Włodarczyk, *Torification and factorization of birational maps*, J. Amer. Math. Soc. **15** (2002), no. 3, 531–572, DOI 10.1090/S0894-0347-02-00396-X. MR1896232

[10] K. Matsuki, *Correction: "A note on the factorization theorem of toric birational maps after Morelli and its toroidal extension" [Tohoku Math. J. (2) **51** (1999), no. 4, 489–537; MR1725624 (2000i:14073)] by D. Abramovich, K. Matsuki, and S. Rashid*, Tohoku Math. J. (2) **52** (2000), no. 4, 629–631, DOI 10.2748/tmj/1178207758. MR1793939

[11] M. Artin, *Some numerical criteria for contractability of curves on algebraic surfaces*, Amer. J. Math. **84** (1962), 485–496, DOI 10.2307/2372985. MR0146182

[12] M. Artin, Néron Models, Chapter VIII of Arithmetic Geometry, edited by G. Cornell and J. H. Silverman, Springer Verlag, 1986. MR0861977

[13] M. F. Atiyah and I. G. Macdonald, *Introduction to commutative algebra*, Addison-Wesley Publishing Co., Reading, Mass.-London-Don Mills, Ont., 1969. MR0242802

[14] M. Auslander, *On the purity of the branch locus*, Amer. J. Math. **84** (1962), 116–125, DOI 10.2307/2372807. MR0137733

[15] M. Auslander and D. A. Buchsbaum, *Unique factorization in regular local rings*, Proc. Nat. Acad. Sci. U.S.A. **45** (1959), 733–734. MR0103906

[16] W. P. Barth, K. Hulek, C. A. M. Peters, and A. Van de Ven, *Compact complex surfaces*, 2nd ed., Ergebnisse der Mathematik und ihrer Grenzgebiete. 3. Folge. A Series of Modern Surveys in Mathematics [Results in Mathematics and Related Areas. 3rd Series. A Series of Modern Surveys in Mathematics], vol. 4, Springer-Verlag, Berlin, 2004. MR2030225

[17] D. Bayer and D. Mumford, *What can be computed in algebraic geometry?*, Computational algebraic geometry and commutative algebra (Cortona, 1991), Sympos. Math., XXXIV, Cambridge Univ. Press, Cambridge, 1993, pp. 1–48. MR1253986

[18] A. Beauville, *Complex algebraic surfaces*, London Mathematical Society Lecture Note Series, vol. 68, Cambridge University Press, Cambridge, 1983. Translated from the French by R. Barlow, N. I. Shepherd-Barron and M. Reid. MR732439

[19] E. Bertini, Introduzione alla geometria proiettiva degli iperspazi, Enrico Spoerri, Pisa, 1907.

[20] C. Birkar, P. Cascini, C. D. Hacon, and J. McKernan, *Existence of minimal models for varieties of log general type*, J. Amer. Math. Soc. **23** (2010), no. 2, 405–468, DOI 10.1090/S0894-0347-09-00649-3. MR2601039

[21] Bhargav Bhatt, Javier Carvajal-Rojas, Patrick Grant, Karl Schwede and Kevin Tucker, Étale fundamental groups of strongly F-regular schemes, arXiv:1611.03884.

[22] E. Bierstone and P. D. Milman, *Canonical desingularization in characteristic zero by blowing up the maximum strata of a local invariant*, Invent. Math. **128** (1997), no. 2, 207–302, DOI 10.1007/s002220050141. MR1440306

[23] N. Bourbaki, *Commutative algebra. Chapters 1–7*, Elements of Mathematics (Berlin), Springer-Verlag, Berlin, 1998. Translated from the French; Reprint of the 1989 English translation. MR1727221

[24] A. M. Bravo, S. Encinas, and O. Villamayor U., *A simplified proof of desingularization and applications*, Rev. Mat. Iberoamericana **21** (2005), no. 2, 349–458, DOI 10.4171/RMI/425. MR2174912

[25] A. Bravo and O. Villamayor U., *Singularities in positive characteristic, stratification and simplification of the singular locus*, Adv. Math. **224** (2010), no. 4, 1349–1418, DOI 10.1016/j.aim.2010.01.005. MR2646300

[26] E. Brieskorn and H. Knörrer, *Plane algebraic curves*, Modern Birkhäuser Classics, Birkhäuser/Springer Basel AG, Basel, 1986. Translated from the German original by John Stillwell; [2012] reprint of the 1986 edition. MR2975988

[27] M. P. Brodmann and R. Y. Sharp, *Local cohomology*, 2nd ed., An algebraic introduction with geometric applications, Cambridge Studies in Advanced Mathematics, vol. 136, Cambridge University Press, Cambridge, 2013. MR3014449

[28] W. Bruns and J. Herzog, *Cohen-Macaulay rings*, Cambridge Studies in Advanced Mathematics, vol. 39, Cambridge University Press, Cambridge, 1993. MR1251956

[29] C. Christensen, *Strong domination/weak factorization of three-dimensional regular local rings*, J. Indian Math. Soc. (N.S.) **45** (1981), no. 1-4, 21–47 (1984). MR828858

[30] I. S. Cohen, *On the structure and ideal theory of complete local rings*, Trans. Amer. Math. Soc. **59** (1946), 54–106, DOI 10.2307/1990313. MR0016094

[31] V. Cossart and O. Piltant, *Resolution of singularities of arithmetic threefolds*, arXiv: 1412.0868.

[32] D. Cox, J. Little, and D. O'Shea, *Ideals, varieties, and algorithms*, 3rd ed., An introduction to computational algebraic geometry and commutative algebra, Undergraduate Texts in Mathematics, Springer, New York, 2007. MR2290010

[33] S. D. Cutkosky, *Local monomialization and factorization of morphisms* (English, with English and French summaries), Astérisque **260** (1999), vi+143. MR1734239

[34] S. D. Cutkosky, *Monomialization of morphisms from 3-folds to surfaces*, Lecture Notes in Mathematics, vol. 1786, Springer-Verlag, Berlin, 2002. MR1927974

[35] S. D. Cutkosky, *Resolution of singularities*, Graduate Studies in Mathematics, vol. 63, American Mathematical Society, Providence, RI, 2004. MR2058431

[36] S. D. Cutkosky, *Local monomialization of transcendental extensions* (English, with English and French summaries), Ann. Inst. Fourier (Grenoble) **55** (2005), no. 5, 1517–1586. MR2172273

[37] S. D. Cutkosky, *Toroidalization of dominant morphisms of 3-folds*, Mem. Amer. Math. Soc. **190** (2007), no. 890, vi+222, DOI 10.1090/memo/0890. MR2356202

[38] S. D. Cutkosky, *Resolution of singularities for 3-folds in positive characteristic*, Amer. J. Math. **131** (2009), no. 1, 59–127, DOI 10.1353/ajm.0.0036. MR2488485

[39] S. D. Cutkosky, *A simpler proof of toroidalization of morphisms from 3-folds to surfaces* (English, with English and French summaries), Ann. Inst. Fourier (Grenoble) **63** (2013), no. 3, 865–922, DOI 10.5802/aif.2779. MR3137475

[40] S. D. Cutkosky, *Counterexamples to local monomialization in positive characteristic*, Math. Ann. **362** (2015), no. 1-2, 321–334, DOI 10.1007/s00208-014-1114-7. MR3343880

[41] S. D. Cutkosky, *Ramification of valuations and local rings in positive characteristic*, Comm. Algebra **44** (2016), no. 7, 2828–2866, DOI 10.1080/00927872.2015.1065845. MR3507154

[42] S. D. Cutkosky, *Asymptotic multiplicities of graded families of ideals and linear series*, Adv. Math. **264** (2014), 55–113, DOI 10.1016/j.aim.2014.07.004. MR3250280

[43] S. D. Cutkosky and O. Piltant, *Ramification of valuations*, Adv. Math. **183** (2004), no. 1, 1–79, DOI 10.1016/S0001-8708(03)00082-3. MR2038546

[44] S. D. Cutkosky and V. Srinivas, *On a problem of Zariski on dimensions of linear systems*, Ann. of Math. (2) **137** (1993), no. 3, 531–559, DOI 10.2307/2946531. MR1217347

[45] S. D. Cutkosky and H. Srinivasan, *Factorizations of birational extensions of local rings*, Illinois J. Math. **51** (2007), no. 1, 41–56. MR2346185

[46] R. Dedekind and H. Weber, *Theorie der algebraischen Functionen einer Veränderlichen* (German), J. Reine Angew. Math. **92** (1882), 181–290, DOI 10.1515/crll.1882.92.181. MR1579901

[47] R. Dedekind and H. Weber, *Theory of algebraic functions of one variable*, translated from the 1882 German original and with an introduction, bibliography and index by John Stillwell, History of Mathematics, vol. 39, American Mathematical Society, Providence, RI; London Mathematical Society, London, 2012. MR2962951

[48] P. Del Pezzo, *Sulle Superficie di ordine n immerse nello spazio di $n+1$ dimensioni*, Rend. Circ. Mat. Palermo **1** (1886).

[49] S. Eilenberg and N. Steenrod, *Foundations of algebraic topology*, Princeton University Press, Princeton, New Jersey, 1952. MR0050886

[50] D. Eisenbud, *Commutative algebra, With a view toward algebraic geometry*, Graduate Texts in Mathematics, vol. 150, Springer-Verlag, New York, 1995. MR1322960

[51] D. Eisenbud and S. Goto, *Linear free resolutions and minimal multiplicity*, J. Algebra **88** (1984), no. 1, 89–133, DOI 10.1016/0021-8693(84)90092-9. MR741934

[52] D. Eisenbud and J. Harris, *On varieties of minimal degree (a centennial account)*, Algebraic geometry, Bowdoin, 1985 (Brunswick, Maine, 1985), Proc. Sympos. Pure Math., vol. 46, Amer. Math. Soc., Providence, RI, 1987, pp. 3–13, DOI 10.1090/pspum/046.1/927946. MR927946

[53] D. Eisenbud and J. Harris, *The geometry of schemes*, Graduate Texts in Mathematics, vol. 197, Springer-Verlag, New York, 2000. MR1730819

[54] S. Encinas and H. Hauser, *Strong resolution of singularities in characteristic zero*, Comment. Math. Helv. **77** (2002), no. 4, 821–845, DOI 10.1007/PL00012443. MR1949115

[55] H. Flenner, L. O'Carroll, and W. Vogel, *Joins and intersections*, Springer Monographs in Mathematics, Springer-Verlag, Berlin, 1999. MR1724388

[56] G. Frey and M. Jarden, *Approximation theory and the rank of abelian varieties over large algebraic fields*, Proc. London Math. Soc. (3) **28** (1974), 112–128, DOI 10.1112/plms/s3-28.1.112. MR0337997

[57] W. Fulton, *Intersection theory*, Ergebnisse der Mathematik und ihrer Grenzgebiete (3) [Results in Mathematics and Related Areas (3)], vol. 2, Springer-Verlag, Berlin, 1984. MR732620

[58] A. M. Gabrièlov, *The formal relations between analytic functions* (Russian), Funkcional. Anal. i Priložen. **5** (1971), no. 4, 64–65. MR0302930

[59] R. Godement, *Topologie algébrique et théorie des faisceaux* (French), Actualités Sci. Ind. No. 1252. Publ. Math. Univ. Strasbourg. No. 13, Hermann, Paris, 1958. MR0102797

[60] S. Goto, K. Nishida and K. Watanabe, *Non-Cohen–Macaulay symbolic blow-ups for space monomial curves and counter-examples to Cowsik's question*, Proc. Amer. Math. Soc. **120** (1994), 383-392. MR1163334 (94d:13005)

[61] H. Grauert, *Über Modifikationen und exzeptionelle analytische Mengen* (German), Math. Ann. **146** (1962), 331–368, DOI 10.1007/BF01441136. MR0137127

[62] P. Griffiths and J. Harris, *Principles of algebraic geometry*, Pure and Applied Mathematics, Wiley-Interscience [John Wiley & Sons], New York, 1978. MR507725

[63] A. Grothendieck, *Sur quelques points d'algèbre homologique* (French), Tôhoku Math. J. (2) **9** (1957), 119–221, DOI 10.2748/tmj/1178244839. MR0102537

[64] A. Grothendieck, *Cohomologie locale des faisceaux cohérents et théorèmes de Lefschetz locaux et globaux (SGA 2)* (French), Augmenté d'un exposé par Michèle Raynaud; Séminaire de Géométrie Algébrique du Bois-Marie, 1962; Advanced Studies in Pure Mathematics, Vol. 2, North-Holland Publishing Co., Amsterdam; Masson & Cie, Éditeur, Paris, 1968. MR0476737

[65] A. Grothendieck and J. Dieudonńe, *Éléments de Géométrie Algébrique* I, Publ. Math. I.H.E.S. **4** (1960).

[66] A. Grothendieck and J. Dieudonńe, *Éléments de Géométrie Algébrique* II, Publ. Math. I.H.E.S. **8** (1961).

[67] A. Grothendieck and J. Dieudonńe, *Éléments de Géométrie Algbŕique* III, Publ. Math. I.H.E.S. **11** (1961) and **17** (1963).

[68] A. Grothendieck and J. Diudonńe, *Éléments de Géométrie Algébrique* IV, Publ. Math. I.H.E.S. **20** (1964), **24** (1965), **28** (1966), **32** (1967).

[69] A. Grothendieck and J. A. Dieudonné, *Eléments de géométrie algébrique. I* (French), Grundlehren der Mathematischen Wissenschaften [Fundamental Principles of Mathematical Sciences], vol. 166, Springer-Verlag, Berlin, 1971. MR3075000

[70] A. Grothendieck, *Revêtements étales et Groupe Fondemental*, Lecture Notes in Math. 224, Springer-Verlag, Heidelberg (1971).

[71] A. Grothendieck and J. P. Murre, *The tame fundamental group of a formal neighbourhood of a divisor with normal crossings on a scheme*, Lecture Notes in Mathematics, Vol. 208, Springer-Verlag, Berlin-New York, 1971. MR0316453

[72] R. Hartshorne, *Local cohomology*, A seminar given by A. Grothendieck, Harvard University, Fall, vol. 1961, Springer-Verlag, Berlin-New York, 1967. MR0224620

[73] R. Hartshorne, *Algebraic geometry*, Graduate Texts in Mathematics, No. 52, Springer-Verlag, New York-Heidelberg, 1977. MR0463157

[74] R. Hartshorne, *Ample subvarieties of algebraic varieties*, Notes written in collaboration with C. Musili, Springer-Verlag, Berlin, Heidelberg, New York, 1970.

[75] H. Hauser, *On the problem of resolution of singularities in positive characteristic (or: a proof we are still waiting for)*, Bull. Amer. Math. Soc. (N.S.) **47** (2010), no. 1, 1–30, DOI 10.1090/S0273-0979-09-01274-9. MR2566444

[76] J. Herzog, *Generators and relations of abelian semigroups and semigroup rings*, Manuscripta Math. **3** (1970), 175–193, DOI 10.1007/BF01273309. MR0269762

[77] J. Herzog, A. Simis, and W. V. Vasconcelos, *On the arithmetic and homology of algebras of linear type*, Trans. Amer. Math. Soc. **283** (1984), no. 2, 661–683, DOI 10.2307/1999153. MR737891

[78] H. Hironaka, *On the theory of birational blowing-up*, ProQuest LLC, Ann Arbor, MI, 1960. Thesis (Ph.D.)–Harvard University. MR2939137

[79] H. Hironaka, *Resolution of singularities of an algebraic variety over a field of characteristic zero*, Ann. of Math. (2) **79** (1964), 109–326. MR0199184

[80] H. Hironaka, *Three key theorems on infinitely near singularities* (English, with English and French summaries), Singularités Franco-Japonaises, Sémin. Congr., vol. 10, Soc. Math. France, Paris, 2005, pp. 87–126. MR2145950

[81] W. V. D. Hodge and D. Pedoe, *Methods of algebraic geometry*, Cambridge, at the University Press; New York, The Macmillan Company, 1947. MR0028055

[82] C. Huneke, *Determinantal ideals of linear type*, Arch. Math. (Basel) **47** (1986), no. 4, 324–329, DOI 10.1007/BF01191358. MR866520

[83] C. Huneke, *On the symmetric and Rees algebra of an ideal generated by a d-sequence*, J. Algebra **62** (1980), no. 2, 268–275, DOI 10.1016/0021-8693(80)90179-9. MR563225

[84] N. Jacobson, *Basic algebra. I*, W. H. Freeman and Co., San Francisco, Calif., 1974. MR0356989

[85] H. W. E. Jung, *Darstellung der Functionen eies algebraishce korrspondenzen und der verallgemeinerte korrespondenprrinzp*, Math. Ann. **28** (1887).

[86] K. Karu, *Local strong factorization of toric birational maps*, J. Algebraic Geom. **14** (2005), no. 1, 165–175, DOI 10.1090/S1056-3911-04-00380-7. MR2092130

[87] H. Kawanoue and K. Matsuki, *Toward resolution of singularities over a field of positive characteristic (the idealistic filtration program) Part II. Basic invariants associated to the idealistic filtration and their properties*, Publ. Res. Inst. Math. Sci. **46** (2010), no. 2, 359–422, DOI 10.2977/PRIMS/12. MR2722782

[88] K. S. Kedlaya, *On the algebraicity of generalized power series*, Beitr. Algebra Geom. **58** (2017), no. 3, 499–527, DOI 10.1007/s13366-016-0325-3. MR3683025

[89] S. L. Kleiman, *Toward a numerical theory of ampleness*, Ann. of Math. (2) **84** (1966), 293–344, DOI 10.2307/1970447. MR0206009

[90] J. Kollár, *Lectures on resolution of singularities*, Annals of Mathematics Studies, vol. 166, Princeton University Press, Princeton, NJ, 2007. MR2289519

[91] J. Kollár and S. Mori, *Birational geometry of algebraic varieties*, with the collaboration of C. H. Clemens and A. Corti, translated from the 1998 Japanese original, Cambridge Tracts in Mathematics, vol. 134, Cambridge University Press, Cambridge, 1998. MR1658959

[92] W. Krull, *Beiträge zur Arithmctik kommutativer Integritätsbereiche. VI. Der allgemeine Diskriminantensatz. Unverzweigte Ringerweiterungen* (German), Math. Z. **45** (1939), no. 1, 1–19, DOI 10.1007/BF01580269. MR1545800

[93] F.-V. Kuhlmann, *Valuation theoretic and model theoretic aspects of local uniformization*, Resolution of singularities (Obergurgl, 1997), Progr. Math., vol. 181, Birkhäuser, Basel, 2000, pp. 381–456. MR1748629

[94] H. Knaf and F.-V. Kuhlmann, *Every place admits local uniformization in a finite extension of the function field*, Adv. Math. **221** (2009), no. 2, 428–453, DOI 10.1016/j.aim.2008.12.009. MR2508927

[95] S. Lang, *Algebra*, 3rd ed., Graduate Texts in Mathematics, vol. 211, Springer-Verlag, New York, 2002. MR1878556

[96] S. Lang, *Abelian varieties*, Springer-Verlag, New York-Berlin, 1983. Reprint of the 1959 original. MR713430

[97] S. Lang and A. Néron, *Rational points of abelian varieties over function fields*, Amer. J. Math. **81** (1959), 95–118, DOI 10.2307/2372851. MR0102520

[98] R. Lazarsfeld, *Positivity in algebraic geometry. I*, Classical setting: line bundles and linear series, Ergebnisse der Mathematik und ihrer Grenzgebiete. 3. Folge. A Series of Modern Surveys in Mathematics [Results in Mathematics and Related Areas. 3rd Series. A Series of Modern Surveys in Mathematics], vol. 48, Springer-Verlag, Berlin, 2004. MR2095471

[99] R. Lazarsfeld and M. Mustaţă, *Convex bodies associated to linear series* (English, with English and French summaries), Ann. Sci. Éc. Norm. Supér. (4) **42** (2009), no. 5, 783–835, DOI 10.24033/asens.2109. MR2571958

[100] J. Lipman, *Rational singularities, with applications to algebraic surfaces and unique factorization*, Inst. Hautes Études Sci. Publ. Math. **36** (1969), 195–279. MR0276239

[101] J. Lipman, *Desingularization of two-dimensional schemes*, Ann. Math. (2) **107** (1978), no. 1, 151–207. MR0491722

[102] S. Łojasiewicz, *Introduction to complex analytic geometry*, translated from the Polish by Maciej Klimek, Birkhäuser Verlag, Basel, 1991. MR1131081

[103] W. S. Massey, *Algebraic topology: An introduction*, Harcourt, Brace & World, Inc., New York, 1967. MR0211390

[104] T. Matsusaka, *The theorem of Bertini on linear systems in modular fields*, Mem. Coll. Sci. Univ. Kyoto Ser. A. Math. **26** (1950), 51–62, DOI 10.1215/kjm/1250778055. MR0041483

[105] H. Matsumura, *Geometric structure of the cohomology rings in abstract algebraic geometry*, Mem. Coll. Sci. Univ. Kyoto. Ser. A Math. **32** (1959), 33–84, DOI 10.1215/kjm/1250776697. MR0137709

[106] H. Matsumura, *Commutative ring theory*, translated from the Japanese by M. Reid, Cambridge Studies in Advanced Mathematics, vol. 8, Cambridge University Press, Cambridge, 1986. MR879273

[107] H. Matsumura, *Commutative algebra*, 2nd ed., Mathematics Lecture Note Series, vol. 56, Benjamin/Cummings Publishing Co., Inc., Reading, Mass., 1980. MR575344

[108] A. Micali, *Sur les algèbres universelles* (French), Ann. Inst. Fourier (Grenoble) **14** (1964), no. fasc. 2, 33–87. MR0177009

[109] J. S. Milne, *Étale cohomology*, Princeton Mathematical Series, vol. 33, Princeton University Press, Princeton, N.J., 1980. MR559531

[110] J. S. Milne, *Abelian varieties*, Arithmetic geometry (Storrs, Conn., 1984), Springer, New York, 1986, pp. 103–150. MR861974

[111] J. S. Milne, *Jacobian varieties*, Arithmetic geometry (Storrs, Conn., 1984), Springer, New York, 1986, pp. 167–212. MR861976

[112] S. Mori, *Projective manifolds with ample tangent bundles*, Ann. of Math. (2) **110** (1979), no. 3, 593–606, DOI 10.2307/1971241. MR554387

[113] S. Mori, *Threefolds whose canonical bundles are not numerically effective*, Ann. of Math. (2) **116** (1982), no. 1, 133–176, DOI 10.2307/2007050. MR662120

[114] S. Mori, *Flip theorem and the existence of minimal models for 3-folds*, J. Amer. Math. Soc. **1** (1988), no. 1, 117–253, DOI 10.2307/1990969. MR924704

[115] D. Mumford, *Algebraic geometry. I. Complex projective varieties*; Grundlehren der Mathematischen Wissenschaften, No. 221, Springer-Verlag, Berlin-New York, 1976. MR0453732

[116] D. Mumford, *The red book of varieties and schemes*, Springer-Verlag, 1980.

[117] D. Mumford, *The topology of normal singularities of an algebraic surface and a criterion for simplicity*, Inst. Hautes Études Sci. Publ. Math. **9** (1961), 5–22. MR0153682

[118] D. Mumford, *Lectures on curves on an algebraic surface*, With a section by G. M. Bergman. Annals of Mathematics Studies, No. 59, Princeton University Press, Princeton, N.J., 1966. MR0209285

[119] D. Mumford, *Abelian varieties*, Tata Institute of Fundamental Research Studies in Mathematics, No. 5, Published for the Tata Institute of Fundamental Research, Bombay; Oxford University Press, London, 1970. MR0282985

[120] J. R. Munkres, *Topology: A first course*, Prentice-Hall, Inc., Englewood Cliffs, N.J., 1975. MR0464128

[121] M. Nagata, *Local rings*, Interscience Tracts in Pure and Applied Mathematics, No. 13, Interscience Publishers a division of John Wiley & Sons New York-London, 1962. MR0155856

[122] M. Nagata, *Lectures on the fourteenth problem of Hilbert*, Tata Institute of Fundamental Research, Bombay, 1965. MR0215828

[123] T. Oda, *Torus embeddings and applications*, based on joint work with Katsuya Miyake, Tata Institute of Fundamental Research Lectures on Mathematics and Physics, vol. 57, Tata Institute of Fundamental Research, Bombay; by Springer-Verlag, Berlin-New York, 1978. MR546291

[124] M. Raynaud, *Anneaux locaux henséliens* (French), Lecture Notes in Mathematics, Vol. 169, Springer-Verlag, Berlin-New York, 1970. MR0277519

[125] D. Rees, *On a problem of Zariski*, Illinois J. Math. **2** (1958), 145–149. MR0095843

[126] M. Reid, *Canonical 3-folds*, Journées de Géometrie Algébrique d'Angers, Juillet 1979/Algebraic Geometry, Angers, 1979, Sijthoff & Noordhoff, Alphen aan den Rijn—Germantown, Md., 1980, pp. 273–310. MR605348

[127] P. C. Roberts, *A prime ideal in a polynomial ring whose symbolic blow-up is not Noetherian*, Proc. Amer. Math. Soc. **94** (1985), no. 4, 589–592, DOI 10.2307/2044869. MR792266

[128] J. J. Rotman, *An introduction to homological algebra*, Pure and Applied Mathematics, vol. 85, Academic Press, Inc. [Harcourt Brace Jovanovich, Publishers], New York-London, 1979. MR538169

[129] J. Sally, *Regular overrings of regular local rings*, Trans. Amer. Math. Soc. **171** (1972), 291–300, DOI 10.2307/1996383. MR0309929

[130] J. D. Sally, *Numbers of generators of ideals in local rings*, Marcel Dekker, Inc., New York-Basel, 1978. MR0485852

[131] P. Samuel, *Algebraic theory of numbers*, translated from the French by Allan J. Silberger, Houghton Mifflin Co., Boston, Mass., 1970. MR0265266

[132] J.-P. Serre, *Faisceaux algébriques cohérents* (French), Ann. of Math. (2) **61** (1955), 197–278, DOI 10.2307/1969915. MR0068874

[133] J.-P. Serre, *Géométrie algébrique et géométrie analytique* (French), Ann. Inst. Fourier, Grenoble **6** (1955), 1–42. MR0082175

[134] J.-P. Serre, *Algebraic groups and class fields*, translated from the French, Graduate Texts in Mathematics, vol. 117, Springer-Verlag, New York, 1988. MR918564

[135] J.-P. Serre, *Local fields*, translated from the French by Marvin Jay Greenberg, Graduate Texts in Mathematics, vol. 67, Springer-Verlag, New York-Berlin, 1979. MR554237

[136] I. R. Shafarevich, *Basic algebraic geometry. Varieties in projective space*; translated from the 1988 Russian edition and with notes by Miles Reid, 2nd ed., Springer-Verlag, Berlin, 1994. MR1328833

[137] D. L. Shannon, *Monoidal transforms of regular local rings*, Amer. J. Math. **95** (1973), 294–320. MR0330154

[138] A. Simis, B. Ulrich, and W. V. Vasconcelos, *Cohen-Macaulay Rees algebras and degrees of polynomial relations*, Math. Ann. **301** (1995), no. 3, 421–444, DOI 10.1007/BF01446637. MR1324518

[139] E. Snapper, *Multiples of divisors*, J. Math. Mech. **8** (1959), 967–992. MR0109156

[140] B. Teissier, *Valuations, deformations, and toric geometry*, Valuation theory and its applications, Vol. II (Saskatoon, SK, 1999), Fields Inst. Commun., vol. 33, Amer. Math. Soc., Providence, RI, 2003, pp. 361–459. MR2018565

[141] M. Temkin, *Inseparable local uniformization*, J. Algebra **373** (2013), 65–119, DOI 10.1016/j.jalgebra.2012.09.023. MR2995017

[142] B. L. van der Waerden, *Modern Algebra*. Vol. I, translated from the second revised German edition by Fred Blum; with revisions and additions by the author, Frederick Ungar Publishing Co., New York, N. Y., 1949. MR0029363

[143] B. L. van der Waerden, *Modern Algebra*. Vol. II, translated from the second revised German edition by Fred Blum; with revisions and additions by the author, Frederick Ungar Publishing Co., New York, N. Y., 1950.

[144] R. J. Walker, *Reduction of the singularities of an algebraic surface*, Ann. of Math. (2) **36** (1935), no. 2, 336–365, DOI 10.2307/1968575. MR1503227

[145] A. Weil, *Foundations of algebraic geometry*, American Mathematical Society, Providence, R.I., 1962. MR0144898

[146] A. Weil, *Variétés Abéliennes et Courbes Algébriques*, Hermann, Paris, 1971.

[147] J. Włodarczyk, *Decomposition of birational toric maps in blow-ups & blow-downs*, Trans. Amer. Math. Soc. **349** (1997), no. 1, 373–411, DOI 10.1090/S0002-9947-97-01701-7. MR1370654

[148] O. Zariski, *Algebraic surfaces*, Second supplemented edition, with appendices by S. S. Abhyankar, J. Lipman, and D. Mumford; Ergebnisse der Mathematik und ihrer Grenzgebiete, Band 61, Springer-Verlag, New York-Heidelberg, 1971. MR0469915

[149] O. Zariski, *The reduction of the singularities of an algebraic surface*, Ann. of Math. (2) **40** (1939), 639–689, DOI 10.2307/1968949. MR0000159

[150] O. Zariski, *Local uniformization on algebraic varieties*, Ann. of Math. (2) **41** (1940), 852–896, DOI 10.2307/1968864. MR0002864

[151] O. Zariski, *Pencils on an algebraic variety and a new proof of a theorem of Bertini*, Trans. Amer. Math. Soc. **50** (1941), 48–70, DOI 10.2307/1989911. MR0004241

[152] O. Zariski, *Foundations of a general theory of birational correspondences*, Trans. Amer. Math. Soc. **53** (1943), 490–542, DOI 10.2307/1990215. MR0008468

[153] O. Zariski, *Reduction of the singularities of algebraic three dimensional varieties*, Ann. of Math. (2) **45** (1944), 472–542, DOI 10.2307/1969189. MR0011006

[154] O. Zariski, *The concept of a simple point of an abstract algebraic variety*, Trans. Amer. Math. Soc. **62** (1947), 1–52. MR0021694

[155] O. Zariski, *A simple analytical proof of a fundamental property of birational transformations*, Proc. Nat. Acad. Sci. U. S. A. **35** (1949), 62–66. MR0028056

[156] O. Zariski, *Theory and applications of holomorphic functions on algebraic varieties over arbitrary ground fields*, Mem. Amer. Math. Soc., **No. 5** (1951), 90. MR0041487

[157] O. Zariski, *On the purity of the branch locus of algebraic functions*, Proc. Nat. Acad. Sci. U.S.A. **44** (1958), 791–796. MR0095846

[158] O. Zariski, *Introduction to the problem of minimal models in the theory of algebraic surfaces*, Publications of the Mathematical Society of Japan, no. 4, The Mathematical Society of Japan, Tokyo, 1958. MR0097403

[159] O. Zariski, *The theorem of Riemann-Roch for high multiples of an effective divisor on an algebraic surface*, Ann. of Math. (2) **76** (1962), 560–615, DOI 10.2307/1970376. MR0141668

[160] O. Zariski and P. Samuel, *Commutative algebra*. Vol. 1, with the cooperation of I. S. Cohen; corrected reprinting of the 1958 edition; Graduate Texts in Mathematics, No. 28, Springer-Verlag, New York-Heidelberg-Berlin, 1975. MR0384768

[161] O. Zariski and P. Samuel, *Commutative algebra*. Vol. II, reprint of the 1960 edition; Graduate Texts in Mathematics, Vol. 29, Springer-Verlag, New York-Heidelberg, 1975. MR0389876

Index

$(\mathcal{L}_1 \cdots \mathcal{L}_t)$, 369
$(\mathcal{L}_1 \cdots \mathcal{L}_t; \mathcal{F})_V$, 368
$(\mathcal{L}_1 \cdots \mathcal{L}_t \cdot W)$, 369
(ϕ), 286
$(\phi)_X$, 286
(f), 241
$(f)_0$, 241
$(f)_\infty$, 242
A-torsion, 208
$B(I)$, 113, 121
$B(\mathcal{I})$, 221–223
$C^*(\underline{U}, \mathcal{F})$, 310
$D(B/A)$, 399
$D(F)$, 71, 103
$D(f)$, 48
$D \cap U$, 242
$D \equiv 0$, 382
$D_1 \geq D_2$, 240
$D_1 \sim D_2$, 242
$D_{L/K}(a_1, \ldots, a_n)$, 398
$F : X_p \to X$, 349
F^a, 71
$Fr(x)$, 349
$G(L, K)$, 7
$G(Y/X)$, 421
$G^i(S/R)$, 440
$G^s(S/R)$, 438
$H^i(A^*)$, 307
$H^i(X, F)$, 308, 309
$H^i_Y(X, \mathcal{F})$, 330
$H^i_I(M)$, 326
$H^0_{\text{Sing}}(U, G)$, 358
$I(S/R)$, 435

$I(Y)$, 29, 33, 69, 70, 103
$I(Z)$, 292
I^{sat}, 66
$I_X(Y)$, 33, 70
$J(S/R)$, 435
K^s, 438
K_X, 286
$M(n)$, 63
$M^\vee$, 243
$M_{(F)}$, 64
$M_{(\mathfrak{p})}$, 64
$N(X)$, 382
$N_{L/K}$, 392
$P(A)$, 243
PG_X, 189
$P_M(z)$, 301
P_Y, 301
$R^i \phi_* \mathcal{F}$, 320
$R^{(d)}$, 64
$S(I; R)$, 56
$S(W)$, 103
$S(X)$, 70
$S(X \times \mathbb{P}^n)$, 105
$S(Y)$, 105
$S(\mathbb{P}^m \times \mathbb{P}^n)$, 102
$S(\mathbb{P}^n)$, 68
$T(A)$, 24
$T_A(M)$, 208
$T_p(X)$, 157
V_ν, 228
W_F, 103
$X \times Y$, 101
$X^{(r)}$, 360

X_F, 71
X_Z, 290
X_f, 48
X_p, 290
Y^H, 422
$Z(A)$, 102
$Z(I)$, 68
$Z(J)$, 70
$Z(T)$, 28, 33
$Z(U)$, 68
$Z(f)$, 28
$Z_1 \cap Z_2$, 290
$Z_X(T)$, 33
$Z_X(U)$, 70
Z_{red}, 289
$[L:K]_i$, 7
$[L:K]_s$, 7
$\mathbb{A}^m \times \mathbb{A}^n$, 101
$\mathbb{A}^n$, 27
$\mathbb{C}\{x_1,\ldots,x_n\}$, 176
Δ, 176
Δ_X, 110
$\Delta_{\mathbb{P}^n}$, 106
$\Gamma_Y(X,\mathcal{F})$, 330
Γ_ν, 228
Γ_ϕ, 106, 107
$\mathbb{N}$, 1
$\Omega^n_{X/k}$, 286
$\Omega^n_{k(X)/k}$, 286
$\Omega_{B/A}$, 280
$\Omega_{X/k}$, 283
$\mathbb{P}^m \times \mathbb{P}^n$, 102
$\mathbb{P}^n_k$, 67
$\mathbb{Z}_+$, 1
$\mathbb{Z}_{\text{an}}$, 359
$\check{H}^p(X,\mathcal{F})$, 312
$\check{H}^p(\underline{U},\mathcal{F})$, 311
$\chi(\mathcal{F})$, 319
$\deg x$, 63
$\deg$, 250
$\deg(D)$, 254
$\deg(Y)$, 301
$\deg(\phi)$, 254, 371, 406
$\dim R$, 19
$\dim X$, 42, 139
$\ell_R(M)$, 9
$\hat{M}$, 409
$\hat{\phi}: \hat{R} \to \hat{S}$, 411
$\lim_\leftarrow A_i$, 184
$\lim_\rightarrow A_i$, 182
$\mathcal{F}(n)$, 210

$\mathcal{F} \otimes \mathcal{G}$, 204, 206
$\mathcal{F} \otimes_{\mathcal{O}_X} \mathcal{G}$, 204, 206
$\mathcal{F}^{\text{an}}$, 359
$\mathcal{G}^*$, 245
$\mathcal{IO}_X$, 205
$\mathcal{I}_Y$, 155, 198, 200
$\mathcal{I}_Z$, 289
$\mathcal{I}_{Y,p}$, 155
$\mathcal{I}_{\text{div}(s)}$, 291
$\mathcal{L}^{-1}$, 269
$\mathcal{O}_D$, 290
$\mathcal{O}_U(V)$, 53
$\mathcal{O}_X$-module, 191
$\mathcal{O}_X$-module homomorphism, 191
$\mathcal{O}_X$-submodule, 192
$\mathcal{O}_X$-torsion, 209
$\mathcal{O}_X(D)$, 242
$\mathcal{O}_X(U)$, 49, 75, 187
$\mathcal{O}_X(n)$, 210
$\mathcal{O}_X^*$, 271
$\mathcal{O}_{U,p}$, 53
$\mathcal{O}_{W,(p,q)}$, 103
$\mathcal{O}_W(U)$, 103
$\mathcal{O}_{X,E}$, 240
$\mathcal{O}_{X,Y}$, 229
$\mathcal{O}_{X,p}$, 50, 75
$\mathcal{T}(\mathcal{F})$, 209
$\mathfrak{p}_i^{(a_i)}$, 245
$\text{Ann}_X(\sigma)$, 208
$\text{Aut}(X)$, 351
$\text{Base}(L)$, 261
$\text{Cl}(X)$, 242
$\text{Cl}^0(X)$, 258
$\text{Cokernel}(\alpha)$, 194
$\text{Der}_A(B,M)$, 279
$\text{Div}(F)$, 249
$\text{Div}(X)$, 240
$\text{Grass}(a,b)$, 73
$\text{Image}(\alpha)$, 194
$\text{Kernel}(\alpha)$, 194
$\text{Map}(U, \mathbb{A}^1)$, 49, 75
$\text{Pic}(X)$, 270
$\text{Pic}^{\text{an}}(X)$, 359
$\text{Proj}(S)$, 67, 297
$\text{Spec}(R)$, 5, 296
$\text{Supp } D$, 240
$\text{bideg}(x)$, 102
$\text{div}(\phi)$, 286
$\text{div}(\phi)_X$, 286
$\text{div}(f)$, 242
$\text{div}(f)_X$, 242

Index

div(s), 291
gcd$(D_0, \ldots, D_n)$, 259
ht(P), 19
reg(M), 328
reg$(\tilde{M})$, 328
trdeg$_K L$, 6
$\mu(I)$, 46
ν_E, 241
$\phi: X \dashrightarrow Y$, 58, 93
ϕ^*, 35, 76
$\phi^* : \text{Cl}(Y) \to \text{Cl}(X)$, 252
$\phi^* \mathcal{M}$, 205
$\phi^{-1}(Z)$, 290
ϕ_L, 260
ϕ_V, 260
Vol$(\mathcal{L})$, 345
$\sqrt{I}$, 3
$\sqrt{\mathcal{I}}$, 289
Tr$_{L/K}$, 392
$\tilde{I}$, 120, 198, 200
$\tilde{J}$, 192
$\tilde{M}$, 196
$\tilde{N}$, 199
$|D|$, 260
$\wedge^n M$, 286
$a^i(M)$, 327
$d(R' : R)$, 443
$d\phi_p$, 157
$e(R)$, 376
$e(S_i/R)$, 396
$e(\nu^*/\nu)$, 397
e_P, 346
$f(S_i/R)$, 396
$f(\nu^*/\nu)$, 397
$f^*\mathcal{G}$, 207
f^h, 71
$f^\#$, 191
$f_* F$, 191
$g(R' : R)$, 443
$g(X)$, 334
$h^i(X, \mathcal{F})$, 319
k, an algebraically closed field, 1
k-linear Frobenius map, 349
$k(U)$, 53, 82
$k(W)$, 103
$k(X)$, 49, 74
$k(p)$, 50, 75, 283
$k[X]$, 33
$k[\mathbb{A}^m \times \mathbb{A}^n]$, 101
$k[\mathbb{A}^n]$, 27
m-regular, 328

m_R, the maximal ideal of a local ring, 1
m_p, 50, 75, 283
$m_{X,p}$, 50, 75
n-fold, 45
p-basis, 425
p-independent, 425
$r(R' : R)$, 443
Aut(L/K), 7
Aut$(k(Y)/k(X))$, 421
Hom$_{\mathcal{O}_X}(\mathcal{F}, \mathcal{G})$, 201
$\mathcal{H}om_{\mathcal{O}_X}(\mathcal{F}, \mathcal{G})$, 204, 206
Num(X), 382
QF(R), 1
QR(A), 291
codim$_X(Y)$, 45
depth$_I M$, 326
res, 34
$gr_{m_R}(R)$, 23

Abel, 360
Abelian variety, 360
Abhyankar, xii, 170, 227, 228, 231, 232, 234, 235, 392, 413, 435
Abhyankar-Jung theorem, 436
Abramovich, 234
Abstract prevariety, 295
Abstract variety, 295
Adjunction, 287, 380
Affine map, 127
Affine scheme, 296
Affine subscheme, 289
Affine subvariety, 34
Affine variety, 55
Algebraic function field, 7
Algebraic local ring, 441
Algebraic set, biprojective, 103
Algebraic set, in $\mathbb{A}^n$, 28
Algebraic set, projective, 68
Algebraic set, quasi-affine, 34
Algebraic set, quasi-biprojective, 103
Ample divisor, 262
Ample invertible sheaf, 270
Analytic implicit function theorem, 176
Artin, 362
Auslander, 25, 433

Base locus, of a linear system, 261
Base point free linear system, 261
Bayer, 330
Bertini, 304, 451
Bézout, 375
Bézout's theorem, 375

Bhatt, 433
Bihomogeneous coordinate ring, 102
Birational equivalence, of affine
  varieties, 60
Birational equivalence, of
  quasi-projective varieties, 94
Birational map, of affine varieties, 60
Birational map, of quasi-projective
  varieties, 94
Birkar, 387
Blow-up, of a subvariety of a projective
  variety, 123
Blow-up, of a subvariety of an affine
  variety, 113
Blow-up, of an ideal of a projective
  variety, 121
Blow-up, of an ideal of an affine variety,
  113
Blow-up, of an ideal sheaf on a
  projective variety, 222
Blow-up, of an ideal sheaf on a
  quasi-projective variety, 223
Blow-up, of an ideal sheaf on an affine
  variety, 221
Buchsbaum, 25

Canonical divisor, 286
Cartier divisor, 245, 273, 291
Carvajal-Rojas, 433
Cascini, 387
Castelnuovo, 328, 383
Castelnuovo's contraction theorem, 383
Chain map, 307
Chevalier, 411
Chinese remainder theorem, 4
Christiansen, 235
Clifford, 342
Clifford's theorem, 342
Closed embedding, of a quasi-affine
  variety, 55
Closed embedding, of a quasi-projective
  variety, 88
Closed embedding, of an affine algebraic
  set, 38
Closed subscheme, 289
Codimension, of a quasi-affine algebraic
  set, 45
Codimension, of a quasi-projective
  algebraic set, 141
Cohen, 410
Coherent $\mathcal{O}_X$-module, 200
Coherent sheaf, 290

Cohomology, 311, 312
Complete abstract variety, 295
Complete linear system, 260
Complex, 307
Complex manifold, 175
Constant sheaf, 189
Coordinate functions, on an affine
  algebraic set, 38
Coordinate ring of $\mathbb{P}^n$, 68
Coordinate ring, of a biprojective
  variety, 103
Coordinate ring, of a projective
  algebraic set, 70
Coordinate ring, of a subvariety of
  $\mathbb{A}^m \times \mathbb{P}^n$, 105
Cossart, 228
Curve, 45
Cutkosky, 235, 236, 345

Dedekind, xi
Degree, of a divisor on a curve, 254
Degree, of a graded module, 301
Degree, of a map of curves, 254
Degree, of a projective subscheme, 301
Degree, of a regular map, 371, 406
Degree, of an invertible sheaf on a
  curve, 259
Del Pezzo, 304
Derivation, 279
Diagonal, 106, 176
Dimension, of a quasi-projective
  algebraic set, 139
Dimension, of a topological space, 42
Direct limit, 181
Discrete valuation, 395
Discrete valuation ring, 395
Discriminant ideal, 398
Divisor, 240
Divisor class group, 242
Divisor of a function, 241
Divisor of a section of an invertible
  sheaf, 273
Divisor of poles of a function, 242
Divisor of zeros of a function, 241
Divisor, local equation, 245
Divisor, of a form, 249
Divisor, of a rational differential
  n-form, 286
Divisor, support, 240
Divisorial valuation, 229
Dominant map, of affine algebraic sets,
  37

Index

Dominant rational map, of affine varieties, 59
Dominant rational map, of quasi-projective varieties, 94
Dominant regular map, of a quasi-projective variety, 82
Dominate, rings, 1
dvr, 395

Effective Cartier divisor, 291
Effective divisor, 240
Eisenbud, 304, 330
Elliptic curve, 341
Étale regular map, 426
Euclidean topology, 175
Euler, 13
Euler characteristic of a sheaf, 319
Excision, 331

Fiber cone, 294
Finite map, 127
Finite map, of affine varieties, 40
First theorem of Bertini, 459
Fractional ideal, 243
Frobenius homomorphism, 349

Gabrièlov, 420
Galois over X, 422
Galois regular map, 422
Generated by global sections, 212
Genus, of a curve, 334
Geometric regularity theorem, 328
Geometrically integral ring, 452
Geometrically integral scheme, 452
Geometrically irreducible ring, 452
Geometrically irreducible scheme, 452
Geometrically reduced ring, 452
Geometrically reduced scheme, 452
Germ, 185
Godement, 181
Going down theorem, 393
Going up theorem, 393
Goto, 324, 330
Graded module, 63
Graded ring, 63
Grant, 433
Graph, of a rational map, 107
Graph, of a regular map, 106
Grauert, 387
Grothendieck, xii, 151, 367, 433

Hacon, 387
Harris, 304

Hartshorne, xii, 181
Heinzer, 395
Herzog, 46, 116
Higher direct images of a sheaf, 320
Hilbert, 11, 29, 301, 376
Hilbert polynomial, of a graded module, 301
Hilbert polynomial, of a projective subscheme, 301
Hilbert's basis theorem, 11
Hilbert's nullstellensatz, 29
Hilbert-Samuel polynomial, 376
Hironaka, 227, 232, 234, 295
Hodge index theorem, 382
Homogeneous coordinates, 90
Homogeneous coordinates, on $\mathbb{P}^n$, 68
Homogeneous ideal, 64
Homogeneous ideal of a subscheme, 292
Homotopy, 308
Huneke, 116
Hurwitz, 348

Ideal sheaf, 120, 205
Ideal sheaf, of a subvariety, 200
Ideal sheaf, on a projective variety, 192
Ideal sheaf, on an affine variety, 192
Ideal transform, 56
Image, of a rational map, 94
Indirect limit, 184
Inertia field, 440
Injective sheaf homomorphism, 194
Integral scheme, 290
Intersection number, 368
Invertible sheaf, 203, 223, 245, 269, 290
Irreducible scheme, 290
Irreducible topological space, 29
Isomorphism, of affine algebraic sets, 38
Isomorphism, of quasi-affine varieties, 54
Isomorphism, of varieties, 76

Jacobi, 360
Jacobian criterion for nonsingularity, 160
Jacobian, of a curve, 360, 361
Jung, 435

Kähler differentials, 280
Kaplansky, 232
Karu, 234, 236
Kollár, 387
Krull, 20

Krull's principal ideal theorem, 20
Künneth, 319
Künneth formula, 319

Lang, 29
Lazarsfeld, 345
Linear equivalence of divisors, 242
Linear subspace, of a projective space, 95
Linear system, 260
Lipman, 227, 231
Local cohomology, 326, 330
Local equations, 155
Local ring, 1
Locally free sheaf, 217
Locally principal ideal sheaf, 223
Locally ringed space, 191

Map of complexes, 307
Matsuki, 234
Matsusaka, 452
Mayer-Vietoris sequence, 331
McKernan, 387
Milne, 362
Mori, 387
Morphism, of affine schemes, 297
Morphism, of locally ringed spaces, 191
Morphism, of schemes, 297
Multiplicity, 376
Mumford, xii, 328, 330
Mustaţă, 345

Nagata, 57, 324, 433
Nakayama, 8
Newton, 437
Nishida, 324
Noether, 17, 134
Noether's normalization lemma, 17
Nondegenerate variety, 299
Nonsingular point, of a variety, 160
Nonsingular, quasi-projective variety, 160
Normal field extension, 394
Normal point, 135
Normal variety, 135
Normalization, in a finite extension, 135
Nullhomotopic, 308
Numerical equivalence of divisors, 382
Numerical polynomial, 300

Oda, 235
Open embedding, of quasi-affine varieties, 55

Open subscheme, 289

Picard group, 270
Piltant, 228
Presheaf, 185
Presheaf, homomorphism, 185
Presheaf, isomorphism, 185
Prime divisor, 240
Primitive element, 401
Primitive element theorem, 7
Principal divisor, 242
Projective algebraic set, 69
Projective Noether normalization, 134
Projective scheme, 297
Projective subscheme, 289
Projective variety, 69
Proper transform, 123
Purity of the branch locus, 433

Quadratic transform, 170
Quasi-affine variety, 34
Quasi-coherent $\mathcal{O}_X$-module, 200
Quasi-coherent sheaf, 290
Quasi-projective algebraic set, 69
Quasi-projective subscheme, 289
Quasi-projective variety, 69
Quotient of Y by H, 422

Ramification index, of a regular map of curves, 346
Ramification locus, in the domain of a finite map, 432, 433
Ramification locus, in the range of a finite map, 407
Ramified extension, of a normal domain, 400
Ramified extension, of local domains, 399
Ramified, map of curves, 346
Rashid, 234
Rational function field, 7
Rational functions, on a biprojective variety, 103
Rational functions, on a projective variety, 74
Rational functions, on a quasi-affine variety, 53
Rational functions, on a quasi-projective variety, 82
Rational functions, on an affine variety, 49
Rational map, of affine varieties, 58

Rational map, of projective varieties, 93
Rational map, of quasi-projective varieties, 93
Rational maps, on a subvariety of $\mathbb{A}^m \times \mathbb{P}^n$, 105
Reduced ramification index, 396
Reduced scheme, 290
Rees, 57
Reflexive rank 1 sheaf, 245
Regular functions, on $\mathbb{A}^n$, 27
Regular functions, on a quasi-affine variety, 49
Regular functions, on an affine algebraic set, 33
Regular functions, on an open subset of a biprojective variety, 103
Regular map, of affine algebraic sets, 36
Regular map, of an affine variety to $\mathbb{A}^n$, 35
Regular map, of quasi-affine varieties, 54
Regular map, of varieties, 76
Regular maps, of biprojective varieties, 103
Regular parameters, 154
Regularity, 328
Relative degree, 396
Resolution of indeterminancy, 225
Resolution of singularities, 225
Riemann, 335, 340
Riemann-Roch inequality, 335
Riemann-Roch problem, 344
Riemann-Roch theorem, 340
Rigidity lemma, 358
Roberts, 324
Roch, 335, 340

Sally, 46, 232
Samuel, 376
Scheme, 289, 297
Scheme-theoretic fiber, 290
Scheme-theoretic intersection, 290
Schwede, 433
Second theorem of Bertini, 458
Section, nonzero divisor, 291
Separable field extension, 452
Separable regular map, 346, 452
Separably generated field, 7
Separated scheme, 297
Separated, variety, 82
Separating transcendence basis, 7
Serre, xii, 301, 319, 335, 339, 468

Serre duality, 319, 339
Shafarevich, xii, 168
Shannon, 232
Sheaf, 186
Sheaf cohomology, 308
Sheaf, axioms, 186
Sheaf, global section, 186
Sheaf, homomorphism, 187
Sheaf, isomorphism, 187
Sheaf, support, 203
Sheafification of a graded module on a projective variety, 199
Sheafification of a module on an affine variety, 196
Sheafification of a presheaf, 187
Short exact sequence of sheaves, 194
Simis, 116
Singular cohomology, 358
Singular locus, 162
Smooth morphism, 456
Snapper, 366
Snapper polynomial, 366
Splitting field, 438
Splitting group, 438
Srinivas, 345
Strict transform, of a projective variety, 123
Strict transform, of an affine variety, 116
Subscheme, 289
Subsheaf, 192
Subvariety, 69
Surface, 45
Surjective sheaf homomorphism, 194
Symbolic power, 245
System of parameters, 142

Tamely ramified, map of curves, 346
Tangent cone, 294
Tangent space, extrinsic definition, 156
Tangent space, intrinsic definition, 157
Torsion sheaf, 366
Transcendence basis, 6
Transcendence degree, 6
Transition functions on an invertible sheaf, 271
Trivialization, of an invertible sheaf, 292
Tucker, 433

Ulrich, 116
Uniformizing parameters, 285
Universal property of blowing up, 223

Unramified extension, of a normal
    domain, 400
Unramified extension, of local domains,
    399

Valuation, 228
Valuation ring, 228
Value group, 228
Variety, biprojective, 103
Variety, quasi-biprojective, 103
Vasconcelos, 116
Veronese subring, 64
Very ample divisor, 262
Very ample invertible sheaf, 270

Walker, 227
Watanabe, 324
Weber, xi
Weierstrass, 176
Weierstrass preparation theorem, 176
Weil, 362
Wildly ramified, map of curves, 346
Włodarczyk, 234

Zariski, xii, 76, 148, 151, 161, 168, 227,
    229–232, 344, 388, 392, 413, 416,
    420, 433, 452, 458
Zariski topology, on $\mathbb{A}^n$, 29
Zariski topology, on $\mathbb{P}^n$, 68
Zariski topology, on a quasi-projective
    algebraic set, 68
Zariski topology, on an affine algebraic
    set, 34
Zariski topology, on the Proj of a
    graded ring, 67
Zariski topology, on the Spec of a ring,
    5
Zariski's connectedness theorem, 151
Zariski's main theorem, 148, 416
Zariski's subspace theorem, 420